Wolf Prize in Mathematics

Volume 4

Wolf Prize in Mathematics

Volume 4

Edited by

Y Sinai
Princeton University, USA

E Stein
Princeton University, USA

World Scientific

NEW JERSEY • LONDON • SINGAPORE • BEIJING • SHANGHAI • HONG KONG • TAIPEI • CHENNAI

Published by

World Scientific Publishing Co. Pte. Ltd.
5 Toh Tuck Link, Singapore 596224
USA office: 27 Warren Street, Suite 401-402, Hackensack, NJ 07601
UK office: 57 Shelton Street, Covent Garden, London WC2H 9HE

British Library Cataloguing-in-Publication Data
A catalogue record for this book is available from the British Library.

The editors and publisher would like to thank the following organisations and publishers of the various journals and books for their assistance and permission to reproduce the selected reprints found in this volume:

Academic Press
American Mathematical Society
Canadian Mathematical Congress
Centre National de la Recherche Scientifique
European Mathematical Society
Gauthier-Villars
Higher Ed. Press
Institut Mittag-Leffler
International Congress of Mathematicians
International Mathematical Union
International Press
The Japan Academy
London Mathematical Society
Mir
Moshe Klein
Plenum Publishing Corp.
Princeton University Press
Scientific Publishers
Shaw Prize Foundation
Springer-Verlag
The Wolf Foundation

While every effort has been made to contact the publishers of reprinted papers prior to publication, we have not been successful in some cases. Where we cold not contact the publishers, we have acknowledged the source of the material. Proper credit will be accorded to the publishers in future editions of this work after permission is granted.

WOLF PRIZE IN MATHEMATICS — Vol. 4
edited by Y. G. Sinai and E. M. Stein

ISBN-13 978-981-4390-29-3
ISBN-10 981-4390-29-1
ISBN-13 978-981-4723-92-3 (pbk)
ISBN-10 981-4723-92-4 (pbk)

List of Wolf Prize Winners

1978	Israel M. Gel'fand	1913–2009
	Carl L. Siegel	1896–1981
1979	Jean Leray	1906–1998
	André Weil	1906–1998
1980	Henri Cartan	1904–2008
	Andrei N. Kolmogorov	1903–1987
1981	Lars V. Ahlfors	1907–1996
	Oscar Zariski	1899–1986
1982	Hassler Whitney	1907–1989
	Mark Grigor'evich Krein	1907–1989
1983/4	Shiing-Shen Chern	1911–2004
	Paul Erdős	1913–1996
1984/5	Kunihiko Kodaira	1915–1997
	Hans Lewy	1904–1988
1986	Samuel Eilenberg	1913–1998
	Atle Selberg	1917–2007
1987	Kiyoshi Itō	1915–2008
	Peter D. Lax	1926
1988	Friedrich Hirzebruch	1927
	Lars Hörmander	1931
1989	Alberto P. Calderón	1920–1998
	John W. Milnor	1931
1990	Ennio de Giorgi	1928–1996
	Ilya Piatetski-Shapiro	1929–2009
1992	Lennart A. E. Carleson	1928
	John G. Thompson	1932

1993	Mikhail Gromov	1943
	Jacques Tits	1930
1994/5	Jürgen K. Moser	1928–1999
1995/6	Robert Langlands	1936
	Andrew J. Wiles	1953
1996/7	Joseph B. Keller	1923
	Yakov G. Sinai	1935
1999	László Lovász	1948
	Elias M. Stein	1931
2000	Raoul Bott	1923–2005
	Jean-Pierre Serre	1926
2001	Vladimir I. Arnold	1937–2010
	Saharon Shelah	1945
2002/3	Mikio Sato	1928
	John T. Tate	1925
2005	Gregory A. Margulis	1946
	Sergei P. Novikov	1938
2006/7	Stephen Smale	1930
	Hillel Fürstenberg	1935
2008	Pierre R. Deligne	1944
	Phillip A. Griffiths	1938
	David Mumford	1937
2010	Shing-Tung Yau	1949
	Dennis P. Sullivan	1941
2012	Michael G. Aschbacher	1944
	Luis A. Caffarelli	1948

Preface

This is the fourth volume in the series dealing with the winners of the Wolf Prize in mathematics. The present volume covers mostly the period of the first decade of this century.

It may be well to recall some facts about the Wolf Prize. It was founded by Dr. Ricardo Wolf and his wife Francisca Subirana Wolf and is given each year in five of the six areas of Agriculture, Chemistry, Mathematics, Medicine, Physics, and the Arts. Laureates receive their awards from the President of the State of Israel in a ceremony at the Knesset (Israel's parliament) in Jerusalem. About this prize, S. S. Chern and F. Hirzebrach said it well when they wrote: "The Fields medal goes to young people ... The Wolf prize often honors the achievements of a whole life."

As in the previous volumes, the documents collected here are varied in nature but are intended to sketch a picture of the work and scientific life of each of the laureates. As such, these include bibliographies and curricula vitae, particularly noteworthy papers or surveys, as well as articles written about the winners, whenever these were available.

It is our pleasure to record here our appreciation to the Wolf foundation. We also wish to express our thanks to World Scientific for their role in publishing this volume, and in particular to the help we have received from Ms. E. H. Chionh.

Yakov G. Sinai
Elias M. Stein

Contents

Mikio Sato

Atle Selberg

Saharon Shelah

John G. Thompson

Curriculum Vitae
David Mumford

Date of Birth: June 11, 1937, Worth, West Sussex, UK

Education:

1957	B.A., Magna Cum Laude, Harvard College
1961	Ph.D., Harvard University

Employment History:

1961–1962	Instructor and Research Fellow in Mathematics Harvard University
1962–1963	Assistant Professor, Harvard University
1962–1963	Member, Institute for Advanced Study, Princeton and Visiting Professor, University of Tokyo
1963–1967	Associate Professor, Harvard University
1967–1977	Professor, Harvard University
1967–1968	Visiting Professor, Tata Institute of Fundamental Research
1970–1971	Nuffield Professor, University of Warwick
1976–1977	Visiting Professor, Institute des Hautes Etudes Scientifiques, Paris
1977–1997	Higgins Professor of Mathematics, Harvard University
1978–1979	Visiting Professor, Tata Institute of Fundamental Research
1981–1984	Chairman, Department of Mathematics, Harvard University
1985–1997	Member, Division of Applied Science, Harvard University
1991–1994	Vice-President, International Mathematical Union
1993(fall)	Rothschild Professor, Isaac Newton Institute, Cambridge University
1995–1998	President, International Mathematical Union
1997–	Professor Emeritus, Harvard University
1996–2007	University Professor, Division of Applied Math. Brown University
1998(fall)	Visiting Professor, Institut Henri Poincare, Paris
2005(spring)	Visiting Professor, Math. Sciences Research Institute, Berkeley
2007–	Professor Emeritus, Brown University

Awards:

1953	Westinghouse Science Talent Search, finalist
1958–1961	Society of Fellows, Harvard University, Junior Fellow
1974	Fields Medal, International Congress of Mathematics, Vancouver
1975	Elected to the National Academy of Sciences

1978	Honorary Fellow, Tata Institute of Fundamental Research
1983	Honorary Degree of Doctor of Science, University of Warwick
1987–1992	MacArthur Foundation Fellow
1991	Elected Foreign Member, Accademia Nazionale dei Lincei, Rome
1995	Elected Honorary Member, London Mathematical Society
1997	Elected American Philosophical Society
2000	Honorary Degree of DSc., Norwegian Univ. of Science and Technology, Trondheim
2001	Honorary Degree of DSc., Rockefeller University
2004	Elected Foreign Member, Norwegian Academy of Science and Letters
2005	IEEE Longuet-Higgins Prize, CVPR 2005, (with Jayant Shah)
2006	The Shaw Prize in Mathematical Sciences, Hong Kong (shared with Wu Wentsun)
2007	Amer. Math. Society Steele Prize for Mathematical Exposition
2008	Wolf Prize in Mathematics (shared with Pierre Deligne and Phillip Griffiths)
2008	Elected Foreign Member of the Royal Society (UK)

List of Publications

1. D. Mumford, What's so baffling about negative numbers? — A cross-cultural comparison, in *Studies in the History of Indian Mathematics*, pp. 113–143, Cult. Hist. Math., Vol. 5 (Hindustan Book Agency, 2010).
2. D. Mumford and A. Desolneux, Pattern theory. The stochastic analysis of real-world signals, in *Applying Mathematics* (A K Peters, Ltd., 2010).
3. D. Mumford, Mathematics in India, *Notices Amer. Math. Soc.* **57** (2010), No. 3, 385–390.
4. A. Ash, D. Mumford, M. Rapoport and Y.-S. Tai, *Smooth Compactifications of Locally Symmetric Varieties*, 2nd edition. With the collaboration of Peter Scholze (Cambridge Univ. Press, 2010).
5. D. Mumford, *Abelian Varieties*. With appendices by C. P. Ramanujam and Yuri Manin. Corrected reprint of the 2nd (1974) edition, Tata Institute of Fundamental Research Studies in Mathematics, Vol. 5 (Hindustan Book Agency, 2008).
6. P. J. Davis and D. Mumford, Henri's crystal ball, *Notices Amer. Math. Soc.* **55** (2008), No. 4, 458–466.
7. L. Younes, P. W. Michor, J. Shah and D. Mumford, A metric on shape space with explicit geodesics, *Atti Accad. Naz. Lincei Cl. Sci. Fis. Mat. Natur. Rend. Lincei* (9) *Mat. Appl.* **19** (2008), No. 1, 25–57.
8. D. Mumford, *Tata Lectures on Theta*, I. With the collaboration of C. Musili, M. Nori, E. Previato and M. Stillman. Reprint of the 1983 edition (Birkhäuser, 2007).
9. P. W. Michor and D. Mumford, An overview of the Riemannian metrics on spaces of curves using the Hamiltonian approach, *Appl. Comput. Harmon. Anal.* **23** (2007), No. 1, 74–113.
10. D. Mumford, *Tata Lectures on Theta*, III. With collaboration of Madhav Nori and Peter Norman. Reprint of the 1991 original (Birkhäuser, 2007).
11. D. Mumford, *Tata Lectures on Theta*, II, *Jacobian Theta Functions and Differential Equations*. With the collaboration of C. Musili, M. Nori, E. Previato, M. Stillman and H. Umemura. Reprint of the 1984 original (Birkhäuser, 2007).
12. P. W. Michor and D. Mumford, Riemannian geometries on spaces of plane curves, *J. Eur. Math. Soc. (JEMS)* **8** (2006), No. 1, 1–48.
13. P. W. Michor and D. Mumford, Vanishing geodesic distance on spaces of submanifolds and diffeomorphisms, *Doc. Math.* **10** (2005), 217–245 (electronic).

14. D. Mumford, *Selected Papers on the Classification of Varieties and Moduli Spaces*. With commentaries by David Gieseker, George Kempf, Herbert Lange and Eckart Viehweg (Springer-Verlag, 2004).
15. D. Mumford, Pattern theory: The mathematics of perception, *Proc. of the International Congress of Mathematicians*, Vol. I (Beijing, 2002), pp. 401–422 (Higher Ed. Press, 2002).
16. D. Mumford, In memoriam: George R. Kempf 1944–2002, *Amer. J. Math.* **124** (2002), No. 6, iii–iv.
17. D. Mumford, C. Series and D. Wright, *Indra's Pearls. The Vision of Felix Klein* (Cambridge Univ. Press, 2002).
18. D. Mumford and B. Gidas, Stochastic models for generic images, *Quart. Appl. Math.* **59** (2001), No. 1, 85–111.
19. D. Mumford, The dawning of the age of stochasticity, Mathematics towards the third millennium (Rome, 1999), *Atti Accad. Naz. Lincei Cl. Sci. Fis. Mat. Natur. Rend. Lincei* (9) *Mat. Appl.* **2000**, Special Issue, 107–125.
20. D. Mumford, The dawning of the age of stochasticity, in *Mathematics: Frontiers and Perspectives*, pp. 197–218 (Amer. Math. Soc., 2000).
21. D. Mamford, The Red Book of Varieties and Schemes, Second, expanded edition. Includes the Michigan lectures (1974) on curves and their Jacobians. With contributions by Enrico Arbarello, Lecture Notes in Mathematics, Vol. 1358 (Springer-Verlag, 1999).
22. P. W. Hallinan, G. G. Gordon, A. L. Yuille, P. Giblin and D. Mumford, *Two- and Three-Dimensional Patterns of the Face* (A K Peters, Ltd., 1999).
23. D. Mumford, Trends in the profession of mathematics, in *Mitt. Dtsch. Math.-Ver.* **1998**, No. 2, 25–29.
24. D. Mumford, Pattern theory: A unifying perspective, in *Fields Medallists' Lectures*, pp. 226–261, World Sci. Ser. 20th Century Math., Vol. 5 (World Scientific, 1997).
25. D. Mumford, Issues in the mathematical modeling of cortical functioning and thought, in *The Legacy of Norbert Wiener: A Centennial Symposium* (Cambridge, MA, 1994), pp. 235–260, Proc. Sympos. Pure Math., Vol. 60 (Amer. Math. Soc., 1997).
26. D. Mumford, Algebraic geometry, I. Complex projective varieties, Reprint of the 1976 edition, in *Classics in Mathematics* (Springer-Verlag, 1995).
27. D. Mumford, Pattern theory: A unifying perspective, *First European Congress of Mathematics*, Vol. I (Paris, 1992), pp. 187–224, Progr. Math., Vol. 119 (Birkhäuser, 1994).
28. D. Mumford, J. Fogarty and F. Kirwan, *Geometric Invariant Theory*, Third edition. Ergebnisse der Mathematik und ihrer Grenzgebiete (2) [Results in Mathematics and Related Areas (2)], Vol. 34 (Springer-Verlag, 1994).

29. F. R. K. Chung and D. Mumford, Chordal completions of planar graphs, *J. Combin. Theory Ser. B* **62** (1994), No. 1, 96–106.
30. D. Mumford, Elastica and computer vision, in *Algebraic Geometry and Its Applications* (West Lafayette, IN, 1990), pp. 491–506 (Springer, 1994).
31. D. Bayer and D. Mumford, What can be computed in algebraic geometry?, in *Computational Algebraic Geometry and Commutative Algebra* (Cortona, 1991), pp. 1–48, Sympos. Math., Vol. XXXIV (Cambridge Univ. Press, 1993).
32. M. Nitzberg, D. Mumford and T. Shiota, *Filtering, Segmentation and Depth*, Lecture Notes in Computer Science, Vol. 662 (Springer-Verlag, 1993).
33. F. R. K. Chung and D. Mumford, Chordal completions of grids and planar graphs, in *Planar Graphs* (New Brunswick, NJ, 1991), pp. 37–40, DIMACS Ser. Discrete Math. Theoret. Comput. Sci., Vol. 9 (Amer. Math. Soc., 1993).
34. D. Mumford, *Tata Lectures on Theta*, III. With the collaboration of Madhav Nori and Peter Norman, Progress in Mathematics, Vol. 97 (Birkhäuser, 1991).
35. D. Mumford and J. Shah, Optimal approximations by piecewise smooth functions and associated variational problems, *Comm. Pure Appl. Math.* **42** (1989), No. 5, 577–685.
36. D. Mumford, *Oscar Zariski and His Work*, A joint AMS-MAA lecture presented in Atlanta, Georgia, January 1988. AMS-MAA Joint Lecture Series (Amer. Math. Soc., 1988).
37. D. Mumford, *The Red Book of Varieties and Schemes*, Lecture Notes in Mathematics, Vol. 1358 (Springer-Verlag, 1988).
38. D. Mumford, *Lektsii o tèta-funktsiyakh* (in Russian) [Lectures on theta-functions], translated from the English by D. Yu. Manin. Translation edited and with a preface by Yu. I. Manin. With appendices by Khirosi Umemura [Hiroshi Umemura] and Takakhiro Shiota [Takahiro Shiota] (Mir, 1988).
39. D. Mumford, S. M. Kosslyn, L. A. Hillger and R. J. Herrnstein, Discriminating figure from ground: The role of edge detection and region growing, *Proc. Nat. Acad. Sci. U.S.A.* **84** (1987), No. 20, 7354–7358.
40. D. Mumford, *Oscar Zariski*: 1899–1986, *Notices Amer. Math. Soc.* **33** (1986), No. 6, 891–894.
41. D. Mumford, *Tata Lectures on Theta*, II. Jacobian theta functions and differential equations. With the collaboration of C. Musili, M. Nori, E. Previato, M. Stillman and H. Umemura, Progress in Mathematics, Vol. 43 (Birkhäuser, 1984).

42. D. Mumford, Towards an enumerative geometry of the moduli space of curves, in *Arithmetic and Geometry*, Vol. II, pp. 271–328, Progr. Math., Vol. 36 (Birkhäuser, 1983).
43. D. Mumford, On the Kodaira dimension of the Siegel modular variety, in *Algebraic Geometry — Open Problems* (Ravello, 1982), pp. 348–375, Lecture Notes in Math., Vol. 997 (Springer-Verlag, 1983).
44. D. Mumford, *Tata Lectures on Theta*, I. With the assistance of C. Musili, M. Nori, E. Previato and M. Stillman. Progress in Mathematics, Vol. 28 (Birkhäuser, 1983).
45. D. Mumford and J. Fogarty, *Geometric Invariant Theory*, 2nd edition, Ergebnisse der Mathematik und ihrer Grenzgebiete [Results in Mathematics and Related Areas], Vol. 34 (Springer-Verlag, 1982).
46. J. Harris and D. Mumford, On the Kodaira dimension of the moduli space of curves. With an appendix by William Fulton, *Invent. Math.* **67** (1982), No. 1, 23–88.
47. D. Mumford, *Algebraic Geometry*, I. Complex projective varieties. Corrected reprint. Grundlehren der Mathematischen Wissenschaften [Fundamental Principles of Mathematical Sciences], Vol. 221 (Springer-Verlag, 1981).
48. D. Mumford, An algebraic surface with K ample, $(K_2) = 9$, $p_g = q = 0$, *Amer. J. Math.* **101** (1979), No. 1, 233–244.
49. D. Mamford, *Algebraicheskaya Geometriya*, I. Komplektsnye proektivnye mnogoobraziya (in Russian) [Algebraic geometry. I. Complex projective varieties] Translated from the English by Ju. I. Manin (Mir, 1979).
50. P. van Moerbeke and D. Mumford, The spectrum of difference operators and algebraic curves, *Acta Math.* **143** (1979), No. 1–2, 93–154.
51. D. Mumford, An algebro-geometric construction of commuting operators and of solutions to the Toda lattice equation, Korteweg deVries equation and related nonlinear equation, in *Proc. of the International Symposium on Algebraic Geometry* (Kyoto Univ., Kyoto, 1977), pp. 115–153 (Kinokuniya Book Store, 1978).
52. D. Mumford, The work of C. P. Ramanujam in algebraic geometry, in *C. P. Ramanujam — A Tribute*, pp. 8–10, Tata Inst. Fund. Res. Studies in Math., Vol. 8 (Springer, 1978).
53. D. Mumford, Some footnotes to the work of C. P. Ramanujam, in *C. P. Ramanujam — A Tribute*, pp. 247–262, Tata Inst. Fund. Res. Studies in Math., Vol. 8 (Springer, 1978).
54. D. Mumford and J. Tate, *Fields Medals*, IV. An instinct for the key idea, *Science* **202** (1978), No. 4369, 737–739.

55. D. Mumford, Hirzebruch's proportionality theorem in the noncompact case, *Invent. Math.* **42** (1977), 239–272.
56. E. Bombieri and D. Mumford, Enriques' classification of surfaces in char. *p*. II, in *Complex Analysis and Algebraic Geometry*, pp. 23–42 (Iwanami Shoten, 1977).
57. D. Mumford, Stability of projective varieties, Lectures given at the Institut des Hautes Études Scientifiques, Bures-sur-Yvette, March-April 1976. Monographie de l'Enseignement Mathématique, No. 24 (L'Enseignement Mathématique, 1977).
58. D. Mumford, Stability of projective varieties, *Enseignement Math.* (2) **23** (1977), No. 1–2, 39–110.
59. E. Bombieri and D. Mumford, Enriques' classification of surfaces in char. *p*. III, *Invent. Math.* **35** (1976), 197–232.
60. D. Mumford, *Algebraic Geometry*, I. Complex projective varieties. Grundlehren der Mathematischen Wissenschaften, No. 221 (Springer-Verlag, 1976).
61. F. F. Knudsen and D. Mumford, The projectivity of the moduli space of stable curves, I. Preliminaries on "det" and "Div", *Math. Scand.* **39** (1976), No. 1, 19–55.
62. D. Mumford, Hilbert's fourteenth problem — The finite generation of subrings such as rings of invariants, in *Mathematical Developments Arising from Hilbert Problems* (Proc. Sympos. Pure Math., Vol. XXVIII, Northern Illinois Univ., De Kalb, Ill., 1974), pp. 431–444 (Amer. Math. Soc., 1976).
63. D. Mumford, A new approach to compactifying locally symmetric varieties, in *Discrete Subgroups of Lie Groups and Applications to Moduli* (Internat. Colloq., Bombay, 1973), pp. 211–224 (Oxford Univ. Press, 1975).
64. D. Mumford, Pathologies IV, *Amer. J. Math.* **97** (1975), No. 3, 847–849.
65. A. Ash, D. Mumford, M. Rapoport and Y. Tai, Smooth compactification of locally symmetric varieties, in *Lie Groups: History, Frontiers and Applications*, Vol. IV (Math. Sci. Press, 1975).
66. A. T. Lascu, D. Mumford and D. B. Scott, The self-intersection formula and the "formule-clef", *Math. Proc. Cambridge Philos. Soc.* **78** (1975), 117–123.
67. D. Mumford, Curves and their Jacobians (The University of Michigan Press, 1975).
68. D. Lieberman and D. Mumford, Matsusaka's big theorem, in *Algebraic Geometry* (Proc. Sympos. Pure Math., Vol. 29, Humboldt State Univ., Arcata, Calif., 1974), pp. 513–530 (Amer. Math. Soc., 1975).

69. D. Mumford, Prym varieties, I, *Contributions to Analysis* (*A Collection of Papers Dedicated to Lipman Bers*), pp. 325–350 (Academic Press, 1974).
70. Ž. D'édonne, Dž. Kerrol and D. Mumford, *Geometricheskaya teoriya invariantov* (in Russian) [Geometric invariant theory] With supplementary material by David Mumford and C. S. Seshadri. Translated from the English by A. N. Paršin. *Izdat* (Mir, 1974).
71. G. Horrocks and D. Mumford, A rank 2 vector bundle on $\mathbf{P}_4$ with 15,000 symmetries, *Topology* **12** (1973), 63–81.
72. D. Mumford, A remark on the paper of M. Schlessinger, in *Complex Analysis*, 1972 (Proc. Conf., Rice Univ., Houston, TX., 1972), Vol. I: Geometry of singularities, *Rice Univ. Studies* **59** (1973), No. 1, 113–117.
73. G. Kempf, F. F. Knudsen, D. Mumford and B. Saint-Donat, Toroidal Embeddings, I. Lecture Notes in Mathematics, Vol. 339 (Springer-Verlag, 1973).
74. D. Mumford and K. Suominen, Introduction to the theory of moduli, in *Algebraic Geometry*, Oslo 1970 (Proc. Fifth Nordic Summer-School in Math.), pp. 171–222 (Wolters-Noordhoff, 1972).
75. D. Mamford, An analytic construction of curves with degenerate reduction over complete local rings (in Russian), translated from the English (Compositio Math. **24** (1972), 129–174), by Ju. I. Manin, *Usp. Mat. Nauk* **27** (1972), No. 6(168), 181–221.
76. D. Mumford, An analytic construction of degenerating abelian varieties over complete rings, *Compositio Math.* **24** (1972), 239–272.
77. D. Mumford, An analytic construction of degenerating curves over complete local rings, *Compositio Math.* **24** (1972), 129–174.
78. M. Artin and D. Mumford, Some elementary examples of unirational varieties which are not rational, *Proc. London Math. Soc.* (3) **25** (1972), 75–95.
79. D. Mumford, The structure of the moduli spaces of curves and Abelian varieties, in *Actes du Congrès International des Mathématiciens* (Nice, 1970), Tome 1, pp. 457–465 (Gauthier-Villars, 1971).
80. D. Mumford, Theta characteristics of an algebraic curve, *Ann. Sci. École Norm. Sup.* (4) **4** (1971), 181–192.
81. D. Mumford, A remark on Mahler's compactness theorem, *Proc. Amer. Math. Soc.* **28** (1971), 289–294.
82. D. Mumford, *Abelian Varieties*, Tata Institute of Fundamental Research Studies in Mathematics, No. 5 (Oxford Univ. Press, 1970).
83. D. Mumford, Varieties defined by quadratic equations, *Questions on Algebraic Varieties* (C.I.M.E., III Ciclo, Varenna, 1969) pp. 29–100 (Edizioni Cremonese, 1970).

84. P. Deligne and D. Mumford, The irreducibility of the space of curves of given genus, *Inst. Hautes* P. Deligne *Études Sci. Publ. Math.* No. 36 (1969), 75–109.
85. D. Mumford, Bi-extensions of formal groups, in *Algebraic Geometry* (Internat. Colloq., Tata Inst. Fund. Res., Bombay, 1968), pp. 307–322 (Oxford Univ. Press, 1969).
86. D. Mumford, Enriques' classification of surfaces in char *p*. I, in *Global Analysis* (Papers in Honor of K. Kodaira), pp. 325–339 (Univ. Tokyo Press, 1969).
87. D. Mumford, A note of Shimura's paper "Discontinuous groups and abelian varieties". *Math. Ann.* **181** (1969), 345–351.
88. T. Matsusaka and D. Mumford, Correction to: "Two fundamental theorems on deformations of polarized varieties". *Amer. J. Math.* **91** (1969), 851.
89. D. Mumford, Rational equivalence of 0-cycles on surfaces, *J. Math. Kyoto Univ.* **9** (1968), 195–204.
90. D. Mumford and P. Newstead, Periods of a moduli space of bundles on curves, *Amer. J. Math.* **90** (1968), 1200–1208.
91. F. Oort and D. Mumford, Deformations and liftings of finite, commutative group schemes, *Invent. Math.* **5** (1968), 317–334.
92. D. Mumford, Abelian quotients of the Teichmüller modular group, *J. Anal. Math.* **18** (1967), 227–244.
93. D. Mumford, On the equations defining abelian varieties, III, *Invent. Math.* **3** (1967), 215–244.
94. D. Mumford, On the equations defining abelian varieties, II, *Invent. Math.* **3** (1967), 75–135.
95. D. Mumford, Pathologies, III, *Amer. J. Math.* **89** (1967), 94–104.
96. D. Mumford, Lectures on curves on an algebraic surface. With a section by G. M. Bergman. Annals of Mathematics Studies, No. 59 (Princeton Univ. Press, 1966).
97. D. Mumford, Families of abelian varieties, in *Algebraic Groups and Discontinuous Subgroups* (Proc. Sympos. Pure Math., Boulder, Colo., 1965), pp. 347–351 (Amer. Math. Soc., 1966).
98. D. Mumford, On the equations defining abelian varieties, I, *Invent. Math.* **1** (1966), 287–354.
99. D. Mumford, *Geometric Invariant Theory*, Ergebnisse der Mathematik und ihrer Grenzgebiete, Neue Folge, Band 34 (Springer-Verlag, 1965).
100. D. Mumford, Picard groups of moduli problems, in *Arithmetical Algebraic Geometry* (Proc. Conf. Purdue Univ., 1963), pp. 33–81 (Harper & Row, 1965).

101. D. Mumford, A remark on Mordell's conjecture, *Amer. J. Math.* **87** (1965), 1007–1016.

102. T. Matsusaka and D. Mumford, Two fundamental theorems on deformations of polarized varieties, *Amer. J. Math.* **86** (1964), 668–684.

103. D. Mumford, Projective invariants of projective structures and applications, in 1963 *Proc. Internat. Congr. Mathematicians* (Stockholm, 1962), pp. 526–530 (Inst. Mittag-Leffler, 1963).

104. D. B. Mumford, Some aspects of the problem of moduli (in Japanese), *Sûgaku* **15** (1963/1964), 155–157.

105. D. Mumford, Further pathologies in algebraic geometry, *Amer. J. Math.* **84** (1962), 642–648.

106. D. Mumford, The topology of normal singularities of an algebraic surface and a criterion for simplicity, *Inst. Hautes Études Sci. Publ. Math.* No. 9 (1961), 5–22.

107. D. Mumford, An elementary theorem in geometric invariant theory, *Bull. Amer. Math. Soc.* **67** (1961), 483–487.

108. D. Mumford, Pathologies of modular algebraic surfaces, *Amer. J. Math.* **83** (1961), 339–342.

Proceedings of the International Congress of Mathematicians
Vancouver, 1974

The Work of David Mumford

J. Tate

It is a great pleasure for me to report on Mumford's work. However I feel there are many people more qualified than I to do this. I have consulted with some of them and would like to thank them all for their help, especially Oscar Zariski.

Mumford's major work has been a tremendously successful multi-pronged attack on problems of the existence and structure of varieties of moduli, that is, varieties whose points parametrize isomorphism classes of some type of geometric object. Besides this he has made several important contributions to the theory of algebraic surfaces. I shall begin by mentioning briefly some of the latter and then will devote most of this talk to a discussion of his work on moduli.

Mumford has carried forward, after Zariski, the project of making algebraic and rigorous the work of the Italian school on algebraic surfaces. He has done much to extend Enriques' theory of classification to characteristic $p > 0$, where many new difficulties appear. This work is impossible to describe in a few words and I shall say no more about it except to remark that our other Field's Medalist, Bombieri, has also made important contributions in this area, and that he and Mumford have recently been continuing their work in collaboration.

We have a good understanding of divisors on an algebraic variety, but our knowledge about algebraic cycles of codimension > 1 is still very meager. The first case is that of 0-cycles on an algebraic surface. In particular, what is the structure of the group of 0-cycles of degree 0 modulo the subgroup of cycles rationally equivalent to zero, i.e., which can be deformed to 0 by a deformation which is parametrized by a line. This group maps onto the Albanese variety of the surface, but what about the kernel of this map? Is it "finite-dimensional"? Severi thought so; but Mumford proved it is not, if the geometric genus of the surface is $\geqq 1$. Mumford's proof uses methods of Severi, and he remarks that in this case the techniques of the classical Italian algebraic geometers seem superior to their vaunted intuition. However, in other cases Mumford has used modern techniques to justify Italian intuition, as in the construction by him and M. Artin of examples of unirational varieties X which are not rational, based on 2-torsion in $H^3(X, \mathbf{Z})$.

Probably Mumford's most famous result on surfaces is his topological characterization of nonsingularity. Let P be a normal point on an algebraic surface V in a complex projective space. Mumford showed that if V is topologically a manifold at P, then it is algebraically nonsingular there. Indeed, consider the intersection K of V with a small sphere about P. This intersection K is 3-dimensional and if V is a manifold at P, then K is a sphere and its fundamental group is trivial. Mumford showed how to compute this fundamental group $\pi_1(K)$ in terms of the diagram of the resolution of the singularity of V at P, and then he showed that $\pi_1(K)$ is not trivial unless the diagram is, i.e., unless V is nonsingular at P. A by-product of this proof is the fact that the Poincaré conjecture holds for the 3-manifolds which occur as K's. Mumford's paper was a critical step between the early work on singularities of branches of plane curves (where K is a torus knot) and fascinating later developments. Brieskorn showed that the analogs of Mumford's results are false in general for V of higher dimension. Consideration of the corresponding problem there led to the discovery of some beautiful relations between algebraic geometry and differential topology, including simple explicit equations for exotic spheres.

Let me now turn to Mumford's main interest, the theory of varieties of moduli. This is a central topic in algebraic geometry having its origins in the theory of elliptic integrals. The development of the algebraic and global aspects of this subject in recent years is due mainly to Mumford, who attacked it with a brilliant combination of classical, almost computational, methods and Grothendieck's new scheme-theoretic techniques.

Mumford's first approach was by the 19th century theory of invariants. In fact, he revived this moribund theory by considering its geometric significance. In pursuing an idea of Hilbert, Mumford was led to the crucial notion of "stable" objects in a moduli problem. The abstract setting behind this notion is the following: Suppose G is a reductive algebraic group acting on a variety V in projective space $\boldsymbol{P}_N$ by projective transformations. Then the action of G is induced by a linear and unimodular representation of some finite covering G^* of G on the affine cone $\boldsymbol{A}^{N+1}$ over the ambient $\boldsymbol{P}_N$. Mumford defines a point $x \in V$ to be *stable* for the action of G on V, relative to the embedding $V \subset \boldsymbol{P}_N$, if for one (and hence every) point $x^* \in \boldsymbol{A}^{N+1}$ over x, the orbit of x^* under G^* is closed in $\boldsymbol{A}^{N+1}$, and the stabilizer of x^* is a finite subgroup of G^*. His fundamental theorem is then that the set of stable points is an open set V_s in V, and V_s/G is a quasi-projective variety.

For example, suppose $V = (\boldsymbol{P}_n)^m$ is the variety of ordered m-tuples of points in projective n-space and G is PGL_n acting diagonally on V via the Segre embedding. Then a point $x = (x_1, x_2, \cdots, x_m) \in V$ is stable if and only if for each proper linear subspace $L \subset \boldsymbol{P}_n$, the number of points $x_i \in L$ is strictly less than $m(\dim L + 1)/(n + 1)$. In case $n = 1$, for example, this means that an m-tuple of points on the projective line is unstable if more than half the points coalesce. The reason such

m-tuples must be excluded is the following: Let $P_t = (tx_1, tx_2, \cdots, tx_r, x_{r+1}, \cdots, x_m)$ and $Q_t = (x_1, \cdots, x_r, t^{-1}x_{r+1}, \cdots, t^{-1}x_m)$, where the x_i are pairwise distinct. Then P_t is in the same orbit as Q_t, for $t \neq 0, \infty$, but $P_0 = (0, \cdots, 0, x_{r+1}, \cdots, x_m)$ is not in the same orbit as $Q_0 = (x_1, \cdots, x_r, \infty, \cdots, \infty)$ unless $m = 2r$, and even then is not in general. Thus if we want a separated orbit space in which $\lim_{t \to 0}$ (Orbit P_t) is unique, we must exclude P_0 or Q_0; and it is natural to exclude the one with more than half its components equal.

Using the existence of the orbit spaces V_s/G, Mumford was able to construct a moduli scheme over the ring of integers for polarized abelian varieties, relative Picard schemes (following a suggestion of Grothendieck), and also moduli varieties for "stable" vector bundles on a curve in characteristic 0. The meaning of stability for a vector bundle is that all proper subbundles are less ample than the bundle itself, if we measure the ampleness of a bundle by the ratio of its degree to its rank. In the special example $V = (\boldsymbol{P}_n)^m$ mentioned above, the results can be proved by explicit computations which work in any characteristic and even over the ring of integers. But in its general abstract form Mumford's theory was limited to characteristic 0 because his proofs used the semisimplicity of linear representations. He conjectures that in characteristic p, linear representations of the classical semisimple groups have the property that complementary to a stable line in such a representation there is always a stable hypersurface (though not necessarily a stable hyperplane which would exist if the representation were semisimple). If this conjecture is true[1] then Mumford's treatment of geometric invariant theory would work in characteristic p. Seshadri has proved the conjecture in case of SL_2. He has also shown recently that the conjecture can be circumvented, by giving different more complicated proofs for the main results of the theory which work in any characteristic.

For moduli of abelian varieties and curves, Mumford has given more refined constructions than those furnished by geometric invariant theory. In three long papers in *Inventiones Mathematicae* he has developed an algebraic theory of theta functions. Classically, over the complex numbers, a theta function for an abelian variety A can be thought of as a complex function on the universal covering space $H_1(A, \boldsymbol{R})$ which transforms in a certain way under the action of $H_1(A, \boldsymbol{Z})$. For Mumford, over any algebraically closed field k, a theta function is a k-valued function on $\prod_{l \in S} H_1(A, \boldsymbol{Q}_l)$ (étale homology) which transforms in a certain way under $\prod_{l \in S} H_1(A, \boldsymbol{Z}_l)$. Here S is any finite set of primes $l \neq \text{char}(k)$, though in treating some of the deeper aspects of the theory Mumford assumed $2 \in S$. In order to get an idea of what these theta functions accomplish let us consider a classical special case. Let A be an elliptic curve over $\boldsymbol{C}$ with its points of order 4 marked. Then we get a canonical embedding $A \subsetneq \boldsymbol{P}^3$ via the theta functions $\theta[{a \atop b}]$; $a, b = 0, 1$. Let 0_A be the origin on A, whose coordinates in $\boldsymbol{P}_3$ are the "theta Nullwerte". Then A is the intersection of all quadric surfaces in $\boldsymbol{P}_3$ which pass through the orbit of 0_A under a certain action of $(\boldsymbol{Z}/4\boldsymbol{Z}) \times (\boldsymbol{Z}/4\boldsymbol{Z})$ on $\boldsymbol{P}_3$. Thus 0_A determines A and

[1](Added during correction of proofs). The conjecture is true; shortly after the Congress, it was proved by W. Haboush in general and by E. Formanek and C. Procesi for $GL(n)$ and $SL(n)$.

can be viewed as a "modulus". Moreover, 0_A lies on the quartic curve $\theta[{0 \atop 0}]^4 = \theta[{0 \atop 1}]^4 + \theta[{1 \atop 0}]^4$ in the plane $\theta[{1 \atop 1}] = 0$, and that curve minus a finite set of points is a variety of moduli for elliptic curves with their points of order 4 marked. Mumford's theory generalizes this construction to abelian varieties of any dimension, with points of any order $\geqq 3$ marked, in any characteristic $\neq 2$. The moduli varieties so obtained have a natural projective embedding, and their closure in that embedding is, essentially, an algebraic version of Satake's topological compactification of Siegel's moduli spaces. Besides these applications to moduli, the theory gives new tools for the study of a single abelian variety by furnishing canonical bases for all linear systems on it.

Next I want to mention briefly p-adic uniformization. Motivated by the study of the boundary of moduli varieties for curves, i.e., of how nonsingular curves can degenerate, Mumford was led to introduce p-adic Schottky groups, and to show how one can obtain certain p-adic curves of genus $\geqq 2$ transcendentally as the quotient by such groups of the p-adic projective line minus a Cantor set. The corresponding theory for genus 1 was discovered by the author, but the generalization to higher genus was far from obvious. Besides its significance for moduli, Mumford's construction is of interest in itself as a highly nontrivial example of "rigid" p-adic analysis.

The theta functions and p-adic uniformization give some insight into what happens on the boundary of the varieties of moduli of curves and abelian varieties, but a much more detailed picture can now be obtained by Mumford's theory of toroidal embeddings. This theory, which unifies ideas that had appeared earlier in the works of several investigators, reduces the study of certain types of varieties and singularities to combinatorial problems in a space of "exponents". The local model for a toroidal embedding is called a torus embedding. This is a compactification $\bar{V}$ of a torus V such that the action of V on itself by translation extends to an action of V on $\bar{V}$. The coordinate ring of V is linearly spanned by the monomials $x^a = x_1^{a_1} x_2^{a_2} \cdots x_n^{a_n}$, $n = \dim V$, with positive or negative integer exponents a_i. Viewing the exponent vectors a as integral points in Euclidean n-space, define a *rational cone* in that space to be a set consisting of r's such that $(r, a) \geqq 0$ for $a \in S$, where S is some finite set of exponent vectors. For each rational cone σ, the monomials x^a such that $(r, a) \geqq 0$ for all $r \in \sigma$ span the coordinate ring of an affine variety $V(\sigma)$ which contains V as an open dense subvariety, if σ contains no nonzero linear subspace of $\boldsymbol{R}^n$. Now if we decompose $\boldsymbol{R}^n$ into the union of a finite number of rational cones σ_α in such a way that each intersection $\sigma_\alpha \cap \sigma_\beta$ is a face of σ_α and σ_β, then the union of the $V(\sigma_\alpha)$ is a compactification of V of the type desired. All such compactifications $\bar{V}$ of V can be obtained in this way and the invariant sheaves of ideals on them can be described in terms of the decomposition into cones. One can also read off whether $\bar{V}$ is nonsingular, and if it is not one can desingularize it by suitably subdividing the decomposition. In short, there is a whole dictionary for translating questions about the algebraic geometry of V and $\bar{V}$ into combinatorial questions about decompositions of $\boldsymbol{R}^n$ into rational cones.

Mumford with the help of his coworkers has used these techniques to prove the following semistable reduction theorem. If a family of varieties X_t over C, in general nonsingular, is parametrized by a parameter t on a curve C, and if X_{t_0} is singular, then one can pull back the family to a ramified covering of C in a neighborhood of t_0 and blow it up over t_0 in such a way that the new singular fiber is of the stablest possible kind, i.e., is a divisor whose components have multiplicities 1 and cross transversally. The corresponding problem in characteristic p is open. For curves in characteristic p the result was proved by Mumford and Deligne and was a crucial step in their proof of the irreducibility of the moduli variety for curves of given genus.

Toroidal embeddings can also be used to construct explicit resolutions of the singularities of the projective varieties $\overline{D/\Gamma}$, where D is a bounded symmetric domain, Γ is an arithmetic group, and the bar denotes the "minimal" compactification of Baily and Borel. The construction of these resolutions is a big step forward. With them one has a powerful tool to analyse the behavior of functions at the "boundary", compute numerical invariants, and, generally, study the finer structure of these varieties.

I hope this report, incomplete as it is, gives some idea of Mumford's achievements and their importance. I heartily congratulate him on them and wish him well for the future!

HARVARD UNIVERSITY
CAMBRIDGE, MASSACHUSETTS 02138, U.S.A.

Some Glimpses of Pure and Applied Mathematics

The University of Hong Kong, China
Sept.14, 2006

David Mumford
Division of Applied Math.,Brown University

I am delighted by the recognition of the intimate partnership between pure and applied mathematics that has been given by the Shaw Prize award this year. My own work has involved both areas and each has its own distinctive charms, its own goals. In this lecture, I will attempt to give some relatively non-technical descriptions of bits of my work in both of these areas. In the first section, I give a somewhat tongue-in-cheek classification of the goals pure mathematicians set themselves, putting myself in what I call the *Linnaean* camp. Linnaeans like to make *maps* and I give some examples. In the second section, I try to sketch what the field of algebraic geometry is all about and what is the moduli space of curves, a space central to most of my research in pure math. In the third section, I switch to applied areas and introduce one of the central problems in analyzing images of the world: the segmentation problem. In the fourth section, I describe the statistical approach to vision problems, called *Pattern Theory*. In the fifth section, I talk about a phenomenon that binds pure and applied maths: the study of self-similar objects. And finally, in the last section, I return to maps and how these have entered computer vision through the study of shape. The original lecture included two brief film clips to illustrate some of the ideas which could not be included in this written version.

I. What sort of goals drive mathematicians?

The work of mathematicians is not as familiar to the public as that of other types of scientists. So I'd like to begin by giving a small taxonomy of the sort of thing we do:

- The *city planner*: these seek to create a web of definitions and thus claim territory – a new area of mathematics.
- The *alchemist*: they like to combine two seemingly unrelated things and *wow*, a startling result comes out.
- The *Linnaean*: their goal is to catalog the wonders of the world, their diversity and beautiful specializations.

Euclid exemplifies the first type of mathematician: in his *Elements,* Euclid codified Greek thought about geometry by systematically defining all the basic concepts and deriving their basic properties in natural order, in thirteen books. Pythagoras's theorem is an excellent example of the second – in this theorem, one combines the geometric of right triangles (or the diagonals of a rectangle) with the algebraic notion of taking squares. The amazing consequence, that the square of the length of the diagonal is the sum of the squares of the lengths of the two perpendicular sides, is the foundation of Cartesian coordinates, of the reduction of geometry to algebra.

This Shaw Prize Lecture is reprinted with permission from the Shaw Prize Foundation.

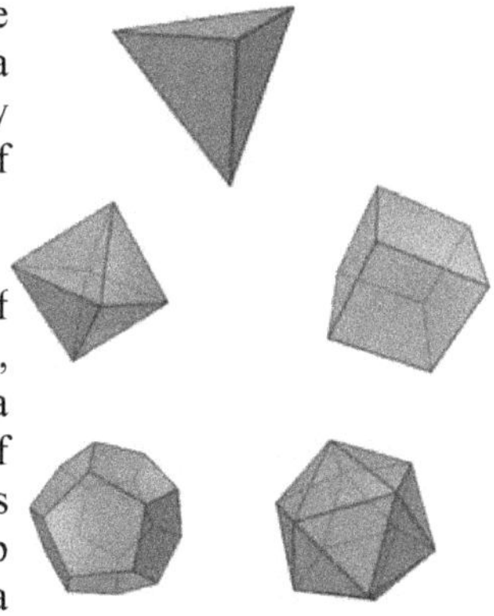

Euclid was also a *Linnaean* in other parts of his treatise. He classified and studied the Platonic solids, the five polyhedra which are fully symmetric: a suitable rotation permutes any vertex with any other, any face with any other. This oldest of mathematical catalogs is shown here on the right.

There are five Platonic solids, a short list, but other types of mathematical objects vary continuously and to 'classify' them, we need not only simple lists but maps. A map is a construction in which each **point** in the map stands for one of the mathematical objects. Below, we have an illustration of this idea using an atlas of the world. Here the **points** in the map stand for actual **places** on the earth. Note that the map itself is a symbolic construction: its paper is not the same thing as the ground on the earth. But it depicts the places in such a way as to preserve various properties of the true earth. When one makes a map like this but now of mathematical objects, the resulting map is itself a new "space", but not one subject to the accidental properties of the universe we live in. Mathematicians like to think of it as a God-given universal object, which they call a "classifying space".

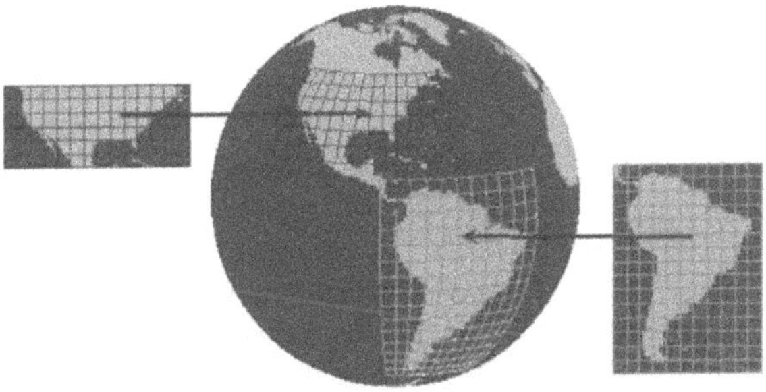

The Linnaean point of view dominated my own work in pure mathematics. After this prelude, we turn to the specific family of maps which I studied, the *moduli space of curves*.

II. Algebraic Geometry

Algebraic geometry is the study of the geometric objects which are defined by the zeros of polynomials. To the right are some plane algebraic curves, zeros of one polynomial in *two* variables. There are, however, two twists. First, one must allow the values of the two variables to be not merely ordinary numbers from the number line, but complex numbers from the number *plane*. As a complex number has both a real and imaginary part, this changes the curves into *surfaces*! Secondly, one must also add certain 'points at infinity' using highest order terms.

These are like the horizon in Renaissance painting: points where parallel lines meet.

To give an example, adding one point at infinity to a line makes it into a circle, and adding its complex points makes it into a sphere as shown below. More complicated curves, after points at infinity and complex solutions are added, become surfaces 'with holes'.

The number of holes is called the *genus* of the surface. Thus, the three surfaces shown here have genus 0, 1 and 2 respectively.

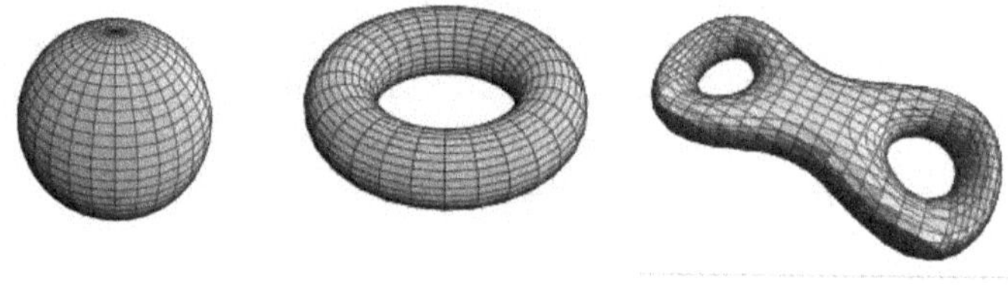

Some curves look very different but their solutions can be related by a *substitution*. An example is the circle and the straight line:

Identify the circle $x^2+y^2=1$ with the line, coord t
via: $x=\frac{t^2-1}{t^2+1}, y=\frac{2t}{t^2+1}, t=\frac{y}{x-1}$

Both of these are both spheres, of genus g=0, when points at infinity and complex solutions are added. But other curves are different: cubic curves with no singularities admit no such substitution because they have genus g=1.

Now we can define our map: we take all curves whose complex solutions form a surface with g 'handles'. We consider as the same two curves which are related by an invertible substitution. Their *map* or *classifying space* is Riemann's moduli space $\mathcal{M}_g$, a fundamental space which has been studied for 150 years. I worked on this space for some 25 years. $\mathcal{M}_g$ is a bit of a tease: there are many ways to play with it, yet it has no really simple description. It hides and one continues to ponder what is it 'really' like. Without giving specifics, here's a list of methods for prying into its nature:

- Geometric invariant theory: the formation of 'quotients' V/G, see [M1]
- Curves have a 'linearization', a product of circles called their *Jacobian*: and these can be described via theta functions for which there is a rich theory.
- Classical analysis, using Kleinian groups and Teichmuller Space
- "Reduction mod p", using fields with only finitely many numbers and counting.
- The "tautological" class method, see [Mad].

Recently, there has been major progress in the theory of this wonderful space and applications of the space to string theory and particle physics.

III. *Models* are the prime goal in applied math

We now turn to applied mathematics, an area I began working in about 1983. No matter how you disguise it, pure math is building 'castles in the air'. But applied math is very different: it seeks parts of the physical world which are amenable to a simplified, abstracted math description. Isaac Newton's theory of gravity and the planets is a quintessential example. In the 17th century, much of the physical world was very hard to understand, e.g. chemistry and its origin in electro-magnetic forces was far in the future. But the exquisite dance of the planets had been studied by every civilization. It was Newton's spectacular insight which finally showed that a very simple mathematical description, the universal law of gravity with a force varying inversely by the square of the distance gave extremely accurate predictions of their motion.

On the other hand, the behavior of the human mind has long resisted being squeezed into *any* mathematical mold. In the twentieth century, Norbert Wiener, with control theory and cybernetics, started anew the quest for models of the mind. This led to the creation of the field of Artificial Intelligence (AI), in the 50's. At first, people in the field attacked the problems that an MIT undergrad wanted to excel in: playing chess, speaking French and solving nasty integrals! But to everyone's surprise, the hard tasks for AI turned out to be things we take for granted! *Vision* is one of these, speech recognition another and common-sense reasoning (e.g. that if you tip a full glass, the water will spill) a third. The realization that vision is subtler than you might think and involves a great deal of reasoning is due to the great Arab scientist *Ibn al Haytham*, who lived around the year 1000 CE.

A first problem in vision is to find the objects in an image. This is called the segmentation problem. Here's the 'state-of-the art' (courtesy of Jitendra Malik and his lab, who have compiled this database [Mal]).

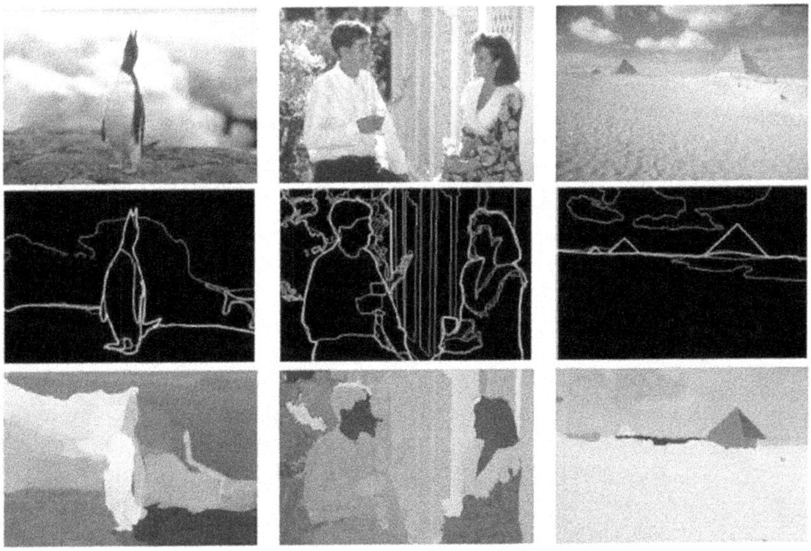

On the top are three black and white images of the world. In the middle are the 'true' segmentations of the images as given by UC Berkeley students, who were asked to draw lines around the prominent objects in each image. On the bottom, the colored regions indicate the segmentation produced by a computer algorithm. Note the failures of the computer algorithm. For example, the computer considers shadows as defining distinct objects and takes the pieces of ice behind the penguin as many distinct objects. Perhaps worse is that it attaches a piece of the background foliage to the man's face, creating a Pinocchio like nose.

My first piece of work in vision was to introduce, with Jayant Shah, a mathematical model, a so-called free-boundary variational method to segment images [MS]. The method uses the fact that images I are typically nearly homogeneous within objects but have large derivatives at their edges. Fixing the image I to be analyzed, the idea is to find a simplified image J and a set of edges of objects Γ which *minimizes* an expression E:

OBSERVED: $I(x, y) =$ brightness of image at point x, y,

INFERRED: $J(x, y) =$ 'cartoon' of image at point x, y

$\Gamma =$ 'edges' of objects/regions in image domain D

METHOD (M-Shah):

minimize $E(J,\Gamma) = \alpha \iint (I-J)^2\,dxdy + \beta \iint_{D-\Gamma} \|\nabla J\|^2\,dxdy + \gamma |\Gamma|$

Although on the right track, I think, our formulation does not work all that well. Much better segmentation techniques are known now, handling texture, some grouping principles, some object shapes, due, among many others, to Andrew Blake, Song-Chun Zhu [ZY], Jitendra Malik and Joachim Buhmann, Ronen Basri et al [BBGS].

IV. Pattern Theory

The above segmentation model for images was based on ideas of classical analysis, dealing with deterministic processes. The proposal of Ulf Grenander [G] was that signals from the world were noisy and incomplete and that it was much better to consider them as incorporating random effects. This was the theoretical perspective which I adopted in my subsequent work in vision and now appearing in the introductory book [DM] with Agnes Desolneux.

More specifically, Grenander's proposal was to construct *generative stochastic models* for all these noisy and incomplete signals from the world, typically based on random graphs, and incorporating random deformations in both range and domain. Given such a messy signal, he suggested using non-parametric Bayesian inference to reconstruct the true patterns. The key to this method is simply Bayes's rule shown below:

$$p(H \mid O) = \frac{p(O \mid H) \cdot p(H)}{p(O)} \propto p(O \mid H) \cdot p(O)$$

O = observations

H = hidden variables

Here one wants to infer things about the real world, the values of the hidden variables, not subject to immediate observation. Each p is a probability and $p(x|y)$ means the probability of an event x, given that another event y occurred. Often one simply infers as likely the value of H which maximize $p(H|O)$, called the *posterior probability* of H when O is observed.

One may contrast traditional AI based on logic and yes/no rules with Pattern Theory, based on likely inferences with the examples in the two boxes below. On the left is a perfectly correct syllogism (from *The Boston Driver's Handbook*) and on the right are the results of a classical experiment of Warren [W]. In the experiment, a sentence was produced mechanically but one consonant was replaced by a loud click. The listener ignored or didn't even notice the click but thought they had heard a complete sentence, replacing the click with the unique consonant which made the sentence make the most sense. Thus some odd shoe might have a tiny wheel on it but every shoe has a heel, etc.

A Parody of Logic

Premises:

a) Tolstoi was a genius,

b) Tolstoi can only be truly appreciated by geniuses,

c) No genius is without some eccentricity,

d) Tolstoi sang the blues,

e) Every eccentric blues singer is appreciated by some half-wit,

f) Eccentrics think they own the road.

⇩ ?????

Consequence: Thcre is always some half-wit who thinks he owns the road.

An actual experiment

ACTUAL SOUND

The ?eel is on the shoe

THE PERCEIVED SOUND

The **h**eel is on the shoe

ACTUAL SOUND

The ?eel is on the car

THE PERCEIVED SOUND

The **wh**eel is on the car

ACTUAL SOUND

The ?eel is on the table

THE PERCEIVED SOUND

The **m**eal is on the table

Bayesian inference is really used!

I find it very convincing that Bayesian inference is indeed the correct mathematical setting for modeling thought. This perspective is now nearly universal in the AI community.

V. Self-similarity

Self-similar processes, processes which exist at a large set of scales with the same appearance at each scale, have been a huge success in twentieth century applied math: phase transitions, turbulence, length of coastlines, etc. all show self-similarity. A big surprise is real world images show self-similarity too: fine detail in blown up images of the real world show *identical statistics* to images depicting large parts of the world. The simplest reason is that images are taken from random viewpoints, hence have no fixed scale. You can see your spouse from a 100 yards away or from a few inches and one image is (ignoring perspective effects) just a larger version of the other. But also the presence of objects of all sizes in typical images creates clutter, which turns out to have statistical self-similarity. The two images below illustrate this: on the left, we see nearly exact self-similarity of the blueberry pickers in the foreground and distance. In the image on the right, there are objects of every size, from the woods and man to tiny twigs and the checks on his shirt.

Basilis Gidas and I developed a stochastic model of these effects in natural images of the world, which used what is called a non-Gaussian renormalization-fixed-point random field based on random wavelet expansions [GM].

Self-similarity is a phenomenon which also occurs in the theory of the moduli space of curves – linking my earlier work with that in computer vision. Benoit Mandelbrot visited Harvard in the late 70's and showed some of the amazing fractals which emerged from very classical questions in complex analysis. That was my first introduction to how powerful and user friendly computers had become at that time. This work led, after many years, to a semi-popular book *Indra's Pearls* written jointly with Dave Wright and Caroline Series [MSW]. The reference to the god Indra was suggested by Caroline, who knew that in the Buddhist tradition, Indra is said to have hung the heavens with a net of pearls, in each of which one sees the reflections of all the others – and the reflections of the reflections, etc. Thus the Buddhist concept that the whole is contained in each of its parts is realized. Or, as Blake said it, *"To see a world in a grain of sand"*.

In my lecture, I showed a movie of Dave's in which a curve is drawn whose length is infinite, which has spirals of every size and thus has Hausdorff dimension nearer to 2 than 1. After the curve is drawn, the movie zooms in a thousand fold showing literally how spirals of every size are present. In lieu of the movie, the full curve is shown below. Metaphorically, we see the struggle of 2 'sibling' genus one curves to carve up the plane. The smooth spirals show some of the symmetries of the curve: the curve reproduces itself exactly after suitable shifts along each of the three spirals present.

VI. Maps of the set of all shapes

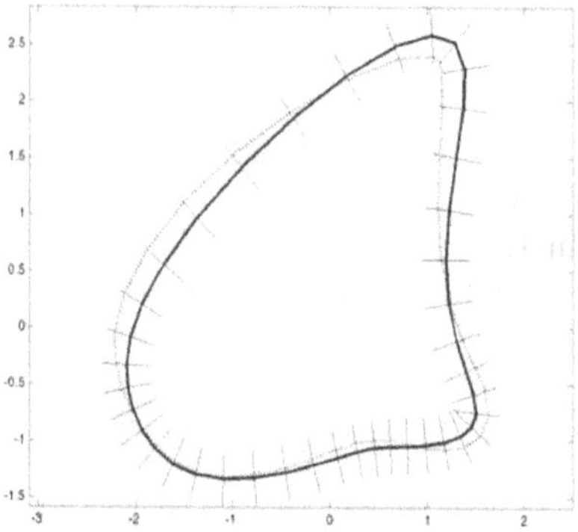

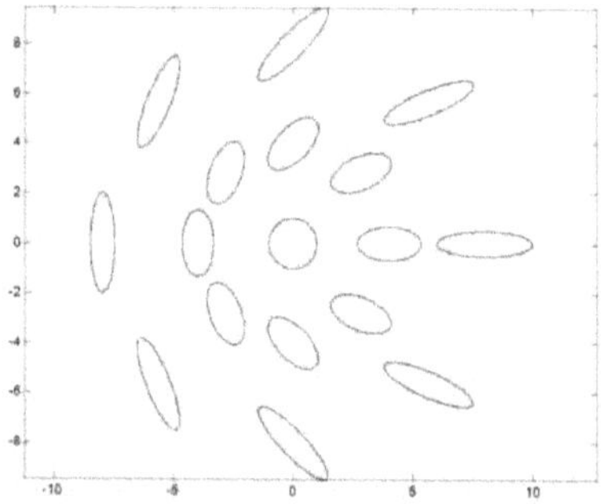

In the final section, we introduce a new and wilder type of map: consider *all* simple closed plane curves (here *simple* means they don't cross themselves as in a figure 8 and *closed* means they are complete loops). We make these the points in our map, i.e. for each such curve, there is to be one point in a new space which we call $\mathcal{S}_2$. The pages of this map are neither flat pieces of paper – 2 dimensional charts – nor are they 3 dimensional, like star charts. They are now infinite-dimensional spaces and all the charts fit together into the infinite dimensional space $\mathcal{S}_2$.

To make the idea clearer, in the top figure, we have a blue curve along which perpendicular 'whiskers' have been drawn. Taking a set of points, one on each whisker, gives a nearby curve. Thus nearby curves are given by functions along the blue curve, and assigning this function to all nearby curves is a coordinate chart. In the bottom figure, the set of all possible ellipses is laid out like a plane: this suggests how inside $\mathcal{S}_2$, the ellipses form a very small 2D subset.
There is a similar theory for surfaces in 3-space leading to $\mathcal{S}_3$; also the group of warpings or diffeomorphisms creates infinite dimensional groups $Diff(\mathcal{R}^2)$, $Diff(\mathcal{R}^3)$.

What is life like in an infinite dimensional world? This question has many sides to it but one thing one can do is investigate Riemannian Geometry on $\mathcal{S}_2$ (see [MM]). That is introduce a measure of distance locally and find first the corresponding geodesics – the shortest paths between two points -- then the curvature of the space – which tells whether geodesics converge or diverge – and finally what diffusion – think of the dissipation of heat -- is like.

The figure below illustrates all of these. On the top left, there is a surface with humps and holes and we have marked one particular geodesic, the shortest path from a point on the bottom to one on the top. Note how it must wind between the humps and holes. In our infinite dimensional space, such a path is the shortest warping of one curve to another. Below on the left are some geodesic paths on $\mathcal{S}_2$ (in one particular metric) between 3 ellipses oriented at 60 degrees to each other. The figure on the top right shows how such geodesic paths may converge or diverge in the 2D setting of humps and holes: this is what is called curvature. We see curvature in the infinite dimensional $\mathcal{S}_2$ also: when the ellipses are small, we get negative curvature; when large, positive curvature. The final figure shows diffusion in the space of simple closed curves in another metric (a translation is added to make the figure readable).

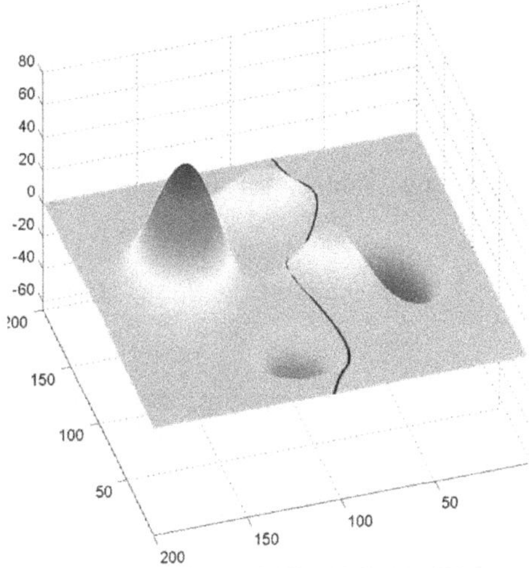

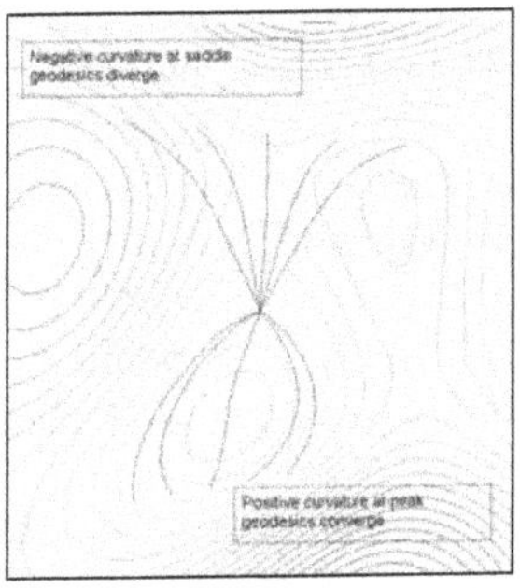

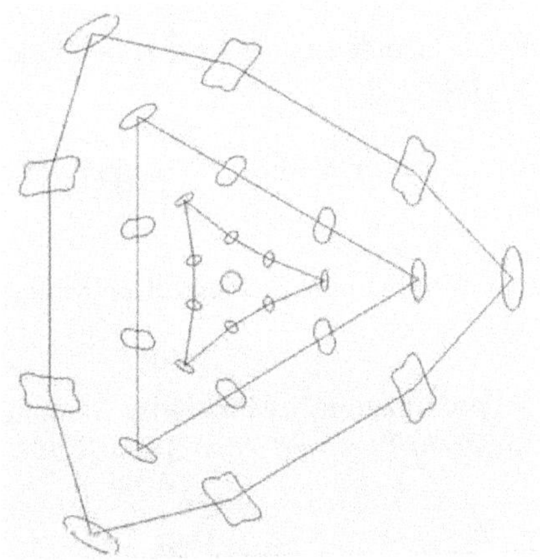

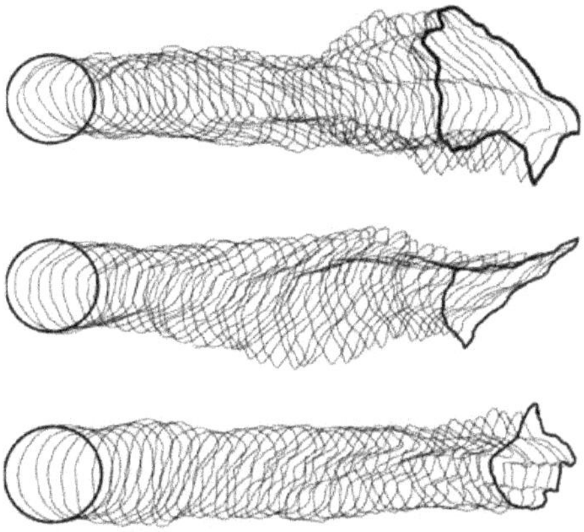

I regret that this sketch of a large set of mathematical ideas has been so brief. But I hope it is sufficient to show you how exciting ideas are brewing in both pure and applied mathematics.

References:

[BBGS] Ronan Basri, Achi Brandt, Meirav Galun and Eitan Sharon, Texture Segmentation by Multiscale Aggregation of Filter Responses and Shape Elements, in *Proceedings of the Ninth IEEE International Conference on Computer Vision*, 2003.

[DM] Agnes Desolneux and David Mumford, *Pattern Theory, the Stochastic Analysis of Real World Signals*, to appear, AKPeters, 2010

[GM] Basilis Gidas and David Mumford, Stochastic Models for Generic Images, *Quarterly Appl. Math.*, **59**, 2001, pp.85-111.

[G] Ulf Grenander, *General Pattern Theory: A Mathematical Study of Regular Structures*, Oxford University Press, 1994.

[Mad] Ib Madsen, Moduli Spaces from a Topological Viewpoint, *Proceedings, International Congress of Mathematicians*, Madrid, 2006.

[Mal] Jitendra Malik et al, The Berkeley Segmentation Dataset and Benchmark, http://www.eecs.berkeley.edu/Research/Projects/CS/vision/bsds/

[MM] Peter Michor and David Mumford, Riemannian Geometries on Spaces of Plane Curves, *J. of the European Math. Society*, **8**, 2006, pp.1-48.

[M1] David Mumford, *Geometric Invariant Theory,* Springer Verlag, 3rd edition with Frances Kirwan and John Fogarty, 1994.

[MSW] David Mumford, Caroline Series and David Wright, *Indra's Pearl,* Cambridge University Press, 2002.

[MS] David Mumford and Jayant Shah, Optimal Approximations of Piecewise Smooth Functions and Associated Variational Problems, *Comm. in Pure and Appl. Math.*, 1989, **42**, pp.577-685.

[W] R.M. Warren, Restoration of missing speech sounds, *Science*, vol. 167, 1970

[ZY] Song-Chun Zhu and Alan Yuille, Region Competition: Unifying Snake/balloon, Region Growing and Bayes/MDL/Energy for multi-band Image Segmentation, *IEEE Trans. on Pattern Analysis and Machine Intelligence*, vol. 18, 1996.

PROJECTIVE INVARIANTS OF PROJECTIVE STRUCTURES AND APPLICATIONS

By DAVID MUMFORD

The basic problem that I wish to discuss is this: if V is a variety, or scheme, parametrizing the set, or functor, of all structures of some type in projective n-space $\mathbf{P}_n$, then the group $PGL(n)$ of automorphisms of $\mathbf{P}_n$ acts on V. Then under what conditions does there exist a quotient or orbit space $V/PGL(n)$, i.e. when can we construct enough "projective invariants" for these structures? For example, let V parametrize the set of hypersurfaces of degree m, with certain types of singularities; or let V parametrize the set of tri-canonical space curves of given genus, or even n-canonical surfaces with at most "negligible singularities" [1]; or let V parametrize the set of 0-cycles of degree m in $\mathbf{P}_n$; or let V parametrize the set of all morphisms of a fixed scheme X into $\mathbf{P}_n$. Moreover, I wish to illustrate how such questions are one essential step in several basic existence and construction problems of algebraic geometry.

One approach to this problem is afforded by the invariant theory of the representations of reductive groups. Here you generalize the problem first: consider an arbitrary action of an algebraic group G on a variety (or scheme) V and seek an orbit space V/G. Then you specialize the problem by (*a*) assuming G is reductive, (*b*) restricting attention to quasi-projective orbit spaces V/G. Of course, if, in particular, the characteristic is 0, $PGL(n)$ is reductive. Now suppose you have a projective embedding $V \subset \mathbf{P}_N$. If V is a normal variety, and if this embedding is defined by a complete linear system on V, it is possible to prove that the action of G on V extends to an action of G on $\mathbf{P}_N$. In any case, if this occurs, I say that G *acts linearly* on $V \subset \mathbf{P}_N$.

Now, if the action of G is linear, its action is induced by a linear and unimodular representation of some finite covering G^* of G on the affine cone $\mathbf{A}^{N+1}$ over the ambient $\mathbf{P}_N$. Then I make the definition:

A point $x \in V$ is *stable* for the action of G on V, relative to the embedding $V \subset \mathbf{P}_N$, if for one (and hence all) homogeneous points $x^* \in \mathbf{A}^{N+1}$ over x, (i) the stabilizer of x^* is a finite subgroup of G^*, and (ii) the orbit of x^* under G^* is closed in $\mathbf{A}^{N+1}$.

Now assume that a reductive algebraic group G acts linearly on $V \subset \mathbf{P}_N$. The fundamental theorem is this:

THEOREM. *The set of stable points forms an open set U in V, and a quasi-projective orbit space U/G exists.*

In case G is semi-simple and of characteristic 0, I can say more. First, let us call the action $\alpha : G \times V \to V$ of G on V a *proper action* if the morphism $\alpha \times Pr_2 : G \times V \to V \times V$ is proper (see [2]). Then:

THEOREM. (1) *A point $x \in V$ is stable if and only if there is a hypersurface section H of V which is invariant under G, and such that $x \notin H$ and the orbit of x is closed in $V - H$.*

(2) *The action of G on the set of stable points U is proper.*

(3) *If G acts properly on V, and if a quasi-projective orbit space V/G exists, then for some projective embedding $V \subset \mathbf{P}_N$, every point of V is stable.*

Finally, if G is any reductive algebraic group in characteristic O, then I can analyze the manner in which stability breaks down in the following way:

THEOREM. *If $x \in V$ is not stable for the action of G, then there is a Borel subgroup $B \subset G$ and a 1-parameter subgroup $\mathfrak{G}_m \subset B$ such that, if $\mathfrak{G}_m$ is any conjugate of $\mathfrak{G}_m$ in B, then x is not stable for the action of $\mathfrak{G}_m$ on V.*

These results are not definitive: I conjecture that they all are valid for semi-simple groups of any characteristic. However, they suggest the kind of answer that should be found for the original question: to every type of structure in $\mathbf{P}_n$, there is a stability condition which is sufficient and, in general, necessary for the existence of enough projective invariants to classify these structures; and, moreover, this stability condition is always of the form: there is no flag in $\mathbf{P}_n$ which has "too high an order of contact" with the given structure.

The following case has been worked out exhaustively by myself and J. Tate in all characteristics, and even over the scheme of integers: let $V = (\mathbf{P}_n)^m$, i.e. V parametrizes ordered 0-cycles of degree m in $\mathbf{P}_n$. Then relative to the Segre embedding of V, a point $x = (x_1, x_2, \ldots, x_m)$ is stable if and only if:

For all linear subspaces $L \subset \mathbf{P}_n$, *then:*

$$\frac{(\text{number of points } x_i \text{ in subspace } L)}{(\text{total number of points } x_i)} < \frac{\dim (L)+1}{n+1}.$$

Then in all characteristics, or even over the scheme of integers, the set of stable 0-cycles forms an open set $U \subset (\mathbf{P}_n)^m$, and a quasi-projective orbit space $U/PGL(n)$ exists. In fact, *U is a principle fibre bundle over $U/PGL(n)$.* Our techniques are entirely elementary (see [3] and [6]).

I would like to illustrate how this simple result can be used to prove the existence of (i) a moduli scheme for polarized abelian varieties, and (ii) (according to a suggestion of Grothendieck) the Picard scheme of any variety.(¹) First, let me fix some notations: if $U \subset (\mathbf{P}_n)^m$ is as above, let $Q_{n,m} = U/PGL(n)$. Second, since U is a principal fibre bundle over $Q_{n,m}$ with group $PGL(n)$, if $PGL(n)$ acts on any scheme X, we can form an associated fibre bundle with fibre X; in particular, if $X = \mathbf{P}_n$, denote the associated fibre bundle by $P_{n,m}$. Let $\pi: P_{n,m} \to Q_{n,m}$ be the bundle morphism. It is easy to see that the bundle $P_{n,m}/Q_{n,m}$ has m distinguished sections $s_i: Q_{n,m} \to P_{n,m}$ associated to the maps which take the m-tuple $(x_1, x_2, \ldots, x_m)$ to its ith factor x_i. Intuitively, regard $P_{n,m}$ plus the collection of sections (s_i) as the universal family of projective n-spaces with given stable 0-cycle of degree m.

To simplify the treatment of the moduli problem, let us consider only the question of finding a quasi-projective variety $M_{g,d}$ whose points parametrize "naturally" the set of all abelian varieties A of dimension g, plus a very ample(²) divisor class D such that $(D^g) = d \cdot g!$ For the question of

(¹) In fact, it can be used to prove much stronger results on the existence of relative Picard schemes: see [4].

(²) in the sense of Grothendieck, i.e. induced via a projective embedding.

making explicit the sense of the word "naturally", and the question of replacing D by its numerical equivalence class, see [6]. Now pick any n such that the characteristic does not divide n, and $n > d \cdot \sqrt{g!}$. Let $\nu = n^{2g}$. Then consider the following set:

Subvarieties $A \subset P_{d-1,\nu}$ such that

(a) $\pi(A)$ is a single point $q \in Q_{d-1,\nu}$,
(b) the degree of A in $\pi^{-1}(q)$ equals $d \cdot g!$, and the dimension of A is g,
(c) A admits a structure of abelian variety such that $\{s_i(q)\}$, for $i = 1, 2, \ldots, \nu$ is the set of points of order n.

Then, on the one hand, this set is parametrized naturally by a locally closed subset of a suitable Chow variety. But, on the other hand, it is isomorphic to the set of abelian varieties A of dimension g, plus a very ample divisor class D such that $(D^g) = d \cdot g!$, *plus* an ordering of the set of points of order n. [Namely, one can prove that the inequality $n > d \cdot \sqrt{g!}$ insures that the set of points of order n form a stable 0-cycle in $\mathbf{P}_{d-1}$ via the embedding defined by the complete linear system $|D|$. Let q be the equivalence class of this 0-cycle mod $PGL(d-1)$; then there is a unique identification of $\pi^{-1}(q)$ with the $\mathbf{P}_{d-1}$ ambient to A, under which $s_i(q)$ corresponds to the ith point of order n. Then A is embedded in $\pi^{-1}(q) \subset P_{d-1,\nu}$ via this identification.] Therefore, the set of all A, plus D alone, is parametrized by the quotient of this set by the group of permutations of ν letters. As such a quotient of a quasi-projective variety is well-known to exist, this solves the problem.

To simplify the treatment of the Picard scheme of a projective variety V, let us consider only the construction of the reduced and connected Picard scheme. The question then may be loosely described as that of finding a variety P whose points parametrize "naturally" the set of all Cartier divisor classes D on V which are algebraically equivalent to 0. But this is equivalent to doing the same for the set of all Cartier divisor classes D on V which are algebraically equivalent to some fixed very ample divisor D_0. But if D_0 is chosen sufficiently ample, and if we associate to each divisor class D the morphism $\phi: V \to \mathbf{P}_n$ defined by the complete linear system $|D|$, then this set will be isomorphic to the set of orbits under the group $PGL(n)$ in the set of morphisms $\phi: V \to \mathbf{P}_n$ which are algebraically equivalent to a fixed $\phi_0: V \to \mathbf{P}_n$. This is a problem of the type originally posed.

It can be reduced to the 0-cycle problem already solved as follows: pick a large number of sufficiently generic points $x_1, x_2, \ldots x_N$ in V, i.e. such that:

For all $\phi: V \to \mathbf{P}_n$ algebraically equivalent to ϕ_0, the set of points $\phi x_1, \phi x_2, \ldots, \phi x_N$ is a stable 0-cycle in $\mathbf{P}_n$.

Then consider the following set:

Morphisms $\phi: V \to P_{n,N}$ such that

(a) $\pi \circ \phi$ maps V to a single point $q \in Q_{n,N}$,
(b) $s_i(q) = \phi(x_i)$,
(c) ϕ, as a morphism from V to $\pi^{-1}(q) \cong \mathbf{P}_n$ is algebraically equivalent to ϕ_0.

Then on the one hand this set is parametrized by a locally closed subset of a suitable Chow variety (i.e. via the graph of ϕ), and on the other hand it is isomorphic to the set of orbits described above.

Here is another example of a stability condition and the resulting quotient theorem. Assume the characteristic is 0, and consider, instead of sequences of points in $\mathbf{P}_n$, sequences of linear subspaces of any dimension. Thus, if $Grass_{k,n}$ stands for the Grassmannian of k-dimensional linear subspaces of $\mathbf{P}_n$, I ask for orbit spaces of the type $(Grass_{k,n})^m/PGL(n)$. Then, in fact, relative to the usual Plücker embedding of the Grassmannian, and to the Segre embedding of this product, it turns out that a point $x=(L_1,L_2,\ldots,L_m)$ of $(Grass_{k,n})^m$ is stable if and only if:

For all linear subspaces $L\subset P_n$, then:

$$\frac{\sum_i (\dim (L\cap L_i)+1)}{\sum_i (\dim L_i+1)} < \frac{(\dim L+1)}{(n+1)}.$$

Then, by the fundamental theorem, the set of stable points forms an open set U, and an orbit space $U/PGL(n)$ exists.

This result can be applied to the problem of classifying vector bundles over a variety in exactly the same way as the result on 0-cycles was applied to the problem of classifying line bundles, i.e. Cartier divisor classes. Of course, it is well-known that the set of vector bundles even over an algebraic curve is not a separated space; in fact, it is not even locally separated, because of the "jump" phenomenon noted by Kodaira and Spencer [5]. However, again, a basic stability condition avoids all these difficulties. For simplicity, let us consider only vector bundles over a fixed-curve C.

DEFINITION. A vector bundle E is *stable* if for all sub-bundles F,

$$\operatorname{Deg} c_1(F) < \operatorname{Deg} c_1(\mathrm{E})\cdot\frac{\operatorname{rank} F}{\operatorname{rank} E},$$

where c_1 denotes the first chern class.

In other words, a vector bundle is stable if all its subbundles are "less ample" than itself. To illustrate the stability condition, let me mention its simplest properties:

(i) If L is a line bundle, then E is stable if and only if $E\otimes L$ is stable; moreover, E is stable if and only if $\check{E}$ is stable.

(ii) If E_1 and E_2 are two vector bundles, $E_1\oplus E_2$ is never stable.

(iii) A line bundle is always stable.

(iv) If a vector bundle E of rank 2 is not stable, then either E is isomorphic to $L_1\oplus L_2$, *or* there is a unique sub-bundle L for which $\geqslant$ holds in the definition and E can be canonically described as an extension.

Then I can prove the following theorem:

THEOREM. *The set of all stable vector bundles of rank* r *over a fixed curve* C *in characteristic* 0 *is "naturally" isomorphic to the set of points of a non-singular quasi-projective variety* $V_r(\mathrm{C})$.

A more complicated example of a stability condition is given by the action of $PGL(2)$ on the variety of plane curves of degree n. There seems to be no simple general rule describing when a plane curve is stable; however, I can prove that if $n\geqslant 3$, then at least every non-singular curve is stable. For low values of n, the precise answer is given by:

n	Stable curves
1, 2	None.
3	Non-singular.
4	No triple points and no tacnodes.
5	No triple points with 3 coincident tangents, or with 2 tangents forming a tacnode.

To conclude, I want to pose a question that seems to me to be the most interesting problem in extending the results discussed above. Moreover, I think this problem is the central one on the road to solving the general problem of algebraic moduli of polarized non-singular varieties. The question is: *when is the Chow point of a non-singular subvariety* $\mathrm{V} \subset \mathbf{P}_n$ *stable* (*in the usual projective embedding of the Chow variety*)? Perhaps more reasonable is the stabilized form of this problem: Given a subvariety $V \subset \mathbf{P}_n$, when is there an n_0 such that if $n \geqslant n_0$, the Chow point of V in the n-ple embedding $V \subset \mathbf{P}_N$ is stable? I have no conjecture to make.

REFERENCES

[1]. ARTIN, M., Some numerical criteria for the contractability of curves on an algebraic surface. *Amer. J. Math.*, 84 (1962), 485.

[2]. PALAIS, R., On the existence of slices for actions of non-compact Lie groups. *Ann. Math.*, 73 (1961), 295.

[3]. MUMFORD, D., An elementary theorem in geometric invariant theory. *Bull. Amer. Math. Soc.*, 67 (1961), 483.

[4]. GROTHENDIECK, A., *Séminaire Bourbaki*, exp. 232 and 236, 1962.

[5]. KODAIRA, K. & SPENCER, D. C., On deformations of complex analytic structures. *Ann. Math.*, 67 (1958), 328.

Most of the material discussed above will be published in:

[6]. MUMFORD, D., Geometric invariant theory. (Forthcoming.)

Actes, Congrès intern. math., 1970. Tome 1, p. 457 à 465.

THE STRUCTURE OF THE MODULI SPACES OF CURVES AND ABELIAN VARIETIES

by DAVID MUMFORD

§ 1. The purpose of this talk is to collect together what seem to me to be the most basic moduli spaces (for curves and abelian varieties) and to indicate some of their most important interrelations and the key features of their internal structure, in particular those that come from the theta functions. We start with abelian varieties. Fix an integer $g \geqslant 1$. To classify g-dimensional abelian varieties, the natural moduli spaces are:

$$\mathscr{A}^{(n)} = \left\{ \begin{array}{l} \text{moduli space of pairs } (X, \lambda),\ X \text{ a } g\text{-dimensional} \\ \text{abelian variety, } \lambda : X \to \hat{X} \text{ a polarization such} \\ \text{that } \deg(\lambda) = n^2 \end{array} \right\}$$

Here and below when we talk of a moduli space, we mean a coarse moduli space in the sense of [11], p. 99 and in all cases these moduli spaces will actually exist as schemes of finite type over Spec $(\boldsymbol{Z})$. This can be proven by the methods of [11], Ch. 7, for instance, which also shows that all the moduli spaces used are quasi-projective at least over every open set Spec $\boldsymbol{Z}\left[\frac{1}{p}\right]$.

The local structure of $\mathscr{A}^{(n)}$ seems quite difficult to work out at some points. However, for every sequence $\delta_1, \ldots, \delta_g$ such that $\delta_1 \mid \ldots \mid \delta_g$, $\prod_{i=1}^{g} \delta_i = n$, let

$$\mathscr{A}^{(\delta)} = \left\{ \begin{array}{l} \text{the open subscheme of } \mathscr{A}^{(n)} \text{ of pairs } (X, \lambda) \\ \text{such that} \\ \qquad \ker(\lambda) \cong \prod_1^g \boldsymbol{Z}/\delta_i \boldsymbol{Z} \times \prod_1^g \mu_{\delta_i} \end{array} \right\}$$

The $\mathscr{A}^{(\delta)}$'s are disjoint and exhaust all of $\mathscr{A}^{(n)}$ except for (X, λ)'s such that char$|$n and ker (λ) contains a subgroup isomorphic to α_p. The local structure of $\mathscr{A}^{(\delta)}$ is not hard to work out (using results of Serre-Tate [20], and Grothendieck and myself on the formal deformation theory of abelian varieties and p-divisible groups, see Oort [17]). In particular all components of $\mathscr{A}^{(\delta)}$ dominate Spec $(\boldsymbol{Z})$. Now I have proven that for all n and all p, the open subset of $\mathscr{A}^{(n)} \times$ Spec $\boldsymbol{Z}/p\boldsymbol{Z}$ of (X, λ)'s such that X is ordinary (*) is *dense* (cf. [14] for a sketch of the proof). Therefore $\bigcup_\delta \mathscr{A}^{(\delta)}$ is dense in $\mathscr{A}^{(n)}$. Since $\mathscr{A}^{(\delta)} \times$ Spec $(\boldsymbol{C})$ is irreducible (see below), it follows that the components of $\mathscr{A}^{(n)}$ are the closures $\overline{\mathscr{A}}^{(\delta)}$ of the $\mathscr{A}^{(\delta)}$ and that all of them dominate Spec $(\boldsymbol{Z})$. It is *not*

(*) i. e. has the maximal number p^g of points of order p.

known, however, whether the geometric fibres of $\mathcal{A}^{(\delta)}$ over finite primes are irreducible or not.

Now these various schemes $\mathcal{A}^{(n)}$ are all related by the isogeny correspondences:

$$Z_{n_1,n_2,k} = \left\{ (X, \lambda)\varepsilon\mathcal{A}^{(n_1)}, (Y, \mu)\varepsilon\mathcal{A}^{(n_2)} \;\middle|\; \begin{array}{l} \exists \text{ an isogeny } \pi : X \to Y \\ \text{of degree } k \text{ such that} \\ n_1{}^2\hat{\pi} \circ \mu \circ \pi = (kn_2)^2\lambda \end{array} \right\}$$

To uniformize all of these, one introduces a second more convenient sequence of moduli spaces. Firstly, over the base scheme $\boldsymbol{Z}[\zeta_n]$, ζ_n a primitive n-th root of 1, let

$$\mathcal{A}_n^* = \left\{ \begin{array}{l} \text{moduli space of triples } (X, \lambda, \alpha),\ X \text{ a } g\text{-dimensional} \\ \text{abelian variety, } \lambda : X \xrightarrow{\approx} \hat{X} \text{ a principal polarization,} \\ \text{and } \alpha : X_n \xrightarrow{\sim} (\boldsymbol{Z}/n\boldsymbol{Z})^g \times \mu_n^g \text{ a symplectic isomorphism} \end{array} \right\}$$

These spaces are normal and irreducible and form a tower with respect to the natural quasi-finite morphisms $\mathcal{A}_{nm}^* \to \mathcal{A}_n^*$ given by $(X, \lambda, \alpha) \mapsto (X, \lambda, \operatorname{res}_{X_n}\alpha)$. Secondly, we enlarge these schemes somewhat by letting $\mathcal{A}_n$ be the normalization of $\mathcal{A}_1$ in the field of rational functions $\boldsymbol{Q}(\mathcal{A}_n^*, \zeta_n)$. Then $\mathcal{A}_n$ is a normal irreducible scheme in which $\mathcal{A}_n^*$ is an open subscheme, and the $\mathcal{A}_n$'s form a tower with respect to finite morphisms $\mathcal{A}_{nm} \to \mathcal{A}_n$. Note that $\mathcal{A}_n = \mathcal{A}_n^*$ except over primes dividing n. Moreover, if $n \geqslant 3$, $\mathcal{A}_n$ is smooth over $\boldsymbol{Z}$ except at non-ordinary abelian schemes in characteristics dividing n. Next, we can uniformize very nearly all of $\mathcal{A}^{(\delta)}$ by the natural morphism:

$$\begin{array}{c} \mathcal{A}_{\delta_g}^* \to \mathcal{A}^{(\delta)} \\ (X, \lambda, \alpha) \mapsto (Y, \mu) \end{array}$$

where Y is the etale covering of X defined by requiring its dual to be the quotient:

$$\hat{Y} = \hat{X}/\alpha^{-1}[(0) \times \Pi\mu_{\delta_i}],$$

and μ is the polarization on Y induced by λ. In the tower $\{\mathcal{A}_n\}$ one now has the Hecke ring of correspondences instead of the isogeny correspondences. These come essentially from 2 types of morphisms:

(*a*) $\boldsymbol{Q}(\mathcal{A}_n, \zeta_n)$ is a Galois extension of $\boldsymbol{Q}(\mathcal{A}_1, \zeta_n)$ with Galois group $\operatorname{Sp}(2g, \boldsymbol{Z}/n\boldsymbol{Z})$, hence $\operatorname{Sp}(2g, \boldsymbol{Z}/n\boldsymbol{Z})$ acts as a group of automorphisms of $\mathcal{A}_n$;

(*b*) the morphism

$$\begin{array}{c} \mathcal{A}_n^* \to \mathcal{A}_1 \\ (X, \lambda, \alpha) \mapsto (Y, \mu) \end{array}$$

(where $Y = X/\alpha^{-1}[(\boldsymbol{Z}/n\boldsymbol{Z})^g \times (0)]$, and if $\pi : X \to Y$ is the natural map, then $\mu : Y \xrightarrow{\sim} \hat{Y}$ is determined by the requirement $\hat{\pi} \circ \mu \circ \pi = n\lambda$). For a discussion on Hecke operators in the classical case, see Shimura [21]. The picture is even clearer when you pass to an inverse limit: e. g. for all n,

$$\varprojlim_k \mathcal{A}_{n^k} \times \operatorname{Spec} R_n$$

where

$$R_n = \boldsymbol{Z}\left[\frac{1}{n}, \zeta_n, \zeta_{n^2}, \ldots\right]$$

exists as a scheme and $\prod_{p|n} \operatorname{Sp}(2g, \boldsymbol{Q}_p)$ acts on it (See [12], § 9 for the case $n = 2$).

Over the complex ground field, these moduli spaces have well-known analytic uniformizations coming from the theory of Siegel modular forms:

$$\mathscr{A}^{(n)} \times \operatorname{Spec}(C) = \coprod_{\delta} \mathscr{A}^{(\delta)} \times \operatorname{Spec}(C)$$
$$\mathscr{A}^{(\delta)} \times \operatorname{Spec}(C) = \mathfrak{H}/\Gamma_\delta$$
$$\mathscr{A}_n \times \operatorname{Spec}(C) = \mathscr{A}_n^* \times \operatorname{Spec}(C) = \mathfrak{H}/\Gamma(n)$$

where

$$\mathfrak{H} = \text{Siegel upper } \frac{1}{2} - \text{plane} = \left\{ Z \,\middle|\, \begin{array}{l} Z = g \times g \text{ complex matrix} \\ {}^t Z = Z, \operatorname{Im} Z > 0 \end{array} \right\}$$

$$\Gamma(n) = \{ A \in \operatorname{Sp}(2g, Z)/(\pm I) \mid A = I_{2g} \bmod n \}$$
$$\Gamma_\delta = \{ A \in GL(2g, Z)/(\pm I) \mid {}^t A . J_\delta . A = J_\delta \},$$

where

$$J_\delta = \left(\begin{array}{c|c} 0 & \begin{matrix} \delta_1 & & \\ & \ddots & \\ & & \delta_g \end{matrix} \\ \hline \begin{matrix} -\delta_1 & & \\ & \ddots & \\ & & -\delta_g \end{matrix} & 0 \end{array} \right)$$

Thus $\mathfrak{H}$ is the analytic " inverse limit " of the $\mathscr{A}_n$'s over Spec C. All the irreducibility assertions made so far are proven by these analytic uniformizations.

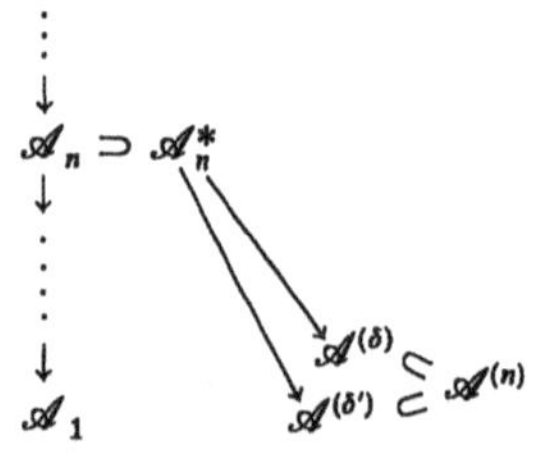

Summary of moduli spaces.

§ 2. The next point is that there is a moduli space intermediate in the tower between $\mathscr{A}_n^*$ and $\mathscr{A}_{2n}^*$ on which there are *canonical coordinates*. Following Igusa, we christen this $\mathscr{A}_{n,2n}^*$ and it is defined as follows in char. $\neq 2$:

$$\mathscr{A}_{n,2n}^* = \left\{ \begin{array}{l} \text{moduli space of triples } (X, L, \alpha),\ X \text{ an abelian} \\ \text{variety of dimension } g,\ L \text{ an ample symmetric} \\ \text{invertible sheaf, } \alpha \text{ a symmetric isomorphism:} \\ \qquad \alpha : \mathscr{G}(L) \xrightarrow{\sim} G_m \times (Z/nZ)^g \times \mu_n^g \\ \text{such that} \\ \quad i) \text{ if } n \text{ even, } e_*^L \equiv 1 \text{ on } X_2, \\ \quad ii) \text{ if } n \text{ odd, } e_*^L \text{ takes the value } +1 \text{ more often} \\ \text{than the value } -1. \end{array} \right\}$$

For definitions of $\mathscr{G}(L)$, e_*^L, etc., see [12], § 1, 2 and [15], § 23. There is an obvious map

$$\mathscr{A}^*_{n,2n} \rightarrow \mathscr{A}^*_n$$
$$(X, L, \alpha) \mapsto \left(X, \frac{1}{n}\varphi_L, \overline{\alpha}\right)$$

where $\overline{\alpha}$ is the induced map from $\mathscr{G}(L)/G_m \cong X_n$ to $(Z/nZ)^g \times \mu_n^g$. There is a not so obvious map $\mathscr{A}^*_{2n} \times \operatorname{Spec} Z\left[\frac{1}{2}\right] \rightarrow \mathscr{A}^*_{n,2n}$ (see [12], § 2). Over C, $\mathscr{A}^*_{n,2n}$ is simply the quotient $\mathfrak{H}/\Gamma(n, 2n)$, where $\Gamma(n, 2n)$ is the subgroup between $\Gamma(n)$ and $\Gamma(2n)$ described by Igusa [9]. Canonical coordinates on $\mathscr{A}^*_{n,2n}$ (where $n \geqslant 2$) are defined as follows:

i) $\mathscr{G}(L)$ and hence $G_m \times (Z/nZ)^g \times \mu_n^g$ acts on $H^0(X, L)$. Write this action as $U_{(\lambda,a,b)}: H^0(X, L) \rightarrow H^0(X, L)$,

ii) there is a section $\sigma \in H^0(X, L)$ unique up to scalars such that $U_{(1,0,c)}\sigma = \sigma$, all $c \in \mu_n^g$,

iii) let $\sigma \rightarrow \sigma(0)$ denote evaluation of sections at $0 \in X$. We obtain a function:

$$(Z/nZ)^g \rightarrow K$$
$$\alpha \mapsto (U_{(1,\alpha,0)}\sigma)(0)$$

unique up to multiplication by a constant, which is never identically zero.

iv) If $N = n^g - 1$, and the homogeneous coordinates of P_N are put in one-one correspondence with the elements of $(Z/nZ)^g$, this defines a morphism:

$$\Theta : \mathscr{A}^*_{n,2n} \rightarrow P_N$$

THEOREM. — If $n \geqslant 4$, Θ is an immersion.

This theorem was proven over C for various n's by Baily [4] and Igusa [9]; in the general case, all the essentials for the proof are in [13]. Over C, Θ is the morphism defined by

$$Z \mapsto \left(\ldots, \theta_{nZ}\begin{bmatrix} 0 \\ \alpha/n \end{bmatrix}(0), \ldots\right)_{\alpha \in (Z/nZ)^g}$$

where $Z \in \mathfrak{H}$, and θ is Riemann's theta function. If $8 \mid n$, one can even find a finite set of homogeneous quartic polynominals—Riemann's theta relations—such that the image of Θ is an open part of the subscheme of P_N defined by these quartics (see [12], § 6).

Even in the char. p case, it is possible to reformulate these canonical coordinates as values of a type of theta function. These theta functions are not functions on the universal covering space of X, but rather on the Tate group.

If $p =$ char. of ground field,

$V(X) =$ group of sequences $\{x_i\}$, $i \geqslant 1$ but $p \nmid i$, where $x_i \in X$, $nx_{in} = x_i$ and x_1 has finite order k prime to p.

Let $T(X) = \{(x_i) \in V(X)$ such that $x_1 = 0\}$.

We get an exact sequence:

$$0 \rightarrow T(X) \rightarrow V(X) \rightarrow (\text{torsion on } X \text{ prime to } p) \rightarrow 0$$

We use the result:

THEOREM. — Let L be an ample symmetric invertible sheaf of degree 1 on an abelian variety X of char. p. If $p \nmid 2n$, then for all $x \in X_n$ for every choice of a point $y \in X$ such that $2ny = x$, there is a canonical isomorphism:

$$L \otimes_{\mathcal{O}_X} k(x) \cong L \otimes_{\mathcal{O}_X} k(0)$$

COROLLARY. — If $\sigma \in \Gamma(L)$, then evaluating σ via the above isomorphisms defines a function

$$\theta : V(X) \to L \otimes_{\mathcal{O}_X} k(0) \cong k$$

such that if $x \in \frac{1}{n} T(X)$, then $\theta(x + y) = \theta(x)$ if $y \in 2nT(X)$.

In fact the functions that we obtain in this way have the following properties:

a) $$\theta(x + a) = e_*\left(\frac{a}{2}\right)e\left(\frac{a}{2}, x\right)\theta(x), \quad x \in V(X),\ a \in T(X)$$

where

$$e_* : \frac{1}{2} T(X) \to \{\pm 1\} \quad \text{and} \quad e : V(X) \times V(X) \to k^*$$

are the functions induced by e_*^L and e_n on $V(X)$.

b) $\theta(-x) = \pm\, \theta(x)$, the sign depending on the Arf invariant of e_*^L.

c) $$\prod_{i=1}^{4} \theta(x_i) = 2^{-g} \sum_{\eta \in \frac{1}{2}T(X)/T(X)} e(y, \eta)\left(\prod_{i=1}^{4} \theta(x_i + y + \eta)\right),$$

where

$$y = -\frac{1}{2}\Sigma x_i$$

d) $$\forall x \in V(X), \quad \exists \eta \in \frac{1}{2} T(X) \quad \text{such that} \quad \theta(x + \eta) \neq 0.$$

e) Up to an elementary linear transformation whose coefficients are roots of 1, the set of values of θ on $\frac{1}{n} T(X)$ is equal to the set of values of the canonical coordinates Θ on the triple (X, L^{n^2}, α) (for any symmetric α).

f) Over $\mathbf{C}$, if Z is a period matrix for X, θ is essentially the function $a \mapsto \theta_Z[a](0)$, $a \in \mathbf{Q}^{2g}$.

g) Moreover, if we restrict the domain to $V_2(X)$, this 2-adic Tate group, all functions $\psi : V_2(X) \to k$ satisfying *a)*, *b)*, *c)* and *d)* arise from a unique principally polarized abelian variety.

(Cf. [12], § 8 through § 12).

§ 3. We turn next to curves. Fix $g \geqslant 2$. Let

$$\mathcal{M} = \left\{ \begin{array}{c} \text{moduli space of non-singular} \\ \text{complete curves } C \text{ of genus } g \end{array} \right\}$$

$\mathcal{M}$ is not only irreducible, but it has irreducible geometric fibres over Spec $(\mathbf{Z})$, cf. [5]. This is proven by introducing a compacification $\overline{\mathcal{M}}$ of $\mathcal{M}$, where

$$\overline{\mathcal{M}} = \left\{ \begin{array}{l} \text{moduli space of stable complete curves } C \\ \text{such that } \dim H^1(o_C) = g \end{array} \right\}$$

and where a *stable curve* is one with at most ordinary double points and such that every non-singular rational component has at least 3 double points on it.

$\overline{\mathcal{M}}$ has recently been proven by F. Knudsen, Seshadri and my self to be a scheme projective over $\mathbf{Z}$

Define:

$$\begin{aligned} t: \mathcal{M} &\rightarrow \mathcal{A}_1 \\ C &\mapsto (\mathrm{Pic}^0(C), \lambda) \end{aligned}$$

where λ is the theta polarization, viz: fixing a base point x_0 on C, we obtain a morphism:

$$\begin{aligned} \phi: C &\rightarrow \mathrm{Pic}^0 C \\ x &\mapsto \text{class of } o_C(x - x_0) \end{aligned}$$

hence

$$\widehat{\mathrm{Pic}^0} C = \mathrm{Pic}^0(\mathrm{Pic}^0 C) \overset{\phi^*}{\underset{\sim}{\rightarrow}} \mathrm{Pic}^0(C)$$

and $\lambda = -(\phi^*)^{-1}$.

According to Torelli's theorem (cf. [1], [10]), t is injective on geometric points. Its image however is not closed since t extends to a morphism on $\tilde{\mathcal{M}}$:

$$\left\{ \begin{array}{l} \text{Stable curves made from} \\ \text{non-singular components} \\ \text{connected together like a} \\ \text{tree} \end{array} \right\} = \begin{array}{ccc} \overline{\mathcal{M}} & & \\ \cup & & \\ \tilde{\mathcal{M}} & \overset{\tilde{t}}{\searrow} & \\ \cup & & \\ \mathcal{M} & \underset{t}{\rightarrow} & \mathcal{A}_1 \end{array}$$

and $\tilde{t}$ can be shown to be a proper morphism taking each stable curve C in $\tilde{\mathcal{M}}$ to $\mathrm{Pic}^0 C$ with a suitable polarization (cf. Hoyt [8]). Let $\mathcal{T} = \overline{t(\mathcal{M})} = \tilde{t}(\tilde{\mathcal{M}})$: this is called the *Torelli locus.* A famous classical problem is to describe $\mathcal{T}$, or its inverse image in some $\mathcal{A}_n$, by explicit equations, e. g. polynomials in the theta-nulls. Partial results on this were obtained in characteristic zero by Riemann [18], Schottky, and Schottky-Jung [19]. Their results have been rigorously established recently by Farkas and Rauch [6], and some interesting generalizations have been given by Fay [7]. A completely different approach to this problem is given in the beautiful paper of Andreotti and Mayer [2]. I want to finish by sketching the key point in Schottky's theory and stating a theorem on what his equations do characterize. We assume char. $\neq 2$.

Let $\Pi: \hat{C} \rightarrow C$ be an etale double covering, and let $\imath: \hat{C} \rightarrow \hat{C}$ be the corresponding involution. If g = genus of C, then $2g - 1$ = genus of $\hat{C}$. The Jacobians $J = \mathrm{Pic}^0 C$, $\hat{J} = \mathrm{Pic}^0 \hat{C}$ are related by 2 homeomorphisms:

$$\hat{J} \underset{\Pi^*}{\overset{Nm}{\rightleftarrows}} J$$

such that $Nm \circ \Pi^* = 2_J$. $\imath$ acts on $\hat{J}$ also. Define:

$$P = \text{locus of points } \{ \imath x - x \} \text{ in } \hat{J}.$$

We get isogenies:

$$\hat{J} \underset{\alpha}{\overset{\beta}{\rightleftarrows}} J \times P$$

$$\alpha(x, y) = \Pi^* x + y$$

$$\beta(z) = (Nmz, z - \iota z)$$

such that $\alpha \circ \beta = \beta \circ \alpha =$ mult. by 2. Next fix γ, a division class on C such that $2\gamma \equiv K_C$, the canonical divisor class, and such that dim $H^0(o(\gamma))$ is even (cf. [16] for this). We get symmetric divisors:

$$\Theta = \{\text{locus of div. classes } \sum_1^{g-1} P_i - \gamma\} \subset J$$

$$\hat{\Theta} = \{\text{locus of div. classes } \sum_1^{2g-2} P_i - \Pi^*\gamma\} \subset \hat{J}$$

representing the standard polarizations of J and $\hat{J}$.

Lemma

a) $\Pi^{*-1}(\hat{\Theta}) = \Theta + \Theta_\kappa$, where $\{0, \kappa\} = \operatorname{Ker}(\Pi^*)$.
b) $\exists$ a symmetric divisor Ξ on P such that $\hat{\Theta}.P = 2\Xi$.
c) $\alpha^{-1}(\hat{\Theta}) \equiv \Theta \times P + \Theta_\kappa \times P + 2J \times \Xi$.

In particular, Ξ has degree 1 and defines a principal polarization on P. Abstractly put now we have a situation with.

i) 3 abelian varieties X, Y, Z of dimensions g, $g-1$, $2g-1$ resp.,

ii) 3 ample degree 1 symmetric divisors $\Theta_X \subset X$, $\Theta_Y \subset Y$, $\Theta_Z \subset Z$, which define as in § 2 theta-functions θ_X on $V(X)$, θ_Y on $V(Y)$ and θ_Z on $V(Z)$,

iii) isogenies $Z \underset{\alpha}{\overset{\beta}{\rightleftarrows}} X \times Y$ such that $\alpha \circ \beta = \beta \circ \alpha =$ mult. by 2. In such a case, $Z \cong X \times Y/H$, where H is a so-called Göpel group, and $V(Z) \cong V(X) \times V(Y)$. Moreover θ_Z can be computed from θ_X and θ_Y by one of the basic theta formulas. But then the lemma, esp. part *a*), implies non-trivial identities on θ_X and θ_Y. In fact, it follows that for a suitable $\eta \in \frac{1}{2}T(X)$ with image κ in X and a suitable homeomorphism:

$$\varphi : \left\{ x \in \frac{1}{2}T(X) \mid e(x, \eta) = 1 \right\} \to \frac{1}{2}T(Y)$$

$$(*) \quad \begin{cases} \dfrac{\theta_X(x).\theta_X(x+\eta)}{\theta_Y(\varphi x)^2} = \dfrac{\theta_X(y).\theta_X(y+\eta)}{\theta_Y(\varphi y)^2} \\ \text{all} \quad x, y \in \frac{1}{2}T(X) \\ \text{with} \quad e(x, \eta) = e(y, \eta) = 1 \end{cases}$$

If we globalize this set-up, we get the following moduli situation: $\mathcal{M}_*$ is to be the normalization of $\mathcal{M}$ is a suitable finite algebraic extension of its function field such that for every point of $\mathcal{M}_*$ there is given rationally not only a curve C of genus g, but (*a*) a (4,8)-structure on J (i. e. a point of $\mathscr{A}_{(4,8)}$ lying over J in $\mathscr{A}_1$), (*b*) a double covering $\Pi : \hat{C} \to C$, (*c*) a (4,8)-structure on P (cf. [5], pp. 104-108 for a precise discussion of such " non-abelian levels "). Thus, if we let $\mathscr{A}$'s (resp. $\mathscr{B}$'s) represent moduli spaces for abelian varieties of dim. g (resp. dim. $g-1$), we have morphisms:

$$\begin{array}{ccc} & & \mathscr{A}_{4,8} \\ & \overset{t_*}{\nearrow} & \\ \mathscr{M}^* & & \\ & \underset{s_*}{\searrow} & \\ & & \mathscr{B}_{4,8} \end{array}$$

and since θ_J on $\frac{1}{2}T(J)$ $\left(\text{resp. } \theta_P \text{ on } \frac{1}{2}T(P)\right)$ are coordinates on $\mathscr{A}_{4,8}$ (resp. $\mathscr{B}_{4,8}$), the identities (*) define a locus $\mathscr{C} \subset \mathscr{A}_{4,8} \times \mathscr{B}_{4,8}$ (the η and φ must be independent of the curve you start with). We find:

THEOREM. — Im (t_*, s_*) is an open subset of one of the components of locus $\mathscr{C}$ of solutions of the Schottky-Jung identites (*) inside the moduli space $\mathscr{A}_{4,8} \times \mathscr{B}_{4,8}$.

REFERENCES

[1] A. ANDREOTTI. — On a theorem of Torelli, *Am. J. Math.*, 80 (1958).
[2] — and A. MAYER. — On period relations for abelian integrals on algebraic curves, *Annali di Scuolo Norm. di Pisa* (1967).
[3] M. ARTIN. — The implicit function theorem in algebraic geometry, in *Algebraic Geometry*, Oxford Univ. Press (1969).
[4] W. BAILY. — On the moduli of abelian varieties with multiplications, *J. Math. Soc. Japan*, 15 (1963).
[5] P. DELIGNE and D. MUMFORD. — The irreducibility of the space of curves of given genus, *Pub. I. H. E. S.*, 36 (1969).
[6] H. FARKAS and RAUCH. — Two kinds of theta constants and period relations on a Riemann Surface, *Proc. Nat. Acad. U. S.*, 62 (1969).
[7] J. FAY. — *Special moduli and theta relations*, Ph. D. Thesis, Harvard (1970).
[8] W. HOYT. — On products and algebraic families of jacobian varieties, *Annals of Math.*, 77 (1963).
[9] J. I. IGUSA. — On the graded ring of theta-constants, *Am. J. Math.*, 86 (1964) and 88 (1966).
[10] T. MATSUSAKA. — On a theorem of Torelli, *Am. J. Math.*, 80 (1958).
[11] D. MUMFORD. — *Geometric Invariant theory*, Springer-Verlag, Heidelberg (1965).
[12] —. — On the equations defining abelian varieties, *Inv. Math.*, Part I: vol. 1 (1966), Parts II and III: vol. 3 (1967).
[13] —. — Varieties defined by quadratic equations, in Questioni sulle varieta algebriche, *Corsi dal C. I. M. E.*, 1969, Edizioni Cremonese, Roma.
[14] —. — *Bi-extensions of formal groups, in Algebraic Geometry*, Oxford Univ. Press (1969).
[15] —. — *Abelian varieties*, Oxford Univ. Press (1970).

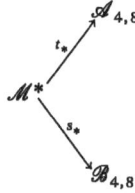

[16] —. — Theta characteristics of an algebraic curve, to appear in *Annales de l'École Normale Supérieure* (1971).

[17] F. Oort. — Finite group schemes, local moduli for abelian varieties and lifting problems, to appear in *Proc. Nordic Summer school of 1970*, Wolters-Noordhoff, Groningen.

[18] B. Riemann. — *Collected works*, Nachtrag IV, Dover edition (1953).

[19] F. Schottky and H. Jung. — Neue Sätze über symmetralfunctionen und die Abelschen funktionen der Riemann'sche theorie, *Sitzungber, Berlin Akad. Wissensch.*, vol. I (1909).

[20] J. P. Serre and J. Tate. — *Proc. of seminar on formal groups at Woods Hole Summer Institute* (1964), mimeographed notes.

[21] G. Shimura. — Moduli of abelian varieties and number theory, in algebraic groups and discontinuous subgroups, *Am. Math. Soc.* (1966).

Harvard University,
Department of Mathematics,
2, Divinity Avenue,
Cambridge, Mass. 02138
(U. S. A.)

ICM 2002 · Vol. I · 401–422

Pattern Theory: The Mathematics of Perception

David Mumford*

Abstract

Is there a mathematical theory underlying intelligence? Control theory addresses the output side, motor control, but the work of the last 30 years has made clear that perception is a matter of Bayesian statistical inference, based on stochastic models of the signals delivered by our senses and the structures in the world producing them. We will start by sketching the simplest such model, the hidden Markov model for speech, and then go on illustrate the complications, mathematical issues and challenges that this has led to.

Keywords and Phrases: Perception, Speech, Vision, Bayesian, Statistics, Inference, Markov.

1. Introduction

How can we understand intelligent behavior? How can we design intelligent computers? These are questions that have been discussed by scientists and the public at large for over 50 years. As mathematicians, however, the question we want to ask is "is there a *mathematical* theory underlying intelligence?" I believe the first mathematical attack on these issues was Control Theory, led by Wiener and Pontryagin. They were studying how to design a controller which drives a motor affecting the world and also sits in a feedback loop receiving measurements from the world about the effect of the motor action. The goal was to control the motor so that the world, as measured, did something specific, i.e. move the tiller so that the boat stays on course. The main complication is that nothing is precisely predictable: the motor control is not exact, the world does unexpected things because of its complexities and the measurements you take of it are imprecise. All this led, in the simplest case, to a beautiful analysis known as the Wiener-Kalman-Bucy filter (to be described below).

But Control Theory is basically a theory of the output side of intelligence with the measurements modeled in the simplest possible way: e.g. linear functions of the

*Division of Applied Mathematics, Brown University, Providence RI 02912, USA. E-mail: David_Mumford@brown.edu

state of the world system being controlled plus additive noise. The real input side of intelligence is perception in a much broader sense, the analysis of all the noisy incomplete signals which you can pick up from the world through natural or artificial senses. Such signals typically display a mix of distinctive patterns which tend to repeat with many kinds of variations and which are confused by noisy distortions and extraneous clutter. The interesting and important structure of the world is thus coded in these signals, using a code which is complex but not perversely so.

1.1. Logic vs. Statistics

The first serious attack on problems of perception was the attempt to recognize speech which was launched by the US defense agency ARPA in 1970. At this point, there were two competing ideas of what was the right formalism for combining the various clues and features which the raw speech yielded. The first was to use logic or, more precisely, a set of 'production rules' to augment a growing database of true propositions about the situation at hand. This was often organized in a 'blackboard', a two-dimensional buffer with the time of the asserted proposition plotted along the x-axis and the level of abstraction (i.e. signal — phone — phoneme — syllable — word — sentence) along the y-axis. The second was to use statistics, that is, to compute probabilities and conditional probabilities of various possible events (like the identity of the phoneme being pronounced at some instant). These statistics were computed by what was called the 'forward-backward' algorithm, making 2 passes in time, before the final verdict about the most probable translation of the speech into words was found. This issue of logic vs. statistics in the modeling of thought has a long history going back to Aristotle about which I have written in [M].

I think it is fair to say that statistics won. People in speech were convinced in the 1970's, artificial intelligence researchers converted during the 1980's as expert systems needed statistics so clearly (see Pearl's influential book [P]), but vision researchers were not converted until the 1990's when computers became powerful enough to handle the much larger datasets and algorithms needed for dealing with 2D images.

The biggest reason why it is hard to accept that statistics underlies all our mental processes — perception, thinking and acting — is that we are not consciously aware of 99% of the ambiguities with which we deal every second. What philosophers call the 'raw qualia', the actual sensations received, do not make it to consciousness; what we are conscious of is a precise unambiguous enhancement of the sensory signal in which our expectations and our memories have been drawn upon to label and complete each element of the percept. A very good example of this comes from the psychophysical experiments of Warren & Warren [W] in 1970: they modified recorded speech by replacing a single phoneme in a sentence by a noise and played this to subjects. Remarkably, the subjects did *not* perceive that a phoneme was missing but believed they had heard the one phoneme which made the sentence semantically consistent:

ACTUAL SOUND	PERCEIVED WORDS
the ¿eel is on the shoe	the *h*eel is on the shoe
the ¿eel is on the car	the *wh*eel is on the car
the ¿eel is on the table	the *m*eal is on the table
the ¿eel is on the orange	the *p*eel is on the orange

Two things should be noted. Firstly, this showed clearly that the actual auditory signal did not reach consciousness. Secondly, the choice of percept was a matter of probability, not certainty. That is, one might find some odd shoe with a wheel on it, a car with a meal on it, a table with a peel on it, etc. but the words which popped into consciousness were the most likely. An example from vision of a simple image, whose contents require major statistical reasoning to reconstruct, is shown in figure 1.

Figure 1: Why is this old man recognizable from a cursory glance? His outline threads a complex path amongst the cluttered background and is broken up by alternating highlights and shadows and by the wrinkles on his coat. There is no single part of this image which suggests a person unambiguously (the ear comes closest but the rest of his face can only be guessed at). No other object in the image stands out — the man's cap, for instance, could be virtually anything. Statistical methods, first grouping contours, secondly guessing at likely illumination effects and finally using probable models of clothes may draw him out. No known computer algorithm comes close to finding a man in this image.

It is important to clarify the role of probability in this approach. The uncertainty in a given situation need not be caused by observations of the world being truly unpredictable as in quantum mechanics or even effectively so as in chaotic phenomena. It is rather a matter of efficiency: in order to understand a sentence being spoken, we do not need to know all the things which affect the sound such as the exact acoustics of the room in which we are listening, nor are we even able to know other factors like the state of mind of the person we are listening to. In other words, we always have incomplete data about a situation. A vast number of physical and mental processes are going on around us, some germane to the meaning of the signal, some extraneous and just cluttering up the environment. In this 'blooming, buzzing' world, as William James called it, we need to extract information and the best way to do it, apparently, is to make a stochastic model in which all the irrelevent events are given a simplified probability distribution. This is not unlike the stochastic approach to Navier-Stokes, where one seeks to replace turbulence or random molecular effects on small scales by stochastic perturbations.

1.2. The Bayesian setup

Having accepted that we need to use probabilities to combine bits and pieces of evidence, what is the mathematical set up for this? We need the following ingredients: a) a set of random variables, some of which describe the observed signal and some the 'hidden' variables describing the events and objects in the world which are causing this signal, b) a class of stochastic models which allow one to express the variability of the world and the noise present in the signals and c) specific parameters for the one stochastic model in this class which best describes the class of signals we are trying to decode now. More formally, we shall assume we have a set $\mathbf{x} = (\mathbf{x}_o, \mathbf{x}_h)$ of observed and hidden random variables, which may have real values or discrete values in some finite or countable sets, we have a set $\boldsymbol{\theta}$ of parameters and we have a class of probability models $\Pr(\mathbf{x} \mid \boldsymbol{\theta})$ on the x's for each set of values of the $\boldsymbol{\theta}$'s. The crafting or learning of this model may be called the first problem in the mathematical theory of perception. It is usual to factor these probability distributions:

$$\Pr(\mathbf{x} \mid \boldsymbol{\theta}) = \Pr(\mathbf{x}_o \mid \mathbf{x}_h, \boldsymbol{\theta}) \cdot \Pr(\mathbf{x}_h \mid \boldsymbol{\theta}),$$

where the first factor, describing the likelihood of the observations from the hidden variables, is called the *imaging model* and the second, giving probabilities on the hidden variables, is called the *prior*. In the full Bayesian setting, one has an even stronger prior, a full probability model $\Pr(\mathbf{x}_h, \boldsymbol{\theta})$, including the parameters.

The second problem of perception is that we need to estimate the values of the parameters $\boldsymbol{\theta}$ which give the best stochastic model of this aspect of the world. This often means that you have some set of measurements $\{\mathbf{x}^{(\alpha)}\}$ and seek the value of $\boldsymbol{\theta}$ which maximizes their likelihood $\prod_\alpha \Pr(\mathbf{x}^{(\alpha)} \mid \boldsymbol{\theta})$. If the hidden variables as well as the observations are known, this is called supervised learning; if the hidden variables are not known, then it is unsupervised and one may maximize, for instance, $\prod_\alpha \sum_{\mathbf{x}_h} \Pr(\mathbf{x}_o^{(\alpha)}, \mathbf{x}_h \mid \boldsymbol{\theta})$. If one has a prior on the $\boldsymbol{\theta}$'s too, one can also estimate them from the mean or mode of the full posterior $\Pr(\boldsymbol{\theta} \mid \{\mathbf{x}^{(\alpha)}\})$.

Usually a more challenging problem is how many parameters $\boldsymbol{\theta}$ to include. At one extreme, there are simple 'off-the-shelf' models with very few parameters and, at the other extreme, there are fully non-parametric models with infinitely many parameters. Here the central issue is how much data one has: for any set of data, models with too few parameters distort the information the data contains and models with too many overfit the accidents of this data set. This is called the *bias-variance dilemma.* There are two main approaches to this issue. One is cross-validation: hold back parts of the data, train the model to have maximal likelihood on the training set and test it by checking the likelihood of the held out data. There is also a beautiful theoretical analysis of the problem due principally to Vapnik [V] and involving the *VC dimension* of the models — the size of the largest set of data which can be split in all possible ways into more and less likely parts by different choices of $\boldsymbol{\theta}$.

As Grenander has emphasized, a very useful test for a class of models is to synthesize from it, i.e. choose random samples according to this probability measure and to see how well they resemble the signals we are accustomed to observing in the world. This is a stringent test as signals in the world usually express layers and layers of structure and the model tries to describe only a few of these.

The third problem of perception is using this machinary to actually perceive: we assume we have measured specific values $\mathbf{x}_o = \widehat{\mathbf{x}}_o$ and want to infer the values of the hidden variables $\mathbf{x}_h$ in this situation. Given these observations, by Bayes' rule, the hidden variables are distributed by the so-called *posterior* distribution:

$$\Pr(\mathbf{x}_h \mid \widehat{\mathbf{x}}_o, \boldsymbol{\theta}) = \frac{\Pr(\widehat{\mathbf{x}}_o \mid \mathbf{x}_h, \boldsymbol{\theta}) \cdot \Pr(\mathbf{x}_h \mid \boldsymbol{\theta})}{\Pr(\widehat{\mathbf{x}}_o \mid \boldsymbol{\theta})} \propto \Pr(\widehat{\mathbf{x}}_o \mid \mathbf{x}_h, \boldsymbol{\theta}) \cdot \Pr(\mathbf{x}_h \mid \boldsymbol{\theta})$$

One may then want to estimate the mode of the posterior, the most likely value of $\mathbf{x}_h$. Or one may want to estimate the mean of some functions $f(\mathbf{x}_h)$ of the hidden variables. Or, if the posterior is often multi-modal and some evidence is expected to available later, one usually wants a more complete description or approximation to the full posterior distribution.

2. A basic example: HMM's and speech recognition

A convenient way to introduce the ideas of Pattern Theory is to outline the simple Hidden Markov Model method in speech recognition to illustrate many of the ideas and problems which occur almost everywhere. Here the observed random variables are the values of the sound signal $s(t)$, a pressure wave in air. The hidden random variables are the states of the speaker's mouth and throat and the identity of the phonemes being spoken at each instant. Usually this is simplified, replacing the signal by samples $s_k = s(k\Delta t)$ and taking for hidden variables a sequence x_k whose values indicate which phone in which phoneme is being pronounced at time $k\Delta t$. The stochastic model used is:

$$\Pr(x_., s_.) = \prod_k p_1(x_k \mid x_{k-1}) p_2(s_k \mid x_k)$$

i.e. the $\{x_k\}$ form a Markov chain and each s_k depends only on x_k. This is expressed by the graph:

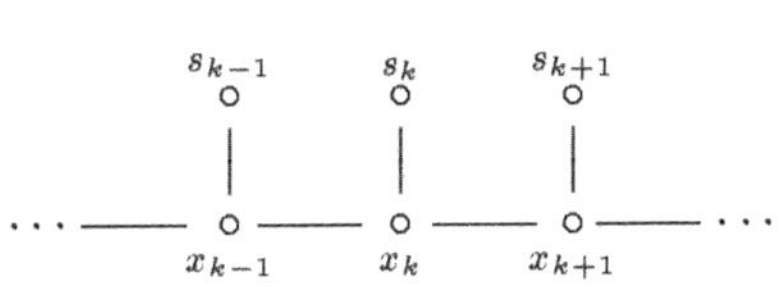

in which each variable corresponds to a vertex and the graphical Markov property holds: if 2 vertices a, b in the graph are separated by a subset S of vertices, then the variables associated to a and b are conditionally independent if we fix the variables associated to S.

This simple model works moderately well to decode speech because of the linear nature of the graph, which allows the ideas of dynamic programming to be used to solve for the marginal distributions and the modes of the hidden variables, given any observations $\widehat{s}_\cdot$. This is expressed simply in the recursive formulas:

$$\begin{aligned}
\Pr(x_k \mid \widehat{s}_{\leq k}) &= \frac{\sum_{x_{k-1}} p_1(x_k \mid x_{k-1}) p_2(\widehat{s}_k \mid x_k) \Pr(x_{k-1} \mid \widehat{s}_{\leq (k-1)})}{\sum_{x_k} \text{numerator}} \\
\text{maxp}(x_k, \widehat{s}_{\leq k}) &\underset{\text{def}}{=} \max_{x_{\leq (k-1)}} \Pr(x_k, x_{\leq k-1}, \widehat{s}_{\leq k}) \\
&= \max_{x_{k-1}} \left(p_1(x_k \mid x_{k-1}) p_2(\widehat{s}_x \mid x_k) \text{maxp}(x_{k-1}, \widehat{s}_{\leq (k-1)}) \right).
\end{aligned}$$

Note that if each x_k can take N values, the complexity of each time step is $O(N^2)$.

In any model, if you can calculate the conditional probabilities of the hidden variables and if the model is of exponential type, i.e.

$$\Pr(x_\cdot \mid \boldsymbol{\theta}_\cdot) = \frac{1}{Z(\boldsymbol{\theta})} e^{\sum_k \theta_k \cdot E_k(x_\cdot)},$$

then there is also an efficient method of optimizing the parameters $\boldsymbol{\theta}$. This is called the *EM algorithm* and, because it holds for HMM's, it is one of the key reasons for the early successes of the stochastic approach to speech recognition. For instance, a Markov chain $\{x_k\}$ is an exponential model if we let the $\boldsymbol{\theta}$'s be $\log(p(a \mid b))$ and write the chain probabilities as:

$$\Pr(x_\cdot) = e^{\sum_{a,b} \log(p(a|b)) |\{k | x_k = a, x_{k-1} = b\}|}.$$

The fundamental result on exponential models is that the $\boldsymbol{\theta}$'s are determined by the expectations $\widehat{E}_k = \text{Exp}(E_k)$ and that any set of expectations $\widehat{E}_k$ that can be achieved in some probability model (with all probabilities non-zero), is also achieved in an exponential model.

2.1. Continuous and discrete variables

In this model, the observations s_k are naturally continuous random variables, like all primary measurements of the physical world. But the hidden variables are discrete: the set of phonemes, although somewhat variable from language to language, is always a small discrete set. This combination of discrete and continuous is characteristic of perception. It is certainly a psychophysical reality: for example experiments show that our perceptions lock onto one or another phoneme, resisting ambiguity (see [L], Ch.8, esp. p.176). But it shows itself more objectively in the low-level statistics of natural signals. Take almost any class of continuous real-valued signals $s(t)$ generated by the world and compile a histogram of their changes $x = s(t+\Delta t) - s(t)$ over some fixed time interval Δt. This empirical distribution will very likely have kurtosis ($= \text{Exp}\left((x-\bar{x})^4\right)/\sigma(x)^4$) greater than 3, the kurtosis of any Gaussian distribution! This means that, compared to a Gaussian distribution with the same mean and standard deviation, x has higher probability of being quite small or quite large but a lower probability of being average. Thus, compared to Brownian motion, $s(t)$ tends to move relatively little most of the time but to make quite large moves sometimes. This can be made precise by the theory of stochastic processes with iid increments, a natural first approximation to any stationary Markov process. The theory of such processes says that (a) their increments always have kurtosis at least 3, (b) if it equals 3 the process is Brownian and (c) if it is greater, samples from the process almost surely have discontinuities. At the risk of over-simplfying, we can say *kurtosis > 3 is nature's universal signal of the presence of discrete events/objects in continuous space-time.*

A classic example of this are stock market prices. Their changes (or better, changes in log(price)) have a highly non-Gaussian distribution with polynomial tails. In speech, the changes in the log(power) of the windowed Fourier transform show the same phenomenon, confirming that $s(t)$ cannot be decently modeled by colored Gaussian noise.

2.2. When compiling full probability tables is impractical

Applying HMM's in realistic settings, it usually happens that N is too large for an exhaustive search of complexity $O(N^2)$ or that the x_k are real valued and, when adequately sampled, again N is too large. There is one other situation in which the HMM-style approach works easily — the *Kalman filter*. In Kalman's setting, each variable x_k and s_k is real vector-valued instead of being discrete and p_1 and p_2 are *Gaussian distributions with fixed covariances and means depending linearly on the conditioning variable.* It is then easy to derive recursive update formulas, similar to those above, for the conditional distributions on each x_k, given the past data $\widehat{s}_{\leq k}$.

But usually, in the real-valued variable setting, the p's are more complex than Gaussian distributions. An example is the tracking problem in vision: the position and velocity x_k of some specific moving object at time $k\Delta t$ is to be inferred from a movie $\widehat{s}_k$, in which the object's location is confused by clutter and noise. It is clear that the search for the optimal reconstruction x_k must be pruned or approximated. A dramatic breakthrough in this and other complex situations has been to adapt the HMM/Kalman ideas by using weak approximations to the marginals $\Pr(x_k \mid \widehat{s}_{\leq k})$ by a finite set of samples, an idea called *particle filtering*:

$$\Pr(x_k \mid \widehat{s}_{\leq k}) \underset{\text{weak}}{\sim} \sum_{i=1}^{N} w_{i,k}\delta_{x_{i,k}}(x_k), \quad \text{that is,}$$

$$\text{Exp}(f(x_k) \mid \widehat{s}_{\leq k}) \approx \sum_{i=1}^{N} w_{i,k} f(x_{i,k}), \text{ for suitable } f.$$

This idea was proposed originally by Gordon, Salmond and Smith [G-S-S] and is developed at length in the recent survey [D-F-G]. An example with explicit estimates of the posterior from the work of Isard and Blake [I-B] is shown in figure 2. They follow the version known as bootstrap particle filtering in which, for each k, N samples x'_l are drawn with replacement from the weak approximation above, each sample is propagated randomly to a new sample x''_l at time $(k+1)$ using the prior $p(x_{k+1} \mid x'_l)$ and these are reweighted proportional to $p(\hat{s}_{k+1} \mid x''_l)$.

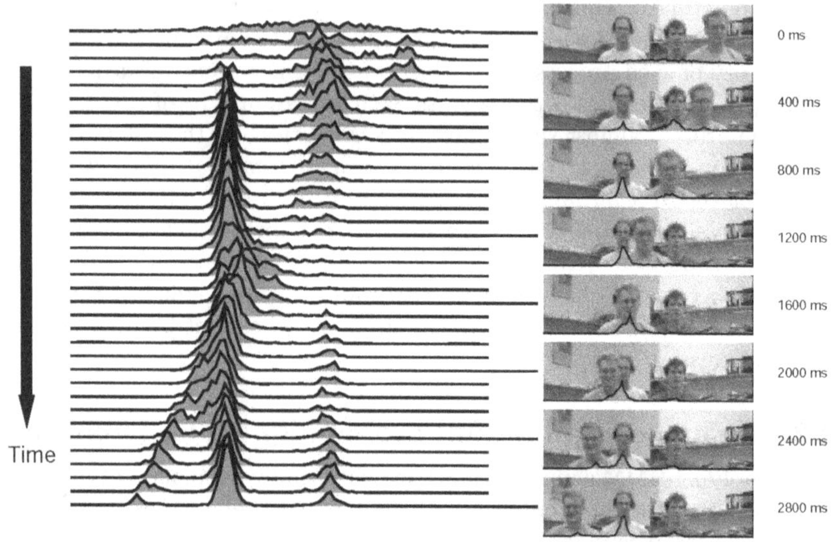

Figure 2: Work of Blake and Isard tracking three faces in a moving image sequence. The curves represent estimates of the posterior probability distributions for faces at each location obtained by smoothing the weighted sum of delta functions at the 'particles'. Note how multi-modal these are and how the tracker recovers from the temporary occlusion of one face by another.

2.3. No process in nature is truly Markov

A more serious problem with the HMM approach is that the Markov assumption is never really valid and it may be much too crude an approximation. Consider speech recognition. The finite lexicon of words clearly constrains the expected phoneme sequences, i.e. if x_k are the phonemes, then $p_1(x_k \mid x_{k-1})$ depends on the current word(s) containing these phonemes, i.e. on a short but variable part of the preceding string $\{x_{k-1}, x_{k-2}, \cdots\}$ of phonemes. To fix this, we could let x_k be a pair consisting of a word and a specific phoneme in this word; then $p_1(x_k \mid x_{k-1})$ would have two quite different values depending on whether x_{k-1} was the last phoneme in the word or not. Within a word, the chain needs only to take into account the variability with which the word can be pronounced. At word boundaries, it should use the conditional probabilities of word pairs. This builds much more of the patterns of the language into the model.

Why stop here? State-of-the-art speech recognizers go further and let x_k be a pair of consecutive words plus a triphone[1] in the second word (or bridging the first and second word) whose middle phoneme is being pronounced at time $k\Delta t$. Then the transition probabilities in the HMM involve the statistics of 'trigrams', consecutive word triples in the language. But grammar tells us that words sequences are also structured into phrases and clauses of variable length forming a parse tree. These clearly affect the statistics. Semantics tells us that words sequences are further constrained by semantic plausibility ('sky' is more probable as the word following 'blue' than 'cry') and pragmatics tells us that sentences are part of human communications which further constrain probable word sequences.

All these effects make it clear that certain parts of the signal should be grouped together into units on a higher level and given labels which determine how likely they are to follow each other or combine in any way. This is the essence of grammar: higher order random variables are needed whose values are subsets of the low order random variables. The simplest class of stochastic models which incorporate variable length random substrings of the phoneme sequence are *probabilistic context free grammars* or PCFG's. Mathematically, they are a particular type of random branching tree.

Definition *A PCFG is a stochastic model in which the random variables are (a) a sequence of rooted trees* $\{\mathcal{T}_n\}$, *(b) a linearly ordered sequence of observations* s_k *and a 1:1 correspondence between the observations* s_k *and the leaves of the whole forest of trees such that the children of any vertex of any tree form an interval* $\{s_k, s_{k+1}, \cdots, s_{k'}\}$ *in time and (c) a set of labels* x_v *for each vertex. The probability model is given by conditional probabilities* $p_1(x_{v_k} \mid x_v)$ *for the labels of each child of each vertex*[2] *and* $p_2(s_k \mid x_{v_k})$ *for the observations, conditional on the label of the corresponding leaf.*

See figure 3 for an example. This has a Markov property if we define the 'extended' state x_k^* at leaf k to be not only the label x_k at this leaf but the whole sequence of labels on the path from this leaf to the root of the tree in which this leaf lies. Conditional on this state, the past and the future are independent.

This is a mathematically elegant and satisfying theory: unfortunately, it also fails, or rather explodes because, in carrying it out, the set of labels gets bigger and bigger. For instance, it is not enough to have a label for noun phrase which expands into an adjective plus a noun. The adjective and noun must agree in number and (in many languages) gender, a constraint that must be carried from the adjective to the noun (which need not be adjacent) via the label of the parent. So we need 4 labels, all combinations of singular/plural masculine/feminine noun phrases. And semantic constraints, such as Pr('blue sky') > Pr('blue cry'), would seem to require even more labels like 'colorable noun phrases'. Rather than letting the label set explode, it is better to consider a bigger class of grammars, which express these relations more succinctly but which are not so easily converted into HMM's: *unification grammars* [Sh] or *compositional grammars* [B-G-P]. The need for grammars of this type is

[1]So-called co-articulation effects mean that the pronunciation of a phoneme is affected by the preceding and suceeding phonemes.

[2]Caution to specialists: our label x_v is the name of the 'production rule' with this vertex as its head, esp. it fixes the arity of the vertex. We are doing it this way to simplify the Markov property.

especially clear when we look at formalisms for expressing the grouping laws in vision: see figure 3. The further development of stochastic compositional grammars, both in language and vision, is one of the main challenges today.

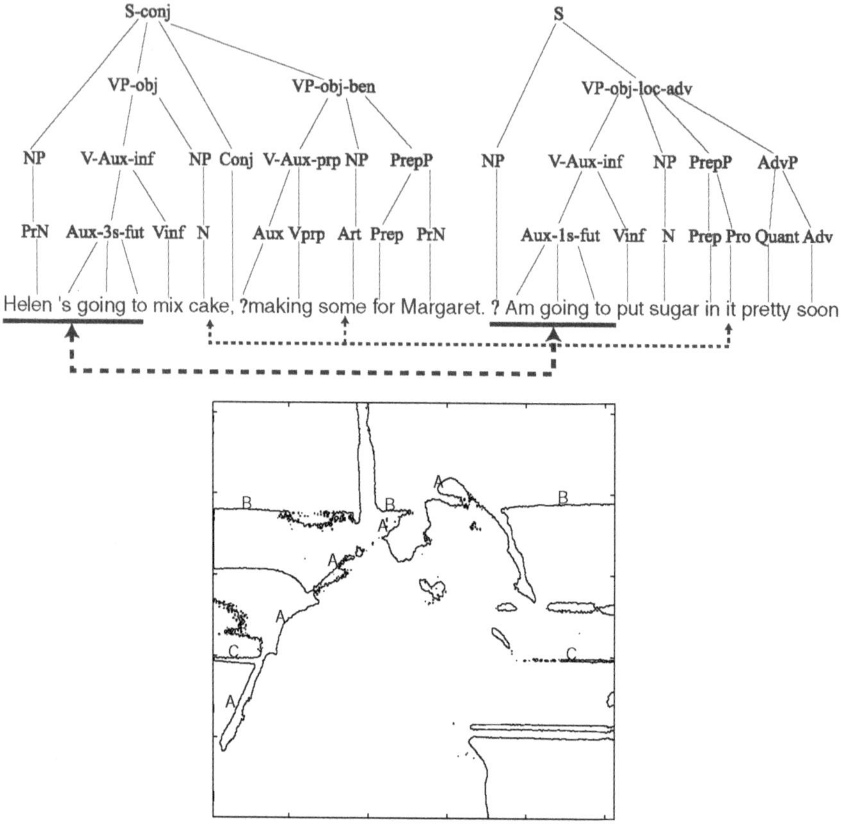

Figure 3: Grouping in language and vision: On top, parsing the not quite grammatical speech of a 2 1/2 year old Helen describing her own intentions ([H]): above the sentence, a context-free parse tree; below it, longer range non-Markov links — the identity 'cake'='some'='it' and the unification of the two parts 'Helen's going to' = '(I) am going to'. On the bottom, 2 kinds of grouping with an iso-intensity contour of the image in Figure 1: note the broken but visible contour of the back marked by 'A' and the occluded contours marked by 'B' and 'C' behind the man.

3. The 'natural degree of generality': MRF's or Graphical Models

The theory of HMM's deals with one-dimensional signals. But images, the signals occurring in vision, are usually two-dimensional — or three-dimensional for

MR scans and movies (3 space dimensions and 2 space plus 1 time dimension), even four-dimensional for echo cardiograms. On the other hand, the parse tree is a more abstract graphical structure and other 'signals', like medical data gathered about a particular patient, are structured in complex ways (e.g. a set of blood tests, a medical history). This leads to the basic insight of Grenander's Pattern Theory [G]: that the variables describing the structures in the world are typically related in a graphical fashion, edges connecting variables which have direct bearing on each other. Finding the right graph or class of graphs is a crucial step in setting up a satisfactory model for any type of patterns. Thus the applications, as well as the mathematical desire to find the most general setting for this theory, lead to the idea of replacing a simple chain of variables by a set of variables with a more general graphical structure. The general concept we need is that of a Markov random field:

Definition *A Markov random field is a graph $G = (V, E)$, a set of random variables $\{x_v\}_{v \in V}$, one for each vertex, and a joint probability distribution on these variables of the form:*

$$\Pr(x_.) = \frac{1}{Z} e^{-\sum_C E_C(\{x_v\}_{v \in C})},$$

where C ranges over the cliques (fully connected subsets) of the graph, E_C are any functions and Z a constant. If the variables x_v are real-valued for $v \in V'$, we make this into a probability density, multiplying by $\prod_{n \in V'} dx_n$. Moreover, we can put each model in a family by introducing a temperature T and defining:

$$\Pr_T(x_.) = \frac{1}{Z_T} e^{-\sum_C E_C(\{x_v\}_{v \in C})/T}.$$

These are also called *Gibbs models* in statistical mechanics (where the E_C are called *energies*) and *graphical models* in learning theory and, like Markov chains, are characterized by their conditional independence properties. This characterization, called the Hammersley-Clifford theorem, is that if two vertices $a, b \in V$ are separated by a subset $S \subset V$ (all paths in G from a to b must include some vertex in S), then x_a and x_b are conditionally independent given $\{x_v\}_{v \in S}$. The equivalence of these independence properties, plus the requirement that all probabilities be positive, with the simple explicit formula for the joint probabilities makes it very convincing that MRF's are a natural class of stochastic models.

3.1. The Ising model

This class of models is very expressive and many types of patterns which occur in the signals of nature can be captured by this sort of stochastic model. A basic example is the Ising model and its application to the image segmentation problem. In the simplest form, we take the graph G to be a square $N \times N$ grid with two layers, with observable random variables $p_{i,j} \in \mathbb{R}, 1 \leq i, j \leq N$ associated to the top layer and hidden random variables $x_{i,j} \in \{+1, -1\}$ associated to the bottom layer. We connect by edges each $x_{i,j}$ vertex to the $p_{i,j}$ vertex above it and to its 4 neighbors $x_{i\pm1,j}, x_{i,j\pm1}$ in the x-grid (except when the neighbor is off the grid) and no others. The cliques are just the pairs of vertices connected by edges. Finally, we take for energies:

$$
\begin{aligned}
E_C &= -x_{i,j} \cdot x_{i',j'}, \text{ when } C = \{(i,j),(i',j')\}, \text{ two adjacent vertices in the } x\text{-grid,} \\
E_C &= -x_{i,j} \cdot y_{i,j}, \text{ when } C \text{ consists of the } (i,j) \text{ vertices in the } x\text{- and } y\text{-grids.}
\end{aligned}
$$

The modes of the posteriors $\Pr_T(x. \mid \widehat{y}.)$ are quite subtle: x's at adjacent vertices try to be equal but they also seek to have the same sign as the correponding $\widehat{y}$. If $\widehat{y}$ has rapid positive and negative swings, these are in conflict. Hence the more probable values of x will align with the larger areas where $\widehat{y}$ is consistently of one sign. This can be used to model a basic problem in vision: the *segmentation problem.* The vision problem is to decompose the domain of an image y into parts where distinct objects are seen. For example, the oldman image might be decomposed into 6 parts: his body, his head, his cap, the bench, the wall behind him and the sky. The decomposition is to be based on the idea that the image will tend to either slowly varying or to be statistically stationary at points on one object, but to change abruptly at the edges of objects. As proposed in [G-G], the Ising model can be used to treat the case where the image has 2 parts, one lighter and one darker, so that at the mode of the posterior the hidden variables x will be $+1$ on one part, -1 on the other. An example is shown in figure 4. This approach makes a beautiful link between statistical mechanics and perception, in which the process of finding global patterns in a signal is like forming large scale structures in a physical material as the temperature cools through a phase transition.

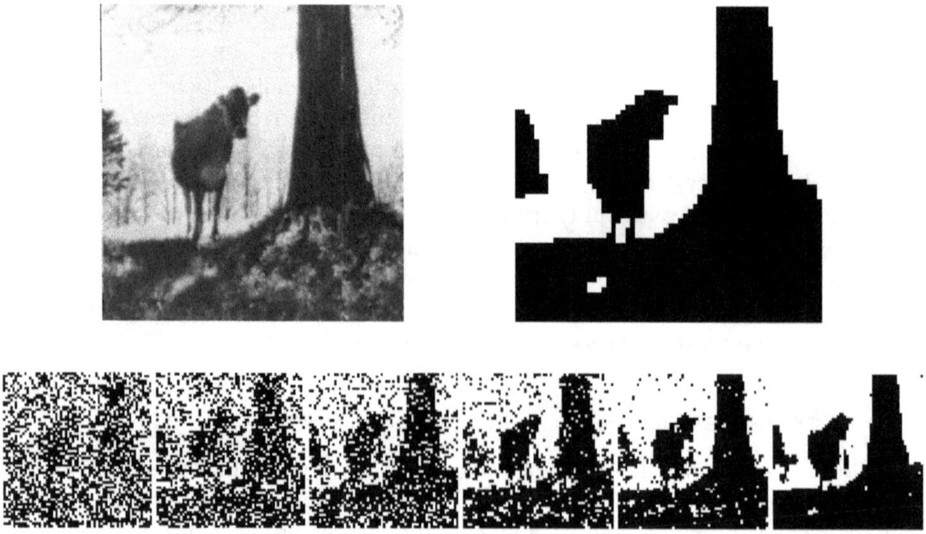

Figure 4: Statistical mechanics can be applied to the segmentation of images. On the top left, a rural scene taken as the external magnetic field, with its intensity scaled so that dark areas are negative, light areas are positive. At the top right, the mode or ground state of the Ising model. Along the bottom, the Gibbs distribution is sampled at a decreasing sequence of temperatures, discovering the global pattern bit by bit.

More complex models of this sort have been used extensively in image analysis, for texture segmentation, for finding disparity in stereo vision, for finding optic flow in moving images and for finding other kinds of groupings. We want to give one example of the expressivity of these models which is quite instructive. We saw above that exponential models can be crafted to reproduce some set of observed expectations but we also saw that scalar statistics from natural signals typically have high kurtosis, i.e. significant outliers, so that their whole distribution and not just their mean needs to be captured in the model. Putting these 2 facts together suggests that we seek exponential models which duplicate the whole distribution of some important statistics $f_{\cdot}$. This can be done using as parameters not just unknown constants but unknown functions:

$$\Pr(x_{\cdot} \mid \phi_{\cdot}) = \frac{1}{Z(\vec{\phi}_{\cdot})} e^{\sum_k \phi_k(f_k(x_{\cdot}))}.$$

If f_k depends only the variables $x_v \in C_k$, for some clique C_k, this is a MRF, whose energies have unknown functions in them. An example of this fitting is shown in Figure 5.

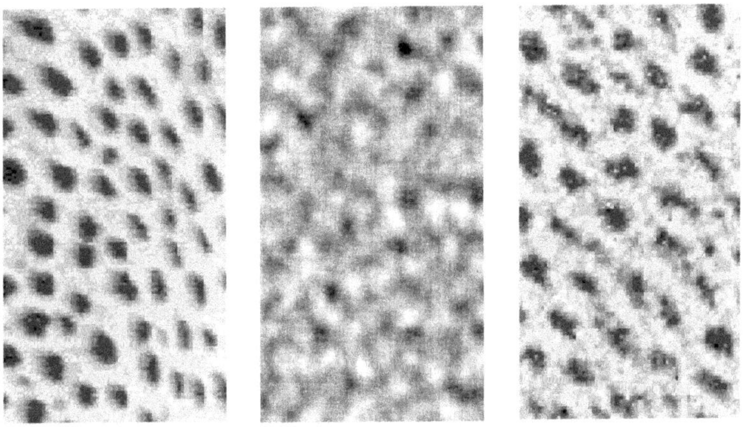

Figure 5: On the left, an image of the texture of a Cheetah's hide, in the middle a synthetic image from the Gaussian model with the same second order statistics, on the right a synthetic image in which the full distribution on 7 filter statistics are reproduced by an exponential model.

3.2. Bayesian belief propagation

However, a problem with MRF models is that the dynamic programming style algorithm used in speech and one-dimensional models to find the posterior mode has no analog in 2D. One strategy for dealing with this, which goes back to Metropolis, is to imitate physics and introduce an artifical dynamics into the state space whose equilibrium is the Gibbs distribution. This dynamics is called a *Monte Carlo Markov Chain* (MCMC) and is how the panels in figure 4 were generated. Letting the temperature converge to zero, we get *simulated annealing* (see [G-G]) and, if we do it slowly enough, will find the mode of the MRF model. Although slow, this can be speeded up by biasing the dynamics (called *importance sampling* — see [T-Z] for a state-of-the-art implementation with many improvements) and is an important tool.

Recently, however, another idea due to Weiss and collaborators (see [Y-F-W]) and linked to statistical mechanics has been found to give new and remarkably effective algorithms for finding these modes. From an algorithmic standpoint, the idea is to use the natural generalization of dynamic programming, called *Bayesian Belief Propagation* (BBP), which computes the marginals and modes correctly whenever the graph is a tree and just use it anyway on an arbitrary graph G! Mathematically, it amounts to working on the universal covering graph $\tilde{G}$, which is a tree, hence much simpler, instead of G. In statistical mechanics, this idea is called the *Bethe approximation*, introduced by him in the 30's.

To explain the idea, start with the *mean field approximation.* The mean field idea is to find the best approximation of the MRF p by a probability distribution in which the variables x_v are all independent. This is formulated as the distribution $\prod_v p_v(x_v)$ which minimizes the Kullback-Liebler divergence $\mathrm{KL}(\prod_v p_v, p)$. Unlike computing the true marginals of p on each x_v which is very hard, this approximation can be found by solving iteratively a coupled set of non-linear equations for the p_v. But the assumption of independence is much too restrictive. The idea of Bethe is instead to approximate p by a $\pi_1(G)$-invariant distribution on $\tilde{G}$.

Such distributions are easy to describe: note that a Markov random field on a *tree* is uniquely determined by its marginals $p_e(x_v, x_w)$ for each edge $e = (v, w)$ and, conversely, if we are given a compatible set of distributions p_e for each edge (in the sense that, for all edges (v, w_k) abutting a vertex v, the marginals of $p_{(v,w_k)}$ give distributions on v independent of k), they define an MRF on G. So if we start with a Markov random field on any G, we get a $\pi_1(G)$-invariant Markov random field on $\tilde{G}$ by making duplicate copies for each random variable $x_v, v \in V$ for each $\tilde{v} \in \tilde{V}$ over v and lifting the edge marginals. But more generally, if we have any compatible set of probability distributions $\{p_e(v, w)\}_{e \in E}$ on G, we also get a $\pi_1(G)$-invariant MRF on $\tilde{G}$. Then the Bethe approximation is that family $\{p_e\}$ which minimizes $\mathrm{KL}(\{p_e\}, p)$. As in the mean field case, there is a natural iterative method of solving for this minimum, which turns out, remarkably, to be identical to the generalization of BBP to general graphs G.

This approach has proved effective in some cases at finding best segmentations of images via the mode of a two-dimensional MRF. Other interesting ideas have been proposed for solving the segmentation problem which we do not have time to sketch: region growing, see esp. [Z-Y]), using the eigenfunctions of the graph-theoretic Laplacian, see [S-M], and multi-scale algorithms, see [P-B] and [S-B-B].

4. Continuous space and time and continuous sets of random variables

Although signals as we measure them are always sampled discretely, in the world itself signals are functions on the continua, time or space or both together. In some situations, a much richer mathematical theory emerges by replacing a countable collection of random variables by random processes and asking whether we can find good stochastic models for these continuous signals. I want to conclude this talk by mentioning three instances where some interesting analysis has arisen when passing to the continuum limit and going into some detail on two. We will not worry about algorithmic issues for these models.

4.1. Deblurring and denoising of images

This is the area where the most work has been done, both because of its links with other areas of analysis and because it is one of the central problems of image processing. You observe a degraded image $I(x, y)$ as a function of continuous variables and seek to restore it, removing simultaneously noise and blur. In the discrete setting, the Ising model or variants thereof discussed above can be applied for this. There are two closely related ways to pass to the continuous limit and reformulate this as a problem in analysis. As both drop the stochastic interpretation and have excellent treatments in the literature, we only mention briefly one of a family of variants of each approach:

Optimal piecewise smooth approximation of I via a variational problem:

$$\min_{J,\Gamma}\left(c_1\iint_D (I-J)^2 dxdy + c_2 \iint_{D-\Gamma} \|\nabla J\|^2 dxdy + c_3|\Gamma|\right)$$

where J, the improved image, has discontinuities along the set of 'edge' curves Γ. This approach is due to the author and Shah and has been extensively pursued by the schools of DeGiorgi and Morel. See [M-S]. It is remarkable that it is still unknown whether the minima to this functional are well behaved, e.g. whether Γ has a finite number of components. Stochastic variants of this approach should exist.

Non-linear diffusion of I:

$$\frac{\partial J}{\partial t} = \operatorname{div}\left(\frac{\nabla J}{\|\nabla J\|}\right) + \lambda(I - J)$$

where J at some future time is the enhancement. This approach started with the work of Perona and Malik and has been extensively pursued by Osher and his coworkers. See [Gu-M]. It can be interpreted as gradient descent for a variant of the previous variational problem.

4.2. Self-similarity of image statistics and image models

One of the earliest discoveries about the statistics of images I was that their power spectra tend to obey power laws

$$\mathrm{Exp}|(\widehat{I}(\xi,\eta)|^2 \approx (\xi^2+\eta^2)^{-\lambda/2},$$

where λ varies somewhat from image to image but clusters around the value 2. This has a very provocative interpretation: this power law is implied by self-similarity! In the language of lattice field theory, if $I(i,j), i,j \in \mathbb{Z}$ is a random lattice field and $\bar{I}$ is the block averaged field

$$\bar{I}(i,j) = \frac{1}{4}\left(I(2i,2j) + I(2i+1,2j) + I(2i,2j+1) + I(2i+1,2j+1)\right),$$

then we say the field is a renormalization fixed point if the distributions of I and of $\bar{I}$ are the same. The hypothesis that natural images of the world, treated as a single large database, have renormalization invariant statistics has received remarkable confirmation from many quite distinct tests.

Why does this hold? It certainly isn't true for auditory or tactile signals. I think there is one major and one minor reason for it. The major one is that the world is viewed from a random viewpoint, so one can move closer or farther from any scene. To first approximation, this scales the image (though not exactly because nearer objects scale faster than distant ones). The minor one is that most objects are opaque but have, by and large, parts or patterns on them and, in turn, belong to clusters of larger things. This observation may be formulated as saying the world is not merely made up of objects but it is cluttered with them.

The natural setting for scale invariance is pass to the limit and model images as random functions $I(x,y)$ of two real variables. Then the hypothesis is that a suitable function space supports a probability measure which is invariant under both translations and scalings $(x,y) \mapsto (\sigma x, \sigma y)$, whose samples are 'natural images'. This hypothesis encounters, however, an infra-red and an ultra-violet catastrophe:
a) The infra-red one is caused by larger and larger scale effects giving bigger and bigger positive and negative swings to a local value of I. But these large scale effects are very low-frequency and this is solved by considering I to be defined only modulo an unknown constant, i.e. it is a sample from a measure on a function space mod constants.
b) The ultra-violet one is worse: there are more and more local oscillations of the signals at finer and finer scales and this contradicts Lusin's theorem that an integrable function is continuous outside sets of arbitrarily small measure. In fact, it is a theorem that *there is no translation and scale invariant probability measure on the space of locally integrable functions mod constants.* This can be avoided by allowing images to be generalized functions. In fact, the support can be as small as the intersection of all negative Sobolev spaces $\bigcap_\epsilon \mathcal{H}^{-\epsilon}$.

To summarize what a good statistical theory of natural images should explain, we have scale-invariance as just described, kurtosis greater than 3 as described in section 2.1 and finally the right local properties:

Hypothesis I A theory of images is a translation and scale invariant probability measure on the space of generalized functions $I(x, y)$ mod constants.

Hypothesis II For any filter F with mean 0, the marginal statistics of $F * I(x, y)$ have kurtosis greater than 3.

Hypothesis III The local statistics of images reflect the preferred local geometries, esp. images of straight edges, but also curved edges, corners, bars, 'T-junctions' and 'blobs' as well as images without geometry, blank 'blue sky' patches.

Hypothesis III is roughly the existence of what Marr, thinking globally of the image called the *primal sketch* and what Julesz, thinking locally of the elements of texture, referred to as *textons*. By scale invariance, the local and global image should have the same elements.

To quantify Hypothesis III, what is needed is a major effort at data mining. Specifically, the natural approach seems to be to take a small filter bank of zero mean local filters $F_1, \cdots, F_k$, a large data base of natural images I_α leading to the sample of points in $\mathbb{R}^k$ given by $(F_1 * I_\alpha(x, y), \cdots, F_K * I_\alpha(x, y)) \in \mathbb{R}^k$ for all α, x and y. One seeks a good non-parametric fit to this dataset. But Hypothesis III shows that this distribution will not be simple. For example Lee et al [L-P-M] have taken $k = 8$, F_i a basis of zero mean filters with fixed 3×3 support. They then make a linear tranformation in $\mathbb{R}^8$ normalizing the covariance of the data to I_8 ('whitening' the data), and to investigate the outliers, map the data with norms in the upper 20% to S^7 by dividing by the norm. The analysis reveals that the resulting data has asymptotic infinite density along a non-linear surface in S^7! This surface is constructed by starting with an ideal image, black and white on the two sides of a straight edge and forming a 3×3 discrete image patch by integrating this ideal image over a tic-tac-toe board of square pixels. As the angle of the edge and the offset of the pixels to the edge vary, the resulting patches form this surface. This is the most concrete piece of evidence showing the complexity of local image statistics.

Are there models for these three hypotheses? We can satisfy the first hypothesis by the unique scale-invariant Gaussian model, called the free field by physicists — but its samples look like clouds and its marginals have kurtosis 3, so neither the second nor third hypothesis is satisfied. The next best approximation seems to be to use infinitely divisible measures, such as the model constructed by the author and B.Gidas [M-G], which we call *random wavelet expansions*:

$$I(x, y) = \sum_i \phi_i(e^{r_i}x - x_i, e^{r_i}y - y_i),$$

where $\{(x_i, y_i, r_i)\}$ is a Poisson process in $\mathbb{R}^3$ and ϕ_i are samples from an auxiliary Levi measure, playing the role of individual random wavelet primitives. But this model is based on adding primitives, as in a world of transparent objects, which causes the probability density functions of its marginal filter statistics to be smooth at 0 instead of having peaks there, i.e. the model does not produce enough 'blue sky' patches with very low constrast.

A better approach are the random collage models, called ***dead leaves models*** by the French school: see [L-M-H]. Here the ϕ_i are assumed to have bounded support, the terms have a random depth and, instead of being simply added, each term occludes anything behind it with respect to depth. This means $I(x,y)$ equals the one ϕ_i which is in front of all the others whose support contains (x,y). This theory has major troubles with both infra-red and ultra-violet limits but it does provide the best approximation to date of the empirical statistics of images. It introduces explicitly the hidden variables describing the discrete objects in the image and allows one to model their preferred geometries.

Crafting models of this type is not simply mathematically satisfying. It is central to the main application of computer vision: object recognition. When an object of interest is obscured in a cluttered badly lit scene, one needs a p-value for the hypothesis test — is this fragment of stuff part of the sought-for object or an accidental conjunction of things occurring in generic images? To get this p-value, one needs a null hypothesis, a theory of generic images.

4.3. Stochastic shapes via random diffeomorphisms and fluid flow

As we have seen in the last section, modeling images leads to objects and these objects have shape — so we need stochastic models of shape, the ultimate non-linear sort of thing. Again it is natural to consider this in the continuum limit and consider a k-dimensional shape to be a subset of $\mathbb{R}^k$, e.g. a connected open subset with nice boundary Γ. It is very common in multiple images of objects like faces, animals, clothes, organs in your body, to find not identical shapes but warped versions. How is this to be modeled? One can follow the ideas of the previous section and take a highly empirical approach, gathering huge databases of faces or kidneys. This is probably the road to the best pattern recognition in the long run. But another principle that Grenander has always emphasized is to take advantage of the group of symmetries of the situation — in this case, the group of all diffeomorphisms of $\mathbb{R}^k$. He and Miller and collaborators (see [Gr-M]) were led to rediscover the point of view of Arnold which we next describe.

Let $\mathcal{G}_n =$ group of diffeomorphisms on $\mathbb{R}^n$ and $\mathcal{SG}_n$ be the volume-preserving subgroup. We want to bypass issues of the exact degree of differentiability of these diffeomorphisms, but consider $\mathcal{G}_n$ and $\mathcal{SG}_n$ as infinite dimensional Riemannian manifolds. Let $\{\theta_t\}_{0\leq t\leq 1}$ be a path in $\mathcal{SG}_n$ and define its length by:

$$\text{length of path} \;=\int\left(\sqrt{\int_{\mathbb{R}^n}\|\frac{\partial\theta_t}{\partial t}(\theta_t^{-1}(x))\|^2 d\vec{x}}\right)dt.$$

This length is nothing but the *right*-invariant Riemannian metric:

$$\text{dist}(\theta,(I+\epsilon\vec{v})\circ\theta)^2=\epsilon^2\int\|\vec{v}\|^2 dx_1\cdot\cdot dx_n,\ \text{ where } \text{div}(\vec{v})\equiv 0.$$

Arnold's beautiful theorem is:

Theorem *Geodesics in $\mathcal{SG}_n$ are solutions of Euler's equation:*

$$\frac{\partial v_t}{\partial t} + (v_t \cdot \nabla)v_t = \nabla p, \text{ some pressure } p.$$

This result suggests using geodesics on suitable infinite dimensional manifolds to model optimal warps between similar shapes in images and using diffusion on these manifolds to craft stochastic models. But we need to get rid of the volume-preserving restriction. The weak metric used by Arnold no longer works on the full $\mathcal{G}_n$ and in [C-R-M], Christensen et al introduced:

$$||\vec{v}||_L^2 = \int < L\vec{v} \cdot \vec{v} > dx_1 \cdots dx_n$$

where v is any vector field and L is a fixed positive self-adjoint differential operator e.g. $(I - \Delta)^m, m > n/2$. Then a path $\{\theta_t\}$ in G has both a velocity:

$$v_t = \frac{\partial \theta_t}{\partial t}(\theta_t^{-1}(x))$$

and a *momentum:* $u_t = Lv_t$ (so $v_t = K * u_t$, K the Green's function of L). What is important here is that the momentum u_t can be a generalized function, even when v_t is smooth. The generalization of Arnold's theorem, first derived by Vishik, states that geodesics are:

$$\frac{\partial u_t}{\partial t} + (v_t \cdot \nabla)(u_t) + \mathrm{div}(v_t)u_t = -\sum_i (u_t)_i \vec{\nabla}((v_t)_i).$$

This equation is a new kind of regularized compressible Euler equation, called by Marsden the template matching equation (TME). The left hand side is the derivative along the flow of the momentum, as a measure, and the right hand side is the force term.

A wonderful fact about this equation is that by making the momentum singular, we get very nice equations for geodesics on the $\mathcal{G}_n$-homogeneous spaces:

(a) $\mathcal{L}_n$ = set of all N-tuples of distinct points in $\mathbb{R}^n$ and

(b) $\mathcal{S}_n$ = set of all images of the unit ball under a diffeomorphism.

In the first case, we have $\mathcal{L}_n \cong \mathcal{G}_n/\mathcal{G}_{n,0}$ where $\mathcal{G}_{n,0}$ is the stabilizer of a specific set $\{P_1^{(0)}, \cdots, P_N^{(0)}\}$ of N distinct points. To get geodesics on $\mathcal{L}_n$, we look for 'particle solutions of the TME', i.e.

$$\vec{u}_t = \sum_{i=1}^{N} \vec{u}_i(t)\delta_{P_i(t)}$$

where $\{P_1(t), \cdots, P_N(t)\}$ is a path in $\mathcal{L}_n$ The geodesics on $\mathcal{G}_n$, which are perpendicular to all cosets $\theta\mathcal{G}_{n,0}$, are then the geodesics on $\mathcal{L}_n$ for the quotient metric:

$$\begin{aligned} \mathrm{dist}(\{P_i\}, \{P_i + \epsilon v_i\})^2 &= \epsilon^2 \inf_{\left(\substack{v \text{ on } \mathbb{R}^n \\ v(P_i)=v_i}\right)} \int < Lv, v > \\ &= \epsilon^2 \sum_{i,j} G_{ij}(v_i \cdot v_j) \end{aligned}$$

where $G = K(||P_i - P_j||)^{-1}$. For these we get the interesting Hamiltonian ODE:

$$\frac{dP_i}{dt} = 2\sum_j K(||P_i - P_j||)\vec{u}_j$$

$$\frac{du_i}{dt} = -\sum_j \nabla_{P_i} K(||P_i - P_j||) \cdot (\vec{u}_i \cdot \vec{u}_j)$$

which makes points traveling in the same direction attract each other and points going in opposite directions repel each other. This space leads to a non-linear version of the theory of landmark points and shape statistics of Kendall [Sm] and has been developed by Younes [Yo].

A similar treatment can be made for the space of shapes $\mathcal{S}_n \cong \mathcal{G}_n/\mathcal{G}_{n,1}$, where $\mathcal{G}_{n,1}$ is the stabilizer of the unit sphere. Geodesics on $\mathcal{S}_n$ come from solutions of the TME for which $\vec{u}_t$ is supported on the boundary of the shape and perpendicular to it. Even though the first of these spaces $\mathcal{S}_2$ might seem to be quite a simple space, it seems to have a remarkable global geometry reflecting the many perceptual distinctions which we make when we recognize a similar shapes, e.g. a cell decomposition reflecting the different possible graphs which can occur as the 'medial axis' of the shape. This is an area in which I anticipate interesting results. We can also use these Riemannian structures to define Brownian motion on $\mathcal{G}_n, \mathcal{S}_n$ and $\mathcal{L}_n$ (see [D-G-M], [Yi]). Putting a random stopping time on this walk, we get probability measures on these spaces. To make the ideas more concrete, in figure 6 we show a simulation of the random walk on $\mathcal{S}_2$.

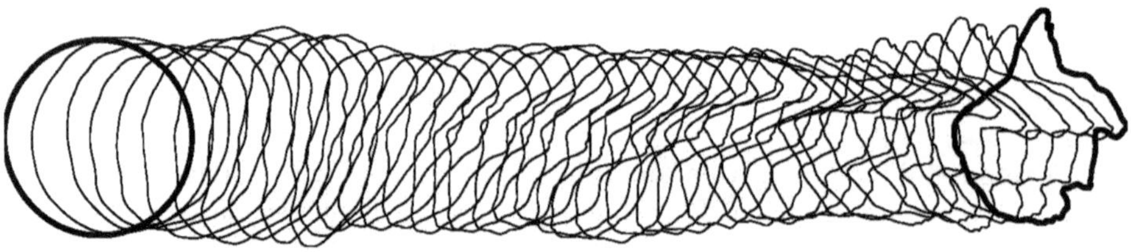

Figure 6: An example of a random walk in the space of 2D shapes $\mathcal{S}_2$. The initial point is the circle on the left. A constant translation to the right has been added so the figures can be distinguished. The operator L defining the metric is $(I - \triangle)^2$

5. Final thoughts

The patterns which occur in nature's sensory signals are complex but allow mathematical modeling. Their study has gone through several phases. At first, 'off-the-shelf' classical models (e.g. linear Gaussian models) were adopted based only on intuition about the variability of the signals. Now, however, two things are happening: computers are large enough to allow massive data gathering to support fully non-parametric models. And the issues raised by these models are driving the study of new areas of mathematics and the development of new algorithms for working with these models. Applications like general purpose speech recognizers and computer driven vehicles are likely in the foreseeable future. Perhaps the ultimate dream is a fully unsupervised learning machine which is given only signals from the world and which finds their statistically significant patterns with no assistance: something like a baby in its first 6 months.

References

[B-G-P] E Bienenstock, S Geman and D Potter, Compositionality, MDL priors and object recognition, *Adv. in Neural Information Proc.*, Mozer, Jordan and Petsche ed., **9**, MIT Press, 1998.

[C-R-M] G Christensen, RD Rabbitt and M Miller, 3D brain mapping using a deformable neuroanatomy, *Physics in Med. and Biol.*, **39**, 1994.

[D-F-G] A Doucet, N. de Freitas and N Gordon editors, *Sequential Monte Carlo Methods in Practice*, Springer, 2001.

[D-G-M] P Dupuis, U Grenander and M Miller, Variational problems on flows of diffeomorphisms for image matching, *Quarterly Appl. Math.*, **56**, 1998.

[G] U Grenander, *General Pattern Theory*, Oxford Univ. Press, 1993.

[G-G] S Geman and D Geman, Stochastic relaxation, Gibbs distr. and Bayesian restoration of images, **PAMI** = *IEEE Trans. Patt. Anal. Mach. Int.*, **6**, 1984.

[Gr-M] U Grenander and M Miller, Computational anatomy, *Quart. of Appl. Math.*, **56**, 1998.

[Gu-M] F Guichard and J-M Morel, *Image Anal. and Partial Diff. Equations*, `http://www.ipam.ucla.edu/publications/gbm2001/gbmtut_jmorel.pdf`.

[G-S-S] N Gordon, D Salmond and A Smith, Novel approach to nonlinear/non-Gaussian Bayesian state estimation, *IEE Proc.-F*, **140**, 1993.

[H] L.C.G. Haggerty, What and two-and-a-half year old child said in one day, *J. Genet. Psych.*, **37**, 1930.

[I-B] M Isard and A Blake, Contour tracking by stochastic propagation of conditional density, *Eur. Conf. Comp. Vis.*, 1996.

[L] P Lieberman, *The Biology and Evolution of Language*, Harvard, 1984.

[L-M-H] A Lee, D Mumford and J Huang, An occlusion model for natural images, *Int. J. Comp. Vis.*, **41**, 2001.

[L-P-M] A Lee, K Pederson and D Mumford, The non-linear statistics of high-contrast patches in natural images, to appear, *Int. J. Comp. Vis.*.

[M] D Mumford, The Dawning of the Age of Stochasticity, in *Mathematics, Frontiers and Perspectives*, ed. by Arnold, Atiyah, Lax and Mazur, AMS, 2000.

[M-G] D Mumford and B Gidas, Stochastic models for generic images, *Quarterly of Appl. Math.* , **59**, 2001.

[M-S] J-M Morel and S Solimini, *Variational Methods in Image Segmentation*, Birkhauser, 1995.

[P] J Pearl, *Probabilistic Reasoning in Int. Systems*, Morgan-Kaufmann, 1997.

[P-B] J Puzicha and J Buhmann, Multiscale annealing for grouping and texture segmentation, *Comp. Vis. and Image Understanding*, **76**, 1999.

[Sh] S Shieber, *Constraint-Based Grammar Formalisms*, MIT Press, 1992.

[Sm] C Small, *The Statistical Theory of Shape*, Springer, 1996.

[S-B-B] E Sharon, A Brandt and R Basri, Segmentation and boundary detection using multiscale intensity measurements, *Proc IEEE Conf. Comp. Vis. Patt. Recognition*, Hawaii, 2001.

[S-M] J Shi and J Malik, Normalized cuts and image segmentation, *PAMI*, **22**, 2000.

[T-Z] -W Tu and S-C Zhu, Image segmentation by data driven Markov chain Monte Carlo, *PAMI*, **24**, 2002.

[V] V Vapnik, *The Nature of Statistical Learning Theory*, Springer, 199?.

[W] R.M. Warren, Restoration of missing speech sounds, *Science*, **167**, 1970.

[Yi] N-K Yip, Stoch. motion by mean curv., *Arch. Rat. Mech. Anal.*, **144**, 1998.

[Yo] L Younes, *Invariance, Déformations et Reconnaissance de Formes*, to appear.

[Y-F-W] J Yedidia, W Freeman, and Y Weiss, Generalized Belief Propagation, *Adv. in Neural Information Proc.*, edited by Leen, Dietterich, Tresp, **13**, 2001.

[Z-Y] S-C Zhu and A Yuille, Region competition, *PAMI*, **18**, 1996.

Curriculum Vitae
Sergei P. Novikov

Date of Birth: March 20, 1938, Nizny Novgorod, USSR

Education:

1955–1960 Moscow State University, Department of Mathematics and Mechanics

1960 Diploma from the Department of Mathematics and Mechanics of Moscow State University (Thesis title: "Homotopy properties of Thom complexes", Prof. M. M. Postnikov — Adviser)

1960–1963 Graduate Studies, the Steklov Institute of Mathematics (Prof. M. M. Postnikov — Adviser)

1964 Candidate of Science (Ph.D.) in Physics and Mathematics (Thesis title: "Differentiable sphere bundles")

1965 Doctor of Science in Physics and Mathematics (Thesis title: "Homotopy equivalent smooth manifolds")

Employment History:

1963–1975 A member of the Steklov Institute of Mathematics, junior researcher till 1965, senior researcher after 1965

1965– A member of the Department of Mathematics and Mechanics of Moscow State University, The Head of the Chair of Differential Geometry, full professor since 1967

1971–1993 Head of the Mathematics Group at the L. D. Landau Institute of Theoretical Physics, Academy of Sciences of the USSR, after 1993 — Principal Researcher in the same Institute

1983 Head of the Chair in Higher Geometry and Topology of Moscow State University

1984 Head of the Group in Geometry and Topology of the Steklov Mathematical Institute of the Academy of Sciences of the USSR

February 1991–August 1991 Research Professor, Laboratory of Theoretical Physics, Ec. Norm. Sup. de Paris, France

1992–1996 University of Maryland at College Park, visiting professor

1996– Full professor IPST and Math Department, University of Maryland, Distinguished Professor since 1997

June 2000, June 2001 and November 2002 Visiting Distinguished Professor of KIAS, (South) Korean Republic, Seoul

1 February–30 March and 10 May–10 June 2009 Newton Institute for Math Sciences, Cambridge, UK: a participant of the Program "Discrete Integrable Systems"

Special Service:

1983–1986 and 2000–2002	Member of Fields Medal Committees of The International Mathematical Union (for the International Mathematical Congresses in Berkeley, 1986 and Beigin, 2002)
1985–1996	President of the Moscow Mathematical Society
1986–1990	Vice-President of the International Association of Mathematical Physics
1986–	Editor-in-Chief of the Journal "Uspekhi Math Nauk" (= "Russia Math Surveys")
1967–1972	Member of Lenin Komsomol Prize Committee for young scientists; Chairman of the Expert Group in Mathematics, Mechanics and Informatics
1983–1988	Member of Expert Committee in Mathematics, Mechanics and Informatics of the Highest Attestation Committee (VAK USSR)
1983–2007	Member of International Lobachevski Prize Committee of USSR/ Russian Academy of Sciences; Chairman after 1991
1985–1991	Member of Expert Group in Mathematics, Mechanics and Informatics of Lenin and State Prize Committee of USSR
1983–2001	Member of Expert Group in Mathematics, Mechanics and Informatics of State Prize Committee of Russian Federation
1984–1991	Head of the Geometry/Topology Problem Committee at the Mathematical Division of the Academy of Sciences of USSR
1994–1996	Member of the Program Committee of the European Math. Society (for the 2nd European Math Congress, Budapest, July 1996)
1995–1998	Member of the Program Committee of the International Mathematical Union (for the International Mathematical Congress, Berlin, August 1998)
1993–1998	Head of the Expert Committee in Mathematics, Mechanics, Informatics in the Russian Foundation for the Fundamental Research (RFFR)
2001–2002	Co-chair (with P.-L. Lions) of the International Program Committee for the European Conference in Applied Mathematics/Applications of Mathematics, AMAM2003 (Nice, 2003)

2006–2007	Member of the Shaw Prize Committee
2008	Member of the Abel Prize Committee

Awards and Honors:

1964	Moscow Math Society Award for young mathematicians
1966–1981	Corresponding member of the Academy of Sciences of the USSR
1967	Lenin Prize
1970	Fields Medal of the International Mathematical Union
1981	Lobachevskii International Prize of the Academy of Sciences of the USSR
1981	Full Member of the Academy of Sciences of the USSR
1987	Honorary Member of the London Math. Society
1988	Honorary Member of the Serbian Academy of Art and Sciences
1988	Doctor Honoris Causa, University of Athens
1991	Foreign Member, "Academia de Lincei", Italy
1993	Academia Europea, member
1994	Foreign associate, National Academy of Sciences, USA
1996	Member, Pontifical Academy of Sciences (Vatican)
1997	Distinguished University Professor, University of Maryland at College Park

1998 Conferences in Honor of 60th birthday:
Solitons, Geometry, Topology: On the Crossroads,

(a) Steklov Math Institute and Landau Institute for Theor Physics, Moscow, Russia, May 26–31, 1998
(b) University of Maryland at College Park, College Park, MD, September 24–26, 1998.

1999	Doctor Honoris Causa, University of Tel Aviv.
2003	Fellow, European Academy of Sciences, Brussels
2005	Wolf Prize in Mathematics
2008	Pogorelov Prize of the Ukranian National Academy of Sciences (NANU)
2009	Bogoliubov Gold Medal of the Russian Academy of Sciences and Dubna Institute for the Nuclear Research

Some Selected Honorable Invited Talks:

1978 Plenary Speaker of the International Mathematical Congress, Helsinki (Theory of Solitons and Algebraic Geometry)

1966 Invited Speaker of the International Mathematical Congress, Moscow, Section of Topology (Presented to the Congress preprint of the lecture "Pontryagin Classes, the Fundamental Group and Some Problems of the Stable Algebra" — Later published in the special edition dedicated to 70th birthday of Georges de Rham; actually made talk in the Cobordism Theory)

1970 Invited Speaker of the International Mathematical Congress, Nice, Section of Topology ("Hermitian Analog of the K-theory and Hamiltonian Formalism"; has not been permitted to attend Congress personally as a punishment for the letters supporting dissidents; the lecture has been read by other person and published in the Materials of the Congress)

1977, 1981, 1986, 1988 Invited Plenary Speaker of the International Congresses in Mathematical Physics in Rome, W.Berlin, Marceille and Swansea

1992 Fermi Lectures, Scuola Normale Superior di Pisa, "Solitons and Geometry" (Cambridge University Press, 1994)

1994 Leonardo da Vinci Lecture, University of Milan, "Algebraic Geometry and Solitons"

2000 Pollack Distinguished Lectures Series, Haifa, Technion, Israel, "2D Schrodinger Operators and Discrete Spectral Symmetries", "Operators on Graphs and Symplectis Geometry", "Topological Phenomena in Normal Metals"

List of Publications

1. S. P. Novikov, Qualitative theory of dynamical systems and foliations in the Moscow mathematical school in the first half of the 1960s (dedicated to the memory of V. I. Arnol'd) (in Russian), Appendix: "Some questions for S. P. Novikov" — An interview conducted by V. M. Bukhshtaber, *Usp. Mat. Nauk* **65** (2010), No. 4(394), 201–207.
2. P. G. Grinevich, A. E. Mironov and S. P. Novikov, The two-dimensional Schrödinger operator: Evolution (2 + 1)-systems and their new reductions, the two-dimensional Burgers hierarchy, and inverse problem data (in Russian), *Usp. Mat. Nauk* **65** (2010), No. 3(393), 195–196; translation in *Russ. Math. Surv.* **65** (2010), No. 3, 580–582.
3. P. G. Grinevich and S. P. Novikov, Singular finite-gap operators and indefinite metrics (in Russian), *Usp. Mat. Nauk* **64** (2009), No. 4(388), 45–72; translation in *Russ. Math. Surv.* **64** (2009), No. 4, 625–650.
4. G. S. Golitsin, R. A. Minlos, S. P. Novikov, V. M. Tikhomirov and M. I. Fortus, Akiva Moiseevich Yaglom (in Russian), *Usp. Mat. Nauk* **63** (2008), No. 2(380), 153–156; translation in *Russ. Math. Surv.* **63** (2008), No. 2, 351–355.
5. Yu. S. Osipov and L. D. Faddeev, Sergeĭ Petrovich Novikov (on the occasion of his seventieth birthday) (in Russian), *Usp. Mat. Nauk* **63** (2008), No. 2(380), 3–4; translation in *Russ. Math. Surv.* **63** (2008), No. 2, 201–203.
6. V. V. Kozlov, The Novikov Day at the Steklov Institute of Mathematics (in Russian), *Usp. Mat. Nauk* **63** (2008), No. 6(384), 3–6; translation in *Russ. Math. Surv.* **63** (2008), No. 6, 995–997.
7. S. P. Novikov, *Dynamical Systems and Differential Forms. Low Dimensional Hamiltonian Systems*, Geometric and probabilistic structures in dynamics, pp. 271–287, Contemp. Math., Vol. 469 (Amer. Math. Soc., 2008).
8. *Geometry, Topology, and Mathematical Physics*. Selected papers from S. P. Novikov's Seminar held in Moscow, 2006–2007, eds. V. M. Buchstaber and I. M. Krichever, American Mathematical Society Translations, Series 2, Vol. 224, Advances in the Mathematical Sciences, Vol. 61 (Amer. Math. Soc., 2008).
9. P. G. Grinevich and S. P. Novikov, Reality problems in the soliton theory, in *Probability, Geometry and Integrable Systems*, pp. 221–239, Math. Sci. Res. Inst. Publ., Vol. 55 (Cambridge Univ. Press, 2008).

10. *Topological Library*. Part 1: *Cobordisms and Their Applications*. Translation by V. O. Manturov, eds. S. P. Novikov and I. A. Taimanov, Series on Knots and Everything, Vol. 39 (World Scientific, 2007).
11. Deformations and contractions in mathematics and physics, Abstracts from the workshop held January 15–21, 2006. Organized by A. Fialowski, M. de Montigny, S. P. Novikov and M. Schlichenmaier. Oberwolfach Reports. Vol. 3, No. 1. *Oberwolfach Rep.* **3** (2006), No. 1, 119–186.
12. S. P. Novikov and I. A. Taimanov, *Modern Geometric Structures and Fields*, translated from the 2005 Russian original by Dimitry Chibisov. Graduate Studies in Mathematics, Vol. 71 (Amer. Math. Soc., 2006).
13. A. Ya. Maltsev and S. P. Novikov, Topology, quasiperiodic functions, and the transport phenomena, in *Topology in Condensed Matter*, pp. 31–59, Springer Ser. Solid-State Sci., Vol. 150 (Springer, 2006).
14. *Topologicheskaya biblioteka.* Tome III. Spektral'nye posledovatel'nosti v topologii (in Russian) [Topology library. Vol. III. Spectral sequences in topology], eds. S. P. Novikov and I. A. Taĭmanov. Institut Komp'yuternykh Issledovaniĭ, Izhevsk, 2005.
15. *Topologicheskaya biblioteka.* Tome II. Kharakteristicheskie klassy i gladkie struktury na mnogoobraziyakh (in Russian) [Topology library. Vol. II. Characteristic classes and smooth structures on manifolds], eds. S. P. Novikov and I. A. Taĭmanov. Institut Komp'yuternykh Issledovaniĭ, Izhevsk, 2005.
16. *Topologicheskaya biblioteka.* Tome I. Kobordizmy i ikh prilozheniya (in Russian) [Topology library. Vol. I. Cobordisms and their applications], eds. S. P. Novikov and I. A. Taĭmanov. Institut Komp'yuternykh Issledovaniĭ, Izhevsk, 2005.
17. S. P. Novikov, Topology of generic Hamiltonian foliations on Riemann surfaces, *Mosc. Math. J.* **5** (2005), No. 3, 633–667, 743.
18. S. P. Novikov, On metric-independent exotic homology (in Russian), *Tr. Mat. Inst. Steklova* **251** (2005), Nelinein. Din., 215–222; translation in *Proc. Steklov Inst. Math.* (2005), No. 4(251), 206–212.
19. S. P. Novikov, The Schrödinger equation and symplectic geometry, in *Surveys in Modern Mathematics*, pp. 203–210, London Math. Soc. Lecture Note Ser., Vol. 321 (Cambridge Univ. Press, 2005).
20. I. A. Dynnikov and S. P. Novikov, Topology of quasiperiodic functions on the plane (in Russian), *Usp. Mat. Nauk* **60** (2005), No. 1(361), 3–28; translation in *Russ. Math. Surv.* **60** (2005), No. 1, 1–26.
21. B. A. Dubrovin, I. M. Krichever and S. P. Novikov, Topological and algebraic geometry methods in contemporary mathematical physics,

in *Classic Reviews in Mathematics and Mathematical Physics*, Vol. 2 (Cambridge Scientific Publishers, 2004).
22. S. P. Novikov, Discrete connections and linear difference equations (in Russian), *Tr. Mat. Inst. Steklova* **247** (2004), Geom. Topol. i Teor. Mnozh., 186–201; translation in *Proc. Steklov Inst. Math.* (2004), No. 4(247), 168–183.
23. S. P. Novikov, Algebraicheskaya topologiya (in Russian) [Algebraic topology] Available electronically at http://www.mi.ras.ru/spm/pdf/004.pdf. Sovremennye Problemy Matematiki [Current Problems in Mathematics], 4. Rossiĭskaya Akademiya Nauk, Matematicheskiĭ Institut im. V. A. Steklova, Moscow, 2004.
24. V. A. Marchenko, S. P. Novikov and I. V. Ostrovskiĭ *et al.*, Mikhail Iosifovich Kadets (on the occasion of his eigthieth birthday) (in Russian), *Usp. Mat. Nauk* **59** (2004), No. 5(359), 183–185; translation in *Russ. Math. Surv.* **59** (2004), No. 5, 1001–1004.
25. S. P. Novikov, Topology in the 20th century: A view from the inside (in Russian), *Usp. Mat. Nauk* **59** (2004), No. 5(359), 3–28; translation in *Russ. Math. Surv.* **59** (2004), No. 5, 803–829.
26. A. Ya Maltsev and S. P. Novikov, Dynamical systems, topology, and conductivity in normal metals, *J. Statist. Phys.* **115** (2004), No. 1–2, 31–46.
27. *Geometry, Topology, and Mathematical Physics*. Selected papers from S. P. Novikov's Seminar held in Moscow, 2002–2003, eds. V. M. Buchstaber and I. M. Krichever, American Mathematical Society Translations, Series 2, Vol. 212, Advances in the Mathematical Sciences, Vol. 55 (Amer. Math. Soc., 2004).
28. V. M. Buchstaber, Interview with S. P. Novikov, in *Geometry, Topology, and Mathematical Physics*, pp. 25–40, Amer. Math. Soc. Transl. Ser. 2, Vol. 212 (Amer. Math. Soc., 2004).
29. S. P. Novikov, The second half of the 20th century and its conclusion: Crisis in the physics and mathematics community in Russia and in the West. Translated from Istor.-Mat. Issled. (2) No. 7(42) (2002), 326–356, 369; by A. Sossinsky. Amer. Math. Soc. Transl. Ser. 2, Vol. 212, *Geometry, Topology, and Mathematical Physics*, pp. 1–24 (Amer. Math. Soc., 2004).
30. *Topology. II. Homotopy and Homology. Classical Manifolds*, eds. S. P. Novikov and V. A. Rokhlin. Encyclopaedia of Mathematical Sciences, Vol. 24 (Springer-Verlag, 2004).
31. V. M. Buchstaber, Yu. S. Ilyashenko, I. M. Krichever, O. K. Sheinman, A. B. Sossinsky and M. A. Tsfasman, Sergey Petrovich Novikov, *Mosc. Math. J.* **3** (2003), No. 4, 1206–1208.

32. I. A. Dynnikov and S. P. Novikov, Geometry of the triangle equation on two-manifolds. Dedicated to Vladimir I. Arnol'd on the occasion of his 65th birthday, *Mosc. Math. J.* **3** (2003), No. 2, 419–438, 742.
33. P. G. Grinevich and S. P. Novikov, Topological phenomena in the real periodic sine–Gordon theory. Integrability, topological solitons and beyond, *J. Math. Phys.* **44** (2003), No. 8, 3174–3184.
34. I. M. Krichever and S. P. Novikov, A two-dimensionalized Toda chain, commuting difference operators, and holomorphic vector bundles (in Russian), *Usp. Mat. Nauk* **58** (2003), No. 3(351), 51–88; translation in *Russ. Math. Surv.* **58** (2003), No. 3, 473–510.
35. A. Ya. Maltsev and S. P. Novikov, Quasiperiodic functions and dynamical systems in quantum solid state physics. Dedicated to the 50th anniversary of IMPA, *Bull. Braz. Math. Soc. (N.S.)* **34** (2003), No. 1, 171–210.
36. Yu. Ilyashenko, I. Krichever, S. P. Novikov, M. A. Tsfasman and V. Vassiliev, Victor Buchstaber, *Mosc. Math. J.* **3** (2003), No. 1, 257.
37. P. G. Grinevich and S. P. Novikov, Topological charge of the real periodic finite-gap sine–Gordon solutions. Dedicated to the memory of *Jürgen K. Moser. Comm. Pure Appl. Math.* **56** (2003), No. 7, 956–978.
38. S. P. Novikov, The second half of the 20th century and its results: The crisis of the society of physicists and mathematicians in Russia and in the West (in Russian), *Istor.-Mat. Issled.* (2) No. 7(42) (2002), 326–356, 369.
39. P. G. Grinevich and S. P. Novikov, Real finite-gap solutions of the sine-Gordon equation: A formula for the topological charge (in Russian), *Usp. Mat. Nauk* **56** (2001), No. 5(341), 181–182; translation in *Russ. Math. Surv.* **56** (2001), No. 5, 980–981.
40. B. A. Dubrovin, I. M. Krichever and S. P. Novikov, Integrable systems. I, in *Dynamical Systems*, IV, pp. 177–332, Encyclopaedia Math. Sci., Vol. 4 (Springer, 2001).
41. *Dynamical Systems*. IV. *Symplectic Geometry and Its Applications*. A translation of *Current Problems in Mathematics*, Fundamental directions, Vol. 4 (in Russian), Akad. Nauk SSSR, Vsesoyuz. Inst. Nauchn. i Tekhn. Inform., Moscow, 1985. Translated by G. Wasserman. Translation edited by V. I. Arnol'd and S. P. Novikov. Second expanded and revised edition, Encyclopaedia of Mathematical Sciences, Vol. 4 (Springer-Verlag, 2001).
42. A. Ya. Maltsev and S. P. Novikov, On the local systems Hamiltonian in the weakly non-local Poisson brackets, *Phys. D* **156** (2001), No. 1–2, 53–80.
43. V. G. Boltyanskiĭ and V. A. Efremovich, *Intuitive Combinatorial Topology*, with an introduction by S. P. Novikov. Translated from the 1982 Russian original by Abe Shenitzer with the editorial assistance of John Stillwell (Springer-Verlag, 2001).

44. S. P. Novikov, Classical and modern topology, Topological phenomena in real world physics. GAFA 2000 (Tel Aviv, 1999). *Geom. Funct. Anal.* (2000), Special Volume, Part I, 406–424.
45. S. P. Novikov, V. V. Sazonov and L. D. Faddeev, Yuriĭ Vasil'evich Prokhorov (on the occasion of his seventieth birthday) (in Russian), *Usp. Mat. Nauk* **55** (2000), No. 5(335), 187–190; translation in *Russ. Math. Surv.* **55** (2000), No. 5, 1009–1013.
46. B. I. Botvinnik, V. M. Bukhshtaber, S. P. Novikov and S. A. Yuzvinskiĭ, Algebraic aspects of multiplication theory in complex cobordisms (in Russian), *Usp. Mat. Nauk* **55** (2000), No. 4(334), 5–24; translation in *Russ. Math. Surv.* **55** (2000), No. 4, 613–633.
47. S. P. Novikov, Pseudo-history and pseudo-mathematics: Fantasy in our life. Comment on: "Linguistics in the sense of A. T. Fomenko" [*Usp. Mat. Nauk* **55** (2000), No. 2, 162–188] by A. A. Zaliznyak (in Russian), *Usp. Mat. Nauk* **55** (2000), No. 2(332), 159–161; translation in *Russ. Math. Surv.* **55** (2000), No. 2, 365–368.
48. S. I. Adyan, E. I. Zel'manov, G. A. Margulis, S. P. Novikov, A. S. Rapinchuk, L. D. Faddeev and V. I. Yanchevskiĭ, Vladimir Petrovich Platonov (on the occasion of his sixtieth birthday) (in Russian), *Usp. Mat. Nauk* **55** (2000), No. 3(333), 197–204; translation in *Russ. Math. Surv.* **55** (2000), No. 3, 601–610.
49. I. M. Krichever and S. P. Novikov, Holomorphic bundles and commuting difference operators. Two-term constructions (in Russian), *Usp. Mat. Nauk* **55** (2000), No. 3(333), 181–182; translation in *Russ. Math. Surv.* **55** (2000), No. 3, 586–588.
50. I. M. Krichever and S. P. Novikov, Holomorphic bundles and difference scalar operators: Single-point constructions (in Russian), *Usp. Mat. Nauk* **55** (2000), No. 1(331), 187–188; translation in *Russ. Math. Surv.* **55** (2000), No. 1, 180–181.
51. S. P. Novikov, Surgery in the 1960s, in *Surveys on Surgery Theory*, Vol. 1, pp. 31–39, Ann. of Math. Stud., Vol. 145 (Princeton Univ. Press, 2000).
52. I. M. Krichever and S. P. Novikov, Trivalent graphs and solitons (in Russian), *Usp. Mat. Nauk* **54** (1999), No. 6(330), 149–150; translation in *Russ. Math. Surv.* **54** (1999), No. 6, 1248–1249.
53. A. D. Aleksandrov, S. S. Kutateladze and S. P. Novikov, Yuriĭ Grigor'evich Reshetnyak (on the occasion of his seventieth birthday) (in Russian), *Usp. Mat. Nauk* **54** (1999), No. 5(329), 191–196; translation in *Russ. Math. Surv.* **54** (1999), No. 5, 1069–1075.

54. S. P. Novikov, Levels of quasiperiodic functions on a plane, and Hamiltonian systems (in Russian), *Usp. Mat. Nauk* **54** (1999), 148; translation in *Russ. Math. Surv.* **54** (1999), No. 5, 1031–1032.
55. S. P. Novikov, Schrödinger operators on graphs and symplectic geometry, in *The Arnol'dfest* (Toronto, ON, 1997), pp. 397–413, Fields Inst. Commun., Vol. 24 (Amer. Math. Soc., 1999).
56. I. Krichever and S. P. Novikov, Periodic and almost-periodic potentials in inverse problems, *Inverse Problems* **15** (1999), No. 6, R117–R144.
57. V. M. Bukhshtaber and A. A. Mal'tsev, From the editors (in Russian), *Tr. Mat. Inst. Steklova* **225** (1999), Solitony Geom. Topol. na Perekrest., 7–10; translation in *Proc. Steklov Inst. Math.* (1999), No. 2 (225), 1–4.
58. K 60-letiyu so dnya rozhdeniya akademika Sergeya Petrovicha Novikova. [Dedicated to Academician Sergeĭ Petrovich Novikov on the occasion of his 60th birthday] (in Russian), eds. V. M. Bukhshtaber and A. A. Mal'tsev, *Tr. Mat. Inst. Steklova* **225** (1999), Solitony Geom. Topol. na Perekrest., pp. 1–400; translation in *Proc. Steklov Inst. Math.* (1999), No. (225), 1–381.
59. V. A. Rokhlin, Izbrannye raboty (in Russian), [Selected works] Vospominaniya o V. A. Rokhline. [Appendix: Reminiscences of V. A. Rokhlin by V. I. Arnol'd, A. M. Vershik, S. P. Novikov and Ya. G. Sinai] With commentaries on Rokhlin's work by N. Yu. Netsvetaev and Vershik. Edited and with a preface by Vershik. Moskovskiĭ Tsentr Nepreryvnogo Matematicheskogo Obrazovaniya, Moscow (1999).
60. S. P. Novikov, The discrete Schrödinger operator (in Russian), *Tr. Mat. Inst. Steklova* **224** (1999), Algebra. Topol. Differ. Uravn. i ikh Prilozh., 275–290; translation in *Proc. Steklov Inst. Math.* (1999), No. 1(224), 250–265.
61. S. P. Novikov (on the occasion of his sixtieth birthday) (in Russian), *Usp. Mat. Nauk* **54** (1999), No. 1(325), 5–10; translation in *Russ. Math. Surv.* **54** (1999), No. 1, 1–7.
62. S. P. Novikov and A. S. Shvarts, Discrete Lagrangian systems on graphs. Symplecto-topological properties (in Russian), *Usp. Mat. Nauk* **54** (1999), No. 1(325), 257–258; translation in *Russ. Math. Surv.* **54** (1999), No. 1, 258–259.
63. S. P. Novikov, Discrete Schrödinger operators and topology. Mikio Sato: A great Japanese mathematician of the twentieth century, *Asian J. Math.* **2** (1998), No. 4, 921–933.
64. N. A. Bobylev, E. A. Gorin, A. Yu. Ishlinskiĭ, S. P. Novikov and V. M. Tikhomirov, Mark Aleksandrovich Krasnosel'skiĭ (in Russian), *Usp. Mat. Nauk* **53** (1998), No. 1(319), 199–201; translation in *Russ. Math. Surv.* **53** (1998), No. 1, 195–198.

65. V. M. Bukhshtaber and S. P. Novikov, On the history of the N. I. Lobachevskiĭ prize (on the centenary of its first awarding in 1897) (in Russian), *Usp. Mat. Nauk* **53** (1998), No. 1(319), 235–238; translation in *Russ. Math. Surv.* **53** (1998), No. 1, 237–241.
66. S. P. Novikov, Role of integrable models in the development of mathematics, in *Fields Medallists' Lectures*, pp. 202–217, World Sci. Ser. 20th Century Math., Vol. 5 (World Scientific, 1997).
67. O. A. Ladyzhenskaya, V. A. Marchenko, Yu. A. Mitropol'skiĭ, S. P. Novikov and A. V. Pogorelov, Evgeniĭ Yakovlevich Khruslov (on the occasion of his sixtieth birthday) (in Russian), *Usp. Mat. Nauk* **52** (1997), No. 6(318), 205–206; translation in *Russ. Math. Surv.* **52** (1997), No. 6, 1351–1353.
68. I. A. Dynnikov and S. P. Novikov, Laplace transformations and simplicial connectivities (in Russian), *Usp. Mat. Nauk* **52** (1997), No. 6(318), 157–158; translation in *Russ. Math. Surv.* **52** (1997), No. 6, 1294–1295.
69. S. P. Novikov, The Schrödinger operator on graphs, and topology (in Russian), *Usp. Mat. Nauk* **52** (1997), No. 6(318), 177–178; translation in *Russ. Math. Surv.* **52** (1997), No. 6, 1320–1321.
70. S. P. Novikov and I. A. Dynnikov, Discrete spectral symmetries of small-dimensional differential operators and difference operators on regular lattices and two-dimensional manifolds (in Russian), *Usp. Mat. Nauk* **52** (1997), No. 5(317), 175–234; translation in *Russ. Math. Surv.* **52** (1997), No. 5, 1057–1116.
71. A. V. Zarelua, A. A. Mal'tsev and S. P. Novikov, Heiner Zieschang (on the occasion of his sixtieth birthday) (in Russian), *Usp. Mat. Nauk* **52** (1997), No. 4(316), 247–250; translation in *Russ. Math. Surv.* **52** (1997), No. 4, 881–885.
72. V. I. Arnol'd, A. A. Bolibrukh, R. V. Gamkrelidze, V. P. Maslov, E. F. Mishchenko, S. P. Novikov, Yu. S. Osipov, Ya. G. Sinai, A. M. Stepin and L. L. Faddeev, Dmitriĭ Viktorovich Anosov (on the occasion of his sixtieth birthday) (in Russian), *Usp. Mat. Nauk* **52** (1997), No. 2(314), 193–200; translation in *Russ. Math. Surv.* **52** (1997), No. 2, 437–445.
73. S. P. Novikov, Mathematical education in Russia: Is it forward-looking? (in Russian), *Voprosy Istor. Estestvoznan. i Tekhn.* (1997), No. 1, 97–106.
74. S. P. Novikov, Mathematics and history (in Russian), *Priroda* (1997), No. 2, 70–7.
75. S. P. Novikov, Algebraic properties of two-dimensional difference operators (in Russian), *Usp. Mat. Nauk* **52** (1997), No. 1(313), 225–226; translation in *Russ. Math. Surv.* **52** (1997), No. 1, 226–227.
76. S. P. Novikov and A. P. Veselov, Exactly solvable two-dimensional Schrödinger operators and Laplace transformations, in *Solitons, Geometry,*

and Topology: On the Crossroad, pp. 109–132, Amer. Math. Soc. Transl. Ser. 2, Vol. 179 (Amer. Math. Soc., 1997).

77. *Solitons, Geometry, and Topology: On the Crossroad*, eds. V. M. Buchstaber [V. M. Bukhshtaber] and S. P. Novikov. American Mathematical Society Translations, Series 2, Vol. 179. Advances in the Mathematical Sciences, Vol. 33 (Amer. Math. Soc., 1997).
78. V. I. Arnol'd, M. I. Vishik, A. S. Kalashnikov, V. P. Maslov, S. M. Nikol'skiĭ, S. P. Novikov, Ol'ga Arsen'evna Oleĭnik (on the occasion of her seventieth birthday) (in Russian), *Tr. Semin. im. I. G. Petrovskogo* No. **19** (1996), 5–25; translation in *J. Math. Sci.* (New York) **85** (1997), No. 6, 2249–2259.
79. O. V. Besov, V. M. Bukhshtaber, A. G. Vitushkin, A. A. Gonchar, S. V. Konyagin, L. D. Kudryavtsev, S. M. Nikol'skiĭ, S. P. Novikov, Yu. S. Osipov, A. Yu. Popov, V. A. Sadovnichiĭ, A. F. Sidorov, Yu. N, Subbotin, S. A. Telyakovskiĭ, P. L. Ul'yanov and N. I. Chernykh, Sergeĭ Borisovich Stechkin (in Russian), *Usp. Mat. Nauk* **51** (1996), No. 6(312), 3–10; translation in *Russ. Math. Surv.* **51** (1996), No. 6, 1007–1014.
80. S. P. Novikov, A correspondence between β-pre-Frattini subalgebras and β-normalizers of multirings (in Russian), *Vestn. Beloruss. Gos. Univ. Ser. 1 Fiz. Mat. Inform.* (1996), No. 1, 46–48, 80.
81. S. P. Novikov, L. A. Bunimovich, A. M. Vershik, B. M. Gurevich, E. I. Dinaburg, G. A. Margulis, V. I. Oseledets, S. A. Pirogov, K. M. Khanin and N. N. Chentsova, Yakov Grigor'evich Sinai (on the occasion of his sixtieth birthday) (in Russian), *Usp. Mat. Nauk* **51** (1996), No. 4(310), 179–191; translation in *Russ. Math. Surv.* **51** (1996), No. 4, 765–778.
82. S. P. Novikov, Theory of the string equation in the double-scaling limit of 1-matrix models, *Int. J. Mod. Phys. B* **10** (1996), Nos. 18–19, 2249–2271.
83. S. P. Novikov, A. I. Aptekarev, E. P. Dolzhenko, V. A. Kalyagin, V. V. Kozlov, Yu, V, Nesterenko, M. K. Potapov, V. N. Sorokin and P. L. Ul'yanov, Evgeniĭ Mikhaĭlovich Nikishin (on the fiftieth anniversary of his birth) (in Russian), *Usp. Mat. Nauk* **51** (1996), No. 2(308), 181–182; translation in *Russ. Math. Surv.* **51** (1996), No. 2, 361–362.
84. S. P. Novikov, Topology, in *Topology*, I, pp. 1–319, Encyclopaedia Math. Sci., Vol. 12 (Springer, 1996).
85. Topology. I. General survey. Translated from *Current Problems in Mathematics, Fundamental Directions*, Vol. 12 (in Russian), *Akad. Nauk SSSR*, Vsesoyuz. Inst. Nauchn. i Tekhn. Inform., Moscow, 1986. Translation by B. Botvinnik and R. G. Burns. Translation edited by S. P. Novikov, Encyclopaedia of Mathematical Sciences, Vol. 12 (Springer-Verlag, 1996).

86. S. P. Novikov, Mathematicians and physicists in the Academy from 1960 to the 1980s (in Russian), *Voprosy Istor. Estestvoznan. i Tekhn.* (1995), No. 4, 55–65, 172.
87. A. P. Veselov and S. P. Novikov, Exactly solvable periodic two-dimensional Schrödinger operators (in Russian), *Usp. Mat. Nauk* **50** (1995), No. 6(306), 171–172; translation in *Russ. Math. Surv.* **50** (1995), No. 6, 1316–1317.
88. O. I. Mokhov, S. P. Novikov and A. K. Pogrebkov, Irina Yakovlevna Dorfman (in Russian), *Usp. Mat. Nauk* **50** (1995), No. 6(306), 151–156; translation in *Russ. Math. Surv.* **50** (1995), No. 6, 1241–1246.
89. N. S. Bakhvalov, S. P. Novikov and A. T. A. T. Fomenko, Ol'ga Arsen'evna Oleĭnik (in Russian), *Usp. Mat. Nauk* **50** (1995), No. 4(304), 177–186; translation in *Russ. Math. Surv.* **50** (1995), No. 4, 837–848.
90. P. G. Grinevich and S. P. Novikov, Nonselfintersecting magnetic orbits on the plane. Proof of the overthrowing of cycles principle, in *Topics in Topology and Mathematical Physics*, pp. 59–82, Amer. Math. Soc. Transl. Ser. 2, Vol. 170 (Amer. Math. Soc., 1995).
91. V. M. Buchstaber and S. P. Novikov, The S. P. Novikov Seminar, in *Topics in Topology and Mathematical Physics*, pp. 1–7, Amer. Math. Soc. Transl. Ser. 2, Vol. 170 (Amer. Math. Soc., 1995).
92. *Topics in Topology and Mathematical Physics*, ed. S. P. Novikov. Translated from the original Russian manuscripts. Translation edited by A. B. Sossinsky [A. B. Sosinskiĭ]. American Mathematical Society Translations, Series 2, Vol. 170. Advances in the Mathematical Sciences, Vol. 27 (Amer. Math. Soc., 1995).
93. Yu. S. Osipov, A. A. Gonchar, S. P. Novikov, V. I. Arnol'd, G. I. Marchuk, P. P. Kulish, V. S. Vladimirov and E. F. Mishchenko, Lyudvig Dmitrievich Faddeev (on the occasion of his sixtieth birthday) (in Russian), *Usp. Mat. Nauk* **50** (1995), No. 3(303), 171–186; translation in *Russ. Math. Surv.* **50** (1995), No. 3, 643–659.
94. V. M. Bukhshtaber and S. P. Novikov, Boris Gershevich Moishezon [1937–1993] (in Russian), *Usp. Mat. Nauk* **50** (1995), No. 3(303), 147–148; translation in *Russ. Math. Surv.* **50** (1995), No. 3, 613–614.
95. S. P. Novikov, The semiclassical electron in a magnetic field and lattice. Some problems of low-dimensional "periodic" topology, *Geom. Funct. Anal.* **5** (1995), No. 2, 434–444.
96. S. P. Novikov, My generation in mathematics (in Russian), *Usp. Mat. Nauk* **49** (1994), No. 6(300), 3–6; translation in *Russ. Math. Surv.* **49** (1994), No. 6, 1–4.

97. P. G. Grinevich and S. P. Novikov, String equation. II. Physical solution (in Russian), *Algebra i Analiz* **6** (1994), No. 3, 118–140; translation in St. Petersburg *Math. J.* **6** (1995), No. 3, 553–574.
98. Yu. M. Berezanskiĭ, V. A. Marchenko and S. P. Novikov, *et al.*, Mikhail Iosifovich Kadets (on the occasion of his seventieth birthday) (in Russian), *Usp. Mat. Nauk* **49** (1994), No. 3(297), 205–206; translation in *Russ. Math. Surv.* **49** (1994), No. 3, 223–225.
99. S. P. Novikov, Solitons and geometry, in *Lezioni Fermiane* [*Fermi Lectures*] Published for the Scuola Normale Superiore, Pisa (Cambridge University Press, 1994).
100. Dynamical systems. VII. Integrable systems, nonholonomic dynamical systems. A translation of *Current Problems in Mathematics*, Fundamental directions, Vol. 16 (Russian), Akad. Nauk SSSR, Vsesoyuz. Inst. Nauchn. i Tekhn. Inform., Moscow, 1987 Translation by A. G. Reyman [A. G. Reĭman] and M. A. Semenov-Tian-Shansky [M. A. Semenov-Tyan-Shanskiĭ]. Translation edited by V. I. Arnol'd and S. P. Novikov. Encyclopaedia of Mathematical Sciences, Vol. 16 (Springer-Verlag, 1994).
101. S. P. Novikov, Differential geometry and hydrodynamics of soliton lattices. Important developments in soliton theory, pp. 242–256, Springer Ser. Nonlinear Dynam. (Springer, 1993).
102. B. A. Dubrovin and S. P. Novikov, Hydrodynamics of soliton lattices, *Sov. Sci. Rev. Sec. C: Math. Phys. Rev.*, Vol. 9, Part 4 (Harwood Academic, 1993).
103. V. A. Zalgaller, S. S. Kutateladze, O. A. Ladyzhenskaya, S. P. Novikov, A. V. Pogorelov and Yu. G. Reshetnyak, Aleksandr Danilovich Aleksandrov (on the occasion of his eightieth birthday) (in Russian), *Usp. Mat. Nauk* **48** (1993), No. 4(292), 239–241; translation in *Russ. Math. Surv.* **48** (1993), No. 4, 257–260.
104. A. A. Gonchar, G. I. Marchuk and S. P. Novikov, Vasiliĭ Sergeevich Vladimirov (on the occasion of his seventieth birthday) (in Russian), *Usp. Mat. Nauk* **48** (1993), No. 1(289), 195–204; translation in *Russ. Math. Surv.* **48** (1993), No. 1, 201–212.
105. S. P. Novikov and A. Ya. Mal'tsev, Liouville form of averaged Poisson brackets (in Russian), *Usp. Mat. Nauk* **48** (1993), No. 1(289), 155–156; translation in *Russ. Math. Surv.* **48** (1993), No. 1, 155–157.
106. S. P. Novikov, Quasiperiodic structures in topology, in *Topological Methods in Modern Mathematics* (Stony Brook, NY, 1991), pp. 223–233 (Publish or Perish, 1993).

107. Topologiya i ee prilozheniya (in Russian) [Topology and its applications] Proceedings of the International Topology Conference held in Baku, October 3–8, 1987, ed. S. P. Novikov. *Trudy Mat. Inst. Steklov*. **193** (1992) (Nauka, 1992).
108. S. P. Novikov, Action-angle variables and algebraic geometry. La Mécanique analytique de Lagrange et son héritage, II (Turin, 1989), *Atti Accad. Sci. Torino Cl. Sci. Fis. Mat. Natur.* **126** (1992), suppl. 2, 139–150.
109. S. P. Novikov, Role of integrable models in the development of mathematics, in *Mathematical Research Today and Tomorrow* (Barcelona, 1991), pp. 13–28, Lecture Notes in Math., Vol. 1525 (Springer, 1992).
110. S. P. Novikov, Various doublings of Hopf algebras. Algebras of operators on quantum groups, complex cobordisms (in Russian), *Usp. Mat. Nauk* **47** (1992), No. 5(287), 189–190; translation in *Russ. Math. Surv.* **47** (1992), No. 5, 198–199.
111. S. P. Novikov, Integrability in mathematics and theoretical physics: Solitons, *Math. Intelligencer* **14** (1992), No. 4, 13–21.
112. S. P. Novikov, Hydrodynamics of soliton lattices: Differential geometry and Hamiltonian formalism. Progress in variational methods in Hamiltonian systems and elliptic equations (L'Aquila, 1990), pp. 144–156, Pitman Res. Notes Math. Ser., Vol. 243 (Longman Sci. Tech., 1992).
113. B. A. Dubrovin, A. T. Fomenko and S. P. Novikov, Modern geometry — Methods and applications. Part I, in *The Geometry of Surfaces, Transformation Groups, and Fields*, 2nd edition. Translated from the Russian by Robert G. Burns, Graduate Texts in Mathematics, Vol. 93 (Springer-Verlag, 1992).
114. Algebra, geometriya i diskretnaya matematika v nelineĭnykh zadachakh (in Russian) [Algebra, Geometry and Discrete Mathematics in Nonlinear Problems], eds. O. B. Lupanov, S. P. Novikov and A. I. Kostrikin (Moskov. Gos. Univ., 1991).
115. S. P. Novikov, Riemann surfaces, operator fields, strings — Analogues of the Fourier–Laurent bases. Common trends in mathematics and quantum field theories (Kyoto, 1990), *Progr. Theor. Phys. Suppl.* No. 102 (1990), 293–300 (1991).
116. S. P. Novikov, On the equation $[L, A] = \varepsilon \cdot 1$. With an appendix by the author and B. A. Dubrovin. Common trends in mathematics and quantum field theories (Kyoto, 1990), *Progr. Theor. Phys. Suppl.* No. 102 (1990), 287–292 (1991).
117. S. P. Novikov and A. T. Fomenko, *Basic Elements of Differential Geometry and Topology*, translated from the Russian by M. V. Tsaplina, Mathematics and its Applications (Soviet Series), Vol. 60 (Kluwer, 1990).

118. I. M. Krichever and S. P. Novikov, Riemann surfaces, operator fields, strings. Analogues of the Fourier–Laurent bases, in *Physics and Mathematics of Strings*, pp. 356–388 (World Scientific, 1990).
119. V. S. Makarov, A. A. Mal'tsev, S. P. Novikov, S. S. Ryshkov, E. S. Tikhomirova and A. S. Shvarts, Vadim Arsen'evich Efremovich (in Russian), *Usp. Mat. Nauk* **45** (1990), No. 6(276), 113–114; translation in *Russ. Math. Surv.* **45** (1990), No. 6, 137–138.
120. S. P. Novikov, Quantization of finite-gap potentials and a nonlinear quasiclassical approximation that arises in nonperturbative string theory (in Russian), *Functional. Anal. i Prilozhen.* **24** (1990), No. 4, 43–53, 96; translation in *Funct. Anal. Appl.* **24** (1990), No. 4, 296–306 (1991).
121. B. A. Dubrovin, A. T. Fomenko and S. P. Novikov, Modern geometry — Methods and applications. Part III, in *Introduction to Homology Theory*, translated from the Russian by Robert G. Burns. Graduate Texts in Mathematics, Vol. 124 (Springer-Verlag, 1990).
122. N. N. Bogolyubov, N. A. Bobylëv, P. P. Zabreĭko, A. Yu. Ishlinskiĭ, V. P. Maslov, Yu. A. Mitropol'skiĭ and S. P. Novikov, Mark Aleksandrovich Krasnosel'skiĭ (on the occasion of his seventieth birthday) (in Russian), *Usp. Mat. Nauk* **45** (1990), No. 2(272), 225–227; translation in *Russ. Math. Surv.* **45** (1990), No. 2, 231–234.
123. A. D. Aleksandrov, S. L. Krushkal, S. S. Kutateladze and S. P. Novikov, Yuriĭ Grigor'evich Reshetnyak (on the occasion of his sixtieth birthday) (in Russian), *Usp. Mat. Nauk* **45** (1990), No. 1(271), 199–204; translation in *Russ. Math. Surv.* **45** (1990), No. 1, 231–238.
124. *Dynamical Systems. IV. Symplectic Geometry and Its Applications.* A translation of Sovremennye problemy matematiki. Fundamental'nye napravleniya, Tome 4, Akad. Nauk SSSR, Vsesoyuz. Inst. Nauchn. i Tekhn. Inform., Moscow, 1985. Translation by G. Wassermann, translation edited by V. I. Arnol'd and S. P. Novikov. Encyclopaedia of Mathematical Sciences, Vol. 4 (Springer-Verlag, 1990).
125. S. P. Novikov, Complex analysis on Riemann surfaces motivated by the operatorial string theory, in Analysis, etc., pp. 501–519 (Academic Press, 1990).
126. V. S. Vladimirov, A. A. Logunov and S. P. Novikov, Nikolaĭ Nikolaevich Bogolyubov (on the occasion of his eightieth birthday) (in Russian), *Usp. Mat. Nauk* **44** (1989), No. 5(269), 4–12; translation in *Russ. Math. Surv.* **44** (1989), No. 5, 1–10.
127. B. A. Dubrovin and S. P. Novikov, Hydrodynamics of weakly deformed soliton lattices. Differential geometry and Hamiltonian theory (in Russian),

Usp. Mat. Nauk **44** (1989), No. 6(270), 29–98, 203; translation in *Russ. Math. Surv.* **44** (1989), No. 6, 35–124.

128. A. D. Aleksandrov, V. A. Marchenko, S. P. Novikov and Yu. G. Reshetnyak, Alekseĭ Vasil'evich Pogorelov (on the occasion of his seventieth birthday) (in Russian), *Usp. Mat. Nauk* **44** (1989), No. 4(268), 245–249; translation in *Russ. Math. Surv.* **44** (1989), No. 4, 217–223.
129. I. M. Krichever and S. P. Novikov, Algebras of Virasoro type, the energy-momentum tensor, and operator expansions on Riemann surfaces (in Russian), *Functional. Anal. i Prilozhen.* **23** (1989), No. 1, 24–40; translated in *Funct. Anal. Appl.* **23** (1989), No. 1, 19–33.
130. Izbrannye voprosy algebry, geometrii i diskretnoĭ matematiki (in Russian) [Selected problems in algebra, geometry and discrete mathematics], eds. A. I. Kostrikin, O. B. Lupanov and S. P. Novikov (Moskov. Gos. Univ., 1988).
131. *Mathematical Physics Reviews*, Vol. 7. Translated from the Russian, eds. S. P. Novikov and Ya. G. Sinai, Sov. Sci. Rev., Sec. C: Math. Phys. Rev., Vol. 7 (Harwood Academic, 1988).
132. I. M. Krichever and S. P. Novikov, Virasoro–Gel'fand–Fuks type algebras, Riemann surfaces, operator's theory of closed strings, *J. Geom. Phys.* **5** (1988), No. 4, 631–661 (1989).
133. L. E. Evtushik, O. B. Lupanov and S. P. Novikov *et al.*, In memory of Anatoliĭ Mikhaĭlovich Vasil'ev (1923–1987) (in Russian), *Vestnik Moskov. Univ. Ser. I Mat. Mekh.* (1988), No. 5, 97–101.
134. O. V. Manturov, S. P. Novikov, L. V. Sabinin, V. V. Trofimov and A. T. Fomenko, In memory of Anatoliĭ Mikhaĭlovich Vasil'ev (1923–1987) (in Russian), *Trudy Sem. Vektor. Tenzor. Anal.* No. **23** (1988), 5–6.
135. B. A. Dubrovin, S. P. Novikov and A. T. Fomenko, Geometria contemporanea. Metodi e applicazioni. Vol. II. (Italian) [*Modern Geometry. Methods and Applications*. Vol. II] Geometria e topologia delle varietà. [Geometry and topology of manifolds] Translated from the second Russian edition by Vitalij Panasenko. Nuova Biblioteca di Cultura. [New Library of Culture] Editori Riuniti, Rome (Mir, 1988).
136. Academician S. P. Novikov (on the occasion of his 50th birthday) (in Russian), *Vestnik Akad. Nauk SSSR* (1988), No. 9, 133–134.
137. S. P. Novikov, Reminiscences about A. N. Kolmogorov (in Russian), *Usp. Mat. Nauk* **43** (1988), No. 6(264), 35–36; translation in *Russ. Math. Surv.* **43** (1988), No. 6, 41–42.
138. N. S. Bakhvalov, V. S. Vladimirov, A. A. Gonchar, L. D. Kudryavtsev, V. I. Lebedev, S. M. Nikol'skiĭ, S. P. Novikov, O. A. Oleĭnik and Yu. G. Reshetnyak, Sergeĭ L'vovich Sobolev (on the occasion of his eightieth

birthday) (in Russian), *Usp. Mat. Nauk* **43**, 3–16; translation in *Russ. Math. Surv.* **43** (1988), No. 5, 1–18.

139. S. P. Novikov, Analytical theory of homotopy groups, in *Topology and Geometry* — Rohlin Seminar, pp. 99–112, Lecture Notes in Math., Vol. 1346 (Springer, 1988).
140. M. A. Akivis, G. Ya. Galin, V. F. Kirichenko, E. A. Morozova, S. P. Novikov and A. S. Fedenko, Anatoliĭ Mikhaĭlovich Vasil'ev (in Russian), *Usp. Mat. Nauk* **43** (1988), No. 4(262), 159–160; translation in *Russ. Math. Surv.* **43** (1988), No. 4, 191–193.
141. S. P. Novikov (on the occasion of his fiftieth birthday) (in Russian), *Usp. Mat. Nauk* **43** (1988), No. 4(262), 3–9; translation in *Russ. Math. Surv.* **43** (1988), No. 4, 1–10.
142. P. G. Grinevich and S. P. Novikov, Inverse scattering problem for the two-dimensional Schrödinger operator at a fixed negative energy and generalized analytic functions, in *Plasma Theory and Nonlinear and Turbulent Processes in Physics*, Vol. 1, 2 (Kiev, 1987), pp. 58–85 (World Scientific, 1988).
143. Yu. F. Borisov, V. A. Zalgaller, S. S. Kutateladze, O. A. Ladyzhenskaya, S. P. Novikov, A. V. Pogorelov, Yu. G. Reshetnyak and S. L. Sobolev, Aleksandr Danilovich Aleksandrov (on the occasion of his seventy-fifth birthday) (in Russian), *Usp. Mat. Nauk* **43** (1988), No. 2(260), 161–167; translation in *Russ. Math. Surv.* **43** (1988), No. 2, 191–199.
144. P. G. Grinevich and S. P. Novikov, A two-dimensional "inverse scattering problem" for negative energies, and generalized-analytic functions. I. Energies lower than the ground state (in Russian), *Functional. Anal. i Prilozhen.* **22** (1988), No. 1, 23–33, 96; translation in *Funct. Anal. Appl.* **22** (1988), No. 1, 19–27.
145. B. A. Dubrovin, S. P. Novikov and A. T. Fomenko, Geometria contemporanea. Metodi e applicazioni. Vol. I. (in Italian) [Modern geometry. Methods and applications. Vol. I] Geometria delle superfici, dei gruppi di trasformazioni e dei campi. [The geometry of surfaces, transformation groups and fields] Translated from the second Russian edition by Vitalij Panasenko. Nuova Biblioteca di Cultura. [New Library of Culture] Editori Riuniti, Rome (Mir, 1987).
146. *Mathematical Physics Reviews*, Vol. 6. Translated from the Russian, eds. S. P. Novikov and Ya. G. Sinai. Soviet Scientific Reviews, Section C: Mathematical Physics Reviews, Vol. 6 (Harwood Academic, 1987).
147. V. I. Arnol'd, M. I. Vishik, Yu. V. Egorov, A. S. Kalashnikov, S. P. Novikov and S. L. Sobolev, Ol'ga Arsen'evna Oleĭnik (on the occasion of her sixtieth birthday) (in Russian), *Trudy Sem. Petrovsk.* No. **12** (1987), 3–21.

148. Yu. M. Berezanskiĭ, N. N. Bogolyubov, Yu. L. Daletskiĭ, P. A. Kuchment, B. Ya. Levin, V. P. Maslov, S. P. Novikov and E. M. Semënov, Selim Grigor'evich Kreĭn (on the occasion of his seventieth birthday) (in Russian), *Usp. Mat. Nauk* **42** (1987), No. 5(257), 223–224.
149. A. G. Vitushkin, A. A. Gonchar, B. S. Kashin, A. I. Kostrikin, S. M. Nikol'skiĭ, S. P. Novikov, P. L. Ul'yanov and L. D. Faddeev, Evgeniĭ Mikhaĭlovich Nikishin (in Russian), *Usp. Mat. Nauk* **42** (1987), No. 5(257), 183–188.
150. I. M. Krichever and S. P. Novikov, Algebras of Virasoro type, Riemann surfaces and strings in Minkowski space (in Russian), *Functional. Anal. i Prilozhen.* **21** (1987), No. 4, 47–61, 96.
151. S. P. Novikov and A. T. Fomenko, Èlementy differentsial'oĭ geometrii i topologii (in Russian) [*Elements of Differential Geometry and Topology*] (Nauka, 1987).
152. S. P. Novikov, Two-dimensional Schrödinger operator and solitons. 3-dimensional integrable systems, *VIII International Congress on Mathe-matical Physics* (Marseille, 1986), pp. 226–241 (World Scientific, 1987).
153. V. V. Avilov, I. M. Krichever and S. P. Novikov, Evolution of the Whitham zone in the Korteweg–de Vries theory (in Russian), *Dokl. Akad. Nauk SSSR* **295** (1987), No. 2, 345–349.
154. I. M. Krichever and S. P. Novikov, Algebras of Virasoro type, Riemann surfaces and the structures of soliton theory (in Russian), *Functional. Anal. i Prilozhen.* **21** (1987), No. 2, 46–63.
155. V. V. Avilov and S. P. Novikov, Evolution of the Whitham zone in KdV theory (in Russian), *Dokl. Akad. Nauk SSSR* **294** (1987), No. 2, 325–329.
156. S. P. Novikov, Topology (in Russian), Current problems in mathematics. Fundamental directions, Vol. 12 (in Russian), pp. 5–252, 322, Itogi Nauki i Tekhniki, Akad. Nauk SSSR, Vsesoyuz. Inst. Nauchn. i Tekhn. Inform., Moscow, 1986.
157. A. N. Kolmogorov, A. A. Mal'tsev and S. P. Novikov, Kirill Aleksandrovich Sitnikov (on the occasion of his sixtieth birthday) (in Russian), *Usp. Mat. Nauk* **41** (1986), No. 6(252), 209–210.
158. V. G. Boltjanskij and V. A. Efremovič, Anschauliche kombinatorische Topologie (in German) [Visual Combinatorial Topology], with a foreword by S. P. Novikov. Translated from the Russian and with a foreword by Detlef Seese and Martin Weese. Mathematische Schülerbücherei [*Mathematical Library for Students*], Vol. 129 (VEB Deutscher Verlag der Wissenschaften, 1986).

159. V. G. Boltjanskij and V. A. Efremovič, Anschauliche kombinatorische Topologie (in German) [Visual Combinatorial Topology] With an appendix by V. P. Mineev. Edited and with a preface by S. P. Novikov. Translated from the Russian and with a preface by Detlef Seese and Martin Weese (Friedr. Vieweg & Sohn, Braunschweig, 1986).
160. B. A. Dubrovin, S. P. Novikov and A. T. Fomenko, Sovremennaya geometriya (in Russian), [*Modern Geometry*] Metody i prilozheniya. [*Methods and Applications*], 2nd edition (Nauka, 1986).
161. A. G. Vitushkin, M. I. Zelikin, N. N. Krasovskiĭ, S. P. Novikov, Yu. S. Osipov and K. A. Sitnikov, Nikolaĭ Terent'evich Tynyanskiĭ (in Russian), *Usp. Mat. Nauk* **41** (1986), No. 4(250), 143–144.
162. S. P. Novikov and M. A. Shubin, Morse inequalities and von Neumann II1-factors (in Russian), *Dokl. Akad. Nauk SSSR* **289** (1986), No. 2, 289–292.
163. V. I. Arnol'd, A. M. Vershik, O. Ya. Viro, A. N. Kolmogorov, S. P. Novikov, Ya. G. Sinai and D. B. Fuks, Vladimir Abramovich Rokhlin (in Russian), *Usp. Mat. Nauk* **41** (1986), No. 3(249), 159–163.
164. S. P. Novikov, Bloch homology. Critical points of functions and closed 1-forms (in Russian), *Dokl. Akad. Nauk SSSR* **287** (1986), No. 6, 1321–1324.
165. S. P. Novikov and A. P. Veselov, Two-dimensional Schrödinger operator: Inverse scattering transform and evolutional equations, Solitons and coherent structures (Santa Barbara, Calif., 1985), *Phys. D* **18** (1986), No. 1-3, 267–273.
166. O. V. Manturov, S. P. Novikov, L. V. Sabinin, V. V. Trofimov and A. T. Fomenko, In memory of Petr Konstantinovich Rashevskiĭ (1907–1983) (in Russian), *Trudy Sem. Vektor. Tenzor. Anal.* No. **22** (1985).
167. A. P. Veselov, I. M. Krichever and S. P. Novikov, Two-dimensional periodic Schrödinger operators and Prym's θ-functions, in *Geometry Today* (Rome, 1984), pp. 283–301, Progr. Math., Vol. 60 (Birkhäuser, 1985).
168. *Mathematical Physics Reviews*, Vol. 5. Translated from the Russian, ed. S. P. Novikov. Soviet Scientific Reviews, Section C: Mathematical Physics Reviews, Vol. 5 (Harwood Academic, 1985).
169. B. A. Dubrovin, I. M. Krichever and S. P. Novikov, Integrable Systems, I (in Russian), Current Problems in Mathematics. Fundamental Directions, Vol. 4, pp. 179–284, 291, Itogi Nauki i Tekhniki, Akad. Nauk SSSR, Vsesoyuz. Inst. Nauchn. i Tekhn. Inform., Moscow, 1985.
170. S. P. Novikov, Algebraic topology at the Steklov Mathematical Institute of the Academy of Sciences of the USSR (in Russian), Topology, ordinary differential equations, dynamical systems, *Trudy Mat. Inst. Steklov.* **169** (1985), 27–49, 254.

171. S. P. Novikov, Differential geometry and the averaging method for field-theoretic systems (in Russian), in *III International Symposium on Selected Topics in Statistical Mechanics*, Vol. II (Dubna, 1984), pp. 106–118, Ob'ed. Inst. Yadernykh Issled. Dubna, D17-84-850, Ob'ed. Inst. Yadernykh Issled. (Dubna, 1985).
172. B. Doubrovine, S. Novikov and A. Fomenko, Géométrie contemporaine, Méthodes et applications, 2e partie (in French) [*Modern Geometry. Methods and Applications.* Part 2] Géométrie et topologie des variétés. [*Geometry and Topology of Manifolds*] Translated from the Russian by Vladimir Kotliar. Reprint of the 1982 translation. Traduit du Russe: Mathématiques. [Translations of Russian Works: Mathematics] (Mir, 1985).
173. B. Doubrovine, S. Novikov and A. Fomenko, Géométrie contemporaine. Méthodes et applications. 1re partie (in French) [Modern Geometry. Methods and Applications. Part 1] Géométrie des surfaces, des groupes de transformations et des champs. [Geometry of Surfaces, Transformation Groups and Fields] Translated from the Russian by Vladimir Kotliar. Reprint of the 1982 translation. Traduit du Russe: Mathématiques. [Translations of Russian Works: Mathematics] (Mir, 1985).
174. V. I. Arnol'd, M. I. Vishik, I. M. Gel'fand, Yu. V. Egorov, A. S. Kalashnikov, A. N. Kolmogorov, S. P. Novikov and S. L. Sobolev, Ol'ga Arsen'evna Oleĭnik (on the occasion of her sixtieth birthday) (in Russian), *Usp. Mat. Nauk* **40** (1985), No. 5(245), 279–293.
175. B. A. Dubrovin, A. T. Fomenko and S. P. Novikov, *Modern Geometry — Methods and Applications.* Part II. The geometry and topology of manifolds. Translated from the Russian by Robert G. Burns, Graduate Texts in Mathematics, Vol. 104 (Springer-Verlag, 1985).
176. S. P. Novikov, Geometry of conservative systems of hydrodynamic type. The averaging method for field-theoretic systems (in Russian), International conference on current problems in algebra and analysis (Moscow-Leningrad, 1984). *Usp. Mat. Nauk* **40** (1985), No. 4(244), 79–89.
177. Yu. V. Egorov, A. Yu. Ishlinskiĭ, A. S. Kalashnikov, A. N. Kolmogorov, O. B. Lupanov and S. P. Novikov, Ol'ga Arsen'evna Oleĭnik (in Russian), *Vestnik Moskov. Univ. Ser. I Mat. Mekh.* (1985), No. 4, 95–104.
178. S. P. Novikov, Analytical homotopy theory. Rigidity of homotopic integrals (in Russian), *Dokl. Akad. Nauk SSSR* **283** (1985), No. 5, 1088–1091.
179. A. A. Balinskiĭ and S. P. Novikov, Poisson brackets of hydrodynamic type, Frobenius algebras and Lie algebras (in Russian), *Dokl. Akad. Nauk SSSR* **283** (1985), No. 5, 1036–1039.

180. A. A. Voronov, D. M. Gvishiani, L. V. Kantorovich, A. N. Kolmogorov and S. P. Novikov, In search of the new (in Russian), *Mat. v Shkole* (1985), No. 2, 68–72.

181. V. G. Boltjanskij and V. A. Jefremovič, Očigledna topologija. (Serbo-Croatian) [*Descriptive Topology*] Edited and with a preface by S. P. Novikov. Translated from the Russian and with a preface by Dušan Adnadjević. Matematička Biblioteka [Mathematical Library], 46 (Zavod za Udžbenike i Nastavna Sredstva, Belgrade, 1984).

182. S. P. Novikov, An averaging method for one-dimensional systems, in *Nonlinear and Turbulent Processes in Physics*, Vol. 3 (Kiev, 1983), pp. 1529–1540 (Harwood Academic, 1984).

183. S. Novikov, S. V. Manakov, L. P. Pitaevskiĭ and V. E. Zakharov, Theory of solitons, in *The Inverse Scattering Method*, translated from the Russian, Contemporary Soviet Mathematics. Consultants Bureau (Plenum, 1984).

184. A. P. Veselov and S. P. Novikov, Finite-gap two-dimensional Schrödinger operators. Potential operators (in Russian), *Dokl. Akad. Nauk SSSR* **279** (1984), No. 4, 784–788.

185. B. A. Dubrovin and S. P. Novikov, Poisson brackets of hydrodynamic type (in Russian), *Dokl. Akad. Nauk SSSR* **279** (1984), No. 2, 294–297.

186. A. P. Veselov and S. P. Novikov, Finite-gap two-dimensional potential Schrödinger operators. Explicit formulas and evolution equations (in Russian), *Dokl. Akad. Nauk SSSR* **279** (1984), No. 1, 20–24.

187. *Mathematical Physics Reviews*. Vol. 4, ed. S. P. Novikov. Papers translated from the Russian by Morton Hamermesh, *Sov. Sci. Rev., Sec. C: Math. Phys. Rev*., Vol. 4 (Harwood Academic, 1984).

188. B. A. Dubrovin, S. P. Novikov and A. T. Fomenko, Sovremennaya geometriya (in Russian) [*Modern Geometry*] Metody teorii gomologiĭ. [*Methods of Homology Theory*] (Nauka, 1984).

189. S. P. Novikov, The analytic generalized Hopf invariant. Multivalued functionals (in Russian), *Usp. Mat. Nauk* **39** (1984), No. 5(239), 97–106.

190. V. G. Dulov, S. P. Novikov, L. V. Ovsyannikov, B. L. Rozhdestvenskiĭ, A. A. Samarskiĭ and Yu. I. Shokin, Nikolaĭ Nikolaevich Yanenko (in Russian), *Usp. Mat. Nauk* **39** (1984), No. 4(238), 85–94.

191. A. P. Veselov and S. P. Novikov, Poisson brackets and complex tori (in Russian), *Algebraic Geometry and Its Applications*, *Trudy Mat. Inst. Steklov*. **165** (1984), 49–61.

192. S. P. Novikov, Critical points and level surfaces of multivalued functions (in Russian), Modern problems of mathematics. Differential equations, mathematical analysis and their applications, *Trudy Mat. Inst. Steklov*. **166** (1984), 201–209.

193. S. P. Novikov, Algebro-topological approach to reality problems. Real action variables in the theory of finite-gap solutions of the sine-Gordon equation (in Russian), Differential geometry, Lie groups and mechanics, VI. *Zap. Nauchn. Sem. Leningrad. Otdel. Mat. Inst. Steklov.* (LOMI) **133** (1984), 177–196.
194. B. A. Dubrovin, A. T. Fomenko and S. P. Novikov, *Modern Geometry — Methods and Applications*. Part I. The geometry of surfaces, transformation groups, and fields. Translated from the Russian by Robert G. Burns. Graduate Texts in Mathematics, Vol. 93 (Springer-Verlag, 1984).
195. S. P. Novikov, Yu. M. Smirnov and A. S. Shvarts, Vadim Arsen'evich Efremovich (in Russian), *Usp. Mat. Nauk* **39** (1984), No. 1(235), 175–176.
196. S. P. Demushkin, A. I. Kostrikin, S. P. Novikov, A. N. Parshin, L. S. Pontryagin, A. N. Tyurin and D. K. Faddeev, Igor' Rostislavovich Shafarevich (in Russian), *Usp. Mat. Nauk* **39** (1984), No. 1(235), 167–174.
197. S. P. Novikov and I. A. Taĭmanov, Periodic extremals of multivalued or not everywhere positive functionals (in Russian), *Dokl. Akad. Nauk SSSR* **274** (1984), No. 1, 26–28.
198. S. P. Novikov, Multivalued functionals in modern mathematical physics. Proceedings of the IUTAM-ISIMM symposium on modern developments in analytical mechanics, Vol. II (Torino, 1982), *Atti Accad. Sci. Torino Cl. Sci. Fis. Mat. Natur.* **117** (1983), suppl. 2, 635–644.
199. Solitony (in Russian) [*Solitons*], eds. R. K. Bullough and P. J. Caudrey, translated from the English by B. A. Dubrovin, I. M. Krichever and S. V. Manakov. Translation edited by S. P. Novikov (Mir, 1983).
200. B. A. Dubrovin and S. P. Novikov, Hamiltonian formalism of one-dimensional systems of the hydrodynamic type and the Bogolyubov-Whitham averaging method (in Russian), *Dokl. Akad. Nauk SSSR* **270** (1983), No. 4, 781–785.
201. S. P. Novikov, Two-dimensional Schrödinger operators in periodic fields (in Russian), in *Current Problems in Mathematics*, Vol. 23, pp. 3–32, Itogi Nauki i Tekhniki, Akad. Nauk SSSR, Vsesoyuz. Inst. Nauchn. i Tekhn. Inform., Moscow, 1983.
202. N. N. Bogolyubov, S. G. Gindikin, A. A. Kirillov, A. N. Kolmogorov, S. P. Novikov and L. D. Faddeev, Izrail' Moiseevich Gel'fand (on the occasion of his seventieth birthday) (in Russian), *Usp. Mat. Nauk* **38** (1983), No. 6(234), 137–152.
203. Topologicheskie i geometricheskie metody v matematicheskoĭ fizike (in Russian) [*Topological and Geometric Methods in Mathematical Physics*], ed. S. P. Novikov. Novoe v Global'nom Analize. [*New Results in Global Analysis*] (Voronezh. Gos. Univ., 1983).

204. A. D. Aleksandrov, S. P. Novikov, A. V. Pogorelov, È. G. Poznyak, P. K. Rashevskiĭ, È. R. Rozendorn, I. Kh. Sabitov and S. B. Stechkin, Obituary: Nikolaĭ Vladimirov Efimov (in Russian), *Usp. Mat. Nauk* **38** (1983), No. 5(233), 111–117.

205. B. Dubrovin, S. Novikov and A. Fomenko, Géométrie contemporaine. Méthodes et applications. II. (in French) [*Modern Geometry. Methods and Applications*. II] Géométrie et topologie des variétés. [*Geometry and Topology of Manifolds*]; translated from the Russian by Vladimir Kotliar (Mir, 1982).

206. B. Dubrovin, S. Novikov and A. Fomenko, Géométrie contemporaine. Méthodes et applications. I. (in French) [*Modern Geometry. Methods and Applications*. I] Géométrie des surfaces, des groupes de transformations et des champs. [*Geometry of Surfaces, Groups of Transformations and Fields*]; translated from the Russian by Vladimir Kotliar (Mir, 1982).

207. S. P. Novikov and P. G. Grinevich, On the spectral theory of commuting operators of rank 2 with periodic coefficients (in Russian), *Functional. Anal. i Prilozhen.* **16** (1982), No. 1, 25–26.

208. B. A. Dubrovin and S. P. Novikov, Algebrogeometric Poisson brackets for real finite-range solutions of the sine–Gordon and nonlinear Schrödinger equations (in Russian), *Dokl. Akad. Nauk SSSR* **267** (1982), No. 6, 1295–1300.

209. A. P. Veselov and S. P. Novikov, Poisson brackets that are compatible with the algebraic geometry and the dynamics of the Korteweg-de Vries equation on the set of finite-gap potentials (in Russian), *Dokl. Akad. Nauk SSSR* **266** (1982), No. 3, 533–537.

210. S. P. Novikov, The Hamiltonian formalism and a many-valued analogue of Morse theory (in Russian), *Usp. Mat. Nauk* **37** (1982), No. 5, 3–49.

211. Yu. M. Berezans'kiĭ, N. N. Bogolyubov, B. Ya. Levin, Yu. A. Mitropol'skiĭ, S. P. Novikov, Ĭ. V. Ostrovs'kiĭ and A. V. Pogorelov, Vladimir Aleksandrovich Marchenko (on the occasion of his sixtieth birthday) (in Russian), *Usp. Mat. Nauk* **37** (1982), No. 6(228), 255–260.

212. V. G. Boltyanskiĭ and V. A. Efremovich, Naglyadnaya topologiya (in Russian) [*Descriptive Topology*], edited and with a preface by S. P. Novikov. Bibliotechka "Kvant" [*Library "Kvant"*], Vol. 21 (Nauka, 1982).

213. *Mathematical Physics Reviews*, Vol. 3, ed. S. P. Novikov, Sov. Sci. Rev., Sec. C: Math. Phys. Rev., Vol. 3 (Harwood Academic, 1982).

214. S. P. Novikov, Hamiltonian formalism and variational-topological methods for finding periodic trajectories of conservative dynamical systems, in

Mathematical Physics Reviews, Vol. 3, pp. 3–51, Sov. Sci. Rev., Sec. C: Math. Phys. Rev., Vol. 3 (Harwood Academic, 1982).
215. S. P. Novikov, Commuting operators of rank with periodic coefficients (in Russian), *Dokl. Akad. Nauk SSSR* **263** (1982), No. 6, 1311–1314.
216. Yu. I. Manin, M. A. Markov, S. P. Novikov, V. I. Ogievetskiĭ, V. Ya. Faĭnberg and E. S. Fradkin, Feliks Aleksandrovich Berezin (in Russian), *Sov. Phys. Usp.* **134** (1981), No. 2, 357–358.
217. *Mathematical Physics Reviews*, Vol. 2, ed. S. P. Novikov, Sov. Sci. Rev., Sec. C: Math. Phys. Rev., Vol. 2 (Harwood Academic, 1981).
218. S. P. Novikov, Variational methods and periodic solutions of equations of Kirchhoff type. II (in Russian), *Functional. Anal. i Prilozhen.* **15** (1981), No. 4, 37–52, 96.
219. S. P. Novikov, Multivalued functions and functionals. An analogue of the Morse theory (in Russian), *Dokl. Akad. Nauk SSSR* **260** (1981), No. 1, 31–35.
220. S. P. Novikov, Shmel'tser, I. Periodic solutions of Kirchhoff equations for the free motion of a rigid body in a fluid and the extended Lyusternik-Shnirel'man-Morse theory. I (in Russian), *Functional. Anal. i Prilozhen.* **15** (1981), No. 3, 54–66.
221. S. P. Novikov, Bloch functions in the magnetic field and vector bundles. Typical dispersion relations and their quantum numbers (in Russian), *Dokl. Akad. Nauk SSSR* **257** (1981), No. 3, 538–543.
222. V. G. Drinfel'd, I. M. Krichever, Yu. I. Manin and S. P. Novikov, Methods of algebraic geometry in contemporary mathematical physics, in *Mathematical Physics Reviews*, Vol. 1, pp. 1–54, Sov. Sci. Rev., Sec. C: Math. Phys. Rev., Vol. 1 (Harwood Academic, 1980).
223. *Mathematical Physics Reviews*, Vol. 1, ed. S. P. Novikov, Sov. Sci. Rev., Sec. C: Math. Phys. Rev., Vol. 1 (Harwood Academic, 1980).
224. I. M. Krichever and S. P. Novikov, Holomorphic bundles over algebraic curves, and nonlinear equations (in Russian), *Usp. Mat. Nauk* **35** (1980), No. 6(216), 47–68, 215.
225. B. A. Dubrovin and S. P. Novikov, Ground states of a two-dimensional electron in a periodic magnetic field (in Russian), *J. Exp. Teor. Phys.* **52** (1980), No. 3, 511–516; translated from *Zh. Eksper. Teor. Fiz.* **79** (1980), No. 3, 1006–1016.
226. B. A. Dubrovin and S. P. Novikov, Fundamental states in a periodic field. Magnetic Bloch functions and vector bundles (in Russian), *Dokl. Akad. Nauk SSSR* **253** (1980), No. 6, 1293–1297.

227. Issledovaniya po metricheskoĭ teorii poverkhnosteĭ (in Russian) [Investigations in the metric theory of surfaces] Papers translated from the English and from the French by I. H. Sabitov, eds. A. N. Kolmogorov and S. P. Novikov. Matematika: Novoe v Zarubezhnoĭ Nauke [Mathematics: Recent Publications in Foreign Science], Vol. 18 (Mir, 1980).
228. V. E. Zaharov, S. V. Manakov, S. P. Novikov and L. P. Pitaevskiĭ, Teoriya solitonov (in Russian) [*Theory of Solitons*], Metod obratnoĭ zadachi [*The Method of the Inverse Problem*] (Nauka, 1980).
229. I. M. Kričever and S. P. Novikov, Holomorphic bundles and nonlinear equations. Finite-gap solutions of rank (in Russian), *Dokl. Akad. Nauk SSSR* **247** (1979), No. 1, 33–37.
230. B. A. Dubrovin, S. P. Novikov and A. T. Fomenko, Sovremennaya geometriya (in Russian), [*Modern Geometry*] Metody i prilozheniya. [*Methods and Applications*] (Nauka, 1979).
231. O. I. Bogojavlenskiĭ and S. P. Novikov, Finite-dimensional oscillatory models in the general theory of relativity and in gas dynamics (in Russian), Boundary value problems of mathematical physics and related questions in the theory of functions, 11. *Zap. Nauchn. Sem. Leningrad. Otdel. Mat. Inst. Steklov*. (*LOMI*) **84** (1979), 7–15, 309–310, 316.
232. V. L. Golo, M. I. Monastyrsky and S. P. Novikov, Solutions to the Ginzburg–Landau equations for planar textures in superfluid, *Comm. Math. Phys.* **69** (1979), No. 3, 237–246.
233. S. P. Novikov, New applications of algebraic geometry to nonlinear equations and inverse problems. Nonlinear evolution equations solvable by the spectral transform (Internat. Sympos., Accad. Lincei, Rome, 1977), pp. 84–96, Res. Notes in Math., Vol. 26 (Pitman, 1978).
234. S. P. Novikov, Periodic solitons and algebraic geometry, in *Mathematical Problems in Theoretical Physics* (Proc. Internat. Conf., Univ. Rome, Rome, 1977), pp. 222–228, Lecture Notes in Phys., Vol. 80 (Springer, 1978).
235. Evklidova kvantovaya teoriya polya (in Russian) [*Euclidean Quantum Field Theory*] Markovskiĭ podkhod. [The Markov approach] Translation from the English edition edited by R. A. Minlos, A. N. Kolmogorov and S. P. Novikov. Matematika: Novoe v Zarubezhnoĭ Nauke [*Mathematics: Recent Publications in Foreign Science*], Vol. 12 (Mir, 1978).
236. I. M. Kričever and S. P. Novikov, Holomorphic vector bundles over Riemann surfaces and the Kadomcev-Petviašvili equation. I (in Russian), *Functional. Anal. i Prilozhen.* **12** (1978), No. 4, 41–52.
237. S. P. Novikov, A method for solving the periodic problem for the KdV equation and its generalizations. Conference on the Theory and

Applications of Solitons (Tucson, Ariz., 1976), *Rocky Mountain J. Math.* **8** (1978), No. 1–2, 83–93.

238. S. P. Novikov and A. T. Fomenko, Petr Konstantinovič Raševskiĭ (on the occasion of his seventieth birthday) (in Russian), *Usp. Mat. Nauk* **32** (1977), No. 5(197), 205–209.
239. O. I. Bogojavlenskiĭ and S. P. Novikov, Homogeneous models in general relativity theory and gas dynamics (in Russian), *Usp. Mat. Nauk* **31** (1976), No. 5(191), 33–48.
240. O. I. Bogojavlenskiĭ, S. P. Novikov, The connection between the Hamiltonian formalisms of stationary and nonstationary problems (in Russian), *Functional. Anal. i Prilozhen.* **10** (1976), No. 1, 9–13.
241. B. A. Dubrovin, V. B. Matveev and S. P. Novikov, Nonlinear equations of Korteweg-de Vries type, finite-band linear operators and Abelian varieties (in Russian), *Usp. Mat. Nauk* **31** (1976), No. 1(187), 55–136.
242. O. I. Bogojavlenskiĭ and S. P. Novikov, Qualitative theory of homogeneous cosmological models (in Russian), *Trudy Sem. Petrovsk. Vyp.* **1** (1975), 7–43.
243. B. A. Dubrovin and S. P. Novikov, A periodic problem for the Korteweg-de Vries and Sturm-Liouville equations. Their connection with algebraic geometry (in Russian), *Dokl. Akad. Nauk SSSR* **219** (1974), 531–534.
244. I. A. Volodin, V. E. Kuznecov and A. T. Fomenko, The problem of the algorithmic discrimination of the standard three-dimensional sphere (in Russian), Appendix by S. P. Novikov. *Usp. Mat. Nauk* **29** (1974), No. 5(179), 71–168.
245. S. P. Novikov, The periodic problem for the Korteweg–de Vries equation (in Russian), *Functional. Anal. i Prilozhen.* **8** (1974), No. 3, 54–66.
246. B. A. Dubrovin and S. P. Novikov, Periodic and conditionally periodic analogs of the many-soliton solutions of the Korteweg-de Vries equation, *Sov. Phys. JETP* **40** (1974), No. 6, 1058–1063; translated from *Ž. Èksper. Teoret. Fiz.* **67** (1974), No. 6, 2131–2144 (Russian).
247. O. I. Bogoyavlenskiĭ and S. P. Novikov, Singularities of the cosmological model of the Bianchi type according to the qualitative theory of differential equations, *Sov. Phys. JETP* **37** (1973), 747–755; translated from *Ž. Èksper. Teoret. Fiz.* **64** (1973), 1475–1494 (Russian).
248. V. M. Buhštaber, A. S. Miščenko and S. P. Novikov, Formal groups and their role in the apparatus of algebraic topology (in Russian), *Usp. Mat. Nauk* **26** (1971), No. 2(158), 131–154.
249. S. P. Novikov, Analogues hermitiens de la-théorie. Actes du Congrès International des Mathématiciens (Nice, 1970), Tome 2, pp. 39–45, (Gauthier-Villars, 1971).

250. M. F. Atiyah, On the work of Serge Novikov. Actes du Congrès International des Mathématiciens (Nice, 1970), Tome 1, pp. 11–13, (Gauthier-Villars, 1971).

251. V. M. Buhštaber and S. P. Novikov, Formal groups, power systems and Adams operators (in Russian), *Mat. Sb.* (*N.S.*) **84**(126) (1971), 81–118.

252. V. I. Arnol'd, I. M. Gel'and, Ju. I. Manin, B. G. Moĭšezon, S. P. Novikov and I. R. Šafarevič, Galina Nikolaevna Tjurina. Obituary (in Russian), *Usp. Mat. Nauk* **26** (1971), No. 1, 207–211.

253. P. S. Aleksandrov, V. I. Arnol'd, I. M. Gel'fand, A. N. Kolmogorov, S. P. Novikov and O. A. Oleĭnik, Ivan Georgievič Petrovskiĭ (on his seventieth birthday) (in Russian), *Usp. Mat. Nauk* **26** (1971), No. 2, 3–24.

254. S. P. Novikov, Algebraic construction and properties of Hermitian analogs of theory over rings with involution from the viewpoint of Hamiltonian formalism. Applications to differential topology and the theory of characteristic classes. I. II. *Math. USSR-Izv.* **4** (1970), 257–292; *ibid.* **4** (1970), 479–505; translated from *Izv. Akad. Nauk SSSR Ser. Mat.* **34** (1970), 253–288; *ibid.* **34** (1970), 475–500.

255. S. P. Novikov, Pontrjagin classes, the fundamental group and some problems of stable algebra. 1970 Essays on Topology and Related Topics (Mémoires dédiés à Georges de Rham), pp. 147–155 (Springer, 1970).

256. S. P. Novikov, Adams operators and fixed points (in Russian), *Izv. Akad. Nauk SSSR Ser. Mat.* **32** (1968), 1245–1263.

257. S. P. Novikov, Pontrjagin classes, the fundamental group and some problems of stable algebra. 1968 Amer. Math. Soc. Translations Ser. 2, Vol. 70: 31 Invited Addresses (8 in Abstract) at the Internat. Congr. Math. (Moscow, 1966) pp. 172–179 (Amer. Math. Soc., 1968).

258. S. P. Novikov, Methods of algebraic topology from the point of view of cobordism theory (in Russian), *Izv. Akad. Nauk SSSR Ser. Mat.* **31** (1967), 855–951.

259. S. P. Novikov, Rings of operations and spectral sequences of Adams type in extraordinary cohomology theories: Cobordism and theory (in Russian), *Dokl. Akad. Nauk SSSR* **172** (1967), 33–36.

260. S. P. Novikov and B. Ju. Sternin, Elliptic operators and submanifolds, *Dokl. Akad. Nauk SSSR* **171**, 525–528 (in Russian); translated as *Sov. Math. Dokl.* **7** (1966), 1508–1512.

261. S. P. Novikov and B. Ju. Sternin, Traces of elliptic operators on submanifolds and theory, *Dokl. Akad. Nauk SSSR* **170**, 1265–1268 (in Russian); translated as *Sov. Math. Dokl.* **7** (1966), 1373–1376.

262. S. P. Novikov, The Cartan–Serre theorem and inner homologies (in Russian), *Usp. Mat. Nauk* **21** (1966), No. 5(131), 217–232.

263. S. P. Novikov, On manifolds with free abelian fundamental group and their application (in Russian), *Izv. Akad. Nauk SSSR Ser. Mat.* **30** (1966), 207–246.
264. S. P. Novikov, The topology of foliations (in Russian), *Trudy Moskov. Mat. Obšč.* **14** (1965), 248–278.
265. S. P. Novikov, Rational Pontrjagin classes, Homeomorphism and homotopy type of closed manifolds. I (in Russian), *Izv. Akad. Nauk SSSR Ser. Mat.* **29** (1965), 1373–1388.
266. S. P. Novikov, Topological invariance of rational Pontrjagin classes (in Russian), *Dokl. Akad. Nauk SSSR* **163** (1965), 298–300.
267. S. P. Novikov, New ideas in algebraic topology. Theory and its applications (in Russian), *Usp. Mat. Nauk* **20** (1965), No. 3(123), 41–66.
268. S. P. Novikov, The homotopy and topological invariance of certain rational Pontrjagin classes (in Russian), *Dokl. Akad. Nauk SSSR* **162** (1965), 1248–1251.
269. S. P. Novikov, Differentiable sphere bundles (in Russian), *Izv. Akad. Nauk SSSR Ser. Mat.* **29** (1965), 71–96.
270. S. P. Novikov, Smooth foliations on three-dimensional manifolds (in Russian), *Usp. Mat. Nauk* **19** (1964), No. 6(120), 89–91.
271. S. P. Novikov, I. I. Pyatezki-Shapiro and I. R. Šafarevič, Fundamental directions in the development of algebraic topology and algebraic geometry (in Russian), *Usp. Mat. Nauk* **19** (1964), No. 6(120), 75–82.
272. S. P. Novikov, Foliations of co-dimension (in Russian), *Dokl. Akad. Nauk SSSR* **157** (1964), 788–790.
273. S. P. Novikov, Foliations of co-dimension on manifolds (in Russian), *Dokl. Akad. Nauk SSSR* **155** (1964), 1010–1013.
274. S. P. Novikov, Homotopically equivalent smooth manifolds. I (in Russian), *Izv. Akad. Nauk SSSR Ser. Mat.* **28** (1964), 365–474.
275. S. P. Novikov, Some properties of manifolds of dimension (in Russian), *Dokl. Akad. Nauk SSSR* **153** (1963), 1005–1008.
276. S. P. Novikov, Differential topology (in Russian), 1963 Itogi Nauki (Algebra. Topology, 1962) pp. 134–160, Akad. Nauk SSSR Inst. Naučn. Informacii, Moscow.
277. S. P. Novikov, Homotopy properties of the group of diffeomorphisms of the sphere (in Russian), *Dokl. Akad. Nauk SSSR* **148** (1963), 32–35.
278. S. P. Novikov, A diffeomorphism of simply connected manifolds (in Russian), *Dokl. Akad. Nauk SSSR* **143** (1962), 1046–1049.
279. S. P. Novikov, Homotopy properties of Thom complexes (in Russian), *Mat. Sb. (N.S.)* **57**(99) (1962), 407–442.

280. S. P. Novikov, Imbedding of simply connected manifolds into Euclidean space (in Russian), *Dokl. Akad. Nauk SSSR* **138** (1961), 775–778.
281. S. P. Novikov, Some problems in the topology of manifolds connected with the theory of Thom spaces. *Dokl. Akad. Nauk SSSR* 132 1031–1034 (Russian); translated as *Sov. Math. Dokl.* **1** (1960), 717–720.
282. S. P. Novikov, Cohomology of the Steenrod algebra (in Russian), *Dokl. Akad. Nauk SSSR* **128** (1959), 893–895.

THE PERIODIC PROBLEM FOR THE KORTEWEG–De VRIES EQUATION

S. P. Novikov

Introduction

The Koeteweg–de Vries equation (KV) arose in the nineteenth century in connection with the theory of waves in shallow water. It is now known (cf., for example, [15]) that this equation also describes the propagation of waves with weak dispersion in various nonlinear media. In reduced form it is written

$$u_t = 6uu_x - u_{xxx}.$$

In 1967 in the well-known work of Gardner, Green, Kruskal, and Miura a remarkable procedure for integrating the Cauchy problem for this equation was discovered for functions u(X) which are rapidly decreasing for $x \to \pm\infty$. This procedure consists in the following: we consider the Schrödinger (Sturm–Liouville) operator $L = -(d^2/dx^2) + u$; let $f(x, k)$ and $g(x, k)$ be such that $Lf(x, k) = k^2 f$, $Lg(x, k) = k^2 g$, whereby $f(x, k) \to e^{-ikx}$ for $x \to -\infty$ and $g(x, k) \to e^{-ikx}$ for $x \to +\infty$. We then have two pairs of linearly independent solutions $f, \bar{f}$ and $g, \bar{g}$ and a transition matrix

$$f(x, k) = a(k)\, g(x, k) + b(k)\, \bar{g}(x, k),$$
$$\bar{f}(x, k) = \bar{b}(k)\, g(x, k) + \bar{a}(k)\, \bar{g}(x, k).$$

If the potential u changes in time according to the KV equation, then the coefficients a and b vary according to the law $\dot{a} = 0$, $\dot{b} = 8ik^3 b$, and for the eigenfunction $f(x, k)$ there is the equation $\dot{f} = Af + \lambda f$, where $A = 4\frac{d^3}{dx^3} - 3\left(u\frac{d}{dx} + \frac{d}{dx}u\right)$, $\lambda = 4ik^3$. If $\lambda_n = +k_n^2 = (i\varkappa_n)^2$ is a point of the discrete spectrum of the potential u(x), then $\dot{\varkappa}_n = 0$ and $\dot{c}_n = +8\varkappa^3 c_n$, where c_n is the natural normalization of the eigenfunction. These formulas make it possible to reduce the Cauchy problem for the KV equation with rapidly decreasing functions to an inverse problem of scattering theory and to use the results of I. M. Gel'fand, B. I. Levitan, V. A. Marchenko, and L. D. Faddeev (cf. [2, 3, 4]). Subsequently, P. Lax [7] discovered that the KV equation is identical to the operator equation $\dot{L} = [A, L]$, where $L = -\frac{d^2}{dx^2} + u$, $A = 4\frac{d^3}{dx^3} - 3\left(u\frac{d}{dx} + \frac{d}{dx}u\right)$, since $\dot{L}$ is the operator of multiplication by $\dot{u}$, while $[A, L]$ is the operator of multiplication by the function $6uu' - u'''$. In particular, it follows from this that the spectrum of the operator L is an integral of the KV equation (this is also true for the periodic problem). This fact reveals the algebraic meaning of the procedure in [5] and is extremely useful for the application of these ideas to other problems. Further, L. D. Faddeev and V. E. Zakharov [8] have shown that the KV equation is a completely integrable Hamiltonian system, where the canonical variables are $\frac{2k}{\pi}\ln|a(k)|$, $\arg b(k)$, $\varkappa_n^2$, $\ln(b_n)$, $b_n = ic_n\left(\frac{da}{dk}\right)_{k=i\varkappa_n}$. In particular, the eigenvalues $\lambda_n = k_n^2$ are commuting integrals of the KV equation also in the periodic problem [and not only in the case of rapidly decreasing functions u(x)]. A number of other equations have subsequently been found which admit the "Lax representation" $\dot{L} = [A, L]$ for a certain pair of operators A, L. P. Lax [7]

V. A. Steklov Mathematics Institute. L. D. Landau Institute of Theoretical Physics, Academy of Sciences of the USSR. Translated from Funktsional'nyi Analiz i Ego Prilozheniya, Vol. 8, No. 3, pp. 54–66, July–September, 1974. Original article submitted February 18, 1974.

and Gardner [6] have shown that the known polynomial integrals $I_n(u) = \int_{-\infty}^{\infty} P_n(u, \ldots, u^{(n)})\,dx$ of the KV equation (the I_n are expressed in terms of the spectrum of the operator L) all determine equations

$$\dot{u} = \frac{d}{dx}\frac{\delta I_n}{\delta u(x)}, \tag{1}$$

admitting the Lax representation $\dot{L} = [A_n, L]$, where $L = (d^2/dx^2) + u$ and the A_n are certain skew-symmetric operators of order $2n+1$,

$$\begin{aligned}
&I_0 = \int u^2\,dx, \quad I_1 = \int\left(\frac{u'^2}{2} + u^3\right)dx,\\
&I_2 = \int\left(\frac{u''^2}{2} - \frac{5}{2}u^2u'' + \frac{5}{2}u^4\right)dx,\\
&A_0 = \frac{d}{dx}, \quad A_1 = 4\frac{d^3}{dx^3} - 3\left(u\frac{d}{dx} + \frac{d}{dx}u\right),\\
&A_2 = \frac{d^5}{dx^5} - \frac{5}{2}u\frac{d^3}{dx^3} - \frac{15}{4}u'\frac{d^2}{dx^2} + \frac{15u^2 - 25u''}{8}\frac{d}{dx} + \frac{15}{8}\left(uu' - \frac{u'''}{2}\right).
\end{aligned} \tag{2}$$

We shall call these equations "higher KV equations." Further, beginning with the papers [9, 10], a number of other important equations were found which admit the Lax representation $\dot{L} = [A, L]$, where L is no longer a Schrödinger operator (and is not always symmetric). In the papers of L. D. Faddeev, V. E. Zakharov, and A. B. Shabat ([10, 11, 12]) the needed generalization of scattering theory was carried out for the new operators which has made it possible to solve the direct and inverse problems and to carry through the "Kruskal" integration of the Cauchy problem for rapidly decreasing (as $x \to \pm\infty$) initial data. A considerable literature has recently been devoted to discovering such new equations and carrying over the Kruskal mechanism to them. However, even for the original KV equation the periodic problem has not moved forward. The only new result in the periodic case, which was obtained by the method of Gardner, Green, Kruskal, and Miura, is the theorem of Faddeev and Zakharov to the effect that the eigenvalues of the operator L are commuting integrals of the KV equation as a Hamiltonian system (we remark that the integrals themselves can be expressed in a one-to-one manner in terms of the previously known polynomial integrals I_n which are thus also involutive). This result has not been used in an essential way until the present work, but here it plays an important role (cf. § 2).

The interaction of simple waves [solutions of the type u(x−ct), usually called "solitons"] are of major interest in the theory of the KV equation. This interaction is described by means of so-called "multisoliton solutions" where $b(k) \equiv 0$ for real k.* These solutions decay into solitons for $x, t \to \pm\infty$ and describe their interaction for finite t. For this case the Gel'fand–Levitan equations are completely solvable (cf., for example [8, 9]). Another method of obtaining multisoliton solutions has been developed in [13]. However, all these results refer to the case of rapidly decreasing functions u(x). In the periodic case the solitons u(x−ct) of the KV equation are of a more complicated structure; there are many more of them and their interaction has not been studied at all. In the present paper we propose a method of studying certain analogs of the "multisoliton" solutions of the KV equation which, generally speaking, are found to be not only periodic, but also conditionally periodic functions u(x) describing the interaction of periodic solitons. Our work is based on certain simple but fundamental algebraic properties of equations admitting the Lax representation which are strongly degenerate in the problem with rapidly decreasing functions (for $x \to \pm\infty$), and have therefore not been noted. Finally, it is essential to note the nonlinear "superposition law for waves" for the KV equation which in the periodic case has an interesting algebraic-geometric interpretation. The superposition law will be discussed in the second part of the work.

In conclusion, we call the attention of the reader to the following circumstance: in classical mechanics and mathematics the appearance of integrals in conservative systems (conservation laws) is almost always related to a Lie symmetry group of the problem in question. Other fundamental algebraic mechanisms of integrability were previously unknown. However, there were several exceptions: for example, the Jacobi case (geodesics on a triaxial ellipsoid) or the case of Kovalevskaya (the problem of the motion of a solid body with a fixed point in a gravitational field). Other exceptional examples are now known. There is not the slightest doubt that all these cases are the manifestation of a Kruskal-type algebraic mechanism based on the possibility of a Lax-type representation for these dynamical systems.

---

*I. M. Gel'fand has informed the author that such potentials u(x) were first considered by Bargmann.

§ 1. The Schrödinger (Sturm – Liouville) Equations with Periodic Coefficients. The Monodromy Matrix

We shall first list systematically simple facts which we shall need.

Let $u(x)$ be a smooth function where $u(x+T) = u(x)$, and let $L = (d^2/dx^2) + u$. We consider on the line the equation $L\psi_k = \lambda\psi_k$, where $\lambda = k^2$ is a real number. We consider the pair of linearly independent solutions $\psi_k(x, x_0)$, $\bar{\psi}_k(x, x_0)$, where $\psi_k(x_0, x_0) = 1$, $\psi_k'(x_0, x_0) = ik$, or the pair φ_k, $\bar{\varphi}_k$, where $\varphi_k(x_0, x_0) = 1$, $\varphi_k'(x_0, x_0) = i$. (The pair ψ_k, $\bar{\psi}_k$ is more convenient but is meaningful only for $k^2 > 0$.) We can define the "monodromy matrix"

$$T(k, x_0) = \begin{pmatrix} a & b \\ \bar{b} & \bar{a} \end{pmatrix}, \quad a = a(k, x_0), \quad b = b(k, x_0),$$

where

$$\psi_k(x+T, x_0) = a\psi_k(x, x_0) + b\bar{\psi}_k(x, x_0),$$
$$\bar{\psi}_k(x+T, x_0) = \bar{b}\psi_k(x, x_0) + \bar{a}\bar{\psi}_k(x, x_0),$$

or the analogous matrix in another basis. In the basis φ_k, $\bar{\varphi}_k$, for example, the monodromy matrix is an entire function of $\lambda = k^2$. The trace of the matrix $\mathrm{Sp}T = a + \bar{a} = 2a_R$ is real, while the determinant is equal to one, $\det T = |a|^2 - |b|^2 = 1$ for all real k, since the Wronskian determinant is conserved.

The eigenvalues of the matrix $T(k, x_0)$ do not depend on the point x_0 and have the form: $\mu_\pm = a_R \pm \sqrt{a_R^2 - 1}$. In particular, we have two cases: $\mu_\pm = e^{\pm ip}$, $|a_R| \leqslant 1$; $\mu_\pm = e^{\pm p}$, $|a_R| \geqslant 1$, where p is a real number ($a_R = \cos p$ for $|a_R| \le 1$). The Bloch eigenfunctions of the operator L are those solutions of the equation $Lf = k^2 f$ such that $f(x+T) = e^{\pm ip} f(x)$, where the number p is called the "quasi-momentum." We have periodic eigenfunctions for $e^{ip} = 1$ or $a_R = 1$, and antiperiodic eigenfunctions $f(x+T) = -f(x)$, where $a_R = -1$. The permitted zones are the regions on the axis $\lambda = k^2$, where $|a_R| \le 1$, and the forbidden zones are the regions on the axis $\lambda = k^2$, where $|a_R| \ge 1$. The boundaries of the permitted and forbidden zones are the points $|a_R| = 1$, where the numbers $\lambda_n = k_n^2$ are the eigenvalues of the periodic or antiperiodic problem. These eigenvalues may be nondegenerate or doubly degenerate. Since $a = a_R + ia_I$, $b = b_R + ib_I$, and $|a|^2 - |b|^2 = 1$, it follows that at points of both spectra, where $|a_R| = 1$, we have $|a_I| = |b|$, $\lambda = \lambda_n = k_n^2$.

In the nondegenerate case $|a_I| = |b| \ne 0$, $\lambda = \lambda_n$. In the degenerate case $|a_I| = |b| = 0$, $\lambda = \lambda_n$ and therefore the forbidden zone in the degenerate case contracts to zero. Thus, the degenerate points of both spectra where $b = 0$ for $\lambda = \lambda_n$ are not boundaries of permitted zones. In the sequel we shall call "essential" only those nondegenerate points of both spectra which are boundaries of Bloch zones and which completely determine the zone structure on the axis $\lambda = k^2$. There is the following simple fact: multiplying the period T by an integer $T \to nT$ does not change the zone structure. Indeed, multiplication of the period $T \to nT$ raises the monodromy matrix to a power $T(k, x_0) \to T^n(k, x_0)$. Thus, the forbidden zones remain forbidden zones $e^{\pm p} \to e^{\pm np}$ and the permitted zones remain permitted zones $e^{\pm ip} \to e^{\pm inp}$, although in the interior of the permitted zones there appear new degenerate ("inessential") levels where $b \equiv 0$ which for increasing n fill out the entire zone in a dense manner.

This fact makes it possible to carry over the definition of the zone structure of a potential to almost periodic functions by approximating them by periodic functions with increasing period. However, in the literature there are no investigations of the convergence of such a process (apparently it has never been studied). It is possible that for linear combinations of periodic functions even with two periods T_1, T_2, the result may depend on the arithmetic of the number T_1/T_2.

We note that the function $\mathrm{Sp}T = 2a_R$ is an entire function of λ in the entire complex plane and can be defined up to a constant multiple as an infinite product (this was indicated to the author by L. D. Faddeev):

$$\frac{1}{2}\mathrm{Sp}\, T(\lambda) = 1 + \mathrm{const} \prod_j (\lambda - \lambda_j),$$

where the λ_j are the points of the spectrum of the periodic problem.

The points of the spectrum of the antiperiodic problem $f(x+T) = -f(x)$ are thus determined in principle by the equation $a_R = -1$ starting from the purely periodic spectrum. However, for this it is necessary to know all the points of the spectrum of the periodic problem, while we do not wish, for example, to use degenerate points of the spectrum. Therefore, we shall continue to work with both spectra.

We defined the monodromy matrix $T(k, x_0)$ by choosing an initial point x_0, although its eigenvalues do not depend on the initial point. Therefore, the matrices $T(k, x_0)$ for fixed k but various x_0 are conjugate. From this it follows immediately that with regard to the dependence of this matrix on the parameter x_0, which we now denote by x, a differential equation of the form

$$\frac{dT}{dx} = [Q, T] \tag{3}$$

is satisfied, where the matrix Q is easily computed as the transition matrix $(1+Qdx)$ from the basis $[\psi_k(x, x_0+dx), \bar{\psi}_k(x, x_0+dx)]$ to the basis $[\psi_k(x, x_0), \bar{\psi}_k(x, x_0)]$. Therefore $(1+Qdx)T(1-Qdx) = T(k, x_0+dx)$ or $T' = [Q, T]$.

In our bases the matrix Q has the form

$$Q = -ik\begin{pmatrix} 1 & 0 \\ 0 & -1 \end{pmatrix} + \frac{iu}{2k}\begin{pmatrix} 1 & -1 \\ 1 & -1 \end{pmatrix} \qquad (\text{basis } \psi_k, \bar{\psi}_k),$$

$$Q = -\frac{i}{2}\begin{pmatrix} 1 & 1 \\ -1 & -1 \end{pmatrix} - \frac{i}{2}(u-k^2)\begin{pmatrix} 1 & -1 \\ 1 & -1 \end{pmatrix} \qquad (\text{basis } \varphi_k, \bar{\varphi}_k). \tag{4}$$

Thus, the monodromy matrix T can be sought as a periodic solution of Eq. (3) which satisfies the condition $\det T = |a|^2 - |b|^2 = 1$. At the nondegenerate points of both spectra $|a_R| = 1$, where $b(k_n) \neq 0$ and $|b| = \pm a_I$, the following equation is obtained from Eq. (3) by direct substitution:

$$\varphi' \pm \frac{u}{k}\sin\varphi = -2k + \frac{u}{k}, \qquad k = k_n, \tag{3'}$$

where $\varphi = \arg b(k, x)$ for $k = k_n$, $a_I = \mp|b| \neq 0$. In principle, the nondegenerate points of the spectrum k_n^2 are determined from the requirement that the latter Eq. (3') should have a periodic solution. In general, this condition may also include "extraneous roots" k_n; however, for the zero potential $u \equiv 0$ we see that $\Delta\varphi = 2\pi n = 2kT$ gives the points of the spectrum exactly (analogously for the case u = const). Formally this equation is applicable only in the nondegenerate case, but in view of the stability of the properties of this equation under small (smooth) variations of the potential u, use can be made of the fact that all the levels become nondegenerate after almost any small perturbation. From this there follows:

PROPOSITION 1.1. 1) For potentials close to zero (or to a constant) the condition of the existence of periodic solutions of Eq. (3') includes all the nondegenerate points of the spectrum k_n and is not solvable for k_n which are not spectral points;

2) If a) the function $\varphi(k, x)$ is known as a function of u for a given potential u for all k, x, where it is meaningful (i.e., on the entire line with the exception of degenerate points of both spectra $a_I = b = 0$), and b) the function $(u/k)\sin\varphi$ extends as a smooth function of the variables x, k to all x, k including the degenerate points of the spectrum $k = k_n$, then each spectral points satisfies the transcendental equation

$$2\pi n = \int_{x_0}^{x_0+T}\left(2k - \frac{u}{k} \pm \frac{u}{k}\sin\varphi\right)dx. \tag{5}$$

Assertion 1) of Proposition 1.1 was proved above, since for a constant potential the condition on the spectral points is exact. For the proof of assertion 2) we remark that the function $\sin\varphi$ at all points x, k where it is well defined (i.e., $b \neq 0$) depends smoothly on the potential u(x). If after an arbitrarily small perturbation δu of the potential u which makes all levels nondegenerate Eq. (5) is not satisfied for given k, then this point k is not a spectral point for the potential u. This implies assertion 2) of Proposition 1.1.

§ 2. Potentials with a Finite Number of Zones and Multisoliton Solutions of the KV Equation

We recall that for an infinite period $T = \infty$ where $u(x) \to 0$ for $x \to \pm\infty$, the monodromy matrix is defined from the transition from $-\infty$ to $+\infty$. The multisoliton solutions $u(x, t)$ of the KV equation are determined from the condition $b \equiv 0$ for real k for these potentials u at any t. In this case a is defined by its zeros (the spectral points) $k_n = i\varkappa_n$ in the upper half plane $\varkappa_n > 0$ by the formula $a = \prod_j \frac{k - i\varkappa_j}{k + i\varkappa_j}$, whereby to a single soliton $u(x-ct)$ there corresponds a single spectral point $k_1 = i\varkappa_1$, where $\varkappa_1^2 = +c/4$ (cf. [5]). What is the right analog of multisoliton solutions in the periodic problem? We suppose that the periodic potential u is such that it has only a finite number of zones; this means that starting from some number $n > n_0$ all points of both spectra $a_R = \pm 1$ are doubly degenerate and that the entire half line $k^2 \geq k_{n_0}^2$ forms a single zone. The degeneracy condition for a spectral point is the condition $b(k_n, x) \equiv 0$ for $k_n^2 > k_{n_0}^2$. On multiplying the period T by an integer $T \to mT$ the entire zone is filled out by degenerate levels $b \equiv 0$ as $m \to \infty$. On the other hand, with correct passage to the period $T \to \infty$, whereby the potential $u_T(x)$ with period T tends to a rapidly decreasing potential for $T \to \infty$, the zones of finite size contract to isolated points of the discrete spectrum. All this indicates that it is natural to suppose that the right analog of multisoliton solutions are the finite-zone potentials. The property that a potential be a finite-zone potential is conserved in time by virtue of the KV equation. How should one seek such solutions of the KV equation? What are finite-zone potentials like?

We consider the "higher KV equations" of order n

$$\dot{u} = \frac{d}{dx}\left(\frac{\delta I_n}{\delta u(x)} + c_1 \frac{\delta I_{n-1}}{\delta u(x)} + \ldots + c_n \frac{\delta I_0}{\delta u(x)}\right), \tag{6}$$

where $I_n = \int P_n(u, u', \ldots, u^{(n)})dx$ is a polynomial integral of the KV equation and $c_1, \ldots, c_n$ are arbitrary constants. Equation (6) has order 2n+1 and by the theorem of Lax and Gardner [6, 7] it admits the Lax representation

$$\dot{L} = [L, A_n + c_1 A_{n-1} + \ldots + c_n A_0], \tag{7}$$

where $L = -(d^2/dx^2) + u$ and the operators A_0, A_1, A_2 are indicated in the introduction, $A_0 = (d/dx)$. We shall now indicate the corollary of the Zakharov–Faddeev theorem mentioned in the introduction which, while not noted by them, is extremely important for our subsequent purposes.

PROPOSITION 2.1. The set of all fixed points (stationary solutions) of any one of the higher KV equations is an invariant manifold also for any other of the higher KV equations (in particular, for the original KV equation) considered as a dynamical system in function space.

Proof. All the higher KV equations are Hamiltonian systems; the Poisson brackets of the integrals are equal to zero $[I_n, I_m] = 0$ for all n, m. Therefore, all the higher KV equations commute as dynamical systems in function space. Therefore, the set of fixed points for one of them is invariant with respect to all the remaining ones. This proves Proposition 2.1.

We have the following basic theorem:

THEOREM 2.1. 1) All the periodic stationary solutions of the higher KV equations

$$\frac{d}{dx}\left(\sum_{i=0}^{i=n} c_i \frac{\delta I_{n-i}}{\delta u(x)}\right) = 0 \tag{8}$$

are potentials $u(x)$, the number of zones of which does not exceed n.

2) The equation

$$\frac{\delta I_n}{\delta u(x)} + \sum_{i=1}^{n} c_i \frac{\delta I_{n-i}}{\delta u(x)} = d \tag{8'}$$

is a completely integrable Hamiltonian system with n degrees of freedom depending on $(n+1)$ parameters $(c_1, \ldots, c_n, d)$, whereby the collection of n commuting integrals of this system and all the parameters $(c_1, \ldots, c_n, d)$ are expressed in terms of $2n+1$ nondegenerate eigenvalues of both spectra of the potentials $u(x)$ which form the boundaries of the zones.

Proof. As is known, in the case of rapidly decreasing functions $u(x)$ from the Lax representation $\dot{L} = [A, L]$ it is easy to derive an equation for the eigenfunctions ψ_k of the operator $L: \dot{\psi}_k = A\psi_k + \lambda\psi_k$. In the periodic case the analogous derivation gives

$$\dot{\psi}_k = A\psi_k + \lambda\psi_k + \mu\bar{\psi}_k, \quad \dot{\bar{\psi}}_k = A\bar{\psi}_k + \bar{\mu}\psi_k + \bar{\lambda}\bar{\psi}_k, \tag{9}$$

where $\lambda + \bar{\lambda} = 0$.

Indeed, $(L-k^2)\psi_k = 0$. Therefore,

$$0 = \dot{L}\psi_k + (L-k^2)\dot{\psi}_k = (AL - LA)\psi_k + (L-k^2)\dot{\psi}_k = (L-k^2)(\dot{\psi}_k - A\psi_k).$$

Since $(L-k^2)\psi_k = (L-k^2)\bar{\psi}_k = 0$, we obtain the desired result with unknown coefficients $\lambda(x_0, t)$, $\mu(x_0, t)$. To determine the coefficients we make use of the fact that $\psi_k(x_0, x_0) = \psi_k'(x_0, x_0) = 0$. From this for $x = x_0$ we have

$$(A\psi_k)_{x=x_0} + \lambda + \mu = 0, \quad (\dot{\psi}_k)_{x=x_0} = 0,$$
$$\left(\frac{d}{dx} A\psi_k\right)_{x=x_0} + ik(\lambda - \mu) = 0, \quad (\dot{\psi}_k')_{x=x_0} = 0 \tag{10}$$

in the basis $\psi_k, \bar{\psi}_k$. (In the basis $\varphi_k, \bar{\varphi}_k$ it is necessary to let $k \mapsto 1$ in the lower equation.)

We consider the matrix $\Lambda = \begin{pmatrix} \lambda & \mu \\ \bar{\mu} & \bar{\lambda} \end{pmatrix}$, where $\lambda + \bar{\lambda} = 0$. This is a matrix of the Lie algebra of the group $SU_{1,1}$ to which the monodromy matrix $T(k, x_0)$ belongs. The matrix Λ, as is evident from Eq. (10), depends on $u(x_0, t), u'(x_0, t), \ldots, u^{(2n)}(x_0, t), k$.

In order to study the time dependence of the monodromy matrix $T(k, x_0)$ by Eq. (6) it is necessary to compute ψ_k and ψ' at the point $x = x_0 + T$, where T is the period. Having done this we obtain $(x = x_0)$

$$\dot{a} + \dot{b} = A(a\psi_k + b\bar{\psi}_k) + \lambda(a\psi_k + b\bar{\psi}_k) + \mu(\bar{b}\psi_k + \bar{a}\bar{\psi}_k),$$
$$ik(\dot{a} - \dot{b}) = \frac{d}{dx} A(a\psi_k + b\bar{\psi}_k) + \lambda(a\psi_k' + b\bar{\psi}_k') + \mu(\bar{b}\psi_k' + \bar{a}\bar{\psi}_k'). \tag{11}$$

Substituting relation (10) into (11) and performing simple computations, we obtain

$$\dot{a} = \mu\bar{b} - b\bar{\mu}, \quad \dot{a} + \dot{\bar{a}} = 2\dot{a}_R = 0, \quad \dot{b} = (\lambda - \bar{\lambda})b + (a - \bar{a})\mu = 2\lambda b + 2ia_I\mu. \tag{11'}$$

PROPOSITION 2.2. On the basis of the higher KV equation the dependence of the monodromy matrix on time is described by the equation

$$\dot{T} = [\Lambda, T], \tag{12}$$

where the matrix Λ is found from formula (10) starting from the time dependence of the eigenfunction basis $(\psi_k, \bar{\psi}_k)$:

$$(\dot{\psi}_k, \dot{\bar{\psi}}_k) = A(\psi_k, \bar{\psi}_k) + \Lambda(\psi_k, \bar{\psi}_k).$$

This proposition has been proved above. It implies:

COROLLARY 2.3. The equation

$$\Lambda' - \dot{Q} = [Q, \Lambda], \tag{13}$$

holds and is the compatibility condition for the two-parameter family of matrices $T(k, x, t)$ satisfying the equations $\dot{T} = [\Lambda, T]$, $T' = [Q, T]$, where the matrix Q is that of formula (4).

Proof. Since $\dot{T} = \dot{T}'$, it follows that $[\Lambda', T] + [\Lambda, T'] = [\dot{Q}, T] + [Q, \dot{T}]$. Using the Jacobi identity, we obtain $[\Lambda' - \dot{Q} - (\Lambda, Q), T] = 0$.

This proves the corollary (it now follows easily from elementary properties of the Lie algebra of the group $SU_{1,1}$).

We now proceed to the basic Theorem 2.1.

If u(x) is a stationary solution of Eq. (6), then the corresponding monodromy matrix T(k, x) is also stationary $\dot{T} \equiv 0$ and $\dot{Q} \equiv 0$. From Corollary 2.3 we obtain the equation

$$\Lambda' = [Q, \Lambda], \tag{14}$$

where $\mathrm{Sp}\,\Lambda = \lambda + \bar{\lambda} = 0$. The eigenvalues of the matrix Λ are integrals of the system (8). Since $\mathrm{Sp}\,\Lambda = 0$, for the eigenvalues of the matrix Λ we have $\alpha_{\pm}(k) = \pm\sqrt{\det\Lambda}$, where $\det\Lambda = |\lambda|^2 - |\mu|^2$.

Further, $\det\Lambda$ is a polynomial in k^2 of degree $2n+1$, as is easily verified from the form of the matrix Λ, whereby the leading coefficient is a nonzero constant. The roots of the polynomial $\det\Lambda = 0$ are numerical integrals of the system $\Lambda' = [Q, \Lambda]$; formally the coefficients of the polynomial $\det\Lambda$ in the variable k^2 are written in the form of polynomial expressions in u(x) and its derivatives with constant coefficients. We remark that the equation $\Lambda' = [Q, \Lambda]$ is equivalent to the original equation $\frac{d}{dx}\left(\sum c_i \frac{\delta I_{n-i}}{\delta u(x)}\right) = 0$ for the function u(x). Thus, we have $2n+1$ integrals of this system (the roots of the equation $\det\Lambda = 0$ or the coefficients of the polynomial $\det\Lambda$), the first $n+1$ of which are formed from the constants $(c_1, \ldots, c_n, d)$.

We shall prove that these integrals are commutative and define all the zone boundaries of the potential u(x). Using Eq. (12) $[\Lambda, T] = 0$, we can obtain the following relations:

$$1)\ (a - \bar{a})\,\mu = 2\lambda b, \quad 2)\ \bar{b}\mu = \bar{\mu} b. \tag{15}$$

From this we immediately obtain

$$e^{2i\varphi} = b/\bar{b} = \mu/\bar{\mu}, \quad \text{where} \quad \varphi = \arg b, \ ia_I/b = \lambda/\mu. \tag{15'}$$

At the (nondegenerate) points of the spectrum $k^2 = k_n$

$$\left|\frac{ia_I}{b}\right| = \left|\frac{\lambda}{\mu}\right| = 1, \quad k = k_n, \tag{16}$$

or $|\lambda|^2 - |\mu|^2 = \det\Lambda = 0$. Thus, the zone boundaries are the zeros of the polynomial $\det\Lambda = 0$. Passage to the limit of infinite period whereby u tends to a rapidly decreasing function shows that the last n coefficients of the polynomial $\det\Lambda$ are algebraically independent integrals which are polynomials in u and its derivatives for almost any values of the first $n+1$ constants $(c_1, \ldots, c_n, d)$. We thus have n independent integrals of the Hamiltonian system (8') with n degrees of freedom. From the previously mentioned theorem of Zakharov and Faddeev it can be seen also that these integrals commute. Indeed, the manifold of functions [the stationary points of Eq. (6)] lies in a complete function space where all $2n+1$ integrals are commutative. On the finite-dimensional manifold of functions in question the symplectic form is degenerate. The integrals $(c_1, \ldots, c_n, d)$ after restriction to the manifold have zero Poisson brackets with all functions on this manifold. It is easy to verify that the symplectic form on this manifold is in fact obtained by restriction of a form from the entire function space. This implies the complete integrability of Eq. (8') and all the assertions of the basic Theorem 2.1.

Remark 1. It is evident from Eqs. (15') that $\varphi = \arg\mu + m\pi$, $\varphi = \arg b$, and thus $\pm\sin\varphi$ is expressed in terms of u and its derivatives. Combining this fact with Proposition 1.1, we find that also the degenerate points of finite-zone potentials u(x) can be obtained from the transcendental equation (5):

$$2\pi n = \int_0^T \left(2k - \frac{u}{k} \pm \frac{u}{k}\sin\varphi\right) dx,$$

where $\pm\sin\varphi = \sin(\arg\mu)$.

Remark 2. Periodic and conditionally periodic solutions are obtained when the level surface of all the commuting integrals found of Eq. (8') for whatever values of the constants $(c_1, \ldots, c_n, d)$ is compact (i.e., is an n-dimensional torus). In general, for randomly chosen eigenvalues or zone boundaries (or, what is the same thing, values of the constants and integrals), periodic functions are obtained with n incommensurable periods. It is thus more natural to solve the inverse scattering problem in almost-periodic functions rather than in periodic functions only.

Remark 3. The matrix Q itself has the form of the matrix Λ for the "zero-order KV equation" $\{\dot{u} = u'\}$, where the operator $A_0 = d/dx$, $[A_0, L] = u'$. The stationary solutions in this case are constants constituting 0-zone potentials. Further, for any n the transformation $u \to u + c$ takes an n-zone potential into an n-zone potential, and by means of this transformation it can be achieved that the function $u = v + \text{const}$ satisfies the equation

$$\frac{\delta I_n}{\delta u(x)} + c_2 \frac{\delta I_{n-2}}{\delta u(x)} + \ldots + c_n \frac{\delta I_0}{\delta u(x)} = d,$$

where $c_1 = 0$ (it is assumed that $c_0 = 1$). By studying the form of the matrix $\Lambda = \Lambda_R + i\Lambda_I$ starting from the formulas (10), it is possible to show that the matrix Λ_R has the form ($n \geq 1$)

$$\Lambda_R = \left(\sum_{q=0}^{n-1} \gamma_q \left(\frac{d}{dx} \frac{\delta I_q}{\delta u(x)}\right) k^{2(n-1-q)}\right) \begin{pmatrix} 0 & 1 \\ 1 & 0 \end{pmatrix},$$

where the constants γ_q are expressed in terms of the constants $c_1, \ldots, c_n$.

Remark 4. The stationary solutions of the KV equation of order q are contained among the stationary solutions of the KV equation of order $n > q$ as degenerate tori of Eq. (8'), where the $n+1$ constants ($c_1, \ldots, c_n, d$) and the values of the n integrals $I_1, \ldots, I_n$ are chosen such that these integrals depend on the entire level surface. We shall now consider stationary solutions of the KV equations for $n = 1$ and $n = 2$.

1) The Case $n = 1$. In this case we have the equation

$$cu' + 6uu' - u''' = 0,$$

$$u'' = 3u^2 + cu + d, \quad u'^2 = 2\left(u^3 + \frac{cu^2}{2} + du + E\right).$$

We obtain the elliptic function

$$x = \int \frac{du}{\sqrt{2u^3 + cu^2 + 2du + 2E}},$$

where $u(x - ct)$ is a solution of the KV equation of the type of a simple wave. According to Theorem 2.1 $u(x)$ is a 1-zone potential. This fact was first proved by E. Ince in 1940 in another language and by another method [14]; the Sturm–Liouville equation with an elliptic potential is a special case of the Lamé equation arising from the Laplace operator on an ellipsoid where $1/2\, n\,(n+1)$-fold elliptic functions also occur in the potential. (As shown in [14], they are n-zone potentials which are a degenerate case of the general n-zone potentials given by Theorem 2.1).

In the basis ψ_k, $\bar{\psi}_k$ the matrix Λ has the form

$$\lambda = \frac{ik}{2k^2}(u'' - 2u^2 + 8k^4) - ick + \frac{icu}{2k},$$

$$\mu = -u' + \frac{ik}{2k^2}(-u'' + 2u^2 + 4k^2 u) - \frac{icu}{2k}, \tag{17}$$

and the characteristic polynomial has the form

$$\det \Lambda = 4k^6 + 2ck^4 + \frac{c^2 + 4d}{4} k^2 + \frac{cd - 2E}{4}. \tag{17'}$$

We see that $k_1^2 + k_2^2 + k_3^2 = -c/2$. If the period $T \to \infty$ and $u(x)$ tends to a rapidly decreasing function, then $E \to 0$, $d \to 0$. Therefore, $k_3 \to 0$, $k_1^2 \to -\varkappa^2$, $k_2^2 \to -\varkappa^2$, and the zone contracts to the eigenvalue $k^2 = -\varkappa^2$ where $\varkappa^2 = c/4$.

We note that the transcendental equation for all the degenerate points of the spectrum follows from Remark 1.

2) The Case n = 2. On the basis of Remark 3 we consider only an equation of the form

$$\frac{d}{dx}\left(\frac{\delta I_2}{\delta u(x)}+8c\frac{\delta I_0}{\delta u(x)}\right)=0, \tag{8''}$$

where $8I_0=\int(8u^2)\,dx$, $I_2=\int\left(\frac{u''^2}{2}-\frac{5}{2}u^2u''+\frac{5}{2}u^4\right)dx$ [we obtain all 2-zone potentials by adding a constant to the solutions of Eq. (8")]. The Lagrangian of the dynamical system (8") has the form

$$L=L_2+8cL_0-du=L(u,u''),\quad 8L_0=8u^2,\quad L_2=\frac{u''^2}{2}-\frac{5}{2}u^2u''+\frac{5}{2}u^4.$$

We denote by q the quantity $q=(\partial L/\partial u'')$. The energy of a system in which the Lagrangian depends on two derivatives has the form

$$E=L-u''q+u'q'.$$

We denote u' by p_q and q' by p_u. Then $E=H(u,q,p_u,p_q)=V(u,q)+p_up_q$, and Eqs. (8") assumes the Hamiltonian form

$$p_u'=-\frac{\partial H}{\partial u},\quad p_q'=-\frac{\partial H}{\partial q},\quad u'=\frac{\partial H}{\partial p},\quad q'=\frac{\partial H}{\partial p_q},$$

where $V=L-u''q=-\frac{q^2}{2}-\frac{5}{2}qu^2-\frac{5}{8}u^4+8cu^2-du$. However, the study of this Hamiltonian system with two degrees of freedom is not so simple. We therefore compute its remaining integral by using the Lax representation $\Lambda'=[Q,\Lambda]$ of this dynamical system. The operator A has here the form

$$A=A_2+16cA_0=\frac{d^5}{dx^5}-\frac{5}{2}u\frac{d^3}{dx^3}-\frac{15}{4}u'\frac{d^2}{dx^2}+\frac{15u^2-25u''}{8}\frac{d}{dx}+\frac{15}{8}\left(uu'-\frac{u'''}{2}\right)+16c\frac{d}{dx}.$$

Computing the matrix Λ by Eqs. (10), we obtain (in the basis $\psi_k,\bar\psi_k$)

$$-16\lambda=\frac{ik}{k^2}\left(-\frac{u^{(4)}}{2}+4uu''+3u'^2-3u^3+2k^2u^2+16k^4\right)+16ick-8\frac{icu}{k},$$

$$16\mu=(6uu'-u'''+4k^2u')+\frac{ik}{k^2}\left(-\frac{u^{(4)}}{2}+4uu''+3u'^2-3u^3+2k^2u''-4k^2u^2-8k^4u\right)-\frac{8icu}{k},$$

$$\det\Lambda=k^{10}+2ck^6-\frac{d}{16}k^4+k^2\left(c^2+\frac{E}{32}\right)+\frac{I+16cd}{16^2}, \tag{18}$$

where

$$I=p_u^2-2up_up_q+(2q-16c)p_q^2+D,$$
$$D=u^5+16cu^3-4uq^2+32cuq-2dq. \tag{19}$$

The integrals I and E = H are commutative, and the Hamiltonian system (8") is thus completely integrable. Let $p_u^2=\alpha_1^2$,

$$(2q-16c)p_q^2=\begin{cases}\alpha_2^2 & \text{for } q\geqslant 8c,\\ -\alpha_2^2 & \text{for } q<8c.\end{cases}$$

Then

$$\alpha_1^2\pm\alpha_2^2=I-D(u,q)+u(E-V),\quad \pm\alpha_1^2\alpha_2^2=(2q-16c)(E-V)^2,$$
$$\pm 2\alpha_j^2=A\pm\sqrt{A^2-4B},\quad j=1,\ 2, \tag{19'}$$

where $A = I-D+2u(E-V)$, $B = (2q-16c)(E-V)^2$. We find that for given values of the constants c, d, E, I we must seek regions of the (q, u)-plane where

$$A \geqslant 0, \quad A^2 - 4B \geqslant 0, \quad q \geqslant 8c. \tag{20}$$

Compact regions satisfying inequality (20) give tori in the phase space (u, q, p_u, p_q). It is evident from the equations that compact regions are possible only in the half space $q \geq 8c$. Indeed, for $q < 8c$ the expression $A \pm \sqrt{A^2-4B}$ always has roots of different signs which coalesce only at the isolated points $A = 0$, $B = 0$.

For these potentials (periodic and conditionally periodic with two periods) the relation $\sum_j k_j^2 = 0$ is satisfied.

As noted in Remark 3, the remaining 2-zone potentials found in Theorem 2.1 are obtained by adding a constant $u \to u + \text{const}$, where this constant is the sum of the eigenvalues – the zone boundaries – of the new potential v(x). The equations giving the singular points of the Hamiltonian system, as is easily seen, reduce to the conditions

$$p_u = p_q = 0, \quad q = -\frac{5}{2}u^2, \quad 10u^3 - 16cu + d = 0$$

or

$$u = u(c, d), \quad q = q(c, d), \quad E = V(c, d), \quad I = D(c, d).$$

In parameter space this is a two-dimensional surface $E = E(c, d)$, $I = I(c, d)$. In general, under these conditions we obtain in phase space compact separated level surfaces E = const, I = const of torus type with one degenerate cycle. The trajectories of the KV equation on these surfaces describe the interaction of a periodic simple wave with a rapidly decreasing soliton. The interaction of two rapidly decreasing solitons (a two-soliton solution of the problem with rapidly decreasing initial data) is obtained [up to an additive constant U(c, d)] if still another condition on the parameters is satisfied under which both cycles on the torus degenerate to a point. This relation has the form of a condition on the parameters under which the polynomial det Λ has three distinct roots, two of which are double roots (the two zones contract to points).

With the exception of this special case, the evolution in time of two-zone potentials according to the Korteweg–de Vries equation is characterized by two constants Δ_1, Δ_2 such that

$$u(x + \Delta_1, t + \Delta_2) = u(x, t).$$

Calculation of these constants Δ_1, Δ_2 in terms of the zone boundaries will be given in the second part of the work.

In the second part of the work we shall study n-soliton solutions in more detail. Does there exist a superposition law synthesizing them from single-soliton solutions – an algebraic function of pairs of solitons (elliptic functions) which contains doubly valued points (roots) and therefore, in general, leads beyond the field of elliptic functions? In terms of the characteristic polynomials this appears as follows: there are two solitons $u_1(x, c_1, d_1, E_1)$, $u_2(x, c_2, d_2, E_2)$ with matrices $\det \Lambda^{(1)} = (k^2-\lambda_1)(k^2-\lambda_2)(k^2-\lambda_3)$, $\det \Lambda^{(2)} = (k^2-\mu_1)(k^2-\mu_2)(k^2-\mu_3)$.

Suppose that the roots λ_j, μ_j are nonmultiple and $\lambda_1 = \mu_1$ (a condition for the possibility of composition); the remaining 4 roots λ_2, λ_3, μ_2, μ_3 are all distinct. The superposition law of solitons is such that the 2-soliton potential $v = F(u_1, u_2)$ has a characteristic polynomial det Λ in the form of the least common multiple of the initial polynomials $\det \Lambda = (k^2-\lambda_2)(k^2-\lambda_3)(k^2-\mu_2)(k^2-\mu_3)(k^2-\lambda_1)$, where $\lambda_1 = \mu_1$. The correct analog of the amplitude a_k for rapidly decreasing functions is here the quantity det Λ. However, this definition of superposition is ineffective. Completely different representations of the superposition law are possible; we shall discuss these in Part II.

Remark 5. V. B. Matveev and L. D. Faddeev have informed the author that in 1961 N. I. Akhiezer essentially formulated and began the solution of the problem of constructing examples of finite-zone potentials starting from the results of [2, 3] on the inverse scattering problem on the half line. In this work [1] N. I. Akhiezer developed an interesting approach to the construction of finite-zone potentials using facts from the theory of hyperelliptic Riemann surfaces. His construction, however, gives for a prescribed zone structure only a finite number of potentials which satisfy specific parity conditions in x. Theorem 2.1 of the present work gives many more periodic and almost-periodic n-zone potentials – they depend on n continuous parameters for a prescribed zone structure. Since the KV equation $\{\dot{u} = 6uu' - u'''\}$ is not invariant under the transformation $x \to -x$, it follows that during evolution in time other potentials will be obtained from those of Akhiezer which are not contained in his construction of n-zone potentials. Judging from the work [1], N. I. Akhiezer was not familiar with the work of E. Ince [14] which actually proved (in another language) that an elliptic function is a 1-zone potential. It is curious to note that the three proofs of this particular fact which follow from the works of E. Ince [14], N. I. Akhiezer [1], and the present work are, in principle, all different.

LITERATURE CITED

1. N. I. Akhiezer, "A continuum analog of orthogonal polynomials in a system of integrals," Dokl. Akad. Nauk SSSR, 141, No. 2, 263-266 (1961).
2. I. M. Gel'fand and B. M. Levitan, "On the determination of a differential operator from its spectral function," Izv. Akad. Nauk SSSR, Ser. Matem., 15, 309-360 (1951).
3. V. A. Marchenko, "Some questions in the theory of one-dimensional differential operators, I," Trudy Mosk. Matem. Obshch. I, 327-420 (1952).
4. L. D. Faddeev, "Properties of the S-matrix of the one-dimensional Schrödinger equation," Trudy Matem. Inst. im. V. A. Steklova, 73, 314-336 (1964).
5. C. Gardner, J. Green, M. Kruskal, and R. Miura, "A method for solving the Kortweg–de Vries equation," Phys. Rev. Lett., 19, 1095-1098 (1967).
6. R. M. Miura, C. S. Gardner, and M. Kruskal, "Kortweg–de Vries equation and generalizations," J. Math. Phys., 9, No. 8, 1202-1209 (1968).
7. P. Lax, "Integrals of nonlinear equations of evolution and solitary waves," Comm. Pure Appl. Math., 21, No. 2, 467-490 (1968).
8. V. E. Zakharov and L. D. Faddeev, "The Kortweg–de Vries equation – a completely integrable Hamiltonian system," Funkt. Analiz., 5, No. 4, 18-27 (1971).
9. V. E. Zakharov, "A kinetic equation for solitons," Zh. Éksp. Teor. Fiz., 60, No. 3, 993-1000 (1971).
10. V. E. Zakharov and A. B. Shabat, "An exact theory of two-dimensional self-focusing and one-dimensional automodulation of waves in nonlinear media," Zh. Éksp. Teor. Fiz., 61, No. 1, 118-134 (1971).
11. A. B. Shabat, "On Kortweg–de Vries equations," Dokl. Akad. Nauk SSSR, 211, No. 6, 1310-1313 (1973).
12. V. E. Zakharov and A. B. Shabat, "On the interaction of solitons in a stable medium," Zh. Éksp. Teor. Fiz., 64, No. 5, 1627-1639 (1973).
13. R. Hirota, "Exact solution of the modified Korteweg–de Vries equation for multiple collisions of solitons," J. Phys. Soc. Japan, 33, No. 5, 1456-1458 (1972).
14. E. L. Ince, "Further investigations into the periodic Lamé functions," Proc. Roy. Soc. Edinburgh, 60, 83-99 (1940).
15. B. B. Kadomtsev and V. I. Karpman, "Nonlinear waves," Usp. Fiz. Nauk, 103, No. 2, 193-232 (1971).
16. V. E. Zakharov and S. B. Manakov, "On the complete integrability of the nonlinear Schrödinger equation," Zh. Matem. i Teor. Fiz., 19, No. 3, 322-343 (1974).

Uspekhi Mat. Nauk 37:5 (1982), 3–49 Russian Math. Surveys 37:5 (1982), 1–56

The Hamiltonian formalism and a many-valued analogue of Morse thoery

S.P. Novikov

CONTENTS

Introduction[(1)]

It is now scarcely a matter of dispute that dynamical systems describing real physical processes are, as a rule, Hamiltonian in one sense or another if the dissipation of energy can be disregarded. However, the Hamiltonian formalism may turn out to be non-classical, that is, it may not originate from a Lagrangian formalism as a result of a Legendre transformation. There may not be global canonical coordinates. This refers in the first place to many systems of hydrodynamic origin. Various aspects of the Hamiltonian formalism will be discussed in greater detail in §§1, 2. Another aim of this survey is to describe topological methods of search for periodic trajectories. The fact is that the overwhelming majority of non-trivial conservative systems are non-integrable even for two degrees of freedom. After stationary points, periodic solutions are the simplest objects of the qualitative theory of dynamical systems; nevertheless, even the problem of the existence of periodic trajectories is often highly non-trivial and requires the use of

(1)This survey was written as a result of reworking and extending the author's contribution to [27], Ch. 1, which was written for the English edition.

topological methods. The Morse and Lyusternik-Shnirel'man (LSM) theory, which combines the calculus of variations with the topology of function spaces consisting of closed contours (curves) on the relevant *configuration space* (see §3), is widely known.

However, the use of the LSM theory necessitates the strict requirement of a positive-definite Lagrangian formalism. From this it is clear that in most general Hamiltonian systems not of Lagrangian origin, this theory, generally speaking, cannot be applied. Variational principles on phase trajectories never give rise to positive-definite functionals. Some very interesting systems, which we call Kirchhoff systems, reduce to a problem, mathematically equivalent to the theory of a charged particle in a magnetic field "the Dirac monopole" (see §4). The following systems are of Kirchhoff type:

(a) the Kirchhoff equation for the motion of a rigid body in an ideal *incompressible fluid* moving under a potential and at rest at infinity;

(b) the equation of motion of a *rigid body* with a fixed point in an axially symmetric strong field;

(c) the Leggett equation for the *magnetic moment* in the low temperature phases of ^{3}He (nuclear magnetic resonance).

In these systems, equations of motion can ultimately be reduced to a principle of extremal action S. But (see §5) from a global point of view the action S turns out to be a "many-valued" functional on the space of closed contours (smooth curves) on the sphere S^2, which after a reduction plays the role of the configuration space. This means that δS is a single-valued quantity (a 1-form or covector) on the space of contours, but the "integrals over cycles" in the space of contours of δS are non-trivial. Therefore, S is a many-valued functional (for example, on a circle $d\varphi$ is a single-valued 1-form, but φ is a many-valued).

One of the purposes of §5 is to extend the topological methods of LSM theory to many-valued functionals. This enables us to establish the existence of a large collection of periodic orbits for systems of Kirchhoff type. The results of §4, 5 are mainly from [1] and [2]. An analogue of Morse theory for many-valued functions (closed 1-forms) on finite-dimensional manifolds is constructed in §6. The results of this section are from [3].

§1. The Hamiltonian formalism. Simplest examples. Systems of Kirchhoff type. Factorization of the Hamiltonian formalism for the B-pahse of ^{3}He

From the contemporary point of view, at the basis of the Hamiltonian formalsim lies the concept of a "*Poisson bracket*". Let y^i be local coordinates on a manifold (the "phase-space"); the Poisson bracket of two functions $f(y)$ and $g(y)$ is given by a tensor field $h^{ij}(y)$:

$$\{f, g\} = h^{ij}(y)\frac{\partial f}{\partial y^i}\frac{\partial g}{\partial y^j}. \tag{1}$$

Here we require the following properties to hold:

(a) bilinearity and skew-symmetry

$$\{f, g\} = -\{g, f\}, \tag{2}$$

(b) the Leibniz identity

$$\{fg, h\} = g\{f, h\} + f\{g, h\}, \tag{3}$$

(c) the Jacobi identity

$$\{f, \{g, h\}\} + \{h, \{f, g\}\} + \{g, \{h, f\}\} = 0. \tag{4}$$

By definition, Hamiltonian systems have the form

$$\dot{f} = \{f, H\}, \tag{5}$$

where f is any function and H is the Hamiltonian.

It can happen that there are non-trivial functions f_q (possibly, defined locally on the manifold) such that

$$\{f_q, g\} = 0 \tag{6}$$

for any function $g(y)$. In this case the Poisson bracket is said to be "degenerate"; the matrix $h^{ij}(y)$ is degenerate. After finding all such quantities $f_l(y)$ then on their common level surface

$$f_l(y) = \text{const} \quad (l = 1, 2, \ldots) \tag{7}$$

the Poisson bracket becomes non-degenerate.

Let z^q be coordinates on the level surface (7). The restriction of the tensor $h^{qt}(z)$ to this surface is non-degenerate, and there is an inverse matrix

$$h_{qp}h^{pt} = \delta_q^t, \tag{8}$$

which determines the 2-form

$$\Omega = h_{qp}(z)\, dz^q \wedge dz^p. \tag{9}$$

From (4) it follows that the form Ω is closed:

$$d\Omega = 0 \leftrightarrow \frac{\partial h_{qp}}{\partial z^t} + \frac{\partial h_{tq}}{\partial z^p} + \frac{\partial h_{pt}}{\partial z^q} = 0. \tag{10}$$

Let us consider the main types of phase spaces.

Type I: The classical Hamiltonian formalism and variational principles. Suppose that $(y) = (x^1, \ldots, x^n, p_1, \ldots, p_n)$ and that the matrix h^{ij} is constant and non-degenerate:

$$h^{ij} = h_{ij} = \begin{pmatrix} 0 & \begin{matrix} 1 & & 0 \\ & \ddots & \\ 0 & & 1 \end{matrix} \\ \begin{matrix} -1 & & 0 \\ & \ddots & \\ 0 & & -1 \end{matrix} & 0 \end{pmatrix} = \text{const}. \tag{11}$$

The equations (6) take the form $\dot{x}^i = \partial H/\partial p_i$, $\dot{p}_i = -\partial H/\partial x^i$.

The coordinates (x, p) are said to be canonical. Locally they can always be found for non-degenerate Poisson brackets (Darboux's theorem).

If $H(x, p)$ is a Hamiltonian, then we have the Lagrangian $L(x, \dot{x})$, where x is the configuration space coordinate, which can be defined by

$$\dot{x}^i = \frac{\partial H}{\partial p^i}(x, p), \quad L = p_i \dot{x}^i - H. \tag{12}$$

We assume that the equation $\dot{x}^i = \partial H/\partial p_i$ can be solved for the variables p_i. The Hamiltonian equations (12) are obtained from the variational principle $\delta S = 0$, where

$$S = \int L(x, \dot{x})\, dt. \tag{13}$$

Type II. The Hamiltonian formalism and Lie algebras.[1]

We consider now the following (second) case in order of complexity, when the tensor h^{ij} is not constant, but depends linearly on the coordinate (y):

$$h^{ij}(y) = C^{ij}_k y^k, \qquad C^{ij}_k = \text{const.} \tag{14}$$

We consider the set L of all linear functions on the phase space, which we denote by L^*. For the basis linear forms (the coordinates y^i) we define the operation of "commutation".

$$[y^i, y^j] = C^{ij}_k y^k = \{y^i, y^j\}. \tag{15}$$

From (2) and (4) it follows that the operation (15) turns the linear space L into a Lie algebra for which the dual space L^* is the phase space for the Poisson bracket (14). A bracket of this kind was first considered by Berezin. It was used by Kirillov and Kostant (in the less convenient language of symplectic manifolds) in the theory of infinite-dimensional representations of Lie groups.

Example 1. A basic example of the Hamiltonian formalism of Type I is the phase space $T^*(M)$: the space of covectors (with subscripts) on the manifold M (the configuration space). The manifold M can be infinite-dimensional (a space of fields $q(x)$ in which x is one of the "indices" in the formulae). In the finite-dimensional case there are the local coordinates x^i and the conjugate momenta p_i with the Poisson brackets

$$\{x^i, x^j\} = \{p_i, p_j\} = 0, \qquad \{x^i, p_j\} = \delta^i_j \tag{16}$$

and the form

$$\Omega_0 = \sum dx^i \wedge dp_i.$$

[1] The third case in order of complexity, when the tensor $h^{ij}(x)$ depends quadratically on x, is also very interesting and has been studied recently (Sklyapin, Faddeev).

In the infinite-dimensional case there are two fields and Poisson brackets of the form

(17) $$\begin{cases} \{q^i(x),\ p_j(y)\} = \delta^i_j \delta(x-y), \\ \{q^i(x),\ q_j(y)\} = \{p_i(x),\ p_j(y)\} = 0. \end{cases}$$

Example 2. It is also useful to consider a Poisson bracket of the form (18) with an additional "external field" $F_{ij}(x)$:

(18) $$\{x^i,\ x^j\} = 0, \quad \{x^i,\ p_j\} = \delta^i_j, \quad \{p_i,\ p_j\} = F_{ij}(x),$$

where the 2-form $F = F_{ij}\, dx^i \wedge dx^j$ is closed:

$$dF = 0.$$

Then we have the 2-form

(19) $$\Omega = \sum dx^i \wedge dp_j + \sum F_{ij}\, dx^i \wedge dx^j = \Omega_0 + F.$$

The equations of motion with a Hamiltonian $H(x, p)$ and a Poisson bracket (18) are (for $n = 2$ or 3) the equations of motion of a charged particle in an external magnetic field F_{ij} (or an electromagnetic field for $n = 4$). In the domain where $F = dA$, (19) reduces to the standard form (16).

Example 3. Rather more general *a priori* but as a rule reducible to the form (18) are Poisson brackets on the space $T^*(M)$ satisfying the following requirement; any pair of functions (f, g) on the base M (independent of the variables p_i on the *fiber* consisting of all covectors with a lower index) has a vanishing Poisson bracket:

(20) $$\{f,\ g\} = 0.$$

We call (20) "variational admissibility" of the Poisson bracket on $T^*(M)$. Clearly, the bracket (18) is variationally admissible. As we know, on sufficiently small domains any (non-degenerate) Poisson bracket reduces to the form (16). Globally this is no longer so; if the form Ω is not exact, then the Poisson bracket does not reduce to the form (16). Variationally admissible Poisson brackets are probably always globally reducible to the form (18), but this has not been proved rigorously; they reduce to the simplest form (16) on any domain when Ω is exact. Let $(x^1, \ldots, x^n, y^1, \ldots, y^n)$ be local coordinates in a domain U_α such that $\{x^i,\ x^j\} = 0$ and let $H(x, y)$ be a Hamiltonian. We consider half of the Hamilton equations

(21) $$\dot{x}^j = \{x^j,\ H\}.$$

We assume that (21) allows us to express the variables (y^i) uniquely in terms of $(x, \dot{x})$:

(22) $$y^i = F^i(x,\ \dot{x}).$$

The other half of the Hamilton equations

$$\dot{y}^j = \{y^j, H\} = G^j(x, y)$$

now reduces by (22) to the second-order system

$$F^j(x, \dot{x})^{\cdot} = G^j(x, \dot{x}). \tag{23}$$

Let us now construct the "phase Lagrangian" $L(x, y)\,dt = -H\,dt + \omega_\alpha$, where $d\omega_\alpha = \Omega$ in U_α. We express L in terms of $(x, \dot{x})$, using (22).

Lemma. *The equations* (23) *are equations of the extremals for the Lagrangian variational principle* $\delta S = 0$, $S = \int L(x, \dot{x})\,dt$.

These are the elementary properties of variationally admissible Poisson brackets.

We now pass on to discuss examples of Poisson brackets of Type II associated with Lie algebras.

Example 1. Let L be the Lie algebra of the group SO_3. The Killing metric is Euclidean and allows us to identify L with L^*. The Poisson bracket of the basis functions M_i on L^* has the form

$$\{M_i, M_j\} = \varepsilon_{ijk} M_k, \quad c_k^{ij} = \varepsilon_{ijk} = \pm 1. \tag{24}$$

The function $M^2 = \sum M_i^2$ is such that

$$\{M^2, M_i\} = 0 \quad (i = 1, 2, 3). \tag{25}$$

Hamiltonian systems on L^* have the form

$$\dot{M}_i = \{M_i, H(M)\}. \tag{26}$$

Let $\Omega^i = \partial H / \partial M_i$; the Killing metric allows us to identify upper and lower indices. The equations (26) reduce to the "Euler equations"

$$\dot{M} = [M, \Omega]. \tag{27}$$

This conclusion holds for all compact (semisimple) Lie groups on which the Killing metric is Euclidean (pseudo-Euclidean) on the Lie algebra and invariant under inner automorphisms

$$L \to gLg^{-1}, \tag{28}$$

where g is an element of the Lie group and L its Lie algebra. Arnol'd calls such systems for the groups SO_N "many-dimensional analogues of a rigid body" if the Hamiltonian is a quadratic form on the space of skew-symmetric matrices $(a_{ij}) = (-a_{ji})$, where $M = (M_{ij})$

$$H(M) = \sum_{i<j} d_{ij} M_{ij}^2, \tag{29}$$

and

$$d_{ij} = q_i + q_j, \quad q_i > 0. \tag{30}$$

Now it is known that all systems of the form (30) on the Lie algebra of SO_N are completely integrable [46]. Moreover, according to [46], a sufficient condition for integrability is[(1)] that

(31) $$d_{ij}=\frac{\bar{a}_i-\bar{a}_j}{\bar{b}_i-\bar{b}_j}.$$

The idea of [46] is as follows. Under the conditions (31) the Euler equation (27) can be represented as a statinary problem for first-order metric systems in the (x, t)-space admitting an "$L-A$"-pair, or the method of the inverse problem (see [47]). In accordance with the formalism of integration of stationary problems [16] there arises the matrix equation:[(2)]

(32) $$\frac{d}{dx}(M-\lambda a)=[M-\lambda a,\ \Omega-\lambda b],$$

$$a=(a_{ij}),\quad b=(b_{ij}),\quad a_{ij}=\bar{a}_i\delta_{ij},\quad b_{ij}=\bar{b}_i\delta_{ij}.$$

The coefficients of the polynomial

(33) $$P(\lambda,\ \mu)=\det(\mu\cdot 1-M-\lambda a)$$

are integrals of (27) "in involution", that is, have zero Poisson brackets between pairs. A complete set of formulae of the motion can be obtained in terms of the θ-function associated with the Riemann surface $P(\lambda, \mu)=0$, starting from the methods of [16] and ending in [49] for first-order matrix systems. We recall that the Poisson bracket (15) is invariant only under the transformations (28). For the classical Euler equations for the free rotation of a rigid body we have

(34) $$G=SO_3,\quad H=\sum\frac{a_iM_i^2}{2},\quad \Omega=\frac{\partial H}{\partial M},$$

where Ω is the angular velocity of the body and M is the angular momentum.

Example 2. Some important systems arising in hydrodynamics are connected with the Lie algebra L of the group $E(3)$ of motions of the Euclidean space $\mathbf{R}^3$. This algebra is no longer semisimple. On the phase space L^* there are 6 coordinates $(M_1, M_2, M_3, p_1, p_2, p_3)$ and the Poisson brackets

(35) $$\{M_i,\ M_j\}=\varepsilon_{ijk}M_k,\quad \{M_i,\ p_j\}=\varepsilon_{ijk}p_k,\quad \{p_i,\ p_j\}=0.$$

---

(1) If $\bar{a}_i=\bar{b}_i^2$, then we have (30), $q_i=\bar{b}$. The Liouville integrability for SO_4 under the condition (30) was first established in [51] and [52]. However, the connection with the method of the inverse problem and the theory of θ-functions of Riemann surfaces remained unknown; for this reason they did not succeed in obtaining an explicit integration, even in this simplest case.

(2) To a reader unfamiliar with the method of the inverse problem (see [47]) the emergence of equations of the type (32) may seem incomprehensible; in this case, to understand what follows he must start directly from (32) as a formal identity whose verification represents no difficulty once it is written down.

The bracket (35) has two independent functions $f_1 = p^2 = \sum p_i^2$, $f_2 = ps = \sum M_i p_i$ such that

$$\{f_q, M_i\} = \{f_q, p_i\} = 0 \quad (q = 1, 2, i = 1, 2, 3). \tag{36}$$

Let $H(M, p)$ be the Hamiltonian. We write $u^i = \partial H/\partial p_i$, $\omega^i = \partial H/\partial M_i$. The Hamilton equations assume the "Kirchhoff" form

$$\dot{p} = [p \times \omega], \quad \dot{M} = [M \times \omega] + [p \times u]. \tag{37}$$

The equations (37) coincide (for a quadratic Hamiltonian H) with the Kirchhoff equations for the motion of a rigid body in an ideal incompressible fluid at rest at infinity [4]. The motion of the liquid itself is assumed to be of potential form. In this case H is the energy, M and p are the total *angular* and *linear momentum* of the system, the body being identified with a moving system of coordinates rigidly attached to it. The energy H is assumed to be positive and quadratic in both variables (M, p). By transformations of the form (28) H can be brought to the form

$$2H = \sum a_i M_i^2 + \sum b_{ij}(p_i M_j + M_i p_j) + \sum c_{ij} p_i p_j. \tag{38}$$

Even in classical hydrodynamics non-trivial integrable cases were discovered of Hamiltonians of the form (38) for the algebra $L = E(3)$. These cases of Clebsch and Steklov do not reduce to an "obvious" group symmetry. We are especially interested in the case of Clebsch, in which the diagonality relations

$$b_{ij} = 0, \quad c_{ij} = \bar{c}_i \delta_{ij} \tag{39}$$

hold as well as the "Clebsch relations"

$$\sum_{i=1}^{i=3} \bar{c}_i a_i (\bar{c}_{i+1} - \bar{c}_{i-1}) = 0 \quad (i + 3 \approx i). \tag{40}$$

The Lie algebra of SO_4 is obtained from $E(3)$ by "deformation" or, conversely, L can be obtained from SO_4 be "retraction". The precise meaning of this is the following: in SO_4 we choose a basis (e_i', e_i'') such that

$$\begin{cases} [e_i', e_j'] = [e_i'', e_j''] = \varepsilon_{ijk} e_k', \\ [e_i', e_j''] = \varepsilon_{ijk} e_k''. \end{cases} \tag{41}$$

Changing to a new basis $\bar{e}_i' = e_i'$, $\bar{e}_i'' = \alpha e_i''$ we obtain

$$\begin{cases} [\bar{e}_i', \bar{e}_j'] = \varepsilon_{ijk} \bar{e}_k', \quad [\bar{e}_i', \bar{e}_j''] = \varepsilon_{ijk} \bar{e}_k'', \\ [\bar{e}_i'', \bar{e}_k''] = \varepsilon_{ijk} \cdot \alpha^2 \bar{e}_j'. \end{cases} \tag{42}$$

Letting $\alpha \to 0$, we obtain from (42) the relations (35), which define the algebra $L = E(3)$.

We now write in (29)

$$\begin{cases} d_{i0} = c_i \quad (i = 1, 2, 3), \\ d_{12} = a_3, \ d_{23} = a_1, \ d_{13} = a_2. \end{cases} \tag{43}$$

The quantities (31) are connected by the following relation for SO_4

$$0=\sum_{\substack{i=1\\ i\approx i+3}}^{i=3} c_i a_i\,(c_{i+1}-c_{i-1}+a_{i+1}-a_{i-1}). \tag{44}$$

Let us complete the retraction of SO_4 to $L = E(3)$ according to (41) and (42); we require that the quadratic form (29) has a finite limit under this limit passage. For this, under the conditions (43) it is then necessary that the a_i are finite as $\alpha \to 0$ and the c_i are of order $c_i \sim \bar{c}_i\alpha^{-2}$. As $\alpha \to 0$, we obtain from (44) precisely the Clebsch relation (40)[1]. So we arrive at the result due to Novikov and Golo:

The recently discovered cases of integrability of systems on SO_4 are deformations of the classical Clebsch case.

For diagonal Hamiltonians of the form (39) on $L = E(3)$ (more precisely, on L^*) that do not satisfy the Clebsch condition, the absence of "superfluous" analytic integrals of motion has recently been proved [50]. Thus, "general" diagonal Hamiltonians on L are non-integrable.

We consider two other applications of (37):

(A) The equations of motion of a rigid body with a fixed point in a strong axially symmetric field with potential $W(z)$ reduce to (37). The corresponding Hamiltonian is

$$H=\sum a_i M_i^2/2+W\,(l^i p_i), \tag{45}$$

where l^i is the vector determined by the position of the centre of mass with respect to the principal axes of inertia at the fixed point. The quantities p_i are dimensionless and cannot be interpreted physically as *momenta.* They are the direction cosines of a unit vector, that is,

$$f_1 = p^2 = 1. \tag{46}$$

(B) The (Leggett) equation of the spin dynamics in the A-phase of the *superfluid* ^{3}He can also be reduced to the form (37); this is the dynamics of the spin variables of the vectors (s, d), where $d^2 = 1$, by analogy with (46). (See the survey by Brinkman and Cross in [5].) On transition to the Leggett equations for nuclear magnetic resonance in the A-phase one must alter the notation (S is the "magnetic moment")

$$M_i \to s_i, \quad p_i \to d_i, \tag{47}$$

---

[1] We note that the coefficients of the "stationary $L-A$-pair" (32) diverge under the retraction $\alpha \to 0$, although the integrals of the motion converge. In this connection, recently in [53] another matrix representation depending on λ of the Kirchhoff equations for the Clebsch case has been constructed. By way of contrast, this representation does not admit a deformation for $\alpha \neq 0$ in any non-trivial way.

and consider a Hamiltonian of the form

$$H=\frac{1}{2}as^2+b\left(\sum s_id_i\right)^2+\lambda\left(\sum s_iH_i\right)+W(d). \tag{48}$$

Here a, b, and λ are constants, H^i is the external magnetic field, and the potential W has the form

$$W(d)=\text{const}\,(l^id_i)^2. \tag{49}$$

By the property (36) of the Poisson bracket (35), $f_2=\sum s_id_i$ is equivalent to a constant in the equations of motion. Therefore, the second term can simply be deleted from the Hamiltonian:

$$H\sim H'=\frac{1}{2}as^2+\lambda\sum s_iH_i+W(d). \tag{50}$$

The quantity d, the spin part of the so-called "order parameter", is a unit vector, $d^2=1$, as was mentioned above.

We consider a very important example (though not associated with Lie algebras).

There is another phase, the B-phase, of ^{3}He, in which the Leggett equation takes a form that is not similar to the classical top (see, for example, the survey by Brinkman and Cross in [5]).

In a state of hydrodynamic equilibrium and with non-zero spin, the state in the B-phase is defined by a pair comprising a rotation matrix $R=(R_{ij})\in SO_3$ and a "*magnetic moment*" $s=(s_i)$ $(i=1,2,3)$.

The variables s_i are coordinates in the dual space of the Lie algebra of SO_3 similar to the angular momentum components M_i. The standard Poisson brackets for $T^*(SO_3)$ in the variables $(s_i,\ R_{jk})$ can be written:

$$\begin{cases}\{s_i,\ s_j\}=\varepsilon_{ijk}s_k, & \{R_{ij},\ R_{kl}\}=0,\\ \{s_i,\ R_{jl}\}=\varepsilon_{ijk}R_{kl}.\end{cases} \tag{51}$$

The Leggett Hamiltonian in the B-phase in an external magnetic field has the form

$$H=\frac{1}{2}as^2+b\sum s_iF_i+V(\cos\Theta), \tag{52}$$

where a and b are constants, $F=(F_i)$ is the external field, and

$$V(\cos\Theta)=\text{const}\left(\frac{1}{2}+2\cos\Theta\right)^2, \tag{53}$$

R_{ij} is the rotation through the angle Θ around the axis n_i, $n^2=1$:

$$\begin{gathered}R_{ij}=\cos\Theta\delta_{ij}+(1-\cos\Theta)\,n_in_j+\sin\Theta\varepsilon_{ijk}n_k,\\ 1+2\cos\Theta=R_{ii}=\operatorname{Sp}R.\end{gathered} \tag{54}$$

After the substitution

$$as_i=\omega_i,\qquad \Omega_{jk}=\varepsilon_{jki}\omega_i=(\dot{R}R^{-1})_{jk} \tag{55}$$

we obtain a Lagrangian system in the variables $(R_{ij}, \dot{R}_{ij})$ on $T^*(SO_3)$ where the kinetic energy is defined by the 2-sided invariant Killing measure, and the potential $V(\cos\Theta)$ is invariant under the inner automorphisms

(56) $$R \to gRg^{-1}, \quad s \to gs, \qquad g \in SO_3.$$

If the field $F = (F_i)$ is constant, then the Lagrangian is invariant under the one-parameter group of transformations (56), where g belongs to the group of rotations around the axis of F. Let $F = (F, 0, 0)$.

When $F = 0$, the system admits the group SO_3 of transformations (56) and is completely integrated in [6]. The transformations (56) generate the conserved vector (when $F = 0$):

(57) $$A = (A_j) = (1-\cos\Theta)\left[n \times \left(\cot\frac{\Theta}{2} S + [n \times S]\right)\right]_j,$$

with the same Poisson brackets as for the usual angular momentum:

(58) $$\begin{cases} \{A_i, A_j\} = \varepsilon_{ijk}A_k, \\ \left\{A_i, \frac{1}{2}as^2 + V(\cos\Theta)\right\} = 0. \end{cases}$$

As Golo has shown [7], when $F = 0$, the variables s^2 and Θ in the Hamiltonian generate a closed algebra of Poisson brackets $\{s^2, s_\parallel, \Theta\}$, where

(59) $$\begin{cases} s_\parallel = \sum s_i n_i, \\ \{s^2, \Theta\} = 2s_\parallel, \quad \{s_\parallel, \Theta\} = 1 \\ \{s^2, s_\parallel\} = \dfrac{1+\cos\Theta}{\sin\Theta}(s^2 - s_\parallel^2). \end{cases}$$

The quantity $A^2 = \sum A_i^2 = (1-\cos\Theta)(s^2 - s_\parallel^2)$ has vanishing Poisson brackets with all generators of this subalgebra

(60) $$\{A^2, s^2\} = \{A^2, s_\parallel\} = \{A^2, \Theta\} = 0.$$

In a non-zero magnetic field $(F, 0, 0)$ there remains only one integral apart from the energy[(1)]

(61) $$\{A_1, H\} = 0.$$

The system becomes non-integrable. In this case it seems to be possible to complete (globally) the procedure of "factorization of the Hamiltonian formalism" and to reduce the system to 2 degrees of freedom.

The integral A_1 generates the group (56), where g is a rotation about the axis $n = (1, 0, 0)$. The invariant variables under this subgroup are

(62) $$s^2, \ s_\parallel, \ \Theta, \ n_1, \ s_1, \quad \tau = s_2 n_3 - n_2 s_3$$

with the purely geometrical constraint

(63) $$s^2\tau^2 = (s^2 - s_1^2)(s^2 - s_\parallel^2) - (s^2 n_1 - s_1 s_\parallel)^2.$$

[(1)] For large fields $F \to \infty$ this system has been studied in [8] with viscosity taken into account.

It is easy to check that the variables (62) form a closed algebra of Poisson brackets, containing the Hamiltonian H (52) and having the functional dimension 5. The quantity A_1 in this algebra has vanishing brackets with all the variables:

$$(64)\quad 0 = \{A_1, s^2\} = \{A_1, s_{\|}\} = \{A_1, \Theta\} = \{A_1, n_1\} = \{A_1, s_1\}.$$

Therefore, by imposing the condition $A_1 = \text{const}$ we can, as before, formally use the formulae for the Poisson brackets of the quantities (62), which arise from (51). Under the condition $A_1 = \text{const}$ we choose as basis the following variables:

$$(65)\qquad (A^2, s_{\|}, \Theta, n_1), \quad n_1 = n.$$

Their brackets have the form

$$(66)\qquad \begin{cases} \{s_{\|}, \Theta\} = 1, \quad \{\Theta, n\} = 0, \\ \{A^2, s^2\} = \{A^2, \Theta\} = \{A^2, s_{\|}\} = 0, \\ \{A^2, n\} = \sqrt{\frac{1}{2}(1-n^2)A^2 - \frac{1}{4}A_1^2}. \end{cases}$$

Thus, the canonical variables can be chosen in the form

$$(67)\qquad \begin{cases} x^1 = \Theta, \quad \xi_1 = p_\Theta = s_{\|}, \\ x^2 = n, \quad \xi_2 = p_n = \sqrt{\frac{2A^2}{1-n^2} - \frac{A_1^2}{(1-n^2)^2}}. \end{cases}$$

The Hamiltonian becomes

$$(68)\qquad H = \frac{1}{2}a\left[p_\Theta^2 + \frac{1-n^2}{2(1-\cos\Theta)}\left(p_n^2 + \frac{A_1^2}{(1-n^2)^2}\right)\right] + \\ + bF\left(np_\Theta + \frac{1-n^2}{2}\sin\Theta p_n + A_1^2\,\frac{2-\sin^2\Theta}{2(1-\cos\Theta)}\right) + V(\cos\Theta).$$

We now introduce the spherical coordinates

$$(69)\qquad \Theta = 2\chi, \quad n = n_1 = \sin\varphi$$

and go over to the Lagrangian formalism. We obtain

$$(70)\qquad L = 2a(\dot\chi^2 + \sin^2\chi\dot\varphi^2) - \tilde{A}_1\dot{y}^1 - \tilde{A}_2\dot{y}^2 - U(y),$$

where $y^1 = \chi,\ y^2 = \varphi$,

$$\tilde{A}_1 = 2b\sin\varphi, \quad \tilde{A}_2 = 8bF\cos\varphi\sin^3\chi\cos\chi,$$

$$(71)\qquad U = V(\cos\Theta) + aA_1^2/4\sin^2\chi\cos^2\varphi + bFA_1(1-\sin^2\chi\cos^2\chi)/2\sin^2\chi - \\ - b^2F^2(\sin^2\varphi + 4\cos^2\varphi\sin^3\chi\cos\chi)/2.$$

Thus, we have obtained a system in a domain in the sphere S^2 with the usual metric, in which there is an effective magnetic field and a scalar potential. When $A_1 \neq 0$, this system cannot be extended to the whole sphere, since it is singular at $\varphi = 0, \pi$.

If $A_1 = 0$, then the system is defined on the whole sphere except at the poles, where it has singularities. We note that the great circle $\varphi = 0, \pi$ corresponds to the axis $n = (\pm 1, 0, 0)$; the rotations around this axis correspond to the group of symmetries of the system.

Finding the stationary points of the Hamiltonian system (68) presents no difficulty. They are given by the equations

$$\frac{\partial H}{\partial \Theta} = \frac{\partial H}{\partial n} = \frac{\partial H}{\partial p_\Theta} = \frac{\partial H}{\partial p_n} = 0. \tag{72}$$

The stationary solutions of (72) are periodic solutions of the original Leggett equations. These exact solutions were not known previously, as Fomin has told the author.

Since we have treated in the text (above) the equations for a magnetic moment under homogeneous nuclear magnetic resonance (NMR) in the superfluid ^3He, it is appropriate to recall the definition of the A- and the B-phases. From the microscopic theory of super-conductivity of Bardeen-Cooper-Schriffer, Bogolyubov, Gor'kii, and Anderson one deduces for a coupling with moment $l = 1$ that the superfluid ^3He can be described macroscopically in the stationary state according to the Ginzburg-Landau scheme by a (3×3) complex matrix $A_{qj}(x, y, z)$, where the index $q = 1, 2, 3$ refers to the (internal) "spin" space, while the index $j = 1, 2, 3$ refers to physical space. The field $A_{qj}(x, y, z)$ is called the "order parameter". It must minimize the free energy functional, which depends on the temperature, the magnetic field, the pressure, and the other external parameters:

$$F\{A\} = \int_{R^3} (F_{\text{grad}} + V)\, d^3x, \tag{73}$$

where

$$F_{\text{grad}} = \gamma_1 (\partial_k \bar{A}_{qj} \partial_k A_{qj}) + \gamma_2 (\partial_k \bar{A}_{qj} \partial_j A_{qk}) + \gamma_3 (\partial_k \bar{A}_{qk})(\partial_i A_{qi}). \tag{74}$$

In the absence of a magnetic field (assuming also that the dipole energy is small) the potential V has the form

$$\begin{aligned} V = {} & \alpha \operatorname{Sp}(\bar{A}^T A) + \beta_1 |\operatorname{Sp} A A^T|^2 + \beta_2 (\operatorname{Sp}(\bar{A}^T A))^2 + \\ & + \beta_3 \operatorname{Sp}[(A^T \bar{A})(\overline{\bar{A} A^T})] + \beta_4 \operatorname{Sp}[(\bar{A}^T A)^2] + \beta_5 \operatorname{Sp}[(\bar{A}^T A)(\overline{\bar{A}^T A})]. \end{aligned} \tag{75}$$

The exact values of the parameters α, β, γ are undetermined and can vary together with the parameters of the system (the temperature etc.).

The concept of "phase" is defined in the spatially homogeneous state $F_{\text{grad}} = 0$ by minimizing the function $V(A_{qj})$. It is difficult to classify the "phases", that is, the minima of V for all values of the parameters α and β are an unsolved problem. In any case, potentials of the form (75) are invariant under the action of the group

$$\begin{cases} G = U_1 \times SO_3 \times SO_3, \\ gA = e^{i\varphi} R_1^{-1} A R_2, \\ g = (e^{i\varphi}, R_1, R_2). \end{cases} \tag{76}$$

Therefore, the minima of the potential are manifolds on which G acts. In the case of "general position" they are homogeneous spaces of G. However, there is an important example of the A_1-phase close to the critical temperature when the pressure and the field are small which is defined by a degenerate minimum (by a non-homogeneous submanifold M_{A_1}, in the matrix space

$$(77) \qquad M_{A_1}=\{A_{qj}=\Delta\cdot(d_q e_j^{(1)}+\bar{d}_q e_j^{(2)})\},\ d_q=d_q'+id_q'',$$
$$e_j^{(\alpha)}=e_j'^{(\alpha)}+e_j''^{(\alpha)},\qquad |e'^{(\alpha)}|^2=|e''^{(\alpha)}|^2=1,\qquad (e'^{(\alpha)},e''^{(\alpha)})=0\qquad (\alpha=1,2).$$

The more popular A- and B-phases are defined by the G-homogeneous matrix manifolds M_A and M_B consisting of matrices of the form

$$(78)\qquad \begin{cases} M_A=\{A_{qj}=2\Delta d_q(e_j'+ie_j'')\}, & |d|^2=1,\\ |e'|^2=|e''|^2=1,\quad (e',e'')=0, & \Delta=\text{const},\\ M_A=(S^2\times SO_3)/\mathbf{Z}_2 \end{cases}$$

or $e^{(1)}=e^{(2)}=e=e'+ie''$, $\quad d=\bar{d}$;

$$(79)\qquad M_B=\{A_{qj}=\Delta/\sqrt{3}\cdot R_{qj}e^{i\varphi}\},\qquad R\in SO_3,$$
$$\Delta=\text{const},\quad M_B=SO_3\times U_1.$$

Passing on to states depending on (x, y, z) we consider "quasi-homogeneous" states, where the deviation of the field $A_{qj}(x, y, z)$, from a spatially-homogeneous state can be disregarded locally, and we may assume that every $A_{qj}(x, y, z)$ lies in a "phase-manifold" M_{A_1}, M_A, M_B, ..., but changes from point to point. Now $F_{\text{grad}}\neq 0$, although the whole field is regarded as having values only in the phase manifold.

The Euler-Lagrange equation $\delta F_{\text{grad}}=0$ for fields with values in the manifolds M_{A_1}, M_A, M_B etc., which define the state of the system, are called the Ginzburg-Landau equations. States depending only on the single variable z, "planar textures", lead for the B-phase to the usual equation of the Euler top (here it is even symmetrical). For the A-phase the equations of planar textures are more complex; they have been fully integrated in [43], where one can find references on ^{3}He (see also [5] and [6]). In a magnetic field, in a state with non-zero spin, the functional of free energy becomes more complex; there arises a new variable, the "magnetic moment" S whose dynamic (see above) is used in the so-called "nuclear magnetic resonance".

The planar textures in the A_1-phase are not known, and it would be interesting to study them. In the manifold M_{A_1} there is a singular submanifold

$$(80)\qquad W: e'^{(1)}\times e''^{(1)}=e'^{(2)}\times e''^{(2)},\quad W\subset W_{A_1}.$$

The submanifold W has codimension 3, although it is given by two equations in the 8-dimensional manifold M_{A_1}. We have become accustomed to the fact that the number of "Goldstone perturbations" is equal to the dimension of degeneration, that is, the dimension of the vacuum manifold. The dimension of M_{A_1} is 8. However, at points of W the number of Goldstone modes turns out to be 9, as Volovik and Fomin have communicated to me. In the given case, the dimension 9 coincides with the dimension of the "tangent space" to M_{A_1} at points of W in the sense of algebraic geometry. Apparently, the number of Goldstone modes always coincides with the dimension of the tangent space of algebraic geometry. In all previously known cases in field theory the vacuum manifold was homogeneous and hence non-singular.

§2. The Hamiltonian formalism of systems of hydrodynamic origin

In this survey we do not discuss any new results on hydrodynamic systems (with the exception of the Kirchhoff system already introduced in §1), and this section is purely methodological in character. The Hamiltonian formalism has already been worked out long ago in the language of the so-called "Clebsch variables" for various types of ideal fluids (see below). However, as will become clear, these field variables cannot always be introduced, and if they can, then frequently only locally. Here the Clebsch variables are extremely unstable under a change of the type of the system: the addition to the system of a superfluous field (for example, even the transition from an incompressible fluid to a weakly compressible one in which the density and entropy are the new field variables) leads to a non-local and by no means small change in the Clebsch variables. Besides, in several cases the number of field variables is odd. In the latter case one must introduce superfluous fictitious degrees of freedom to define the Clebsch variables; they represent a system with a large, complicated and poorly understood "calibrated" freedom. Consequently, an invariant exposition of the Hamiltonian formalism of hydrodynamic systems is useful. In the incompressible case such an invariant exposition can be found in [19], but its language (that of "symplectic manifolds") seems to be artificially complicated and inconvenient compared with the language of "Poisson brackets". The situation is more complex for compressible fluids and further systems (see below).

The underlying Lie algebra L for hydrodynamic systems over which the subsequent superstructure will be erected, is the algebra of vector fields (we do not yet specify the domain of definition). For vector fields $v^i(x)$, $w^i(x)$ in an n-dimensional space the commutator is

$$[v, w]^i(x) = v^j \frac{\partial w^i}{\partial x^j} - w^j \frac{\partial v^i}{\partial x^j}. \tag{1}$$

Here the pairs (x, i) (the point x and the index i) act as a single "index". The operation must be expressed in terms of the structure constants in the form

$$[v, w]^i(x) = \int dy\, dz C^i_{jk}(x, y, z)\, v^j(y)\, w^k(z). \tag{2}$$

Comparing (1) and (2) we obtain

(3) $$C^i_{jk}(x, y, z) = \delta^i_j \delta(z-x)\, \partial^{(y)}_k (y-z) - \delta^i_k \delta (y-x)\, \partial^{(z)}_j \delta(z-y),$$
$$\partial^{(x)}_j = \frac{\partial}{\partial x^j}, \quad \int \partial^{(z)}_j \delta(z-x) f(z) = -\frac{\partial f}{\partial z^j}(x).$$

The variables $p_i(x)$ *conjugate* to the velocity components on the dual space L^* to the vector fields $v^i(x)$ must be such that

(4) $$\int p_i(x) v^i(x) d^n x$$

is scalar under change of variables. This means that the variables $p_i(x)$ are densities of covectors, which under changes of variables are additionally multiplied by the Jacobian (we call them momentum densities). According to § 1 (14), the Poisson brackets are of the form

(5) $$\{p_j(y), p_k(z)\} = \int C^i_{jk}(x, y, z)\, p_i(x)\, d^n x =$$
$$= p_k(y)\, \partial^{(y)}_j \delta(y-z) - p_j(z)\, \partial^{(z)}_k \delta(z-y).$$

Here is an important example, the case $n = 1$. Then we obtain

(6) $$\{p(y), p(z)\} = p(y)\delta'(y-z) - p(z)\delta'(z-y).$$

By making the substitution $p = u^2$ we arrive at the standard Poisson bracket (Gardner, Zakharov, Fadeev) occurring in the theory of the KdV (Korteweg-de Vries) equation, see [9], [10], [11]:

(7) $$\{u(x), u(y)\} = \delta'(x-y).$$

For in the KdV theory it is precisely the quantity $I_0 = \int u^2\, dx$ that plays the role of the momentum (see [11]). The Poisson bracket of two functionals has the form

(8) $$\{J, I\} = \int \frac{\delta J}{\delta u(x)} \frac{\partial}{\partial x} \frac{\delta I}{\delta u(x)}\, dx.$$

Since the operator $\partial/\partial x$ has the constants as non-trivial kernel, there is a quantity $I_{-1} = \int u\, dx$ such that

(9) $$\{J, I_{-1}\} = 0$$

for any functional J. The KdV equation itself is given by the Hamiltonian

(10) $$I_1 = H = \int \left(\frac{u'^2}{2} + u^3\right) dx,$$
$$\dot{u} = \frac{\partial}{\partial x} \frac{\delta H}{\delta u(x)} = 6uu_x - u_{xxx}.$$

It is curious that one of the phenomena of the integrability of the KdV equation by the *method of the inverse problem* is the presence of another local Poisson bracket of the two functionals [12] and even of a family of brackets with the operators $A+\lambda\partial/\partial x$;

$$(11)\qquad \begin{cases} A=-\dfrac{\partial^3}{\partial x^3}+2\left(u\dfrac{\partial}{\partial x}+\dfrac{\partial}{\partial x}u\right), \\ \{J,\ I\}_2=\displaystyle\int\frac{\delta J}{\delta u(x)}A\frac{\delta I}{\delta u(x)}\,dx. \end{cases}$$

The operators $A+\lambda\partial/\partial x$ are obtained from A by the substitution $u\to u+\text{const}$. The KdV equation itself has the following form in the new Hamiltonian structure:

$$(12)\qquad \dot{u}=A\frac{\delta I_0/2}{\delta u(x)},\qquad I_0=\int u^2\,dx.$$

A further investigation of systems that are Hamiltonian for a family of Poisson brackets can be found in [15].

Note 1 (Adler, Manin, Lebedev). We mention (although this adds nothing new to the construction of solutions of non-linear equations by the inverse problem method) that from the purely algebraic point of view the integrable systems can be interpreted, starting from standard properties of "transformation operators" [48], as systems on phase spaces of type L^* for Lie algebras of Volterra integral ("upper triangular") operators L with the corresponding Hamiltonian formalism (see [13], [14]); the set of Poisson brackets arising here was already known earlier [11]). However, this algebraic interpretation does not completely cover the algebraic essence of the Hamiltonian formalism in the method of the inverse problem.

Note 2 (Bogoyavlenskii, Novikov). It is appropriate to note here another interesting phenomenon arising in KdV theory: the connection between the Hamiltonian formalisms of stationary and non-stationary problems for Hamiltonian systems given by the Poisson bracket (7); suppose that we are given a system

$$(13)\qquad u_t=\frac{\partial}{\partial x}\frac{\delta H}{\delta u(x)},\qquad H=\int P(u,\ u_x,\ \ldots,\ u^{(n)})\,dx,$$

where the Hamiltonian H has the form (13) and P is a polynomial with constant coefficients, and an integral of it, that is, a functional of the same form $J=\int Q(u,\ u_x,\ \ldots,\ u^{(m)})\,dx$ such that $\{J,\ H\}=0$. We consider the stationary equation $u_t=0$ or

$$(14)\qquad \frac{\delta(H+\lambda I_{-1})}{\delta u(x)}=0.$$

Since $\{J,\ H\}=0$, we have for any function $u(x)$ the identity

$$(15)\qquad \left(\frac{\partial}{\partial x}\frac{\delta J}{\delta u(x)}\right)\left(\frac{\delta(H+\lambda I_{-1})}{\delta u(x)}\right)=\frac{\partial}{\partial x}T_\lambda(u,\ u_x,\ \ldots).$$

Therefore, T_λ is an integral of (14). For the translation group $J = I_0$ the integral T_λ is the energy for (14).

We consider a *flow* with the Hamiltonian J in the Poisson bracket (8) that commutes with the initial flow (13):

$$u_\tau = \frac{\partial}{\partial x}\frac{\delta J}{\delta u(x)}. \tag{16}$$

Proposition. *The restriction of the flow* (16) *to the finite-dimensional phase space of the stationary system* (14) *is also Hamiltonian in the new bracket and is generated in Hamiltonian fashion by* T_λ *(see* [11], [16], [17]).

Apparently this is true for a wide class of Poisson brackets (see [18] and a number of other papers quoted there).

We return to systems of hydrodynamic type. In the algebra of vector fields L in a Euclidean space (in which are distinguished the Euclidean metric and the element of volume, namely, the mass density, which is assumed to be constant) we specify the subalgebra of *divergence-free* fields $L_0 \subset L$ by

$$\partial_i v^i = 0. \tag{17}$$

As is easy to see, the dual space L_0^* is obtained by

$$L_0^* = L^*/(\partial_i \varphi). \tag{18}$$

By (17), the momentum densities $p_i(x)$ give trivial linear forms on L_0 if $p_i = \partial_i\varphi$:

$$0 = \int p_i v^i\, d^n x = \int v^i\, \partial_i\varphi\, d^n x = -\int \varphi\, \partial_i v^i\, d^n x. \tag{19}$$

The hydrodynamical Euler equation for an ideal incompressible fluid, as a Hamiltonian system, can be written (see [19]) in the space $L_0^* = L^*/\partial_i\varphi$ with a Hamiltonian of the form

$$H = \int \rho \frac{v^2}{2}\, d^n x, \quad \rho = \text{const}, \quad \partial_i v^i = 0, \quad p_i = \rho v^i \tag{20}$$

and the Poisson brackets (5). These equations are always written on the complete space L^*, which in this case is equivalent to the space of velocities:

$$\begin{cases} \rho v_t^i = \{p_i, H\} + \partial_i p, \\ \partial_i v^i = 0. \end{cases} \tag{21}$$

The terms $\partial_i p$ come from the transition from L_0^* to L^*, where quantities of the form $\partial_i\varphi$ are equivalent to zero. Here, the pressure p is in principle determined by $\partial_i p$ only. On the space L_0^* we can write the Poisson bracket in the form

$$\begin{cases} \{v_i(x), v_j(y)\} = \frac{1}{\rho}(\partial_i v_j - \partial_j v_i)\,\delta(x-y), \\ p_i = \rho v_i, \end{cases} \tag{22}$$

where $\rho = \text{const}$.

In fact, in the presence of boundary conditions the velocity $v_j(x)$ of an incompressible fluid is determined by the *vortex* $\Omega_{ij} = \partial_i v_j - \partial_j v_i$ under the condition $\partial_i v_i = 0$.

Example. The case $n = 2$. When $n = 2$ the vortex Ω_{12} reduces to a single scalar function $\Omega_{12} = f(x)$. Thus, for $n = 2$ the Poisson bracket (22) reduces to a Poisson bracket for scalar functions $f(x)$. It has the form

$$\{f(x), f(y)\} = \partial_1 f \partial_2 \delta(x-y) - \partial_2 f \partial_1 \delta(x-y). \tag{23}$$

However, a Hamiltonian H of the form (20) becomes complicated in the "vortex" variables. For the "finite-dimensional" case we have a set of discrete *vortices* ($x = x^1$, $y = x^2$):

$$\Omega_{12} = f(x, y) = \sum_{\alpha=1}^{N} q_\alpha \delta(x - x_\alpha)\delta(y - y_\alpha). \tag{24}$$

For such f, assuming the q_α to be constant, we obtain from (23) the usual $2N$-dimensional phase space with the canonical variables

$$x_1, \ldots, x_N, \quad p_1 = y_1, \ldots, p_N = y_N \tag{25}$$

(that is, the coordinates x and y are canonically conjugate in the plane). The Hamiltonian of the system of vortices has the form

$$H = \sum_{\alpha>\beta} q_\alpha q_\beta \log \sqrt{(x_\alpha - x_\beta)^2 + (y_\alpha - y_\beta)^2}. \tag{26}$$

For a 3-dimensional incompressible fluid one can introduce the canonical "Clebsch variables" locally, starting from the representation

$$\begin{cases} p_i = \psi \partial_i \varphi \quad (\mathrm{mod}\ \rho \partial_i f), \\ p_i = \rho v_i, \quad \rho = \mathrm{const}, \\ \frac{1}{\rho} d\psi \wedge d\varphi = \Omega_{ij}\, dx^i \wedge dx^j = d(v_i\, dx^i), \\ \{\psi(x), \psi(x')\} = \{\varphi(x), \varphi(x')\} = 0, \\ \{\psi(x), \varphi(x')\} = \delta(x - x'). \end{cases} \tag{27}$$

Since $\partial_i p_j - \partial_j p_i = \rho \Omega_{ij}$, we find that the vortex lines are given by

$$\varphi = \mathrm{const}, \quad \psi = \mathrm{const}, \tag{28}$$

because

$$\rho \Omega_{ij}\, dx^i \wedge dx^j = d\psi \wedge d\varphi.$$

Thus, we arrive at the conclusion: the canonical Clebsch variables can be introduced globally if (and only if) the form Ω_{ij} can be decomposed into the product of two 1-forms; the decomposition gives a mapping of the domain under investigation into a two-dimensional space (for example, into a domain of a plane, a sphere, or a torus) such that the vortex lines are *inverse images* of points. Hence we conclude:

if the vortex lines are entangled and form a complex dynamical system, then the Clebsch variables cannot be introduced globally.

We now return to the Lie algebra L of all smooth vector fields. We consider a simple example: the simplest Hamiltonian on the Lie algebra L of all vector fields and the phase space L^* without derivative of momenta is the Hamiltonian of non-interacting sound waves, of the "background gas"

$$H = \int c(x) \mid p \mid d^n x. \tag{29}$$

In this case the Hamilton equations are easily integrated; for c = const the solution is given by the standard substitution

$$v(x, t) = v_0(x - tv(x, t)), \tag{30}$$

where

$$v(x, t) = cp/\mid p \mid, \quad v_0 = v(x, 0).$$

The solution (30) means that the particles conserve momentum and their motion is free and rectilinear; if $c(x) \neq$ const, then the motion is also free, but proceeds along a geodesic of the metric $g_{ij} = c(x)\delta_{ij}$, similarly to *Fermat's principle.*

In spite of its evident meaning, the formula (30) contains topologically non-trivial possibilities, if we wish to know the solution $v(x, t)$. For any x and t there is the mapping $F_{x,t} : S^{n-1} \to S^{n-1}$ given by $F_{x,t}(m) = v_0(x - mt)$ for a unit vector m. The fixed points of

$$m = v_0(x - mt) = F_{x,t}\ (m) \tag{31}$$

also give the solution $m = v(x, t)$. If $p(x)$ vanishes nowhere for $t = 0$, then the degree deg $F_{x,t}$ is always 0.

The Hamiltonian formalism for an ideal compressible fluid cannot be realized on the algebra L; this is a special case of the Hamiltonian formalism for fluids with internal degrees of freedom. Two of the more complicated systems of this kind are the magnetohydrodynamics, where the magnetic field is "frozen" into the particles of the fluid [21], and also the superfluid ^{4}He, which has an internal degree of freedom of quantum provenance [22]. A number of more complicated systems are now known (spin glasses, rigid bodies with dislocations and disclinations, and anisotropic phases of the superfluid ^{3}He; see [23]-[25]). Of course, in real systems in addition to the Hamiltonian part there are "viscous" terms in the equations. However, even when these are large, the approximate Hamiltonian formalism enables us to predict correctly (we hope) the structure of the equations of motion themselves for which in certain cases, for example ^{3}He, there is as yet no alternative.

Example 1. *A classical compressible fluid.* However, we are now interested in the fact that even an ordinary compressible fluid has such internal degrees of freedom: the mass density ρ and the entropy density s, and if we wish to include them, we have to extend the Lie algebra of vector fields. To the vector fields v^i we add another pair of fields v^ρ and v^s with commutators of the form

$$(32)\quad [(v, v^\rho, v^s), (w, w^\rho, w^s)] = \\ = ([v, w], v^i\partial_i w^\rho - w^i\partial_i v^\rho, v^i\partial_i w^s - w^i\partial_i v^s).$$

We denote the algebra (32) by $L_{\rho,s}$ and the variables in the dual space $L^*_{\rho,s}$ by ρ (mass density) and s (entropy density) with Poisson brackets (the velocities are here the covectors $v_i = p_i\rho^{-1}$):

$$(33)\quad \begin{cases} \{p_i(x), \rho(y)\} = \rho(x)\partial_i\delta(y-x), \\ \{p_i(x), s(y)\} = s(x)\partial_i\delta(y-x), \\ \{\rho(x), \rho(y)\} = \{s(x), s(y)\} = \{\rho(x), s(y)\} = 0, \\ \{v_i(x), v_j(y)\} = \frac{1}{\rho}\Omega_{ij}(x)\delta(x-y). \end{cases}$$

Let $H = \int(1/2\rho p^2 + \varepsilon(\rho, s))\, d^n x$ be this energy. The Euclidean metric contained in the Hamiltonian permits us to identify upper and lower indices.

The quantities $M = \int \rho\, d^n x$ and $S = \int s\, d^n x$ have vanishing Poisson brackets with all functionals ("trivial" conservation laws). The Poisson brackets (33) were chosen essentially so that that mass and entropy are transported with the particles, in contrast to the energy, which is conserved only globally. For $n = 2$ we can introduce the canonical "Clebsch variables" (evidently globally):

$$(34)\quad \begin{cases} p_i = \rho\partial_i\varphi + s\partial_i\psi, \\ \{\rho, s\} = \{\varphi, \psi\} = \{\rho, \psi\} = \{s, \varphi\} = 0, \\ \{\rho(x), \varphi(y)\} = \{s(x), \psi(y)\} = \delta(x-y). \end{cases}$$

For $n = 3$ there are three cases:

(a) An irrotational barotropic *flow*, where the vortex is zero and the entropy is redundant as a field variable. The Clebsch variables are

$$p_i = \rho\partial_i\varphi,$$

$$(35)\quad \{\rho, \rho\} = \{\varphi, \varphi\} = 0,\ \{\rho(x), \varphi(y)\} = \delta(x-y).$$

(b) A barotropic flow (the entropy is not a field variable)

$$(36)\quad p_i = \rho\,\partial_i\varphi + \alpha\,\partial_i\beta,\ \Omega_{ij} = d(\alpha\rho^{-1}) \wedge d\beta,$$

where α is conjugate to β and φ to ρ. For the same reasons as above (see (28)) a global introduction of Clebsch variables is, in general, not possible.

(c) General flows. Here the canonical Clebsch variables contain the "redundant" field variable

$$(37)\quad \begin{cases} p_i = \rho\partial_i\varphi + s\partial_i\psi + \alpha\partial_i\beta, \\ \Omega_{ij}\,dx^i \wedge dx^j = d\left(\frac{s}{\rho}\right) \wedge d\psi + d\left(\frac{\alpha}{\rho}\right) \wedge d\beta. \end{cases}$$

It would be useful to calculate the degree of many-valuedness of the representation (37) and to clarify the extent to which it holds globally. Let us consider a simpler example.

We recall that in two-dimensional barotropic flow the Clebsch variables also contain the redundant field variable

$$(38)\qquad \begin{cases} p_i = \rho\partial_i\varphi + \alpha\partial_i\beta, \\ \Omega_{12}\,dx^1 \wedge dx^2 = d\left(\frac{\alpha}{\beta}\right) \wedge d\beta. \end{cases}$$

For this reason, on the space of field variables $\alpha/\rho(x)$, and $\beta(x)$ there acts a "calibrating group" (that is, a group of transformations of the plane $\mathbf{R}^2$ depending on $\alpha\rho^{-1}$ and conserving their exterior product: the element of area (or 2-form)). This group preserves the representations (38).

Example 2. *The superfluid* ^{4}He. The equations of hydrodynamics (without dissipation far from the *point* of transition) for the superfluid ^{4}He can be written down in the same variables p_i, ρ, and s together with the "superfluid velocity" $v_{si} = \partial_i\varphi$ (see [22]). The Poisson brackets have the form (33), where additionally the brackets for all quantities with the variable φ are:

$$(39)\qquad \begin{cases} \{\varphi(x),\ s(y)\} = 0, \quad \{\varphi(x),\ \rho(y)\} = \delta(x-y), \\ \{p_i(x),\ \rho\varphi(y)\} = \rho\varphi(x)\,\partial_i\delta(y-x), \quad \{\varphi,\ \varphi\} = 0. \end{cases}$$

As previously, the energy acts as Hamiltonian. The Hamiltonian is given in the form

$$(40)\qquad H = \int\left[\frac{\rho v_s^2}{2} + p_{0i}v_s^i + \varepsilon_0(\rho,\ s,\ p_0)\right]d^nx.$$

It is assumed that p_0 is proportional to $v_n - v_s$:

$$(41)\qquad \begin{cases} p_0 = \rho_n(v_n - v_s) = p - \rho v_s, \\ v_n^i - v_s^i = \dfrac{\partial\varepsilon_0}{\partial p_{0i}}, \quad v_s^i = \partial_i\varphi. \end{cases}$$

The quantities ρ_n and ρ_s are called the densities of the normal and of the superfluid components of the fluid. The momentum is $p = \rho_n v_n + \rho_s v_s$. We introduce the "Clebsch variables" as usually,

$$p_i = \rho\,\partial_i\varphi + s\,\partial_i\psi + \alpha\,\partial_i\beta.$$

It would make sense to investigate the question of global impediments to the introduction of Clebsch variables in more detail.

Various more complicated versions of equations of "superfluid" systems and other anisotropic fluids can be found in the surveys [24] and [25].

Note. By analogy with §1, (14), we can write down the Poisson brackets for the algebra of vector fields in an "external" magnetic field given by a 2-form $F = F_{ij}\,dx^i \wedge dx^j$. These brackets are defined by the extended Lie algebra L_F, in which the commutator of the basic vector fields $e_i = \partial/\partial x^i$ is given in the form (of an e-extension)

$$(42) \qquad [e_i, e]=0, \quad [e_i, e_j]=F_{ij}(x)\, e.$$

For the fields $v=v^i e_i+\psi e$ and $w=w^i e_i+\varphi e$ we obtain

$$(43) \qquad [v, w]=(v^j\partial_j w^i-w^j\partial_j v^i)\, e_i+(F_{ij}w^i v^j+w^i\partial_i\psi+v^i\partial_i\varphi)\, e.$$

To this algebra there correspond in L_i^* the conjugate variables (p_i, q) and the Poisson brackets

$$(44) \qquad \begin{cases} \{p_i(x), q(y)\}=q(x)\,\partial_i\delta(y-x), \\ \{q(x), q(y)\}=0, \\ \{p_i(x), p_i(y)\}=p_j(x)\,\partial_i\delta(y-x)- \\ \qquad\qquad -p_i(y)\,\partial_j\delta(x-y)+F_{ij}(x)\,\delta(x-y). \end{cases}$$

Example 3. As already mentioned, in magneto hydrodynamics the magnetic field is not external but is "frozen" into the particles of the fluid (as always, the Hamiltonian coincides with the energy, including the magnetic energy); the Poisson brackets have another form: the Poisson bracket of the momentum densities conserve the form (5), while that of the magnetic field itself with momenta are such that the field is transported by the particles, as with ρ and s. This means that the flux through an arbitrary fluid surface remains unchanged (only the surface itself is transported) (see [21]). Such brackets with momenta can be introduced for differential forms of any rank: if $H_{i_1,\ldots,i_k}(x)$ is a skew-symmetric tensor of any rank (the k-form $H = H_{i_1,\ldots,i_k}\, dx^{i_1}\wedge\ldots\wedge dx^{i_k}$), then the bracket has the form

$$(45) \qquad \{H(x), H(y)\} = 0,$$

$$\{p_i(x), H(y)\} = H(x)\,\partial_i\delta(y-x) - \partial_i\wedge(H(x)\delta(x-y)),$$

where the operation $\partial_i\wedge\ldots$ on skew-symmetric tensors (forms) of rank k gives a skew-symmetric tensor of rank $k+1$. For example,

(1) $\{p_i(x), \varphi(y)\} = \varphi(x)\,\partial_i\delta(y-x) - \partial_i(\varphi\delta) = (-\partial_i\varphi)\delta(x-y)$ where is φ a scalar;

$$(46) \qquad 2)\ \{p_i(x), \rho(y)\} = \rho(x)\,\partial_i(y-x),$$

where ρ is a form of rank n (scalar density);

3) $\{p_i(x), A_j(y)\} = A_j(x)\,\partial_i\delta(y-x) - \partial_i(A_j\delta) + \partial_j\delta(A_i\delta)$, where A_j is a covector;

4) $\{p_i(x), H_{jk}(y)\} = H_{jk}(x)\,\partial_i\delta(y-x) - \partial_i(H_{jk}\delta) + \partial_j(H_{ik}\delta) - \partial_k(H_{ij}\delta)$, $H_{jk} = -H_{kj}$.

§3. What is Morse (LSM) theory?

The general Morse theory [26] deals with the solution of the following problem: given a finite- or infinite-dimensional space M (*manifold*) on which there is given a function (functional) $S\colon\ M \to R$.

Fundamental problem of Morse theory. How are the stationary points $dS = 0$ (or $\delta S = 0$ for functionals) connected with the topology of the manifold M?

If the critical points are non-degenerate, that is if $\delta^2 S$ is non-degenerate at critical points (as one says, there are no "zero modes"), then the "index" (the Morse index) is the number of negative squares of the form $\delta^2 S$ if this is finite.

Morse theory (in its classical version) is constructed under the following assumptions:

(a) all the critical points are non-degenerate, and the Morse indices are finite;

(b) all the domains $S \leqslant$ const for the function S are relatively compact (the Arzélà principle); this means that a sequence of points x_i such that $S(x_i) < C$ has a limit point in M.

Under these hypotheses the following inequality is established: the number $M_i(S)$ of critical points of index i is not less than the Betti number (the rank of the *homology group*) of M:

$$M_i(S) \geqslant b_i(M). \tag{1}$$

The mechanism by which this inequality arises is very simple. Each critical point x of index k has a "*surface of most rapid discharge*", that is, a map of the disk D^k (the open ball of dimension k, where $\sum_{\alpha=1}^{k} (y^\alpha)^2 < 1$):

$$f\colon D^k \to M.$$

A function S that is bounded on the disk D^k can have only one critical point: one maximum at the centre 0, where $f(0) = x$. The map f should be continued "downwards" through the levels of the function in such a way that the image of the boundary $f(\delta D^k)$ falls into the union of "surfaces of fastest discharge" of the various critical points x_q, where a) $S(x_q) < S(x)$ and b) the indices of all the points x_q are less than k.

Thus, the function S generates a *cell partition* of the manifold M, where the number of *cells* of dimension k is equal to the number of critical points of index k for S.

Since all k-dimensional cycles can be formed from the cells of a cell partition, the rank of the group of cycles (and of the homology group: its factor group by the boundary) does not exceed the number of cells:

$$M_k(S) \geqslant b_k(M).$$

If M is finite-dimensional, then the "Poincaré-Morse theorem" holds:

$$\sum_{i\geqslant 0}(-1)^i M_i(S) = \sum(-1)^i b_i(M) = \chi(M),$$

where $\chi(M)$ is the Euler-Poincaré characteristic and $M_i(S)$ the number of cells.

In passing through the critical value $c = c_k$ the level surface $V_c\{S = c\}$ and the domain $W_c\{S \leqslant c\}$ undergo the operations of *reconstruction* (it is assumed that there is only one critical point on $S = c_k$):

(a) $W_{c_k+\varepsilon} = W_{c_k-\varepsilon}$ plus "a *handle* of index k",

(b) $V_{c_k+\varepsilon}$ is the "Morse reconstruction" of the manifold $V_{c_k-\varepsilon}$.

The operations of "*attaching a handle*" and of "*Morse reconstruction*" have great significance in topology itself. (There are manifold invariants that are finer than the Betti numbers, which enable us to give a lower bound for the number of critical points of S, even when δS is a degenerate form. These are the so-called "Lyusternik-Shnirel'man" categories. We do not define these invariants here (see [28]).) In the case of "general position" all the critical points are non-degenerate. Also useful is the case (which arises quite frequently, especially when there is symmetry),

$$Q_h \subset M, \quad \delta S = 0.$$

Suppose that a) l_k is the dimension of the critical manifold Q_k; b) that the form $\delta^2 S$ is non-degenerate on planes normal to the submanifolds Q_k, and that it has a finite number k of negative squares (the Morse index). Then there is an inequality for the numbers determined by the homology of the set of critical points

$$M_j(S) = \sum_k b_{j-k}(Q_k) \geqslant b_j(M), \tag{2}$$

where b_j is the Betti number in the homology mod 2 (under certain hypotheses of orientability this is also true for the ranks of the homology groups with arbitrary coefficients).

Such is "Morse theory" on compact or open manifolds *without boundary*. For manifolds with an *edge* Morse theory can be extended naturally when the whole boundary is a level surface $S = c$ and near the boundary $S < c$.

Example 1 (This observation is apparently due to Maxwell). In a mountainous island the number of peaks minus the numbers of colls plus the number of depressions is 1 (the peaks, colls, and depressions are critical points of the function "height").

For $\chi(D^2) = 1$, D^2 being the island whose boundary is the sea, that is, a level surface of the height function $g \geqslant 0$.

Example 2. On a closed orientable surface of genus $g \geqslant 0$ a function always has at least one minimum and at least one maximum. The number of critical points of saddle type is not less than $2g$ if they are all non-degenerate. If the degeneration is resolved, then for $g > 0$ the function may have in all three critical points: a minimum, a maximum, and a "degenerate saddle". A non-constant function on a closed surface (with $g > 0$) cannot have fewer than three critical points.

We do not discuss here the various purely topological applications of Morse theory in the theory of finite-dimensional smooth manifolds: in the problem of calculating the *homotopy groups* of Lie groups [26], [29], in techniques used in the classification of wide classes of manifolds: in the first place, of manifolds of spherical type [30]-[32], then of arbitrary *simply-connected manifolds* [33], [34], and also of some non-simply connected manifolds [35]-[37].

Initially we are interested in a functional $S(\gamma)$ on some class or another of contours on a finite-dimensional manifold M^n, say, without boundary. The classical Poincaré-Birchhoff-Lyusternik-Shnirel'man-Morse theory considers in the first place the positive functional consisting of length in the Riemannian metric $g_{ij}(x)$:

$$l(\gamma)=\int_\gamma dl=\int_\gamma \sqrt{g_{ij}\dot{x}^i\dot{x}^j}\,dt \tag{3}$$

or the more general positive functional of *"Finsler" length*

$$l_F(\gamma)=\int_\gamma F(x,\dot{x})\,dt>0, \tag{4}$$

which gives rise to a Banach space structure on each n-dimensional tangent plane, where $F(x, \lambda\dot{x}) = \lambda F(x, \dot{x})$ for $\lambda > 0$. Here the x^i are local coordinates on M^n, and the curve γ has the form $x^i(t)$. Frequently the *action functional* of a mechanical system occurs

$$s(\gamma) = \int_\gamma \left[\frac{1}{2} g_{ij}\dot{x}^i\dot{x}^j - U(x)\right] dt. \tag{5}$$

We recall the *"Maupertuis-Fermat"* principle [40]: a functional of length, depending on the energy

$$l^E(\gamma)=\int \sqrt{(E-U(x))\,g_{ij}\dot{x}^i\dot{x}^j}\,d\tau, \tag{6}$$

has *extremals* that coincide trajectorially with (5). The metric $(E - U)g_{ij}$ is *non-singular* when $E > \max U(x)$. In what follows we require the metric in question to be non-singular and *complete*. For completeness it is sufficient that M^n is compact.

Under these conditions the Arzélà principle holds (the set of curves joining two points and having length $\leqslant c$ is relatively compact; similarly for closed curves of length $\leqslant c$ on a compact manifold M^n). From this there follows the theorem (of Hilbert) that there is a geodesic joining an arbitrary pair of points on a complete Riemannian manifold. The Morse theorem on the finiteness of the number of negative squares and the finiteness of the degrees of degeneracy of the form $\delta^2 S$ for the extremals of a γ-periodic form or one joining a pair of points $x_0, x_1 \in M^n$ holds. Subsequently it will be important for us that this theorem is valid for an arbitrary functional of the form

$$S(\gamma) = \int L(x, \dot{x})\, dt, \quad \frac{\partial^2}{\partial \dot{x}^i \partial \dot{x}^j} \xi^i \xi^j > 0. \tag{7}$$

All functionals of the form (7) have well-defined "level surfaces" S = const and "lines of steepest descent" (along the gradient ∇S), although not all have the Arzélà principle. The Arzélà principle holds for length (3) functionals on a complete Riemannian manifold. Also Finsler (positive) metrics (4).

Consequently, for functionals of the form (4) the Morse inequalities (1) hold, where M is either a space $\Omega(x_0, x_1)$ of contours, joining two points of M, or the space of closed contours $M = \Omega$.

If the value of $S(\gamma)$ depends on the direction of the curve γ, then we consider the spaces of directed closed curves Ω^+. One must distinguish two cases:

a) M^n is simply-connected. In this case the homotopy groups are

$$\begin{cases} 1. \ \pi_{i-1}(\Omega(x_0, x_1)) = \pi_i(M^n), \\ 2. \ \pi_i(\Omega^+) = \pi_{i+1}(M^n) + \pi_i(M^n). \end{cases} \tag{8}$$

These equalities are established starting from the two *fibrations* (Serre):

1. $E(x_0) \xrightarrow[p_1]{\Omega(x_0, x_1)} M^n$, where $E(x_0)$ is the contractible space of all paths with origin $\gamma(0) = x_0$, $p_1(\gamma) = \gamma(1) = x_1$.

2. $\Omega^+ \xrightarrow[p_2]{\Omega(x, x)} M^n$, where $p_2(\gamma) = x = \gamma(0) = \gamma(1)$.

There is a *section* ψ: $M^n \to \Omega^+$ consisting of one-point curves in Ω^+. In the case of the sphere $M^n = S^n$ the Betti numbers are

$$b_i(\Omega(x_0, x_1)) = \begin{cases} 1, & i = k(n-1), \\ 0, & i \neq k(n-1). \end{cases} \tag{9}$$

b) M^n is not simply-connected (for example, all surfaces M^2 except for the sphere S^2).

In this case the space $\Omega(x_0, x_1)$ splits into the union

$$\Omega(x_0, x_1) = \bigcup_\alpha \Omega_\alpha(x_0, x_1) \tag{10}$$

over all homotopy classes $\alpha \in \pi_1(M^n)$ in each of which the functional must have a minimum.

For closed contours we also have the splitting

$$\Omega^+ = \bigcup_\beta \Omega^+_\beta, \tag{11}$$

where β is a homotopy class of closed paths, that is, a class of conjugate elements in $\pi_1(M^n)$. The minima in each class β correspond to conjugacy classes in π_1. For example, for manifolds M^n with a complete Riemannian metric for which the curvature of any element of area is non-positive

$$R_{ijkl}\xi^i\eta^j\xi^k\eta^l \leqslant 0,$$

the situation is as follows: all stationary points of the length functional $l(\gamma)$ are minima, both for the problem with two ends and for the periodic problem; all the spaces $\Omega_\alpha(x_0, x_1)$, Ω^+_β, $\beta \neq 1$ are *contractible* (homotopically trivial), and each contains one minimum for the length l.

We call attention to certain peculiarities (important in what follows, see §4) of the periodic problem. We consider the space Ω^+_0 of curves homotopic to zero ($\Omega^+_0 = \Omega^+$ in simply-connected manifolds). Then the minimum of the functional l is achieved on one-point curves

$$\psi(M^n) = M^n \subset \Omega^+_0.$$

As a consequence of this, not all stationary points can be non-degenerate in the strict sense of the word (see above): we may require all except the single-point extremals of the functional to be non-degenerate. The Morse inequalities (1) must take the following form:

$$M_i(S) \geqslant b_i(\Omega^+_0, M^n) \tag{12}$$

in the relative homology modulo the single-point curves.

However, here yet another difficulty arises: it is not a priori excluded that all closed extremals except one are multiples of each one of them. This means that we may find only one periodic extremal from Morse theory other than the one-point one.

For $n = 2$ and $M^2 = S^2$ this difficulty was overcome by Lyusternik and Shnirel'man in 1930 (see [28]), who were able to show that for $n = 2$ the number $M^n_i(S^2)$ of *non-self-intersecting* periodic extremals can be estimated from below by the homology (and other topological invariants) of the Lyusternik-Shnirel'man subspace $\hat{\Omega}^+$ of closed non-self-intersecting curves in S^2 completed by the one-point curves (the sign + denotes directed curves)

$$\begin{cases} S^2_+ \cup S^2_- \subset \hat{\Omega}^+(S^2) \subset \Omega^+(S^2), \\ M^{\wedge}_i(S) \geqslant b_i(\hat{\Omega}^+, S^2_+ \cup S^2_-). \end{cases} \tag{13}$$

The space of non-self-intersecting directed curves on the sphere (completed by the one-point curves) contracts modulo single-point curves to the subset of plane sections of the sphere S^2 having the form $S^2 \times I$, where I is the interval $-1 \leqslant \tau \leqslant +1$, and the boundary is formed by the one-point curves

$$(14) \qquad \begin{cases} S^2_+ \cup S^2_- \subset S^2 \times I \subset \hat{\Omega}^+ \subset \Omega^+, \\ b_i\,(S^2 \times I,\ S^2_+ \cup S^2_-) = \begin{cases} 1, & i=1,\ 3, \\ 0, & i \neq 1,\ 3. \end{cases} \end{cases}$$

In the classical papers only the functional of Riemannian length (3) is considered, independently of the choice of direction (there is an invariance $t \to -t$).

Therefore, the subject of study are the spaces of directed closed curves

$$S^2 \subset \hat{\Omega} \subset \Omega(S^2),$$

where $\hat{\Omega}$ are the non-self-intersecting curves. In this case the Betti numbers mod 2 have the form

$$(15) \qquad b_i(\hat{\Omega},\ S^2) = 1 \quad (i = 1,\ 2,\ 3)$$

(here the Lyusternik-Shnirel'man category turns out also to be 3).

In this case one can extract from the methods of the LSM theory no less than three closed non-intersecting geodesics (without taking the direction into account; with direction there would be 6).

For functionals of type (4) without the invariance $t \to -t$ the LSM theory gives from (14) the existence of two non-self-intersecting closed extremals (in the neighbourhood of which, by the Poincaré-Birkhoff-Kolmogorov-Arnol'd-Moser perturbation theory for conservative systems with 2 degrees of freedom, there is, in general position, an infinite number of self-intersecting periodic extremals, if the initial system is elliptic [38]). On spheres of dimension $n \geqslant 3$ these arguments no longer work. At present the only rigorously established result is that in general position there is at least one further periodic extremal that is not a multiple of the first [39]. On manifolds on which the Betti numbers of the space of paths $b_i(\Omega^+)$ increases as $i \to \infty$, the matter is far simpler: the number of critical points of the functional $l = S$ is much greater than the number of periodic geodesics that could be multiples of any finite number of "basic" geodesics (see [44]). However, this argument fails for the sphere S^n.

Unfortunately, topological methods are, as a rule, not applicable to all natural functionals whose domain of definition has dimension >1 (that is, the Euler-Lagrange equations involve partial derivatives). In some examples the minima that arise naturally in modern geometry (or in the apparatus of modern physics) form non-degenerate critical manifolds in each *connected component* of the function space, and their neighbourhoods are of "good" structure (see [40], II, Ch. 6). However, the theory of critical points of saddle type and the Morse theory no longer hold here, as a rule.

§4. Equations of Kirchhoff type and the Dirac monopole

Systems of Kirchhoff type were discussed in §2. These are systems on the phase space L^* of the Lie algebra L of the group $E(3)$ of motions of $\mathbf{R}^3$. Among them are: a) the Kirchhoff equations for the motion of a rigid body in an ideal fluid (without vortices); b) the motion of a top in a gravitational field; c) the (Leggett) system for the spin dynamics of the superfluid $^3\mathrm{He}-A$.

The phase-variables are (M_i, p_i) $(i = 1, 2, 3)$, the Poisson bracket is given by §1, (35). The Kirchhoff integrals are $f_1 = p^2$ and $f_2 = ps = \sum M_i p_i$ such that $\{f_q, M_i\} = \{f_q, p_i\} = 0$ $(q = 1, 2, i = 1, 2, 3)$. The Poisson bracket on the level surface $p^2 = \text{const} \neq 0$ and $ps = \text{const}$ can be found from the same formulae (35) in §1.

It is easy to see that the level surface f_1, f_2 for $f_1 = p^2 \neq 0$ is topologically equivalent to the tangent manifold $T^*(S^2)$ of the two-dimensional sphere S^2 given by the equation $p^2 = \text{const}$. The variables in the tangent space are given by

$$\sigma_i = M_i - \gamma p_i, \qquad \gamma = s/p, \tag{1}$$

so that

$$\sum \sigma_i p_i = 0. \tag{2}$$

According to (35) of §1, the coordinates p_i have zero Poisson bracket $\{p_i, p_j\} = 0$ on S^2. Therefore, the Poisson bracket on $T^*(S^2)$ turns out to be variationally admissible (see §1) and must reduce to the form (18) of §1. The corresponding change (see [1]) has the form

$$\begin{cases} -\pi/2 \leqslant \Theta \leqslant \pi/2, \quad 0 \leqslant \psi \leqslant 2\pi, \\ p_1 = p\cos\Theta\cos\psi, \quad p_2 = p\cos\Theta\sin\psi, \quad p_3 = p\sin\Theta, \\ \sigma_1 = p_\psi \tan\Theta\cos\psi - p_\Theta\sin\psi, \quad \sigma_3 = -p_\psi, \\ \sigma_2 = p_\psi\tan\Theta\sin\psi + p_\Theta\cos\psi, \quad \sigma_i = M_i - sp^{-1}p_i. \end{cases} \tag{3}$$

It is easy to verify that from (3) it follows that

$$\begin{cases} \{\Theta, \psi\} = \{p_\Theta, \psi\} = \{p_\psi, \Theta\} = 0, \\ \{\Theta, p_\Theta\} = \{\psi, p_\psi\} = 1, \\ \{p_\Theta, p_\psi\} = \cos\Theta. \end{cases} \tag{4}$$

The corresponding 2-form is

$$\begin{cases} \Omega = d\Theta \wedge dp_\Theta + d\psi \wedge dp_\psi + s\cos\Theta\, d\Theta \wedge d\psi, \\ x^1 = \Theta, \quad x^2 = \psi, \quad \xi_1 = p_\Theta, \quad \xi_2 = p_\psi, \\ \Omega = dx^\alpha \wedge d\xi_\alpha + \cos x^1\, dx^1 \wedge dx^2. \end{cases} \tag{5}$$

Thus, the Poisson bracket is explicitly reduced to the form (18) of §1 where the ξ_α for $\alpha = 1$ and 2 are the momenta.

In these variables the Hamiltonian $H(M, p)$ of equations of Kirchhoff type (see (38) and (45) of § 1) has the form

$$H=\frac{1}{2}g^{\alpha\beta}\xi_\alpha\xi_\beta+A^\alpha\xi_\alpha+V(x^1, x^2) \tag{6}$$

for a rigid body in a fluid. Here

$$\left\{\begin{array}{l}\sum a_i\sigma_i^2=g^{\alpha\beta}\xi_\alpha\xi_\beta>0, \quad \sigma_i=M_i-sp^{-1}p_i, \\ A^\alpha\xi_\alpha=s(\sum a_ip_ip^{-1}\sigma_i)+p(\sum_{i,j} b_{ij}(\sigma_ip_jp^{-1}+\sigma_jp_ip^{-1})), \\ 2V=s^2(\sum a_ip_i^2p^{-2})+2ps(\sum_{ij} b_{ij}p_ip_jp^{-2})+p^2(\sum c_{ij}p_ip_jp^{-2}).\end{array}\right. \tag{7}$$

By virtue of homogeneity, the Hamiltonian H depends only on sp^{-1}.

By substituting for σ_i and p_i in the expressions (3); $\Theta = x^1$, $\psi = x^2$, $p_\Theta = \xi_1$, $p_\psi = \xi_2$, we obtain the final formulae

$$H=\frac{1}{2}g^{\alpha\beta}\xi_\alpha\xi_\beta+A'^\alpha\xi_\alpha+V' \tag{8}$$

for the top. Here the $g^{\alpha\beta}(x)$ are the same, $g_{\alpha\beta}g^{\beta\gamma}=\delta_\alpha^\gamma$. For A' and V' we have

$$\left\{\begin{array}{l}A'^\alpha\xi_\alpha=s(\sum a_i\sigma_ip_ip^{-1}), \\ 2V'=s^2(\sum a_ip_i^2p^{-2})+2W(l^ip_i).\end{array}\right. \tag{9}$$

In addition, $p^2 = 1$. Therefore, the Hamiltonian depends only on the level $f_2 = S$

$$H=\frac{1}{2}a(\dot\Theta^2+\cos^2\Theta\dot\psi^2)+A''^\alpha\xi_\alpha+V'' \tag{10}$$

for the Leggett equations in $^3\text{He}-A$. Here, always $p^2 = 1$ (p is given by d);

$$\left\{\begin{array}{l}A''^\alpha\xi_\alpha=\lambda(\sum\sigma_iH_i), \\ V''=\lambda(s\sum p_iH_i)+W(p_1, p_2, p_3), \qquad p^2=1.\end{array}\right. \tag{11}$$

The Hamiltonian H depends on the parameter $s = f_2$.

Thus we reach the following conclusion.

Conclusion. Equations of Kirchhoff type reduce to a system mathematically equivalent to a classical charged particle moving on a sphere S^2 with Riemannian metric $g_{\alpha\beta}(x)$ in a potential field $U(x)$, and also in an effective magnetic field $F_{12}(x)$. In spherical coordinates (Θ, ψ) this magnetic field has the form

$$U(x)=V(x)-\frac{1}{2}g_{\alpha\beta}A^\alpha A^\beta$$

$$F_{12}=s\cos\Theta+\partial_1A_2-\partial_2A_1, \quad A_\alpha=g_{\alpha\beta}A^\beta, \quad s=f_2f_1^{-1/2}. \tag{12}$$

We note that the form $A_\alpha dx^\alpha$ is defined globally on the sphere S^2. For the flow we obtain

$$\iint_{S^2}F_{12}\,d\Theta\wedge d\psi=\int_{-\pi/2}^{\pi/2}\int_0^{2\pi}s\cos\Theta\,d\Theta\wedge d\psi=4\pi s. \tag{13}$$

Thus, when $s \neq 0$, the effective magnetic field is always non-zero and represents a (non quantized) "Dirac monopole". Here s is the level of the Kirchhoff integrals. For $s = 0$ the "magnetic field", if it does not vanish, has zero flow through S^2.

Note. When $s = 0$ for the top (45) of §1, there arises a mechanical system of traditional type on the sphere S^2 (with a non-zero effective magnetic field).(1) This result was obtained by another method somewhat earlier by Kozlov and Kharlaniov (see [41], Ch. 6): these authors then turned to the classical LSM theory to find periodic solutions. As indicated in §3, here there arises the Maupertuis-Fermat functional of type (6) in §3, which for $s = 0$ and $E > \max U$ is the length in a certain Riemannian metric. Since the length functional does not depend on the direction, it follows from LSM theory that there exist no fewer than 6 periodic motions (geometrically 3 curves) that are non-self-intersecting on the sphere S^2.

§5. Many-valued functionals and an analogue of Morse theory. The periodic problem for equations of Kirchhoff type. Chiral fields in an external field.

We have reduced equations of Kirchhoff type to the theory of a charged particle on the sphere S^2 (with some metric) in a scalar potential field and an effective "magnetic field" with non-zero total flow $4\pi s$, that is, a "Dirac monopole". The magnetic field $F = F_{12}\, dx^1 \wedge dx^2$ is a closed, but not necessarily exact 2-form on S^2 (for $s \neq 0$) (see [1], [2]).

It is useful to generalize this situation: let M^n, $n > 1$, be a manifold with a metric $g_{\alpha\beta}$, let U be a scalar function (potential) and F a 2-form (magnetic field), not necessarily exact. We consider a domain $Q \subset M^n$ such that F is exact on Q:

$$(1) \qquad \begin{cases} F = d\omega_Q = d(A^Q_\alpha\, dx^\alpha), \\ -F_{\alpha\beta} = \partial_\beta A^Q_\alpha - \partial_\alpha A^Q_\beta. \end{cases}$$

Let γ be a curve located entirely in Q. Then we can define the action for it:

$$(2) \qquad S_Q(\gamma) = \int_\gamma \left[\frac{1}{2} g_{\alpha\beta}\dot{x}^\alpha \dot{x}^\beta - A^Q_\alpha \dot{x}^\alpha - U\right] dt.$$

By the Maupertuis-Fermat principle, these same trajectories of motion (up to parametrization) can be obtained for a fixed energy from the functional

$$(3) \qquad l^E_Q(\gamma) = \int_\gamma \left[\sqrt{(E-U)\, g_{\alpha\beta}\dot{x}^\alpha \dot{x}^\beta} - A^Q_\alpha \dot{x}^\alpha\right] d\tau.$$

(1)In its physical meaning the problem of a top in an axially symmetrical field is not associated with the Lie algebra $L = L(3)$. This problem is naturally depicted as a Lagrangian system on SO_3. Its factorization and transition to $T^*(S^2)$ with a certain symplectic structure is discussed, though not investigated further, in [45]. As the author of the book [45] has communicated to me, the erroneous assertion that the resulting symplectic structure is equivalent to the standard structure on $T^*(S^*)$ has been removed in the English translation.

This can be done for any domain Q in which $F = d\omega_Q$. We fix a 1-form ω_Q for all possible domains Q on which the 2-form F is exact. If γ lies entirely in both domains Q_1 and Q_2, then:(1)

$$F = d\omega_{Q_1} = d\omega_{Q_2},$$
$$S_{Q_1}(\gamma) - S_{Q_2}(\gamma) = l^E_{Q_1}(\gamma) - l^E_{Q_2}(\gamma) = \int_\gamma (\omega_{Q_1} - \omega_{Q_2}). \tag{4}$$

The value of the integral remains unchanged under any deformation γ_λ of the curve $\gamma = \gamma_0$, assuming that γ is periodic (or under any deformation γ_λ of γ_0 with the same end-points if γ has such):

$$0 = \frac{d}{d\lambda} \int_{\gamma_\lambda} (\omega_{Q_1} - \omega_{Q_2}) = \frac{d}{d\lambda} [l^E_{Q_1}(\gamma_\lambda) - l^E_{Q_2}(\gamma_\lambda)], \tag{5}$$

since the form $\omega_{Q_1} - \omega_{Q_2}$ is closed.

From this we obtain the following conclusion.

Conclusion. The set of local actions l^E_Q (or S_Q) for all domains Q defines a "many-valued functional" on the function spaces: a) of the closed contours (the directed curves Ω^+); b) of the paths joining two points, $\Omega(x_0, x_1)$. Here we assume that $E > \max U(x)$. This means that the infinite-dimensional 1-form δl^E is everywhere uniquely determined and closed, but its "path integral", in general, determines a many-valued function on Ω^+ or $\Omega(x_0, x_1)$.

Near any extremal this function may be assumed to be unique. The Morse index theorem and all other "good" local properties hold for the functionals (3) in so far as the condition (7) of §3 is satisfied. For example, it is clear that one-point curves give a local minimum of the functionals (3). This is a very important fact for our purposes.

The many-value function (functional) l^E becomes single-valued after transition to a certain covering with infinitely many sheets:

$$\hat{\Omega} \to \Omega^+, \quad \tilde{\Omega} \to \Omega(x_0, x_1), \tag{6}$$

by defining on the covering space a single-valued function $l^E(S)$, running through all the values

$$-\infty < l^E < 0, \quad -\infty < S < \infty$$

(on $\hat{\Omega}$ or $\tilde{\Omega}$). Of course, no analogue of the "Arzélà principle", of §3 can hold.

When the magnetic field F_{ij} is an exact 2-form, then the functional S or l^E is defined everywhere ($Q = M^n$) and is therefore single-valued. Nevertheless, this functional can turn out to be non-positive. In this case also the Arzélà principle fails for $-\infty < S < \infty$, $-\infty < l^E < \infty$.

(1)This situation essentially arose in connection with arguments (see [42]) for the construction of the quantum amplitude $\exp\{iS\}$ as a single-valued functional under the condition that the *flow* of the magnetic field of the "Dirac monopole" is integer-valued.

Example 1. Let $\mathbf{R}^2(x, y)$ be a plane with the Euclidean metric, and $F \neq 0$ a homogeneous magnetic field, directed along the z-axis $\perp \mathbf{R}^2$. All orbits of motion of a charged particle are circles (with a definite direction, depending on the sign of F). The radius of the (Larmor) orbits has the form

$$r^2 = \text{const}\cdot E/F^2 \tag{7}$$

(the constant involves the charge, the mass, and c). From this it follows that if the distance between x_0 and x_1 is sufficiently large, then there is no extremal in $\Omega(x_0, x_1)$. The reason is that the functional l^E on curves γ with large area is not positive (although l^E is single-valued).

Example 2. Let S^2 be the sphere with the standard metric and F a magnetic field invariant under all the motions from SO_3. For fixed energy E and large fields F the Larmor radius $r^2 \sim EF^{-2}$ becomes arbitrarily small. The problem is exactly integrable: as for the plane $\mathbf{R}^2$ all the orbits are closed. By arguments similar to those of Example 1 we arrive at the conclusion that there is a pair of points x_0, x_1 on S^2 such that the many-valued functional l^E has no extremal in $\Omega(x_0, x_1)$.

We note that in Example 2 the manifold (the sphere S^2) is compact, but the functional is many-valued.

The periodic problem of the variational calculus in this case differs strongly, on the whole, from the problem with two fixed end-points. In the periodic case, the Maupertuis-Fermat functional l^E when δl^E is an everywhere defined 1-form on Ω^+ always has "trivial" critical points: these are the one-point curves, which form a submanifold of local minima $M^n \subset \Omega^+(M^n)$. On any *sheet of the covering* $\hat{\Omega}_q \to \Omega^+$ the complete inverse image

$$q^{-1}(M^n) = \bigcup_j M_j^n = M_0 \cup M_1 \cup M_{-1} \cup \cdots \tag{8}$$

gives a manifold of local minima of l^E that is single-valued on $\hat{\hat{\Omega}}$.

Let us join by a homotopy two manifolds of local minima, say, M_0^n and M_1^n; that is, we construct a map of the cylinder (I is the interval $0 \leqslant \tau \leqslant 1$)

$$f\colon M^q \times I \to \hat{\hat{\Omega}} \tag{9}$$

over any q-dimensional cycle $M^q \subset M^n$.

At the boundary we impose the condition

$$\begin{cases} f(x, 0) = M_0^q \subset M_0^n, \\ f(x, 1) = M_1^q \subset M_1^n. \end{cases} \tag{10}$$

In particular, for $q = 0$ we obtain a map of the interval $I \to \hat{\hat{\Omega}}$; for $q = n$ we obtain a map $M^n \times I \to \hat{\hat{\Omega}}$ of the cylinder over M^n, provided that it is compact and closed.

When we restrict the functional l^E to $M^q \times I$ and begin to move the map f "downwards" along the gradient ∇l^E, then the ends $f\,|_{\tau=0}$ and $f\,|_{\tau=1}$ do not move (they even occur in the local minima); we see that somewhere "in the middle" l^E has a maximum (x_f, τ_f) on $M^q \times I$ for any map f; we have the obvious inequality

$$l^E(x_f, \tau_f) > l^E(x, \tau)\,|_{\tau=0, 1}.$$

Since the values on the boundaries do not depend on f, we arrive at the following conclusion:

Theorem. *In general position, every basic cycle in the homology group $H_q(M^n)$ generates a critical point of the many-valued functional l^E of Morse index $i = q+1$. For arbitrary energy E, l^E has at least one critical point (the non-trivial periodic extremal) that is not a singleton. It is assumed that the Maupertuis metric $(E - U)g_{ij}$ has no singularities $E >$ max $U(x)$ and is complete on the compact manifold M^n (this is always so, for example, when $M^n = S^2$).*

The same arguments can also be applied to the case of the two-ended problem, for a many-valued functional l^E on $\Omega(x_0, x_1)$. Here one has to assume that either $x_0 = x_1$ is any fixed point $x_0 \in M^n$, or that x_0 and x_1 are so close that there is a unique "short" extremal (locally minimal) from x_0 to x_1 that can be denoted by $[x_0, x_1]$. In this case the manifold of local minima consists of the single point $[x_0, x_1] \in \Omega(x_0, x_1)$. So we obtain the result: in addition to $[x_0, x_1]$ there is also a "long" extremal of index 1.

Now let $x_0 = x_1$. We have obtained a "long" extremal $\gamma(x_0)$ with an "angle" at x_0. We consider the scalar function $l^E(\gamma(x_0)) = \psi(x_0)$ (which is easily seen to be single-valued) on M^n, where $\gamma(x_0)$ is the "short" extremal.

Proposition. *If ψ is smooth, then its critical points on M^n are also periodic extremals.*

Conjecture. *The periodic extremals obtained from cycles on M^n of one-point curves cannot all concide geometrically (that is, they cannot all be multiples of one of them).*

This condition is easily met in the dimension $n = 2$ on the sphere S^2, by using the Lyusternik-Shnirel'man space of smooth non-self-intersecting curves $\hat{\Omega}^+(S^2)$, completed by the one-point curves. Since $\hat{\Omega}^+$ is simply-connected, the many-valued functional l^E on this space reduces to a single-valued one; however, this functional extends up to the "boundary" (the set of one-point curves) with two different values (the set of single-pointed curves "bifurcates" into two pieces). As was shown above (§3, (14)), homotopically we have

$$\hat{\Omega}^+ \sim S^2 \times I, \quad -1 \leqslant \tau \leqslant +1;$$

$S^2_+ \cup S^2_-$ are the one-point curves (the boundary).

More precisely, we must consider the subspace $\hat{\hat{\Omega}}^+$ in the covering space $\hat{\hat{\Omega}} \to \Omega^+$ where $\hat{\Omega}^+ \subset \hat{\hat{\Omega}}$. Two copies of the one-point curves $(S^2_+ \cup S^2_-)$ are contained in the closure of $\hat{\Omega}^+$. We normalize the value of l^E so that

$$l^E(S^2_+) = 0, \quad l^E(S^2_-) \geqslant 0. \qquad (11)$$

For equations of Kirchhoff type we obtain from (4.14) that

$$l^E(S^2_-) = 4\pi \mid s \mid.$$

Even among the non-self-intersecting curves we obtain two extrema of indices 1 and 3 by using two saddles as in (9) and (10) above:

$$(12) \qquad \begin{cases} f\colon I \to \Omega^+, & f(0) \in S^2_+, \quad f(1) \in S^2_-, \\ f\colon S^2 \times I \to \hat{\Omega}^+, & f(x, 0) = S^2_+, \quad f(x, 1) = S^2_-, \\ & 0 \leqslant \tau \leqslant 1. \end{cases}$$

Thus, in this case we obtain two non-self-intersecting periodic extremals γ_1 and γ_2 of Morse indices 1 and 3, respectively.

For a single-valued functional l^E in the presence of a magnetic field we have $s = 0$; all the results remain valid.

Thus, we have obtained the following result:

Theorem. *For all the values of the parameters $f_1 = p^2 \neq 0$ and $f_2 = ps$ for which the Hamiltonians are defined of systems of Kirchhoff type* (4), (6)–(10), *reduced to the sphere S^2, and for all energies $E >$* max *U the system has at least two periodic orbits (non-self-intersecting in p), depending on the parameters E, s, and p. If they are non-degenerate, then their Morse indices are* 1 *and* 3, *respectively. (In fact, for the motion of a rigid body in a fluid because of the homogeneity of the Hamiltonian this dependence reduces to the variables E and sp^{-1}. For a top and the Leggett system we must set $p =$* 1.)[1]

As Arnol'd has told the author, from Poincaré's paper of 1905 (for the quotation, see [28]) one can extract an idea whose natural development makes it possible to prove the existence of a closed extremal in a magnetic field in a number of cases. Poincaré's arguments, when translated into modern language, have the following meaning: we consider the sphere S^2 with some Riemannian metric g_{ab}. Among all curves γ bounding a given area A we look for the shortest (the isoperimetric problem). We denote this[1] by γ_A^*. It is easy to see from arguments with Lagrangian multipliers that γ_A^* is a closed extremal of a (formal) charged particle in some magnetic field F proportional to the element of area $F = \lambda d^2\sigma$ of the metric g_{ab} with some (so far undetermined) $\lambda = \lambda(A)$. When A is increased from zero to the

---

(1) As is shown in [2], §5, for a broad class of Hamiltonians there is a $\mathbf{Z}_2$-symmetry enabling us to find periodic motions, unknown classically, among plane sections of a sphere, by looking for the extremum of action as a function of a single variable.

(2) Anosov has pointed out that although the isoperimetric problem is present in Poincaré's work, Arnold's idea is lacking; that is, the further arguments with the change of the parameter A from 0 to the area of the sphere; it would be useful to make this argument rigorous.

whole area of the sphere S^2, then $\lambda(A)$ increases from $-\infty$ to $+\infty$, as is easy to verify. Hence, by continuity, $\lambda(A_0) = 0$ for some A_0. From this we also obtain the fact that in any "constant" magnetic field (that is, $F = \lambda d^2\sigma$ for all λ) there is at least one non-self-intersecting extremal. By a trivial generalization of this argument, one has to consider for any given 2-form ω on S^2 the problem of finding the shortest $\gamma^{\min}(A, \omega)$ in the set of non-self-intersecting curves γ for which $\int_U \omega = A$, $\gamma = \partial U$. By changing A one can apparently obtain closed extremals in the magnetic field $\lambda\omega$ for any $-\infty < \lambda < \infty$. In any case this is true for forms like $f(x)d^2\sigma$, where $f > 0$ and $d^2\sigma$ is the element of area.

However, it is more convenient, as Arnol'd has suggested, to act dually. We fix the length of a curve $l(\gamma) = L$ homotopic to zero in any complete Riemannian metric on a simply-connected manifold M^n, and we fix a closed 2-form Ω. We look for a curve γ such that the quantity $I(\gamma) = \int_\sigma \Omega$ has a maximum or a minimum (a local maximum or even a stationary point if $H_2(M^n, R) \neq 0$ and the homology class $[\Omega] \neq 0$, although here it is already difficult to prove an existence theorem). If the class $[\Omega] \neq 0$, then the whole of this construction must be carried out on the space of pairs $(\gamma, n) \in \hat{P}_L$, where γ is a closed null-homotopic curve of fixed length L, and n is the homotopy class of the membrane σ, $\partial\sigma = \gamma$. The space of pairs $\hat{P}_L$ is an infinitely-sheeted covering of the space P_L of curves γ of the given length, $\hat{P}_L \to P_L$, but for small lengths L the covering is trivial;

$$\hat{P}_L = P_L \times \mathbf{Z} \quad \text{as} \quad L \to 0;$$

as $L \to \infty$, this covering becomes non-trivial.

The space P_L itself is compact. Therefore, if the class $[\Omega] = 0$ or the covering $\hat{P}_L$ is trivial, then there is always a maximum of $I(\gamma)$. Suppose that it is attained at $\gamma_L^* \in P_L$. Using the previous arguments, and changing L from 0 to ∞, we obtain a closed periodic extremal in any magnetic field proportional to Ω. Things are more complicated if the homology class is $[\Omega] \neq 0$. For a certain "critical" $L = L_0$ the covering $\hat{P}_L$ becomes non-trivial.[(1)] However, on the two-dimensional sphere S^2 one can use non-self-intersecting curves γ which bound only two membranes $\partial\sigma_1 = \gamma$, $\partial\sigma_2 = \gamma$, $\sigma_1 \cup \sigma_2 = S^2$.

In this case there is always a maximum; with a change of parameter the previous arguments reduce to the theorem on a single periodic extremal.

[(1)]In this case the critical value $L = L_0$ is a stationary point of the functional $L(\gamma)$ of index 1^0; it is appropriate to conjecture that: the minimal (maxima) γ_L^*, generate "short" closed orbits in the magnetic field $\lambda^{-1}(L)\Omega$ where $\lambda(L)$ runs through *all* the values from 0 to ∞ as L changes from 0 to L_0.

The "finite-dimensional" model of the present arguments with the space of closed curves and its subspaces of curves of length L is as follows: given a manifold M (open, of large dimension); suppose that on it there are given

a) a smooth function $l(x) \geqslant 0$ such that the domains $l \leqslant L$ and the level surfaces $l = L$ are all compact,

b) a closed 1-form $\hat{\omega}$, $d\hat{\omega} = 0$.

We investigate the critical points of the form $\omega = dl + \hat{\omega}$. For this purpose we consider the family of forms $\omega_L = \omega$ on the level $l = L$. If the form ω_L on the level $l = L$ is exact: $\omega_L = d\varphi_L$, then we consider the maximum γ_L^* of φ_L on the level $l = L$. By varying L, $\infty > L \geqslant 0$, we find the critical points of the form $\lambda(L)dl + \hat{\omega}$. We recall that for $L = 0$ we must obtain not an isolated minimum, but a whole non-degenerate manifold of minima since in the case of curves we obtain all the one-point curves. All other critical points may be assumed to be non-degenerate. Moreover, in the domains $l \geqslant L$ for small $L \sim \varepsilon$ the form $\omega = dl + \hat{\omega}$ must be (locally) the gradient of a function $f = l + \varphi$, where $\varphi \sim \varepsilon^2$, that is, f has a local minimum on the whole set $L = 0$ of minima of $l(x)$. As $L \to 0$, we see that $\lambda(L) \to 0$.

We now assume that the functional l^E of the Kirchhoff problem (6)–(10) of §4 on the space of non-self-intersecting curves, normalized by the conditions (11), turns out to be everywhere positive.

According to Stokes' formula, the magnetic part of the functional l^E for a non-self-intersecting curve γ is always bounded above by the area of the domains on the sphere which it bounds. Hence, l^E is always semi-bounded on the subspace $\hat{\Omega}^+(S^2)$:

$$l^E(\gamma) > \text{const} > -\infty. \tag{13}$$

If there is a curve γ such that $l^E(\gamma) < 0$ (under the conditions (11) for one-point curves), then there is a minimum $\gamma_{\min} \in \hat{\Omega}^+$, that is, one more "superfluous" periodic extremal. Apart from the minimum there is also a "superfluous" saddle γ_1 where $l^E(\gamma_1) > 0$ in the space $\hat{\Omega}^+$.

Example. We consider the Kirchhoff problem (6)–(10) of §4; at energies close to the maximum of the potential $E \sim \max U$, we look for "small" test curves γ_ε surrounding the maximum point x_0, $U(x_0) = \max$, such that

$$l^E(\gamma_\varepsilon) < 0, \quad \varepsilon \to 0. \tag{14}$$

Let $x^1 = \Theta$ and $x^2 = \psi$ be the coordinates (5) of §4, where the Lagrangian and the magnetic field have the form (6)–(10) of §4 and $g_{\alpha\beta}$ is a Riemannian metric.

If the effective magnetic field $F = F_{12}$ exceeds a certain "threshold", then there are small test curves γ_ε such that $l^E(\gamma_\varepsilon) < 0$ for energies E sufficiently close to the critical energy:

$$9{,}2\ \lambda_{\max} < 4\ |F_{12}(x_0)|\ \det g^{\alpha\beta}(x), \tag{15}$$

where $\lambda_{\max}$ is the largest eigenvalue of the form $(-\partial^2 U/\partial x^\alpha \partial x^\beta)$:

$$\det\left(-\frac{\partial^2 U}{\partial x^\alpha \partial x^\beta} - \lambda_{\max} g_{\alpha\beta}\right)_{x=x_0} = 0 \tag{16}$$

(see [2]). The conclusion that there is a saddle when inequalities of the type (15) hold is valid in all dimensions, in contrast to the existence of a minimum.

For purely methodological purposes it is useful to consider an example that is unrelated to equations of Kirchhoff type. Suppose that in the (x, y)-plane $\mathbf{R}^2$ there is a magnetic field $F(x, y) = F(x+T_1, y) = F(x, y+T_2)$ with two periods, directed along the z-axis. We consider the classical motion of a charged particle in this field (a generalization of Larmor orbits). Suppose that the average field is non-zero

$$\overline{F} = \frac{1}{T_1 T_2} \int_0^{T_1}\int_0^{T_2} F\, dx \wedge dy \neq 0. \tag{17}$$

We form the Maupertuis-Fermat functional, when $Q = \mathbf{R}^2$. We have a single-valued functional $l^E(\gamma)$ on the space $\hat{\Omega}_0^+$ of all smooth closed non-self-intersecting curves, directed in the same sense as the motion along a Larmor orbit in the homogeneous field $\overline{F}$. The functional l^E is non-positive (for $F = \overline{F}$ this was shown in Example 1 above). For circles γ_r of radius r we have

$$\begin{cases} l^E(\gamma_2) < 0, & r \to \infty, \\ l^E(\gamma_2) > 0, & r \to 0. \end{cases} \tag{18}$$

We consider the function $\psi(r) = l^E(\gamma_r)$ on the half-line. Using the periodicity of F, we identify the curves $\gamma_1 \sim \gamma_2$ if they differ by a shift through a vector of the lattice (mT_2, nT_2). Then the one-point curves form a torus. We apply the same arguments as above (see (11), (12)) taking instead of the interval I a map of the half-line

$$f\colon M^q \times R^+ \to \hat{\Omega}_0^+.$$

Here M^q is a cycle on the torus T^2 $(q = 0, 1, 2)$. We have four cycles: a point, two neighbourhoods, and the whole torus. The map f is subject to the boundary conditions:

(a) $f(x, 0) = M_0^q \subset T^2$ are one-point curves;

(b) for large $\tau \to \infty$ the images $f(x, \tau)$ consist of curves γ such that $l^E(\gamma) < 0$.

By analogy with the preceding we establish in the case of general position the existence of four "Larmor" orbits for any energy $E > 0$ with Morse indices (1, 2, 2, 3).

Other examples and the development of an analogue of Morse theory for closed 1-form can be found in [1]-[3].

Let us now introduce the class of *chiral fields* [3] among the generalized "external fields"; it is natural to correct "many-valued functionals" with these fields, not unlike those arising on the space of contours for the Dirac monopole.

The definition of a non-linear chiral field is as follows (see, for example, [40], II, Ch. 6): let N^q and M^n be arbitrary Riemannian manifolds; let $S_0(f)$ be a functional defined on the map $f\colon N^q \to M^n$. Usually, $N^q = \mathbf{R}^q$ or $N^q = S^q$. If $N^q = \mathbf{R}^q$, then we require that at infinity the field $f(x)$ tends to a constant, $f(x) \to y_0 \in M^n$ as $|x| \to \infty$. Here $S_0(f)$ has the form of a Dirichlet functional that is quadratic in the derivatives of f, possibly with some additions. The principal chiral fields arise when $M^n = G$ is a Lie group. In field theory one considers the case when the metric on G is invariant on both sides (the Killing metric); the metrics on $\mathbf{R}^q$ or S^q are also assumed to be standard. The standard "chiral Lagrangian" has the form

$$S_0(f) = \int_{N^q} \mathrm{Sp}\,(A_\mu A^\mu)\sqrt{g}\,d^q x, \tag{19}$$

where $g_{\mu\nu}$ is a metric on N^q (we note that in the theory of the Ginzburg-Landau equation for the superfluid $^3\mathrm{He}-A$ or $^3\mathrm{He}-B$, more complex Lagrangians arise for chiral fields; for references, see [43]).

Let Ω be an additional closed $(q+1)$-form on M^n (the "external field"):

$$d\Omega = 0. \tag{20}$$

We take a *covering* of M^n by domains

$$M^n = \bigcup_\alpha W_\alpha$$

(with continuously many domains) such that

1. Ω is exact, $\Omega = d\psi_\alpha$, on each W_α;
2. the image of each map $f\colon N^q \to M^n$ lies entirely in some domain W_α.

The "local action functionals" are defined by

$$S_\alpha(f) = S_0(f) + \int_{(N^q,\,f)} \psi_\alpha. \tag{21}$$

In the intersections, if

$$f(N^q) \subset W_\alpha \cap W_\beta,$$

then

$$S_\alpha - S_\beta = \int_{(N^q,\,f)} (\psi_\alpha - \psi_\beta), \tag{22}$$

where $\psi_\alpha - \psi_\beta$ is a closed q-form in $W_\alpha \cap W_\beta$. By analogy with §5, where $q = 1$, $N^q = S^1$, $M^n = S^2$ (and $M^n = \mathbf{R}^2$) we obtain the following lemma:

Lemma. *The set of functionals $S_\alpha(f)$ defines a closed 1-form δS on the functional space F of admissible map $f\colon N^q \to M^n$. This 1-form δS determines a many-valued functional S, which is single-valued on an infinitely-sheeted covering $\hat{F} \to F$.*

In the quantization of such fields one has to require that the quantity ("amplitude") $\exp(iS)$ is a single-valued functional on F. From this it follows that Ω has integer-valued integrals along $(q+1)$-cycles in M^n.

Example. Let $q = 2$, $N^q = \mathbf{R}^2$ or S^2 with the standard metric, and let M^n be a compact Lie group with the Killing metric. The functional $S_0(f)$ is taken in the form (19); Ω is a two-sided invariant 3-form on the group $G = M^n$ (such a form always exists; when $G = SU_2$ then Ω is the volume element).

Problem. Does the Euler-Lagrange equation $\delta S = 0$ have a solution by the inverse problem method (see [11], Ch. III for *ordinary* chiral fields with $\Omega = 0$)?

Remark. We obtain a curious example for the so-called relativistic "strings" in Minkowsky space $M^n = \mathbf{R}^{3,1}$ or Euclidean space $M^4 = \mathbf{R}^4$, where $q = 2$, $N^2 = \mathbf{R}^2$ (or S^2 or D^2) and $f\colon N^2 \to \mathbf{R}^4$. Let $S_0(f) = \iint_{N^2} \sqrt{g}\, d^2x$ where $g = \det g_{ij}$, g_{ij} being the metric on N^2 induced by the embedding f. For any closed 3-form on some domain in $\mathbf{R}^4$ we again obtain the "many-valued functional"

$$S(f) = S_0(f) + \iint_{(N^2,\, f)} \psi, \qquad d\psi = \Omega. \tag{23}$$

If Ω is defined everywhere in $\mathbf{R}^4$ except for isolated singularities at the points $(x_1, \ldots, x_k)$, then there is a set of integrals over small spheres S_i^3 $(i = 1, \ldots, k)$ around these points

$$\varkappa_i = \int_{S_i^3} \Omega. \tag{24}$$

From the requirement that $\exp(iS)$ is single-valued it follows that $(2\pi)^{-1}\varkappa_i$ is an integer. This situation is similar to the "Dirac monopole", but here it is more natural to call the singular points x_i "instantons", since they are localized in $\mathbf{R}^4$. If $\Omega \to 0$ sufficiently rapidly as $|x| \to \infty$, then $\sum \varkappa_i = 0$.

The class of many-valued functionals for chiral fields introduced above can be extended naturally. Let $E \xrightarrow{p} N^q$ be a smooth fibration (or a direct product) with fiber M^n. In the case of a direct product, $E = M^n \times N^q$. Let $S_0(f)$ be a single-valued functional on the sections $f\colon N^q \to E$, $p \circ f = 1$, and let Ω be a closed $(q+1)$-form on the manifold E, $d\Omega = 0$. The subsequent definition of the "many-valued functional" $S\{f\} = S_0\{f\} + \int_{N^q f} d^{-1}(\Omega)$ is a word-for-word repetition of (20)–(22) above with the obvious change that the W_α are domains on E. In the special case mentioned above Ω was a form on the fiber M^n, and the W_α where domains on M^n, which naturally generate "cylindrical" domains and forms on $E = M^n \times N^q$, independent of the basis N^q. The following interesting problem was solved in [54], [55]:

let $P(f) = 0$ be a differential equation of the sections of the fibration E, f: $N^q \xrightarrow{p} E$, $p \circ f = 1$, such that the "local" expression $P(f)$ is formally the variational derivative of some functional. When is the operator $P(f)$ globally the variational derivative of a functional (which, of course, is assumed to be single-valued), that is, $P(f) = \delta S/\delta f$?

In [54], [55] an "obstruction" $\alpha[P] \in H^{q+1}(E, R)$, $\alpha[P_1 + P_2] = \alpha_1 + \alpha_2$ was constructed such that $\alpha = 0$ is equivalent to the global existence of $S\{f\}$. By comparing this with our construction of "many-valued functionals", we obtain the following proposition: all locally Lagrangian systems of differential equations reduce to the variational derivatives of many-valued functionals of the form $S_0\{f\} + \int_{(N^q, f)} d^{-1}(\Omega)$, $d\Omega = 0$.

For the direct product $E = N^q \times M^n$ the simplest natural class of examples of such "external fields", that is, of $(q+1)$-forms Ω, is given by products of closed forms of the base and the fiber: $x^1, \ldots, x^q$ are local coordinates in the base N^q, and $\varphi^1, \ldots, \varphi^n$ in the fibre M^n;

$$\Omega_{k,l} = \Omega = \omega'_k \wedge \omega''_l, \qquad k + l = q + 1, \qquad d\omega'_k = d\omega''_l = 0, \tag{25}$$

where ω'_k is a form on N^q and ω''_l a form on M^n.

Example 1. Let $l = 1$, $k = q$, $\omega''_1 = dU(\varphi)$ (locally), and let $\omega'_k = \sqrt{g}\, dx^1 \wedge \ldots \wedge dx^q$ be the volume element on N^q. Then

$$S = S_0\{f\} + \lambda \int_{(N^q, f)} U(\varphi) \sqrt{g}\, d^n x. \tag{26}$$

An external field $\Omega_{q,1}$ of this kind reduces to a potential $U(\varphi)$ on M^n or the covering $\hat{M} \to M^n$.

Example 2. Let $k = 2$, $l = q - 1$, $\omega'_2 = dA$, where $A = A_\alpha(x)dx^\alpha$ is a vector potential. In this case

$$S\{f\} = S_0\{f\} + \lambda \int_{N^q} (A_\alpha\, dx^\alpha) \wedge f^* \omega''_{q-1}.$$

The field $\Omega'_{2, q-1}$ can represent a pair: the "magnetic field" $H_{\alpha\beta} = \partial_\alpha A_\beta - \partial_\beta A_\alpha$ for $q = 2, 3$ and another field ω''_{q-1} of φ, interacting with the chiral field f: $N^q \to M^n$.

Example 3. Let $k = 1$, $l = q$, $\omega'_1 = E_\alpha dx^\alpha = dU(x)$. In this case E_α can be an electrical field.

In the simplest interesting cases we have

a) $N^q = \mathbf{R}^q$ (the field f tends to f_0 as $|x| \to \infty$) or $N^q = T^q$ (periodic boundary conditions), the metric is $g_{\alpha\beta} = \delta_{\alpha\beta}$.

b) $M^n = S^1$ (the field f of "sine-Gordon" type) or, more generally, $M^n = S^n$. Let ω''_{q-1} be the element of area.

The magnetic field in Example 2 interacts with the chiral field of unit tangent vectors for $N^q = \mathbf{R}^q$, T^q. The basic examples of the functionals S_0 are as follows:

I. The Dirichlet integral on the sections $(x, f(x)) \subset N^q \times M^n$.
Let $g_{\alpha\beta}$ be the metric in N^q and G_{ab} in M^n. We put

$$(27) \qquad S_0\{f\} = \int_{N^q} \left[\lambda + g^{\alpha\beta}(x)\, G_{ab}(f(x))\, \frac{\partial\varphi^a}{\partial x^\alpha}\frac{\partial\varphi^b}{\partial x^\beta}\right]\sqrt{g}\, d^n x.$$

More generally, the Dirichlet integral plus a potential:

$$(28) \qquad S_0\{f\} = \int_{N^q} \left[g^{\alpha\beta} G_{ab}\frac{\partial\varphi^a}{\partial x^\alpha}\frac{\partial\varphi^b}{\partial x^\beta} + V(\varphi)\right]\sqrt{g}\, d^n x,$$

where $V(\varphi)$ is a single-valued scalar (in Example 1 it can happen that $dV(\varphi)$ is closed, but not an exact 1-form).

II. Sectional volume.
In the same notation

$$(29) \qquad S_0\{f\} = \int_{N^q} \sqrt{h}\, d^n x, \qquad h = \det h_{\alpha\beta}(x),$$

where $h_{\alpha\beta} = g_{\alpha\beta} + G_{ab}(f(x))\dfrac{\partial\varphi^a}{\partial x^\alpha}\dfrac{\partial\varphi^b}{\partial x^\beta}$ is the induced metric on the section $\tilde{f}\colon N^q \to N^q \times M^n$ generated by the map $f\colon N^q \to M^n$.

For functionals $S_0\{f\}$ of type I and II on the set of null-homotopic maps $N^q \to M^n$ the following proposition holds: if $f_0 = \text{const}$ is a local minimum of the functional $S_0\{f\}$, then f_0 is also a local minimum of the functional $S\{f\} = S_0\{f\} + \lambda \int_{(N^q, f)} d^{-1}(\Omega)$. Let $S\{f_0\} = 0$. If $S\{f\}$ is essentially many-valued or single-valued but non-positive (that is, $S\{f\} < 0$ for some null-homotopic f, then $S\{f\}$ has a "saddle" extremal. The proof of this statement is similar to that of the corresponding theorem for closed curves, where $q = 1$.

§6. Many-valued functions on finite-dimensional manifolds. An analogue of Morse theory.

On the manifold M^n we specify a closed 1-form ω; there is an (infinitely-sheeted) covering $\hat{M} \xrightarrow{p} M^n$ such that the form $p^*\omega$ is the differential of a function (the simplest example is $\omega = d\varphi$ on $\mathbf{R}^2 \setminus 0 = M^n$, where $\hat{M}$ is the Riemann surface of the logarithm):

$$(1) \qquad p^*\omega = dS.$$

We call S a "many-valued function" on M^n. In fact, we consider only the case when all the critical points are either non-degenerate or form non-degenerate critical manifolds (see §3). We also assume that S has a well-defined "gradient discharge"; that is, on M any compact space under descent along the gradient ∇S either approaches a critical point or passes successively "downwards" through all levels of S.

Problem. To construct an analogue of Morse theory for an estimate of the number of stationary points of a many-valued function S (that is, of a closed 1-form ω) of any Morse index i. We denote the number of stationary points of Morse index i by $m_i(S)$ (or $m_i(\omega)$), $p^*\omega = dS$.

In the group $H_1(M, \mathbf{Z})$ we can choose a basis $\gamma_1, \ldots, \gamma_k, \gamma_{k+1}, \ldots, \gamma_N$ such that

$$\oint_{\gamma_j} \omega = \begin{cases} 0, & j \geqslant k+1, \\ \varkappa_j \neq 0, & j \leqslant k, \end{cases} \tag{2}$$

and the numbers $\varkappa_j$ for $j = 1, \ldots, k$ are rationally (or integrally) independent. The number $k-1$ is called the "degree of irrationality" of ω. The monodromy group of the minimal covering $p\colon \hat{M} \to M^n$, turning ω into a differential of a single-valued function $dS = p^*\omega$ is precisely equal to $\mathbf{Z}^k$, the free Abelian group with k generators $t_1, \ldots, t_k$ acting by shifts on $\hat{M}$:

$$t_j\colon \hat{M} \to \hat{M}.$$

In fact, the irrationality exponent is a point of the projective space

$$\varkappa = (\varkappa_1 : \varkappa_2 : \ldots : \varkappa_k) \in \mathbf{RP}^{k-1}.$$

A particularly simple and interesting case is $k = 1$, when ω (possibly after multiplication by a factor) gives an element of the first integral cohomology group $[\omega] \in H^1(M^n, \mathbf{Z})$. In this case, $\exp(2\pi i S)$ is a well-defined complex-valued function of modulus 1, that is,

$$f = \exp(2\pi i S)\colon M^n \to S^1. \tag{3}$$

The problem of constructing an analogue of Morse theory for the critical points of such maps appears to be absolutely classical; this problem has never previously (up to 1981) been studied in the literature. We consider the following case. If there are no critical points, then the map f defines a fibration with base $B = S^1$. A cyclic $\mathbf{Z}$-covering $\hat{M} \xrightarrow{p} M^n$ is constructed as follows: we realize the cycle $D[\omega] \in H_{n-1}(M^n, \mathbf{Z})$ by the submanifold N^{n-1}, where D is the Poincaré duality operator. By cutting the manifold along the cycle N^{n-1} we obtain a membrane W^n with two edges $\partial W = N^{n-1} \cup N_1^{n-1}$, diffeomorphic to N^{n-1}. We take infinitely many of copies of this membrane $W \approx W_i$ with boundaries $\partial W_i = N_{i0} \cup N_{i,1}$, diffeomorphic to N^{n-1}. We paste them to each other along the boundary and according to the number of components of the boundary

$$\hat{M} = \bigcup W_i, \quad N'_{i+1,0} = N_{i,1}, \quad -\infty < i < \infty. \tag{4}$$

The manifold $N^{n-1} = N_0^{n-1}$ may be assumed to be a level surface of the function S (or the complete inverse image of a point under the map $f = \exp(2\pi i S)$). The monodromy operator acts as follows:

$$t\colon W_i \to W_{i+1},\ N_{i,0} \to N_{i,1} = N_{i+1,0}. \tag{5}$$

In accordance with general principles, S must generate a cell complex (see §3). However, in our case the most important requirement on which the usual Morse theory is based is not satisfied: this theory requires that the domains of lesser values $S \leqslant a$ are relatively compact, both in the finite-dimensional and infinite-dimensional case. In our case this is not true. However, in our case from each critical point of index i the "surface of most rapid descent" (or, if necessary, its smallest displacement) which can naturally be regarded as a "cell", emerges "downwards" through the levels. However, this "cell" can be pulled through the levels of S as far as $-\infty$; infinitely many such "cells" of dimension $i-1$ can be contained in its algebraic boundary. Under the shift $t\colon \hat{M} \to \hat{M}$ the functions S goes over into itself with the addition of a constant, taking critical points into critical points. Thus, we conclude that a) every critical point determines a free generator in the complex in question; b) the boundary of a cell can be an infinite linear combination of cells of this complex, lying "lower" in the levels of S, that is, emanating from ∞ only to one side in $\hat{M}$; c) all the "cells" are obtained from finitely many of all possible base shift through elements t^m of $\mathbf{Z}$ acting on $\hat{M}$.

We introduce the ring of Laurent series of the form

$$\sum_{j>\text{const}>-\infty} m_j t^j \in \hat{K} \tag{6}$$

with integer coefficients m_j that vanish for sufficiently large negative j. We denote this ring by $K = \hat{\mathbf{Z}}^+[t, t^{-1}]$. We regard the cell complex generated by a many-valued function on the manifold M^n or a function S on the covering $\hat{M} \to M^n$ as a free complex of finitely generated K-modules C (since the number of critical points is finite). The complex C has the form

$$0 \to C_n \xrightarrow{\partial} C_{n-1} \xrightarrow{\partial} \dots \xrightarrow{\partial} C_1 \xrightarrow{\partial} C_0 \to 0,$$

where ∂ is a K-module homomorphism. We note that in contrast to the usual Morse theory it can happen that $C_0 = C_n = 0$. Furthermore, on any manifold M^n there is a closed 1-form of any non-trivial cohomology class $[\omega] \in H^1(M^n, \mathbf{R})$ such that there are no local minima and maxima at all (that is, $C_0 = C_n = 0$).

For the skew products of M^n with the base S^1 there is a form ω without critical points, that is, $C_n = C_{n-1} = \dots = C_1 = C_0 = 0$.

Lemma. *The homology of the complex of K-modules C, generated by any smooth closed 1-form ω is homotopy invariant.*

Without proving this simple lemma, we see that the invariants of these homology groups can be used to obtain analogues of the Morse inequalities for the case of many-valued functions generating maps into the circle

$$\exp(2\pi i S)\colon M^n \to S^1.$$

The ring K is homologically one-dimensional (if the coefficients m_j of the series are elements of a field, then the corresponding analogue of K is also a field). Consequently, submodules of free modules are always free. This enables us to choose free bases in the groups (modules) of "cycles" $\mathbf{Z}_k = \ker \partial \subset C_n$ and "boundaries" $B_k = \operatorname{Im} \partial \subset C_n$. The difference in rank of these modules is called the "Betti number" and is denoted by $b_k(M^n, a)$ where $a = [\omega] \in H^1(M^n, \mathbf{Z})$.

The analogues of the torsion numbers $q_k(M^n, a)$ are defined as follows: we can choose free bases of the module $\mathbf{Z}_k(e_1, \ldots, e_N)$ and the submodule $B_k(e'_1, \ldots, e'_L)$, where $N - L = b_k$, such that:

$$e'_j = \Bigl(n_j + \sum_{k \geqslant 1} n_{jk} t^k\Bigr) e_j + \sum_{i > L} q_{ij}(t) e_i, \tag{7}$$

moreover:

1) the number n_j is divisible by n_{j+1};

2) the degrees of all the terms $q_{ij}(t)$ of the series are non-negative;

3) the numbers $q_{ij}(0) \neq 0$ are also divisible by n_j for all i and j (if the series does not vanish identically).

The total number of indices j such that $n_j \neq 1$ is called the torsion number and is denoted by $q_k(M^n, a)$. The number $q_k + b_k$ coincides with the minimal number of generators of the module $H_k = \mathbf{Z}_k / B_k$.

Theorem. *The following analogues of the Morse inequalities hold for the numbers $m_i(S)$ (or $m_i(\omega)$) of critical points of index i for a map in the neighbourhood of $\exp(2\pi i S)$ or for a closed 1-form ω, where $[\omega] = a \in H^1(M^n, \mathbf{Z})$:*

$$m_i(S) \geqslant b_i(M^n, a) + q_i(M^n, a) + q_{i-1}(M^n, a). \tag{8}$$

The proof of this theorem is easily deduced from the preceding.

We note that these analogues of the Morse inequalities are similar to the classical ones, but the topological invariants in them have a more complex geometrical meaning.

For manifolds with $\pi_1(M^n) = \mathbf{Z}$ it makes sense to ask when there is equality in (8), which resembles the familiar Smale theorem on single-valued functions on simply-connected manifolds. One can construct without difficulty a level surface N^{n-1} that is dual to the class $a = [\omega] \in H^{-1}(M^n)$ and is connected and simply-connected (in any case, for $n \geqslant 6$). Next, by using the Smale function on the membrane W^n with two boundaries $\partial W = N_0^{n-1} \cup N_1^{n-1}$, which is obtained from M^n by a section, a level surface N^{n-1} can be "minimally" continued (by using the Smale function on W) to the whole manifold M^n resulting in a form on M^n and a function S on the covering $\hat{M}$. However, this form (or many-valued function) can be far from minimal in its number of critical points. The construction of a minimal 1-form ω requires the choice of an initial manifold $N^{n-1} \subset M^n$ that is "minimal" in a certain sense if this choice is at all possible. It would be interesting to analyse to the end this problem for manifolds with the group $\pi_1 = \mathbf{Z}$.

We make a few remarks concerning the more complicated case $k > 1$, that is, when the form ω has at least two rationally independent integrals over one-dimensional cycles $\varkappa_i = \oint_{\gamma_i} \omega$, $\gamma_1, \ldots, \gamma_k$, where $\gamma_{k+1}, \ldots, \gamma_N$ is a basis of $H_1(M^n, \mathbf{Z})$, $\varkappa_i \neq 0$, $i \leqslant k$, $\sum m_i \varkappa_i \neq 0$, and the m_i are arbitrary integers. Here we have the covering $\hat{M} \xrightarrow[p]{} M^n$, where $p\omega = dS$, and the monodromy group is free Abelian. We introduce the ring $K_\varkappa$ of series $b \in K_\varkappa$ with integer coefficients

$$(9) \qquad b = \sum_{m=(m_1, \ldots, m_k)} b_m t_1^{m_1} \cdot \ldots \cdot t_k^{m_k}.$$

Here

1. $b_m = 0$ if $\sum m_i \varkappa_i$ is sufficiently large in modulus and negative.

2. "Stability", that is, for any series b there are numbers $\varepsilon > 0$ and N such that $b_m = 0$ if

$$(10) \qquad \sum m_i \varkappa_i^* < -N, \quad \sum |\varkappa_i^* - \varkappa_i| < \varepsilon.$$

The closed 1-form ω defines a cell complex, regarded as a complex of $K_\varkappa$-modules. The homology of this complex is homotopy invariant and can serve as a basis for constructing inequalities of Morse type. It is interesting to study the way in which the complexes and homology that arise here depend on $\varkappa$ if ω is altered slightly and the critical points remain essentially as before. If ω has no critical points at all, then the manifold M^n has the form

$$M^n = \hat{M}/\mathbf{Z}^k = (\hat{N} \times R)/\mathbf{Z}^k,$$

where $\hat{N}$ is a typical fiber of the fibration $\omega = 0$. All the fibers in this case are identical. From an approximation of ω by closed forms $\omega_j \to \omega$ with rational integrals over cycles without critical points it is clear that M^n is a skew product with circular base. The fibers of these skew products are compact manifolds N_j^{n-1} that are factors of $\hat{N}$:

$$\hat{N} \to N_j^{n-1},$$

That is, $\hat{N}$ is a regular covering over N_j with monodromy group $\mathbf{Z}^{k-1}$.

Any Riemannian metric on M^n together with a form ω without critical points generate a discrete "dynamical system", namely, the action ρ of $\mathbf{Z}^k$ on M^n, conserving the fiber: the transformations $\rho(t_j) \in \mathbf{Z}^k$ on $\hat{M}$ are constructed by inversely transferring to the initial fiber $\hat{N} \subset \hat{M}$ the image $t_j(\hat{N})$ of the monodromy map t_j: $\hat{M} \to \hat{M}$ along the normals to the fibers in the given metric. In the group $\rho(\mathbf{Z}^k)$ there are subgroups isomorphic to $\mathbf{Z}^{k-1}$ that act discretely on the fibers with a compact factor; their action can be everywhere dense on M^n, as shown by the very simple example $M^2 = T^2$, $\omega = \varkappa_1\, d\varphi_1 + \varkappa_2\, d\varphi_2$, $\varkappa_1/\varkappa_2$ is irrational. The whole group $\rho(\mathbf{Z}^k)$ acts non-discretely even on the fibers. What are the ergodic properties of the action of $\rho(\mathbf{Z}^k)$ on M^n? Do they depend on the Riemannian metric? Can we assert that in the "typical" case the orbits of $\rho(\mathbf{Z}^k)$ are everywhere dense on the fibers? The optimal Riemannian metric that can be used here naturally must be such that the distance between neighbouring fibers is constant.

For forms with critical points the geometrical picture becomes substantially more complicated. Let us assume that all the critical points are non-degenerate, therefore, that there are finitely many of them. For simplicity we may assume that the form has no local minima and maxima (such closed forms are always in any cohomology class). Apparently, in the "typical" case the non-singular fibers are everywhere dense. It is an interesting problem to describe the topological properties of non-singular fibers. In a certain natural sense it is a "quasi-periodic manifold".

The simplest non-trivial case is $k = 2$; here the minimal covering $\hat{M} \xrightarrow{p} M^n$ that turns the form ω into the differential of a single-valued function $p^*\omega = dS$ has the monodromy group $G = \mathbf{Z}^2$; the fibers (that is, the surfaces S = const) in the covering $\hat{M}$ are in a certain sense similar to $\mathbf{Z}$-coverings over compact $(n-1)$-manifolds. In any case, these fibers extend to infinity in two directions, topologically speaking, they have two "ends" $(\pm\infty)$ if they are connected around $\pm\infty$. The simplest model of a quasi-periodic manifold that can occur is as follows: there is a finite set of manifolds W_i^{n-1} with boundaries $\partial W_i^{n-1} = V_{0i}^{n-2} \cup V_{1i}^{n-2}$ $(i = 1, 2, \ldots, m)$. Let $\alpha = (\ldots, i_{-2}, i_{-1}, i_0, i_1, i_2, \ldots)$ be a doubly infinite sequence of m symbols. If this sequence is "admissible", then we can construct an open manifold W_α in the following natural way;

(11) $$W_\alpha = (\ldots \cup W_{i_{-2}} \cup W_{i_{-1}} \cup W_{i_0} \cup W_{i_1} \cup \ldots)$$

with the pastings

(12) $$V_{1, i_j}^{n-2} = V_{0, i_{j+1}}^{n-2}, \qquad -\infty < j < \infty.$$

Admissibility indicates that the pastings (12) are possible.

For $k = 2$ the proposition is that the non-singular fibers $\omega = 0$ (the level surfaces of the many-valued function) can all be obtained by this construction. The singular points, of which there are finitely many, can also be constructed, but in the pasting (11) one of the elements W_{i_s} is replaced in just one place by a manifold with the simplest Morse singularity of index i.

On the covering $\hat{M}$ the fibers S = const are singular for a countable everywhere dense set of values (in the "typical" case), and on each singular fiber there is only one critical point of S. Let $S = c$, $c + \varepsilon$ be non-singular fibers and $\varepsilon > 0$ sufficiently small; we can achieve that all the non-singular fibers are pasted together from one and the same elements W_i, but possibly relative to distinct sequences $\alpha = \alpha_c$, $S = c$. The transition $c \to c + \varepsilon$ in the set α_c gives rise to a change in the subset of indices $\gamma \subset \alpha_c$ at the expense of the Morse reconstruction of critical points in the domain $(c, c + \varepsilon)$. The sequence $\gamma \subset \alpha_i$ has "on average" finitely many elements $n(\gamma)$ on an interval of order $1/\varepsilon$. The relevant "density" $\bar{n}(\gamma) = \varepsilon n(\gamma)$ determines the average number of reconstructions of a level surface in the interval $(c, c + \varepsilon)$. It is natural to introduce the quantities $\bar{n}_i(\gamma)$: the densities of the number of Morse reconstructions of a given index i in the interval $(c, c + \varepsilon)$, $\bar{n}(\gamma) = \sum_{i=1}^{n-1} \bar{n}_i(\gamma)$ (we recall that by assumption there are no minima and maxima). Thus, for all non-singular fibers $S = c$ the function α_c, together with an indication of the places and type of replacements in an everywhere dense countable set of singular fibers c_j, determines a series of quantities that characterize the family of level surfaces of a many-valued function for $k = 2$, a function that has so far only been studied on the covering $\hat{M}$. The transition to M^n, that is, the factorization over $\mathbf{Z}^2$ with the generators (t_1, t_2) causes new difficulties. We can achieve that the representations of fibers in the form (11) are consistent with a single shift t_1. This has the following significance: the whole manifold $\hat{M}$ is constructed, beginning with a single fiber represented in the form (11), by the sequence of Morse reconstructions of the pieces W_j, according to the combinatorial scheme described above. Here the partition can be made so that under the shift $t_1^{q_1}: \hat{M} \to \hat{M}$ for some integer $q_1 \neq 0$ ($\hat{N}_c \to \hat{N}_{c+q_1\varkappa_1}$) the representation of the fiber $\hat{N}_{q_1\varkappa_1+c}$ in the form (11) is obtained from a representation of the fiber $\hat{N}_c$ by some shift of the sequences $\alpha = \alpha_c$ by an integer s_1:

$$\alpha_{c+q_1\varkappa_1} = s_1(\alpha_c), \qquad i_j \to i_j + s_1. \tag{13}$$

Into the topological arbitrariness of this construction there enters yet another elementwise diffeomorphism $\psi^{(1)}: \hat{N}_c \to \hat{N}_c$, where $\psi_m^{(1)}: W_{j_m} \to W_{j_m}$ and the diffeomorphisms $\psi_m^{(1)}$ are compatible at the boundaries. One can choose another partition of the fiber $\{W_l'\}$, where $\hat{N}_c = W'_\beta$, $\beta = (\ldots, l_{-1}, l_0, l_1, \ldots)$, and construct the whole manifold $\hat{M}$ by analogy with preceding but adapted to the shift t_2: for some integer $q_2 \neq 0$ the fiber $\hat{N}_{c+q_2\varkappa_2}$ is obtained after a series of reconstructions from a representation of the fiber $\hat{N}_c$ in the form

$$\hat{N}_{c+q_2\varkappa_2} = W'_{s_2(\beta)} = W'_{\beta_{c+q_2\varkappa_2}}, \tag{14}$$

and $t_2^{q_1}$: $\hat{N}_c \to \hat{N}_{c+q_2\varkappa_2}$, where $s_2(\beta_c)$ is a shift of the sequences $\beta = \beta_c$ by an integer and possibly an elementwise diffeomorphism $\psi^{(2)}$:

$$\psi^{(2)}: \hat{N}_c \to \hat{N}_c, \quad \psi^{(2)} = \bigcup \psi_m^{(2)}, \ \psi_m^{(2)}: W'_{l_m} \to W'_{l_m}.$$

The fact that these two partitions can be consistent is plausible, but not proved: can we choose them so that $W_j = W_j'$? Undoubtedly of great interest is the question when the "quasi-periodic manifold" (11) can be realized as a non-singular everywhere dense fiber: the level surface of a closed 1-form is on the compact manifold M^n. All the constructions easily generalized to the case $k > 2$: the manifolds W_j^{n-1} must be given together with maps into the cube φ_j: $W_j^{n-1} \to I^{k-1}$ in such a way that the complete inverse images of the faces of the cube give the corresponding partition of the boundary $\partial W_j^{n-1} = \varphi_j^{-1}(\partial I^{k-1})$. The sequences of symbols α must be "admissible" functions on a $(k-1)$-dimensional lattice $j(n_1, \ldots, n_{k-1})$ with value in the set of symbols (j) numbering the manifolds W_j^{n-1}. The pasting

$$(15) \quad W_\alpha = \bigcup_{(n_1, \ldots, n_{k-1})} W^{n-1}_{j(n_1, \ldots, n_{k-1})}, \quad \alpha = \{j(n_1, \ldots, n_{k-1})\},$$

can be defined as in (11), if α is an admissible distribution of indices (of course, the complete inverse images φ_j^{-1} are pasted together from the adjacent faces of cubes according to the numbering on the lattice). All the problems raised here naturally become more complicated for $k > 2$.

By analogy with the forms without critical points (above), for forms ω with Morse singularities one can also define "almost everywhere" the action of a somewhat smaller group $\rho(\mathbf{Z}^k)$ as follows. Let a and b be integers such that

$$|a\varkappa_1 + b\varkappa_2| < \varepsilon,$$

where $\varepsilon \to 0$ is sufficiently small. The transformation $t_1^a t_2^b$ is such that the image of any fiber $\hat{N}_c$ on $\hat{M}$ turns out to be equal to $\hat{N}_{c+a\varkappa_1+b\varkappa_2}$ and so is uniformly distributed close to $\hat{N}_c$ for any c everywhere on $\hat{M}$. The map $\rho(t_1^a t_2^b)$: $\hat{N}_c \to \hat{N}_c$ is constructed by means of the composition $t_1^a t_2^b$ with a translation along the normals to the fibers in the given Riemannian metric on M^n. This map is not defined on the set of measure zero consisting of the intersection of fibers "of surfaces of most rapid descent" emerging on both sides from critical points lying between the fibers $\hat{N}_c$ and $\hat{N}_{c+a\varkappa_1+b\varkappa_2}$ (We recall that by assumption the form ω has no minima and maxima, therefore, the set where the map and its inverse are not defined is of measure zero on the fibers; moreover, the critical points between these fibers are "sectionally" distributed as $\varepsilon \to 0$.) The map $\rho(t_1^a t_2^b)$ and its inverse are defined on M^n and are everywhere smooth, except at critical points and those parts of their "surfaces of most rapid volume decrease" that fall into ε-neighbourhoods of critical points. Of course, to define the whole group $\rho(\mathbf{Z}^k)$ one would have to eliminate entirely all these surfaces from M^n.

A closer investigation of the family of level surfaces of closed 1-forms seems to the author to be an extremely interesting (purely topological) problem. Incidentally, in the Hamiltonian formalism, as we have seen, the Hamiltonian is not necessarily single-valued on the symplectic manifold on which the Poisson brackets are defined, but only a closed 1-form. The simplest example of this is the motion of a classical particle in space under the influence of a periodic potential (for some lattice Γ) plus a constant strong field. In this case the Hamiltonian

$$(16)\qquad \begin{cases} H=\frac{p^2}{2m}+eU(x)+eE_i x^i, \\ E_i=\text{const}, \quad U(\vec{x}+\vec{\Gamma}_i)=U(\vec{x}), \end{cases}$$

$$\Gamma=(n_1\vec{\Gamma}_1+n_2\vec{\Gamma}_2+n_3\vec{\Gamma}_3)$$

is a 1-form on $T^*(T^3)=\mathbf{R}^3_{(p)}\times T^3_{(x)}$. If p are the so-called "quasi-momenta", periodic with the inverse lattice Γ^* and $\varepsilon(p)$ is the "dispersion law", then the Hamiltonian, after the inclusion of a ("weak") external electric field, can have the form

$$(17)\qquad H=\varepsilon(p)+eE_i x^i+eU(x),$$

where $U(x)$ is periodic with the lattice Γ or one of its sublattices $\Gamma'\subset\Gamma$, and $\varepsilon(p)$ is periodic for Γ^*. Here $M^n=T^3_{(p)}\times T^3_{(x)}$. Of course, these examples are somewhat artificial, but we have given them to illustrate the fact that the study of the level surfaces of 1-forms can be useful from various points of view. It is also possible to include a weak magnetic field where the Hamiltonian of the form (17) is modified and becomes

$$(18)\qquad H=\varepsilon\left(p-\frac{e}{c}A(x)\right)+e(E_i x^i+U(x)),$$

where $A_j(\ldots, x_i+\Gamma_i, \ldots)=A_j(x)+\partial_j f_i$ the f_i being a single-valued function. It is easy to see that the term $\varepsilon\left(p-\frac{e}{c}A\right)$ determines a single-valued function on the torus T^6 (or on a finite covering of it) if and only if the flows of the magnetic field $H_{ij}=\partial_i A_j-\partial_j A_i$ across all two-dimensional elementary cells of the lattice Γ are rational multiples of the unit of magnetic flow $2\pi ce^{-1}(\hbar=1)$. The additional distorted potential $U(x)$ is not, as a rule, considered. The "electrical part" of the Hamiltonian $E_i x^i$ gives a closed, but not-exact 1-form dH on the torus $T^3\times T^3$, which in general has 3 non-commensurable non-zero periods:

$$(19)\qquad dH=\frac{\partial\varepsilon}{\partial p_i}dp_i+e\left(E_i+\frac{\partial U}{\partial x^i}-\frac{1}{c}\frac{\partial\varepsilon}{\partial p_j}\frac{\partial A_j}{\partial x^i}\right)dx^i.$$

The symplectic 2-form has the usual shape $\Omega=dp_i\wedge dx^i$ and determines the Poisson brackets and the Hamiltonian systems. A classical dynamical system with Hamiltonian (18), where $\varepsilon(p)$ is derived from quantum theory, can in principle be of considerable interest in solid state theory, whereas the system with the Hamiltonian (16) is interesting only after quantization, where its properties are non-trivial even in the one-dimensional case because of the presence of the 1-form $E_i dx^i$ (of "constant force").

Remark. In 3-dimensional space, in the absence of a constant electric field and weak external periodic potential $E_i = 0$, $U = 0$ we have the semiclassical Hamiltonian $H = \varepsilon\left(p - \frac{e}{c}A\right)$ of a particle with momentum $p = p' + \frac{e}{c}A$, $p' \in T^3_{(p')}$ in a constant (homogeneous) magnetic field $\bar{H}$: strictly speaking, the field $\bar{H}$ is a 2-form

$$\bar{H} = H_1\, dy \wedge dz + H_2\, dz \wedge dx + H_3\, dx \wedge dy$$

such that $d\bar{H} = 0$. In the absence of local currents we have rot $\bar{H} = 0$ or $d(*\bar{H}) = 0$. Of course, this is satisfied for a homogeneous field. The semiclassical "motion" of a particle occurs in the p'-space T^3 under the influence of the field $\bar{H}$ by virtue of the equation $\dot{p}' = \frac{e}{c}[v \times \bar{H}]$ or $\dot{p}'_j = \frac{e}{c} v^k H_{kj}$, $v = \dot{x} = \partial\varepsilon/\partial p'$; where $p' = p - \frac{e}{c}A$. This is the "old momentum" to within inclusion of the field $\bar{H}$; the particle moves along the surfaces $\varepsilon(p') = \text{const}$ orthogonally to $\bar{H}$. Thus, the motion is along the level surfaces of the 1-form ($\omega = 0$) on the 2-dimensional manifolds $\varepsilon(p') = \text{const}$, provided that the vector H has two or three pairwise incommemsurable coefficients of inclination with respect to the initial lattice in which $\omega = H_1\, dp'_1 + H_2\, dp'_2 + H_3\, dp'_3$. Although the fiber $\omega = 0$ is, in general, not connected, transitions are possible from one trajectory to another on the same fiber $\omega = 0$ near the critical points. Thus, it is natural to regard the fiber $\omega = 0$ as a single integer and to study (in the irrational case) its quasiperiodic structure discussed above.

Here we have the following proposition:

If the Fermi surface $\varepsilon(p) = \varepsilon_0$ in the torus T^3 has the genus $g \leqslant 1$, then the form ω has the degree of irrationality $k - 1 \leqslant 0$; if the genus is $g = 2$, then $k \leqslant 2$; if the genus is $g \geqslant 3$, then $k \leqslant 3$. Here ω is the restriction of the form $*\bar{H}$ to the Fermi surface and k is the rank of the monodromy group of the covering $\hat{M} \xrightarrow{p} M^2$ such that $p^*\omega = dS$.

The proof uses the face that $\varepsilon(p)$ is a smooth single-valued function on the torus T^3. Therefore, any Fermi surface $\varepsilon(p) = \varepsilon_0$ bounds a membrane in T^3: and so, the inclusion j: $M^2 \to T^3$ is such that $j_*[M^2] = 0$. The covering over M^2 is induced by a $\mathbf{Z}^3$-covering over T^3. The restriction of the cohomology j^*: $H^1(T^3, \mathbf{Z}) \to H^1(M^2)$ has an image on which the multiplication of cohomology is trivial, since $j_*[M^2] = 0$. Hence, the rank of the image $j^*H^1 \subset H^1(M^2)$ does not exceed the genus g (but also never exceeds 3), since $\omega = j^*(*\bar{H})$, and the proof is complete.

Therefore, if the genus of the Fermi-surface is $\leqslant 1$, then the trajectories of a semiclassical motion are always closed (compact) in $\mathbf{R}^3$ for the quasi-momenta, although for $g = 1$ the covering itself over the Fermi-surface in $\mathbf{R}^3$ may be an open covering of cylinder type (it cannot be $\mathbf{R}^2$). If the genus is $g = 2$, then $k = 2$ is possible. In this case the level surfaces $\omega = 0$ (orthogonal to the field $\bar{H}$, as a vector in $\mathbf{R}^3$) may be "quasi-periodic" manifolds and extend to $\pm\infty$ in two "asymptotic" directions. We recall that according to Morse theory the function $\varepsilon(p)$ on the torus T^3 must have no less than $\binom{l}{3} = b_l(T^3)$ critical points of index l (if they are non-degenerate), and their number increases in the pressence of a non-trivial finite group of symmetries, which (in general) occurs in crystals. Consequently, Fermi-surfaces of high genus are possible.

References

[1] S.P. Novikov and I. Shmel'tser, Periodic solutions of the Kirchhoff equations for the free motion of a rigid body in an ideal fluid and the extended Lyusternik-Shnirel'man-Morse theory. I, Funktsional. Anal. i Prilozhen. **15**:3 (1981), 54-66. MR **83a**:58026a.
= Functional Anal. Appl. **15** (1981), 197-207.

[2] ———, Variational methods and periodic solutions of equations of Kirchhoff type. II, Funktsional. Anal. i Prilozhen. **15**:4 (1981), 37-52. MR **83a**:58026b.
= Functional Anal. Appl. **15** (1981), 263-274.

[3] ———, Many-valued functions and functionals. An analogue of Morse theory, Dokl. Akad. Nauk SSSR **260** (1981), 31-35. MR **83a**:58025.
= Soviet Math. Dokl. **24** (1981), 222-226.

[4] L.M. Milne-Thomson, Theoretical hydrodynamics, 4th ed., Macmillan, London-New York 1960. MR **22** # 3286.
Translation: *Teoreticheskaya gidrodinamika,* Mir, Moscow 1964.

[5] W.F. Brinkman and M.C. Cross, Spin and orbital dynamics of superfluid ^{3}He, in: Progress in low temperature physics, ed. D.F. Brewer, North Holland, Amsterdam 1978, 105.

[6] K. Maki and H. Ebisawa, Exact magnetic ringing solutions in superfluid ^{3}He – B, Phys. Rev. B **13** (1976), 2924-2930. PA 76-61577.

[7] V.L. Golo, Non-linear regimes in spin dynamics of superfluid ^{3}He, Lett. Math. Phys. **5** (1981), 155-159. MR **82g**:81123.

[8] I.A. Fomin, Solution of spin dynamics equations for ^{3}He superfluid phases in a strong magnetic fluid, J. Low Temp. Phys. **31** (1978), 509-526. PA 78-56010.

[9] C. Gardner, Korteweg-de Vries equation and generalizations. IV, The Korteweg-de Vries equation as a Hamiltonian system, J. Math. Phys. **12** (1971), 1548-1551. MR **44** # 3615.

[10] V.E. Zakharov and L.D. Faddeev, The Korteweg-de Vries equation is a fully integrable Hamiltonian system, Funktsional. Anal. i Prilozhen. **5**:4 (1971), 18-27. MR **46** # 2270.
= Functional Anal. Appl. **5** (1971), 280-287.

[11] I.M. Gel'fand and L.A. Dikii, Fractional powers of operators, and Hamiltonian systems, Funktsional. Anal. i Prilozhen. **10**:4 (1976), 13-29. MR **55** # 6484.
= Functional Anal. Appl. **10** (1976), 259-273.

[12] F. Magri, A simple model of the integrable Hamiltonian equation, J. Math. Phys. **19** (1978), 1156-1162. MR **80a**:35112.

[13] M. Adler, On a trace functional for formal pseudo-differential operators and the symplectic structure of the Korteweg-de Vries type equations, Invent. Math. **50** (1979), 219-248. MR **80i**:58026.
[14] D.R. Lebedev and Yu.I. Manin, The Gel'fand-Dikii Hamiltonian operator and the coadjoint representation of the Volterra group, Funktsional. Anal. i Prilozhen. **13**:4 (1979), 40-46. MR **81a**:58029.
= Functional Anal. Appl. **13** (1979), 268-273.
[15] I.M. Gel'fand and I.Ya. Dorfman, Hamiltonian operators and infinite-dimensional Lie algebras, Funktsional. Anal. i Prilozhen. **15**:3 (1981), 23-40. MR 82j:58045.
= Functional Anal. Appl. **15** (1981), 173-187.
[16] B.A. Dubrovin, V.B. Matveev, and S.P. Novikov, Non-linear equations of the Korteweg-de Vries type, finite-zone linear operators, and Abelian manifolds, Uspekhi Mat. Nauk **31**:1 (1976), 55-136. MR **55** # 899.
= Russian Math. Surveys **31**:1 (1976), 59-146.
[17] O.I. Bogoyavlenskii and S.P. Novikov, The connection between the Hamiltonian formalisms of stationary and non-stationary problems, Funktsional. Anal. i Prilozhen. **10**:1 (1976), 9-11. MR **57** # 7666.
= Functional Anal. Appl. **10** (1976), 8-11.
[18] A.P. Veselov, Lagrangian and Hamiltonian formalism for Novikov-Krichever equations, Funktsional. Anal. i Prilozhen. **13**:1 (1979), 1-7. MR **80i**:58028.
= Functional Anal. App. **13** (1979), 1-6.
[19] V.I. Arnold, The Hamiltonian nature of the Euler equations in the dynamics of a rigid body and of an ideal fluid, Uspekhi Mat. Nauk **24**:3 (1969), 225-226. MR **43** # 2900.
[20] H. Lamb, Hydrodynamics, CUP, 6th ed., Cambridge 1932.
Translation: *Gidrodinamika,* Mir, Moscow 1947.
[21] A.G. Kulikovskii and G.A. Lyubimov, *Magnitnaya gidrodinamika* (Magnetohydrodynamics), Fizmatgiz, Moscow 1962.
[22] I.M. Khalatnikov, *Teoriya sverkhtekuchesti* (Superfluidity theory), Nauka, Moscow 1971.
[23] ——— and V.V. Lebedev, Canonical equations of hydrodynamics of quantum liquids, J. Low Temp. Phys. **32** (1978), 789-801 (in English). PA 78-91497.
[24] ——— and ———, Hamiltonian equations of hydrodynamics of anisotropic superfluid liquid ^{3}He − A, Phys. Lett. **61A** (1977), 319-320 (in English). PA 77-69138.
[25] I.E. Dzyaloshinskii and G.E. Volovick, Poisson brackets in condensed matter physics, Ann. Physics **125** (1980), 67-97 (in English). PA 80-54296.
[26] J.W. Milnor, Morse theory, Princeton Univ. Press, Princeton, NJ 1963. MR **29** # 634.
Translation: *Teoriya Morsa,* Mir, Moscow 1965. MR **31** # 6249.
[27] B.A. Dubrovin, I.M. Krichever, and S.P. Novikov, Topological and algebraic-geometrical methods in mathematical physics. II, in: Soviet Scientific Rev. (ed. S.P. Novikov), Phys. Rev., vol. 3, Harwood Acad. Publishers, New York 1982.
[28] L.A. Lyusternik and L.G. Shnirel'man, Topological methods in variational problems and their application to the differential geometry of surfaces, Uspekhi Mat. Nauk 2:1 (1947), 166-217. MR **10**-624.
[29] R. Bott, The stable homotopy of the classical groups, Ann. of Math. **70** (1959), 313-337. MR **22** # 987.
[30] J.W. Milnor, A procedure for killing the homotopy groups of different manifolds, Proc. Symp. Pure Maths. III, 39-55, American Math. Soc. 1961. MR **24** # A556.
[31] ——— and M. Kervaire, Groups of homotopy spheres, Ann. of Math. (2) **77** (1963), 504-537. MR **26** # 5584.

[32] S. Smale, On the structure of manifolds, Amer. J. Math. **84** (1962), 387-399. MR **27** # 2991.

[33] S.P. Novikov, Homotopically equivalent smooth manifolds. I, Izv. Akad. Nauk SSSR Ser. Mat. **28** (1964), 365-464. MR **28** # 5445.

[34] W. Browder, Surgery on simply-connected manifolds, Springer, New York-Heidelberg 1972. MR **50** # 11272.

[35] S.P. Novikov, On manifolds with a free Abelian fundamental group and their applications, Izv. Akad. Nauk SSSR Ser. Mat. **30** (1966), 207-246. MR **33** #4951.

[36] C.T.C. Wall, Surgery on compact manifolds, Academic Press, London-New York 1970. MR **55** # 4217.

[37] S.P. Novikov, Pontrjagin classes, the fundamental groups and some problems of stable algebra. Essays on topology and related topics (Mémoires dédiés à Georges de Rham), 147-155, Springer, New York 1970 (in English). MR **42** # 3804.

[38] V.I. Arnol'd, Small denominators and the problem of stability of motion in classical and celestial mechanics, Uspekhi Mat. Nauk **18**:6 (1963), 91-192. MR **30** # 943.
= Russian Math. Surveys **18**:6 (1963), 85-191.

[39] A.I. Fet, On a periodicity problem in the calculus of variations, Dokl. Akad. Nauk SSSR **160** (1965), 287-289. MR **36** # 3381.
= Soviet Math. Dokl. **6** (1965), 85-88.

[40] B.A. Dubrovin, S.P. Novikov, and A.T. Fomenko, *Sovremmenaya geometriya. Geometriya i topologiya mnogoobrazii,* (Modern geometry. The geometry and topology of manifolds), Nauka, Moscow 1979.

[41] V.V. Kozlov, *Metody kachestvennogo analiza v dinamike tverdogo tela* (Methods of qualitative analysis in the dynamics of a rigid body), Moskov. Gos. Univ., Moscow 1980. MR **82e**:70002.

[42] Tai Tsun Wu and Chen Ning Yang, Dirac's monopoles without strings: Classical Lagrangian theory, Phys. Rev. D **14** (1976), 437-445.

[43] V.L. Golo, M.I. Monastyrskii, and S.P. Novikov, Solutions to the Ginsberg-Landau equations for planar textures in superfluid ^{3}He, Commun. Math. Phys. **69** (1979), 237-246 (in English). PA 80-10729.

[44] D. Gromall, W. Klingenberg, and W. Meyer, Riemannsche Geometrie im Grossen, Lecture Notes in Math. **55** (1968, 2nd ed. 1975). MR **37** # 4751.
Translation: *Rimanova geometriya v tselom,* Mir, Moscow 1971.

[45] V.I. Arnol'd, *Matematicheskie metody klassicheskoi mekhaniki,* Nauka, Moscow 1974. MR **57** # 14032.
Translation: Mathematical methods of classical mechanics, Springer, Berlin-New York 1979.

[46] S.V. Manakov, A remark on the integration of the Euler equations for the dynamics of an n-dimensional rigid body, Funktsional. Anal. i Prilozhen. **10**:4 (1976), 93-94. MR **56** # 13272.
= Functional Anal. Appl. **10** (1976), 328-329.

[47] ———, V.E. Zakharov, S.P. Novikov, and L.P. Pitaevskii, *Teoriya solitonov. Metod obratnoi zadachi* (Theory of solitons. The inverse problem method), Nauka, Moscow 1980. MR **81g**:35112.

[48] V.E. Zakharov and A.B. Shabat, A plan for integrating the non-linear equations of mathematical physics by the method of the inverse scattering problem. I, Funktsional. Anal. i Prilozhen. 8:3 (1974), 43-53. MR **58** # 1768.
= Functional Anal. Appl. **8** (1974), 226-235.

[49] B.A. Dubrovin, Fully integrable Hamiltonian systems associated with matrix operators, and Abelian manifolds, Funktsional. Anal. i Prilozhen. **11**:4 (1977), 28-41. MR **58** # 31219.
= Functional Anal. Appl. **11** (1977), 265-277.

[50] V.V. Kozlov and D.A. Onishchenko, Non-integrability of the Kirchhoff equations, Uspekhi Math. Nauk **37**:4 (1982), 110.
[51] A.S. Mishchenko, Integrals of geodesic flows on Lie groups, Funktsional. Anal. i Prilozhen. **4**:3 (1970), 73-77. MR **43** # 649.
= Functional Anal. Appl. **4** (1970), 232-235.
[52] L.A. Dikii, Remarks on Hamiltonian systems associated with the rotation group, Funktsional. Anal. i Prilozhen. **6**:4 (1972), 83-84. MR **47** # 1084.
= Functional Anal. Appl. **6** (1972), 326-327.
[53] A.M. Perelomov, Some remarks about the integrability of the equations of motion of a rigid body in an ideal fluid, Funktsional. Anal. i Prilozhen. **15**:2 (1981), 83-85. MR **82j**:58066.
= Functional Anal. Appl. **15** (1981), 144-146.
[54] F. Takens, A global version of the inverse problem of the calculus of variations, J. Differential Geom. **14** (1979), 543-562. MR **83b**:58028.
[55] A.M. Vinogradov, A spectral sequence associated with a non-linear differential equation and the algebraic-geometrical foundations of Lagrangian field theory with connections, Dokl. Akad. Nauk SSSR **238** (1978), 1028-1031. MR **81d**:58060.
= Soviet Math. Dokl. **19** (1978), 144-148.

Translated by R.L. and G. Hudson

The Landau Institute of Theoretical Physics

Received by the Editors 22 April 1982

Doklady 1965
Tom 163, No.2

TOPOLOGICAL INVARIANCE OF RATIONAL PONTRJAGIN CLASSES

S. P. NOVIKOV

The object of this paper is to prove the following theorem:

Theorem 1. *Let M_1^n and M_2^n be two smooth manifolds, or PL manifolds, and let $h: M_1^n \to M_2^n$ be a continuous homeomorphism. Then $h^* p_i(M_2^n) = p_i(M_1^n)$, where $p_i(M^n)$ are the Pontrjagin classes, with rational or real coefficients, of a manifold M^n.*

In this paper we give a sketch of the proof of this theorem for smooth simply-connected manifolds M_1^n and M_2^n; the non-simplyconnected case reduces to the simply-connected case. We remark that the method used in a previous paper [2] by the present author, where this proposition was proved in a special case, differs from that presented here, although the ideas in the two papers are closely related.

Theorem 1 follows from the following fundamental lemma:

Fundamental lemma. *Suppose that we are given a smooth structure on the Cartesian product $M^{4k} \times R^m$, making $M^{4k} \times R^m$ into a smooth open manifold W (here M^{4k} is a closed compact simply-connected manifold). Then the following formula holds:*

$$(L_k(W), [M^{4k}] \otimes 1) = \tau(M^{4k}).$$

Here L_k is the Hirzebruch polynomial in the Pontrjagin classes of the manifold W, and $\tau(M^{4k})$ is the signature of the manifold M^{4k}.

The proof is divided into several steps.

Step 1. We take the open subset $T^{m-1} \times R \subset R^m$, where T^{m-1} is the $(m-1)$-dimensional torus. We consider the open submanifold

$$W_1 = M^{4k} \times T^{m-1} \times R \subset W.$$

The following lemma is not difficult.

Lemma 1. *There exists a smooth closed submanifold $V_1 \subset W_1$ representing the cycle $[M^{4k} \times T^{m-1}] \otimes 1 \in H_{4k+m-1}(W_1)$, such that the natural projection $p: V_1 \to M^{4k} \times T^{m-1}$ induces isomorphisms of homotopy and homology groups in dimensions $< 2k + [(m-1)/2]$.*

Similarly, let W be a smooth manifold having the homotopy type of $M^{4k} \times T^m$, and let W_1' be a covering having the homotopy type of $M^{4k} \times T^{m-1}$. Then there exists a natural projection $f: W_1' \to M^{4k} \times T^{m-1}$ of degree +1; and $H_{4k+m-1}(W_1') = Z$.

Lemma 1'. *There exists a smooth submanifold $V_1' \subset W_1'$ representing the fundamental cycle of the group $H_{4k+m-1}(W_1') = Z$, such that the natural projection $f: V_1' \to M^{4k} \times T^{m-1}$ induces isomorphisms of homotopy and homology groups in dimensions $< 2k + [(m-1)/2]$.*

The proofs of the two lemmas are similar. They are accomplished by performing successive Morse surgeries on the kernel of the inclusion $V_1 \subset W_1$ or $V_1' \subset W_1'$, similarly, for example, to [1,3]. Here we make the additional remark that, since the group ring of an abelian group is noetherian, all homotopy kernels of the mapping f are finitely generated as $Z(\pi_1)$-modules, where $\pi_1 = Z + \cdots + Z$.

Reprinted with permission from the American Mathematical Society, Topological Invariance of Rational Pontrjagin Classes, Sov. Math. Dokl. **6** (1965) 921–923.

Thus in the dimensions indicated the homotopy kernels of the mapping f can be killed by surgery. We note that the mapping of universal coverings $f\colon V_1 \to M^{4k} \times R^{m-1}$ has degree +1 and is a proper map. Therefore the formula $f_* \hat{D} f^* \hat{D} = 1$ holds, where D stands for Poincaré duality. Moreover, in this case we have the following equality, which is not hard to prove:

$$\operatorname{Ker} \hat{f}_* (\pi_i) / Z_0 (\pi_1) \operatorname{Ker} \hat{f}_* (\pi_i) = \operatorname{Ker} f_* (H_i)$$

(by the Hurewicz theorem) where $\operatorname{Ker} \hat{f}_* (\pi_j) = \operatorname{Ker} f_* (\pi_j) = 0$ for $j < i$, and $\operatorname{Ker} f_* (\pi_i) = \operatorname{Ker} f_* (H_i)$. This is enough to give Lemmas 1 and 1′.

Step 2. The investigation of dimension $i = [2k + (m-1)/2]$ is more difficult. For this purpose, we make some remarks on intersection numbers. Let $P\colon \hat{M} \to M$ be a covering with monodromy group $\pi : \hat{M} \to \hat{M}$. The group π acts on the groups $H_i(\hat{M})$. We consider two cycles $x, y \in H_*(\hat{M})$ and all elements $\alpha_i \in \pi$. Then the formula $(p_* x) \circ (p_* y) = \Sigma_{\alpha_i \in \pi} x \circ (\alpha_i y)$ holds, where the symbol $\circ$ stands for taking intersection numbers in M and $\hat{M}$ respectively. Suppose that, using Lemmas 1 and 1′, we have manifolds $V_1 \subset W_1$ (or $V_1' \subset W_1'$) together with a projection $W_1 \to M_{4k} \times T^{m-1}$, having degree +1 on V_1, which is a homotopy equivalence on W_1 and a homotopy equivalence in dimensions $< [2k + (m-1)/2]$ on V_1. If we let N be the kernel $\operatorname{Ker} f_* (\pi_l)$ for $l = [2k + (m-1)/2]$, then $\operatorname{Ker} f_* (H_l) = N / Z_0 (\pi_1) N$, where $\pi_1 = Z + \cdots + Z$. An analogous formula holds for mappings of all intermediate coverings of V_1 and of $M^{4k} \times T^{m-1}$ which correspond to subgroups $\pi' \subset \pi_1$. If π' is of finite index in π_1, by the properties of mappings of degree +1 the kernels of the induced homomorphisms of homology groups of the corresponding coverings satisfy the usual Poincaré duality.

As we already remarked, intersection numbers in the homotopy kernels of π'-coverings can be obtained from intersection numbers in N by the following formula:

$$(qx) \circ (qy) = \sum_{\alpha_i \in \pi'} x \circ (\alpha_i y), \quad q : N \to N / Z_0 (\pi') N.$$

It is also important to note that in this case the kernels of the homotopy homomorphisms for all the coverings split into the sum of the "right kernel" and the "left kernel" from V_1 to W_1 (or from V_1' to W_1').

We now proceed as follows:

1. For odd values of the integer $2l = 4k + m - 1$, by performing surgery on V_1 inside W_1 (only to one side of V_1) operating on a $Z(\pi_1)$ basis of the "right kernel" in homotopy, we can get the "right kernel" in homotopy to be trivial in V_1 in dimension $l = [2k + (m-1)/2]$. Then, (similarly to [1]) the homotopy kernels in this dimension will be free on all coverings with finitely many sheets.

2. For even values of $2l = 4k + m - 1$ one can show that $N/N_\infty = F \oplus \ldots \oplus F$, where N_∞ is the kernel of projection onto all coverings with finitely many sheets and $F = \Sigma_{\alpha_i \in \pi'} \alpha_i Q$, $\dim Q = 2$; the intersection matrix for Q is of the form $\begin{pmatrix} 0 & 1 \\ \pm 1 & 0 \end{pmatrix}$ and $\alpha_i Q \cdot \alpha_j Q = 0$ for $\alpha_i \neq \alpha_j$, where π' is a subgroup of finite index in π_1.

3. We continue in both cases ($4k + m - 1$ even and odd) by performing surgery on the basis elements of the $Z(\pi_1)$-module N to get the kernels of the homotopy mappings on all coverings with finitely many sheets to be trivial, without changing the homotopy in dimensions $< l$.

4. However, at this stage the surgeries are already being carried out not inside the manifold W_1 (or W_1'), but abstractly on a π'-covering.

It is easily seen that after these operations the scalar product of the class $L_k(V_2)$ with the cycle $f_*^{-1}([M^{4k}] \otimes 1)$ is still the same as the scalar product $(L_k(W_1), f_*^{-1}[M^{4k}] \otimes 1)$, since this cycle lies in V_1 and V_1 has a trivial normal bundle in W after the last modification.

Step 3. The following fact is of the greatest importance: after the last modification, we obtain $N = \operatorname{Ker} f_*(\pi_l) = 0$.

This is not obvious, since all we knew was that $N/N_\infty = 0$, where N_∞ is the kernel of all projections on coverings with finitely many sheets. Since the ring $Z(\pi_1)$ is noetherian, the module N is finite-dimensional over $Z(\pi_1)$. By the Hurewicz theorem, the module N_∞ can be defined here in a purely algebraic manner. By an easy argument the problem can be reduced to one-dimensional modules in the case $\pi_1 = Z$, when it takes the following form: there exists a polynomial $P(x)$, with integer coefficients, such that $P(0) \neq 0$ and $P(1) = 1$ and for any root of unity $\zeta^i = 1$ the value $P(\zeta)$ is invertible among the integral elements of the corresponding cyclotomic field. One must show that $P(x) \equiv 1$. This proposition is proved using elementary algebraic number theory.* A similar assertion holds for polynomials mod p.

It follows that, after carrying out modifications on V_1 (or V_1'), we obtain $\operatorname{Ker} f_*(\pi_l) = 0$.

Therefore the modified mapping $f\colon V_1 \to M^{4k} \times T^{m-1}$ is a homotopy equivalence in dimensions $\leq [2k + (m-1)/2]$. Using Poincaré duality, the Hurewicz theorem and the proposition on polynomials mod p stated above, one can show that the mapping f is a homotopy equivalence in all dimensions.

Step 4. We are now ready for the fundamental lemma.

We consider $W = M^{4k} \times T^{m-1} \times R$ with some smoothing, and using the preceding work we construct a manifold V_1 having the homotopy type of $M^{4k} \times T^{m-1}$ such that its class $L_k(V_1)$ is the same as $L_k(W)$.

Furthermore, we consider the covering of V_1 with group Z: $V_1 = W_2 \to V_1$, and a mapping f_2: $V_1 \to M^{4k} \times T^{m-2}$. Using our preceding work, from $V_1 = W_2$ we can obtain a similar manifold V_2 having the same class $L_k(V_2)$ together with a mapping $V_2 \to M^{4k} \times T^{m-2}$ of degree $+1$, and so forth, until we reach dimension $4k$. We obtain a manifold V_m of dimension $4k$ having the homotopy type of M^{4k} and whose class $L_k(V_m)$ is the same as $L_k(W)$; more precisely, the scalar product of the class L_k with the corresponding cycle is the same. Using Hirzebruch's formula, we conclude that $L_k(V_m) = (L_k(W), f_*^{-1}([M^{4k}] \otimes 1)) = \tau(M^{4k})$. This gives the fundamental lemma.

Remark 1. The author's method does not, so far, give a means of defining Pontrjagin classes of topological microbundles and of topological manifolds, since the whole argument was based on a smooth structure; how such a definition can be given constitutes the most immediate problem suggested by the results of this paper.

Remark 2. The proof of invariance seems somewhat artificial to the author, since the surgeries were obviously not essential. It should become clearer when the connection between the classes L_k and coverings is explained; this is shown in a special case by the proof in [2], and by hitherto unpublished results concerning these problems which have been obtained by the present author and by V. A. Rohlin.

V. A. Steklov Mathematical Institute
Academy of Sciences of the USSR

Received 19/APR/65

BIBLIOGRAPHY

[1] W. Browder, Preprint, Princeton University, Princeton, N.J., 1964.
[2] S. P. Novikov, Dokl. Akad. Nauk SSSR 162 (1965), 1248 = Soviet Math. Dokl. 6 (1965), 854.
[3] ———, Izv. Akad. Nauk SSSR Ser. Mat. 28 (1964), 365; English transl., Amer. Math. Soc. Transl. (2) 48 (1965), 271. MR 28 #5445

Translated by: C. Wasiutynski

*The author expresses his deep gratitude to S. P. Demuškin, who proved this assertion at the author's request.

Curriculum Vitae
Mikio Sato

Date of Birth: April 18, 1928, Tokyo, Japan

Education:

Apr. 1949	Entered Univ. of Tokyo
Mar. 1952	Graduated from Univ. of Tokyo
Degree:	Doctor of Science, Univ. of Tokyo

Previous Positions (permanent):

Apr. 1958–Jan. 1960	Assistant, Univ. of Tokyo, Math. Dept.
Feb. 1960–Mar. 1963	Lecturer, Tokyo Univ. of Education, Math. Dept.
Apr. 1963–Aug. 1966	Professor, Osaka Univ., Math. Dept.
Apr. 1968–Jun. 1970	Professor, Univ. of Tokyo, Faculty of General Education
Jun. 1970–Mar. 1992	Professor, RIMS, Kyoto Univ., Japan
Jan. 31, 1987–Jan. 30, 1991	Director, RIMS, Kyoto Univ., Japan
Apr. 1992	Professor Emeritus, Kyoto Univ., Japan
Nov. 1984	Culture Contributor of Japan
Apr. 1993	Foreign Associates, Nat. Acad. of Sci. of the U.S.A.

Visiting Positions:

Sep. 1960–Jun. 1962	Member, IAS, Princeton
Sep. 1964–Jun. 1966	Visiting Professor, Columbia Univ., Math. Dept.
Sep. 1972–Jun. 1973	Visiting Professor, Nice Univ., Math. Dept.
Sep. 1992–Dec. 1992	Visiting Professor (Eirenberg Chair), Columbia Univ.

Awards:

1969	Asahi Shimbun Culture Prize (with Prof. H. Komatsu)
1976	Japan AcademyPrize
1987	Fujiwara Prize
1997	Rolf SchockPrize
2003	Wolf Prize

Honorary Degree:

Sep. 1993 Docteur Honoris Causa, Université Paris

List of Publications

1. M. Sato, Algebraic analysis and I (in Japanese), in *Algebraic Analysis and Number Theory* (Kyoto, 1992). *Sūrikaisekikenkyūsho Kōkyūroku*, No. 810 (1992), pp. 164–217.
2. M. Sato, Theory of prehomogeneous vector spaces (algebraic part) — The English translation of Sato's lecture from Shintani's note. Notes by Takuro Shintani; translated from the Japanese by M. Muro, *Nagoya Math. J.* **120** (1990), 1–34.
3. M. Sato, *D*-modules and nonlinear integrable systems, in *Algebraic Analysis, Geometry, and Number Theory* (Baltimore, MD, 1988), pp. 325–339 (Johns Hopkins Univ. Press, 1989).
4. M. Sato, $\mathcal{D}$-modules and nonlinear systems, in *Integrable Systems in Quantum Field Theory and Statistical Mechanics*, pp. 417–434, Adv. Stud. Pure Math., Vol. 19 (Academic Press, 1989).
5. M. Sato, The KP hierarchy and infinite-dimensional Grassmann manifolds, in *Theta Functions — Bowdoin 1987*, Part 1 (Brunswick, ME, 1987), pp. 51–66, Proc. Sympos. Pure Math., Vol. 49, Part 1 (Amer. Math. Soc., 1989).
6. M. Sato, M. Kashiwara and T. Kawai, Microlocal analysis of theta functions, in *Group Representations and Systems of Differential Equations* (Tokyo, 1982), pp. 267–289, Adv. Stud. Pure Math., Vol. 4 (North-Holland, 1984).
7. M. Sato and Y. Sato, Soliton equations as dynamical systems on infinite-dimensional Grassmann manifold, in *Nonlinear Partial Differential Equations in Applied Science* (Tokyo, 1982), pp. 259–271, North-Holland Math. Stud., Vol. 81 (North-Holland, 1983).
8. M. Sato, M. Kashiwara and T. Kawai, Linear differential equations of infinite order and theta functions, *Adv. in Math.* **47** (1983), No. 3, 300–325.
9. M. Jimbo, T. Miwa and M. Sato, Supplement to: "Holonomic quantum fields. IV" [*Publ. Res. Inst. Math. Sci.* **15** (1979), No. 3, 871–972]. *Publ. Res. Inst. Math. Sci.* **17** (1981), No. 1, 137–151.
10. M. Jimbo, T. Miwa, Y. Môri and M. Sato, Density matrix of an impenetrable Bose gas and the fifth Painlevé transcendent, *Phys. D* **1** (1980), No. 1, 80–158.
11. M. Sato, M. Kashiwara, T. Kimura and T. Ōshima, Microlocal analysis of prehomogeneous vector spaces, *Invent. Math.* **62** (1980/81), No. 1, 117–179.
12. M. Sato, T. Miwa and M. Jimbo, Holonomic quantum fields, V, *Publ. Res. Inst. Math. Sci.* **16** (1980), No. 2, 531–584.
13. M. Jimbo, T. Miwa, M. Sato and Y. Môri, Holonomic quantum fields. The unanticipated link between deformation theory of differential equations and

quantum fields, in *Mathematical Problems in Theoretical Physics* (Proc. Internat. Conf. Math. Phys., Lausanne, 1979), pp. 119–142, Lecture Notes in Phys., 116 (Springer, 1980).

14. M. Sato, T. Miwa and M. Jimbo, Aspects of holonomic quantum fields. Isomonodromic deformation and Ising model, in *Complex Analysis, Microlocal Calculus and Relativistic Quantum Theory* (Proc. Internat. Colloq., Centre Phys., Les Houches, 1979), pp. 429–491, Lecture Notes in Phys., Vol. 126 (Springer, 1980).
15. M. Jimbo, T. Miwa, Y. Môri and M. Sato, Studies on holonomic quantum fields, XVI. Density matrix of impenetrable Bose gas, *Proc. Japan Acad. Ser. A Math. Sci.* **55** (1979), No. 9, 317–322.
16. M. Jimbo, T. Miwa and M. Sato, Studies on holonomic quantum fields, XV. Double scaling limit of one-dimensional *XY* model, *Proc. Japan Acad. Ser. A Math. Sci.* **55** (1979), No. 8, 267–272.
17. M. Sato, T. Miwa and M. Jimbo, Holonomic quantum fields, II. The Riemann–Hilbert problem, *Publ. Res. Inst. Math. Sci.* **15** (1979), No. 1, 201–278.
18. M. Sato, T. Miwa and M. Jimbo, Studies on holonomic quantum fields, XII, *Proc. Japan Acad. Ser. A Math. Sci.* **55** (1979), No. 3, 73–77.
19. M. Sato, T. Miwa and M. Jimbo, Studies on holonomic quantum fields, XI, *Proc. Japan Acad. Ser. A Math. Sci.* **55** (1979), No. 1, 6–9.
20. M. Sato, T. Miwa and M. Jimbo, Holonomic quantum fields, IV, *Publ. Res. Inst. Math. Sci.* **15** (1979), No. 3, 871–972.
21. M. Sato, T. Miwa and M. Jimbo, Holonomic quantum fields, III, *Publ. Res. Inst. Math. Sci.* **15** (1979), No. 2, 577–629.
22. M. Sato, T. Miwa and M. Jimbo, Studies on holonomic quantum fields, X, *Proc. Japan Acad. Ser. A Math. Sci.* **54** (1978), No. 10, 309–313.
23. M. Sato, T. Miwa and M. Jimbo, Studies on holonomic quantum fields, VIII, *Proc. Japan Acad. Ser. A Math. Sci.* **54** (1978), No. 8, 221–225.
24. M. Sato, T. Miwa and M. Jimbo, Holonomic quantum fields, I, *Publ. Res. Inst. Math. Sci.* **14** (1978), No. 1, 223–267.
25. M. Sato, T. Miwa and M. Jimbo, Studies on holonomic quantum fields, VII, *Proc. Japan Acad. Ser. A Math. Sci.* **54** (1978), No. 2, 36–41.
26. M. Sato, T. Miwa and M. Jimbo, Studies on holonomic quantum fields, VI, *Proc. Japan Acad. Ser. A Math. Sci.* **54** (1978), No. 1, 1–5.
27. M. Sato, T. Miwa and M. Jimbo, Studies on holonomic quantum fields, V, *Proc. Japan Acad. Ser. A Math. Sci.* **53** (1977), No. 7, 219–224.
28. M. Sato, T. Miwa and M. Jimbo, Studies on holonomic quantum fields, IV, *Proc. Japan Acad. Ser. A Math. Sci.* **53** (1977), No. 6, 183–185.

29. M. Sato, T. Miwa and M. Jimbo, Studies on holonomic quantum fields, III, *Proc. Japan Acad. Ser. A Math. Sci.* **53** (1977), No. 5, 153–158.
30. M. Sato, T. Miwa and M. Jimbo, Studies on holonomic quantum fields, II, *Proc. Japan Acad. Ser. A Math. Sci.* **53** (1977), No. 5, 147–152.
31. M. Sato, T. Miwa and M. Jimbo, Studies on holonomic quantum fields, I, *Proc. Japan Acad. Ser. A Math. Sci.* **53** (1977), No. 1, 6–10.
32. M. Satō, T. Miwa and M. Jimbō, Studies in holonomic quantum fields (in Japanese), in *Research on Microlocal Analysis* (Proc. Sympos., Res. Inst. Math. Sci., Kyoto Univ., Kyoto, 1977). *Sûrikaisekikenkyûsho Kókyûroku*, No. 295 (1977), 77–87.
33. M. Sato and T. Kimura, A classification of irreducible prehomogeneous vector spaces and their relative invariants, *Nagoya Math. J.* **65** (1977), 1–155.
34. M. Kashiwara, M. Sato, T. Miwa and M. Muro, Erratum: "The Fourier transform in the case of an imaginary Lagrangian" (Sûrikaisekikenkyûsho Kókyûroku, No. 248 (1975), 212–260) (in Japanese), *Various Problems in Algebraic Analysis* (Proc. Sympos., Res. Inst. Math. Sci., Kyoto Univ., Kyoto, 1975), *Sûrikaisekikenkyûsho Kókyûroku* No. 266 (1976), 235.
35. M. Sato, T. Miwa, M. Jimbo and T. Ōshima, Holonomy structure of Landau singularities and Feynman integrals, *Proc. of the Oji Seminar on Algebraic Analysis and the RIMS Symposium on Algebraic Analysis* (Kyoto Univ., Kyoto, 1976), *Publ. Res. Inst. Math. Sci.* **12** (1976/77), suppl, 387–439.
36. M. Sato, T. Miwa and M. Jimbo, Dimension formula for the Landau singularity, in *Various Problems in Algebraic Analysis* (Proc. Sympos., Res. Inst. Math. Sci., Kyoto Univ., Kyoto, 1975), *Sûrikaisekikenkyûsho Kókyûroku* No. 266 (1976), 91–107.
37. M. Sato, Recent development in hyperfunction theory and its application to physics (microlocal analysis of S-matrices and related quantities), in *International Symposium on Mathematical Problems in Theoretical Physics* (Kyoto Univ., Kyoto, 1975), pp. 13–29, Lecture Notes in Phys., Vol. 39 (Springer, 1975).
38. M. Kashiwara, M. Sato, T. Mikio and M. Muro, The Fourier transform in the case of an imaginary Lagrangian (in Japanese), in *Distributions and Linear Differential Equations* (Proc. Fourth Sympos., Res. Inst. Math. Sci., Kyoto Univ., Kyoto, 1975), *Sûrikaisekikenkyûsho Kókyûroku*, No. 248 (1975), pp. 212–260.
39. M. Sato, Reduced *b*-functions (in Japanese), Notes by Tamaki Yano. Conference on Singular Points of Hypersurfaces and *b*-Functions (Res. Inst. Math. Sci., Kyoto Univ., Kyoto, 1973), *Sûrikaisekikenkyûsho Kókyûroku*, No. 225 (1975), pp. 1–15.

40. M. Satō, M. Kashiwara and T. Miwa, Microlocal study of infinite systems (in Japanese), *Distributions and Linear Differential Equations* (Proc. Fourth Sympos., Res. Inst. Math. Sci., Kyoto Univ., Kyoto, 1975), *Sûrikaisekikenkyûsho Kókyûroku*, No. 248 (1975), pp. 261–292.
41. M. Sato and M. Kashiwara, The determinant of matrices of pseudo-differential operators, *Proc. Japan Acad.* **51** (1975), 17–19.
42. M. Sato and T. Shintani, On zeta functions associated with prehomogeneous vector spaces, *Ann. of Math.* (2) **100** (1974), 131–170.
43. M. Sato, T. Kawai and M. Kashiwara, The theory of pseudodifferential equations in the theory of hyperfunctions (in Japanese), Collection of articles on hyperfunctions, *Sûgaku* **25** (1973), 213–238.
44. M. Sato, Pseudo-differential equations and theta functions, in *Colloque International CNRS sur les Équations aux Dérivées Partielles Linéaires* (Univ. Paris-Sud, Orsay, 1972), pp. 286–291. Asterisque, 2 et 3 (Soc. Math. France, 1973).
45. M. Sato, Microlocal structure of a single linear pseudodifferential equation, in *Séminaire Goulaouic-Schwartz 1972–1973: Équations aux Dérivées Partielles et Analyse Fonctionnelle*, Exp. No. 18, 9 pp. (Centre de Math., 1973).
46. M. Sato, T. Kawai and M. Kashiwara, Microfunctions and pseudo-differential equations, in *Hyperfunctions and Pseudo-differential Equations* (Proc. Conf., Katata, 1971; dedicated to the memory of André Martineau), pp. 265–529, Lecture Notes in Math., Vol. 287 (Springer, 1973).
47. M. Sato, T. Kawai and M. Kashiwara, On the structure of single linear pseudo-differential equations, *Proc. Japan Acad.* **48** (1972), 643–646.
48. M. Sato and T. Shintani, On zeta functions associated with prehomogeneous vector spaces, *Proc. Nat. Acad. Sci. U.S.A.* **69** (1972), 1081–1082.
49. M. Sato, Regularity of hyperfunctions solutions of partial differential equations, in *Actes du Congrès International des Mathématiciens* (Nice, 1970), Tomc 2, pp. 785–794 (Gauthier-Villars, 1971).
50. M. Sato, Theory of hyperfunctions, II, *J. Fac. Sci. Univ. Tokyo Sect. I* **8** (1960), 387–437.
51. M. Sato, Theory of hyperfunctions, I, *J. Fac. Sci. Univ. Tokyo. Sect. I* **8** (1959), 139–193.
52. M. Sato, On a generalization of the concept of functions, II, *Proc. Japan Acad.* **34** (1958), 604–608.
53. M. Sato, On a generalization of the concept of functions, *Proc. Japan Acad.* **34** (1958), 126–130.

Interview with Mikio Sato

ceived his Ph.D. from the University of Tokyo in 1963. He was a professor at Osaka University and the University of Tokyo before moving to the Research Institute for Mathematical Sciences (RIMS) at Kyoto University in 1970. He served as the director of RIMS from 1987 to 1991. He is now a professor emeritus at Kyoto University. Among Sato's many honors are the Asahi Prize of Science (1969), the Japan Academy Prize (1976), the Person of Cultural Merit Award of the Japanese Education Ministry (1984), the Fujiwara Prize (1987), the Schock Prize of the Royal Swedish Academy of Sciences (1997), and the Wolf Prize (2003).

This interview was conducted in August 1990 by the late Emmanuel Andronikof; a brief account of his life appears in the sidebar. Sato's contributions to mathematics are described in the article "Mikio Sato, a visionary of mathematics" by Pierre Schapira, in this issue of the *Notices*.

Andronikof prepared the interview transcript, which was edited by Andrea D'Agnolo of the Università degli Studi di Padova. Masaki Kashiwara of RIMS and Tetsuji Miwa of Kyoto University helped in various ways, including checking the interview text and assembling the list of papers by Sato. The *Notices* gratefully acknowledges all of these contributions.

—Allyn Jackson

Learning Mathematics in Post-War Japan

Andronikof: *What was it like, learning mathematics in post-war Japan?*

Sato: You know, there is a saying that goes like this: in happy times lives are all the same, but sorrows bring each individual a different story. In other words, I can tell of my hardships, but this will not answer your general question. Besides, I think the reader's interest should lie in the formation of the ideas of hyperfunctions, microlocal analysis, and so forth. It is true that in my young age I encountered some difficulties, but I don't think I should put emphasis on such personal matters.

Andronikof: *Still, I think we could start from a personal level. We could mix up journalism with mathematics, and go from one to the other. After all, you might not have become a—I would say, such a—mathematician without the experience of these hard times.*

Sato: Let me tell you this. In pre-war Japan, school was organized like the old German system. Elementary school ranged from the age of six to twelve, then followed middle school from twelve to seventeen, then three years of high school before entering university, where you graduated after three years. After the War, the system was changed to the American one: the five years of middle school were replaced by three years of junior high school and three years of high school. In order to become a graduate student, one then has to attend university for four years.

When I entered the middle school in Tokyo in 1941, I was already lagging behind: in Japan, the school year starts in early April, and I was born in late April 1928. The system was rigid, and thus I had to wait one year before getting in. Actually, it did not really matter, since I was not a quick boy. On the contrary, when I was a child, say, four like my son is now[1], I was called *bonchan*, which means a boy who is very slow in responding, very inadequate. I think I am very much the same now, ha! ha! Anyway, I turned thirteen right after entering middle school. In December of that year, Japan entered the war against the allied forces: U.S., UK, Holland, and China.

Andronikof: *Hectic times?*

Sato: Not so much in the beginning, as Japan was in a winning position. After Pearl Harbor, the British fleet was destroyed in the Far East, Singapore was occupied, and so on. Things looked favorable for Japan. But soon after, a year or so later, things started changing.

This was the beginning of my hard experiences. My regular courses in middle school lasted for only two years, and the rest of my school life was total chaos. The war in the Pacific ended on May 15, 1945. The first atomic bomb was dropped on August 6, 1945, after which the USSR declared war on Japan in order to secure the Kurilsk and Sachalin islands. At that time I was fifteen. Being a teenager, I had to work in factories. From 1943 to 1945, I had to carry coal. Very hard work… bad food… In late 1944, the systematic bombings of

[1] *That is, in August 1990.*

civilian targets by the U.S. started, after the fall of Micronesia, which then served as a base. In early 1945, Tokyo was a target. The first attack on Tokyo was on March 10, 1945, and some 80,000 were killed that night, but my family was spared. This was a short respite, since a month later there was a second attack and another broad area of Tokyo was burned down, including our house. I narrowly escaped the fire. We lost everything, but the family was safe. Due to the smoke, I partially lost my eyesight for a couple of weeks. Who cares about such details, anyway? Well, Japan had been rough on some people elsewhere, like the occupation army in China, and now it was hard for Japan. It was hard for many.

But I didn't intend to get into such detail... Isn't it tiresome?

Andronikof: *The tape recorder has no opinion, besides, I am personally interested.*

Sato: General Tōjō, Japan's military strong man, had taken power in Japan and conducted the whole war. He remained as prime minister at the time, and he, and the government, decided to move the schools to the countryside—they were practically closed, anyway. We could not find a place or job outside Tokyo. In a way, our family collapsed. We did not have any relatives in the countryside we could stay with, and we had no house. Pupils with family or friends in the countryside were supposed to go there, and those without such advantages had to join a party led by a schoolteacher. My father was a lawyer, but in 1941 or 1942 he fell ill and could not work as such anymore. Still, he thought that he could provide for his family. But then there was a sharp devaluation of the yen, by a factor of 100 (a yen was nearly the equivalent to a dollar before the war). Soon, the money my father had left from his work was down to practically nothing. We could not live on that anymore, and we nearly starved.

But let me talk about my formation. In elementary school we learned some arithmetic, coming from traditional Japanese mathematics. This is arithmetic for small boys. A typical example is the "counting of tortoises and cranes". Say there are several tortoises and several cranes. The total number is 7 and the total number of legs is 18. Then, how many tortoises are there? We had to manage to follow the reasoning without using equations.

I became very interested in mathematics at the age of twelve, I guess. This is when I moved from elementary school to middle school, and I had my first experience with algebra. We learned how to handle x's and y's, so things were solved very systematically. I was charmed by the simplicity of it. I was amazed by algebra, and in two years I made very quick progress and learned, for example, about complex numbers and their use in trigonometrical formulas, like the Euler formula. Of course, this was not taught at school. I remember saving my money systematically, by walking for hours instead of taking buses and streetcars to central Tokyo, and then spending hours in the bookstore to find the book that would give me the best value for my money. I remember an expository book by Fujimori on the theory of complex numbers where power series expansion and Taylor series were touched upon. The conformal transformation for the airplane wings was given as an example of the usefulness of complex numbers in the war industry... Some kind of propaganda book! Another book I read at that age was a book by Iwata on projective geometry: Desargues' theorem, Pascal's theorem, Brianchon's theorem dual to Pascal's...

My teacher of mathematics at that time was Mr. Ohashi, who is still alive. I actually met him again six years ago, after more than forty years. He contacted me after seeing my name in a newspaper or on television. I was lucky to have a good teacher. Of course I don't mean that he taught me such mathematical things during my first year. At that time I was rather timid, not talkative at all, and I did not try to consult him on my choice of books. But still, he was very encouraging: he always pushed me and taught me. That gave me the feeling that indulging in mathematics was not a game. This was very important, because on many occasions—at least in the Japanese educational system—pupils are supposed to follow the lead of the teacher and should not get off the track which he sets, whereas I was running completely off the track of the educational system. So I was just feeling it as a gift, so to speak, I was enjoying a kind of permitted pleasure.

Andronikof: *Were the conditions such that you had plenty of time on your own?*

Sato: Well, in the daytime we were supposed to stay at school, whether we had any classes or not. Besides, the only mathematical library was in Tokyo, and not a big library. Unbelievable! It was the contrary of reality in Japan today.

Andronikof: *So, you have always been interested in mathematics.*

Sato: Since twelve years of age, yes, and also a little in physics. But you see, my interest in physics grew much later, around 1945, when I read a university textbook which I had the opportunity to borrow from some graduate student.

In a way, the first three years of middle school were very fruitful for me. Afterwards, I developed my way of mathematical thinking through reading books and making calculations, solving interesting problems and so on. But I did not receive any good education after that, ha! ha! I was lucky enough to have had a chance of awakening my ability in mathematics at an early age. For the rest, I was sort of a dull boy. In the Japanese school system there are such subjects as geography, where memorizing names and years and so on is important, and in this field my performance was extremely low. That gave me the feeling that studying at school was a kind of unpleasant job. Doing my own mathematics—that is, not school mathematics, but reading those books I mentioned—was like watching television would be for a present-day boy. See, I was probably indulging in such things to forget the unpleasant school courses. A way to escape the school system. During and after the war things became harder and harder and I went deeper and deeper into mathematics, so to speak, like another would dive into alcohol.

Sato at blackboard, around 1972.

After middle school—though we had not completed it... I was admitted to high school. At that time, high school was rather elitist, more like École Polytechnique or École Normale Supérieure in France. The high school that I entered was called the First High School, and was closely attached to Tōdai[2] (Imperial University at the time). Both were national, i.e., non-private. The First High School is considered to be the top of elite schools, and I was lucky enough to skip the entrance examinations, because of the war. Well, there was a kind of test, but just to check some ability in mathematics: if they had tested my knowledge, then I couldn't have entered. Today, there are entrance examinations at many universities, including Tokyo or Kyoto: a bad test... But this is not interesting.

After the war, chaos occurred again—or rather, persisted. As I said, because of the devaluation of the yen, my family was starving. My father was sick, and I had a younger sister (by nine years) and a younger brother (by five years). I had to support them, so, in 1948, after three years of the First High School, I immediately started to work as a full-time teacher at the *new* high school, just when the school system was changed and middle school was cut by half. Housing and food conditions were extremely bad at the time, as you can imagine: like in Eastern Europe or Southeast Asia now. I entered Tōdai in 1949, having failed to enter in 1948. I had very little time to get prepared, then.

These hard times as schoolteacher lasted ten years, from 1948 to 1958. In 1958 I published the theory of hyperfunctions, in order to get a job at the university. I was an old student at the time, but it was like today: finishing university is sort of automatic, provided you succeed in getting in, where the competition is very tough.

Andronikof: *I read in your CV that you got a BSc in physics after your BSc in mathematics at Tōdai.*

Sato: You see, in Japan teaching depends on each professor, and one of my professors was very strict. At the time I was to graduate, he called me up, and told me that my term paper was very good but I had not attended the mandatory exercise sessions—not even once. This was an obligation that I didn't know of. Remember, that's why I was called *bonchan* when I was a little boy, and I'm still very much that way now. So, he said: "I cannot give you the points, so you cannot graduate". Then, he remained silent and watched me for a good minute. He opened his mouth again and said: "Okay, I'll give you the lowest points, so you can just graduate. Your paper is the top one". But this barred me from getting a position at the university as an assistant, which is customary for top students. Being assistant in Japan is a tenured position. The second-best student may also get some special position, and hence is assured of some top financial support. Anyway, I lost that kind of chance then. Since at the time I had also become interested in theoretical physics, I just moved to physics for two years under the new, American-style university system. I was still teaching full time in high school, so in physics I ran into the same academic problems as in mathematics. After two years at the Tōdai Physics Department, I moved to the graduate school of another university, Tokyo School of Education, where Professor Tomonaga taught theoretical physics. I stayed there until 1958.

This was the end of my twenties. At the time I was undergoing some kind of crisis in physical strength. Since by then my younger brother and sister were able to support themselves, my duty to them was sort of accomplished. I was able to return to my own life, so to speak, and go back to mathematics.

The Birth of Hyperfunctions and Microfunctions

Andronikof: *So you decided to go back to mathematics, rather than physics?*

Sato: Yes, and it was a good decision since competition in physics seemed stiffer. See, after these tough years I was beginning to feel physically tired, and my youth was leaving me. Even if I wasn't a man of quick response, I nevertheless understood that I had to face real life, so to speak, and to try to show what I could do in mathematics.

[2] *University of Tokyo.*

Names Mentioned in the Interview

Below is a list of names of mathematicians mentioned during the interview.

Armand Borel
Élie Cartan
Etsurō Date
Pierre Deligne
Jean Ecalle
Leon Ehrenpreis
Daisuje Fujiwara
Roger Godement
Alexandre Grothendieck
Ryōgo Hirota
Sin Hitotumatu
Yasutaka Ihara
Kenkichi Iwasawa
Shōkichi Iyanaga
Michio Jimbo
Akira Kaneko
Masaki Kashiwara
Yukiyosi Kawada
Takahiro Kawai
Tatsuo Kimura
Hikosaburo Komatsu
Serge Lang
Jean Leray
André Martineau (1930–1972)
Yozō Matsushima
Barry M. McCoy
Tetsuji Miwa
Mitsuo Morimoto
Atsushi Nakayashiki
Yousuke Ohyama
Kiyoshi Oka
Frédéric Pham
Yasuko Sato
Pierre Schapira
Jean-Pierre Serre
Goro Shimura
Takuro Shintani (1943–1980)
Masuo Suzuki
Teiji Takagi
Kanehisa Takasaki
Shunichi Tanaka
Junichi Uchiyama
André Voros
Kōsaku Yosida
André Weil

My advisor at the mathematics department was Professor Iyanaga, and I wanted to show him what I was capable of. High school teachers had a forty-day vacation between the first and second semester. So, during the summer of 1957 I tried to prepare something that I could show him, and that was hyperfunctions. I worked out hyperfunction series and outlined the theory for several variables—though the complete theory was finished later, since it required a generalization of cohomology theory. In December of that year, I went to see Professor Iyanaga, after an interruption of some years, and told him about it. Professor Iyanaga showed interest in my work and persuaded Professor Kōsaku Yosida to offer me a position as an assistant.

Actually, it seems that Iyanaga was one of the professors who wanted me to get a position as an assistant already when I had graduated, but he was not a senior professor at the time, so his opinion did not prevail. I was glad that Professor Iyanaga showed interest in hyperfunction theory. I was lucky: if the professor had not been Iyanaga—if he were a specialist in analysis, for example—perhaps he would have told me: "You are doing nothing". Fortunately, Iyanaga was a generous and open-minded mathematician, open enough to appreciate what I was doing. He was a student of Professor Teiji Takagi, the founder of class field theory and number theory.

Of course, coming from ten years as a schoolteacher, this new position made me lose some financial advantages. But finally, after the hard times, I could have the pleasure of doing solely mathematics. What I wanted to do was to organize mathematics.

You see, being a high school teacher then was not like today. We had to work very hard. So, I couldn't do difficult things, but only general things like sheaf theory, category theory, and so on. I just tried to organize my own mathematics and studied a lot, always keeping my interest alive. During that period, I had the opportunity to see many papers published in *Sūgaku*[3]. In particular, my construction of relative cohomology was inspired by a report on complex variables by Hitotumatu, which contained a short account on sheaf theory and cohomology of sheaves, perhaps in three pages. *Sūgaku* was a very useful journal, which has been central to me: a compact publication covering every branch of mathematics. There were many journals that I could not get, so *Sūgaku* was essentially the only publication I was reading. And this, usually, when I was commuting on crowded public transportation, during rush hour. Well, some people read newspapers, anyway.

I believe that after ten years I had quite good ideas about theoretical physics and diverse branches of mathematics. Like a spider, I went on spinning my web for ten years, extending it by attaching it to different places. Some things get stuck in it, some go through. You go to some

[3] *A journal published in Japanese by the Mathematical Society of Japan. Recent volumes are translated into English and published by the AMS under the title* Sugaku Expositions.

things, keeping some others for later. But you don't let go.

Anyway, I was assistant to Professor Yosida in Tōdai for two years, and in 1960 I moved to Tokyo University of Education as a lecturer—a higher position than assistant. Professor Iyanaga sent a copy of my work to André Weil who showed some interest, and suggested that I come to the Institute for Advanced Study (IAS). I visited the IAS from the fall of 1960 to 1962. But before leaving for Princeton, in June 1960, I gave a talk at the Extended Colloquium of Tōdai. This was a periodic meeting, organized by the Tōdai Mathematics Department twice a year. There, I had the opportunity to present my program in analysis. I explained how a manifold is the geometric counterpart of a commutative ring, and vector bundles are the counterpart of modules over that ring, and if you go to the non-commutative case you can treat linear and nonlinear differential equations. From this point of view, linear equations are defined to be $\mathcal{D}$-modules, and if you write $\mathcal{D}$ in a more general form, you can consider nonlinear systems.

At that time, that method was only built to establish the algebraic theory of Picard-Vessiot. This has not much contact with pure analysis, like the study of hyperbolic equations. In that field, the more geometrical methods of Élie Cartan, based on the theory of differential systems, were considered more effective. Nonetheless, as I have told several people, like Masaki Kashiwara or Pierre Schapira, I already had the feeling that Cartan's methods were not the right ones to build a general nonlinear theory. But it was only after 1970 that I—how should I say—I became determined to throw away exterior differential methods and stick to this new point of view.

Anyway, in my 1960 talk I clearly stated the setting of $\mathcal{D}$-module theory in the linear case, the notion of maximally overdetermined systems (a name that we later changed to *holonomic systems*), and the important role that holonomic systems play even in the study of overdetermined systems, through elementary solutions. For example, the Riemann function of a hyperbolic equation usually satisfies a holonomic system. I also explained my program using homological methods: $\mathcal{H}om$ describing the homogeneous solutions to the system, $\mathcal{E}xt^1$ the obstruction to solvability, and things like that.

Andronikof: *Did that talk contain maximally overdetermined systems? That is, did you have an idea of the involutivity theorem[4]?*

Sato: Mmm... Not exactly. It was only ten years later, after the establishment of microlocal analysis, that everything was clarified. Actually, I already had some vague ideas before 1960, but then I had only spent two years on these problems. I then moved to Princeton, and there I had to change my subject. I'll tell you about that later.

The first half of my talk at the colloquium in June 1960 was recorded in the notes by Hikosaburo Komatsu.

Andronikof: *What happened to the second half?*

Sato: Well, I kept going on and went into the nonlinear systems, and so on and so on... and he just couldn't keep up, ha! ha! You don't have to tell him this. Anyway, I am certainly indebted to Komatsu because this is a very rare case of a record of my talks. He was introduced to me in 1958 by Professor Yosida, of whom he was the best student. Komatsu was the first man who really understood hyperfunction theory, and he was just a graduate student!

At that time I gave a lot of talks at different occasions: for example, I gave a talk about derived categories. I needed that kind of theory because the notion of spectral sequence by Jean Leray was inconvenient in my theory. So I wanted to improve it and produced derived categories between 1958 and 1960.

Andronikof: *Were derived categories included in your 1960 colloquium talk?*

Sato: No, not at the colloquium talk. Professor Kawada, a number theorist at Tōdai, was very kind to me and organized some sort of seminar in 1959—small periodic meetings—to offer me the chance to expose my theory in a systematic way. There were a few lectures in his office, for a limited audience. But I found that no one understood me. So after a few times, I don't remember the exact number, this collapsed. Actually, Professor Kawada himself wondered daily how long it would last, ha! ha! He couldn't get the participants to understand.

This was before the summer vacation of 1959/1960. During the vacation, I had to attend some English training, before moving to Princeton, to be able to get a Fulbright grant. This is the kind of competition I'm not too good at... Then, I remember crossing the ocean on a 10,000-ton ship and crossing the USA by train: a very interesting trip. The first one for me.

Andronikof: *So you left behind some people pondering over your theories?*

Sato: Mmm... Well, I do a poor expository job in general. Most of the audience gets lost on the way.

Andronikof: *That might be due to the contents, which are always new.*

Sato: You see, it seems I cannot adapt to the audience. I just expose my ideas according to my way of thinking and pay little attention to how the audience receives them.

[4] *This theorem is a kind of analogue in the theory of differential equations to the uncertainty principle in quantum mechanics. It asserts that characteristic (co)directions constitute a variety of low codimension in the phase space.*

Let me tell you one more thing that happened before leaving for Princeton. It concerns the birth of microfunctions. On one occasion, a summer school I think, Professor Hitotumatu—who was just four years my senior—talked about the edge-of-the-wedge theorem in the theory of several complex variables. Of course, this was not done in the framework of hyperfunctions. So, I just mentioned that that concept could be better handled in the hyperfunction category: I had the cotangent nature of boundary values in mind.

Andronikof: *You knew at once that it should be the cotangent bundle*[5] *and not the tangent bundle?*

Sato: Yes, because of Cousin-like theorems: a holomorphic function can be decomposed into greater domains. At that time, I clearly felt the existence of some kind of microlocal structure underlying hyperfunctions, but I didn't consider it seriously. Anyway, that impression didn't leave me for a long time. It stayed with me.

For nine years, after my active period of 1958-1960, there was a long intermission for hyperfunctions. From 1960 to 1969 I didn't work in that field at all. Then, in April 1969, Professor Yosida held an international symposium on functional analysis at RIMS[6], and I was asked to give a talk. I think I was told this in December 1968. So I tried to organize those ideas I had ten years before and started to compute the theory. I found that my thought was correct. This time I wanted to show that the decomposition of hyperfunctions into cotangent components is actually important in analysis. I wanted to show that it would be useful in order to establish some results in mathematical physics. I wrote this down in less than a month, and this was the start of microfunction theory.

Andronikof: *The interruption in hyperfunction theory was due to your stay in Princeton?*

Sato: Yes. It seemed to me that even André Weil did not like my way of putting things in terms of cohomology very much. In fact, I learned that he was very much against cohomology. I got the idea that hyperfunctions were not taken very seriously, and since I was sort of a little boy (though I was 32!) I just wanted to show the usefulness of my ideas. My general program, which I had expanded at the colloquium talk of 1960, was just a very formal kind of general nonsense. I then wanted to give some concrete example of it in the analysis of differential equations. But my knowledge of that field was very poor. I'm not a reader of big books and specialized papers. I live for practical examples and, just as now, I was very slow in developing my deeper thoughts.

But let me go back to the talk I had occasion to give in 1969. I found that microlocal analysis could explain several interactions with classical analysis. This time I wanted to show that my theory was not just nonsense, but could be applied to explicit problems. I was confident that microlocal analysis, once organized, could persuade people of the usefulness of hyperfunction theory.

Andronikof: *Microfunctions were devised to help hyperfunction theory—so to speak!*

Sato: Yes, so that hyperfunction theory could at least be perceived as a method in analysis. $\mathcal{D}$-modules and things like that were not accepted at the time. The only one who really appreciated $\mathcal{D}$-modules was Masaki Kashiwara, who became interested in developing a survey of the theory. It was in 1969 or in 1970 that he did it, to get his master's degree. In Japan, this degree is considered of primary importance in order to obtain a position. So he published a very nice paper on $\mathcal{D}$-modules.

Andronikof: *You mean Kashiwara's master's thesis, handwritten in pencil, that can be found in Tōdai library*[7].

Sato: Yes, precisely. That was completely his own work. He had a very good background in general nonsense. In the lectures before the colloquium, I already developed as much general nonsense as I thought was needed for hyperfunction and $\mathcal{D}$-module theory. But it was rather informal, because I didn't do it in a systematic way, whereas Kashiwara started his mathematical career in a clear way from the very beginning. To perceive some of the difficulties of his task, recall that it was a kind of hunting age for such general nonsense. He learned Bourbaki, Grothendieck, and these things when he was eighteen or nineteen years old. He studied it by himself, with no teacher, when he was only—what do you call it?—a senior mathematics student. Yes, he is very ingenious. The best young boy I ever met.

Andronikof: *But who put him onto these subjects?*

Sato: After the first two years at Tōdai, he moved to Hongo campus. There, he first learned of hyperfunction theory in 1968, at a lecture by Komatsu. I think this lecture is fundamental in the history of algebraic analysis. It was published in the Seminar Notes of the University of Tokyo after notes taken by student participants like Kawai, Uchiyama, ...

Andronikof: *How did $\mathcal{D}$-modules come to Kashiwara?*

Sato: In 1968 Komatsu and I organized a weekly seminar on algebraic analysis at the Tōdai Mathematics Department, where I gave several talks—very disorganized, as my talks always are. It wasn't

---

[5] *The phase space of classical mechanics.*

[6] *Research Institute for the Mathematical Sciences, which is part of Kyoto University.*

[7] *Kashiwara's thesis has been translated into English by Andrea D'Agnolo and Jean-Pierre Schneiders: Masaki Kashiwara,* Algebraic study of systems of partial differential equations, *Mém. Soc. Math. France (N.S.) (1995), no. 63, xiv+72.*

an official seminar in Tōdai, but rather a kind of "Jacobin Club". Among the participants, there were many very eager young students, including Kawai and Kashiwara. I met them there for the first time, and the group of Kawai, Kashiwara, and myself was formed that year.

In spring 1969 some old friends of mine in Komaba, which is part of the Faculty of General Education of the University of Tokyo, arranged to have me go there as a professor. I stayed in Komaba for two years.

***Andronikof:** And when did you come to RIMS?*

Sato and Emmanuel Andronikof, 1990.

Sato: It was in June 1970. Actually, in Tokyo I had a great understanding with Komatsu and other seniors, as well as with many young mathematicians who gathered at our seminar. Among the participants, besides Kashiwara and Kawai who were extremely active, there were Morimoto, Kaneko, Fujiwara, Shintani, Uchiyama, and some others. So, I could supervise a lot of people who were very eager to study mathematics with me, and I thought I should better stay at Tōdai than come to RIMS. Anyway, Professor Kôsaku Yosida, who was director of the Institute from 1969 to 1972, and of whom I was once an assistant, put great pressure on me to come to RIMS. He had already asked on the occasion of the seminar he had organized in 1969. But since I had a position at Komaba, that was delayed until 1970. I was unhappy when I had to move to Kyoto because it meant I would be separated from this group: I could bring Kawai and Kashiwara to Kyoto, but I had to leave others behind.

The Katata Conference and S-K-K

***Andronikof:** As for the "milestones" in the birth of hyperfunction and microfunction theory, can you comment on the famous Katata conference in fall 1971?*

Sato: Actually, what I said at that conference was sort of completed quite early, just after 1969. I have already told you how microfunctions originated in preparing the talk I gave at the international symposium at RIMS in April 1969. I had planned to present some of the things I had in mind, like the cotangential decomposition of hyperfunctions, so I had to check whether my ideas were working or not. I started to check this in the three-hour *shinkansen*[8] trip, commuting from Tokyo to Kyoto (or vice versa, I don't remember) to attend a pre-symposium meeting at RIMS. You could say that the basic part of the theory was conceived during these three hours. But later I checked it in detail, and it was completed at the international symposium. The final touch was a proof of microlocal regularity for elliptic systems[9]. At the time, I employed Fritz John's method of plane wave decomposition. Of course, the idea went back to 1960, when I attended Professor Hitotumatu's talk on the edge-of-the-wedge theorem.

***Andronikof:** When were the famous S-K-K[10] proceedings written?*

Sato: The basic structure of the paper hinges on my talk at the Katata conference, but the manuscript was completely prepared by Kawai and Kashiwara. Let us say I presented the whole story, but did not prove every detail. For example, concerning the notion of microdifferential operators, I worked out some cohomological constructions, but then Kawai and Kashiwara gave a better, more direct presentation, by which the proof of the invertibility for microelliptic operators, instead of using Fritz John's plane wave method, reduced to a kind of abstract nonsense. Kawai and Kashiwara must have taken a lot of effort to complete every detail.

The work was done between 1969 and 1971: surely the golden age of microfunctions. At the time, the three of us were working together, in the same places. In 1969 we were in Tokyo, then we moved to RIMS in 1970. Kawai came here as an assistant, while Kashiwara had only a kind of grant since he was very young at the time. He became assistant in 1971. I think the main part of the job was finished prior to the Katata conference, and was already presented in my talk at the Nice Congress [International Congress of Mathematicians] in 1970. To be precise, in the Nice talk the structure theorem for microdifferential systems was not yet finished. It was presented at the summer school on partial differential equations at Berkeley in 1971. I also prepared a kind of preprint, which did not appear in the proceedings of the Berkeley summer school, though it was distributed. There, I stated the structure theorem, asserting that all microdifferential systems are—at least generically—classified into three categories, the most important being what we called Lewy-Mizohata type system. The proof of this reduced to some simple nonlinear equations

[8] *A high-speed Japanese train.*

[9] *Now known as Sato's theorem.*

[10] *M. Sato, T. Kawai, and M. Kashiwara,* Microfunctions and pseudo-differential equations, *In Komatsu (ed.),* Hyperfunctions and pseudo-differential equations, *Proceedings Katata 1971, Lecture Notes in Mathematics, no. 287, Springer, 1973, pp. 265–529.*

describing some geometrical transformations[11]. I clearly remember discussing it with Kawai and Kashiwara. Actually, at Berkeley this was stated only for the simply characteristic case. The multiple characteristic case was done the following month, and all this was finished for the Katata conference. I must also say that at Katata I was given plenty of time, while at Berkeley my work was not taken into great consideration. I was sort of neglected there.

Andronikof: *So, when was the S-K-K manuscript ready?*

Sato: It was ready in 1971, but then Komatsu asked me to write a preface. I kept putting it off, so it was my fault that the publication was delayed by more than a year. I always behave in that silly way.

Andronikof: *Concerning another important result in S-K-K, could you comment on what is now called the* Fourier-Sato transformation*? When was it cooked up?*

Sato: As I told you, during my years as a schoolteacher I began to develop my own ideas in mathematics. Since the beginning, I had chosen the subject of generalized functions: both as one of the building blocks of my mathematical world and as a way to present some of my mathematical results, in order to get a job. It seemed to me that this subject could be quite easily appreciated. I estimated that the classical functional analytic approach, considering dual spaces to Banach spaces or to locally convex spaces, was not satisfactory at all. I was determined to throw away all these things, and to construct a theory which completely relied upon algebraic methods, like cohomology. What I wanted was to build something in the spirit of Cartan-Serre's presentation of Oka's theory of complex variables, or of Godement's work on sheaf theory. That was the idea of algebraic analysis, around 1960. You see, in this context the Fourier transform was not such a drastic idea to me, but rather a natural consequence of my kind of thinking. Of course, the actual theory of the Fourier transform was made later, in the course of systematizing microfunction theory, after 1969. At first, I constructed it using a quite straightforward method, considering the edge-of-the-wedge and the microlocal decomposition. But then I tried to describe it in a more algebraic way. That is when I arrived at the formulation of the Fourier transform as it appears in the S-K-K proceedings, with the three exact sequences: one on the base space, and two in the tangent and cotangent bundles.

[11] *What they called* quantized contact transformations.

The Theory of Prehomogeneous Vector Spaces

Andronikof: *Between 1960 and 1967, what happened to the broad program in mathematics that you devised? You said that you had stopped working in microfunction theory.*

Sato: Things are not so linear, of course. Let us say that I was working on microfunction theory in the background, while considering several other problems in parallel. In my twenties, my thoughts were scattered. While trying to organize analysis on an algebraic basis, I was also very much interested in subjects such as special functions via concrete examples, class field theory, automorphic forms, and of course quantum field theory. As I told you, after mathematics I had switched to the physics department for six years, from 1952 to 1958. Since I had to work as a high school teacher, I couldn't participate in research activities at the university, but nevertheless physics was one of my major subjects of interest.

Anyway, when I moved to Princeton in 1960, I wasn't acquainted with the mathematical society. I always pursued mathematics my own way, which was not the academic one. I was a kind of amateur in mathematics, certainly not a professional. As a consequence, whenever I tried to explain my ideas, I did not know which parts were well known or which were new. Most of the time people did not understand me. In Princeton there were a number of eminent and active mathematicians, but to them I just seemed a strange man with some very strange ideas. So I decided that in order to connect with some audience, I should devise some applications of my theory to differential equations. Of course, I then thought of developing the theory of differential equations from scratch. In my talk in Tokyo in 1960, I had already pointed out how holonomic systems are a powerful tool in understanding other differential equations, and how they give an algebraic way of defining special functions. But gradually, I understood that fundamental solutions may not satisfy holonomic systems in the usual sense: to understand the situation better you have to go to some wider class of equations: nonlinear, of infinite degree, or with an infinite number of independent variables.

To develop the theory of differential equations from the beginning, I thought I had to get acquainted with differential equations through concrete examples. As you know, the differential equations one usually encounters are mainly of second order, like the Laplace equation, the wave equation, the heat equation, or some modifications of these where the coefficients are variable. So, I tried to get a good example that was beyond the second order but that was still manageable through a special function. That is how, in spring

1961 I think, I devised the theory of prehomogeneous vector spaces (PHVS). Though the name is not a good one, I wanted it to mean a vector space that is not exactly homogeneous, but homogeneous except for a rather inessential subspace. PHVS gave a good generalization of the Laplacian, whose principal part is a quadratic form. A quadratic form is connected with orthogonal transformations, and PHVS generalizes this to the case of arbitrary groups. In this framework, instead of a quadratic form we can consider other interesting polynomials, associated to differential equations of higher order, whose principal part is very symmetrical, like a determinant or a discriminant. Moreover, I immediately realized that PHVS could also be useful in the theory of the Fourier transform, or to get generalized ζ-functions, and so on. Throughout this month of 1961, I was engaged in this theory, and in connection with this concept I introduced the notion of the a-function and b-function. The most important is the b-function[12], since it makes it possible to define a generalization of hypergeometric functions, and the a-function is nothing but the principal symbol of the b-function. The reason for "a" and "b" is that a is easier but b can also be developed from this concept. In late 1961, I explained these things at a seminar at Princeton conducted by Armand Borel. I gave two talks there, but again I think the subject was not fashionable at the time. It was something people were quite unfamiliar with, out of the focus of contemporary mathematicians, and so they just ignored or forgot it.

Because of this situation, the following year I tried to do something fashionable. So, I took up something else which I had in mind, about number theory. I was quite well acquainted with algebraic number theory and especially class-field theory and ζ-functions or automorphic forms. You know, Japan had a strong tradition in algebraic number theory, since Takagi and its school: Iwasawa, Shimura... It was one of my favorite subjects, and I started investigating it more seriously. I was interested in the Ramanujan conjecture, and I began working on it during the summer of 1962. Actually, Shimura was also working very actively in that field at that time. My term in Princeton expired that summer, and just before my departure I succeeded in establishing the relation between Ramanujan-type functions and the Dirichlet series. I won't go into detail about this, since I think we should stay within microlocal analysis. We cannot focus on too many subjects at a time.

In August 1962 I returned to Japan. At that time, mathematics departments were very small. Including all universities like Tokyo, Osaka, and Kyoto, there were only five positions for professors. A professor of analysis from Osaka University was retiring, and Matsushima invited me very strongly to go there. So I moved to Osaka University in spring 1963. There, I had a couple of good young students working with me. Once again, what I then did remained unpublished, with the possible exception of a talk that I gave at Tōdai, and of which Ihara wrote a report: some record of it exists in the form of handwritten notes in Japanese, in a journal for young mathematicians. Later, Ihara wrote a paper about this work of mine, but the story is rather complicated since there is a relation with later work by Deligne, and if I present it briefly I risk being imprecise. So just forget about it.

The following year, in 1964, I went to Columbia University, where I stayed as a visiting professor for two years. Serge Lang invited me there. You see, he came to Japan and was impressed by my work: that is, my work in number theory! Ha!, ha!

Andronikof: *Let me return to PHVS. You said the theory came about through your interest in high order PDEs.*

Sato: Yes, but at the same time it has a relation with algebra, geometry, number theory, θ-functions, generalized ζ-functions, and so on. It is a very interesting subject, related to later developments of microlocal calculus. But again, this is another subject, and so I won't go into detail. Let me just mention that a very good mathematician, Takuro Shintani, who unfortunately died some years ago, wrote a paper which is the basis for the theory of PHVS applied to ζ-functions. After a seminar he gave in Tokyo, a number of mathematicians from Japan, the USA, and Europe began to work on that subject. So, although I am not active in that field at the moment, the theory of PHVS is far from being buried, and instead is a rapidly developing subject.

This is more or less how my time was spent from 1960 to 1967. In 1967 I really started considering PDEs through algebraic analysis, i.e., in the category of hyperfunctions and microfunctions.

Andronikof: *What happened just after 1966, when you came back from Columbia?*

Sato: I resigned from Osaka in fall 1966 and was out of the university for a year and a half, when I became professor at Komaba. In fact, I had been living in Tokyo since early 1967. There, I met Komatsu, and we started the Tokyo seminar on algebraic analysis I told you about. It lasted until Kawai, Kashiwara, and I left for Kyoto in 1970.

Toward Mathematical Physics

Andronikof: *And after the congress in Nice, you went to France.*

Sato: Yes, Kawai, Kashiwara, and I stayed in Nice from September 1972 until the next aca-

[12] *Now called* Bernstein-Sato polynomial.

Emmanuel Andronikof

Pierre Schapira

Emmanuel Andronikof passed away on September 15, 1995, from a brain tumor discovered two years earlier, just after he had been named professor at the University of Nantes. He faced his illness with exceptional courage, never complaining about his misfortune or about the unfairness of this world. Perhaps his character was a product of his strict orthodox and aristocratic Georgian education. An excellent sportsman, especially in boxing and swimming, he had a unique vision of life and was able to spend long days in a forest with nothing but a book of Russian poetry. Needless to say, he was not interested in the games of power, not even in his own career. While he was devoted to science, he looked at the scientific world as a theater with very few actors and many jokers, considering himself to be an extra.

Nevertheless, his contribution to mathematics is highly significant. In his thesis, he succeeded in making a synthesis of the microlocalization functor of M. Sato and the temperate cohomology of M. Kashiwara, defining the functor of temperate microlocalization. This opened the way to the linear analysts to study microlocally (i.e., in the cotangent bundle) distributions, or more generally temperate cohomology classes, with the tools of homological algebra, sheaf theory, and $\mathcal{D}$-module theory. In other words, this was the first step towards a "temperate microlocal algebraic analysis", and Emmanuel Andronikof began to apply his functor to handle various problems. He provided an illuminating proof of the Nilsson theorem on the integrals of holomorphic functions of "Nilsson class", he proved that the C^∞-wave front set of a distribution solution of a regular holonomic system coincides with its analytic wave front set, and he was the first to give a microlocal version of the Riemann-Hilbert correspondence, this last work being at the origin of further important developments (see [7] and [6]). He also had many projects that he was unfortunately unable to carry out to completion.

The departing of Emmanuel Andronikof leaves us with the memory of an excellent man, both rigorous and open, tough and gentle.

References

[1] E. Andronikof, Intégrales de Nilsson et faisceaux constructibles, *Bull. Soc. Math. France* **120** (1992), p. 51-85.

[2] ——, On the C^∞-singularities of regular holonomic distributions, *Ann. Inst. Fourier (Grenoble)* **42** (1992), p. 695-705.

[3] ——, The Kashiwara conjugation and wave front set of regular holonomic distributions on complex manifolds, *Invent. Math.* **111** (1993), p. 35-49.

[4] ——, Microlocalisation tempérée, *Mémoires Soc. Math. France* **57** (1994), 176 pp.

[5] ——, A microlocal version of the Riemann-Hilbert correspondence, *Topol. Methods Nonlinear Anal.* **4** (1994), p. 417-425.

[6] S. Gelfand, R. MacPherson, and K. Vilonen, Microlocal perverse sheaves, arXiv math.AG/0509440.

[7] I. Waschkies, Microlocal Riemann-Hilbert correspondence, *Publ. Res. Inst. Math. Sci.* **41** (2005), p. 37-72.

Pierre Schapira is professor of mathematics at the Université Pierre et Marie Curie, Institut de Mathématiques, Paris, France. His contact addresses are: schapira@math.jussieu.fr *and* http://www.math.jussieu.fr/~schapira/. *Extracted and adapted from* La Gazette des Mathématiciens **66** *(1995).*

demic year. There, I became acquainted with a number of mathematical physicists who Frédéric Pham had invited from different parts of Europe. Moreover, Pham stimulated my interest in studying a program in mathematical physics. He introduced me to the momentum space structure of an S-matrix and Green functions in quantum field theory. This was the first time I became interested in applying microlocal analysis to such subjects. Later, Pham worked in that field for quite a long time.

Ten Papers by Mikio Sato

Theory of hyperfunctions. I and II, *J. Fac. Sci. Univ. Tokyo. Sect. I*, **8**, 1959 and 1960, 139–193 and 387–437.

Hyperfunctions and partial differential equations, *Proc. Internat. Conf. on Functional Analysis and Related Topics, 1969*, Univ. Tokyo Press, Tokyo, 1970, 91–94.

Regularity of hyperfunctions solutions of partial differential equations, *Actes Congrès Internat. Math., 1970*, Gauthier-Villars, Paris, 1971, **2**, 785–794.

(with T. Kawai and M. Kashiwara), Microfunctions and pseudo-differential equations, *Hyperfunctions and pseudo-differential equations, Proc. Conf., Katata, 1971*, 265–529, Lecture Notes in Math., Vol. 287, Springer, 1973.

(with T. Shintani), On zeta functions associated with prehomogeneous vector spaces, *Ann. of Math.* **100**, 1974, 131–170.

(with T. Kimura), A classification of irreducible prehomogeneous vector spaces and their relative invariants, *Nagoya Math. J.*, **65**, 1977, 1–155.

(with T. Miwa and M. Jimbo), *Holonomic quantum fields. I–V*, Publ. RIMS, Kyoto Univ., **14** (1978) 223–267 (I), **15** (1979) 201–278 (II), 577–629 (III), 871–972 (IV), **16** (1980) 531–584 (V), 137–151 (IV Suppl.), **17** (1981).

(with T. Miwa, M. Jimbo, and Y. Mōri), Density matrix of an impenetrable Bose gas and the fifth Painlevé transcendent, *Phys. D*, **1**, 1980, no. 1, 80–158.

(with M. Kashiwara, T. Kimura, and T. Oshima), Microlocal analysis of prehomogeneous vector spaces, *Invent. Math.*, **62**, 1980, no. 1, 117–179.

(with Yasuko Sato), Soliton equations as dynamical systems on infinite-dimensional Grassmann manifold, *Nonlinear partial differential equations in applied science (Tokyo, 1982)*, North-Holland, Amsterdam, 1983, pp. 259–271.

Then I returned to Japan in the spring of 1973. In April of that year, Tetsuji Miwa came to Kyoto and obtained a position as an assistant at RIMS. He was a very good student of Komatsu at Tōdai and had studied hyperbolic equations with him. The following year, in 1974, Michio Jimbo had just finished his undergraduate courses and was preparing to enter the graduate school of Kyoto. He spent two years there and then got a position as an assistant at RIMS in 1976.

Miwa, Jimbo, and I worked on the subject in mathematical physics in which I'd become interested through Pham and his friends. We began studying the momentum space structure. In order to have a clearer idea than from the general quantum field theory, I intended to use some very concrete examples. Working in that direction, I just took up the Ising model. In fact, when I was a student in physics I was very much interested in quantum field theory, of course, but also in statistical physics and statistical mechanics. At that time, Dr. K. Ito, a theoretical physicist who was then a graduate student, mentioned a paper by a young Russian mathematician on correlation functions in the Ising model that we found very interesting. Later, Professor Masuo Suzuki from Tōdai mentioned the work of Wu-McCoy-Tracy-Barouch ([1]) on the two-point function of the Ising model. It was probably published in 1976, but I learned later from Barry McCoy that their work had been conceived much earlier, around 1973. The remarkable point is that their work contained a correct expression by means of Painlevé transcendents. It was a pleasant surprise to me that such special functions actually appeared in concrete problems of theoretical physics, especially in one of my favorite subjects, the Ising model. So, we developed this much further with Miwa and Jimbo. We started doing this around 1976, having spent the previous three years doing much more elementary things, in order to get acquainted with the subject. Everything worked very effectively, and very quickly, in 1977, we finished working in that direction. 1976–1978 was a very fruitful time. Since then, Jimbo and Miwa have done quite a lot of things.

Andronikof: *What about the KP hierarchy?*

Sato: We became acquainted with many people in soliton physics. Date was a student of Professor Tanaka, who was in Osaka at that period and then moved to Kyushu University. Tanaka and Date were pioneer mathematicians, who introduced soliton theory in Japan, including Krichever's work, Jacobian variety, θ-function, KdV, and KP equations. We learned about these things through them during that period. Of course, there were many people in physics departments or engineering departments who were interested in soliton theory, but in the world of pure mathematics in Japan, it was Tanaka who introduced such modern developments in connection with soliton theory. I also learned a lot from Hirota. He is a very important name in soliton theory, although he is not a mathematician in the strict sense. He is very unique, in that he has his own system of mathematics, so to speak, and has devised quite a unique method of analysis of soliton theory. And this inspired me. You see, most of the time, when I meet some new subject in mathematics, I don't find it very strange. I'm thinking about the traditional systems of mathematics, like those of Newton, Leibniz, or even Descartes, Gauss, Riemann, and so on. But Hirota's mathematics seemed to me quite strange, so I just wanted to understand it in the language of modern mathematics. In this way I finally arrived at the concept of infinite-dimensional Grassmannian manifold around Christmas 1980, in collaboration with my wife. We performed lots of computations, using BASIC programs. Looking at them day after day, I finally realized that in the background of the whole story stood a Grassmannian manifold of infinite dimension. Again, once I arrived at this concept of Grassmannian manifold and Plücker coordinates, things developed quite quickly, and in a very short period—less than a month—everything was almost finished.

Andronikof: *Who works on KP nowadays here?*

Sato: I don't think that anyone is working seriously on it now. But at the time Miwa and Jimbo and others developed the theory in connection with Kac-Moody algebras. I worked to generalize this to higher dimensions. You see, KP is a kind of one-dimensional theory, using microdifferential operators in one variable, so we tried to generalize it to higher dimensions and did it in part with my graduate students Ohyama and Nakayashiki. But I am not yet satisfied with this. I gave some talks on this subject at the AMS summer institute organized by Ehrenpreis in 1988. They have been written down by Takasaki([4]).

Andronikof: *It seems that many subjects have been lying dormant inside like still waters, like a volcano waiting to erupt. Where did you turn your attention to after KP theory?*

Sato: Life is not limited to mathematics, and one is forced to engage in other things. So, unwillingly of course, at some times I was prevented from focusing in mathematics. Starting from about 1980, I was not able to concentrate on mathematics for several years, and I was rather unhappy. As I told you, mathematics give me relief from everyday life, so to speak. Anyway, it has only been in the last two or three years that I have been able to work again. I'm now trying to reorganize my own mathematics these days.

Sato's School

Andronikof: *Which mathematicians played a role as supporters of your ideas, as propagators of the theory?*

Sato: At the early stage the most important names were Iyanaga, Komatsu, Martineau and, in a sense, André Weil. After 1970 I became acquainted with a lot of French people, like Schapira, who is a student of Martineau, or Pham, as I told you, and Jean Leray.

Andronikof: *Did you create a school?*

Sato: This may be partly true, but in fact many of the active people were students of Komatsu at Tōdai. Kawai graduated with him in 1968 and came to RIMS in 1970 as an assistant. In 1968 I also met Kashiwara, who was only a junior student at Tōdai. Since I was at Komaba at the time, he formally registered with Kodaira, but in practice he was with me. Miwa was also a student of Komatsu and came to RIMS in 1973 as an assistant. Jimbo was my only graduate student among these people. As concerns PHVS, Shintani was the most important person. He was not a permanent member of the Tokyo seminar, because Komatsu was not interested in PHVS. He was a very talented mathematician, and if he were alive he would now be a major figure in the world of mathematics, like Kashiwara. Kimura was in the same generation as Miwa. Though not in a formal sense, he was also my student and often came to Kyoto to study PHVS—in which he is an important figure.

Andronikof: *For nonlinear equations you had Takasaki, Ohyama, Nakayashiki?*

Sato: Yes. Actually, Takasaki was a student of Komatsu, and he joined me rather late. After finishing his course at Tōdai, he formally stayed in Tokyo for three more years, but actually came to Kyoto. Ohyama and Nakayashiki are from Kyoto University and came as graduate students to RIMS, so they are really my students.

Andronikof: *Do you have any students working on your ideas on number theory?*

Sato: Mmmhh... Not many.

Andronikof: *Tōdai has been a good source of students for you.*

Five Papers About Mikio Sato

S. Iyanaga, Three personal reminiscences, *Algebraic analysis*, Vol. I, Academic Press, Boston, MA, 1988, 9-11.

Masaki Kashiwara and Takahiro Kawai, The award to Mikio Sato of a medal for distinguished services to culture, *Sūgaku* **37**, 1985, no. 2, 161-163.

_____, Introduction: Professor Mikio Sato and his work, *Algebraic analysis*, Vol. I, Academic Press, Boston, MA, 1988, 1-7.

Gian-Carlo Rota, Mikio Sato: impressions, *Algebraic analysis*, Vol. I, Academic Press, Boston, MA, 1988, 13-15.

Kōsaku Yosida, Sato, a perfectionist, *Algebraic analysis*, Vol. I, Academic Press, Boston, MA, 1988, 17-18.

Sato: True. Remember that when I came here twenty years ago I didn't want to leave Tokyo because I didn't want to sever ties with the young mathematicians who were eager to work with me, and it was with some regret that I came to Kyoto. In spring 1992 I'll be retiring from the University of Kyoto, and I will have no other choice but to work here, ha! ha!

Andronikof: *Is there such a thing as retirement in mathematics?*

Sato: Well, I'm not retiring from mathematics at all, but I have less time to work now. I am not young anymore, and my brain is not as fast...

Research Now and Future[13]

Andronikof: *What is left of your colonies of ideas? What has not been exploited yet of the program you exposed at the colloquium in 1960?*

Sato: In that sense: nonlinear equations. I have been keeping it in mind since that time, and it's still not worked out very well. I now want to include singular perturbation in my framework, but all this is not enough.

Andronikof: *So you are not yet satisfied with the form you gave to nonlinear equations?*

Sato: It is just a starting point, as you see... In 1960 I proposed the $\mathcal{D}$-module concept as the most natural way to deal with linear equations, so everything can be worked out by means of homological algebra, and I also introduced what are now called holonomic systems. But in the case of nonlinear equations, the only application of differential algebra was to a kind of Galois theory for equations of the Picard-Vessiot type. It is a rather minor branch in analysis, the major stream in analysis being, say, Hadamard's theory of hyperbolic equations.

At the time, the major battlefield in nonlinear PDEs was mathematical physics, as described for example in Courant-Hilbert's famous book, and differential geometry, with the Monge-Ampère, Hamilton-Jacobi, and Einstein equations, and contact transformation theories. These concepts naturally fitted in Élie Cartan's scheme, and I was not sure that my approach by means of the nonlinear filtered differential algebra $\mathcal{D}$ could work as effectively... But later I became more and more confident that this algebraic method was the only natural way to develop a general theory, because it represents the structure of differential equations without depending on how you write down the equations. Anyway, in my 1960 talk I just touched upon two main methods to develop the general theory of nonlinear equations, without sorting out either of them.

Andronikof: *In the one-dimensional case what do you think of Jean Ecalle's approach?*

[13] *That is, in August 1990.*

Sato: My way of understanding it is that this kind of problem is close to what Pham noticed at a meeting in Greece three years ago and is related to my oldest problem in connection with PDEs. In my talk of 1960 I explained that linear nonholonomic systems can be related to holonomic ones by considering elementary solutions: what Riemann called singular functions. Suppose you have a PDE on a manifold X. Then its fundamental solution can be expected to satisfy some holonomic system, not on X but on $X \times X$. In that way, one could hope to control the whole theory of PDEs through holonomic systems.

However, it soon appeared that this program could not be carried out, because in certain cases the fundamental solution may not be holonomic in a strict sense. For example, some particular fundamental solutions may contain a factor of the form x^y, with both x and y being variables (of course, x^α, α constant, is a solution to a holonomic system). This happens quite often, and this means that you have to further extend the concept of holonomic system by considering equations of infinite order, or by involving an infinite number of independent variables. This is now a quite natural thing in modern mathematical physics. The work of Pham, Ecalle, André Voros, as well as some earlier works by mathematical physicists, seem to have close connections with such difficulties. This is also related to singular perturbation theory. There, an equation involves a parameter and at some value—which could be a coupling constant, or Planck's constant—the system becomes singular, so convergence is lost, although it is summable by some kind of Borel resummation. This kind of resummation is not unique—contrary to the convergent case—but depends on the chosen direction. If you move the direction it may suddenly change, yielding a kind of Stokes phenomenon. And this again is related to concepts like instantons or tunneling, which are also quite common in mathematical physics and are related to monodromy groups and so on. It is clear that such quite general concepts should be handled systematically in analysis.

In my talk at Okayama[14], I explained that singular perturbation can be dealt with by considering solutions of nonlinear equations which are neither holomorphic nor hyperfunctions but rather a kind of modified microfunctions, living in a filtered formal function space. For example, microfunctions supported at the origin are generated by the Dirac δ function together with its derivatives. If you apply microdifferential operators of order zero you get something like a Heaviside function times analytic functions, and this is the kind of special microfunction whose

[14] *On the occasion of ICM-90 Satellite Conference on Special Functions.*

support is the origin. In several variables solutions have to be holomorphic in some variables but in the remaining ones it can be like a δ function or Heaviside function. Such mixed solutions can give a singular perturbation situation. This is what I tried to explain in Okayama—though I did not develop my talk in a very organized way. Anyway, this is how I see the work of Ecalle, Voros, or others. I think these things are very important and very natural in many problems of mathematical physics.

Andronikof: *Like shock waves?*

Sato: Yes, shock waves or diffraction, which are described by subdominant terms. The magnitude of the backward wave is exponentially small, so that the C^∞-theory cannot take it into account. But the analytic theory can single it out. This subdominant effect is related to Ecalle's notion of resurgency. In quantum physics, tunneling is also related to subdominant or Stokes phenomena. This aspect of analysis should be exploited to reach a consistent understanding, a correct general theory of such phenomena. I think this should be done in coming years.

During the congress [15] it should have clearly appeared that mathematical physics is very important, not only in applied, but also pure mathematics, including analysis, geometry, algebraic geometry, and number theory. But there are actually deeper connections with mathematical physics. This is not fully exploited yet. Ecalle's, Voros's, or earlier works, like Balian-Bloch ([2]) and Bender-Wu ([3]), are instances of applications of analysis to mathematical physics. But, in my opinion, this is still very primitive. There is work for the next decade in order to settle things so that analysis can really be applied to mathematical physics. While methods of mathematical physics in quantum field theory have profited various branches of mathematics (topology, braid theory, number theory, geometry), the converse is not necessarily true. Today, mathematical physicists mostly use number theory or algebraic geometry. Mathematical physics is receptive only to higher developed areas of mathematics, some of which are exploited in superstring theory, though not to its full extent. Mathematics has not succeeded in providing a more effective way of computation than perturbation expansions. Of course, there are some primitive methods of computation, like the Monte-Carlo method. All these are kind of brute force computations, not refined mathematics, surely not refined enough for the problems physics is now confronted with, like determining the mass of particles or quarks. All these things are discussed on a very abstract level, not on a quantitative level. So I think that mathematical analysis should be developed much further to match the reality of physics.

Andronikof: *What could remain of these problems in the twenty-first century?*

Sato: I cannot say: a good mathematician can appear in each branch. If one of these branches attracts a good young mathematician, this branch might develop quite rapidly. Look for instance at the past: if there were no Grothendieck or Deligne, things in algebraic geometry would have developed in a different way. But you can imagine how it is going to evolve in the next few years, because a number of very active mathematicians at this congress gave a number of very interesting talks and would give you an idea of what mathematicians in the 1990s will look like.

Andronikof: *My last question is about Japanese mathematics—the fact that in modern mathematics Japan has risen from nowhere to its current outstanding place.*

Sato: I'm not a specialist in Japanese mathematical history, but as I have explained it, Japan has a good background for modern mathematics. About two or three hundred years ago, at the time of Leibniz and Newton, we had Seki, one of the founders of Japanese mathematics. He and some of his great followers developed a kind of algebra. Some elementary mathematics, related to counting, was introduced from China. But Seki developed a major system of algebra and even infinitesimal calculus. He developed a theory for algebraic equations, with Chinese characters instead of x and y. He dealt with linear equations in several variables, with higher-order algebraic equations, though he did nothing like Galois theory or Cardano formulas for cubic equations: instead he developed approximate solutions akin to Newton-Horner methods. Also, he knew how to solve linear equations using the Cramer method, since he had the notion of determinant, earlier than in Europe. He used derivatives to determine maxima and minima of a polynomial in several variables. But the biggest defect of his system was that they did not see at the time that integration is the converse of differentiation. So they computed integrals the way Archimedes did, obtaining volumes and areas using a kind of slicing procedure. One of the typical integrals appearing in connection with computing area or volume for which they had such a formula, is $\int_0^1 (\sqrt{1-x^2})^n dx$, for n odd. This is just one type of accomplishment of Japanese mathematics in the seventeenth and eighteenth century.

Andronikof: *Did it go on?*

Sato: Although we had this tradition, mathematics declined to a kind of hobby. The government in the Meiji era decided to adopt Western mathematics, because Japanese mathematics is just a kind of empirical thing. They did not pay

[15] *Kyoto, ICM 1990.*

much attention to logic. So even in geometry, they did not develop a demonstration. Just a discussion is given, but not a completely logical inference.

Andronikof: *What about RIMS*[16]*?*

Sato: When the Russians sent the Sputnik into orbit in 1957, it was a big shock in the USA. This certainly motivated them to develop their science. At that time, the number of professors in scientific branches, including mathematics, simply doubled. The same thing took place in Japan, a little later. Scientists had a good excuse to get more money. A number of institutes of technology and physics, but also of pure sciences, were built at that time. Many senior mathematicians, like Yosida and Iyanaga, persuaded our government to found a new institute for mathematics. The Institute was created in 1959, when I had just started my career.

Tōdai had already too many research institutes independent of the departments and was reluctant to have a new one. So it was finally decided that the new mathematical institute should be opened within Kyoto University, which supported the idea.

Andronikof: *And from the start it was devoted to pure mathematics?*

Sato: It was devoted to mathematical sciences. The fact is that its emphasis is on pure mathematics, but it also includes a few mathematical physicists and a few computer scientists. We are even trying to put emphasis on applied mathematics. This balance has now changed. We have to recover this. The importance of computer science in mathematics is recognized, as we can see by the Nevanlinna Prize, dedicated to that discipline.

References

[1] T. T. Wu, B. M. McCoy, C. A. Tracy, and E. Barouch, Spin-spin correlation functions for the two dimensional Ising model: Exact results in the scaling region, *Phys. Rev. B* **13** (1976), 316–374.

[2] R. Balian and C. Bloch, Solutions of the Schrödinger equation in terms of classical paths, *Ann. Phys.* **85** (1974), 514-545.

[3] C. M. Bender and T. T. Wu, Anharmonic oscillator, *Phys. Rev.* **184** (1969), 1231-1260.

[4] M. Sato, The KP hierarchy and infinite-dimensional Grassmann manifolds, *Theta Functions, Bowdoin 1987, Proceedings of Symposia in Pure Mathematics*, vol. 49, Part 1, Amer. Math. Soc., Providence, RI, 1989, pp. 51–66,

[16] *Sato was the director of RIMS at the time of this interview.*

Proceedings of Symposia in Pure Mathematics
Volume **49** (1989), Part 1

The KP Hierarchy and Infinite-Dimensional Grassmann Manifolds

MIKIO SATO[1]

First Lecture

1.1. Finite-dimensional Grassmann manifolds. Let $V(N)$ be an N-dimensional vector space over a commutative field $\mathscr{C}$, $N = m + n$ a partition of N into two positive integers. The Grassmann manifold,

$$\begin{aligned}\mathrm{GM}(m; V(N)) := \{&m\text{-tuples } (\xi^{(0)}, \dots, \xi^{(m-1)}) \text{ of} \\ &\text{linearly independent vectors in } V(N)\}/\mathrm{GL}(m),\end{aligned}$$

gives a parameter space of all m-dimensional vector subspaces $\{U \subset V(N)\colon \dim U = m\}$. The correspondence between a subspace and a frame is

$$U = \mathscr{C}\xi^{(0)} + \cdots + \mathscr{C}\xi^{(m-1)}.$$

Such an m-tuple $(\xi^{(0)}, \dots, \xi^{(m-1)})$ is called an *m-frame*. The action of $\mathrm{GL}(m)$ defining the quotient is

$$(\xi^{(0)}, \dots, \xi^{(m-1)}) \mapsto (\xi^{(0)}, \dots, \xi^{(m-1)})h, \qquad h \in \mathrm{GL}(m).$$

Replacing $\mathrm{GL}(m)$ with $\mathrm{SL}(m)$ one can also construct the quotient manifold

$$\widetilde{\mathrm{GM}}(m; V(N)) := \{m\text{-frames}\}/\mathrm{SL}(m),$$

which becomes a line bundle over $\mathrm{GM}(m; V(N))$ with the canonical projection $\pi\colon \widetilde{\mathrm{GM}}(m; V(N)) \to \mathrm{GM}(m; V(N))$, whose fibers are isomorphic to $\mathrm{GL}(1)$. This line bundle gives rise to a projective embedding of the Grassmann manifold as follows.

An element of $\widetilde{\mathrm{GM}}(m; V(N))$ is in one-to-one correspondence with the exterior product $\xi^{(0)} \wedge \cdots \wedge \xi^{(m-1)}$. Thus one obtains an embedding of $\widetilde{\mathrm{GM}}(m; V(N))$ into $\bigwedge^m V(N) - \{0\}$, and accordingly an embedding of $\mathrm{GM}(m; V(N))$ into $\mathbf{P}(\bigwedge^m V(N))$. These are called the *Plücker embeddings*.

1980 *Mathematics Subject Classification* (1985 *Revision*). Primary 58F07, 14K25, 14M15.
[1]This note is written by his coworker, Dr. Kanehisa Takasaki, RIMS, Kyoto University.

To give a more explicit picture, choose a basis $e_0, \dots, e_{N-1}$ of $V(N)$ and write the members of the m-frames as:

$$\xi^{(j)} = \sum_{0 \le i < N} \xi_{ij} e_i, \qquad \xi_{ij} \in \mathscr{C}.$$

Then

$$\xi^{(0)} \wedge \cdots \wedge \xi^{(m-1)} = \sum_{0 \le l_0 < \cdots < l_{m-1} < N} \xi_{l_0 \cdots l_{m-1}} e_{l_0} \wedge \cdots \wedge e_{l_{m-1}}$$

where

$$\xi_{l_0 \cdots l_{m-1}} := \det(\xi_{l_i j})_{0 \le i,j < m}.$$

The last determinants are called *Plücker coordinates.* The Plücker coordinates are well-defined functions on $\widetilde{\mathrm{GM}}(m; V(N))$, and nothing else than the coordinates of the image of the Plücker embedding under the identification

$$\bigwedge^m V(N) \cong \mathscr{C}^{d_{mn}}, \qquad d_{mn} := \binom{m+n}{m}.$$

The image of the Plücker embedding is a nonsingular algebraic variety defined by the *Plücker relations*

$$\sum_{0 \le i \le m} (-1)^i \xi_{l_0 \cdots l_{m-2} l'_i} \xi_{l'_0 \cdots l'_{i-1} l'_{i+1} \cdots l'_m} = 0,$$

where $l_0, \dots, l_{m-2}$ and $l'_0, \dots, l'_m$ range over all possible values.

1.2. Nesting of finite-dimensional Grassmann manifolds. To go forward to the construction of an infinite-dimensional limit called the *universal Grassmann manifold,* we start from the case where $V(N) = V(m+n) = \mathscr{C}^{m+n}$ (the vector space of column vectors of size $m+n$). For convenience let us renumber the entries of column vectors with $\{-m, -m+1, \dots, n-1\}$ instead of $\{0, 1, \dots, m+n-1\}$ and write the corresponding Grassmann manifold and the line bundle as $\mathrm{GM}(m,n)$ and $\widetilde{\mathrm{GM}}(m,n)$.

In this setting an m-frame is nothing else than a matrix ξ of the form

$$\xi = (\xi_{ij})_{-m \le i < n, -m \le j < 0}, \qquad \xi_{ij} \in \mathscr{C}, \quad \operatorname{rank} \xi = m.$$

Thus

$$\mathrm{GM}(m,n) = \{\xi;\ \text{as above}\}/\mathrm{GL}(m),$$
$$\widetilde{\mathrm{GM}}(m,n) = \{\xi;\ \text{as above}\}/\mathrm{SL}(m).$$

From each m-frame we define the Plücker coordinates as

$$\xi_{l_{-m} \cdots l_{-1}} = \det(\xi_{l_i j})_{-m \le i,j < 0}$$

where the indices $l_{-m}, \dots, l_{-1}$ range over $-m, \dots, n-1$.

We find a nesting structure among these finite-dimensional Grassmann manifolds as follows. If $m \le m'$ and $n \le n'$, $\mathrm{GM}(m,n)$ (resp., $\widetilde{\mathrm{GM}}(m,n)$)

can be embedded into $\mathrm{GM}(m', n')$ (resp., $\widetilde{\mathrm{GM}}(m', n')$) through the embedding of frames as

$$\xi \mapsto \xi' := \begin{pmatrix} 1_{m'-m} & 0_{(m'-m)\times m} \\ 0_{(m+n)\times(m'-m)} & \xi \\ 0_{(n'-n)\times(m'-m)} & 0_{(n'-n)\times m} \end{pmatrix},$$

where 1_* and $0_{*\times *}$ denote the unit and null matrices of size indicated therein. The Plücker coordinates for both Grassmann manifolds are related as

$$\begin{cases} \xi'_{-m'\cdots -m-1 l_{-m}\cdots l_{-1}} = \xi_{l_{-m}\cdots l_{-1}} \quad (-m \le l_{-m} < \cdots < l_{-1} < n), \\ \text{other Plücker coordinates of } \xi' = 0. \end{cases}$$

1.3. Universal Grassmann manifold. In a limit as $m, n \to \infty$, thus, we will arrive at an infinite-dimensional Grassmann manifold in which all $\mathrm{GM}(m, n)$ are embedded. This is the universal Grassmann manifold

$$\mathrm{UGM} := \{\text{all } \mathbf{N}^c\text{-frames}\}/\mathrm{GL}(\mathbf{N}^c),$$

where an $\mathbf{N}^c$-*frame* is an infinite matrix whose rows and columns are indexed by $\mathbf{Z} = \{0, \pm 1, \pm 2, \dots\}$ and $\mathbf{N}^c = \{-1, -2, \dots\}$ as $\xi = (\xi_{ij})_{i\in\mathbf{Z}, j\in\mathbf{N}^c}$ and which has a form as indicated below:

$$\xi = \left(\begin{array}{ccc|ccc} \ddots & & & & & \\ & 1 & & & 0 & \\ & & 1 & & & \\ 0 & & & 1 & & \\ \hline ** & ** & ** & * & * & * \\ ** & ** & ** & & m\text{-frame} & \\ ** & ** & ** & * & * & * \\ \hline ** & ** & ** & ** & ** & ** \\ ** & ** & ** & ** & ** & ** \end{array}\right) \begin{array}{l} \vdots \\ -m-1 \\ -m \\ \vdots \\ n-1 \\ \vdots \end{array}$$

$$\begin{array}{cccc} \cdots\cdots & -m & \dots & -1 \end{array}$$

where m and n are positive integers depending on ξ. $\mathrm{GL}(\mathbf{N}^c)$ is accordingly made up of square matrices $h = (h_{ij})_{i,j\in\mathbf{N}^c}$ of the following form:

$$h = \left(\begin{array}{ccc|ccc} \ddots & & & & & \\ & 1 & & & 0 & \\ & & 1 & & & \\ 0 & & & 1 & & \\ \hline ** & ** & ** & * & * & * \\ ** & ** & ** & & \mathrm{GL}(l) & \\ ** & ** & ** & * & * & * \end{array}\right) \begin{array}{l} \vdots \\ -l-1 \\ -l \\ \vdots \\ -1 \end{array},$$

$$\begin{array}{cccc} \cdots\cdots & -l & \dots & -1 \end{array}$$

l being likewise a positive integer depending on h.

UGM carries a line bundle,

$$\widetilde{\text{UGM}} := \{\text{all } \mathbf{N}^c\text{-frames}\}/\text{SL}(\mathbf{N}^c),$$

where $\text{SL}(\mathbf{N}^c)$ is the subgroup of $\text{GL}(\mathbf{N}^c)$ obtained by replacing the block at the lower right corner with a unimodular (i.e., $\text{SL}(l)$) matrix. Plücker coordinates are defined as functions on $\widetilde{\text{UGM}}$; we use an infinite sequence $(l_i)_{i \in \mathbf{N}^c}$ of integers with $l_i = i$ for almost all $i \in \mathbf{N}^c$ (i.e., allowing at most a finite number of exceptions) to label them as

$$\xi_{\cdots l_{-2} l_{-1}} := \det(\xi_{l_i j})_{i,j \in \mathbf{N}^c},$$

where the determinant is understood as the limit of the truncated one, $\det(\xi_{l_i j})_{-m \le i, j < 0}$ as $m \to \infty$. This limit is meaningful because the truncated determinant, as a consequence of the above assumption on ξ, becomes independent of m for all m greater than a finite bound. Plücker coordinates thus defined are totally antisymmetric with regard to permutations of the indices. Hence only those with strictly increasing indices are sufficient to recover all information, the others either vanishing (if there is an overlap of indices) or being equal to those with strictly increasing indices except for a possible difference of the sign. Such strictly increasing index sequences have a particular property that they are in one-to-one correspondence with Young diagrams, e.g.,

$$(\cdots \ell_{-2} \ell_{-1}) \longleftrightarrow Y = \begin{vmatrix} \rule{3cm}{0pt} \\ \rule{3cm}{0pt} \\ \end{vmatrix} \begin{matrix} \ell_{-1} + 1 \\ \ell_{-2} + 2. \\ \vdots \end{matrix}$$

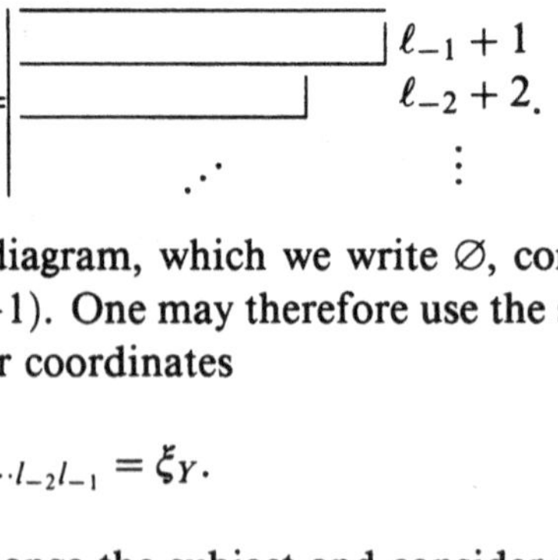

In particular, the empty Young diagram, which we write $\varnothing$, corresponds to the index sequence $(\ldots, -3, -2, -1)$. One may therefore use the set of Young diagrams $\{Y\}$ as labels of Plücker coordinates

$$\xi_{\cdots l_{-2} l_{-1}} = \xi_Y.$$

1.5. KP hierarchy. We now change the subject and consider the KP hierarchy. Let us start with a review of the notion of micro-differential operators. From a purely algebraic point of view, a micro-differential operator in one variable x is a linear combination of both positive and negative powers of d/dx with an upper bound

$$A = \sum_{i=-\infty}^{m} a_i(x) \left(\frac{d}{dx}\right)^i,$$

the coefficients being taken from, say, the ring $\mathscr{E}[[x]]$ of formal power series of x. The leading exponent m with $a_m \neq 0$ is called the *order* of A, and if

$a_m = 1$, A is said to be *monic*. The rules of addition and multiplication are

$$\sum a_i \left(\frac{d}{dx}\right)^i + \sum b_i \left(\frac{d}{dx}\right)^i := \sum (a_i + b_i) \left(\frac{d}{dx}\right)^i,$$

$$\sum a_i \left(\frac{d}{dx}\right)^i \cdot \sum b_j \left(\frac{d}{dx}\right)^j := \sum c_k \left(\frac{d}{dx}\right)^k$$

$$\text{where } c_k := \sum_{i \in \mathbf{Z}, l \geq 0} \binom{i}{l} a_i \left(\frac{d}{dx}\right)^l b_{k-i+l},$$

which are consistent with the calculus of differential operators. Thus the set of micro-differential operators $\mathscr{E} = \mathscr{E}_{\mathscr{C}[[x]]}$ gives a ring extension of the ring $\mathscr{D} = \mathscr{D}_{\mathscr{C}[[x]]}$ of differential operators. In fact one can replace $\mathscr{C}[[x]]$ above with any differential ring $\mathscr{O}$ with a derivation d/dx to define the rings $\mathscr{D}_{\mathscr{O}}$ and $\mathscr{E}_{\mathscr{O}}$ of differential and micro-differential operators with coefficients in $\mathscr{O}$. A basic property, which does not depend on a particular choice of the coefficient ring $\mathscr{O}$, is that a micro-differential operator has an inverse if and only if its leading coefficient has an inverse within $\mathscr{O}$.

The KP hierarchy admits several equivalent expressions. The first one is a system of *Lax equations* to a first-order micro-differential operator of the form

$$L = \frac{d}{dx} + u_2(x,t) \left(\frac{d}{dx}\right)^{-1} + u_3(x,t) \left(\frac{d}{dx}\right)^{-2} + \cdots$$

where t is a collection of an infinite number of *time* variables $(t_1, t_2, \ldots)$. The system of Lax equations takes the form of evolution equations

$$\frac{\partial L}{\partial t_n} = [B_n, L], \qquad n = 1, 2, \ldots,$$

where B_n are the differential operators

$$B_n := (L^n)_+,$$

$(\ )_+$ being the projection onto the differential operator part of micro-differential operators. The second one is made up of the system of the following form:

$$\left[\frac{\partial}{\partial t_m} - B_m, \frac{\partial}{\partial t_n} - B_n\right] = 0, \qquad m, n = 1, 2, \ldots.$$

In the third expression the central role is played by a monic 0th order micro-differential operator of the form

$$W = 1 + w_1(x,t) \left(\frac{d}{dx}\right)^{-1} + w_2(x,t) \left(\frac{d}{dx}\right)^{-2} + \cdots,$$

which is introduced to represent L as

$$L = W \cdot \frac{d}{dx} \cdot W^{-1}.$$

Under this relation the system of equations

$$\frac{\partial W}{\partial t_n} = B_n W - W\left(\frac{d}{dx}\right)^n, \qquad n = 1, 2, \ldots,$$

reproduces the Lax system, and conversely if L satisfies the Lax system, one can find W with the above properties by solving linear differential equations recursively for the coefficients of W. Since

$$B_n = \left(W \cdot \left(\frac{d}{dx}\right)^n \cdot W^{-1}\right)_+$$

under the relation above, the last system of equations closes in itself as a system of evolution equations for W. One can regard this as defining commuting dynamical flows in the space of 0th order monic micro-differential operators $\{W\}$, each flow being attached with a time parameter in $\{t_n\}$.

1.6. Dynamical system on universal Grassmann manifold. There is a one-to-one correspondence between the set of W's and the universal Grassmann manifold. With this correspondence the KP hierarchy gives rise to dynamical flows on the universal Grassman manifold. The latter turns out to be governed by a very simple law, i.e., the dynamical motion is, roughly speaking, caused by the left action by the matrix-valued exponential

$$\xi \mapsto \exp\left(\sum_{n=1}^{\infty} t_n \Lambda^n\right)\xi, \qquad \Lambda^n := (\delta_{i,j-n})_{i,j\in\mathbf{Z}}.$$

To be precise, the last statement above includes a subtlety: in general, the action of the exponential produces from an *admissible* $\mathbf{N}^c$-frame (i.e., an infinite matrix $\xi = (\xi_{ij})_{i\in\mathbf{Z},j\in\mathbf{N}^c}$ of the form as mentioned above) a *nonadmissible* one. A remedy for this difficulty is to multiply the resultant nonadmissible $\mathbf{N}^c$-frame with a compensating matrix from the right so as to render it again into an admissible one. This is a natural extension of the quotient process by $\mathrm{GL}(\mathbf{N}^c)$ in the definition of UGM, corresponding to the change of bases within a vector subspace. The only difference is that the prescription above is accompanied with *infinite sums* of vectors, in contrast to the case of $\mathrm{GL}(\mathbf{N}^c)$ where a base change is done within finite sums of vectors.

Suppose, for example, that ξ has a *normalized* form:

$$\xi = \left(\begin{array}{cccc} \ddots & & & \\ & 1 & & \\ & & 1 & \\ 0 & & & 1 \\ \hline * & * & \cdots & * \\ * & * & \cdots & * \\ * & * & \cdots & * \end{array}\right), \qquad \text{i.e., } \xi_{ij} = \delta_{ij} \text{ for } i, j \in \mathbf{N}^c.$$

(This is a *generic* situation, i.e., such ξ's form an open dense subset, $\mathrm{UGM}^{\varnothing}$, of UGM. The entries in the lower half, ξ_{ij} for $i \in \mathbf{N}$ and $j \in \mathbf{N}^c$, then become affine coordinates on it.) Then

$$h(t) := \text{ the } \mathbf{N}^c \times \mathbf{N}^c \text{ part of } \exp\left(\sum_{n=1}^{\infty} t_n \Lambda^n\right)\xi$$

turns out to be an invertible matrix (but not an element of $\mathrm{GL}(\mathbf{N}^c)$), and

$$\xi(t) := \exp\left(\sum_{n=1}^{\infty} t_n \Lambda^n\right)\xi \cdot h(t)^{-1}$$

becomes an admissible $\mathbf{N}^c$-frame (in fact, it is normalized as ξ is) whose entries are formal power series in t.

Another prescription, which is actually equivalent to the above one, is to define the action in the language of Plücker coordinates. To this end, let us use Young diagrams $\{Y\}$ as labels of Plücker coordinates. This is simply for notational convenience. We then define

$$\xi_Y(t) := \sum_{Y':\text{all}} \chi_{Y'/Y}(t)\xi_{Y'},$$

where

$$\begin{aligned} \chi_{Y'/Y}(t) &:= \det(p_{l'_i - l_j}(t))_{i,j\in\mathbf{N}^c}; \\ p_n(t) &:= \sum_{\nu_1+2\nu_2+3\nu_3+\cdots=n} t_1^{\nu_1} t_2^{\nu_2} \cdots / (\nu_1!\nu_2!\cdots); \\ p_0(t) &:= 1. \end{aligned}$$

In particular, we write $\chi_{Y/\phi}$ simply as χ_Y, which are nothing else than the *Schur functions*:

$$\chi_Y(t) = \det(p_{l_i - j}(t))_{i,j\in\mathbf{N}^c}.$$

The infinite determinants in the definition above are understood to be of the same meaning as in the definition of Plücker coordinates. These $\xi_Y(t)$ indeed satisfy all Plücker relations (one can easily check this fact, reducing it to an analogous fact in $\mathrm{GM}(m,n)$), hence this certainly defines a dynamical motion in $\widetilde{\mathrm{UGM}}$. Its projection onto UGM agrees with the previous one.

The KP hierarchy and the dynamical flows on the universal Grassmann manifold are connected by the following relation:

$$\begin{aligned} w_n(x,t) &= -\frac{\xi_{(\ldots,-n-1,0,-n+1,\ldots,-1)}(x+t)}{\xi_{(\ldots,-3,-2,-1)}(x+t)} \\ &= (-1)^n \frac{\xi_{\Delta_{n,1}}(x+t)}{\xi_{\emptyset}(x+t)} \end{aligned}$$

where $x+t := (x+t_1, t_2, t_3, \ldots)$ and $\Delta_{n,m}$ denotes the $n \times m$ rectangular Young diagram.

A remarkable fact is that the basic Plücker coordinate

$$\xi_{\emptyset}(t) = \sum_{Y:\text{all}} \chi_Y(t)\xi_Y$$

can reproduce all the other Plücker coordinates as:

$$\xi_Y(t) = \chi_Y(\partial_t)\xi_\varnothing(t), \qquad \partial_t := \left(\frac{\partial}{\partial t_1}, \frac{1}{2}\frac{\partial}{\partial t_2}, \frac{1}{3}\frac{\partial}{\partial t_3}, \dots\right).$$

$\xi_\varnothing(t)$ is nothing else than the τ (tau) function $\tau(t,\xi)$. Thus the relation above connecting W and ξ can be rewritten in the language of τ functions.

Second Lecture

The presentation of the previous lecture is yet not enough refined in the sense that it fully relies on a particular choice of bases and associated matrix representations. Such an expression is eventually very useful in applications, but from a theoretical point of view a more intrinsic formulation should also be pursued.

2.1. Micro-differential operators. We now attempt to reconstruct the whole theory on the basis of the notion of $\mathscr{D}$- and $\mathscr{E}$-modules, where $\mathscr{D}$ and $\mathscr{E}$ are the rings of differential and micro-differential operators, respectively. In the present context only the case with one variable x is relevant; a multi-dimensional extension will be discussed in the final lecture.

As already mentioned in the first lecture, $\mathscr{D}$ and $\mathscr{E}$ in one variable are defined as

$$\mathscr{D} := \left\{ A = \sum_{0 \le i \ll \infty} a_i(x) \left(\frac{d}{dx}\right)^i : a_i \in \mathscr{O} \right\};$$

$$\mathscr{E} := \left\{ A = \sum_{-\infty < i \ll \infty} a_i(x) \left(\frac{d}{dx}\right)^i : a_i \in \mathscr{O} \right\};$$

where $\mathscr{O}$ is a differential $\mathscr{C}$-algebra on which the derivation d/dx acts (e.g., $\mathscr{O} = \mathscr{C}[[x]]$ = formal power series in x) and "$i \ll \infty$" means that the summation ranges over the indicated domain with a *finite upper bound* as "$i \le m$" for some integer m that may depend on A. Addition and multiplication in $\mathscr{E}$ are introduced along just the same way as mentioned in the first lecture, and $\mathscr{E}$ gives a ring extension of $\mathscr{D}$. For an element $A = \sum a_i(d/dx)^i$ of $\mathscr{D}$ or $\mathscr{E}$ its *order* is by definition the maximal exponent of d/dx in A:

$$\operatorname{ord}(A) := \max\{i; a_i \ne 0\}.$$

The order of the zero operator is understood to be $-\infty$. With this notion one can introduce the filtrations

$$\mathscr{D}^{(i)} := \{A \in \mathscr{D}; \operatorname{ord}(A) \le i\},$$
$$\mathscr{E}^{(i)} := \{A \in \mathscr{E}; \operatorname{ord}(A) \le i\}.$$

2.2. Infinite-dimensional vector space. We now introduce an infinite-dimensional vector space as

$$\mathbf{V} := \mathscr{E}/\mathscr{E}x.$$

This is primarily a left $\mathscr{E}$-module with a single generator, say 1 mod $\mathscr{E}x$, but through the embedding $\mathscr{C} \hookrightarrow \mathscr{E}$ this naturally acquires the structure of a $\mathscr{C}$-vector space. As a $\mathscr{C}$-vector space

$$\mathbf{V} \simeq \mathscr{E}_{\text{const}} := \left\{ \sum_{-\infty<i\ll\infty} a_i \left(\frac{d}{dx}\right)^i ; a_i \in \mathscr{C} \right\},$$

equipped with the induced filtration

$$\mathbf{V}^{(i)} \simeq \mathscr{E}^{(-i-1)}_{\text{const}} := \mathscr{E}_{\text{const}} \cap \mathscr{E}^{(-i-1)}, \qquad i \in \mathbf{Z}.$$

One can regard this filtration as a fundamental system of neighborhoods of the origin to define a uniform topology in $\mathbf{V}$, which accordingly becomes a locally linearly compact vector space in the sense of Lefschetz. A standard basis of $\mathbf{V}$ is

$$e_i := \left(\frac{d}{dx}\right)^{-i-1} \bmod \mathscr{E}x, \qquad i \in \mathbf{Z}.$$

With this basis one can identify $\mathbf{V}$ with the vector space of infinite column vectors as

$$v = \sum v_i e_i \leftrightarrow (v_j)_{j\in\mathbf{Z}} = \begin{pmatrix} \vdots \\ v_{-1} \\ v_0 \\ \vdots \end{pmatrix}.$$

The action of x and d/dx on the members of this basis, which reflects the $\mathscr{E}$-module structure of $\mathbf{V}$, reads:

$$xe_i = (i+1)e_{i+1},$$
$$\frac{d}{dx}e_i = e_{i-1};$$

therefore, in the language of infinite column vectors as above, these operators have the following matrix representations:

$$x \leftrightarrow K := (i\delta_{i,j+1})_{i,j\in\mathbf{Z}},$$
$$\frac{d}{dx} \leftrightarrow \Lambda := (\delta_{i,j-1})_{i,j\in\mathbf{Z}}.$$

2.3. Intrinsic definition of universal Grassmann manifold. The universal Grassmann manifold UGM can be redefined as follows:

$$\text{UGM} := \{\text{vector subspaces } U \subset \mathbf{V} \text{ such that } \dim(U \cap \mathbf{V}^{(0)}) = \dim(\mathbf{V}/(U + \mathbf{V}^{(0)})) < \infty\}.$$

Roughly speaking, UGM is the set of all vector subspaces of $\mathbf{V}$ of the "same size" as the reference subspace

$$U_\varnothing := \mathscr{D}e_{-1} = \left\{ \sum_{\nu<0} c_\nu e_\nu ; c_\nu \in \mathscr{C} \right\},$$

which we refer to as the *origin* of UGM.

A generic case is such that U and $\mathbf{V}^{(0)}$ are complementary to each other in the sense

$$\mathbf{V} = U \oplus \mathbf{V}^{(0)}.$$

This is equivalent to the condition

$$\xi_\varnothing \neq 0,$$

under which one can choose an $\mathbf{N}^c$-frame of the following form:

$$\xi = \begin{pmatrix} \ddots & & \\ & 1 & \\ & & 1 \\ 0 & & & 1 \\ \hline *&*&\cdots&* \\ *&*&\cdots&* \\ *&*&\cdots&* \end{pmatrix}.$$

The set of these generic points forms an open dense subset $\mathrm{UGM}^\varnothing$ of UGM. It is actually an affine open set, ξ_{ij} $(i \geq 0, j < 0)$ being affine coordinates under the normalization condition $\xi_{ij} = \delta_{ij}$ $(i, j < 0)$. For any point of UGM, likewise, there is a unique minimal Young diagram Y for which $\xi_Y \neq 0$. UGM is accordingly stratified as

$$\mathrm{UGM} = \bigsqcup_Y \mathrm{UGM}^Y \quad \text{(disjoint union)},$$

$$\mathrm{UGM}^Y := \{\xi_{Y'} = 0 \text{ for any } Y' \not\supset Y, \text{ and } \xi_Y \neq 0\}.$$

2.4. Family of $\mathscr{D}$-modules. We next consider an intrinsic way of understanding the KP hierarchy on the basis of the notion of "deformations of $\mathscr{D}$-modules". What plays a central role is the family of left $\mathscr{D}$-submodules $\{\mathscr{J}\}$ of $\mathscr{E}$ (i.e., $\mathscr{D}\mathscr{J} \subset \mathscr{J}$) under the following condition:

$$\mathscr{E} = \mathscr{J} \oplus \mathscr{E}^{(-1)}$$

(i.e., $\mathscr{E} = \mathscr{J} + \mathscr{E}^{(-1)}$, $\{0\} = \mathscr{J} \cap \mathscr{E}^{(-1)}$). The last condition should be compared with the characterization of $\mathrm{UGM}^\varnothing$ above. In the following let us focus on the case where $\mathscr{D} = \mathscr{D}_{\mathscr{C}[[x]]}$ and $\mathscr{E} = \mathscr{E}_{\mathscr{C}[[x]]}$, for simplicity.

It is not hard to see that any $\mathscr{D}$-submodule $\mathscr{J} \subset \mathscr{E}$ with the above property is *cyclic*, i.e., generated by a single element W as

$$\mathscr{J} = \mathscr{D} W,$$

the generator W being a monic 0th order micro-differential operator

$$W = 1 + w_1 \left(\frac{d}{dx}\right)^{-1} + w_2 \left(\frac{d}{dx}\right)^{-2} + \cdots.$$

One can check this with the following simple observation. From the assumption, any element of $\mathscr{E}$ should be expressed as the sum of a member of $\mathscr{J}$

and a member of $\mathscr{E}^{(-1)}$. In particular, for any $i \geq 0$,

$$\left(\frac{d}{dx}\right)^i = W_i + \sum_{j<0} w_{ij} \left(\frac{d}{dx}\right)^j$$

for some $W_i \in \mathscr{I}$ and $w_{ij} \in \mathscr{E}[[x]]$. Thus one obtains a system of generators of $\mathscr{I}$, $W_i = (d/dx)^i - \sum_{j<0} w_{ij}(d/dx)^j$, whose $\mathscr{E}[[x]]$-linear combinations span the whole $\mathscr{I}$, i.e., $\mathscr{I} = \sum_{i\geq 0} \mathscr{E}[[x]]W_i$. The requirement that $\mathscr{I}$ be a left $\mathscr{D}$-module yields a set of relations among W_i's of the form:

$$W_{i+1} - \left(\frac{d}{dx}\right) \cdot W_i - w_{i,-1} W_0 = 0 \quad \text{for} \quad i \geq 0.$$

From these relations it follows that $\mathscr{I} = \mathscr{D}W$ with $W = W_0 = 1 - \sum_{j<0} w_{0,j}(d/dx)^j$.

2.5. Link between $\mathscr{D}$-modules and vector spaces. Left $\mathscr{D}$-submodules $\{\mathscr{I}\}$ of $\mathscr{E}$ as above are in one-to-one correspondence with points $\{U\}$ of UGM. To be precise, only the open dense subset $\text{UGM}^{\varnothing}$ is relevant to the above situation. The correspondence from each side to the other is as follows:

$$\text{(From } \mathscr{I} \text{ to } U) \quad U = W^{-1} U_{\varnothing} = \{v \in \mathbf{V}; \mathscr{I} v \subset U_{\varnothing}\};$$
$$\text{(From } U \text{ to } \mathscr{I}) \quad \mathscr{I} = \{A \in \mathscr{E}; AU \subset U_{\varnothing}\}.$$

In the language of Plücker coordinates the coefficients of W can be written

$$w_i(x) = \frac{(-1)^i \xi_{\Delta_{i,1}}(x)}{\xi_{\varnothing}(x)} = -\frac{\xi_{(\ldots,-i-1,0,-i+1,\ldots,-1)}(x)}{\xi_{\varnothing}(x)},$$

where $\xi_Y(x) := \xi_Y(t_1 = x, t_2 = 0, t_3 = 0, \ldots)$. Since $\xi_{\varnothing}(x) = \xi_{\varnothing} + \xi_{\square} x + O(x^2)$, the coefficients $w_i(x)$ will have poles if U goes outside $\text{UGM}^{\varnothing}$. In that case the setting above must be appropriately modified.

2.6. KP hierarchy as deformations of $\mathscr{D}$-modules. In the language of W the KP hierarchy is the system of differential equations

$$\frac{\partial W}{\partial t_n} = B_n W - W \left(\frac{d}{dx}\right)^n, \qquad B_n = \left(W \cdot \left(\frac{d}{dx}\right)^n \cdot W^{-1}\right)_+.$$

One can rewrite them as

$$\frac{\partial W}{\partial t_n} + W \cdot \left(\frac{d}{dx}\right)^n \in \mathscr{I} = \mathscr{D} W.$$

Note that the somewhat involved terms $B_n W$ are absorbed into the $\mathscr{D}$-module structure of $\mathscr{I}$. Nothing is lost in the passage to the second expression, which we now regard as defining deformations of $\mathscr{D}$-modules with multi-dimensional parameters. This is an "infinitesimal" expression. Its finite form should read

$$\mathscr{I}_t = \mathscr{I}_{t=0} \exp\left(-\sum t_n \left(\frac{d}{dx}\right)^n\right).$$

The interpretation above is of quite universal nature, and the essence does not change even in a multi-component or multi-dimensional case.

Third Lecture

Up to here we have confined ourselves to the case of the KP hierarchy and related structures. We now change the subject to the issue of the multi-dimensional extensions.

3.1. Multi-dimensional micro-differential operators. As explained in the last lecture, the KP hierarchy can be interpreted as a kind of deformation of left $\mathscr{D}$-submodules of the ring $\mathscr{E}$ of micro-differential operators in one variable. A possible extension will be, naturally, to consider a similar situation in multi-dimensions.

For simplicity let us adopt the ring $\mathscr{C}[[x]]$ of formal power series in r variables $x = (x_0, x_1, \dots, x_{r-1})$ as a basic differential algebra, the derivatives being $\partial_i = \partial/\partial x_i$, $0 \le i < r$. The ring $\mathscr{D} = \mathscr{D}_{\mathscr{C}[[x]]}$ of differential operators is

$$\mathscr{D} := \left\{ \sum_{|\alpha| \ll \infty} a_\alpha(x)\partial^\alpha ; a_\alpha(x) \in \mathscr{C}[[x]], \alpha \in \mathbf{N}^r \right\},$$

where $\mathbf{N} := \{0, 1, \dots\}$ (all nonnegative integers), $\partial^\alpha := (\partial_0)^{\alpha_0} \cdots (\partial_{r-1})^{\alpha_{r-1}}$ and $|\alpha| := \alpha_0 + \cdots + \alpha_{r-1}$ for every "multi-index" $\alpha = (\alpha_0, \dots, \alpha_{r-1})$.

In multi-dimensions, in contrast to the one-dimensional case, the notion of micro-differential operators is indeed *microlocal*, i.e., depends on the *co-direction* (= ray in the cotangent space). If one chooses $dx_0 = (1, 0, \dots, 0)$ as such a co-direction, an algebraic definition of the ring $\mathscr{E} = \mathscr{E}_{\mathscr{C}[[x]]}$ of micro-differential operators becomes

$$\mathscr{E} := \left\{ \sum_{|\alpha| \ll \infty} a_\alpha(x)\partial^\alpha ; a_\alpha(x) \in \mathscr{C}[[x]], \alpha \in \mathbf{Z} \times \mathbf{N}^{r-1} \right\}.$$

Note that the leading entry α_0 of $\alpha = (\alpha_0, \dots, \alpha_{r-1})$ now ranges over both positive and negative integers. $\mathscr{E}$ is thus an extension of $\mathscr{D}$ allowing negative powers of ∂_0. The rules of addition and multiplication are

$$\sum a_\alpha \partial^\alpha + \sum b_\alpha \partial^\alpha := \sum (a_\alpha + b_\alpha)\partial^\alpha,$$
$$\sum a_\alpha \partial^\alpha \cdot \sum b_\beta \partial^\beta := \sum c_\gamma \partial^\gamma,$$
$$\text{where } c_\gamma := \sum_{\alpha \in \mathbf{Z} \times \mathbf{N}^{r-1}, \nu \in \mathbf{N}^r} \binom{\alpha}{\nu} a_\alpha \partial^\nu (b_{\gamma - \alpha + \nu})$$

where

$$\binom{\alpha}{\nu} := \binom{\alpha_0}{\nu_0} \cdots \binom{\alpha_{r-1}}{\nu_{r-1}}.$$

With these structures $\mathscr{E}$ becomes a ring extension of $\mathscr{D}$.

The maximal degree of derivatives included in an operator is called the *order* of that operator. This gives rise to a filtration in both $\mathscr{D}$ and $\mathscr{E}$:

$$\mathscr{E}^{(m)} := \{A \in \mathscr{E}; \mathrm{ord}(A) \leq m\} \subset \mathscr{E},$$
$$\mathscr{D}^{(m)} := \{A \in \mathscr{D}; \mathrm{ord}(A) \leq m\} \subset \mathscr{D}.$$

3.2. Families of $\mathscr{D}$-modules and vector spaces. Left $\mathscr{D}$-modules of $\mathscr{E}$ to be deformed are those that satisfy the condition

$$\mathscr{E} = \mathscr{I} \oplus \mathscr{E}_0,$$

where $\mathscr{E}_0$ is a free left $\mathscr{C}[[x]]$-submodule of $\mathscr{E}$. The setting for the KP hierarchy is such that $r = 1$ and $\mathscr{E}_0 = \mathscr{E}^{(-1)}$. In a multi-dimensional case there is a priori no standard choice of $\mathscr{E}_0$; there are rather an infinite number of possibilities. The condition above is retained under small perturbations of $\mathscr{I}$.

As a source vector space for constructing a Grassmann manifold, we take

$$\mathbf{V} := \mathscr{E}/(\mathscr{E}x_0 + \mathscr{E}x_1 + \cdots + \mathscr{E}x_{r-1}),$$

which also has an $\mathscr{E}$-module structure. As a left $\mathscr{E}$-module, $\mathbf{V}$ is cyclic with generator $1 \bmod \mathscr{E}x_0 + \cdots + \mathscr{E}x_{r-1}$, which may be identified with the delta function $\delta(x) = \delta(x_0, \ldots, x_{r-1})$. Its derivatives $\partial^\alpha \bmod \mathscr{E}x_0 + \cdots + \mathscr{E}x_{r-1}$ $(= \delta^{(\alpha)}(x))$, $\alpha \in \mathbf{Z} \times \mathbf{N}^{r-1}$, give a basis as a $\mathscr{C}$-vector space. Left $\mathscr{D}$-submodules $\{\mathscr{I}\}$ of $\mathscr{E}$ as above are in one-to-one correspondence with vector subspaces $\{U\}$ of $\mathbf{V}$ of the "same size" as some reference subspace $U_\emptyset$. One can give a more stringent formulation of "being the same size" just as we discussed in the second lecture.

We now illustrate the ideas above in a somewhat special case, which is however very interesting in itself. It is a multi-dimensional extension of the notion of "quasi-periodic solutions" of soliton equations, i.e., those which are written in a closed form with theta functions and Abelian integrals.

3.3. Special solutions in KP hierarchy. Let us recall how such special solutions are characterized in the theory of KP hierarchy. General (or generic) solutions of the KP hierarchy are parametrized by generic points of UGM. Special solutions, accordingly, will live in a submanifold of UGM which is stable under the dynamical motions of the KP hierarchy. To classify these stable submanifolds, we consider $\mathscr{E}_{\mathrm{const}} := \{\sum c_i (d/dx)^i \in \mathscr{E}; c_i \in \mathscr{C}\}$ and its subrings $\{\mathscr{A}\}$ under the following condition:

$$\mathscr{A} \cap \mathscr{E}^{(0)}_{\mathrm{const}} = \mathscr{C},$$

where $\mathscr{E}_{\mathrm{const}}$ is understood to have a filtration $\{\mathscr{E}^{(n)}_{\mathrm{const}}; n \in \mathbf{Z}\}$ induced from that of $\mathscr{E}$. With the correspondence

$$\left(\frac{d}{dx}\right)^{-1} \leftrightarrow z$$

one has an isomorphism $\mathscr{E}_{\text{const}} \simeq \mathscr{C}((z)) = \mathscr{C}[[z]][z^{-1}]$ (the ring of formal Laurent series in a formal variable z), which is to be used to give a geometric interpretation of the present construction. The condition on $\mathscr{A}$ above then reads

$$\mathscr{A} \cap \mathscr{C}[[z]] = \mathscr{C}.$$

Since $\mathscr{E}_{\text{const}}$ gives rise to infinitesimal action on UGM (at the level of matrices, $\mathscr{A}(d/dx)^n$ corresponds to Λ^n), one may consider the subset UGM $\subset$ UGM consisting of points which are fixed under the action of $\mathscr{A}$. In the language of vector subspaces of $\mathbf{V}$,

$$\text{UGM}^{\mathscr{A}} = \left\{ U \in \text{UGM}; f(\Lambda)U \subset U \quad \text{for } \forall f\left(\frac{d}{dx}\right) \in \mathscr{A} \right\}.$$

Evidently this is stable under the infinitesimal action of Λ^n for $n \in \mathbf{Z}$ which, for $n > 0$, are infinitesimal generators of "time evolutions" of the KP dynamical system on UGM. A number of interesting families of solutions can be singled out through this construction. (Note that we also take into account the infinitesimal actions of nonpositive powers of Λ; this becomes crucial in the treatment of quasi-periodic solutions.)

For example, if $\mathscr{A} = \mathscr{C}[z^{-2}], \mathscr{C}[z^{-3}], \ldots$, $\text{UGM}^{\mathscr{A}}$ can be accordingly identified with the solution spaces of the KdV equation, the Boussinesq equation, etc. These are examples of the case where $\mathscr{A}$ is not "very large". If $\mathscr{A}$ is "large" in the sense that $g := \dim \mathscr{C}((z))/(\mathscr{A} + \mathscr{C}[[z]])$ is finite, $\mathscr{A}$ becomes the affine ring $\mathscr{O}(X \backslash p_\infty)$ of an algebraic curve (defined over $\mathscr{C}$) of "genus" g, p_∞ being a point "at infinity". $\text{UGM}^{\mathscr{A}}$ then parametrizes quasi-periodic solutions (if X is a nonsingular algebraic curve over $\mathscr{C} = \mathbf{C}$), soliton or rational solutions (if X is singular and its normalization is rational), etc. A typical case is elliptic solutions of the KP hierarchy, which one can obtain in the following situation:

$$\mathscr{A} = \mathbf{C}[\wp(z), \wp'(z)], \mathscr{C} = \mathbf{C}, \wp(z) = \text{Weierstrass } \wp \text{ function}.$$

3.4. Multi-dimensional analogues. The basic program as above will remain valid in multi-dimensional cases as well. One starts again from the ring $\mathscr{E}_{\text{const}}$ of micro-differential operators with constant coefficients, which is now written

$$\mathscr{E}_{\text{const}} = \mathscr{C}[[\partial_0^{-1}, \partial_0^{-1}\partial_1, \ldots, \partial_0^{-1}\partial_{r-1}]][\partial_0].$$

With the correspondence

$$\partial_0^{-1} \leftrightarrow z_0, \qquad \partial_0^{-1}\partial_i \leftrightarrow z_i \quad (1 \le i < r)$$

one obtains an isomorphism

$$\mathscr{E}_{\text{const}} \simeq \mathscr{C}[[z]][z_0^{-1}], \qquad z = (z_0, \ldots, z_{r-1}).$$

This isomorphism also respects filtrations; the filtration in $\mathscr{E}_{\text{const}}$ is the one induced from $\mathscr{E}$ as:

$$\mathscr{E}_{\text{const}}^{(n)} := \mathscr{E}_{\text{const}} \cap \mathscr{E}^{(n)},$$

whereas the corresponding one in $\mathscr{C}[[z]][z_0^{-1}]$ measures the order of poles or zeros along $z_0 = 0$:

$$\mathscr{C}[[z]][z_0^{-1}]^{(n)} := \mathscr{C}[[z]]z_0^{-n}.$$

Filtrations play much more important roles than in the one-dimensional case.

What one has to do next is to choose a subring $\mathscr{A} \subset \mathscr{E}_{\text{const}}$ to form the stable subset $\text{UGM}^{\mathscr{A}} \subset \text{UGM}$. Actually, up to now, this program has been worked out only in such a very limited case as

$$(*) \qquad \begin{cases} \mathscr{A} = \mathscr{O}(X\backslash H), \\ X = r\text{-dimensional nonsingular projective variety}, \\ H = \text{ample divisor in } X. \end{cases}$$

A typical example is the principally polarized Abelian variety, H being the theta divisor. This can be thought of as a multi-dimensional analogue of the elliptic case mentioned above.

In such a case $X\backslash H$ becomes an affine algebraic variety, $\mathscr{A}$ being a coordinate ring. $\mathscr{A}$ has a natural filtration

$$\mathscr{A}^{(n)} := \{f \in \mathscr{A};\ f \text{ has poles of order } \leq n \text{ along } H\},$$

which carries information on the compactification (completion) of this affine variety. In the multi-dimensional theory such a filtration plays in general a far more crucial role than in the one-dimensional case. We shall see that even $\mathscr{D}$-modules to be deformed and associated points of some Grassmann manifold must respect a certain filtration structure. From the assumption, evidently, $\mathscr{A}^{(0)} = \mathscr{C}$ and $\chi(n) := \dim_{\mathscr{C}} \mathscr{A}^{(n)} < \infty$ for every n. It is well known that in an appropriate situation, $\chi(n)$ for $n \geq n_0$ (n_0 is a nonnegative integer) agree with the values of a polynomial called the "Hilbert polynomial". This is one of the most basic invariants for such a filtered algebra.

In addition to these global data, one has to choose several local data, which consist of a regular point p_∞ of H and local coordinates $z = (z_0, \ldots, z_{r-1})$ around p_∞ for which H is defined by the equation $z_0 = 0$ in a neighborhood of p_∞. With these local data one can embed $\mathscr{A}$ into $\mathscr{E}_{\text{const}}$ as follows:

$$\mathscr{A} \hookrightarrow \mathscr{E}_{\text{const}} \simeq \mathscr{C}[[z]][z_0^{-1}],$$

assigning to each element f of $\mathscr{A}$ its Laurent expansion along H in a neighborhood of p_∞. This embedding also preserves the filtration structure.

Having obtained $\mathscr{A}$ as a subring of $\mathscr{E}_{\text{const}}$, one can now consider $\text{UGM}^{\mathscr{A}}$. Through the natural isomorphism

$$\mathbf{V} \simeq \mathscr{E}_{\text{const}}, \qquad \partial^\alpha \bmod \mathscr{E}x_0 + \cdots + \mathscr{E}x_{r-1} \leftrightarrow \partial^\alpha$$

one can regard $\mathscr{A}$ as a $\mathscr{C}$-vector subspace of $\mathbf{V}$, which evidently belongs to $\text{UGM}^{\mathscr{A}}$. In the following we understand $\text{UGM}^{\mathscr{A}}$ as the set of vector subspaces of $\mathbf{V}$ which are invariant under the infinitesimal action of $\mathscr{A}$ and which have the "same size" as $\mathscr{A}$ itself. An important family of such vector subspaces are those consisting of "projective" $\mathscr{A}$-modules $\{U\}(U \subset \mathbf{V} \simeq \mathscr{E}_{\text{const}})$

of rank one. Geometrically, this is equivalent to considering line bundles over X. Thus we are naturally led to the deformations of line bundles.

For example, if X is a principally polarized Abelian variety defined over $\mathscr{C} = \mathbf{C}$, one can exploit the theory of theta functions to carry out more detailed computations. If H is defined by the theta divisor $\theta = 0$, both $\mathscr{A}$ and projective $\mathscr{A}$-modules $\{U\}$ of rank one can be explicitly constructed as theta quotients, the denominators being the powers θ^n, $n \geq 0$. The order of poles at $H\colon \theta = 0$ gives rise to a filtration $\{U^{(n)}; n \geq 0\}$ of U. Again the integer sequence $\dim U^{(n)}$ includes basic information on the size of U.

As elements of an infinite-dimensional Grassmann manifold, thus, one has to consider only those which carry some filtration as an additional structure. Admissible dynamical motions should accordingly take place only under the condition that the filtration structure be retained throughout. The presence of such a filtration appears to be unavoidable to put the whole deformation theory under good control.

A detailed analysis in the case of principally polarized Abelian varieties, as a by-product, also suggests an interesting fact. Namely it turns out that in a good situation (such as the case of principally polarized Abelian varieties) any projective $\mathscr{A}$-module, $\mathscr{A}U \subset U$, of rank one becomes a $\mathscr{D}_{\mathscr{A}}$-module, $\mathscr{D}_{\mathscr{A}}U \subset U$, where $\mathscr{D}_{\mathscr{A}}$ denotes the ring of differential operators in $\mathscr{A}$ generated by $\mathscr{A}$ and $\mathrm{Der}_{\mathscr{C}}\mathscr{A} = \{\text{derivatives}\colon \mathscr{A} \to \mathscr{A}\}$. In such a case U is effectively a "holonomic" $\mathscr{D}_{\mathscr{A}}$-module, and this implies a possibility of determining the structure of solutions explicitly with the aid of differential equations that the $\mathscr{D}$-module structure represents. The case of polarized Abelian varieties indeed provides an example advocating such a hope. This method appears to be useful in general cases as well.

REFERENCES

1. M. Sato, *Soliton Equations as Dynamical Systems on an Infinite Dimensional Grassmann Manifold*, RIMS Kokyuroku **439** (1981), 30–46.

2. M. Sato and Y. Sato, *Soliton Equations as Dynamical Systems on Infinite Dimensional Grassmann Manifold*, in Proc. U.S.-Japan seminar, *Nonlinear Partial Differential Equations in Applied Science*, Tokyo 1982, P. D. Lax and H. Fujita, eds., pp. 259–271, North-Holland/Kinokuniya, 1982.

3. M. Sato, T. Kawai, and M. Kashiwara, *Microfunctions and pseudo-differential equations*, Lecture Notes in Math. **287**, pp. 265–529, Springer, 1973.

RESEARCH INSTITUTE FOR MATHEMATICAL SCIENCES, KITASHIRAKAWA, SAKYO-KU, KYOTO-SHI 606, JAPAN

No. 1] Proc. Japan Acad., **51** (1975) 17

5. *The Determinant of Matrices of Pseudo-differential Operators*

By Mikio SATO*) and Masaki KASHIWARA**)

(Comm. by Kôsaku YOSIDA, M. J. A., Jan. 13, 1975)

The purpose of this paper is to give a definition of the determinant of matrices of pseudo-differential operators (of finite order) and to establish some of its properties. Let X be a complex manifold, and P^*X (resp. T^*X) be its cotangent projective (resp. vector) bundle. The projection from T^*X-X onto P^*X is denoted by γ.

Our result is the following.

Theorem. *For every matrix* $A(x,D)=(A_{ij}(x,D))_{1\le i,j\le N}$, *whose entries* $A_{ij}(x,D)$ *are pseudo-differential operators defined on an open set* $U\subset P^*X$, *one can canonically associate* $\det A(x,D)$, *which is a homogeneous holomorphic function defined on* $\gamma^{-1}(U)$, *and possesses the following properties*

a) $\det A(x,D)B(x,D)=\det A(x,D)\cdot\det B(x,D)$

b) $\det(A(x,D)\oplus B(x,D))=\det A(x,D)\cdot\det B(x,D)$

c) *if there are integers* m_i *and* n_j *such that order* $A_{ij}(x,D)\le m_i+n_j$ *and* $\det(\sigma_{m_i+n_j}(A(x,D))$ *does not vanish identically, then*

$$\det A(x,D)=\det(\sigma_{m_i+n_j}(A_{i,j})),$$

where $\sigma_{m_i+n_j}(A_{ij})$ *denotes the principal symbol of* A_{ij} (*which is* 0 *if* A_{ij} *is of the order* $\le m_i+n_j-1$). *In particular, our determinant reduces to the concept of the principal symbol, if the size* N *is* 1.

d) $A(x,D)$ *is invertible if and only if* $\det A(x,D)$ *vanishes nowhere.*

e) *if* $P(x,D)$ *is a pseudo-differential operator such that* $[P,A]=0$, *then* $\{\sigma(P),\det A\}=0$.

Corollary. *If* $A(x,D)$ *is a matrix of differential operators, then* $\det A(x,D)$ *is a homogeneous polynomial on the fiber coordinate* ξ.

Corollary is an immediate consequence of Theorem. In fact, by adding an auxiliary parameter t, one can regard $A(x,D)$ as a pseudo-differential operator defined on a (t,x)-space $C\times X$. Therefore, $\det A(x,D)$ is defined all over T^*X, which implies $\det A(x,D)$ is a polynomial on ξ.

In order to prove Theorem, we prepare the following lemma.

Lemma (see [2]). *Let* K *be a (not necessarily commutative) field,* $K=\bigcup_{m\in Z}K_m$ *be a filtration of* K *satisfying*

*) Research Institute of Mathematical Sciences, Kyoto University.
**) Department of Mathematics, Faculty of Sciences, Nagoya University.

1) *The intersection of all* K_m *is zero,*
2) $K_m \subset K_{m+1}$, *and* $K_1 \neq K_0$,
3) K_m *is closed under addition,*
4) $K_{m_1}K_{m_2} \subset K_{m_1+m_2}$,
5) $[K_{m_1}, K_{m_2}] \subset K_{m_1+m_2-1}$,
6) *If* α *does not belong to* K_m, *then* α^{-1} *belongs to* K_{-1-m}.

Then $k=K_0/K_{-1}$ *is a commutative field and* $L=K_1/K_0$ *is a vector space over* k *of dimension* 1 *and* $K_m/K_{m-1}=L^{\otimes m}$. *The canonical homomorphism from* K_m *to* $L^{\otimes m}$ *is denoted by* σ_m. *Then, there is a map* $\det : M(n:K) \to \oplus_{m\in \mathbf{Z}} L^{\otimes m}$ *satisfying*

a) $\det(AB)=\det A \det B$.

b) $\det(A\oplus B)=\det A \det B$.

c) *If there are integers* m_i *and* n_j *such that* $A_{ij} \in K_{m_i+n_j}$ *and* $\det(\sigma_{m_i+n_j}(A_{ij}))$ *is non zero, then* $\det(A_{ij})=\det(\sigma_{m_i+n_j}(A_{ij}))$.

d) $\det A \neq 0$ *if and only if* A *is invertible.*

e) *If* $\alpha \in K$ *centralizes a matrix* A, *then* $\sigma(\alpha)$ *centralizes* $\det A$.

Since this is a purely algebraic lemma, we omit its proof.

Now, let p be a point in P^*X. Let K be a quotient field of a stalk $\mathcal{P}^f_{X,p}$ of $\mathcal{P}^f_X$ at p. Then the canonical filtration of $\mathcal{P}^f_X$ defined by order induces a filtration of K. Then $k=K_0/K_{-1}$ is a field of germs of meromorphic functions at p, and L is a set of germs of homogeneous meromorphic functions of order 1 at p. Thus, we can define $\det A(x, D)$ as a homogeneous *meromorphic* functions defined on $\gamma^{-1}(U)$.

Proposition. $\det A(x, D)$ *is a holomorphic function.*

Proof. We will prove this by the induction on the size of $A(x, D)$. Levi's theorem says that a meromorphic function is holomorphic if it is holomorphic except on a 2-codimensional analytic set. Therefore, it suffices to prove that $\det A(x, D)$ is holomorphic outside a 2-codimensional set. We may assume $\det(A(x, D))$ is holomorphic except on a non singular hypersurface $f=0$, and $(\sum \xi_i dx_i)|_{f^{-1}(0)} \neq 0$. By a quantized contact transformation, we can set $f=\xi_1$. Let r_{ij} be a multiplicity of $\sigma(A_{ij})$ at $\{\xi_1=0\}$. Set $r=\min(r_{ij})$. We prove the proposition by the induction of r.

Without loss of generality, we may assume $r_{11}=r$, and $\sigma(A_{11})/\xi_1^r$ never vanishes by Levi's theorem.

By Späth's theorem, A_{ij} has the form

$$A_{1j}=A_{11}Q_j+R_j$$

where

$$R_j=\sum_{\nu<r} R_{j,\nu}(x, D')D_1^\nu \quad (\text{where } D'=(D_2, \cdots, D_n)).$$

Therefore

$$A(x,D)=\begin{bmatrix} A_{11}, R_2, \cdots\cdots\cdots R_N \\ A_{21}, A_{22}-A_{21}Q_2, \cdots \\ \cdots\cdots\cdots\cdots\cdots\cdots \\ \cdots\cdots\cdots\cdots\cdots\cdots \end{bmatrix}\begin{bmatrix} 1 & Q_2 & \cdots & Q_N \\ & 1 & & \\ & & \ddots & \\ & & & 1 \end{bmatrix}.$$

Setting the first matrix of the right hand side $\tilde{A}(x,D)$, we have $\det \tilde{A}(x,D)=\det A(x,D)$. If one of R_j is non zero, the multiplicity of some $\sigma(R_j)$ at $\{\xi_1=0\}$ is strictly less than r. Therefore the hypothesis of the induction implies $\det \tilde{A}(x,D)=\det A(x,D)$ is holomorphic. If all R_j are zero, $\det \tilde{A}$ is the product of $\sigma(A_{11})$ and the determinant of an $(N-1)\times(N-1)$ matrix of pseudo-differential operators. In this case, also, the hypothesis of induction on the size again implies that $\det A(x,D)$ is holomorphic. q.e.d.

Since Property (d) is proved in the same argument, we omit its proof.

Example.

$$A(x,D)=\begin{bmatrix} xD+\alpha(x) & D^2+\beta(x)D+\gamma(x) \\ x^2 & xD+\delta(x) \end{bmatrix}$$

In this case,

$$\det A(x,D)=\begin{cases} (\alpha+\beta-1-x\gamma)\delta & \text{if it is not zero} \\ \alpha\beta-2\beta+x\beta'-x^2\delta & \text{if } \alpha+\beta-1-x\gamma=0. \end{cases}$$

In fact,

$$A(x,D)=\begin{bmatrix} 1 & \\ & x^2 \end{bmatrix}\begin{bmatrix} 1 & xD+\alpha \\ & 1 \end{bmatrix}\begin{bmatrix} 0 & Q \\ 1 & xD+\beta+2 \end{bmatrix}\begin{bmatrix} 1 & \\ & x^{-2} \end{bmatrix},$$

where $Q=(1-\alpha-\beta+x\gamma)xD+(2+2\gamma x+x^2\delta-\alpha\beta-2\alpha-x\beta')$.

Example. Let $A=(A_{ij})_{1<i,j\leq 2}$. If $A_{21}\neq 0$ $\det A=\sigma(A_{21})\sigma(A_{11}A_{21}^{-1}A_{22}-A_{12})$. If $A_{11}\neq 0$ $\det A=\sigma(A_{11})\sigma(A_{22}-A_{21}A_{11}^{-1}A_{12})$.

References

[1] M. Sato, T. Kawai, and M. Kashiwara: Microfunctions and Pseudo-differential Equations. Lecture note in Math., No. 287, Springer, Berlin-Heidelberg-New York, pp. 265–529 (1973).

[2] E. Artin: Geometric Algebra. Interscience (1957).

Advanced Studies in Pure Mathematics 19, 1989
Integrable Systems in Quantum Field Theory and Statistical Mechanics
pp. 417–434

$\mathcal{D}$-Modules and Nonlinear Systems

M. Sato

§1. Tschirnhaus transformations for algebraic systems

From the start of my research in analysis, I made some programs about how to organize such concepts like functions, generalized functions, differential equations, both linear and nonlinear, and all about that. My first systematic talk about this subject was given in july 1960. Since then, as is well known, much progress was made, at least, in the field of a general theory of linear partial differential equations by means of the concept of $\mathcal{D}$-modules and specialized concept of holonomic systems. They have several applications made by Kashiwara, Kawai and others. But my original program was just to develop the theory of nonlinear equations in the same spirit. So I shall give a brief sketch about it.

First recall the special case of algebraic geometry, that is, the concept of manifolds, vector bundles and things like that which live on a manifold. All these things are presented and studied systematically by means of a commutative ring and modules over it. As already pointed out by René Descartes, geometrical objects like curves, surfaces and others are described by means of algebraic equations like

$$f_i(x) = 0, \qquad 0 \le i \le n,$$

where x denotes points of the ambient linear space. In particular in the case of one variable we have the equation

$$f(x) = 0. \tag{1}$$

This is an algebraic equation in one indeterminate. Many studies were done for such equations in the past. Especially a first systematic study was made by Tschirnhaus in 17-th century. He is a friend of Spinoza. He

Received January 30, 1989.

*Lecture delivered at RIMS, Kyoto, October 18, 1988. Notes taken by A. Nakayashiki.

gave the following general idea to study such an algebraic equation, say cubic equation, quartic equation and other more complicated equations. Introduce a new variable y by

$$y = \varphi(x), \tag{2}$$

where φ may be a polynomial or can be a rational function in x. If this is a rational function, we simply multiply the equation by the denominator. Then we get an equation which is a polynomial in x and whose coefficients are linear in y. Then regarding equations (1) and (2) as a system of two algebraic relations in two variables x and y, we eliminate x from these. Then the resultant equation is of the form

$$g(y) = 0. \tag{3}$$

In this way equation (1) is transformed to (3). This process of transforming an algebraic equation is called Tschirnhaus transformation. If we can find a suitable Tschirnhaus transformation so that the resultant equation–the algebraic equation in y–is simpler than the original one, then this means that we achieved some progress in solving equation (1). Here let me just change the view point. Suppose that in particular, not only equation (2) of y but we have an expression like

$$x = \psi(y), \tag{4}$$

where ψ is a rational function in y. I do not mean that we can always do this. Then, in that case, by eliminating y from equations (3) and (4), we obtain the original equation (1). So in that sense, the sets of equations (1), (2) and (3), (4) are mutually equivalent. Now let us make a trivial generalization of this and go to the case of several variables. Then we have not a single variable but a set of variables, say $x = (x_0, \cdots, x_{n-1})$. And we have a number of equations of the form

$$f_i(x) = 0, \qquad 0 \leq i < N. \tag{5}$$

In the case of single variable (1), the equation means a finite number of points in a geometrical picture. But here these equations represent some algebraic variety. Now introduce a number of new indeterminates by the following rational expressions:

$$y_\mu = \varphi_\mu(x), \qquad 0 \leq \mu < m. \tag{6}$$

Then by making the elimination process, which is always possible, we find a number of equations of the form:

$$g_j(y) = 0, \qquad j = 0, 1, 2, \cdots. \tag{7}$$

Here the number of equations are always finite by Hilbert basis theorem. Suppose we have an expression of the form

$$x_\nu = \psi_\nu(y), \qquad 0 \leq \nu < n. \tag{8}$$

Then the equations (5) and (7) are mutually equivalent. The same algebraic variety is defined by means of equations (5) and (7). The only difference is the ambient space. In the equation (5) it is x-space and y-space in (7). In other words, the variety is the same and only the embedding is different. This means that we should consider the commutative ring

$$A = \mathcal{C}[x_0, \cdots, x_{n-1}]/\mathcal{I}.$$

Here $\mathcal{C}$ is a commutative ring or field which represents constants, and $\mathcal{I}$ is an ideal in A generated by $f_i(x)$'s:

$$\mathcal{I} = (f_0(x), \cdots).$$

This ring A represents a set of all regular functions on the algebraic variety in consideration. So this is the algebraic representation of the variety. The variety is a geometrical object and this is equivalently represented by a commutative ring A. They are mutually contragredient categories, that is, a morphism in the side of a variety is represented by the homomorphism in the opposite direction in the ring side. In this way, we have a one-to-one correspondence between rings and geometrical objects.

If we choose any number of elements $y_0, \cdots, y_{m-1}$ of A, then this will certainly generate a subring B of A. Of course B is not necessarily free and can be written as

$$B = \mathcal{C}[z_0, \cdots, z_{m-1}]/\mathcal{I}',$$

where $\mathcal{I}' = (g_0, \cdots)$. Now the special situation where $\mathcal{C}[x_0, \cdots, x_{n-1}]/\mathcal{I}$ and B are equivalent, is the case where $y_0, \cdots, y_{m-1}$ can generate the whole ring A, that is, $A = B$. So this means that Tschirnhaus transformation simply means the change of generators within a commutative ring. Let us just use the term "Tschirnhaus transformation" in this wider sense, that is, a Tschirnhaus transformation is just a change of generators for a given algebraic system. So suppose we have some object consisting of elements for which we can perform algebraic operations like polynomial operations. They are of course a kind of algebraic systems. Then we can talk about the change of generators which we shall call Tschirnhaus transformation in that system. So the same can be applied, for instance, to differential equations. In linear or non-linear differential equations we have a lot of systems of equations. But this is just one representation of some single entity.

First we consider the case of linear differential equations. We define

$$\mathcal{D} = \Big\{ \sum_\alpha a_\alpha(x) \Big(\frac{\partial}{\partial x}\Big)^\alpha \Big| \alpha = (\alpha_0, \cdots, \alpha_{n-1}) \in \mathbf{N}^n \Big\}$$

$$= \{\text{linear differential operators}\},$$

$$\Big(\frac{\partial}{\partial x}\Big)^\alpha = \Big(\frac{\partial}{\partial x_0}\Big)^{\alpha_0} \cdots \Big(\frac{\partial}{\partial x_{n-1}}\Big)^{\alpha_{n-1}},$$

where coefficients $a_\alpha(x)$'s belong to analytic functions or other classes of functions for which differentiations make sense. Then a single linear differential equation can be written as

$$Pu = 0$$

with $P \in \mathcal{D}$, where u is an unknown function. I am just writing it in an explicit form but, of course, we can as well define the same concept of differential operators in a more abstract context like differential operators on an algebraic or an analytical manifold.

In a more general situation we have

$$\sum_{0 \le \kappa < k} P_{i\kappa} u_\kappa = 0, \qquad 0 \le i < l, \tag{9}$$

where $\{u_\kappa\}$ is a system of unknown functions. This is a system of partial differential equations for several unknown functions. Again here we have the concept of Tschirnhaus transformation. We now introduce

$$v_\lambda = \sum_{0 \le \kappa < k} A_{\lambda\kappa} u_\kappa, \qquad A_{\lambda\kappa} \in \mathcal{D}, \quad 0 \le \lambda < N. \tag{10}$$

This is a change of unknown functions. If we define new unknown functions by the above equations, then, by the process of elimination, we get some linear differential equations satisfied by these new variables:

$$\sum_{0 \le \kappa < k} Q_{j\lambda} v_\lambda = 0, \qquad 0 \le j < l'. \tag{11}$$

Suppose now that, in particular, opposite process of solving these equations with respect to u_κ's is possible, that is,

$$u_\kappa = \sum_{0 \le \lambda < N} B_{\kappa\lambda} v_\lambda, \qquad B_{\kappa\lambda} \in \mathcal{D}, \quad 0 \le \kappa < k. \tag{12}$$

In this case, again this means that (9) and (10) are mutually equivalent. These two equations are apparently different expressions for the same entity of the equations in the intrinsic sense. This entity can most naturally be called a $\mathcal{D}$-module. Now I explain what this is like. We first construct

$$\mathcal{D}u_0 \oplus \cdots \oplus \mathcal{D}u_{k-1} \simeq \mathcal{D}^k,$$

the direct sum of k-copies of $\mathcal{D}$, which we consider a left $\mathcal{D}$-module. Dividing this by a left $\mathcal{D}$-module $\mathcal{I}$, where $\mathcal{I}$ is defined as

$$\mathcal{I} = \mathcal{D} \sum_{0 \le \kappa < k} P_{0\kappa} u_\kappa + \mathcal{D} \sum_{1 \le \kappa < k} P_{0\kappa} u_\kappa + \cdots,$$

we obtain a left $\mathcal{D}$-module $\mathcal{M}$:

$$\mathcal{M} = \mathcal{D}u_0 \oplus \cdots \oplus \mathcal{D}u_{k-1} / \mathcal{I}.$$

Conversely whenever we have a left $\mathcal{D}$-module $\mathcal{M}$ which is finitely generated over $\mathcal{D}$, then we can choose a finite number of generators to be

$$\mathcal{M} = \mathcal{D}v_0 \oplus \cdots \oplus \mathcal{D}v_{k'-1}.$$

Therefore it must be a quotient of a free module $\mathcal{D}^{k'}$ by a certain submodule. This submodule represents fundamental relations between generators $v_0, \cdots, v_{k'-1}$. So unknown functions are just generators of a $\mathcal{D}$-module and linear differential equations are just fundamental relations between generators. The Tschirnhaus transformation, in this case, is again a change of generators of $\mathcal{D}$-module $\mathcal{M}$. Consequently this means the change of fundamental relations between generators. Of course the same can be applied to non-linear equations. To go to that, let us remind that any geometrical object should be represented by the algebraic concept, like a commutative ring or a C*-algebra in Gel'fand's representation theory, etc., corresponding to various situations. Anyway likewise, here I just explained that the concept of linear partial differential equations is conveniently and naturally be represented by $\mathcal{D}$-modules. All concepts related linear differential equations can be interpreted in terms of $\mathcal{D}$-modules. For example, solving an equation means finding a homomorphism of $\mathcal{M}$ into a certain natural well known $\mathcal{D}$-module such as the $\mathcal{D}$-module of functions, generalized functions, etc.

I have already sketched how the concept of an algebraic system, like a commutative ring or a $\mathcal{D}$-module, represents an algebraic variety or a differential equation. In the commutative algebra case, consider a commutative ring A which represents a variety. Now a quite natural object attached to A is the category of A-modules. The totality of A-modules constitutes a so-called Abelian category and the study of A itself can be carried out by means of the study of this category. In particular if we have a special element in this category, say a finite projective A-module, then it represents a vector bundle on the algebraic variety corresponding to A. Even if the module, say $\mathcal{M}$, is not necessarily projective but coherent, that is, $\mathcal{M}$ is finitely generated and has finite relations, then it may also represent some geometrical object. It is generally a vector bundle but, on certain subvarieties, is degenerate. And again on that subvarieties, generically it represents a vector bundle and so on. So we have some hierarchy of vector bundles attached, in a rather complicated way, to such subvarieties. But anyway, such concept as A-module represents some linear object attached to that non-linear object, that is, the algebraic variety. Here I first discussed about linear equations, but this should rather be compared with $\mathcal{M}$ or A-modules. If A is $\mathcal{C}$ itself, then this is 0-dimensional and this represents just a single point. If we take an algebraic extension to $\mathcal{C}$, then it represents a finite number of points. In this 0-dimensional case, a $\mathcal{C}$-module is nothing but a vector space, so is just a linear algebra. Therefore algrbraic geometry, especially the category of A-modules, is just a generalization of usual linear algebra of vector spaces. It represents just a deformation of linear algebra along the variety. Here I am just talking about $\mathcal{D}$-modules which just correspond to such $\mathcal{C}$-modules, that is, vector spaces. I will now explain how non-linear partial differential equations are described by means of algebraic concepts and how associated linear systems can be viewed as a deformation of linear differential equations, that is, $\mathcal{D}$-modules.

§2. Non-linear equations as non-commutative algebras

Now we start from a non-commutative associative algebra R over certain field of constants $\mathcal{C}$. A non-commutative ring, if no restriction is imposed, is usually a very, very wild concept. To make it more tame, so to speak, we asuume the following conditions:

$$(13)_1 \qquad R = \bigcup_{m \in \mathbf{N}} R^{(m)}, \qquad R^{(m)} = 0 \quad \text{if } m < 0,$$

$$(13)_2 \qquad R^{(m)} : \mathcal{C}\text{-subspace of } R,$$

$$(13)_3 \qquad R^{(0)} \subset R^{(1)} \subset R^{(2)} \subset \cdots,$$

$$(14) \qquad R^{(m)} R^{(n)} \subset R^{(m+n)},$$

$$(15) \qquad [R^{(m)}, R^{(n)}] \subset R^{(m+n-1)}.$$

The conditions $(13)_1$–$(13)_3$ and (14) mean that R is filtered. The condition (14) means that, if we call elements of $R^{(m)}$ the operators of order at most m, the product of two operators of order at most m and n is an operator of order at most $m+n$. The condition (15) means that the products PQ and QP of operators P and Q, are approximately the same, that is, the highest order parts are the same. So R is quasi-commutative, so to speak. These conditions make our non-commutative ring rather tame. As a special case, the condition (15) means

$$[R^{(1)}, R^{(1)}] \subset R^{(1)}, \tag{16}$$

$$[R^{(1)}, R^{(0)}] \subset R^{(0)}, \tag{17}$$

$$[R^{(0)}, R^{(0)}] \subset R^{(-1)} = 0. \tag{18}$$

Hence $R^{(1)}$ has a Lie algebra structure, acts on $R^{(0)}$ through a commutator and $R^{(0)}$ is a commutative subalgebra of R. Imagine the case of a ring of usual linear differential operators $\mathcal{D}$. It satisfies the above conditions $(13)_1$–(15). There $\mathcal{D}^{(0)}$ represents a commutative ring of functions, and $\mathcal{D}^{(1)}$ represents differential operators of the first order, that is, derivations and functions. So this is a way of constructing differential equations in a similar way to algebraic geometry.

Now to give a more concrete picture of the situation, let us consider the special case. Since $R^{(0)}$ is a commutative ring, let us denote it by A. We define Θ by

$$\Theta = R^{(1)}/R^{(0)}.$$

This Θ is also a Lie algebra and acts on A because of the condition (15):

$$[\Theta, \Theta] \subset \Theta,$$
$$[\Theta, A] \subset A.$$

Now we start from a commutative ring A, Lie algebra Θ which is an A-module. We assume that Θ acts on A i0n a natural way as derivations, that is,

$$X(fg) = fX(g) + X(f)g \quad \text{for } f, g \in A \text{ and } X \in \Theta. \tag{19}$$

Further we assume the usual rule of calculation between Θ and A:

$$[X, f] = X(f) \qquad \text{for } f \in A \text{ and } X \in \Theta. \tag{20}$$

Then we can define the universal enveloping algebra $\mathcal{D}$ of Θ in a natural way and $\mathcal{D}$ acts on A, that is, A has a structure of a left $\mathcal{D}$-module. Consider, for example, the case where A is just the space of analytical functions or just of functions on some manifold and Θ is the totality of vector fields or derivations on that manifold. Then Θ is certainly a Lie algebra and also a left A-module. In fact for any $X, X' \in \Theta$, we can construct $aX + a'X' \in \Theta$, where $a, a' \in \Theta$. But Θ is not necessarily a right A-module. Now if we are given a parameter representation of functions (=elements of A), say $a = f(x) \in A$ etc., then any element X in Θ may be written as

$$X = \sum_{i=0}^{n-1} f_i \frac{\partial}{\partial x_i},$$

where f_i's are functions in n variables $x = (x_0, \cdots, x_{n-1})$. This means that we can construct a non-commutative associative algebra $\mathcal{D}$ of linear differential operators with coefficient in A as

$$\mathcal{D} = \Big\{ P = \sum_{\alpha} a_\alpha(x) \Big(\frac{\partial}{\partial x}\Big)^\alpha \mid a_\alpha(x) \in A, \quad \alpha = (\alpha_0, \cdots, \alpha_{n-1}) \in \mathbf{N}^n \Big\}.$$

Let us return to the situation where A is simply a commutative ring on which Θ acts and consequently $\mathcal{D}$ acts, and the assumptions (19) and (20) are satisfied. This is a new starting point of our story. In this construction we can show that $\mathcal{D}$ is filtered and satisfies the conditions (14) and (15). First $\mathcal{D}$ is filtered, because

$$(21)_1 \qquad \mathcal{D}^{(0)} = A,$$

$$(21)_2 \qquad \mathcal{D}^{(1)} = A \oplus \Theta,$$

$$(21)_3 \qquad \mathcal{D}^{(m)} = \underbrace{\mathcal{D}^{(1)} \cdots \mathcal{D}^{(1)}}_{m \text{ times}} \qquad \text{for } m \geq 2.$$

Then by using these expressions, we can show the conditions (14) and (15) by induction on m. So this is a completely well known construction of such models. But there are certainly cases where the given R is not obtained in such a way. For this situation to be the case, it is necessary and sufficient that our ring R is generated by $R^{(1)}$ as a non-commutative ring. If R is finitely generated over A, that is, A and a finite number of elements in R generate R (this is quite a natural assumption), and in addition if these generators can be chosen from $R^{(1)}$, then this is the case. Anyway there are some slight discrepancies between the general formulation of R and this some more specialized formulation. And we have natural cases where such formulation is not adequate. In that case we should start from the above general formulation for R. But the discrepancies are not very big one. So in the following we shall sketch our theory in this specialized framework.

Anyway we start from a commutative ring A which more or less represents a ring of functions. But I want to emphasize that A is not a space of functions in the ordinary sense. It is the space of functions containing both known and unknown functions, so to speak. That is the main difference. Remember that, in the case of algebraic geometry, the simplest case is just $\mathcal{C}$ itself — the field of constants —, in this case it represents one point. If we are given some more complicated ring which is finitely generated over $\mathcal{C}$, then this represents some higher dimensional object containing an infinite number of points. The case $\mathcal{C}$ corresponds to $\mathcal{D}$ in the usual sense, that is, $\mathcal{D}$ which is a ring of differential operators on the given manifold V, whose coefficients are known functions. But our situation is something more complicated. Here A is just a space of functions, both known and unknown. And hence $\mathcal{D}$ is a ring of differential operators whose coefficients contain both known and unknown functions, so to speak. I am just talking in an intuitive language, but mathematically they are just presented by simple assumptions $(13)_1$–(15), (19) and (20).

Now how does (A, Θ), hence $\mathcal{D}$, represent non-linear differential equations ? If A does not have null divisors, we can construct a field of quotient K. If the ground field is of characteristic 0, there is no problem about constructing the enveloping algebra etc., in connection with the extension of A to K. Then it has a meaning to ask the dimensionality of Θ. Set

$$\Theta_K = K \otimes_A \Theta.$$

Now we assume that this is finite dimensional:

$$\dim_K \Theta_K = n < +\infty.$$

This n represents the number of independent variables. So suppose

$$\Theta_K = K\delta_0 \oplus \cdots \oplus K\delta_{n-1}, \tag{22}$$

where δ_i's are in Θ. Assume that Θ acts faithfully on A, we see that there are elements $f_0, \cdots, f_{n-1}$ in A such that

$$\det(\delta_i(f_j)) \neq 0.$$

This means that $\delta_0, \cdots, \delta_{n-1}$ are linearly independent as derivations acting on A. Since $\det(\delta_i(f_j))$ is a non-zero element in K which is a commutative field, we can construct the inverse matrix of $(\delta_i(f_j))$. By performing a linear transformation given by the matrix $(\delta_i(f_j))^{-1}$, we now go to a new basis $\{\delta'_0, \cdots, \delta'_{n-1}\}$. Then in this new basis of Θ_K we see that

$$\delta'_i(f_j) = \begin{cases} 1 & \text{if } i = j, \\ 0 & \text{if } i \neq j. \end{cases}$$

This means that by regarding $\{f_j\}$ as independent variables chosen from A, we can imagine as

$$x_j = f_j,$$

which is a more intuitive notation than f_j. In the following we assume that A is a field, that is, $A = K$, for simplicity. Since the condition $\det(\delta_i(f_j)) \neq 0$ is generic, a general set of n elements can always be chosen to be independent variables. Accordingly we can find derivations which serve as derivations with respect to independent variables. In this picture it does not mean that K is generated by $x_0, \cdots, x_{n-1}$. Let A' denote the subring of K generated by $x_0, \cdots, x_{n-1}$:

$$A' = \mathcal{C}[x_0, \cdots, x_{n-1}].$$

But this is not unique at all. In fact we can choose any non-special set of n elements from K and we can take the subring generated by these elements. Since we already assume that $\mathcal{D}$ is finitely generated, we can find $y_0, \cdots, y_{n-1} \in K$ such that A' and $y_\rho^{(\alpha)}$'s generate K, where

$$y_\rho^{(\alpha)} = \delta_0'^{\alpha_0} \cdots \delta_0'^{\alpha_{n-1}}(y_\rho), \qquad 0 \leq \rho < r, \quad \alpha = (\alpha_0, \cdots \alpha_{n-1}) \in \mathbf{N}^n,$$
$$\delta'_\nu = \frac{\partial}{\partial x_\nu}.$$

So K is not finitely generated in the usual sense but is only finitely generated by admitting an application of an infinite number of derivations as mentioned above. So A is generated by independent variables, dependent variables and their derivatives in a suitable sense. It is usually possible that x_i's, y_j's and $y_\rho^{(\alpha)}$'s are not algebraically independent. This means that there are a lot of relations such that

$$(23) \quad F_i(x_0, \cdots, x_{n-1}, y_0, \cdots, y_{r-1}, \cdots, y_\rho^{(\alpha)} \cdots) = 0, \quad i = 0, 1, 2, \cdots,$$

where F_i's are rational functions in a finite number of indeterminates. These are just defining relations of K. These, as we see, are nothing but non-linear partial differential equations. So a system of partial differential equations is just a representation of defining relations between generators of our structure. Suppose we choose another set of independent variables $x'_0, \cdots, x'_{n-1}$ which can be just arbitrary as far as they satisfy $\det(\delta'_i(x'_j)) \neq 0$. Then accordingly we can choose some additional dependent variables $y'_0, \cdots, y'_{s-1}$. Here n is a fixed number

($=\dim_K \Theta_K$), but s and r can be different. Then we have fundamental relations of these generators:

$$(24)\quad G_i(x_0,\cdots,x_{n-1},y_0,\cdots,y_{r-1},\cdots y_\rho^{(\alpha)}\cdots)=0,\quad i=0,1,2,\cdots,$$

where $y'^{(\alpha)}_\rho = \delta_0''^{\alpha_0}\cdots\delta_0''^{\alpha_{n-1}}(y_\rho)$ and $\delta''_\nu=\frac{\partial}{\partial x'_\nu}$. To go from (23) to (24) we should observe that we can write

$$(25)_1\qquad x'_\nu=\varphi_\nu(x_0,\cdots,x_{n-1},y_0,\cdots,y_{r-1},\cdots y_\rho^{(\alpha)}\cdots),$$

$$(25)_2\qquad y'_\sigma=\pi_\sigma(x_0,\cdots,x_{n-1},y_0,\cdots,y_{r-1},\cdots y_\rho^{(\alpha)}\cdots).$$

Since, of course, δ''_ν is an element of Θ, from the expression (22) we have

$$\delta''_\nu=\sum_{i=0}^{n-1}\psi_{\nu i}\delta'_i.$$

Here we see that the coefficients $\psi_{\nu i}$'s or the functions φ_ν's and π_σ's are not functions in x_i's and y_ρ's only, but they also contain the derivatives of y_ρ's. So it is consistent with the transformation theory in linear or non-linear differential equations, such as contact transformations, canonical transformations, Bäcklund transformations, etc. They all fall within this category — transformations involving both independent and dependent variables and their derivatives. They are just changes of generators, that is, Tschirnhaus transformations of our algebraic structure. So new equations (24) in terms of new variables are obtained by eliminating old variables from equations (23) and (25) — just a process of elimination which we encountered in the case of algebraic equations. This is just a kind of the most general forms of transformation. Whenever the equation is solved in one expression (23), then the other equations (24) can be solved immediately by substituting (25) by the solution. Conversely if the $x'_\nu, y'_\sigma, y'^{(\alpha)}_\sigma$ are again generators, then the original variables should again be expressed by means of new quantities:

$$x_\nu=\varphi'_\nu(x'_0,\cdots,x'_{n-1},y'_0,\cdots,y'_{r-1},\cdots y'^{(\alpha)}_\sigma\cdots),$$

$$y_\rho=\varphi'_\rho(x'_0,\cdots,x_{n-1},y'_0,\cdots,y'_{r-1},\cdots y'^{(\alpha)}_\sigma\cdots).$$

So (23) and (24) are completely equivalent under the above assumption. Hence our structure, or more generally a filtered non-commutative ring satisfying the conditions specified above $(13)_1$–(15), represents a system of non-linear partial differential equations in an intrinsic sense. In the case of linear differential equations, the concept of $\mathcal{D}$-module gives an intrinsic understanding of linear differential equations. Quite similarly we now understand non-linear partial differential equations in an intrinsic language. Now whenever we are given a commutative or a non-commutative associative algebra, we should naturally consider the category of modules over that given ring. Of course, in our case, the corresponding modules should also be filtered. Anyway we have a good category of $\mathcal{D}$-modules. Now what such $\mathcal{D}$-modules should represent ? In the commutative case, a general ring, if it is finitely generated over the constant field $\mathcal{C}$, then represents some algebraic geometrical object ,i.e., an algebraic variety. An A-module, say $\mathcal{M}$, represents some complicated towering of vector bundles, so to speak. At each point, it represents a vector space. At a generic point it represents a vector space of a fixed dimension, but at a certain subvariety, the structure changes. So we have some stratified, what we call, structure on that variety. Therefore $\mathcal{M}$ is a vector bundle in a generalized sense and this is just a collection of vector spaces which are living on each point of V. So the algebraic variety serves as a moduli space or a parameter space for such vector spaces. Likewise our non-linear partial differential equation can also be a parameter space, so to speak, to deformation families of linear partial differential equations, that is $\mathcal{D}$-modules. So now we should consider the category of $\mathcal{D}$-modules in our sense.

It represents a deformation family of linear partial differential equations depending on solutions to non-linear differential equations. This is a very familiar situation which we encounter in various theories like gauge field theory, soliton theory or even in monodromy preserving deformation theory, etc. Anyway we consider such situations where linear equations contain some unknown functions so that these unknown functions satisfy non-linear equations and solutions to these equations serve as a parameter space to the linear equations. This is just a very general situation. Now I explain what is a solution to non-linear equations or to our $\mathcal{D}$-module. Recall that, in the case of commutative algebra, a morphism between commutative algebras gives rise to the opposite morphism between corresponding varieties:

$$\begin{array}{ccccl} A & \longrightarrow & A' & & \text{: rings} \\ \updownarrow & & \updownarrow & & \\ V & \longleftarrow & V' & & \text{: varieties.} \end{array}$$

Consider the situation where $A = \mathcal{C}[x_0, \cdots, x_{n-1}]/\mathcal{I}$ is a ring and V is the corresponding variety. If we have a point on V, then we have a morphism of A to $\mathcal{C}'$, where $\mathcal{C}'$ is some extension field of the constant field $\mathcal{C}$:

$$A \longrightarrow \mathcal{C}'$$

$$x_i \longrightarrow c_i.$$

Then we are dealing with a $\mathcal{C}'$-rational point on the variety V. This means that we just find some value of $a \in A$ in $\mathcal{C}'$, which satisfies equations

$$f_i(x) = 0, \qquad 0 \leq i < m, \tag{26}$$

where f_i's are generators of the ideal $\mathcal{I}$. Because f_i's are 0 in A and 0 should go to 0. So this means that finding homomorphism is just solving the equation (26). Likewise, if we consider $\mathcal{D}$, then a morphism from such a structure to a simple structure, where A is just a ring of regular functions, rational functions, holomorphic or meromorphic functions over the given manifold. Each of these rings does not contain unknown functions, in other words, elements of these rings are known functions on the given manifold X and $\mathcal{D}$ is $\mathcal{D}_X$. Certainly $\mathcal{D}_X$ is a known object. So if we are given a morphism

$$\mathcal{D} = A'[\delta_0, \cdots, \delta_{n-1}](y_0, \cdots, y_{r-1}) \longrightarrow \mathcal{D}_X,$$

then this means that we solved the equation (23), that is, unknown functions y_ρ's are now represented by elements of known functions. Since the above morphism corresponds to a point, so to speak, now instead of a variety, we have, in general, as a totality of solutions, an infinite dimensional manifold like the universal Grassmann manifold. So we have, in general, an infinite dimensional manifold which is a parameter space to a system of non-linear partial differential equations. Whenever we specify a point on that infinite dimensional manifold, that is, we specify a special solution to that equation, we can substitute unknown functions by that known functions and the $\mathcal{D}$-module now contains only known functions. Then the $\mathcal{D}$-module $\mathcal{M}$ is replaced by $\mathcal{M}_X$ which is some $\mathcal{D}_X$-module on the given manifold X. So this is just a usual linear partial differential equation as has been already explained. So this means that we have an infinite dimensional manifold consisting of solutions to the non-linear equations, and at each point of it we have such $\mathcal{M}_X$. So the situation is completely the same as commutative algebraic geometry. There are a lot of things to mention, but here we just add two additional points. One thing, which is central, is a construction of some special $\mathcal{D}$-module a priori. In the case of commutative algebraic geometry, we have a special class of A-modules, that is, vector bundles. For instance the simplest one may be just tangent bundles, cotangent bundles and its tensor bundles. Similar things can be applied to our situation. Whenever we are given such a genearl non-linear equations as (23), we can construct some special $\mathcal{D}$-module a priori.

§3. Linearization

Now we already know that

$$\Theta = K\delta_0 \oplus \cdots \oplus K\delta_{n-1},$$

where we omit the prime of δ_i in the former notation for simplicity. We recall that

$$K = \mathcal{C}(x_0, \cdots, x_{n-1}, y_0, \cdots, y_{r-1}, \cdots y_\rho^{(\alpha)} \cdots).$$

Then formal differentiation of $f \in K$ with respect to x_ν is

$$\delta_\nu(f) = \frac{\partial f}{\partial x_\nu} + \sum_{\rho,\alpha} y_\rho^{(\alpha+\epsilon_\nu)} \frac{\partial f}{\partial y_\rho^{(\alpha)}},$$

where $\epsilon_\nu = (0, \cdots, \overset{\nu}{\check{1}}, \cdots, 0) \in \mathbf{N}^n$. I am now going to give a brief account of how to find a priori a linearization of the original equation. Since A contains the $x_\nu, y_\rho, {y_\rho}^{(\alpha)}$, there are an infinite number of indeterminates. Algebraically they are just subject to only such equations as

$$F_\kappa(x_0, \cdots, x_{n-1}, y_0, \cdots, y_{r-1}, \cdots y_\rho^{(\alpha)} \cdots) = 0, \quad \kappa = 0, 1, \cdots, \tag{27}$$

and their formal differentiations with respect to x_ν's. This is a structure known as differential algebra. It is studied long years ago by the people in Columbia University associated to Ritt, their students such as Kolchin and others. They introduce the concept of differential algebra to construct a kind of Galois theory to some class of differential equations. But now we see that the same concept of differential algebra should be viewed as a basic concept to intrinsically describe non-linear partial differential equations.

Geometrically the above A represents a kind of an infinite dimensional manifold. We do not mean that this infinite dimensional manifold is the spectra of A. They are completely different. We mean the manifold which is a collection of solutions to the nonlinear equations. Anyway here we consider the spectra of A, which is infinite dimensional. Then the corresponding 1-form on Spec$A$ is also infinite dimensional. We denote it by Ω^1. This is just

$$\Omega^1 = \sum_\nu A dx_\nu + \sum_{\rho,\alpha} A dy_\rho^{(\alpha)}. \tag{28}$$

Let me mention that although the sum of the second term in the right hand side of (28) is infinite, an actual application of the operation d to individual $f \in A$ gives rise to a finite expression. Because each $f \in A$ contains a finite number of arguments. So each element of Ω^1 can be written as a finite sum. Now we have to incorporate an analytical structure. This is done in the following way. We can define, in a natural way, a pairing between Θ and Ω^1 by $< \delta, df >= \delta(f)$ for $f \in K$ and $\delta \in \Theta$. Since Θ is finite dimensional, we see that we can find some subspace $\mathcal{M}$ which is orthogonal to Θ. That is, $\mathcal{M}$ is a totality of 1-forms which is perpendicular to Θ:

$$\mathcal{M} = \Theta^{\perp} = \{\omega \in \Omega^1 | < \delta, \omega >= 0 \text{ for all } \delta \in \Theta\}.$$

I am just talking, for the sake of simplicity, by concrete expressions. But all these things can be defined in a purely intrinsic way. Now $\mathcal{M}$ acquires a natural $\mathcal{D}$-module structure. First we can introduce into $\mathcal{M}$ a Θ-module structure by $\delta(\omega) =< \delta, d\omega >$ for $\delta \in \Theta$ and $\omega \in \Omega^1$, where $< \delta, \eta \wedge \eta' >=< \delta, \eta > \eta' - \eta < \delta, \eta' >$ for η, $\eta' \in \Omega^1$. Whenever Θ acts in a natural way as a Lie algebra, the action can be generalized to the action of $\mathcal{D}$. So finally $\mathcal{M}$ acquires a $\mathcal{D}$-module structure. This $\mathcal{M}$ is a kind of conormal bundle, so to speak, with respect to solutions embedded in the infinite dimensional manifold Spec$A$. In this way $\mathcal{M}$ is intrinsically obtained a $\mathcal{D}$-module structure from the given structure from which we started. This $\mathcal{M}$ represents infinitesimal deformations of a given solution to the original equations. Suppose we have a solution to the original non-linear equations, which can be a special or a kind of generic solutions. Since $\mathcal{M}$ is a $\mathcal{D}$-module, this represents linear partial differential equations. Now we transform

$$x_\nu \longrightarrow x_\nu,$$
$$y_\rho^{(\alpha)} \longrightarrow y_\rho^{(\alpha)} + \epsilon u_\rho^{(\alpha)} \bmod \epsilon^2.$$

If $\{x_\nu,\ y_\rho^{(\alpha)}\}$ is a solution, then the condition that $\{x_\nu,\ y_\rho^{(\alpha)} + \epsilon u_\rho^{(\alpha)}\}$ should be a solution to the original non-linear equations modulo ϵ^2, is given by means of linear equations for u_ρ's. These linear equations just coincide with $\mathcal{M}$. In this way $\mathcal{M}$ represents an infinitesimal deformation of the given solution to the original non-linear equations. This is the meaning of $\mathcal{M}$. Now I can go further, but here I am not going to that any more.

§4. Microlocalization

I should talk about another important point which I announced at the beginning. That is the microlocalization. Let me first explain a localization in the commutative case. Let A be a commutative ring. We take some multiplicative subset in A. This represents a set of denominators which we want to introduce. Then we can construct something like $S^{-1}A$. For instance, if we take a prime ideal $\mathfrak{p}$ in A, then the complementary set $A - \mathfrak{p}$ is closed under the multiplication. Then $S^{-1}A$ represents a localization at $\mathfrak{p}$. If $\mathfrak{p}$ is a maximal ideal, then it represents local functions at the point corresponding to the maximal ideal. In general it represents local functions along the subvariety corresponding to $\mathfrak{p}$. In this way we can introduce the concept of localization. It represents functions in a neighborhood of a given point or some generalized concept of a point.

To go further we may construct a completion of $S^{-1}A$. Let $A_\mathfrak{p}$ denote the ring $S^{-1}A$ if $S = A - \mathfrak{p}$. Then $A_\mathfrak{p}$ has a filtration defined by powers of $\mathfrak{p}$:

$$A_\mathfrak{p} \supset A_\mathfrak{p}\mathfrak{p} \supset A_\mathfrak{p}\mathfrak{p}^2 \supset \cdots.$$

Then we can construct the projective limit of $A_\mathfrak{p}/A_\mathfrak{p}\mathfrak{p}^n$:

$$A_\mathfrak{p} \subset \varprojlim A_\mathfrak{p}/A_\mathfrak{p}\mathfrak{p}^n.$$

This is a kind of formal power series, so to speak. In general we take something between $A_\mathfrak{p}$ and $\varprojlim A_\mathfrak{p}/A_\mathfrak{p}\mathfrak{p}^n$ according to each purpose. Anyway all these are viewed to be serving as local functions. So each of these is a local ring giving rise to some sheaf structure to the given variety V. Thus a localization, roughly speaking, means introducing denominators which do not vanish at a given point.

This process of forming a ring of quotients with denominators within given S, can be done as well for non-commutative rings satisfying a suitable condition known as Ore condition. This is the following. Suppose we are given an associative algebra A which is not necessarily commutative and are given a multiplicative subset of A. Then this pair is said to satisfy the Ore condition if the following is satisfied. For any given $a \in A$ and $s \in S$, there exist $a' \in A$ and $s' \in S$ such that

$$s'a = a's.$$

The meaning of this is the following. We just want to construct something like $S^{-1}A$ which may be called a ring of left quotients. For a non-commutative case, the problem is the following. We want to construct

$$S^{-1}A = \{s^{-1}a | (s, a) \in S \times A\} = S \times A / \sim,$$

where $\sim$ is some equivalence condition and $s^{-1}a$ means the equivalence class of (s, a). If the Ore condition is satisfied, we can define addition and multiplication within $S^{-1}A$. For instance, for the product $s_1^{-1}a_1 \cdot s_2^{-1}a_2$, the problem is the part $a_1 s_2^{-1}$. By the Ore condition there exists $(s_3, a_3) \in S \times A$ such that

$$s_3 a_1 = a_3 s_2,$$

that is, $a_1 s_2^{-1} = s_3^{-1} a_3$. Then

$$s_1^{-1}a_1 \cdot s_2^{-1}a_2 = (s_3 s_1)^{-1} \cdot a_3 a_2.$$

So $S^{-1}A$ is closed under the multiplication. Likewise we can show that it is also closed under the addition as well. So if the Ore condition is satisfied, we can construct a ring of quotient from the left hand side. This construction is a process known as the Ore condition. We can show that, in our setting, $\mathcal{D}$ certainly satisfies the Ore condition where S can be quite arbitrary. For instance, S can be chosen as the totality of non-zero elements in $\mathcal{D}$ if $\mathcal{D}$ is without zero divisors, so that every differential operators with unknown functions within coefficients can have the inverse. In this way we can construct a localization. Before explaining microlocalization, I briefly mention about the principal symbol of an element of $\mathcal{D}$.

Since $\mathcal{D}$ is a filtered ring, we can construct its gradation ring gr$\mathcal{D}$ which is defined as

$$\mathrm{gr}\mathcal{D} = \bigoplus_n \mathcal{D}^{(n)}/\mathcal{D}^{(n-1)}.$$

This is a commutative ring. Hence we can construct a projective variety Proj(gr$\mathcal{D}$). Therefore it has a meaning to talk about a principal symbol or a characteristic variety. So we can construct a quotient ring, where S is chosen to be the set of operators whose principal symbols do not vanish at a given point. The process of microlocalization is nothing but such a process which admits taking the inverse of a differential operator whose principal symbol does not vanish at a given microlocal point. Here a microlocal point means a point of Proj(gr$\mathcal{D}$). So the well known process of microlocalization in the theory of microfunctions and microdifferential operators can be applied to this general situation. Here if we go to such microlocalization, we must make a completion of $S^{-1}\mathcal{D}$. Then it is some formal object which we can deal with in the analytical category as well. Anyway we can construct various classes of microdifferential operators. It is again a filtered ring $\mathcal{E} = \bigcup_{m \in \mathbf{Z}} \mathcal{E}^{(m)}$. But here m belongs to $\mathbf{Z}$ not to $\mathbf{N}$. In this way we have the concept of microlocalization and consequently we also have the concept of characteristic variety etc.

Research Institute for Mathematical Sciences
Kyoto University
Kyoto 606, Japan

Curriculum Vitae

Atle Selberg

Date of Birth: June 14, 1917, Langesund, Norway
Died: August 6, 2007, Princeton, NJ, USA

Education:

1943 Ph.D. Oslo University
1972 Hon. Dr. University of Trondheim

Employment History:

1942–1947	Research fellow, Oslo University
1947–1948	Institute for Advanced Study, Princeton
1948–1949	Associate Professor, Syracuse University
1949	Permanent member, Institute for Advanced Study, Princeton

Honors and Awards:

Fields Medal (1950)
Honorary Fellow, Tata Institute of Fundamental Research, Mumbai
Norwegian Academy of Sciences and Letters
Royal Norwegian Society of Science
Royal Danish Academy of Sciences and Letters
Royal Swedish Academy of Sciences
American Academy of Arts and Sciences (1960)
Indian National Science Academy
Honorary Member, London Mathematical Society (1985)
Wolf Prize (1986)

List of Publications

1. A. Selberg, Reflections around the Ramanujan centenary, in *Ramanujan: Essays and Surveys*, pp. 203–213, Hist. Math., Vol. 22 (Amer. Math. Soc., 2001).
2. A. Selberg, Old and new conjectures and results about a class of Dirichlet series, in *Proc. of the Amalfi Conference on Analytic Number Theory* (Maiori, 1989), pp. 367–385 (Univ. Salerno, 1992).
3. A. Selberg, *Collected Papers*, Vol. II. With a foreword by K. Chandrasekharan (Springer-Verlag, 1991).
4. A. Selberg, Remarks on the distribution of poles of Eisenstein series, in *Festschrift in Honor of I. I. Piatetski-Shapiro on the Occasion of His Sixtieth Birthday*, Part II (Ramat Aviv, 1989), pp. 251–278, Israel Math. Conf. Proc., Vol. 3 (Weizmann, 1990).
5. A. Selberg, Linear operators and automorphic forms, in *Number Theory and Related Topics* (Bombay, 1988), pp. 203–216, Tata Inst. Fund. Res. Stud. Math., Vol. 12 (Tata Inst. Fund. Res., 1989).
6. A. Selberg, *Collected Papers*, Vol. I. With a foreword by K. Chandrasekharan (Springer-Verlag, 1989).
7. A. Selberg, The history of Rademacher's formula for the partition function $p(n)$ (in Norwegian), *Normat* **37** (1989), No. 4, 141–146, 176.
8. A. Selberg, Reflections around the Ramanujan centenary (in Norwegian), *Normat* **37** (1989), No. 1, 2–7, 43.
9. A. Selberg, Sifting problems, sifting density, and sieves, in *Number Theory, Trace Formulas and Discrete Groups* (Oslo, 1987), pp. 467–484 (Academic Press, 1989).
10. A. Selberg, Remarks on multiplicative functions, in *Number Theory Day* (Proc. Conf., Rockefeller Univ., New York, 1976), pp. 232–241, Lecture Notes in Math., Vol. 626 (Springer, 1977).
11. A. Selberg, Remarks on sieves, in *Proc. of the Number Theory Conference* (Univ. Colorado, Boulder, Colo., 1972), pp. 205–216 (Univ. Colorado, 1972).
12. A. Selberg, Sieve methods, in 1969 *Number Theory Institute* (Proc. Sympos. Pure Math., Vol. XX, State Univ. New York, Stony Brook, N.Y., 1969), pp. 311–351 (Amer. Math. Soc., 1971).
13. A. Selberg, Recent developments in the theory of discontinuous groups of motions of symmetric spaces, in 1970 *Proc. of the Fifteenth Scandinavian Congress* (Oslo, 1968), Lecture Notes in Mathematics, Vol. 118, pp. 99–120 (Springer, 1970).

14. A. Selberg and S. Chowla, On Epstein's zeta-function, *J. Reine Angew. Math.* **227** (1967), 86–110.
15. A. Selberg, On the estimation of Fourier coefficients of modular forms, in 1965 *Proc. Sympos. Pure Math.*, Vol. VIII, pp. 1–15 (Amer. Math. Soc., 1965).
16. A. Selberg, Discontinuous groups and harmonic analysis, in 1963 *Proc. Internat. Congr. Mathematicians* (Stockholm, 1962), pp. 177–189 (Inst. Mittag-Leffler, 1963).
17. A. Selberg, On discontinuous groups in higher-dimensional symmetric spaces, in 1960 *Contributions to Function Theory* (Internat. Colloq. Function Theory, Bombay, 1960), pp. 147–164 (Tata Institute of Fundamental Research, 1960).
18. A. Selberg, Harmonic analysis and discontinuous groups in weakly symmetric Riemannian spaces with applications to Dirichlet series, *J. Indian Math. Soc.* (*N.S.*) **20** (1956), 47–87.
19. A. Selberg, Note on a paper by L. G. Sathe, *J. Indian Math. Soc.* (*N.S.*) **18** (1954), 83–87.
20. A. Selberg, On elementary methods in primenumber-theory and their limitations, in *Den 11te Skandinaviske Matematikerkongress* (Trondheim, 1949), pp. 13–22 (Johan Grundt Tanums Forlag, 1952).
21. A. Selberg, An elementary proof of the prime-number theorem, *Ann. Math.* (2) **50** (1949), 305–313.

22.

An elementary proof of the prime-number theorem

Annals of Mathematics, Vol. 50 (1949), No. 2, 305–313

(Received October 14, 1948)

1. Introduction

In this paper will be given a new proof of the prime-number theorem, which is elementary in the sense that it uses practically no analysis, except the simplest properties of the logarithm.

We shall prove the prime-number theorem in the form

$$\lim_{x\to\infty} \frac{\vartheta(x)}{x} = 1 \tag{1.1}$$

where for $x > 0$, $\vartheta(x)$ is defined as usual by

$$\vartheta(x) = \sum_{p \leqq x} \log p, \tag{1.2}$$

p denoting the primes.

The basic new thing in the proof is a certain assymptotic formula (2.8), which may be written

$$\vartheta(x) \log x + \sum_{p \leqq x} \log p \, \vartheta\left(\frac{x}{p}\right) = 2x \log x + O(x). \tag{1.3}$$

From this formula there are several ways to deduce the prime-number theorem. The way I present §§2–4 of this paper, is chosen because it seems at the present to be the most direct and most elementary way.[1] But for completeness it has to be mentioned that this was not my first proof. The original proof was in fact rather different, and made use of the following result by P. Erdös, that for an arbitrary, positive fixed number δ, there exist a $K(\delta) > 0$ and an $x_0 = x_0(\delta)$ such that for $x > x_0$, there are more than

$$K(\delta)\, x/\log x$$

primes in the interval from x to $x + \delta x$.

My first proof then ran as follows: Introducing the notations

$$\underline{\lim} \frac{\vartheta(x)}{x} = a, \qquad \overline{\lim} \frac{\vartheta(x)}{x} = A,$$

one can easily deduce from (1.3), using the well-known result

$$\sum_{p \leqq x} \frac{\log p}{x} = \log x + O(1), \tag{1.4}$$

[1] Because it avoids the concept of lower and upper limit. It is in fact easy to modify the proof in a few places so as to avoid the concept of limit at all, of course (1.1) would then have to be stated differently.

that

$$a + A = 2. \tag{1.5}$$

Next, taking a large x, with

$$\vartheta(x) = ax + o(x),$$

one can deduce from (1.3) in the modified form

$$(\vartheta(x) - ax)\log x + \sum_{p \le x} \log p \left(\vartheta\left(\frac{x}{p}\right) - A\,\frac{x}{p}\right) = O(x), \tag{1.6}$$

that, for a fixed positive number δ, one has

$$\vartheta\left(\frac{x}{p}\right) > (A - \delta)\,\frac{x}{p}, \tag{1.7}$$

except for an exceptional set of primes $\leqq x$ with

$$\sum \frac{\log p}{p} = o\,(\log x).$$

Also one easily deduces that there exists an x' in the range $\sqrt{x} < x' < x$, with

$$\vartheta(x') = Ax' + o(x').$$

Again from (1.6) with a and A interchanged, and x' instead of x, one deduces that

$$\vartheta\left(\frac{x'}{p}\right) < (a + \delta)\,\frac{x'}{p}, \tag{1.8}$$

except for an exceptional set of primes $\leqq x'$ with

$$\sum \frac{\log p}{p} = o\,(\log x).$$

From Erdös' result it is then possible to show that one can chose primes p and p', not belonging to any of the exceptional sets, with

$$\frac{x}{p} < \frac{x'}{p'} < (1 + \delta)\,\frac{x}{p}.$$

Then we get from (1.7) and (1.8) that

$$(A - \delta)\,\frac{x}{p} < \vartheta\left(\frac{x}{p}\right) \leqq \vartheta\left(\frac{x'}{p'}\right) < (a + \delta)\,\frac{x'}{p'} < (a + \delta)(1 + \delta)\,\frac{x}{p},$$

so that

$$A - \delta < (a + \delta)(1 + \delta).$$

or making δ tend to zero

$$A \leqq a.$$

Hence since also $A \geqq a$ and $a + A = 2$ we have $a = A = 1$, which proves our theorem.

Erdös' result was obtained without knowledge of my work, except that it is based on my formula (2.8); and after I had the other parts of the above proof. His proof contains ideas related to those in the above proof, at which related ideas he had arrived independently.

The method can be applied also to more general problems. For instance one can prove some theorems proved by analytical means by Beurling, but the results are not quite as sharp as Beurlings.[2] Also one can prove the prime-number theorem for arithmetic progressions, one has then to use in addition ideas and results from my previous paper on Dirichlets theorem.[3]

Of known results we use frequently besides (1.4) also its consequence

$$\vartheta(x) = O(x). \tag{1.9}$$

Throughout the paper p, q and r denote prime numbers. $\mu(n)$ denotes Möbius' number-theoretic function, $\tau(n)$ denotes the number of divisors of n. The letter c will be used to denote absolute constants, and K to denote absolute positive constants. Some of the more trivial estimations are not carried out but left to the reader.

2. Proof of the basic formulas

We write, when x is a positive number and d a positive integer,

$$\lambda_d = \lambda_{d,x} = \mu(d) \log^2 \frac{x}{d}, \tag{2.1}$$

and if n is a positive integer,

$$\theta_n = \theta_{n,x} = \sum_{d/n} \lambda_d . \tag{2.2}$$

Then we have

$$\theta_n = \begin{cases} \log^2 x, \text{ for } n = 1, \\ \log p \log x^2/p, \text{ for } n = p^\alpha, \alpha \geqq 1, \\ 2 \log p \log q, \text{ for } n = p^\alpha q^\beta, \alpha \geqq 1, \beta \geqq 1, \\ 0, \text{ for all other } n. \end{cases} \tag{2.3}$$

The first three of these statements follow readily from (2.2) and (2.1), the fourth is easily proved by induction. Clearly it is enough to consider n square-free, then if $n = p_1 p_2 \cdots p_k$,

$$\theta_{n,x} = \theta_{n/p_k,x} - \theta_{n/p_k,x/p_k} .$$

From this the remaining part of (2.3) follows.

---

[2] A. Beurling: *Analyse de la loi asymptotique de la distribution des nombres premiers généralisés*, Acta Math., vol. 68, pp. 255–291 (1937).

[3] These Annals this issue, pp. 297–304.

Now consider the expression

$$\begin{aligned}\sum_{n\le x}\theta_n &= \sum_{n\le x}\sum_{d/n}\lambda_d = \sum_{d\le x}\lambda_d\left[\frac{x}{d}\right] = x\sum_{d\le x}\frac{\lambda_d}{d} + O\,(\sum_{d\le x}|\lambda_d|)\\ &= x\sum_{d\le x}\frac{\mu(d)}{d}\log^2\frac{x}{d} + O\left(\sum_{d\le x}\log^2\frac{x}{d}\right) = x\sum_{d\le x}\frac{\mu(d)}{d}\log^2\frac{x}{d} + O(x).\end{aligned}\tag{2.4}$$

This on the other hand is equal to, by (2.3),

$$\begin{aligned}\sum_{n\le x}\theta_n = \log^2 x &+ \sum_{p^\alpha\le x}\log p\log\frac{x^2}{p}\\ &+ 2\sum_{\substack{p^\alpha q^\beta\le x\\ p<q}}\log p\log q = \sum_{p\le x}\log^2 p\\ &+ \sum_{pq\le x}\log p\log q + O\left(\sum_{p\le x}\log p\log\frac{x}{p}\right)\\ &+ O\,(\sum_{\substack{p^\alpha\le x\\ \alpha>1}}\log^2 x) + O\,(\sum_{\substack{p^\alpha q^\beta\le x\\ \alpha>1}}\log p\log q)\\ &+ \log^2 x = \sum_{p\le x}\log^2 p + \sum_{pq\le x}\log p\log q + O(x).\end{aligned}\tag{2.5}$$

The remainder term being obtained by use of (1.4) and (1.9). Hence from (2.4) and (2.5),

$$\sum_{p\le x}\log^2 p + \sum_{pq\le x}\log p\log q = x\sum_{d\le x}\frac{\mu(d)}{d}\log^2\frac{x}{d} + O(x).\tag{2.6}$$

It remains now to estimate the sum on the right-hand-side. To this purpose we need the formulas

$$\sum_{\nu\le z}\frac{1}{\nu} = \log z + c_1 + O(z^{-\frac{1}{2}}),\tag{2.7}$$

and

$$\sum_{\nu\le z}\frac{\tau(\nu)}{\nu} = \tfrac{1}{2}\log^2 z + c_2\log z + c_3 + O(z^{-\frac{1}{2}})\tag{2.7$'$}$$

where the c's are absolute constants, (2.7) is well known, and (2.7′) may be easily derived by partial summation from the well-known result

$$\sum_{\nu\le z}\tau(\nu) = z\log z + c_4 z + O(\sqrt{z}).$$

From (2.7) and (2.7′) we get

$$\log^2 z = 2\sum_{\nu\le z}\frac{\tau(\nu)}{\nu} + c_5\sum_{\nu\le z}\frac{1}{\nu} + c_6 + O(z^{-\frac{1}{2}}).$$

By taking here $z = x/d$, we get

$$\sum_{d\leq x}\frac{\mu(d)}{d}\log^2\frac{x}{d} = 2\sum_{d\leq x}\frac{\mu(d)}{d}\sum_{\nu\leq x/d}\frac{\tau(\nu)}{\nu} + c_5\sum_{d\leq x}\frac{\mu(d)}{d}\sum_{\nu\leq x/d}\frac{1}{\nu}$$

$$+ c_6\sum_{d\leq x}\frac{\mu(d)}{d} + O(x^{-\frac{1}{2}}\sum_{d\leq x}d^{-\frac{1}{2}}) = 2\sum_{d\nu\leq x}\frac{\mu(d)\tau(\nu)}{d\nu}$$

$$+ c_5\sum_{d\nu\leq x}\frac{\mu(d)}{d\nu} + c_6\sum_{d\leq x}\frac{\mu(d)}{d} + O(1)$$

$$= 2\sum_{n\leq x}\frac{1}{n}\sum_{d/n}\mu(d)\tau\left(\frac{n}{d}\right) + c_5\sum_{n\leq x}\frac{1}{n}\sum_{d/n}\mu(d)$$

$$+ O(1) = 2\sum_{n\leq x}\frac{1}{n} + c_5 + O(1) = 2\log x + O(1).$$

We used here that $\sum_{d/n}\mu(d)\tau(n/d) = 1$, and the well-known $\sum_{d\leq x}(\mu(d))/d = O(1)$. Now (2.6) yields

$$\sum_{p\leq x}\log^2 p + \sum_{pq\leq x}\log p\log q = 2x\log x + O(x). \tag{2.8}$$

This formula may also be written in the form given in the introduction

$$\vartheta(x)\log x + \sum_{p\leq x}\log p\,\vartheta\left(\frac{x}{p}\right) = 2x\log x + O(x), \tag{2.9}$$

by noticing that

$$\sum_{p\leq x}\log^2 p = \vartheta(x)\log x + O(x).$$

By partial summation we get from (2.8)

$$\sum_{p\leq x}\log p + \sum_{pq\leq x}\frac{\log p\log q}{\log pq} = 2x + O\left(\frac{x}{\log x}\right). \tag{2.10}$$

This gives

$$\sum_{pq\leq x}\log p\log q = \sum_{p\leq x}\log p\sum_{q\leq x/p}\log q = 2x\sum_{p\leq x}\frac{\log p}{p}$$

$$- \sum_{p\leq x}\log p\sum_{qr\leq x/p}\frac{\log q\log r}{\log qr} + O\left(x\sum_{p\leq x}\frac{\log p}{p\left(1+\log\frac{x}{p}\right)}\right)$$

$$= 2x\log x - \sum_{qr\leq x}\frac{\log q\log r}{\log qr}\vartheta\left(\frac{x}{qr}\right) + O(x\log\log x).$$

Inserting this for the second term in (2.8) we get

$$\vartheta(x)\log x = \sum_{pq\leq x}\frac{\log p\log q}{\log pq}\vartheta\left(\frac{x}{pq}\right) + O(x\log\log x). \tag{2.11}$$

Writing now

$$\vartheta(x) = x + R(x) \qquad , \text{(2.9) easily gives}$$

$$R(x) \log x = -\sum_{p \leq x} \log p \, R\left(\frac{x}{p}\right) + O(x), \tag{2.12}$$

and (2.11) yields in the same manner

$$R(x) \log x = \sum_{pq \leq x} \frac{\log p \log q}{\log pq} R\left(\frac{x}{pq}\right) + O(x \log \log x), \tag{2.13}$$

since

$$\sum_{pq \leq x} \frac{\log p \log q}{pq \log pq} = \log x + O(\log \log x),$$

which follows by partial summation from

$$\sum_{pq \leq x} \frac{\log p \log q}{pq} = \tfrac{1}{2} \log^2 x + O(\log x),$$

which again follows easily from (1.4).

The (2.12) and (2.13) yield

$$2 \mid R(x) \mid \log x \leq \sum_{p \leq x} \log p \left| R\left(\frac{x}{p}\right) \right| + \sum_{pq \leq x} \frac{\log p \log q}{\log pq} \left| R\left(\frac{x}{pq}\right) \right| + O(x \log \log x).$$

From this, by partial summation,

$$2 \mid R(x) \mid \log x \leq \sum_{n \leq x} \left\{ \sum_{p \leq n} \log p + \sum_{pq \leq n} \frac{\log p \log q}{\log pq} \right\} \cdot \left\{ \left| R\left(\frac{x}{n}\right) \right| - \left| R\left(\frac{x}{n+1}\right) \right| \right\} + O(x \log \log x),$$

or by (2.10)

$$\begin{aligned} 2 \mid R(x) \mid \log x &\leq 2 \sum_{n \leq x} n \left\{ \left| R\left(\frac{x}{n}\right) \right| - \left| R\left(\frac{x}{n+1}\right) \right| \right\} \\ &\quad + O\left(\sum_{n \leq x} \frac{n}{1 + \log n} \left| R\left(\frac{x}{n}\right) - R\left(\frac{x}{n+1}\right) \right| \right) + O(x \log \log x) \\ &= 2 \sum_{n \leq x} \left| R\left(\frac{x}{n}\right) \right| + O\left(\sum_{n \leq x} \frac{n}{1 + \log n} \left\{ \vartheta\left(\frac{x}{n}\right) - \vartheta\left(\frac{x}{n+1}\right) \right\} \right) \\ &\quad + O\left(x \sum_{n \leq x} \frac{1}{n(1 + \log n)} \right) + O(x \log \log x) = 2 \sum_{n \leq x} \left| R\left(\frac{x}{n}\right) \right| \\ &\quad + O\left(\sum_{n \leq x} \frac{1}{1 + \log n} \vartheta\left(\frac{x}{n}\right) \right) + O(x \log \log x) \\ &= 2 \sum_{n \leq x} \left| R\left(\frac{x}{n}\right) \right| + O(x \log \log x), \end{aligned}$$

or

$$(2.14) \qquad |R(x)| \leqq \frac{1}{\log x} \sum_{n \leqq x} \left| R\left(\frac{x}{n}\right) \right| + O\left(x \frac{\log \log x}{\log x}\right),$$

which is the result we will use in the following.[4]

3. Some properties of $R(x)$

From (1.4) we get by partial summation that

$$\sum_{n \leqq x} \frac{\vartheta(n)}{n^2} = \log x + O(1),$$

or

$$\sum_{n \leqq x} \frac{R(n)}{n^2} = O(1).$$

This means there exists an absolute positive constant K_1, so that for all $x > 4$ and $x' > x$,

$$(3.1) \qquad \left| \sum_{x \leqq n \leqq x'} \frac{R(n)}{n^2} \right| < K_1 .$$

Accordingly we have, if $R(n)$ does not change its sign between x and x', that there is a y in the interval $x \leqq y \leqq x'$, so that

$$(3.2) \qquad \left| \frac{R(y)}{y} \right| < \frac{K_2}{\log \dfrac{x'}{x}}, \qquad K_2 \geqq 1.$$

This is easily seen to hold true if $R(n)$ changes the sign also.[5]

Thus for an arbitrary fixed positive $\delta < 1$ and $x > 4$, there will exist a y in the interval $x \leqq y \leqq e^{K_2/\delta} x$, with

$$(3.3) \qquad |R(y)| < \delta y.$$

From (2.10) we see that for $y < y'$,

$$0 \leqq \sum_{y < p \leqq y'} \log p \leqq 2(y' - y) + O\left(\frac{y'}{\log y'}\right),$$

from which follows that

$$|R(y') - R(y)| \leqq y' - y + O\left(\frac{y'}{\log y'}\right).$$

[4] Apparently we have here lost something in the order of the remainder-term compared to (2.8). Actually we could instead of (2.14) have used the inequality

$$|R(x)| \leqq \frac{2}{\log^2 x} \sum_{n \leqq x} \frac{\log n}{n} \left| R\left(\frac{x}{n}\right) \right| + O\left(\frac{x}{\log x}\right),$$

which can be proved in a similar way.

[5] Because there will then be a $|R(y)| < \log y$.

Hence, if $y/2 \leqq y' \leqq 2y$, $y > 4$,

$$| R(y') - R(y) | \leqq | y' - y | + O\left(\frac{y'}{\log y'}\right),$$

or

$$| R(y') | \leqq | R(y) | + | y' - y) + O\left(\frac{y'}{\log y'}\right).$$

Now consider an interval $(x, e^{K_2/\delta} x)$, according to (3.3) there exists a y in this interval with

$$| R(y) | < \delta y.$$

Thus for any y' in the interval $y/2 \leqq y' \leqq 2\, y$, we have

$$| R(y') | \leqq \delta y + | y' - y | + \frac{K_3 y'}{\log x},$$

or

$$\left| \frac{R(y')}{y'} \right| < 2\delta + \left| 1 - \frac{y'}{y} \right| + \frac{K_3}{\log x}.$$

Hence if $x > e^{K_3/\delta}$ and $e^{-(\delta/2)} \leqq y'/y \leqq e^{\delta/2}$, we get

$$\left| \frac{R(y')}{y'} \right| < 2\delta + (e^{\delta/2} - 1) + \delta < 4\delta.$$

Thus for $x > e^{K_3/\delta}$ the interval $(x, e^{K_2/\delta} x)$ will always contain a sub-interval $(y_1, e^{\delta/2} y_1)$, such that $| R(z) | < 4\delta z$ if z belongs to this sub-interval.

4. Proof of the prime-number theorem

We are now going to prove the

THEOREM.

$$\lim_{x \to \infty} \frac{\vartheta(x)}{x} = 1.$$

Obviously this is equivalent to

$$\lim_{x \to \infty} \frac{R(x)}{x} = 0. \tag{4.1}$$

We know that for $x > 1$,

$$| R(x) | < K_4 x. \tag{4.2}$$

Now assume that for some positive number $\alpha < 8$,

$$| R(x) | < \alpha x, \tag{4.3}$$

holds for all $x > x_0$. Taking $\delta = \alpha/8$, we have according to the preceding section (since we may assume that $x_0 > e^{K_2/\delta}$), that all intervals of the type $(x, e^{K_2/\delta}x)$ with $x > x_0$, contain an interval $(y, e^{\delta/2}y)$ such that

$$|R(z)| < \alpha z/2, \tag{4.4}$$

for $y \leqq z \leqq e^{\delta/2}y$.

The inequality (2.14) then gives, using (4.2),

$$|R(x)| \leqq \frac{1}{\log x} \sum_{n \leqq x} \left| R\left(\frac{x}{n}\right) \right| + O\left(\frac{x}{\sqrt{\log x}}\right)$$

$$< K_4 \frac{x}{\log x} \sum_{(x/x_0) < n \leqq x} \frac{1}{n} + \frac{x}{\log x} \sum_{n \leqq (x/x_0)} \frac{1}{n} \left| \frac{n}{x} R\left(\frac{x}{n}\right) \right| + O\left(\frac{x}{\sqrt{\log x}}\right),$$

writing now $\rho = e^{K_2/\delta}$, we get further, using (4.3) and (4.4),

$$|R(x)| < \frac{\alpha x}{\log x} \sum_{n \leqq (x/x_0)} \frac{1}{n} - \frac{\alpha x}{2 \log x} \sum_{1 \leqq \nu \leqq (\log (x/x_0)/\log \rho)}$$

$$\sum_{\substack{y_\nu \leqq n \leqq y_\nu e^{(\delta/2)} \\ \rho^{\nu-1} < y_\nu \leqq \rho^\nu e^{-(\delta/2)}}} \frac{1}{n} + O\left(\frac{x}{\sqrt{\log x}}\right) = \alpha x - \frac{\alpha x}{2 \log x} \sum_{1 \leqq \nu \leqq (\log (x/x_0)/\log \rho)} \frac{\delta}{2}$$

$$+ O\left(\frac{x}{\sqrt{\log x}}\right) = \alpha x - \frac{\alpha \delta}{4 \log \rho} x + O\left(\frac{x}{\sqrt{\log x}}\right)$$

$$= \alpha\left(1 - \frac{\alpha^2}{256K_2}\right) x + O\left(\frac{x}{\sqrt{\log x}}\right) < \alpha\left(1 - \frac{\alpha^2}{300K_2}\right) x,$$

for $x > x_1$. Since the iteration-process

$$\alpha_{n+1} = \alpha_n \left(1 - \frac{\alpha_n^2}{300K_2}\right),$$

obviously converges to zero if we start for instance with $\alpha_1 = 4$ (one sees easily that then $\alpha_n < K_5/\sqrt{n}$), this proves (4.1) and thus our theorem.

FINAL REMARK. As one sees we have actually never used the full force of (2.8) in the proof, we could just as well have used it with the remainder term $o(x \log x)$ instead of $O(x)$. It is not necessary to use the full force of (1.4) either, if we have here the remainder-term $o(\log x)$ but in addition knowing that $\vartheta(x) > Kx$ for $x > 1$ and some positive constant K, we can still prove the theorem. However, we have then to make some change in the arguments of §3.

THE INSTITUTE FOR ADVANCED STUDY
AND
SYRACUSE UNIVERSITY

HARMONIC ANALYSIS AND DISCONTINUOUS GROUPS IN WEAKLY SYMMETRIC RIEMANNIAN SPACES WITH APPLICATIONS TO DIRICHLET SERIES

By A. SELBERG

[Received May 2, 1956]

IN THE following lectures we shall give a brief sketch of some representative parts of certain investigations that have been undertaken during the last five years. The center of these investigations is a general relation which can be considered as a generalization of the so-called Poisson summation formula (in one or more dimensions). This relation we here refer to as the " trace-formula."

1. Let S be a Riemannian space, whose points we denote by x and the (local) coordinates by $x^1, x^2, \ldots, x^n$, with a positive definite metric

$$ds^2 = \sum g_{ij}\, dx^i\, dx^j.$$

We shall assume the g_{ij} to be analytic in the coordinates. Further we assume that we have a locally compact group G of isometries of S (not necessarily the full group of isometries), whose elements we denote by m, and that G acts transitively on S so that given x and y in S, there exists an $m \in G$ such that $x = my$. We shall be concerned with the linear operators on functions $f(x)$ defined on S, which have the property that the operators are invariant under G, or otherwise expressed, linear operators that commute with the isometries m in G. We restrict ourselves here to the class of linear operators that are differential operators of finite order, integral operators of the form $\int_S k(x, y)\, f(y)\, dy$ (where dy denotes the invariant element of volume derived from the metric), or any

This is a summary of the results presented by the author to the International Colloquium on Zeta-functions held at the Tata Institute of Fundamental Research, Bombay, on February 14-21, 1956.

finite combination (by addition or multiplication) of such. This class evidently forms a ring.

Turning first to the integral operators, one observes that in order that the operator

$$\int_S k(x, y) f(y)\, dy$$

should be invariant, it is necessary and sufficient that the kernel satisfy the relation

$$k(mx, my) = k(x, y), \tag{1.1}$$

for all x, y in S and all m in G. We shall refer to such a kernel as a "point-pair invariant". If we consider such a "point-pair invariant" $k(x, y)$ as a function of one of the arguments, say x, keeping the other point y fixed, we see that $k(x, y)$ is invariant under the subgroup of G that leaves y fixed. This subgroup we denote by Ry and call it the rotation group of y. We express this property of k by saying that it has as a function of x rotational symmetry around the point y. Let x_0 be a chosen fixed point in S and R^0 with elements r^0 the rotation group of x_0. R^0 is isomorphic to a compact (or possibly finite) subgroup of the orthogonal group of n elements. Norming the bi-invariant Haar measure on R^0 so that $\int_{R^0} dr^0 = 1$, we can define for a function $f(x)$ a symmetrized function

$$f(x\,; x_0) = \int_{R^0} f(r^0 x)\, dr^0\,; \tag{1.2}$$

$f(x; x_0)$ clearly has rotational symmetry around the point x_0. Furthermore, if we have a function $f(x\,; x_0)$ with rotational symmetry around x_0, we can define a point-pair invariant $k(x, y)$ by the relation

$$k(x, y) = f(mx\,; x_0), \quad \text{where} \quad my = x_0,$$

this definition is seen not to depend on the particular choice of m if there is more than one m satisfying the relation $my = x_0$. Therefore the study of point pair invariants is equivalent to the study of functions with rotational symmetry around some point x_0.

We observe also the following facts, before turning to the consideration of differential operators. Because G acts transitively on S, an invariant operator, say L, of our class is completely characterized by its action at one point, say x_0. By this we mean that, introducing the notation $[Lf(x)]_{x=x_0}$ to denote the value of the function $Lf(x)$ at the point $x = x_0$, we can for an arbitrary point x_1 express $[Lf(x)]_{x=x_1}$ by means of the relation

$$[L f(x)]_{x=x_1} = [L f(mx)]_{x=x_0},$$

where m is a solution of $mx_0 = x_1$. Conversely if we have an operator $\mathscr{L}$ (not necessarily invariant), we can from its action at x_0, construct an invariant operator L by the relation

$$[L f(x)]_{x=x_0} = [\mathscr{L} f(x)]_{x=x_0},$$

provided

$$[\mathscr{L} f(r^0 x)]_{x=x_0} = [\mathscr{L} f(x)]_{x=x_0},$$

for every element r^0 in the rotation group R^0 of x_0. Finally if $\mathscr{L}$ does not have this property we may define

$$[L f(x)]_{x=x_0} = [\mathscr{L} f(x\,;\,x_0)]_{x=x_0},$$

where $f(x\,;\,x_0)$ is the symmetrized function of f around x_0 defined by (1.2), because $f(r^0x)$ and $f(x)$ have the same symmetrized function around x_0. If $\mathscr{L}$ is invariant then $L = \mathscr{L}$.

Furthermore, one observes that an invariant operator applied to a function with rotational symmetry around a point, gives a function which again is rotationally symmetric around the same point. Also an invariant operator applied to a point-pair invariant as a function of say the first point, gives as result again a point-pair invariant.

Consider now the class of invariant differential operators of finite order, and let for simplicity the local co-ordinates around x_0 be chosen such that the matrix (g_{ij}) at $x = x_0$ is the identity matrix E_n. Let D be an invariant differential operator, its action at the point $x = x_0$ is identical to that of a differential operator $D^{(0)}$ with constant coefficients,

$$D^{(0)} = \sum a_{i_1,i_2,\dots,i_n} \left(\frac{\partial}{\partial x^1}\right)^{i_1} \left(\frac{\partial}{\partial x^2}\right)^{i_2} \cdots \left(\frac{\partial}{\partial x^n}\right)^{i_n}.$$

By the highest homogeneous part of $D^{(0)}$ we mean the aggregate of terms in the above sum, where $a_{i_1,\dots,i_n} \neq 0$ and $i_1 + i_2 + \dots + i_n$ attains its maximal value; we denote this by $\overline{D}^{(0)}$, and write

$$\overline{D}^{(0)} = p_D \left(\frac{\partial}{\partial x^1}, \frac{\partial}{\partial x^2}, \dots, \frac{\partial}{\partial x^n}\right),$$

where p_D is a homogeneous polynomial. The rotation group R^0 induces on the tangent space of S at x_0 a subgroup R of the orthogonal group $\mathcal{O}_n$, and the polynomial $p_D(u_1, u_2, \dots, u_n)$ is seen to be invariant under this group R of orthogonal transformations. Conversely, if we have a homogeneous polynomial $p(u_1 \dots u_n)$ which is invariant under the group R, we may define an invariant differential operator D_p by the relation

$$[D_p f(x)]_{x=x_0} = \left[p\left(\frac{\partial}{\partial x^1}, \frac{\partial}{\partial x^2}, \dots, \frac{\partial}{\partial x^n}\right) f(x; x_0) \right]_{x=x_0}$$

It should be observed that whereas $p_{(D_p)} = p$, all one can say about $D_{(p_D)} - D$ is that it is an invariant operator of lower order than D. One also easily shows that if p_1 and p_2 are two such homogeneous polynomials invariant under R, we have that $D_{p_1 p_2} - D_{p_1} D_{p_2}$ is an operator of lower order than $D_{p_1 p_2}$. Using these facts, and a well-known result by Hilbert which says that the polynomials p have a finite basis of homogeneous polynomials $p_1, p_2, \dots, p_l$, $1 \leqslant l$, such that every homogeneous polynomial p can be written as a polynomial (not necessarily in a unique way) of $p_1, p_2, \dots, p_l$, with constant coefficients, one obtains the result that $D_{p_1}, D_{p_2}, \dots, D_{p_l}$ generate the ring of the invariant differential operators in the sense that any invariant differential operator D can be written as a finite expression

$$D = \sum A_{\nu_1,\nu_2,\dots,\nu l} \; D_{p_1}^{\nu_1} D_{p_2}^{\nu_2} \dots D_{pl}^{\nu l}, \tag{1.3}$$

where the A's are constants. Writing $D_{p_i} = D_i$ for $i = 1, 2, \dots, l$, we shall call $D_1, D_2, \dots, D_l$ a set of fundamental operators and we may assume that it is so chosen that l is minimal.

The fundamental operators in general do not commute, and as commutativity is essential for our later considerations, we shall make an additional assumption about G and S, which will imply commutativity (as we do not know, however, whether this assumption is necessary for commutativity, we should note that it is only the commutativity that is really necessary for the following developments).

We assume that there is a fixed isometry μ of S (possibly not in G), such that $\mu G \mu^{-1} = G$, $\mu^2 \in G$, and that for any pair of points x and y in S, there exists an m in G for which $mx = \mu y$ and $my = \mu x$. We may call a space for which there is some group of isometries G with these properties (if that is the case then the full group of all isometries will have these properties too) a " weakly symmetric " Riemannian space. This concept is more general than E. Cartan's concept of a symmetric space, as symmetric implies weakly symmetric, whereas it can be shown by examples that weakly symmetric does not imply symmetric.

Under this assumption we can prove that all the invariant operators commute. We first show that they commute when applied to point-pair invariants $k(x, y)$ considered as functions of the first point x. We first notice that if L is an invariant operator then so is also $\tilde{L}$ defined by

$$\tilde{L} f(x) = [L f(\mu^{-1} x)\,]_{x \to \mu x}.$$

Also from our assumption about G follows that for any point-pair invariant $k(x, y)$ we have

$$k(\mu y, \mu x) = k(mx, my) = k(x, y).$$

Denoting by a subscript the argument (x or y) that the operator is to act on, we have

$$L_x k(x, y) = k'(x, y),$$

where $k'(x, y)$ again is a point-pair invariant. Now we have

$$\begin{aligned}\widetilde{L}_y k(x, y) &= \widetilde{L}_y k(\mu y, \mu x)\\ &= [L_y k(y, \mu x)]_{y \to \mu y} = [k'(y, \mu x)]_{y \to \mu y}\\ &= k'(\mu y, \mu x) = k'(x, y).\end{aligned}$$

Thus

$$L_x k(x, y) = \widetilde{L}_y\, k(x, y),$$

so that we may shift the operator from the first to the second argument by replacing it with $\widetilde{L}$. If we now have two operators $L^{(1)}$ and $L^{(2)}$ we may write

$$\begin{aligned}L_x^{(1)} L_x^{(2)} k(x, y) &= L_x^{(1)} \widetilde{L}_y^{(2)}\, k(x, y)\\ &= \widetilde{L}_y^{(2)} L_x^{(1)} k(x, y) = L_x^{(2)} L_x^{(1)}\, k(x, y),\end{aligned}$$

(since the operators clearly may be interchanged when they act on different arguments). Thus we have commutativity when our operators are applied to point-pair invariants. Therefore we have also commutativity if our operators are applied to a function with rotational symmetry around a point, say x_0. For a function without rotational symmetry we notice that

$$[L^{(1)} L^{(2)} f(x)]_{x=x_0} = [L^{(1)} L^{(2)} f(x; x_0)]_{x=x_0},$$

where $f(x; x_0)$ is the function with rotational symmetry defined by (1.2). From this follows

$$[L^{(1)} L^{(2)} f(x)]_{x=x_0} = [L^{(2)} L^{(1)} f(x)]_{x=x_0},$$

or what is the same

$$L^{(1)} L^{(2)} f(x) = L^{(2)} L^{(1)} f(x),$$

that is, the operators commute.

It can be shown that the operator $\tilde{\bar{L}}$ where the bar denotes conjugation, is the formal adjoint of the operator L.

Returning to (1.3) we may now write

$$D = P\,(D_1, D_2, \ldots, D_l), \tag{1.4}$$

where P is a polynomial with constant coefficients. It should be noted that though our fundamental operators were chosen so that l was minimal, there may sometimes still be algebraic relations between them, so that the representation (1.4) may not necessarily be unique. Further it can be shown that one can always choose a set of fundamental operators with minimal l, such that each of them is self-adjoint.

Now let $f(x)$ be a function which is an eigenfunction of all our fundamental operators D_i so that

$$D_i\, f(x) = \lambda_i\, f(x), \quad i = 1, 2, \ldots, l, \tag{1.5}$$

where the λ_i are constants; because of (1.4) it will then be an eigenfunction of all the invariant differential operators, and in particular of the Laplace operator derived from the metric, therefore $f(x)$ will be analytic in the coordinates. If we take a point x_0 such that $f(x_0) \neq 0$, and form $f(x\,;\,x_0)$ defined by (1.2), this will again satisfy the equations (1.5) and will not vanish identically in x since $f(x_0\,;\,x_0) = f(x_0) \neq 0$. We now write

$$f(x\,;\,x_0) = f(x_0)\, \omega_\lambda(x, x_0), \tag{1.6}$$

where the subscript λ is an abbreviation for the l-tuple $(\lambda_1, \lambda_2, \ldots, \lambda_l)$ so that $\omega_\lambda(x_0, x_0) = 1$. We call this the "normed" eigenfunction with rotational symmetry around x_0, and shall show that it is unique, that is to say a function with rotational symmetry around x_0 which takes the value 1 at the point x_0 and which satisfies the equations (1.5) is identical with $\omega_\lambda(x, x_0)$. To prove this we observe that for such a function $g(x)$, we have, because $g(x) = g(x\,;\,x_0)$,

$$\left[\left(\frac{\partial}{\partial x^1}\right)^{\nu_1}\left(\frac{\partial}{\partial x^2}\right)^{\nu_2}\cdots\left(\frac{\partial}{\partial x^n}\right)^{\nu_n} g(x)\right]_{x=x_0}$$

$$=\left[\left(\frac{\partial}{\partial x^1}\right)^{\nu_1}\left(\frac{\partial}{\partial x^2}\right)^{\nu_2}\cdots\left(\frac{\partial}{\partial x^n}\right)^{\nu_n} g(x;\, x_0)\right]_{x=x_0} = [D\, g(x)]_{x=x_0},$$

where D is an invariant differential operator depending only on $(\nu_1, \nu_2, \ldots, \nu_n)$. Because of (1.4) and since $g(x)$ satisfies the equations (1.5) we thus get on using $g(x_0) = 1$,

$$\left[\left(\frac{\partial}{\partial x^1}\right)^{\nu_1}\left(\frac{\partial}{\partial x^2}\right)^{\nu_2}\cdots\left(\frac{\partial}{\partial x^n}\right)^{\nu_n} g(x)\right]_{x=x_0} = P(\lambda_1, \lambda_2, \ldots, \lambda_l),$$

where P is a polynomial depending only on $(\nu_1, \nu_2, \ldots, \nu_n)$. This shows that all the partial derivatives of $g(x)$ at the point x_0 are uniquely determined by the l-tuple $(\lambda_1, \lambda_2, \ldots, \lambda_l)$ and so since $g(x)$ is analytic in the coordinates, $g(x)$ is unique, that is, it coincides with $\omega_\lambda(x, x_0)$. We may from $\omega_\lambda\,(x, x_0)$ construct the point-pair invariant $\omega_\lambda(x, y)$ which will, because of the relation

$$D_x\, \omega_\lambda(x, y) = \tilde{D}_y\, \omega_\lambda\,(x, y),$$

be a normed eigenfunction also in y with rotational symmetry around the point x. Therefore we must have

$$\omega_\lambda(x, y) = \omega_{\tilde{\lambda}}\,(y, x), \tag{1.7}$$

where $\tilde{\lambda}$ denotes an l-tuple $(\tilde{\lambda}_1, \tilde{\lambda}_2, \ldots, \tilde{\lambda}_l)$ not necessarily identical to the original one. $\omega_\lambda(x, y)$ is now easily seen to be an eigenfunction (considered as a function of x) of our whole class of invariant operators for the reason that

$$L_x\, \omega_\lambda(x, y)$$

because of the commutativity of L and the D_i, $i = 1, 2, \ldots, l$, again satisfies the equations (1.5), and furthermore it is again a function with rotational symmetry around y, and differs therefore only by a factor independent of x (and hence since the factor is a point pair invariant it is independent of y also) from $\omega_\lambda(x, y)$, that is to say

$$L_x\, \omega_\lambda(x, y) = \Lambda\, \omega_\lambda(x, y),$$

where Λ is a constant depending on L and the l-tuple λ only.

We can now show that any function which satisfies the equations (1.5) will be an eigenfunction of our class of invariant operators, namely we have

$$\begin{aligned}[L\, f(x)]_{x=x_0} &= [L\, f(x\,;\, x_0)]_{x=x_0}\\ &= [L\, f(x_0)\, \omega_\lambda\, (x, x_0)]_{x=x_0}\\ &= \Lambda\, f(x_0)\, \omega_\lambda\, (x_0, x_0) = \Lambda\, f(x_0).\end{aligned}$$

Since this holds for any point x_0, we have

$$L\, f(x) = \Lambda\, f(x),$$

and we see that the eigenvalue Λ does depend only on L and the l-tuple λ, but not on the particular function $f(x)$.

Thus for an integral operator we may write

$$\int_S k(x, y)\, f(y)\, dy = h(\lambda)\, f(x), \tag{1.8}$$

where $h(\lambda) = h(\lambda_1, \lambda_2, \ldots, \lambda_l)$ depends on k and λ only. In order to get an expression for $h(\lambda)$ it is therefore enough to produce a " representative " set of eigenfunctions, that is, one that exhausts all the possibilities for the l-tuple $(\lambda_1, \ldots, \lambda_l)$, that is, l-tuples for which there really do exist functions satisfying the equations (1.5).

In a number of cases that are of particular interest for applications, such a set can be obtained from the following lemma:

Let T with elements t be a subgroup of G which is simply transitive on S, that is, such that the equation $x = tx_0$, where x is any point in S and x_0 a chosen fixed point, always has one and only one group element t as a solution. Further suppose that we have a continuous non-vanishing function $\phi(t)$ on T that satisfies the relation

$$\phi(t_1 t_2) = \phi(t_1)\,\phi(t_2),$$

for all t_1 and t_2 in T. If we now define $f(x) = f(tx_0) = \phi(t)$, where $tx_0 = x$, then $f(x)$ is an eigenfunction of our operators, because

$$L\,f(x) = [L\,f(tx)]_{x=x_0} = \phi(t)\,[L\,f(x)]_{x=x_0} = [L\,f(x)]_{x=x_0}\,f(x).$$

If we have several such multiplicatively independent functions $\phi_1(t)$, $\phi_2(t),\ldots,\phi_\kappa(t)$, then

$$\phi_1^{s_1}(t)\,\phi_2^{s_2}(t)\,\ldots\,\phi_\kappa^{s_\kappa}(t)$$

will also be one (where, if T or what is the same S, is not simply connected the exponents s_1, $s_2,\ldots,s_\kappa$ have to be chosen such that the resulting function is single-valued). It is of course not always so that different choices of the κ-tuple $s_1,\ldots,s_\kappa$ necessarily lead to different l-tuples $\lambda_1,\ldots,\lambda_l$. In many cases one gets all possibilities for which eigenfunctions exist covered by this construction.

The nature of the set of possible λ's may differ from the completely discrete set that would occur if S is compact,† to the situation for many non-compact spaces where the set of all l-tuples of complex numbers $\lambda_1, \ldots, \lambda_l$, which satisfy the possible algebraic relations between the D_i, $i = 1, 2, \ldots, l$, does occur. Intermediary situations can of course also occur. In the case when the set of all l-tuples λ of complex numbers satisfying the algebraic relations between the D_i's does occur, it is easily shown that $\omega_\lambda(x, y)$ as a function of λ is an analytic function on the algebraic variety defined by these relations, which is regular whenever all λ_i's are finite.

As an illustration we may for instance consider the space of n by n positive definite symmetric matrices $Y = (y_{ij})$ with the metric

$$ds^2 = \sigma\,(Y^{-1}\,dY\,Y^{-1}\,dY),‡$$

† Because we require our functions to be regular globally, if one admits " local " eigenfunctions (that cannot be continued everywhere in S, or that by such continuation would not be single-valued) the situation is different as shown by the examples of the surface of a sphere or the periphery of a circle.

‡ σ here and in the following denotes the trace.

where $dY = (dy_{ij})$, and the group G may be taken as the group of all non-singular real n by n matrices A, the isometries being

$$Y \to A\,Y\,A'$$

(A' is the transposed of A); finally the isometry μ may be taken as

$$Y \to Y^{-1}.$$

It is then easily established that all our requirements are satisfied. The point-pair invariants are easily seen to be of the form that $k(Y_1, Y_2)$ is a symmetric function of the n eigenvalues of the matrix $Y_2\,Y_1^{-1}$, or if one prefers it, $k(Y_1, Y_2)$ is a function of the n arguments $\sigma((Y_2\,Y_1^{-1})^\nu)$, $\nu = 1, 2, \ldots, n$. Conversely any such function is a point-pair invariant.

A set of fundamental operators can be obtained as follows: let $\frac{\partial}{\partial Y}$ denote the matrix $\left(\frac{1+\delta_{i,j}}{2}\,\frac{\partial}{\partial y_{ij}}\right)$, where $\delta_{i,j}$ is the Kronecker symbol; then the operators

$$D_i = \sigma\left(\left(Y\frac{\partial}{\partial Y}\right)^i\right), \quad i = 1, 2, \ldots, n \tag{1.9}$$

are a set of fundamental operators, and they are algebraically independent.

To obtain a representative set of eigenfunctions, consider the subgroup of G formed by the "triangular" matrices $T = (t_{ij})$ with $t_{ij} = 0$ for $i < j$ and $t_{ii} > 0$ for $i = 1, 2, \ldots, n$. This group acts simply transitive on our space, and for any complex n-tuple $s = (s_1, s_2, \ldots, s_n)$ the function

$$\phi_s(T) = \prod_{i=1}^{n} t_{ii}{}^{2s_i + i - (n+1)/2} \tag{1.10}$$

is single-valued and continuous on this group and has the property

$$\phi_s(T_1)\,\phi_s(T_2) = \phi_s\,(T_1\,T_2).$$

Thus defining for $Y = TT'$

$$f_s(Y) = \phi_s(T), \tag{1.11}$$

this is an eigenfunction. One can show that

$$D_i\, f_s(Y) = \lambda_i\,(s)\, f_s(y),$$

where $\lambda_i(s) = \lambda_i(s_1, s_2, \ldots, s_n)$ is a polynomial in the s_j of degree i which is symmetric in the s_j and of the form

$$\lambda_i(s_1, s_2, \ldots, s_n) = s_1^{\,i} + s_2^{\,i} + \ldots + s_n^{\,i} + \text{terms of lower degree.}$$

From this one sees that λ_i are a basis for the symmetric polynomials of the s_j, so that the s_j are determined as roots of an algebraic equation of nth degree whose coefficients are rational in the λ_i, so they are determined up to a permutation of the s_j. From this it follows that we may by suitable choice of the s_j make the λ_i any n-tuple of finite complex numbers. One also can show that

$$\tilde{\lambda}_i(s_1, s_2, \ldots, s_n) = \lambda_i(-s_1, -s_2, \ldots, -s_n).$$

To find an expression for the $h(\lambda)$ defined in (1.8),

$$\int\limits_S k(Y_1, Y)\, f_s(Y)\, dY = h(\lambda)\, f_s(Y_1),$$

where dY is the invariant element of volume

$$dY = \frac{2^{\frac{1}{2}n(n-1)}}{|Y|^{\frac{1}{2}(n+1)}} \prod_{i \leqslant j} dy_{ij},$$

we may write

$$k(Y_1, Y) = k(\sigma\,(YY_1^{-1}),\, \sigma(\,(YY_1^{-1})^2), \ldots, \sigma(\,(YY_1^{-1})^n\,)\,),$$

and take $Y_1 = E_n$ the identity matrix so that $f_s(Y_1) = 1$; further we may introduce the t_{ij}, $i \geqslant j$, in $Y = TT'$ as new coordinates in our space, the element of volume then becomes $\dfrac{2^{n(n+3)/4}}{t_{11}\, t_{22}^2 \ldots t_{nn}^n} \prod\limits_{i \geqslant j} dt_{ij}$ and the relation becomes

$$2^{\frac{1}{4}n(n+3)}\int k(\,\sigma(TT'),\,\sigma(\,(\,TT')^2),\,\ldots,\,\sigma(\,(TT')^n\,)\,)\times$$
$$\times\prod_{i=1}^{n} t_{ii}^{\,2s_i-\frac{1}{2}(n+1)}\prod_{i\geqslant j} dt_{ij}=h(\lambda),$$

where the integration is carried from 0 to ∞ over the t_{ii} and from $-\infty$ to ∞ over the t_{ij} with $i>j$. For special forms of k these integrations can be carried out explicitly, for instance if

$$k(Y_1,Y)=|\,YY_1^{-1}\,|^{\alpha}\,e^{-\beta\sigma(YY_1^{-1})},$$

where the real part of β is positive, and the real part of $2s_i+2\alpha>\frac{1}{2}(n-1)$ for $i=1,2,\ldots,n$; the integral then becomes

$$2^{\frac{1}{4}n(n+3)}\int\left\{\exp\left(-\beta\sum_{i\geqslant j}t_{ij}^{\,2}\right)\right\}\prod_{i=1}^{n}t_{ii}^{\,2s_i+2\alpha-(n+1)/2}\prod_{i\geqslant j}dt_{ij},$$

and splits into a product of $\dfrac{n(n+1)}{2}$ simple integrals, each of which is expressible in terms of Gamma functions.[§]

2. Let now Γ be a discrete subgroup of G which acts properly discontinuous on the space S, and let there be given a representation of Γ by unitary ν by ν matrices $\chi(M)$, where we denote the elements of Γ by M. Consider function vectors $F(x)$, that are column vectors, whose ν components are scalar functions of the point x, and which furthermore satisfy the relation

$$F(M\,x)=\chi(M)\,F(x),\tag{2.1}$$

for all x in S and M in Γ. Such a function $F(x)$ is then of course fully determined by its values on a fundamental domain $\mathscr{D}$ of Γ in S. Applying one of the invariant integral operators to such a function $F(x)$ one sees that

$$\int_S k(x,y)\,F(y)\,dy=\int_{\mathscr{D}}K(x,y;\chi)\,F(y)\,dy,$$

§ For this special choice of k, the resulting form of formula (1.8) has in the meantime been derived by different means by H. Maass, *Journal of the Indian Math. Soc.* 19 (1955), 1-24.

where the kernel K is a matrix given by

$$K(x, y; \chi) = \sum_{M \in \Gamma} \chi(M)\, k(x, My). \tag{2.2}$$

Considering now the Hilbert space defined by the inner product

$$(F_1, F_2) = \int_{\mathscr{D}} \bar{F}_1'(x)\, F_2(x)\, dx,$$

where $\bar{F}_1'$ is the conjugate transposed of F_1, one sees easily that the operator

$$\int_{\mathscr{D}} K(x, y; \chi)\, F(y)\, dy \tag{2.3}$$

is normal since the adjoint operator has a kernel that is derived from the right-hand side of (2.2) by replacing $k(x, y)$ by $\tilde{k}(x, y) = \overline{k(y, x)}$, and thus it commutes with the operator (2.3). The invariant differential operators are also seen to be normal.

We have not up to now put any restrictions on our point-pair invariants $k(x, y)$, but always only assumed that the kernel and the function that the operator acted on were such that the integral also existed if absolute values were taken of the integrands.

It is now time to impose conditions that will enable us to make definite statements about the absolute convergence of the series on the right-hand side of (2.2) and also about the behavior of $K(x, y; \chi)$.

We make the following assumptions :

$k(x, y)$ should have a majorant,

$k_1(x, y)$ such that (a) $\int_S k_1(x, y)\, dy < \infty$, (b) $k_1(x, y)$ is of regular growth; that is to say, there should exist positive constants δ and A such that for all x and y,†

† One can relax this, and permit kernels with, for instance, a singularity at $x = y$ by requiring (b′) to be fulfilled only if the smallest geodesic distance $d(x, y)$ exceeds some fixed number.

$$k_1(x, y) \leqslant A \int\limits_{d(y,y')<\delta} k_1(x, y')\, dy', \tag{b'}$$

where $d(y, y')$ denotes the smallest geodesic distance between y and y'. Under these assumptions the above series for $K(x, y; \chi)$ converges absolutely for x and y in S, and uniformly if x and y are in some compact subregion of S.‡

We also make the assumption that the fundamental domain $\mathscr{D}$ of Γ in S is compact. Then $K(x, y; \chi)$ will be uniformly bounded for x and y in $\mathscr{D}$ (and therefore also for all x and y in S). Therefore also the expression

$$\int\limits_{\mathscr{D}} \int\limits_{\mathscr{D}} \sigma\,(K(x, y\,;\chi)\,\overline{K(x, y\,;\chi)'})\,dx\,dy$$

is finite ($\overline{K}'$ denotes the conjugate transposed of the matrix K) so that the integral operator is of the Hilbert-Schmidt class, and the classical methods from the theory of integral equations can be applied.

Consider now the functions $F(x)$ satisfying (2.1) which are eigenfunctions of our fundamental operators D_i for $i = 1, 2, \ldots, l$. We can then show from the preceding results about our integral and differential operators, that there exist an orthonormal system of eigenfunctions $F_i(x)$, which is complete in our Hilbert space, and such that if we write

$$D_j\, F_i(x) = \lambda_j^i\, F_i(x) \tag{2.4}$$

for $j = 1, 2, \ldots, l$; the l-tuples $\lambda^i = (\lambda_1^i, \lambda_2^i, \ldots, \lambda_l^i)$ have no finite point of accumulation in l-dimensional space. The completeness in particular follows from the easily established fact that the system of all admissible kernels $K(x, y; \chi)$ is complete.

About the eigenvalues, the l-tuples λ^i, one could at once make statements based upon the fact that if the kernel of an integral

‡ Thus in particular $K\,(x, y\,; \chi)$ is continuous if $k\,(x, y)$ is.

operator is Hermitian (which is the case if $k(x, y) = \overline{k(y, x)}$), the eigenvalues $h(\lambda^i)$ must be real; also, by looking at the differential operators, if we have chosen the fundamental operators self-adjoint, as one always can, the λ_j^i for $j = 1, 2, \ldots, l$ have to be real, and for the elliptic ones the sign of the eigenvalue is also given. In terms of the corresponding normed rotationally symmetric eigenfunctions, it follows that $\omega_{\lambda^i}(x, y) = \overline{\omega_{\lambda^i}(y, x)}$, and $|\omega_{\lambda^i}(x, y)| \leqslant 1$ for all x and y in S.

Formally we have the expansion of $K(x, y; \chi)$ in terms of the eigenfunctions F_i,

$$\sum_{M \in \Gamma} \chi(M)\, k(x, My) = \sum_i h(\lambda^i)\, F_i(x)\, \bar{F}_i'(y).^{\S} \tag{2.5}$$

The absolute convergence of the right-hand side and the equality of the two sides could be proved under suitable additional assumptions about $k(x, y)$. However, since the eigenfunctions themselves occur in (2.5), our attention here will instead centre on the trace of the integral operators, where the eigenfunctions do not anymore occur.

We may formally compute the trace of the integral operator in two ways, namely on the one hand as

$$\sum_i h(\lambda^i), \tag{2.6}$$

and on the other hand as

$$\int_{\mathscr{D}} \sigma(K(x, x; \chi))\, dx = \sum_{M \in \Gamma} \sigma(\chi(M)) \int_{\mathscr{D}} k(x, Mx)\, dx. \tag{2.7}$$

We leave aside for the moment the question whether the series (2.6) is convergent or only summable in some sense and also the

§ This formula in the case $\chi(M)$ identically 1, can be used for estimation of the number of points Mx in large regions with rotational symmetry about the point y.

question whether the sum is actually equal to the expression (2.7), and turn our attention first to the latter expression. Under our assumption on $\mathscr{D}$ and $k(x, y)$ the series on the right-hand side of (2.7) actually is absolutely convergent, even when we take absolute values under the integral signs.

We shall rearrange the series on the right-hand side of (2.7) by combining the terms in a suitable way. For this purpose we introduce some notations.

Two elements M_1 and M_2 in Γ are said to be conjugate within Γ if there exists $M_3 \in \Gamma$ such that $M_1 = M_3 M_2 M_3^{-1}$; we call the class of all elements in Γ which are conjugate to a given M the conjugate class of M in Γ, and denote it by the symbol $\{M\}_\Gamma$. The subgroup of Γ formed by the elements which commute with M we call Γ_M and denote its elements by N_M. Similarly we define conjugacy within G, and denote by $\{m\}_G$ the class of all elements in G conjugate to an m in G. Clearly $\{M\}_\Gamma$ is contained in $\{M\}_G$. Also the subgroup of G formed by the elements of G which commute with m we call G_m and denote its elements by n_m; clearly Γ_M is contained in G_M.

We now group together the terms on the right-hand side of (2.7), where M belongs to the same conjugacy class in Γ. The factor $\sigma(\cdot\chi(M))$ has the same value for all elements M belonging to the same conjugacy class in Γ. Therefore we consider the sum

$$\sum_{M \in \{M_0\}_\Gamma} \int_{\mathscr{D}} k(x, Mx)\, dx. \tag{2.8}$$

The terms here are of the form

$$\begin{aligned} \int_{\mathscr{D}} k(x, M_1^{-1} M_0 M_1 x)\, dx &= \int_{\mathscr{D}} k(M_1 x, M_0 M_1 x)\, dx \\ &= \int_{M_1 \mathscr{D}} k(x, M_0 x)\, dx, \end{aligned}$$

with $M_1\mathscr{D}$ denoting the image of $\mathscr{D}$ under the transformation M_1. Two M_1 give the same $M_1^{-1}M_0M_1$, if and only if they differ on the left by an element of Γ_{M_0}. Thus the expression (2.8) becomes

$$\int_{\mathscr{D}_{M_0}} k(x, M_0\, x)\, dx,$$

where the domain of integration is given by $\mathscr{D}_{M_0} = \sum^*_{M \in \Gamma} M\mathscr{D}$, Σ^* indicating that the summation is carried over a complete set of elements M such that no two differ on the left by an element of Γ_{M_0}. It is easily seen that $\mathscr{D}_{M_0}$ is actually a fundamental domain of the discontinuous group Γ_{M_0} in S. Thus we may rewrite the right-hand side of (2.7) as

$$\sum_{\{M\}_\Gamma} \sigma(\chi\,(M)\,) \int_{\mathscr{D}_M} k(x, Mx)\, dx, \tag{2.9}$$

where the summation is extended over one representative for each conjugacy class in Γ. We shall transform the expression

$$\int_{\mathscr{D}_M} k(x, M\, x)\, dx$$

still further. We introduce on G_M with elements n_M the Haar measure dn_M which is invariant with respect to multiplication on the right. We construct some function $p(x)$ which is everywhere on S real and non-negative, and for which

$$\int_{G_M} p(n_M x)\, dn_M = 1, \quad \text{for all } x \text{ in } S.$$

This can be done by constructing first a function $q(x) \geqslant 0$, everywhere on S, for which the integral

$$\int_{G_M} q(n_M x)\, dn_M = q_1(x)$$

exists and is positive for every x in S. This can be done for instance by defining

$$q(x) = \begin{cases} 1 + \rho(x) - d(x), & \text{for } d(x) < 1 + \rho(x), \\ 0, & \text{for } d(x) \geqslant 1 + \rho(x), \end{cases}$$

where $d(x)$ denotes the smallest geodesic distance from x to some fixed point x_0, and $\rho(x) = \min_{n_M \in G_M} d(n_M\, x)$. Then $p(x) = q(x)/q_1(x)$ is seen to satisfy the above requirements. The group Γ_M acting on the right of G_M is discontinuous and we may denote by G_M/Γ_M a fundamental domain of Γ_M in G_M; we then get

$$\begin{aligned}
&\int\limits_{\mathscr{D}_M} k(x, Mx)\, dx \\
&= \int\limits_{G_M} \int\limits_{\mathscr{D}_M} k(x, Mx)\, p(n_M x)\, dx\, dn_M \\
&= \sum_{N_M \in \Gamma_M} \int\limits_{G_M/\Gamma_M} \int\limits_{\mathscr{D}_M} k(x, Mx)\, p(n_M\, N_M\, x)\, dx\, dn_M \\
&= \sum_{N_M \in \Gamma_M} \int\limits_{G_M/\Gamma_M} \int\limits_{\mathscr{D}_M} k(N_M\, x, M\, N_M\, x)\, p\, (n_M\, N_M x)\, dx\, dn_M \\
&= \sum_{N_M \in \Gamma_M} \int\limits_{G_M/\Gamma_M} \int\limits_{N_M \mathscr{D}_M} k(x, Mx)\, p(n_M x)\, dx\, dn_M \\
&= \int\limits_{G_M/\Gamma_M} \int\limits_{S} k(x, M\, x)\, p\, (n_M x)\, dx\, dn_M \\
&= \int\limits_{G_M/\Gamma_M} \int\limits_{S} k(n_M x, M n_M x)\, p(n_M x)\, dx\, dn_M \\
&= \int\limits_{G_M/\Gamma_M} \int\limits_{S} k(x, Mx)\, p(x)\, dx\, dn_M \\
&= \int\limits_{G_M/\Gamma_M} dn_M \int\limits_{S} k(x, Mx)\, p(x)\, dx,
\end{aligned}$$

where we repeatedly have used the fact that $k(x, Mx)$ is invariant under the group G_M, that the measure dn_M is right-invariant, that the measure dx is invariant, and also that

$$G_M = \sum_{N_M \in \Gamma_M} (G_M/\Gamma_M) \cdot N_M \text{ and } S = \sum_{N_M \in \Gamma_M} N_M D_M.$$

Writing now

$$\int_{G_M/\Gamma_M} dn_M = \mu(G_M/\Gamma_M),$$

this factor measures the volume of the fundamental domain of Γ_M in G_M, and does not in any way depend on $k(x, y)$. For the other factor we write

$$\int_S k(x, Mx)\, p(x)\, dx = g(\{M\}_G),$$

and observe that this factor only depends on $k(x, y)$ and on the conjugacy class $\{M\}_G$ of M in G. Combining our results we may now write

$$\int_{\mathscr{D}} \sigma(K(x, x; \chi))\, dx = \sum_{\{M\}_\Gamma} \sigma(\chi(M))\, \mu(G_M/\Gamma_M)\, g(\{M\}_G). \quad (2.10)$$

We now turn to the question 'when and in what sense are the two expressions (2.6) and (2.10) equal?' We can at first say that the series (2.6) converges absolutely and is equal to (2.10) if $k(x, y)$ can be written in the form

$$\int_S k_1(x, z)\, \overline{k_1(y, z)}\, dz, \quad (2.11)$$

where k_1 is a point-pair invariant satisfying our conditions (a) and (b). From this we get next that the same conclusion holds if k can be written in the form

$$\int_S k_1(x, z)\, k_2(z, y)\, dz, \quad (2.11')$$

if k_1 and k_2 both satisfy the conditions (a) and (b), since (2.11′) can be written as a linear combination of expressions of the form (2.11).

Introducing the notation, for $\epsilon > 0$,

$$\kappa_\epsilon(x, y) = \begin{cases} C_\epsilon, & \text{for } d(x, y) < \epsilon, \\ 0, & \text{for } d(x, y) \geqslant \epsilon, \end{cases}$$

where $d(x, y)$ is the smallest geodesic distance between x and y, and where C_ϵ is a constant depending on ϵ, chosen such that

$$\int_S \kappa_\epsilon(x, y)\, dy = 1,$$

(the integral clearly is independent of x), $\kappa_\epsilon(x, y)$ is a point-pair invariant satisfying (a) and (b). Writing

$$\theta_\epsilon(\lambda) = \int_S \kappa_\epsilon(x, y)\, \omega_\lambda(y, x)\, dy,$$

we have

$$\lim_{\epsilon \to 0} \theta_\epsilon(\lambda) = 1,$$

and that for the λ^i, in addition $|\theta(\lambda^i)| \leqslant 1$.†

Now let $k(x, y)$ satisfy (a) and (b) and in addition be continuous; considering the class

$$k_\epsilon(x, y) = \int_S \kappa_\epsilon(x, z)\, k(z, y)\, dz,$$

for $0 < \epsilon < 1$, we get that the class k_ϵ satisfy our conditions (a) and (b) uniformly, and that $\lim_{\epsilon \to 0} k_\epsilon(x, y) = k(x, y)$, uniformly for x and y in any compact subregion of S. Using this we can show that the "trace formula"

$$\sum_i h(\lambda^i) = \sum_{\{M\}_\Gamma} \sigma(\chi(M))\, \mu(G_M/\Gamma_M)\, g(\{M\}_G) \qquad (2.12)$$

† With equality only if $\omega_{\lambda^i}(x, y) = 1$.

is valid if we give the left-hand side the interpretation

$$\lim_{\epsilon\to 0} \sum_i h(\lambda^i)\, \theta_\epsilon(\lambda^i).$$

In particular (2.12) holds whenever k satisfies (a) and (b) and is continuous and the left-hand side of (2.12) converges absolutely.

Various types of sufficient conditions for absolute convergence can be given,‡ for instance that $K(x, y; \chi)$ have partial derivatives up to the order $[n/2] + 1$, which is the case if $k(x, y)$ has partial derivatives up to this order which are such that (2.2) can be differentiated term by term and the resulting series converges absolutely.

The trace formula (2.12) may be used on the one hand to investigate the distribution of the l-tuples λ^i and on the other hand also to investigate the distribution of the conjugate classes $\{M\}_\Gamma$, the latter in the following sense : The conjugate classes in G can be characterized by a certain number of numerical parameters and so with each $\{M\}_\Gamma$ can be associated the numerical parameters that characterize $\{M\}_G$; it is the distribution of these numerical parameters that can be investigated by means of (2.12).

We shall mention briefly a certain generalization of (2.12) which is of interest in connection with the so-called Hecke-operator for the classical modular group and their analogues.

Let us have given in connection with our group Γ and the representation $\chi(M)$, a subset Γ^* of elements M^* of G with the following properties : The set Γ^* (it does not need to be a group) and the

‡ Actually in the case of a particular G and S, the more convenient such conditions are those that can be expressed in terms of $h(\lambda)$ only. This involves expressing $k(x, y)$ in the form $\int h(\lambda)\, \omega_\lambda(x, y) d\lambda$ where $d\lambda$ is a certain measure, and seeing what properties of $h(\lambda)$ are sufficient to ensure that $k(x, y)$ is continuous and satisfies (a) and (b), then determining enough about the asymptotic distribution of the $\lambda^{(i)}$ to see what additional condition should be imposed to ensure the absolute convergence of $\sum_i h(\lambda^i)$.

elements M^*, are such that with M^* the inverse M^{*-1} is also in Γ^*, further for M^* in Γ^* and M in Γ the element M^*M is also in Γ^*, and there should be a finite set of "left-representatives" M_1^*, M_2^*, $\ldots$, M_κ^* such that $\Gamma^* = \sum_{i=1}^{\kappa} M_i^*\Gamma$, or otherwise expressed, every M^* can in a unique way be represented as M_i^*M with $M \in \Gamma$. Further let there be associated with each M^* a ν by ν matrix $\chi(M^*)$ (not necessarily unitary) such that $\chi(M^*M) = \chi(M^*)\chi(M)$ for M^* in Γ^* and M in Γ, and such that

$$\chi(M^{*-1}) = \overline{\chi(M^*)'}.$$

Defining now the operator T^* by

$$T^*F(x) = \sum_{i=1}^{\kappa} \chi(M_i^*)\; F(M_i^{*-1}x), \tag{2.13}$$

one establishes that $T^*F(x)$ again satisfies (2.1). T^* is seen to be self-adjoint in our Hilbert space and further to commute with our invariant integral operators (2.3) and with the fundamental differential operators. Therefore our complete orthonormal system of eigenfunctions $F_i(x)$ may be chosen such that they are also eigenfunctions of T^*; writing then

$$T^*F_i(x) = \lambda_*^i F_i(x),$$

it can be shown by multiplying the T^* with an operator of the form (2.3), which gives us an integral operator with the kernel

$$K^*(x, y; \chi) = \sum_{M^* \in \Gamma^*} \chi(M^*)\; k(x, M^*y),$$

and computing the trace of this integral operator in a similar way that

$$\sum_i h(\lambda^i)\,\lambda_*^i = \sum_{\{M^*\}_\Gamma} \sigma(\chi(M^*))\; \mu(G_{M^*}/\Gamma_{M^*})\; g(\{M^*\}_G), \tag{2.14}$$

where the conjugacy classes $\{M^*\}_\Gamma$ are defined by conjugacy with respect to Γ (that is M_1^* and M_2^* belong to the same conjugacy class, if and only if there exists $M \in \Gamma$ such that $M_1^* = MM_2^* M^{-1}$), and Γ_{M^*} is the subgroup of Γ that commute with M^*. What was said about the validity of (2.12) also holds for (2.14).

If for some l-tuple Λ it happens that $\omega_\Lambda(x, y)$ satisfies the condition

$$\int_S |\omega_\Lambda(x, y)| \, dy < \infty,$$

then one can show that $\omega_\Lambda(x, y)$ satisfies both our conditions (a) and (b), and it can therefore be used as a $k(x, y)$ in our trace formula. Since it is seen that for the $h(\lambda)$ corresponding to $\omega_\Lambda(x, y)$ one has $h(\lambda) = 0$ for $\lambda \neq \Lambda$, and

$$h(\Lambda) = \int_S |\omega_\Lambda (x, y)|^2 \, dy,$$

we get on the left-hand side of the formula (2.12) simply $N(\Lambda)\, h(\Lambda)$, where $N(\Lambda)$ is the number of the l-tuples λ^i that are equal to Λ. I conjecture, but have only so far been able to verify this conjecture for special types of spaces, that in this case $g\,\{M\}_G$† $= 0$ for all M which do not belong to some compact subgroup of G so that (as one easily establishes) the number of terms on the right-hand side which are not zero is finite. This would imply that one gets a finite expression for $N(\Lambda)$. As will be indicated later this has interesting applications to the problem of determining the number of linearly independent regular analytic automorphic forms of a given dimension, in one or more complex variables.‡

We have so far assumed that the fundamental domain $\mathscr{D}$ of our group Γ is compact. If we relax this condition and only require that D have finite volume, the situation changes somewhat. While the kernel $K(x, y; \chi)$ will behave as before as long as at least one

† Of course the special g that is derived from $\omega_\Lambda(x, y)$.

‡ Similar remarks apply to formula (2.14), which is of interest for the theory of Hecke-operators, as applied to the analytic modular forms.

of the points x and y is restricted to a compact subregion of $\mathscr{D}$ (or of S for that matter), the kernel may exhibit a singular behavior as both points tend simultaneously towards the " non-compact boundary " of $\mathscr{D}$, such that the integral

$$\int_{\mathscr{D}} \int_{\mathscr{D}} \sigma(K(x, y; \chi) \overline{K(x, y; \chi)}') \, dx \, dy \tag{2.15}$$

does not exist. If, as it may happen for some χ, the kernels K behave well enough at the " non-compact boundary " for (2.15) to exist, the situation is not significantly changed, the spectrum of l-tuples λ for which there are eigenfunctions F is still discrete and the eigenfunctions are in our Hilbert space, and one may in specific cases by showing special care with the transformations M that leave some " part " of the " non-compact boundary " fixed (namely by grouping together those that have the same Γ_M), prove a trace formula that is not essentially different in form from (2.12), only that some terms on the right-hand side will no longer correspond to a single conjugacy class $\{M\}_\Gamma$, but to an aggregate of conjugacy classes.

If however χ is such that (2.15) does not exist, there are in general continuous spectra (which may even be multi-dimensional) besides the discrete spectrum. In some of the simpler cases, where these continuous spectra have been studied, it is possible to remove them by replacing the kernel $K(x, y; \chi)$ with a modified kernel which retains only the eigenfunctions from the discrete spectrum and with unchanged eigenvalues $h(\lambda^i)$, the computation of the trace of this modified integral operator leads then to a trace formula, which however besides terms of the type occurring on the righthand side of (2.12) will contain terms of a radically new nature.

3. We shall in the following give some explicit illustrations of the formulas in the case of some simpler spaces S and groups G satisfying our conditions.

First we consider the case when S is the hyperbolic plane for which we use the model represented by the upper complex half-plane $z = x + iy$, $y > 0$, with the metric $ds^2 = \dfrac{dx^2 + dy^2}{y^2}$. Our group G may be taken as the group formed by all motions $mz = \dfrac{az + b}{cz + d}$ where $ad - bc = 1$, a, d, b, and c real.† The Laplacian corresponding to the metric $y^2\Delta = y^2\left(\dfrac{\partial^2}{\partial x^2} + \dfrac{\partial^2}{\partial y^2}\right)$ is the only fundamental operator, and the point-pair invariants are seen to be all of the form

$$k(z, z') = k\left(\frac{|z - z'|^2}{yy'}\right).$$

A representative set of eigenfunctions is given by y^s since

$$y^2\Delta\, y^s = \lambda\, y^s$$

with $\lambda = -s(1 - s)$. Writing $s = \frac{1}{2} + ir$, we shall use for convenience the r instead of the λ as parameter. The connection between $k(z, z')$ and $h(r)$ is given by the relations

$$\left.\begin{aligned} &\int_w^\infty \frac{k(t)}{\sqrt{(t - w)}}\, dt = Q(w), \quad k(t) = -\frac{1}{\pi}\int_t^\infty \frac{dQ(w)}{\sqrt{(w - t)}}, \\ &Q(e^u + e^{-u} - 2) = g(u), \\ &h(r) = \int_{-\infty}^{\infty} e^{iru}\, g(u)\, du, \quad g(u) = \frac{1}{2\pi}\int_{-\infty}^{\infty} e^{-iru}\, h(r)\, dr. \end{aligned}\right\} \qquad (3.1)$$

Regarding now $h(r)$ as the primary function, we see that if $h(r)$ satisfies the conditions§ :

(1) $h(r) = h(-r)$,

(2) $h(r)$ is regular analytic in a strip $|\mathrm{Im}\, r| < \frac{1}{2} + \epsilon$, where $\epsilon > 0$,

and

§ This is not the full group of isometries, since this also contains the elements $\dfrac{a\bar{z} + b}{c\bar{z} + d}$, with $ad - bc = -1$. However, we shall for simplicity assume that our discontinuous group Γ has only true motions as elements.

§ The conditions (2) and (3) could be somewhat weakened.

(3) $h(r) = O((1 + |r|^2)^{-1-\epsilon}$ in this strip;

then k will exist and satisfy our conditions (a) and (b).

The elements of G can, as it is well known, be divided into four types, of which the first consists only of the identity element, while the others are respectively the hyperbolic, the elliptic and the parabolic elements. For a hyperbolic element m there is always a representative of the conjugacy class $\{m\}_G$ of the form $z \to \rho z$, where ρ is real and > 1. We call ρ the norm of m, and also the norm of the hyperbolic conjugacy class $\{m\}_G$ and denote it by $N\{m\}$, leaving the subscript G out. An elliptic element has always one (and only one) fixed point in the space and represents a rotation of the plane around this point, by an angle which we may count positive in the counter-clockwise direction; we call this the rotation angle of the elliptic element and also of the elliptic conjugacy class in G represented by the element. Finally if an element is parabolic it belongs to one of the two parabolic conjugacy classes represented by $z \to z + 1$ and $z \to z - 1$ respectively.

In Γ we shall call a hyperbolic element P primitive, if it is not a power with exponent > 1 of any other element in the group Γ, correspondingly we say that the conjugacy class $\{P\}_\Gamma$ is primitive. For the elliptic elements of Γ, those with the same fixed point form a finite group generated by a single element, and the one that has the smallest positive rotation angle we call primitive and denote it by R and call the corresponding class a primitive elliptic conjugacy class $\{R\}_\Gamma$ in Γ. Finally a parabolic element of Γ which is not a power with exponent > 1 of any other element in Γ, and which belongs to the first of the two parabolic conjugacy classes in G, we call a primitive parabolic element of Γ and, denoting it by S, the corresponding class $\{S\}_\Gamma$ a primitive parabolic class. It should be mentioned that if the area of the fundamental domain $\mathscr{D}$ of Γ is finite, that is to say

$$A(\mathscr{D}) = \int_{\mathscr{D}} \frac{dx\,dy}{y^2} < \infty,$$

there are only a finite number of elliptic and primitive parabolic conjugacy classes in Γ, and if $\mathscr{D}$ is compact there are no parabolic ones. The primitive hyperbolic classes $\{P\}_\Gamma$ on the other hand are always present in infinite number.

Assuming first that $\mathscr{D}$ is compact, the trace formula takes the form

$$\sum_i h(r_i) = \frac{A(\mathscr{D})\,\nu}{2\pi} \int_{-\infty}^{\infty} r\,\frac{e^{\pi r} - e^{-\pi r}}{e^{\pi r} + e^{-\pi r}}\,h(r)\,dr +$$

$$+ \sum_{\{R\}_\Gamma} \sum_{k=1}^{m-1} \frac{\sigma(\chi^k(R))}{M \sin k\pi/m} \int_{-\infty}^{\infty} \frac{e^{-2\pi rk/m}}{1 + e^{2-\pi r}}\,h(r)\,dr +$$

$$+ 2 \sum_{\{P\}_\Gamma} \sum_{k=1}^{\infty} \frac{\sigma(\chi^k(P)) \log N\{P\}}{(N\{P\})^{k/2} - (N\{P\})^{-k/2}}\,g(k \log N\{P\}). \quad (3.2)$$

Here the r_i are the values for which there is a solution of the equation

$$y^2 \Delta F(z) = \lambda F(z), \quad \lambda = -\left(\tfrac{1}{4} + r^2\right)$$

with $F(z)$ in our Hilbert space ; since we count both values of r that give the same λ (and if $\lambda = -\tfrac{1}{4}$, $r = 0$ with double multiplicity) our formula actually represents twice the trace of the integral operator. $A(\mathscr{D})$ is the area of the fundamental domain. $m = m(R)$ represents the order of the primitive elliptic element R, and the summations $\sum_R$ and $\sum_P$ are taken over one representative from each primitive elliptic and each primitive hyperbolic class respectively. The r_i have to be such that $\tfrac{1}{4} + r^2$ is real and non-negative, so that the r_i are either real, or they are purely imaginary with absolute value $\leqslant \tfrac{1}{2}$.† The formula (3.2) can now on the one hand be used for determining the asymptotic distribution of the r_i, and on the other hand the asymptotic distribution of the norms of the primitive hyperbolic classes in Γ. Under our assumptions on $h(r)$ all infinite series occurring in (3.2) converge absolutely.

† These latter could of course only occur in finite number, but one can show that their number for suitable Γ and χ may become arbitrarily large, although it can be shown to be less than a certain constant times $\nu\,A(\mathscr{D})$.

(3.2) has a rather striking analogy to certain formulas that arise in analytic number theory from the zeta - and L-functions of algebraic number fields. This leads us to introduce the function defined by

$$Z_\Gamma(s;\chi) = \prod_{\{P\}_\Gamma} \prod_{k=0}^{\infty} | E_\nu - \chi(P)\,(N\{P\})^{-s-k} | , \quad (3.3)$$

for real part of $s > 1$, when the product converges absolutely. E_ν is here the ν by ν identity matrix, and $| \ldots |$ denotes the determinant.

From (3.2) one can derive the following facts about this analytic function of s :

(A). $Z_\Gamma(s;\chi)$ is an integral function of s of order 2, except in the case when the genus of the fundamental domain $\mathscr{D}$ is zero, in this case there may be a pole at $s = 0$ of order at most ν.‡

(B). $Z_\Gamma(s;\chi)$ has "trivial" zeros at the integers $-k$ for $k \geqslant 0$, whose multiplicity can be explicitly given in terms of k, ν, $A(\mathscr{D})$ (or the genus of the fundamental domain if one prefers), and the $m(R)$, the orders of the primitive elliptic classes, and the traces $\sigma(\chi^i(R))$ for $i = 1, 2, \ldots m(R) - 1$. In the particular case that there are no elliptic classes in Γ one has that the multiplicity of the trivial zero at $-k$ is $(2k+1)(2p-2)$, where p is the genus§ (which is in this case always > 1).

(C). $Z_\Gamma(s;\chi)$ satisfies a functional equation which relates the value of $Z_\Gamma(1-s;\chi)$ to that of $Z_\Gamma(s;\chi)$. The form of this functional equation depends on the quantities ν, $A(\mathscr{D})$, and the orders $m(R)$ of the primitive elliptic classes and the traces $\sigma(\chi^i(R))$ for $i = 1, 2, \ldots,$ $m(R) - 1$. In the particular case that there are no elliptic classes in Γ this functional equation has the form

$$Z_\Gamma(1-s;\chi) = Z_\Gamma(s;\chi)\exp\left\{ -\nu A(\mathscr{D}) \int_0^{s-\frac{1}{2}} v \operatorname{tg} \pi v \, dv \right\}. \quad (3.4)$$

‡ If one assumes the representation $\chi(M)$ to be irreducible, this pole only occurs for $\chi(M)$ identically equal to 1, and is then a simple pole.

§ In this particular case $p - 1 = \dfrac{A(D)}{4\pi}$.

(D). The zeros of $Z_\Gamma(s, \chi)$ which are not mentioned under (B), are the numbers $\frac{1}{2} + ir_i$, and have thus real part equal to $\frac{1}{2}$, with the possible exception of a finite number of zeros that are real and lie in the interval $0 \leqslant s \leqslant 1$.

As one sees from (D) the analog of the Riemann hypothesis is true for our $Z_\Gamma(s; \chi)$ with the slight modification that real zeros may occur in the interval $0 \leqslant s \leqslant 1$.

If we only require $A(\mathscr{D})$ to be finite, there will, if $\mathscr{D}$ is not compact, always be at least one primitive parabolic class $\{S\}_\Gamma$. If $\{S_i\}_\Gamma$ for $i = 1, 2, \ldots, \kappa$, are the different primitive parabolic classes in Γ, the situation will depend on the matrices $\chi(S_i)$; if $\chi(S_i)$ has μ_i eigenvalues equal to 1, we say that χ is singular of degree μ_i with respect to the class $\{S_i\}_\Gamma$, and singular of degree $\mu = \sum_{i=1}^{\kappa} \mu_i$ with respect to Γ. If $\mu = 0$, that is if χ is non-singular with respect to Γ, the situation is only slightly altered from the compact case. The spectrum is still discrete and in our trace-formula (3.2), will occur on the right-hand side the new term

$$- 2\, g(0) \sum_{i=1}^{\kappa} \log \| E_\nu - \chi(S_i) \| . \tag{3.5}$$

This new term does not essentially alter the statements (A), (B), (C) and (D) about $Z_\Gamma(s; \chi)$. If $\mu \geqslant 1$ however the situation is very much altered, in that we have then for our eigenvalue problem, besides the discrete spectrum, also a continuous spectrum of multiplicity μ. As mentioned in the previous section we have then first to investigate the eigenfunctions in the continuous spectra and then to remove their contribution to the kernel K and develop a trace formula for the modified kernel. As a description of the general case is rather complicated, we shall here only briefly indicate the results in the simplest case when there is only one parabolic class $\{S\}_\Gamma$ with respect to which χ is singular, and further that χ is one dimensional, so that $\chi(S) = 1$.

We may assume for simplicity that one representative S of the class $\{S\}_\Gamma$ is $Sz = z + 1$. Forming for real part of s greater than 1 the function

$$\left.\begin{aligned} E(z, s; \chi) &= \sum_{M \in S/\Gamma} \overline{\chi(M)}\,(\operatorname{Im} M z)^s = \sum_{M \in S/\Gamma} \overline{\chi(M)}\,\frac{y^s}{|cz + d|^{2s}} \\ M z &= \frac{a z + b}{c z + b}, \end{aligned}\right\} \tag{3.6}$$

where $M \in S/\Gamma$ means that M runs over a complete set of elements of Γ that do not differ by a power of S on the left, one establishes that this series is absolutely convergent for $\sigma > 1$, $s = \sigma + it$. Further one has

$$E(Mz, s; \chi) = \chi(M)\, E(z, s; \chi),$$

for M in Γ, and

$$y^2 \Delta\, E(z, s; \chi) = -\, s(1 - s)\, E(z, s; \chi).$$

It can then be proved that $E(z, s; \chi)$ is a meromorphic function of s in the whole s-plane, and that the poles are all in the region $\sigma < \frac{1}{2}$, with the possible exception of a finite number of simple poles which are real and lie in the interval $\frac{1}{2} < s \leqslant 1$; these poles are independent of z, and $E(z, s; \chi)$ may be written as a quotient of two integral functions in s, each of which is at most of order 2 and where the denominator is independent of z. Further $E(z, s; \chi)$ satisfies a functional equation, which may be described as follows :

We write for $\sigma > 1$,

$$\phi(s, \chi) = \frac{\pi^{\frac{1}{2}}\,\Gamma(s - \frac{1}{2})}{\Gamma(s)} \sum_{c \neq 0} \sum_{0 \leqslant d < |c|} \frac{\overline{\chi(M)}}{|c|^{2s}}; \tag{3.7}$$

then one can show that $\phi(s, \chi)$ is meromorphic in the whole s-plane and regular for $\sigma \geqslant \frac{1}{2}$ with the possible exception of a finite number of simple poles in the interval $\frac{1}{2} < s \leqslant 1$, and can be written as a quotient of two integral functions at most of order 2. Further one has the functional equation

$$\phi(s, \chi)\, \phi(1-s, \chi) = 1.^{\dagger} \tag{3.8}$$

Then the functional equation of $E(z, s; \chi)$ is

$$E(z, s; \chi) = \phi(s, \chi)\; E(z, 1-s; \chi). \tag{3.9}$$

Forming now the kernel

$$H(z, z'; \chi) = \frac{1}{4\pi} \int_{-\infty}^{\infty} h(r)\, E(z, \tfrac{1}{2}+ir; \chi)\, \overline{E(z', \tfrac{1}{2}+ir; \chi)}\, dr, \tag{3.10}$$

one can show that the kernel

$$K^*(z, z'; \chi) = K(z, z'; \chi) - H(z, z'; \chi), \tag{3.11}$$

where

$$K(z, z'; \chi) = \sum_{M \in \Gamma} \chi(M)\, k(z, Mz'),$$

has the property that it retains only the discrete spectrum (that is all eigenfunctions which are not in our Hilbert space are annihilated by the integral operator with kernel K^*), and this is retained with unchanged eigenvalues $h(r_i)$. The evaluation of the trace of this modified integral operator then gives us a trace formula which differs from the earlier in that on the right-hand side we have the new terms

$$\frac{1}{2\pi} \int_{-\infty}^{\infty} h(r) \frac{\phi'}{\phi}(\tfrac{1}{2}+ir, \chi)\, dr - \frac{1}{\pi} \int_{-\infty}^{\infty} h(r) \frac{\Gamma'}{\Gamma}(1+ir)\, dr -$$

$$- 2 \log 2.\ g(0) + \tfrac{1}{2}\,(1 - \phi(\tfrac{1}{2}, \chi))\, h(0). \tag{3.12}$$

These new terms make a rather drastic change in the nature of $Z_\Gamma(s, \chi)$, in particular $Z_\Gamma(s; \chi)$ will have simple poles at $s = -1/2, -3/2, -5/2, \ldots$; because of the second term in (3.12), the last term produces a simple pole at $s = \frac{1}{2}$ if $\phi(\frac{1}{2}, \chi) = -1$ (this pole may however be cancelled by a zero if one or more of the r_i equals zero), so that $Z_\Gamma(s; \chi)$ is no longer an integral function. Furthermore in addition to the non-trivial zeros at the points $\frac{1}{2} + ir_i$, namely wherever $\phi(s, \chi)$ has a pole in the region $\sigma < \frac{1}{2}$, $Z_\Gamma(s, \chi)$ will have a

† Since the coefficients of the Dirichlet series of (3.7) are actually real, this implies $|\phi(\frac{1}{2} + ir, \chi)| = 1$.

a zero of the same multiplicity. The functional equation is also correspondingly modified in that besides simple factors also the function $\phi(s, \chi)$ occurs in it.

In the general case one has a system of μ series like (3.6), with similar properties : when s is replaced by $1 - s$, this system transforms by a matrix $\phi(s, \chi)$ whose elements are of a similar nature as (3.7) ; the determinant of this matrix will essentially then play the role that $\phi(s, \chi)$ does in the former case.

In the 3-dimensional hyperbolic space, the situation is similar but in some respects simpler. One can introduce also there a $Z_\Gamma(s; \chi)$‡ which although it will be a function of order 3, has a functional equation which is essentially simpler than in the case of the hyperbolic plane. For general n-dimensional hyperbolic space the explicit computations are somewhat complicated by the fact that the groups Γ_M and G_M now may not always be abelian when M is different from the identity element; this complicates the form of the trace formula, which however is always in a certain sense simpler when n is odd than when n is even. The non-compact case with finite volume of $\mathscr{D}$ can in all these cases be treated satisfactorily.

For groups acting simultaneously on the product of a finite number of such spaces,§ the situation can also be handled even in the non-compact case as long as the " non-compact boundaries " of $\mathscr{D}$ are point-like.

For other higher dimensional spaces, as for instance the space of positive definite, n by n symmetric matrices with determinant 1, the situation, for $n > 2$, is not so simple. The continuous spectra that may occur in the non-compact case at present cannot be handled properly. One will also here try to obtain them by analytic continuation of certain Dirichlet series, like we did for the hyperbolic plane; only these Dirichlet series are more complicated and in the case of spectra that have a dimension > 1, they are Dirichlet series in several

‡ Defined by a somewhat more complicated product than (3.3).

§ Like the so-called Hilbert group acting on a product of hyperbolic planes.

variables. This problem of analytic continuation cannot be handled at present, except for special groups that arise from arithmetic, where one may be able to utilize this to effect the continuation. As an example could be mentioned the case when $n = 3$ for the above space, and the group Γ is the group of 3 by 3 matrices with determinant 1 and integral rational elements, and χ identical to 1; when one is led to consider the series

$$\zeta_Y(s, s') = \sum_{X'Z=0} (X' Y X)^{-s} (Z' Y^{-1} Z)^{-s'},$$

where the summation is carried over all pairs of column vectors X and Z, with integral rational components which satisfy the conditions $X'X > 0$, $Z'Z > 0$, $X'Z = 0$. The series converges absolutely for $\sigma > 1, \sigma' > 1$, where $s = \sigma + it$, $s' = \sigma' + it'$. One can in this case show that

$$(s-1)\ (s'-1)\ (s+s'-3/2)\ \zeta(2s+2s'-1)\ \zeta_Y(s, s'),$$

where $\zeta(2s + 2s' - 1)$ is the ordinary Riemann zeta-function, is an integral function in the two complex arguments s and s'. Further if one writes

$$\xi_Y(s, s') = \pi^{-2s-2s'}\, \Gamma(s)\, \Gamma(s')\, \Gamma(s+s'-\tfrac{1}{2})\ \zeta(2s+2s'-1)\ \zeta_Y(s, s'),$$

then the function $\xi_Y(s, s')$ remains invariant by replacing (s, s') by any of the following pairs of complex arguments $(s + s' - 1/2, 1 - s')$, $(1 - s, s + s' - 1/2)$, $(3/2 - s - s', s)$, $(s', 3/2 - s - s')$ and $(1 - s', 1 - s)$, so that it has a larger number of functional equations than the zeta-functions in one variable. It should be noted that the group under which $\xi_Y(s, s')$ is invariant is isomorphic to the permutation group of 3 elements, as the three quantities $4s + 2s' - 3$, $2s' - 2s$, $-4s' - 2s + 3$, undergo permutations. $\xi_Y(1/2 + it, 1/2 + it')$ is here connected with the two-dimensional continuous spectrum. Besides this there is a denumerably infinite sequence of Dirichlet series in one complex variable that are connected with one-dimensional continuous spectra.

Similar Dirichlet series in up to $(n - 1)$ complex variables, can be defined for general n, by looking at the definite forms in $(n - 1)$

variables that can be represented by the quadratic form with matrix Y, then the $(n-2)$ forms that can be represented by the $(n-1)$ form, and so on down to a form in one variable, and forming a product of the determinants of the $(n-1)$, $(n-2)$, ..., 1 form raised to complex exponents $-s_{n-1}, -s_{n-2}, \ldots, -s_1$ respectively, and summing over all such " descending " series of forms that are inequivalent in a certain sense. In the case $n = 3$ this would lead to a function which differs only by a simple factor (which is independent of Y) from $\zeta_Y(s_1, s_2)$ as it was defined above. The general study of these series has not yet been undertaken, but it is conceivable that it may prove of value for the theory of quadratic forms.

4. We shall finally give some applications to more classical problems. We go back to the hyperbolic plane $z = x + iy$, $y > 0$, and add a third coordinate ϕ, where we will identify ϕ and $\phi + 2\pi$. On this space consisting now of points (z, ϕ), we take the following group G with elements m_α, where m is a real matrix $\begin{pmatrix} a & b \\ c & d \end{pmatrix}$ with determinant 1, and α a real number, and let it act on the space (z, ϕ) in the way that

$$m_\alpha(z, \phi) = \left(\frac{az+b}{cz+d},\ \phi + \arg(cz+d) + \alpha\right).$$

Further we define μ such that

$$\mu(z, \phi) = (-\bar{z}, -\phi).$$

One then establishes that the two differential forms

$$\frac{dx^2 + dy^2}{y^2} \text{ and } d\phi - \frac{dx}{2y},$$

have the property that they both are invariant under G, the first one is also invariant under μ whereas the second only changes sign. We may therefore take for instance

$$ds^2 = \frac{dx^2 + dy^2}{y^2} + \left(d\phi - \frac{dx}{2y}\right)^2$$

as our invariant metric. We have two fundamental differential operators in this case, namely $\frac{\partial}{\partial \phi}$ and the Laplacian derived from the metric. The point-pair invariants are seen to be all functions of the two real arguments

$$\frac{|z-\bar{z}'|^2}{4\,yy'} \text{ and } \phi - \phi' + \arg \frac{z-\bar{z}'}{2i},$$

where (z, ϕ) and (z', ϕ') are the two points.

If we now have a group Γ which is discontinuous in the hyperbolic plane, it is seen that the group $\bar{\Gamma}$ obtained from Γ by, for each transformation $Mz = \frac{az+b}{cz+d}$ in Γ, counting both $M = \begin{pmatrix} a & b \\ c & d \end{pmatrix}$ and $-M = \begin{pmatrix} -a, & -b \\ -c, & -d \end{pmatrix}$ as different elements of $\bar{\Gamma}$, has the property that when $\bar{\Gamma}$ acts on our space (z, ϕ) in the way

$$M(z, \phi) = \left(\frac{a\,z+b}{c\,z+d},\ \phi + \arg(c\,z+d)\right),$$

the group $\bar{\Gamma}$ is discontinuous in this space, and if the fundamental domain $\mathscr{D}$ in the hyperbolic plane of Γ is compact, then so is the fundamental domain $\bar{\mathscr{D}}$ of $\bar{\Gamma}$, and if $\mathscr{D}$ has finite area $\bar{\mathscr{D}}$ has finite volume. The converse is also true.

Similarly a representation χ of Γ can be extended to $\bar{\Gamma}$ by letting both M and $-M$ correspond to the same χ as the transformation Mz in Γ. There may however also be other representations χ of $\bar{\Gamma}$ where the two elements M and $-M$ correspond to different χ.

If we have such a representation $\chi(M)$ of Γ one now sees that the eigenfunctions of our operators, because of the presence of the fundamental operator $\frac{\partial}{\partial \phi}$ and the identification of (z, ϕ) and $(z, \phi + 2\pi)$ must be of the form $e^{-ik\phi}$ times a function of the point z, where k is an integer. The eigenfunctions $F(z, \phi)$ which satisfy the relation

$$F(M(z, \phi)) = \chi(M)\ F(z, \phi)$$

can therefore be written in the form

$$F(z, \phi) = y^{k/2}\ F(z)\ e^{-ik\phi}, \tag{4.1}$$

and the former relation takes the form

$$F(Mz) = \chi(M)\ (cz + d)^k\ F(z). \tag{4.2}$$

Thus we see that we have the same type of transformation law as (in the case of one-dimensional χ) is known from the theory of the analytic automorphic forms.†

Instead of studying the general eigenfunction, and the general form of the trace-formula, which can be carried out without serious difficulties, we shall here only study a particular type that is associated with certain eigenfunctions with rotational symmetry which have the property that they satisfy our conditions (a) and (b) and so can be used as point-pair invariants in forming our kernels K.

It can be established that

$$\omega_k\ (z, \phi\,;\, z', \phi') = \frac{(y\ y')^{k/2}}{\{(z - \bar{z}')/2i\}^k}\ e^{-ik(\phi - \phi')}, \tag{4.3}$$

for any integer k is an eigenfunction in (z, ϕ) which is a point-pair invariant in the two points (z, ϕ) and (z', ϕ'); further that for $k > 2$ the conditions (a) and (b) are satisfied. As a consequence the integral operator with the kernel

$$K_k\ (z, \phi\,;\, z', \phi'\,;\, \chi) = \sum_{M \in \bar{\Gamma}} \chi(M)\ \omega_k\ \{z, \phi\,;\, M(z', \phi')\}, \tag{4.4}$$

can be shown to have only eigenfunctions§ of the form (4.1) where $F(z)$ is an analytic function of z regular in the interior of the upper half plane and satisfying the condition that

† We get here only integral dimensions, k; if one wants to study arbitrary real dimension, one has to give up the identification $(z, \phi) = (z, \phi + 2\pi)$, also $\bar{\Gamma}$ has to be defined in a different way.

§ That is corresponding to an eigenvalue different from zero,

$$y^{k/2}\,F(z)$$

is uniformly bounded throughout this region, and every such eigenfunction corresponds to the same eigenvalue given by

$$\int_S |\omega_k(z, \phi; z', \phi')|^2\, d(z, \phi)$$

where the integral is taken over our whole (z, ϕ)-space and $d(z, \phi)$ is the invariant element of volume. The trace-formula for this particular kernel gives us then the number N_k of regular analytic forms $F(z)$ satisfying (4.2)‡ as a finite expression depending on k, ν, the area $A(\mathscr{D})$ of the fundamental domain, the elliptic primitive classes $\{R\}_\Gamma$ and the eigenvalues of the χ's that correspond to them, and the primitive parabolic classes and the eigenvalues of the χ's that correspond to them. The hyperbolic classes give no contribution at all. For $k = 2$ it is possible to obtain a similar result by replacing ω_2 with $\omega_2 \left(\dfrac{(y\,y')^{1/2}}{|(z-\bar{z}')/2i|} \right)^\delta$ where $\delta > 0$, and in the trace formula for this kernel letting δ tend to zero.

If we consider the classical modular group Γ with elements $\begin{pmatrix} a & b \\ c & d \end{pmatrix}$, where $ad - bc = 1$, and a, d, b, c are rational integers, and the representation χ identical to 1, Hecke has introduced certain operators T_n for each positive integer n and studied their action on the regular modular forms, in connection with his theory about Dirichlet series with functional equations (of a certain type) and Euler products.

These T_n are of the type (2.13), associated with the set of transformations $M^{(n)} = \dfrac{1}{\sqrt{n}} \begin{pmatrix} a & b \\ c & d \end{pmatrix}$, where a, b, c, d are rational integers with $ad - bc = n$, in the way that T^* was associated with the set M^*. The generalized trace formula (2.14) gives then applied to the

‡ If D is not compact, but has finite area, the condition that $y^{k/2}\,F(z)$ is uniformly bounded, implies that we are only counting the so-called cusp-forms here,

point-pair invariant eigenfunction ω_k for $k > 2$,§ the following formula for the trace of the Hecke operator T_n acting on the space of cusp-forms of dimension $-k$.

$$\sigma_k(T_n) = -\frac{1}{2} \sum_{-2\sqrt{n}<m<2\sqrt{n}} H(4n - m^2) \frac{\eta_m^{k-1} - \bar{\eta}_m^{k-1}}{\eta_m - \bar{\eta}_m} - \\ - \sum_{\substack{d/n \\ d \leqslant \sqrt{n}}}' d^{k-1} + \delta(\sqrt{n}) \frac{k-1}{12} n^{k/2-1}. \quad (4.5)$$

Here the $H(d)$ denotes the number of inequivalent positive definite forms $ax^2 + bxy + cy^2$ with $4ac - b^2 = d$, counted in the usual way that a form equivalent to $a(x^2 + y^2)$ is counted with the weight $\frac{1}{2}$ and one equivalent to $a(x^2 + xy + y^2)$ with weight $1/3$. Further

$$\eta_m = \frac{m + i(4n - m^2)^{\frac{1}{2}}}{2}.$$

$\delta(x)$ is defined as 1 if x is an integer and zero otherwise, and Σ' means that if $d = \sqrt{n}$ the corresponding term is counted with weight $\frac{1}{2}$. For $k = 2$ one can again by a limit process arrive at a similar formula which however will contain one new term, and turns out (since there are no cusp forms of dimension -2 for the modular group, so that $\sigma_2(T_n) = 0$), to be identical with the so-called Kronecker class number relation. For $k = 4, 6, 8, 10$ and 14 there are again no cusp forms, so that the left-hand side of (4.5) is zero, which gives five new class number relations, while for $k = 12$, for instance the left-hand side is identical to the number theoretical function $\tau(n)$ of Ramanujan, so that one gets an explicit (admittedly rather complicated) formula for this. While the results about the number of regular analytic forms of a given dimension $-k$ and representation χ of $\bar{\Gamma}$ are classical† and previously were derived from the Riemann-Roch formula, the evaluation of the trace of

§ k will here be even, since with χ identical to one there are no non-vanishing functions satisfying (4.2), for k odd, since the left-hand side remains the same by replacing M by $-M$ whereas the right-hand side changes sign.

† Although as far as I know only the case of one-dimensional χ occurs in the literature.

the Hecke operator has not yet been accomplished by other means. From our point of view these expressions are finite elementary cases of the general trace formulas (2.12) and (2.14).

It is of interest to note that the method sketched above carries immediately over to the analytic automorphic forms in higher dimensional spaces, as for instance a product space formed by a finite number of hyperbolic planes or the general symplectic space, which can all be handled in a similar way without any essential difficulties occurring as long as the discontinuous group Γ has compact fundamental domain. For the symplectic space for instance, one can introduce in a similar way as before a space (Z, ϕ) and define the group G acting on the space with elements M_α, where the symplectic matrix $M = \begin{pmatrix} A & B \\ C & D \end{pmatrix}$, with $MZ = (AZ + B)(CZ + D)^{-1}$, and we define

$$M_\alpha(Z, \phi) = (MZ,\ \phi + \arg|CZ + D| + \alpha),$$

and as before

$$\mu(Z, \phi) = (-\bar{Z}, -\phi).$$

In this space again the point-pair invariant of the two points (Z, ϕ) and (Z^*, ϕ^*) which has the form

$$\omega_k(Z, \phi;\ Z^*, \phi^*) = \frac{|Y|^{k/2}\,|Y^*|^{k/2}}{\left|\dfrac{Z - \bar{Z}^*}{2i}\right|^k} e^{-ik(\phi - \phi^*)},$$

where $Z = X + iY$, is an eigenfunction for every integer k and for k positive and large enough‡ it will again satisfy our requirements (a) and (b).

We shall finally briefly indicate the most general result that we at present can obtain along these lines. Let there be in our space S a sequence of l-tuples $\lambda^{(k)}$, $k = 1, 2, 3 \ldots$, with the property that we have the relation

‡ If we consider the symplectic space of dimension $n^2 + n$, this takes place for $k > 2n$.

$$\omega_{\lambda^{(k)}}(x, y) = (\omega_{\lambda^{(1)}}(x, y))^k$$

for all positive integers k, where as before $\omega_\lambda(x, y)$ denotes the eigenfunction in x that corresponds to the l-tuple λ, and has rotational symmetry around the point y and is normed so as to take the value 1 for $x = y$. Further assume that for k sufficiently large and positive,

$$\int_S |\omega_{\lambda^{(k)}}(x, y)| \, dy < \infty;$$

then $\omega_{\lambda^{(k)}}(x, y)$ can be seen to satisfy both conditions (a) and (b). If we now have a discontinuous group Γ whose fundamental domain is compact, with a representation by unitary matrices χ, in our space, and denote the number of eigenfunctions corresponding to the eigenvalue $\lambda^{(k)}$ by N_k, then one can show that for k sufficiently large, N_k is given by a finite expression,

$$N_k = P_0(k) + \Sigma \, \epsilon_i^k \, P_i(k), \tag{4.6}$$

where P_0 is a polynomial and the P_i certain polynomials in general of lower degree† and the ϵ_i are certain roots of unity, such that if q_i is the smallest positive integer for which $\epsilon_i^{q_i} = 1$, the number q_i divides the order of some element§ in Γ which is of finite order.

The Institute for Advanced Study
Princeton, N. J., U.S.A.

† The only case when some of them can be of the same degree as P_0 is when Γ contains other elements than the identity which commute with the whole group G.

§ Different from the identity.

REMARKS ON MULTIPLICATIVE FUNCTIONS

Atle Selberg
Institute for Advanced Study, Princeton, New Jersey 08540

My principal reason for choosing this rather elementary topic is to draw attention to the uses of multiplicative functions in more than one variable.

1. We begin by recalling the standard definition of a multiplicative function of one variable defined on the positive integers: it is a function satisfying the conditions

$$f(m)\, f(n) = f(mn) \quad \text{for} \quad (m,n) = 1, \tag{1.1}$$

and

$$f(1) = 1. \tag{1.2}$$

I have never been very satisfied with this definition, and would prefer to define a multiplicative function as follows:

Write

$$n = \prod_p p^a, \tag{1.3}$$

where the product extends over all primes (so that all but a finite number of the a are zero). Let there be defined for each p a function $f_p(a)$ on the non-negative integers such that $f_p(0) = 1$ except for at most finitely many p. Then

$$f(n) = \prod_p f_p(a) \tag{1.4}$$

defines a multiplicative function.

This definition is clearly more general than the previous one*). We call $f(n)$ singular if $f(1) = 0$, otherwise we call $f(n)$ regular. If finally $f(1) = 1$, we say that $f(n)$ is normal. The class of multiplicative functions

*) It should be noted that it permits $f(n)$ to vanish identically.

defined by the standard definition coincides with the class of normal multiplicative functions according to our new definition.

With the new definition it remains true for instance that if $f(n)$ and $g(n)$ are multiplicative, then so is the convolution

$$f * g(n) = \sum_{d|n} f(d)\, g\left(\frac{n}{d}\right),$$

it also remains true that if $f(n)$ is multiplicative and a a positive integer then $f((a,n))$ is multiplicative. However, with our new definition, $f(an)$ and $f([a,n])$*) are also multiplicative, something which is not necessarily true with the standard definition.

Another advantage is that the new definition can be used without change to define multiplicative functions of several variables.

If we denote by $\{n\}_r$ an r-tuple of positive integers $n_1,\dots,n_r$ and write

$$\{n\}_r = \prod_p p^{\{a\}_r} \tag{1.5}$$

to denote that

$$n_i = \prod_p p^{a_i} \qquad \text{for } i = 1,2,\dots,r,$$

we say that a function $f(n_1,\dots,n_r) = f(\{n\}_r)$ is multiplicative if we can write it in the form

$$f(\{n\}_r) = \prod_p f_p(\{a\}_r), \tag{1.6}$$

where the functions $f_p(\{a\}r)$ satisfy the following conditions.

For each p, $f_p(a_1,\dots,a_r)$ is defined on the r-tuples of nonnegative integers, $f_p(0,\dots,0) = 1$ except for at most finitely many p.

Again, writing $\{1\}_r$ for the r-tuple all of whose entries are 1, we say that $f(\{n\}_r)$ is singular if $f(\{1\}_r) = 0$, regular if $f(\{1\}_r) \neq 0$, and normal if $f(\{1\}_r) = 1$.

It is easily seen that if one keeps some of the variables fixed in a multi-

*) We use $[a,n]$ to denote the least common multiple of a and n.

plicative function one gets a function which is multiplicative in the remaining variables.

Let us finally mention that in case of functions of one variable the class of multiplicative functions defined by (1.4) could also be defined by the requirements:

$$f(m)\ f(n) = f([m,n])\ f((m,n)) \tag{1.7}$$

for all positive integers m and n. This is, in spite of its simplicity, not as practical as the constructive definition (1.4). Also one meets complications when trying to adapt it to the case of several variables.

2. We shall now concentrate on functions of two positive integral variables, though as of yet we shall not necessarily assume them to be multiplicative. We say that a function $f(m,n)$ is symmetric if $f(m,n) = f(n,m)$, lower triangular if $f(m,n) = 0$ for $n > m$, and finally normal lower triangular if $f(n,n) = 1$ for all n.

If $t(m,n)$ is normal lower triangular and we have two sequences x_m and y_m connected by the relations

$$x_m = \sum_n t(m,n)\ y_n \tag{2.1}$$

then there exists a unique normal lower triangular function $t^*(m,n)$ such that

$$y_m = \sum_n t^*(m,n)\ x_n. \tag{2.2}$$

t and t^* are connected by the relations

$$\sum_\ell t(m,\ell)\ t^*(\ell,n) = \delta_{m,n}, \tag{2.3}$$

where $\delta_{m,n}$ is the Kronecker symbol, or, alternatively we have

$$\sum_\ell t^*(m,\ell)\ t(\ell,n) = \delta_{m,n}. \tag{2.3'}$$

If we assume that $t(m,n)$ is multiplicative, it follows immediately that $t(m,n) = 0$ unless $n|m$. It is not hard to see then that $t^*(m,n)$ is also multiplicative. Namely, let us define $\tilde{t}(p^r,p^s)$ for $r \geq s$ by the relations

(2.4) $$\sum_{s\le t\le r} t(p^r,p^t)\,\tilde{t}(p^t,p^s) = \delta_{r,s}.$$

Constructing now a multiplicative function $\tilde{t}(m,n) = \prod_p \tilde{t}(p^r,p^s)$ where $m = \prod_p p^r$, $n = \prod_p p^s$, we see that

(2.7) $$\sum_{\ell} t(m,\ell)\,\tilde{t}(\ell,n) = \delta_{m,n}$$

since the left hand side of (2.7) arises by multiplying together the left hand sides of (2.4) for all p. Thus $\tilde{t}(m,n) = t^*(m,n)$ which is therefore multiplicative.

When $t(m,n)$ is multiplicative as well as normal lower triangular (2.1) and (2.2) take the forms

(2.8) $$x_m = \sum_{d|m} t(m,d)\,y_d$$

and

(2.8') $$y_m = \sum_{d|m} t^*(m,d)\,x_d.$$

This generalizes the usual inversion formulae.

We have, of course, also the dual set of formulae: if $t(m,n)$ is not assumed multiplicative but is normal lower triangular and if the sequences x_m and y_m are connected by the relations

(2.9) $$x_n = \sum_m t(m,n)\,y_m,$$

then we have that

(2.9') $$y_n = \sum_m t^*(m,n)\,x_m.$$

Here, since the sums on the right hand side are infinite, one has to assume that, say, the y_m are such that the sums occurring converge absolutely. This, for instance, is the case if we assume that the y_m (and as a consequence also the x_m) vanish for m sufficiently large. For multiplicative $t(m,n)$ we get that if

(2.10) $$x_d = \sum_{d|m} t(m,d)\,y_m,$$

then

(2.10') $$y_d = \sum_{d|m} t^*(m,d)\, x_m.$$

We call a symmetric function $f(m,n)$ positive definite if the quadratic form

(2.11) $$Q(x) = \sum_{m,n} f(m,n)\, x_m\, x_n$$

satisfies $Q > 0$ for all real sequences x_m with at least one and at most finitely many non-zero elements.

For $f(m,n)$ positive definite, we can always find functions $g(n)$ and normal lower triangular $t(m,n)$ such that

(2.12) $$f(m,n) = \sum_{\ell} g(\ell)\, t(m,\ell)\, t(n,\ell),$$

these functions are uniquely determined and can be expressed rationally in terms of the $f(m,n)$.

If we, in addition, require $f(m,n)$ to be multiplicative, it is easily seen that both $g(n)$ and $t(m,n)$ will also be multiplicative*). Namely we define $\tilde{g}(p^r)$ and $\tilde{t}(p^r,p^s)$ by the relations $f(p^r,p^s) = \sum_t \tilde{g}(p^t)\, \tilde{t}(p^r,p^t)\, \tilde{t}(p^s,p^t)$ for all $r,s \geq 0$, $\tilde{t}(p^r,p^r) = 1$, and $\tilde{t}(p^r,p^s) = 0$ for $s > r$. For each p this determines $\tilde{g}(p^r)$ and $\tilde{t}(p^r,p^s)$ uniquely for all $r,s \geq 0$.

We now construct the multiplicative functions $\tilde{g}(m)$ and $\tilde{t}(m,n)$.

Writing $m = \prod p^r$; $n = \prod p^s$ we then have $f(m,n) = \prod_p f(p^r,p^s) = \prod_p \left(\sum_t \tilde{g}(p^t)\, \tilde{t}(p^r,p^t)\, \tilde{t}(p^s,p^t) \right) = \sum_\ell \tilde{g}(\ell)\, \tilde{t}(m,\ell)\, \tilde{t}(n,\ell)$. Thus $\tilde{g}$ and $\tilde{t}$ must be identical with g and t which are therefore multiplicative.

For multiplicative positive definite $f(m,n)$ (2.12) therefore assumes the form

(2.13) $$f(m,n) = \sum_{\substack{d|m \\ d|n}} g(d)\, t(m,d)\, t(n,d).$$

3. Suppose that we wish to determine the minimum of Q under the side conditions $x_n = 0$ for $n > N$ and $x_1 = 1$. Writing $f(m,n)$ in the form given by (2.12) we obtain

(3.1) $$Q(x) = \sum_\ell g(\ell) \left\{ \sum_m t(m,\ell)\, x_m \right\}^2 .$$

*) We assume for simplicity in this argument that $f(1,1)$ (and therefore also $g(1)$) equals 1. This is no restriction since we could otherwise divide by $f(1,1)$ which is positive.

Writing further

(3.2) $$y_n = \sum_m t(m,n)\, x_m,$$

so that also $y_n = 0$ for $n > N$, we get

(3.3) $$Q(x) = \sum_{n \le N} g(n)\, y_n^2 .$$

(3.2) gives

(3.4) $$x_n = \sum_m t^*(m,n)\, y_m.$$

In particular the condition $x_1 = 1$ takes the form

(3.5) $$\sum_{m \le N} t^*(m,1)\, y_m = 1.$$

The minimum is then by standard procedures found to be

(3.6) $$Q_{min} = \frac{1}{\sum_{m \le N} \frac{t^*(m,1)^2}{g(m)}} ,$$

and the minimizing x_n are given by

(3.7) $$x_n = Q_{min} \cdot \sum_{m \le N} \frac{t^*(m,n)\, t^*(m,1)}{g(m)} .$$

Finally, if we assume $f(m,n)$ to be multiplicative, then so are $g(m)$, $t(m,n)$ and $t^*(m,n)$. These functions have to be determined by computing the $g(p^r)$, $t(p^r,p^s)$ and $t^*(p^r,p^s)$ from the $f(p^r,p^s)$ for $r,s \ge 0$. This may be simple or not, depending on the nature of the $f(p^r,p^s)$.

4. We shall consider an application of the preceding to a sieve problem.

Let us denote by $p^r \| n$ that $p^r | n$ and $\left(\frac{n}{p^r}, p\right) = 1$, we say that p^r is an exact prime divisor of n.

Assume now that we have a set of primes p. For each p^r with $r > 0$ we designate a set of $\omega(p^r)$ residue classes modulo p^r to be removed or excluded. We assume that for fixed p the residue classes removed modulo p^r and modulo p^s are disjoint if $r \ne s$. We introduce the notation $nX(p^r)$ to denote that the integer n lies in one of the $\omega(p^r)$ residue classes excluded modulo p^r. We define $nX(1)$ for all integers n, and further for $d > 1$, $nX(d)$ if $nX(p^r)$ whenever $p^r \| d$.

Let there now be given a set of integers n with each of which there is associated a weight $w_n \geq 0$. We assume that $W = \sum w_n < \infty$. Writing $\lambda_1 = 1$; $\lambda_d = 0$ for $d > Z$ and leaving the other λ_d as free real variables, we form the expression

$$(4.1) \qquad Q(\lambda) = \sum_n w_n \left\{ \sum_{nX(d)} \lambda_d \right\}^2 .$$

Clearly $Q(\lambda)$ is always an upper bound for the sum of the weights of the integers which remain after we have removed those that lie in any of the $\omega(p^r)$ excluded residue classes modulo each p^r. Under rather general assumptions about the set of weights w_n the quadratic form $Q(\lambda)$ can be written in the form

$$Q(\lambda) = Q_1(\lambda) + R$$

where

$$Q_1(\lambda) = W \sum_{d,d'} f(d,d') \lambda_d \lambda_{d'}$$

and $f(d,d')$ is a symmetric multiplicative function (positive definite, of course) and R is a remainder term generally bounded by a simple quadratic form in the $|\lambda_d|$. The machinery from the previous section then applies, one can minimize $Q_1(\lambda)$ subject to the side conditions on the λ's, the choice of the parameter Z is then determined by the requirement that R should be small enough not to spoil the result.

We shall apply this technique to the case of an interval I_X of length X (so that we assign the values $w_n = 1$, for $n \in I_X$ and $w_n = 0$ outside I_X. We introduce a symmetric multiplicative function $\varepsilon(d,d')$ by defining $\varepsilon(p^r,p^s) = 1$ if $r = s$ or if $rs = 0$, otherwise we define $\varepsilon(p^r,p^s) = 0$. We say that d and d' are compatible if $\varepsilon(d,d') = 1$, otherwise they are said to be incompatible. We now get

$$Q(\lambda) = \sum_{n\in I_X} \left\{ \sum_{n X(d)} \lambda_d \right\}^2 \leq X \sum_{d,d'} f(d,d') \lambda_d \lambda_{d'} \tag{4.2}$$

$$+ \sum_{d,d'} |\lambda_d| \, |\lambda_{d'}| \, \omega(d,d') \, \varepsilon(d,d'),$$

where

$$f(d,d') = \frac{\omega([d,d'])}{[d,d']} \varepsilon(d,d'). \tag{4.3}$$

Here $\omega(d)$ is the multiplicative function defined by $\omega(1) = 1$ and

$$\omega(d) = \prod_{p^r \| d} \omega(p^r).$$

An alternate form of the upper bound for the interval I_X can be obtained as follows*). Consider a function $w(u)$ defined on the real line, such that $w(u) \geq 1$ for u in I_X and $w(u) \geq 0$ always. We furthermore require that its fourier transform

$$\hat{w}(v) = \int_{-\infty}^{\infty} w(u)\, e^{2\pi iuv}\, du$$

should vanish identically for $|v| > \frac{1}{z^2}$. We then have

$$Q(\lambda) \leq \sum_{-\infty}^{\infty} w(n) \left\{ \sum_{n X(d)} \lambda_d \right\}^2 = \hat{w}(0) \sum_{d,d'} f(d,d') \lambda_d \lambda_{d'}. \tag{4.4}$$

It can be shown that we can choose $w(u)$ satisfying our conditions and such that $\hat{w}(0) \leq X + z^2$. Thus we get

$$Q(\lambda) \leq (X+z^2) \sum_{d,d'} f(d,d') \lambda_d \lambda_{d'}. \tag{4.4'}$$

To use the results of the previous section to minimize the quadratic form on the right hand side of (4.4'), we observe that we have $f(p^r,p^s) = 0$ for $(r-s)rs \neq 0$, and $f(p^r,p^r) = f(p^r,1) = f(1,p^r) = \frac{\omega(p^r)}{p^r}$. Writing

$$\theta(p^r) = 1 - \sum_{1\leq s\leq r} \frac{\omega(p^s)}{p^s}, \tag{4.5}$$

we have $g(1) = 1$, and for $r > 1$

*) See Selberg [1].

$$(4.6)\qquad g(p^r) = (\theta(p^{r-1}) - \theta(p^r))\,\frac{\theta(p^r)}{\theta(p^{r-1})}\,.$$

For $r > s$, we find

$$(4.7)\qquad t(p^r,p^s) = \begin{cases} \theta(p^{r-1})-\theta(p^r), & \text{if } s = 0 \\ -\dfrac{\theta(p^{r-1})-\theta(p^r)}{\theta(p^s)}, & \text{if } 0 < s < r\,. \end{cases}$$

Similarly for $r > s$

$$(4.8)\qquad t^*(p^r,p^s) = \begin{cases} -\dfrac{\theta(p^{r-1})-\theta(p^r)}{\theta(p^{r-1})}, & \text{if } s = 0 \\ \dfrac{\theta(p^{r-1})-\theta(p^r)}{\theta(p^{r-1})}, & \text{if } 0 < s < r, \end{cases}$$

Thus

$$(4.9)\qquad \frac{t^*(p^r,1)^2}{g(p^r)} = \left(\frac{1}{\theta(p^r)} - \frac{1}{\theta(p^{r-1})}\right)\,.$$

Thus the minimum of

$$\sum f(d,d')\,\lambda_d\,\lambda_{d'}$$

is

$$\frac{1}{\displaystyle\sum_{d\le Z}\prod_{p^r\| d}\left(\frac{1}{\theta(p^r)} - \frac{1}{\theta(p^{r-1})}\right)}\,.$$

From (4.4') we thus get the upper bound of the number of integers that the sieve leaves in I_X as

$$(4.10)\qquad \frac{X + Z^2}{\displaystyle\sum_{d\le Z}\prod_{p^r\| d}\left(\frac{1}{\theta(p^r)} - \frac{1}{\theta(p^{r-1})}\right)}\,.$$

A result, identical in form with (4.10), has been given earlier by Gallagher and Johnsen*), who derived it from the large sieve and under the rather restrictive assumption that for each $r > 1$ the $\omega(p^r)$ residue classes excluded modulo p^r are

*) See Gallagher [2].

equidistributed among the $p^{r-1}\,\theta(p^{r-1})$ residue classes that remain modulo p^{r-1} after the earlier exclusions modulo $p,\ldots,p^{r-1}$. This restriction implies, for instance, that $\dfrac{\omega(p^r)}{p^{r-1}\,\theta(p^{r-1})}$ is always an integer (or, otherwise expressed, $p\cdot\dfrac{\theta(p^r)}{\theta(p^{r-1})}$ is always an integer)[*)].

References

1. A. Selberg: Remarks on sieves. Proceedings of the 1972 Number Theory Conference, University of Colorado, 1972, pp. 205-216.

2. P. X. Gallagher: Sieving by prime powers. Proceedings of the 1972 Number Theory Conference, University of Colorado, 1972, pp. 95-99.

[*)] When I first saw Gallagher's paper [2], I thought that the unnecessary restriction was due to faulty technique in the use of the large sieve inequality. As I, however, was unable to derive (4.10) from the large sieve without imposing the same restriction, it remains an open question whether the more general result given here is implied by the large sieve inequality or not.

Curriculum Vitae
Saharon Shelah

Date of Birth: July 3, 1945, Jerusalem, Israel

Education:

October 1962–June 1964	B.Sc. with distinction, Tel Aviv University
October 1965–July 1968	M.Sc. with distinction, Hebrew University (M.Sc. Thesis in Mathematics under the supervision of Professor H. Gaifman, On Hanf numbers)
October 1967–July 1969	Ph.D. in Mathematics, Hebrew University, Jerusalem (under the supervision of Professor M. O. Rabin)
March 16, 1970	Ph.D. Summa Cum Laude

Employment History:

October 1967–March 1969	Teaching assistant in the Institute of Mathematics, The Hebrew University of Jerusalem
April–September 1969	Instructor, Institute of Mathematics, The Hebrew University of Jerusalem
September 1969–July 1970	Lecturer in the Mathematics Department, Princeton University
July 1970–July 1971	Assistant Professor (tenure track), University of California Los Angeles
October 1, 1971–April 30, 1972	Assistant Professor in the Institute of Mathematics, The Hebrew University of Jerusalem
May 1, 1972–December 1, 1974	Associate Professor, Institute of Mathematics, The Hebrew University of Jerusalem
December 1, 1974–	Professor, Institute of Mathematics, The Hebrew University of Jerusalem
October 1986–	Distinguished Visiting Professor, Rutgers University

Honorary Positions:

October 1980–September 1981	Head of the Model Theory group in the Institute for Advanced Studies, Jerusalem

October 1992–	Member of the General Editorial Board of Israel Journal of Mathematics and Journal D'Analyse Math´ematique
January 1994–	Editor of Fundamenta Mathematicae
January 1994–	Honorary Editor of Mathematica Japonica
January 1996–	Editor of Journal of Applied Analysis
January 1998–	Editor of Asian Journal of Mathematics

Awards:

1977	Awarded Erdős Prize
1978	A. Robinson Chair for Mathematical Logic
1982	Rothschild Prize in Mathematics
1983	C. Karp prize
1988	Academy of Science and Humanities
1991	Foreign Honorary Member of the American Academy of Arts and Sciences
1991	Awarded SIAM 1991 George Polya Prize in Applications of Combinatorial Mathematics
1996	Gödel Lecture in Annual Meeting of ASL, Madison, Wisconsin
1998	Israel Prize in Mathematics
1999	JAMS Prize
2000	Janos Bolyai prize by the Hungarian Academy of Sciences
2001	Wolf Prize

List of Publications

1. S. Shelah, On long increasing chains modulo flat ideals, *MLQ Math. Log. Q.* **56** (2010), No. 4, 397–399.
2. P. Larson, I. Neeman and S. Shelah, Universally measurable sets in generic extensions, *Fund. Math.* **208** (2010), No. 2, 173–192.
3. D. Herden and S. Shelah, κ-fold transitive groups, *Forum Math.* **22** (2010), No. 4, 627–640.
4. S. Shelah and B. Tsaban, On a problem of Juhász and van Mill, *Topology Proc.* **36** (2010), 385–392.
5. S. Shelah, Large continuum, oracles, *Cent. Eur. J. Math.* **8** (2010), No. 2, 213–234.
6. S. Shelah and L. Strüngmann, Filtration-equivalent $\aleph_1$-separable abelian groups of cardinality $\aleph_1$, *Ann. Pure Appl. Logic* **161** (2010), No. 7, 935–943.
7. S. Shelah, Diamonds, *Proc. Amer. Math. Soc.* **138** (2010), No. 6, 2151–2161.
8. J. Kellner and S. Shelah, A sacks real out of nowhere, *J. Symbolic Logic* **75** (2010), No. 1, 51–76.
9. M. Machura, S. Shelah and B. Tsaban, Squares of Menger-bounded groups, *Trans. Amer. Math. Soc.* **362** (2010), No. 4, 1751–1764.
10. S. Shelah, *Classification Theory for Abstract Elementary Classes*, Vol. 2, Studies in Logic (London), Vol. 20, Mathematical Logic and Foundations (College Publications, 2009).
11. S. Shelah, *Classification Theory for Elementary Abstract Classes*, Studies in Logic (London), Vol. 18, Mathematical Logic and Foundations (College Publications, 2009).
12. S. Shelah, Dependent first order theories, continued, *Israel J. Math.* **173** (2009), 1–60.
13. T. Eisworth and S. Shelah, Successors of singular cardinals and coloring theorems, II, *J. Symbolic Logic* **74** (2009), No. 4, 1287–1309.
14. I. Kaplan and S. Shelah, The automorphism tower of a centerless group without choice, *Arch. Math. Logic* **48** (2009), No. 8, 799–815.
15. R. Göbel, D. Herden and S. Shelah, Skeletons, bodies and generalized $E(R)$-algebras, *J. Eur. Math. Soc. (JEMS)* **11** (2009), No. 4, 845–901.
16. J. T. Baldwin, A. Kolesnikov and S. Shelah, The amalgamation spectrum, *J. Symbolic Logic* **74** (2009), No. 3, 914–928.
17. S. Shelah and L. Strüngmann, Large indecomposable minimal groups, *Q. J. Math.* **60** (2009), No. 3, 353–365.

18. H. Mildenberger and S. Shelah, Specializing Aronszajn trees and preserving some weak diamonds, *J. Appl. Anal.* **15** (2009), No. 1, 47–78.
19. R. Göbel and S. Shelah, $\aleph_n$-free modules with trivial duals, *Results Math.* **54** (2009), No. 1–2, 53–64.
20. I. Juhász, S. Shelah and L. Soukup, Resolvability vs. almost resolvability, *Topology Appl.* **156** (2009), No. 11, 1966–1969.
21. H. Ardal, J. Maňuch, M. Rosenfeld, S. Shelah and L. Stacho, The odd-distance plane graph, *Discrete Comput. Geom.* **42** (2009), No. 2, 132–141.
22. P. Larson and S. Shelah, Splitting stationary sets from weak forms of choice, *MLQ Math. Log. Q.* **55** (2009), No. 3, 299–306.
23. D. Herden and S. Shelah, An upper cardinal bound on absolute *E*-rings, *Proc. Amer. Math. Soc.* **137** (2009), No. 9, 2843–2847.
24. S. Shelah, A comment on "$p < t$". *Canad. Math. Bull.* **52** (2009), No. 2, 303–314.
25. S. Shelah, Model theory without choice? Categoricity, *J. Symbolic Logic* **74** (2009), No. 2, 361–401.
26. A. Dow and S. Shelah, More on tie-points and homeomorphism in $N*$. *Fund. Math.* **203** (2009), No. 3, 191–210.
27. S. Shelah, What majority decisions are possible, *Discrete Math.* **309** (2009), No. 8, 2349–2364.
28. J. Kellner and S. Shelah, Decisive creatures and large continuum, *J. Symbolic Logic* **74** (2009), No. 1, 73–104.
29. S. Shelah, The Erdös–Rado arrow for singular cardinals, *Canad. Math. Bull.* **52** (2009), No. 1, 127–131.
30. H. Mildenberger and S. Shelah, The near coherence of filters principle does not imply the filter dichotomy principle, *Trans. Amer. Math. Soc.* **361** (2009), No. 5, 2305–2317.
31. I. Juhász and S. Shelah, Hereditarily Lindelöf spaces of singular density, *Studia Sci. Math. Hungar.* **45** (2008), No. 4, 557–562.
32. S. Shelah, Theories with Ehrenfeucht–Fraïssé equivalent non-isomorphic models, *Tbil. Math. J.* **1** (2008), 133–164.
33. A. Blass and S. Shelah, Basic subgroups and freeness, a counterexample, in *Models, Modules and Abelian Groups*, pp. 63–73 (Walter de Gruyter, 2008).
34. S. Shelah and A. Usvyatsov, More on SOP_1 and SOP_2. *Ann. Pure Appl. Logic* **155** (2008), No. 1, 16–31.
35. J. Kennedy, S. Shelah and J. Väänänen, Regular ultrafilters and finite square principles, *J. Symbolic Logic* **73** (2008), No. 3, 817–823.
36. J. T. Baldwin and S. Shelah, Examples of non-locality. *J. Symbolic Logic* **73** (2008), No. 3, 765–782.

37. S. Shelah, *EF*-equivalent not isomorphic pair of models, *Proc. Amer. Math. Soc.* **136** (2008), No. 12, 4405–4412.
38. S. Shelah, Vive la différence, III, *Israel J. Math.* **166** (2008), 61–96.
39. A. Dow and S. Shelah, Tie-points and fixed-points in N_*. *Topology Appl.* **155** (2008), No. 15, 1661–1671.
40. S. Shelah, Groupwise density cannot be much bigger than the unbounded number, *MLQ Math. Log. Q.* **54** (2008), No. 4, 340–344.
41. T. Bartoszynski and S. Shelah, On the density of Hausdorff ultrafilters, in *Logic Colloquium 2004*, pp. 18–32, Lecture Notes Log., Vol. 29 (*Assoc. Symbol. Logic,* 2008).
42. A. Rosłanowski and S. Shelah, Generating ultrafilters in a reasonable way, *MLQ Math. Log. Q.* **54** (2008), No. 2, 202–220.
43. P. B. Larson and S. Shelah, The stationary set splitting game, *MLQ Math. Log. Q.* **54** (2008), No. 2, 187–193.
44. S. Garti and S. Shelah, On depth and depth$_+$ of Boolean algebras, *Algebra Universalis* **58** (2008), No. 2, 243–248.
45. S. Geschke and S. Shelah, The number of openly generated Boolean algebras, *J. Symbolic Logic* **73** (2008), No. 1, 151–164.
46. S. Shelah, Reflection implies the SCH, *Fund. Math.* **198** (2008), No. 2, 95–111.
47. S. Shelah, The spectrum of characters of ultrafilters on ω, *Colloq. Math.* **111** (2008), No. 2, 213–220.
48. S. Shelah, Minimal bounded index subgroup for dependent theories, *Proc. Amer. Math. Soc.* **136** (2008), No. 3, 1087–1091 (electronic).
49. J. Kellner, M. Pauna and S. Shelah, Winning the pressing down game but not Banach–Mazur, *J. Symbolic Logic* **72** (2007), No. 4, 1323–1335.
50. S. Shelah and M. Doron, Relational structures constructible by quantifier free definable operations, *J. Symbolic Logic* **72** (2007), No. 4, 1283–1298.
51. H. Mildenberger and S. Shelah, Increasing the groupwise density number by c.c.c. forcing, *Ann. Pure Appl. Logic* **149** (2007), No. 1–3, 7–13.
52. E. D. Farjoun, R. Göbel, Y. Segev and S. Shelah, On kernels of cellular covers, *Groups Geom. Dyn.* **1** (2007), No. 4, 409–419.
53. S. Garti and S. Shelah, Two cardinal models for singular μ, *MLQ Math. Log. Q.* **53** (2007), No. 6, 636–641.
54. S. Shelah and J. Steprāns, Possible cardinalities of maximal abelian subgroups of quotients of permutation groups of the integers, *Fund. Math.* **196** (2007), No. 3, 197–235.
55. S. Shelah and L. Shtryungman, On the *p*-rank of ExtZ(G, Z) in certain models of ZFC (in Russian), *Algebra Logika* **46** (2007), No. 3, 369–397; 403–404; translation in *Algebra Logic* **46** (2007), No. 3, 200–215.

56. S. Shelah, $\aleph_n$-free abelian group with no non-zero homomorphism to Z, *Cubo* **9** (2007), No. 2, 59–79.
57. A. Rosłanowski and S. Shelah, Sheva–Sheva–Sheva: Large creatures, *Israel J. Math.* **159** (2007), 109–174.
58. C. Havlin and S. Shelah, Existence of EF-equivalent non-isomorphic models, *MLQ Math. Log. Q.* **53** (2007), No. 2, 111–127.
59. A. Rosłanowski and S. Shelah, Universal forcing notions and ideals, *Arch. Math. Logic* **46** (2007), No. 3–4, 179–196.
60. G. Cherlin and S. Shelah, Universal graphs with a forbidden subtree, *J. Combin. Theory Ser. B* **97** (2007), No. 3, 293–333.
61. S. Shelah, Power set modulo small, the singular of uncountable cofinality, *J. Symbolic Logic* **72** (2007), No. 1, 226–242.
62. R. Göbel and S. Shelah, Absolutely indecomposable modules, *Proc. Amer. Math. Soc.* **135** (2007), No. 6, 1641–1649 (electronic).
63. S. Shelah and L. Strüngmann, A characterization of $\mathrm{Ext}(G, Z)$ assuming $(V=L)$, *Fund. Math.* **193** (2007), No. 2, 141–151.
64. H. Mildenberger, S. Shelah and B. Tsaban, The combinatorics of τ-covers, *Topology Appl.* **154** (2007), No. 1, 263–276.
65. A. Rosłanowski and S. Shelah, Reasonably complete forcing notions, in *Set Theory: Recent Trends and Applications*, pp. 195–239, Quad. Mat., Vol. 17 (Seconda Univ., 2006).
66. A. Blass and S. Shelah, Disjoint non-free subgroups of abelian groups, in *Set Theory: Recent Trends and Applications*, pp. 1–24, Quad. Mat., Vol. 17 (Seconda Univ., 2006).
67. G. M. Bergman and S. Shelah, Closed subgroups of the infinite symmetric group, Special issue dedicated to Walter Taylor, *Algebra Universalis* **55** (2006), No. 2–3, 137–173.
68. S. Shelah, The combinatorics of reasonable ultrafilters, *Fund. Math.* **192** (2006), No. 1, 1–23.
69. R. Göbel and S. Shelah, Generalized E-algebras via λ-calculus, I, *Fund. Math.* **192** (2006), No. 2, 155–181.
70. M. Kojman and S. Shelah, Almost isometric embedding between metric spaces, *Israel J. Math.* **155** (2006), 309–334.
71. M. Gitik, R. Schindler and S. Shelah, PCF theory and Woodin cardinals, *Logic Colloquium '02*, pp. 172–205, Lecture Notes Log., Vol. 27 (Assoc. Symbol. Logic, 2006).
72. F. Niedermeyer, S. Shelah and K. Steffens, The f-factor problem for graphs and the hereditary property, *Arch. Math. Logic* **45** (2006), No. 6, 665–672.
73. S. Fuchino, N. Greenberg and S. Shelah, Models of real-valued measurability, *Ann. Pure Appl. Logic* **142** (2006), No. 1–3, 380–397.

74. S. Shelah, Non-Cohen oracle C.C.C, *J. Appl. Anal.* **12** (2006), No. 1, 1–17.
75. M. C. Laskowski and S. Shelah, Decompositions of saturated models of stable theories, *Fund. Math.* **191** (2006), No. 2, 95–124.
76. R. Göbel and S. Shelah, Torsionless linearly compact modules, in *Abelian Groups, Rings, Modules, and Homological Algebra*, pp. 153–158, Lecture Notes Pure Appl. Math., Vol. 249 (Chapman & Hall/CRC, 2006).
77. S. Shelah, More on the revised GCH and the black box, *Ann. Pure Appl. Logic* **140** (2006), No. 1–3, 133–160.
78. H. Mildenberger, S. Shelah and B. Tsaban, Covering the Baire space by families which are not finitely dominating, *Ann. Pure Appl. Logic* **140** (2006), No. 1–3, 60–71.
79. S. Shelah and A. Usvyatsov, Banach spaces and groups — Order properties and universal models, *Israel J. Math.* **152** (2006), 245–270.
80. S. Shelah and J. Väänänen, Recursive logic frames, *MLQ Math. Log. Q.* **52** (2006), No. 2, 151–164.
81. A. Rosłanowski and S. Shelah, Measured creatures, *Israel J. Math.* **151** (2006), 61–110.
82. G. Sági and S. Shelah, On weak and strong interpolation in algebraic logics, *J. Symbolic Logic* **71** (2006), No. 1, 104–118.
83. S. Shelah, On long *EF*-equivalence in non-isomorphic models, in *Logic Colloquium '03*, pp. 315–325, Lecture Notes Log., Vol. 24 (Assoc. Symbol. Logic, 2006).
84. M. Džamonja and S. Shelah, On properties of theories which preclude the existence of universal models, *Ann. Pure Appl. Logic* **139** (2006), No. 1–3, 280–302.
85. A. Rosłanowski and S. Shelah, How much sweetness is there in the universe?, *MLQ Math. Log. Q.* **52** (2006), No. 1, 71–86.
86. S. Shelah and M. Shioya, Nonreflecting stationary sets in $P_\kappa\lambda$, *Adv. Math.* **199** (2006), No. 1, 185–191.
87. P. Matet, C. Péan and S. Shelah, Confinality of normal ideals on $P_\kappa(\lambda)$, II, *Israel J. Math.* **150** (2005), 253–283.
88. S. Shelah, The depth of ultraproducts of Boolean algebras, *Algebra Universalis* **54** (2005), No. 1, 91–96.
89. T. Eisworth and S. Shelah, Successors of singular cardinals and coloring theorems, I, *Arch. Math. Logic* **44** (2005), No. 5, 597–618.
90. S. Shelah, Middle diamond, *Arch. Math. Logic* **44** (2005), No. 5, 527–560.
91. S. Shelah and M. Doron, A dichotomy in classifying quantifiers for finite models, *J. Symbolic Logic* **70** (2005), No. 4, 1297–1324.
92. S. Shelah, P. Väisänen and J. Väänänen, On ordinals accessible by infinitary languages, *Fund. Math.* **186** (2005), No. 3, 193–214.

93. R. Göbel and S. Shelah, On Crawley modules, *Comm. Algebra* **33** (2005), No. 11, 4211–4218.
94. S. Kuhlmann and S. Shelah, κ-bounded exponential-logarithmic power series fields, *Ann. Pure Appl. Logic* **136** (2005), No. 3, 284–296.
95. P. Matet, A. Rosłanowski and S. Shelah, Cofinality of the nonstationary ideal, *Trans. Amer. Math. Soc.* **357** (2005), No. 12, 4813–4837 (electronic).
96. S. Shelah, Two cardinals models with gap one revisited, *MLQ Math. Log. Q.* **51** (2005), No. 5, 437–447.
97. R. Göbel and S. Shelah, How rigid are reduced products?, *J. Pure Appl. Algebra* **202** (2005), No. 1–3, 230–258.
98. S. Shelah, Zero-one laws for graphs with edge probabilities decaying with distance, II, *Fund. Math.* **185** (2005), No. 3, 211–245.
99. J. Kellner and S. Shelah, Preserving preservation. *J. Symbolic Logic* **70** (2005), No. 3, 914–945.
100. S. Shelah, The pair $(\aleph_n, \aleph_0)$ may fail $\aleph_0$-compactness, *Logic Colloquium '01*, pp. 402–433, Lecture Notes Log., Vol. 20 (Assoc. Symbol. Logic, 2005).
101. T. Hyttinen, O. Lessmann and S. Shelah, Interpreting groups and fields in some nonelementary classes, *J. Math. Log.* **5** (2005), No. 1, 1–47.
102. S. Shelah, On nicely definable forcing notions, *J. Appl. Anal.* **11** (2005), No. 1, 1–17.
103. A. Blass and S. Shelah, Ultrafilters and partial products of infinite cyclic groups, *Comm. Algebra* **33** (2005), No. 6, 1997–2007.
104. M. Goldstern and S. Shelah, Clones from creatures, *Trans. Amer. Math. Soc.* **357** (2005), No. 9, 3525–3551 (electronic).
105. G. Sági and S. Shelah, On topological properties of ultraproducts of finite sets, *MLQ Math. Log. Q.* **51** (2005), No. 3, 254–257.
106. L. Kramer, S. Shelah, K. Tent and S. Thomas, Asymptotic cones of finitely presented groups, *Adv. Math.* **193** (2005), No. 1, 142–173.
107. P. Komjáth and S. Shelah, Finite subgraphs of uncountably chromatic graphs, *J. Graph Theory* **49** (2005), No. 1, 28–38.
108. S. Shelah and J. Steprāns, Comparing the uniformity invariants of null sets for different measures, *Adv. Math.* **192** (2005), No. 2, 403–426.
109. B. Baizhanov, J. T. Baldwin and S. Shelah, Subsets of superstable structures are weakly benign, *J. Symbolic Logic* **70** (2005), No. 1, 142–150.
110. S. Shelah and J. Väänänen, A note on extensions of infinitary logic, *Arch. Math. Logic* **44** (2005), No. 1, 63–69.
111. S. Shelah, On the Arrow property, *Adv. Appl. Math.* **34** (2005), No. 2, 217–251.

112. J. C. Kennedy and S. Shelah, More on regular reduced products, *J. Symbolic Logic* **69** (2004), No. 4, 1261–1266.
113. A. Soifer and S. Shelah, How the axiom of choice can affect the chromatic number of distance graphs: Three examples on the plane, in *Proceedings of the 35th Southeastern International Conference on Combinatorics, Graph Theory and Computing, Congr. Numer.* **166** (2004), 5–9.
114. S. Shelah, Properness without elementaricity, *J. Appl. Anal.* **10** (2004), No. 2, 169–289.
115. S. Shelah, Two cardinal invariants of the continuum ($d < a$) and FS linearly ordered iterated forcing, *Acta Math.* **192** (2004), No. 2, 187–223.
116. M. Džamonja and S. Shelah, On the existence of universal models, *Arch. Math. Logic* **43** (2004), No. 7, 901–936.
117. S. Shelah, Quite complete real closed fields. *Israel J. Math.* **142** (2004), 261–272.
118. A. Rosłanowski and S. Shelah, Sweet & sour and other flavours of ccc forcing notions, *Arch. Math. Logic* **43** (2004), No. 5, 583–663.
119. M. Kojman, W. Kubiś and S. Shelah, On two problems of Erdős and Hechler: New methods in singular madness, *Proc. Amer. Math. Soc.* **132** (2004), No. 11, 3357–3365 (electronic).
120. I. Juhász, S. Shelah, L. Soukup and Z. Szentmiklóssy, Cardinal sequences and Cohen real extensions, *Fund. Math.* **181** (2004), No. 1, 75–88.
121. P. C. Eklof, S. Shelah and J. Trlifaj, On the cogeneration of cotorsion pairs, *J. Algebra* **277** (2004), No. 2, 572–578.
122. S. Shelah, Spectra of monadic second order sentences, Special issue on set theory and algebraic model theory, *Sci. Math. Jpn.* **59** (2004), No. 2, 351–355.
123. S. Shelah, Classification theory for elementary classes with the dependence property — A modest beginning, Special issue on set theory and algebraic model theory, *Sci. Math. Jpn.* **59** (2004), No. 2, 265–316.
124. S. Shelah, Anti-homogeneous partitions of a topological space, Special issue on set theory and algebraic model theory, *Sci. Math. Jpn.* **59** (2004), No. 2, 203–255.
125. H. Mildenberger and S. Shelah, On needed reals, *Israel J. Math.* **141** (2004), 1–37.
126. U. Abraham and S. Shelah, Ladder gaps over stationary sets, *J. Symbolic Logic* **69** (2004), No. 2, 518–532.
127. R. Göbel, S. Shelah and L. Strüngmann, Generalized *E*-rings, in *Rings, Modules, Algebras, and Abelian Groups*, pp. 291–306, Lecture Notes in Pure and Appl. Math., Vol. 236 (Dekker, 2004).

128. R. Göbel and S. Shelah, Uniquely transitive torsion-free abelian groups, in *Rings, Modules, Algebras, and Abelian Groups*, pp. 271–290, Lecture Notes in Pure and Appl. Math., Vol. 236 (Dekker, 2004).
129. S. Shelah, Characterizing an $\aleph_\varepsilon$-saturated model of superstable NDOP theories by its $L_{\infty,\aleph_\varepsilon}$-theory, *Israel J. Math.* **140** (2004), 61–111.
130. S. Shelah, Forcing axiom failure for any $\lambda > \aleph_1$, *Arch. Math. Logic* **43** (2004), No. 3, 285–295.
131. A. Soifer and S. Shelah, Axiom of choice and chromatic number: Examples on the plane, *J. Combin. Theory Ser. A* **105** (2004), No. 2, 359–364.
132. V. Kanovei and S. Shelah, A definable nonstandard model of the reals, *J. Symbolic Logic* **69** (2004), No. 1, 159–164.
133. S. Shelah, On nice equivalence relations on λ_2, *Arch. Math. Logic* **43** (2004), No. 1, 31–64.
134. M. Džamonja and S. Shelah, On $\lhd_*$-maximality, *Ann. Pure Appl. Logic* **125** (2004), No. 1–3, 119–158.
135. S. Shelah and J. Zapletal, Games with creatures, *Comment. Math. Univ. Carolin.* **44** (2003), No. 1, 9–21.
136. P. Komjáth and S. Shelah, A partition theorem for scattered order types. Special issue on Ramsey theory, *Combin. Probab. Comput.* **12** (2003), No. 5–6, 621–626.
137. R. Göbel, S. Shelah and S. L. Wallutis, On universal and epi-universal locally nilpotent groups, Special issue in honor of Reinhold Baer (1902–1979). *Illinois J. Math.* **47** (2003), No. 1–2, 223–236.
138. P. C. Eklof and S. Shelah, On the existence of precovers, Special issue in honor of Reinhold Baer (1902–1979), *Illinois J. Math.* **47** (2003), No. 1–2, 173–188.
139. P. Larson and S. Shelah, Bounding by canonical functions, with CH, *J. Math. Log.* **3** (2003), No. 2, 193–215.
140. S. Shelah, The null ideal restricted to some non-null set may be $\aleph_1$-saturated, *Fund. Math.* **179** (2003), No. 2, 97–129.
141. S. Shelah and B. Tsaban, Critical cardinalities and additivity properties of combinatorial notions of smallness, *J. Appl. Anal.* **9** (2003), No. 2, 149–162.
142. J. Brendle and S. Shelah, Evasion and prediction, IV, Strong forms of constant prediction, *Arch. Math. Logic* **42** (2003), No. 4, 349–360.
143. T. Bartoszynski and S. Shelah, Strongly meager sets of size continuum, *Arch. Math. Logic* **42** (2003), No. 8, 769–779.
144. T. Bartoszynski, S. Shelah and B. Tsaban, Additivity properties of topological diagonalizations, *J. Symbolic Logic* **68** (2003), No. 4, 1254–1260.

145. H. Mildenberger and S. Shelah, Specialising Aronszajn trees by countable approximations, *Arch. Math. Logic* **42** (2003), No. 7, 627–647.
146. S. Shelah, On ultraproducts of Boolean algebras and irr, *Arch. Math. Logic* **42** (2003), No. 6, 569–581.
147. S. Geschke and S. Shelah, Some notes concerning the homogeneity of Boolean algebras and Boolean spaces, *Topology Appl.* **133** (2003), No. 3, 241–253.
148. T. Eisworth, P. Nyikos and S. Shelah, Gently killing S-spaces, *Israel J. Math.* **136** (2003), 189–220.
149. S. Shelah, Not collapsing cardinals $\leq\kappa$ in ($<\kappa$)-support iterations, *Israel J. Math.* **136** (2003), 29–115.
150. L. Fuchs and S. Shelah, On a non-vanishing Ext, *Rend. Sem. Mat. Univ. Padova* **109** (2003), 235–239.
151. S. Shelah and A. Soifer, Axiom of choice and chromatic number of the plane, *J. Combin. Theory Ser. A* **103** (2003), No. 2, 387–391.
152. J. Nešetřil and S. Shelah, On the order of countable graphs, *European J. Combin.* **24** (2003), No. 6, 649–663.
153. R. Göbel, S. Shelah and L. Strüngmann, Almost-free E-rings of cardinality $\aleph_1$, *Canad. J. Math.* **55** (2003), No. 4, 750–765.
154. F.-V. Kuhlmann, S. Kuhlmann and S. Shelah, Functorial equations for lexicographic products, *Proc. Amer. Math. Soc.* **131** (2003), No. 10, 2969–2976 (electronic).
155. I. Juhász and S. Shelah, Generic left-separated spaces and calibers, *Topology Appl.* **132** (2003), No. 2, 103–108.
156. S. Shelah and A. Soifer, Chromatic number of the plane, III, Its future, *Geombinatorics* **13** (2003), No. 1, 41–46.
157. W. Kubiś and S. Shelah, Analytic colorings, *Ann. Pure Appl. Logic* **121** (2003), No. 2–3, 145–161.
158. S. Shelah and L. Strüngmann, It is consistent with ZFC that B_1-groups are not B_2, *Forum Math.* **15** (2003), No. 4, 507–524.
159. M. Džamonja and S. Shelah, Universal graphs at the successor of a singular cardinal, *J. Symbolic Logic* **68** (2003), No. 2, 366–388.
160. S. Shelah, A partition relation using strongly compact cardinals, *Proc. Amer. Math. Soc.* **131** (2003), No. 8, 2585–2592 (electronic).
161. S. Shelah, Successor of singulars: Combinatorics and not collapsing cardinals $\leq\kappa$ in ($<\kappa$)-support iterations, *Israel J. Math.* **134** (2003), 127–155.
162. J. Kennedy and S. Shelah, On embedding models of arithmetic of cardinality $\aleph_1$ into reduced powers, *Fund. Math.* **176** (2003), No. 1, 17–24.

163. S. Shelah and L. Strüngmann, Kulikov's problem on universal torsion-free abelian groups, *J. London Math. Soc.* (2) **67** (2003), No. 3, 626–642.
164. S. Shelah, Logical dreams, *Bull. Amer. Math. Soc.* (*N.S.*) **40** (2003), No. 2, 203–228 (electronic).
165. R. Göbel and S. Shelah, Characterizing automorphism groups of ordered abelian groups, *Bull. London Math. Soc.* **35** (2003), No. 3, 289–292.
166. I. Juhász, S. Shelah, L. Soukup and Z. Szentmiklóssy, A tall space with a small bottom, *Proc. Amer. Math. Soc.* **131** (2003), No. 6, 1907–1916 (electronic).
167. S. Shelah, More Jonsson algebras, *Arch. Math. Logic* **42** (2003), No. 1, 1–44.
168. M. C. Laskowski and S. Shelah, Karp complexity and classes with the independence property, *Ann. Pure Appl. Logic* **120** (2003), No. 1–3, 263–283.
169. M. Kojman and S. Shelah, van der Waerden spaces and Hindman spaces are not the same, *Proc. Amer. Math. Soc.* **131** (2003), No. 5, 1619–1622 (electronic).
170. M. Džamonja and S. Shelah, Weak reflection at the successor of a singular cardinal, *J. London Math. Soc.* (2) **67** (2003), No. 1, 1–15.
171. R. Göbel and S. Shelah, Philip Hall's problem on non-abelian splitters. *Math. Proc. Cambridge Philos. Soc.* **134** (2003), No. 1, 23–31.
172. S. Shelah, A countable structure does not have a free uncountable automorphism group, *Bull. London Math. Soc.* **35** (2003), No. 1, 1–7.
173. S. Shelah and A. Tsuboi, Definability of initial segments, *Notre Dame J. Formal Logic* **43** (2002), No. 2, 65–73.
174. S. Shelah, Martin's axiom and maximal orthogonal families, *Real Anal. Exchange* **28** (2002/03), No. 2, 477–480.
175. S. Shelah and Z. Spasojević, Cardinal invariants b_κ and t_κ, *Publ. Inst. Math.* (*Beograd*) (*N.S.*) **72**(86) (2002), 1–9.
176. S. Shelah and L. Strüngmann, Cotorsion theories cogenerated by $\aleph_1$-free abelian groups, *Proc. of the Second Honolulu Conference on Abelian Groups and Modules* (Honolulu, HI, 2001), *Rocky Mountain J. Math.* **32** (2002), No. 4, 1617–1626.
177. S. Shelah, Zero-one laws for graphs with edge probabilities decaying with distance, I, *Fund. Math.* **175** (2002), No. 3, 195–239.
178. T. Hyttinen, S. Shelah and J. Väänänen, More on the Ehrenfeucht-Fraïssé game of length ω_1, *Fund. Math.* **175** (2002), No. 1, 79–96.
179. A. Hellsten, T. Hyttinen and S. Shelah, Potential isomorphism and semi-proper trees, *Fund. Math.* **175** (2002), No. 2, 127–142.

180. S. Shelah, You can enter Cantor's paradise!, in *Paul Erdős and His Mathematics, II* (Budapest, 1999), pp. 555–564, Bolyai Soc. Math. Stud., Vol. 11 (János Bolyai Math. Soc., 2002).
181. S. Shelah and J. Zapletal, Duality and the PCF theory, *Math. Res. Lett.* **9** (2002), No. 5–6, 585–595.
182. M. Goldstern and S. Shelah, Antichains in products of linear orders, *Order* **19** (2002), No. 3, 213–222.
183. S. Shelah and P. Väisänen, Almost free groups and Ehrenfeucht-Fraïssé games for successors of singular cardinals, *Ann. Pure Appl. Logic* **118** (2002), No. 1–2, 147–173.
184. S. Shelah and P. Väisänen, The number of $L_{\infty\kappa}$-equivalent nonisomorphic models for κ weakly compact, *Fund. Math.* **174** (2002), No. 2, 97–126.
185. J. Kennedy and S. Shelah, On regular reduced products, *J. Symbolic Logic* **67** (2002), No. 3, 1169–1177.
186. A. Blass, Y. Gurevich and S. Shelah, On polynomial time computation over unordered structures, *J. Symbolic Logic* **67** (2002), No. 3, 1093–1125.
187. R. Göbel, J. L. Rodriguez and S. Shelah, Large localizations of finite simple groups, *J. Reine Angew. Math.* **550** (2002), 1–24.
188. S. Shelah and P. Väisänen, On equivalence relations second order definable over $H(\kappa)$, *Fund. Math.* **174** (2002), No. 1, 1–21.
189. S. Shelah, A partition theorem, *Sci. Math. Jpn.* **56** (2002), No. 2, 413–438.
190. K. Eda and S. Shelah, The non-commutative Specker phenomenon in the uncountable case, *J. Algebra* **252** (2002), No. 1, 22–26.
191. T. Bartoszynski and S. Shelah, Perfectly meager sets and universally null sets, *Proc. Amer. Math. Soc.* **130** (2002), No. 12, 3701–3711 (electronic).
192. S. Shelah, More constructions for Boolean algebras, *Arch. Math. Logic* **41** (2002), No. 5, 401–441.
193. R. Göbel, A. T. Paras and S. Shelah, Groups isomorphic to all their non-trivial normal subgroups, *Israel J. Math.* **129** (2002), 21–27.
194. S. Shelah, Superatomic Boolean algebras: maximal rigidity, in *Set Theory* (Piscataway, NJ, 1999), pp. 107–128, DIMACS Ser. Discrete Math. Theoret. Comput. Sci., Vol. 58 (Amer. Math. Soc., 2002).
195. T. Hyttinen and S. Shelah, Forcing a Boolean algebra with predesigned automorphism group, *Proc. Amer. Math. Soc.* **130** (2002), No. 10, 2837–2843 (electronic).
196. S. Shelah, PCF and infinite free subsets in an algebra, *Arch. Math. Logic* **41** (2002), No. 4, 321–359.
197. U. Abraham and S. Shelah, Coding with ladders a well ordering of the reals, *J. Symbolic Logic* **67** (2002), No. 2, 579–597.

198. T. Bartoszyński and S. Shelah, Strongly meager and strong measure zero sets, *Arch. Math. Logic* **41** (2002), No. 3, 245–250.
199. S. Shelah, On the existence of large subsets of $[\lambda]^{<\kappa}$ which contain no unbounded non-stationary subsets, *Arch. Math. Logic* **41** (2002), No. 3, 207–213.
200. S. Shelah, Weak diamond, *Sci. Math. Jpn.* **55** (2002), No. 3, 531–538.
201. Y. Matsubara and S. Shelah, Nowhere precipitousness of the non-stationary ideal over $P_\kappa\lambda$, *J. Math. Log.* **2** (2002), No. 1, 81–89.
202. M. Droste and S. Shelah, Outer automorphism groups of ordered permutation groups, *Forum Math.* **14** (2002), No. 4, 605–621.
203. P. C. Eklof and S. Shelah, Whitehead modules over large principal ideal domains, *Forum Math.* **14** (2002), No. 3, 477–482.
204. M. Goldstern and S. Shelah, Clones on regular cardinals, *Fund. Math.* **173** (2002), No. 1, 1–20.
205. S. Shelah and J. Steprāns, Martin's axiom is consistent with the existence of nowhere trivial automorphisms, *Proc. Amer. Math. Soc.* **130** (2002), No. 7, 2097–2106 (electronic).
206. H. Mildenberger and S. Shelah, The relative consistency of $g < \mathrm{cf}(\mathrm{Sym}(\omega))$, *J. Symbolic Logic* **67** (2002), No. 1, 297–314.
207. H. Kikyo and S. Shelah, The strict order property and generic automorphisms, *J. Symbolic Logic* **67** (2002), No. 1, 214–216.
208. R. Göbel and S. Shelah, Constructing simple groups for localizations, *Comm. Algebra* **30** (2002), No. 2, 809–837.
209. H. Mildenberger and S. Shelah, The splitting number can be smaller than the matrix chaos number, *Fund. Math.* **171** (2002), No. 2, 167–176.
210. P. C. Eklof and S. Shelah, The structure of Ext(A, Z) and GCH: Possible co-Moore spaces, *Math. Z.* **239** (2002), No. 1, 143–157.
211. R. Göbel and S. Shelah, Radicals and Plotkin's problem concerning geometrically equivalent groups, *Proc. Amer. Math. Soc.* **130** (2002), No. 3, 673–674 (electronic).
212. J. T. Baldwin and S. Shelah, Model companions of T_{Aut} for stable T, *Notre Dame J. Formal Logic* **42** (2001), No. 3, 129–142.
213. A. Rosłanowski and S. Shelah, Iteration of λ-complete forcing notions not collapsing λ+, *Int. J. Math. Math. Sci.* **28** (2001), No. 2, 63–82.
214. S. Shelah, Categoricity of an abstract elementary class in two successive cardinals, *Israel J. Math.* **126** (2001), 29–128.
215. S. Shelah, Categoricity of theories in $L_{\kappa*,\omega}$, when $\kappa*$ is a measurable cardinal, II. Dedicated to the memory of Jerzy Łoś. *Fund. Math.* **170** (2001), No. 1–2, 165–196.

216. A. Halko and S. Shelah, On strong measure zero subsets of κ_2, *Fund. Math.* **170** (2001), No. 3, 219–229.
217. M. Gitik and S. Shelah, On some configurations related to the Shelah weak hypothesis, *Arch. Math. Logic* **40** (2001), No. 8, 639–650.
218. S. Shelah and L. Strüngmann, The failure of the uncountable non-commutative Specker phenomenon, *J. Group Theory* **4** (2001), No. 4, 417–426.
219. S. Shelah and L. J. Stanley, Forcing many positive polarized partition relations between a cardinal and its powerset, *J. Symbolic Logic* **66** (2001), No. 3, 1359–1370.
220. T. Hyttinen and S. Shelah, Main gap for locally saturated elementary submodels of a homogeneous structure, *J. Symbolic Logic* **66** (2001), No. 3, 1286–1302.
221. M. Gitik and S. Shelah, More on real-valued measurable cardinals and forcing with ideals, *Israel J. Math.* **124** (2001), 221–242.
222. S. Shelah and M. Shioya, Nonreflecting stationary sets in $P_\kappa\lambda$, axiomatic set theory (in Japanese) (Kyoto, 2000), *Sūrikaisekikenkyūsho Kōkyūroku* No. 1202 (2001), 61–65.
223. S. Fuchino and S. Shelah, Models of real-valued measurability, axiomatic set theory (in Japanese) (Kyoto, 2000), *Sūrikaisekikenkyūsho Kōkyūroku* No. 1202 (2001), 38–60.
224. T. Bartoszynski and S. Shelah, Continuous images of sets of reals, *Topology Appl.* **116** (2001), No. 2, 243–253.
225. A. Blass, Y. Gurevich and S. Shelah, Addendum to: "Choiceless polynomial time" [*Ann. Pure Appl. Logic* **100** (1999), No. 1–3, 141–187], *Ann. Pure Appl. Logic* **112** (2001), No. 1, 117.
226. A. Rosłanowski and S. Shelah, Historic forcing for depth, *Colloq. Math.* **89** (2001), No. 1, 99–115.
227. A. Rosłanowski and S. Shelah, Forcing for hL and hd, *Colloq. Math.* **88** (2001), No. 2, 273–310.
228. U. Abraham and S. Shelah, Lusin sequences under CH and under Martin's axiom, *Fund. Math.* **169** (2001), No. 2, 97–103.
229. S. Fuchino, S. Geschke, S. Shelah and L. Soukup, On the weak Freese-Nation property of complete Boolean algebras, *Ann. Pure Appl. Logic* **110** (2001), No. 1–3, 89–105.
230. R. Göbel and S. Shelah, Some nasty reflexive groups, *Math. Z.* **237** (2001), No. 3, 547–559.
231. S. Shelah and J. Trlifaj, Spectra of the Γ-invariant of uniform modules, *J. Pure Appl. Algebra* **162** (2001), No. 2–3, 367–379.

232. L. Halbeisen and S. Shelah, Relations between some cardinals in the absence of the axiom of choice. *Bull. Symbolic Logic* **7** (2001), No. 2, 237–261.
233. T. Bartoszynski and S. Shelah, Strongly meager sets do not form an ideal, *J. Math. Log.* **1** (2001), No. 1, 1–34.
234. M. Kojman and S. Shelah, Fallen cardinals, Dedicated to Petr Vopěnka, *Ann. Pure Appl. Logic* **109** (2001), No. 1–2, 117–129.
235. S. Shelah, Consistently there is no nontrivial ccc forcing notion with the Sacks or Laver property, in Paul Erdös and His Mathematics (Budapest, 1999), *Combinatorica* **21** (2001), No. 2, 309–319.
236. A. Błaszczyk and S. Shelah, Regular subalgebras of complete Boolean algebras, *J. Symbolic Logic* **66** (2001), No. 2, 792–800.
237. S. Shelah and J. Steprāns, The covering numbers of Mycielski ideals are all equal, *J. Symbolic Logic* **66** (2001), No. 2, 707–718.
238. R. Göbel, S. Shelah and S. L. Wallutis, On the lattice of cotorsion theories, *J. Algebra* **238** (2001), No. 1, 292–313.
239. R. Göbel and S. Shelah, Reflexive subgroups of the Baer–Specker group and Martin's axiom, in *Abelian Groups, Rings and Modules* (Perth, 2000), pp. 145–158, Contemp. Math., Vol. 273 (Amer. Math. Soc., 2001).
240. S. Shelah, Strong dichotomy of cardinality, *Results Math.* **39** (2001), No. 1–2, 131–154.
241. R. Göbel and S. Shelah, Decompositions of reflexive modules, *Arch. Math. (Basel)* **76** (2001), No. 3, 166–181.
242. S. Shelah, Constructing Boolean algebras for cardinal invariants, *Algebra Universalis* **45** (2001), No. 4, 353–373.
243. S. Shelah, Non-existence of universal members in classes of abelian groups, *J. Group Theory* **4** (2001), No. 2, 169–191.
244. M. C. Laskowski and S. Shelah, The Karp complexity of unstable classes, *Arch. Math. Logic* **40** (2001), No. 2, 69–88.
245. R. Göbel and S. Shelah, An addendum and corrigendum to: "Almost free splitters" [*Colloq. Math.* No. 2, 193–221], *Colloq. Math.* **88** (2001), No. 1, 155–158.
246. P. C. Eklof amd S. Shelah, A non-reflexive Whitehead group, *J. Pure Appl. Algebra* **156** (2001), No. 2–3, 199–214.
247. S. Shelah and P. Väisänen, On the number of $L_{\infty\omega_1}$-equivalent non-isomorphic models, *Trans. Amer. Math. Soc.* **353** (2001), No. 5, 1781–1817.
248. T. Jech and S. Shelah, Simple complete Boolean algebras, *Proc. Amer. Math. Soc.* **129** (2001), No. 2, 543–549.
249. A. Rosłanowski and S. Shelah, The yellow cake, *Proc. Amer. Math. Soc.* **129** (2001), No. 1, 279–291.

250. S. Shelah, Cellularity of free products of Boolean algebras (or topologies). Saharon Shelah's anniversary issue, *Fund. Math.* **166** (2000), No. 1–2, 153–208.
251. S. Shelah, On a problem of Steve Kalikow, Saharon Shelah's anniversary issue, *Fund. Math.* **166** (2000), No. 1–2, 137–151.
252. S. Shelah, Covering of the null ideal may have countable cofinality, Saharon Shelah's anniversary issue, *Fund. Math.* **166** (2000), No. 1–2, 109–136.
253. S. Shelah, Strong covering without squares, Saharon Shelah's anniversary issue, *Fund. Math.* **166** (2000), No. 1–2, 87–107.
254. S. Shelah, Embedding Cohen algebras using pcf theory, Saharon Shelah's anniversary issue, *Fund. Math.* **166** (2000), No. 1–2, 83–86.
255. S. Shelah, On what I do not understand (and have something to say), I. Saharon Shelah's anniversary issue, *Fund. Math.* **166** (2000), No. 1–2, 1–82.
256. O. Shafir and S. Shelah, More on entangled orders, *J. Symbolic Logic* **65** (2000), No. 4, 1823–1832.
257. S. Shelah, Applications of PCF theory, *J. Symbolic Logic* **65** (2000), No. 4, 1624–1674.
258. S. Shelah, A space with only Borel subsets, *Period. Math. Hungar.* **40** (2000), No. 2, 81–84.
259. K. Ciesielski and S. Shelah, Category analogue of sup-measurability problem, *J. Appl. Anal.* **6** (2000), No. 2, 159–172.
260. S. Shelah and O. Kolman, Infinitary axiomatizability of slender and cotorsion-free groups, *Bull. Belg. Math. Soc. Simon Stevin* **7** (2000), No. 4, 623–629.
261. M. Džamonja and S. Shelah, Erratum: "♣ does not imply the existence of a Suslin tree" [*Israel J. Math.* **113** (1999), 163–204], *Israel J. Math.* **119** (2000), 379.
262. S. Shelah and O. Spinas, On incomparability and related cardinal functions on ultraproducts of Boolean algebras, *Math. Japon.* **52** (2000), No. 3, 345–358.
263. S. Shelah and J. Väänänen, Stationary sets and infinitary logic, *J. Symbolic Logic* **65** (2000), No. 3, 1311–1320.
264. S. Shelah, On quantification with a finite universe, *J. Symbolic Logic* **65** (2000), No. 3, 1055–1075.
265. S. Shelah, Sierpiński right?, IV, *J. Symbolic Logic* **65** (2000), No. 3, 1031–1054.
266. H. Mildenberger and S. Shelah, Changing cardinal characteristics without changing ω-sequences or confinalities, *Ann. Pure Appl. Logic* **106** (2000), No. 1–3, 207–261.

267. P. Komjáth and S. Shelah, Two consistency results on set mappings, *J. Symbolic Logic* **65** (2000), No. 1, 333–338.
268. S. Shelah and P. Väisänen, On inverse γ-systems and the number of $L_{\infty\lambda}$-equivalent, non-isomorphic models for λ-singular, *J. Symbolic Logic* **65** (2000), No. 1, 272–284.
269. S. Shelah and L. J. Stanley, Filters, Cohen sets and consistent extensions of the Erdös-Dushnik-Miller theorem, *J. Symbolic Logic* **65** (2000), No. 1, 259–271.
270. T. Bartoszyński, A. Rosłanowski and S. Shelah, After all, there are some inequalities which are provable in ZFC, *J. Symbolic Logic* **65** (2000), No. 2, 803–816.
271. S. Shelah, The generalized continuum hypothesis revisited, *Israel J. Math.* **116** (2000), 285–321.
272. M. Kojman and S. Shelah, The PCF trichotomy theorem does not hold for short sequences, *Arch. Math. Logic* **39** (2000), No. 3, 213–218.
273. M. Rabus and S. Shelah, Covering a function on the plane by two continuous functions on an uncountable square — The consistency, *Ann. Pure Appl. Logic* **103** (2000), No. 1–3, 229–240.
274. T. Hyttinen and S. Shelah, Strong splitting in stable homogeneous models, *Ann. Pure Appl. Logic* **103** (2000), No. 1–3, 201–228.
275. A. Rosłanowski and S. Shelah, More on cardinal invariants of Boolean algebras, *Ann. Pure Appl. Logic* **103** (2000), No. 1–3, 1–37.
276. S. Shelah and O. Spinas, The distributivity numbers of $P(\omega)$/fin and its square, *Trans. Amer. Math. Soc.* **352** (2000), No. 5, 2023–2047 (electronic).
277. A. Hajna, I. Juhász and S. Shelah, Strongly almost disjoint families, revisited, *Fund. Math.* **163** (2000), No. 1, 13–23.
278. S. Shelah, On what I do not understand (and have something to say), model theory, *Math. Jpn.* **51** (2000), No. 2, 329–377.
279. M. Džamonja and S. Shelah, On versions of ♣ on cardinals larger than $\aleph_1$, *Math. Jpn.* **51** (2000), No. 1, 53–61.
280. J. T. Baldwin and S. Shelah, On the classifiability of cellular automata, *Theoret. Comput. Sci.* **230** (2000), No. 1–2, 117–129.
281. Z. T. Balogh, S. W. Davis, W. Just, S. Shelah and P. J. Szeptycki, Strongly almost disjoint sets and weakly uniform bases, *Trans. Amer. Math. Soc.* **352** (2000), No. 11, 4971–4987.
282. R. Göbel and S. Shelah, Cotorsion theories and splitters, *Trans. Amer. Math. Soc.* **352** (2000), No. 11, 5357–5379.
283. T. Jech and S. Shelah, On reflection of stationary sets in $P_\kappa\lambda$, *Trans. Amer. Math. Soc.* **352** (2000), No. 6, 2507–2515.

284. M. Gitik and S. Shelah, Cardinal preserving ideals, *J. Symbolic Logic* **64** (1999), No. 4, 1527–1551.
285. K. Ciesielski and S. Shelah, A model with no magic set, *J. Symbolic Logic* **64** (1999), No. 4, 1467–1490.
286. J. T. Baldwin, R. Grossberg and S. Shelah, Transfering saturation, the finite cover property, and stability, *J. Symbolic Logic* **64** (1999), No. 2, 678–684.
287. T. Hyttinen and S. Shelah, Constructing strongly equivalent nonisomorphic models for unsuperstable theories, Part C, *J. Symbolic Logic* **64** (1999), No. 2, 634–642.
288. P. C. Eklof and S. Shelah, Absolutely rigid systems and absolutely indecomposable groups, in *Abelian Groups and Modules* (Dublin, 1998), pp. 257–268, Trends Math. (Birkhäuser, 1999).
289. O. Kolman and S. Shelah, Almost disjoint pure subgroups of the Baer–Specker group, in *Abelian Groups and Modules* (Dublin, 1998), pp. 225–230, Trends Math. (Birkhäuser, 1999).
290. W. Just, S. Shelah and S. Thomas, The automorphism tower problem revisited, *Adv. Math.* **148** (1999), No. 2, 243–265.
291. M. Goldstern and S. Shelah, There are no infinite order polynomially complete lattices, after all, *Algebra Universalis* **42** (1999), No. 1–2, 49–57.
292. Z. Shami and S. Shelah, Rigid $\aleph_\varepsilon$-saturated models of superstable theories, *Fund. Math.* **162** (1999), No. 1, 37–46.
293. S. Shelah, Special subsets of $\mathrm{cf}(\mu)\mu$, Boolean algebras and Maharam measure algebras, in *8th Prague Topological Symposium on General Topology and its Relations to Modern Analysis and Algebra*, Part II (1996), *Topology Appl.* **99** (1999), No. 2–3, 135–235.
294. M. Džamonja and S. Shelah, ♣ does not imply the existence of a Suslin tree, *Israel J. Math.* **113** (1999), 163–204.
295. S. Shelah, On T_3-topological space omitting many cardinals, *Period. Math. Hungar.* **38** (1999), No. 1–2, 87–98.
296. A. Blass, Y. Gurevich and S. Shelah, Choiceless polynomial time, *Ann. Pure Appl. Logic* **100** (1999), No. 1–3, 141–187.
297. R. Göbel and S. Shelah, Almost free splitters, *Colloq. Math.* **81** (1999), No. 2, 193–221.
298. A. Mekler, A. Rosłanowski and S. Shelah, On the p-rank of Ext, *Israel J. Math.* **112** (1999), 327–356.
299. S. Shelah, Borel Whitehead groups, *Math. Jpn.* **50** (1999), No. 1, 121–130.
300. S. Fuchino, H. Mildenberger, S. Shelah and P. Vojtáš, On absolutely divergent series, *Fund. Math.* **160** (1999), No. 3, 255–268.

301. S. Lifsches and S. Shelah, Random graphs in the monadic theory of order, in *Logic Colloquium '95* (Haifa), *Arch. Math. Logic* **38** (1999), No. 4–5, 273–312.

302. S. Shelah, Categoricity for abstract classes with amalgamation, *Ann. Pure Appl. Logic* **98** (1999), No. 1–3, 261–294.

303. S. Shelah and J. Zapletal, Canonical models for $\aleph_1$-combinatorics, *Ann. Pure Appl. Logic* **98** (1999), No. 1–3, 217–259.

304. J. Cummings and S. Shelah, Some independence results on reflection, *J. London Math. Soc.* (2) **59** (1999), No. 1, 37–49.

305. J. Brendle and S. Shelah, Ultrafilters on ω — Their ideals and their cardinal characteristics, *Trans. Amer. Math. Soc.* **351** (1999), No. 7, 2643–2674.

306. M. Džamonja and S. Shelah, Similar but not the same: Various versions of $\clubsuit$ do not coincide, *J. Symbolic Logic* **64** (1999), No. 1, 180–198.

307. G. Cherlin, S. Shelah and N. Shi, Universal graphs with forbidden subgraphs and algebraic closure, *Adv. Appl. Math.* **22** (1999), No. 4, 454–491.

308. M. Kojman and S. Shelah, Regressive Ramsey numbers are Ackermannian, *J. Combin. Theory Ser. A* **86** (1999), No. 1, 177–181.

309. S. Shelah and J. K. Truss, On distinguishing quotients of symmetric groups, *Ann. Pure Appl. Logic* **97** (1999), No. 1–3, 47–83.

310. S. Shelah and A. Villaveces, Toward categoricity for classes with no maximal models, *Ann. Pure Appl. Logic* **97** (1999), No. 1–3, 1–25.

311. S. Shelah, On full Suslin trees, *Colloq. Math.* **79** (1999), No. 1, 1–7.

312. S. Shelah, Borel sets with large squares, *Fund. Math.* **159** (1999), No. 1, 1–50.

313. R. Jin and S. Shelah, Possible size of an ultrapower of ω, *Arch. Math. Logic* **38** (1999), No. 1, 61–77.

314. A. Rosłanowski and S. Shelah, Norms on possibilities, I. Forcing with trees and creatures, *Mem. Amer. Math. Soc.* **141** (1999), No. 671.

315. S. Shelah and O. Spinas, On tightness and depth in superatomic Boolean algebras, *Proc. Amer. Math. Soc.* **127** (1999), No. 12, 3475–3480.

316. M. Rabus and S. Shelah, Topological density of ccc Boolean algebras — Every cardinality occurs, *Proc. Amer. Math. Soc.* **127** (1999), No. 9, 2573–2581.

317. K. Ciesielski and S. Shelah, Uniformly antisymmetric functions with bounded range, *Real Anal. Exchange* **24** (1998/99), No. 2, 615–619.

318. S. Shelah, Covering numbers associated with trees branching into a countably generated set of possibilities, *Real Anal. Exchange* **24** (1998/99), No. 1, 205–213.

319. S. Shelah, There may be no nowhere dense ultrafilter, in *Logic Colloquium '95 (Haifa)*, pp. 305–324, Lecture Notes Logic, Vol. 11 (Springer, 1998).
320. M. Magidor and S. Shelah, Length of Boolean algebras and ultraproducts, *Math. Jpn.* **48** (1998), No. 2, 301–307.
321. O. Kolman and S. Shelah, A result related to the problem CN of Fremlin, *J. Appl. Anal.* **4** (1998), No. 2, 161–165.
322. R. Jin and S. Shelah, Compactness of Loeb spaces, *J. Symbolic Logic* **63** (1998), No. 4, 1371–1392.
323. R. Göbel and S. Shelah, Endomorphism rings of modules whose cardinality is cofinal to omega, in *Abelian Groups, Module Theory, and Topology (Padua, 1997)*, pp. 235–248, Lecture Notes in Pure and Appl. Math., Vol. 201 (Dekker, 1998).
324. S. Shelah, The lifting problem with the full ideal, *J. Appl. Anal.* **4** (1998), No. 1, 1–17.
325. T. Hyttinen and S. Shelah, On the number of elementary submodels of an unsuperstable homogeneous structure, *MLQ Math. Log. Q.* **44** (1998), No. 3, 354–358.
326. S. Shelah and O. Spinas, The distributivity numbers of finite products of $P(\omega)$/fin, *Fund. Math.* **158** (1998), No. 1, 81–93.
327. R. Göbel and S. Shelah, Indecomposable almost free modules — The local case, *Canad. J. Math.* **50** (1998), No. 4, 719–738.
328. M. Goldstern and S. Shelah, Order polynomially complete lattices must be large, *Algebra Universalis* **39** (1998), No. 3–4, 197–209.
329. M. Gitik and S. Shelah, On densities of box products, *Topology Appl.* **88** (1998), No. 3, 219–237.
330. J. D. Hamkins and S. Shelah, Superdestructibility: A dual to Laver's indestructibility, *J. Symbolic Logic* **63** (1998), No. 2, 549–554.
331. J. T. Baldwin and S. Shelah, DOP and FCP in generic structures, *J. Symbolic Logic* **63** (1998), No. 2, 427–438.
332. S. Shelah, *Proper and Improper Forcing*, Second edn., Perspectives in Mathematical Logic (Springer-Verlag, 1998).
333. S. Shelah, Erdös and Rényi conjecture, *J. Combin. Theory Ser. A* **82** (1998), No. 2, 179–185.
334. S. Shelah and R. Grossberg, On cardinalities in quotients of inverse limits of groups, *Math. Japon.* **47** (1998), No. 2, 189–197.
335. M. Balcerzak, A. Rosłanowski and S. Shelah, Ideals without ccc, *J. Symbolic Logic* **63** (1998), No. 1, 128–148.
336. S. Lifsches and S. Shelah, Uniformization and Skolem functions in the class of trees, *J. Symbolic Logic* **63** (1998), No. 1, 103–127.

337. S. Shelah and E. C. Milner, A tree-arrowing graph, in *Set Theory* (Curaçao, 1995, Barcelona, 1996), pp. 175–182 (Kluwer, 1998).

338. S. Shelah, A polarized partition relation and failure of GCH at singular strong limit, *Fund. Math.* **155** (1998), No. 2, 153–160.

339. A. Rosłanowski and S. Shelah, Cardinal invariants of ultraproducts of Boolean algebras, *Fund. Math.* **155** (1998), No. 2, 101–151.

340. M. Kojman and S. Shelah, A ZFC Dowker space in $\aleph_{\omega+1}$: An application of PCF theory to topology, *Proc. Amer. Math. Soc.* **126** (1998), No. 8, 2459–2465.

341. I. Juhász and S. Shelah, On the cardinality and weight spectra of compact spaces, II, *Fund. Math.* **155** (1998), No. 1, 91–94.

342. P. C. Eklof and S. Shelah, The Kaplansky test problems for $\aleph_1$-separable groups, *Proc. Amer. Math. Soc.* **126** (1998), No. 7, 1901–1907.

343. S. Shelah, Non-existence of universals for classes like reduced torsion free abelian groups under embeddings which are not necessarily pure, in *Advances in Algebra and Model Theory* (Essen, 1994, Dresden, 1995), pp. 229–286, Algebra Logic Appl., Vol. 9 (Gordon and Breach, 1997).

344. M. Gilchrist and S. Shelah, The consistency of $\mathrm{ZFC} + 2^{\aleph_0} > \aleph_\omega + \mathcal{I}(\aleph_2) = \mathcal{I}(\aleph_\omega)$, *J. Symbolic Logic* **62** (1997), No. 4, 1151–1160.

345. A. Rosłanowski and S. Shelah, Simple forcing notions and forcing axioms, *J. Symbolic Logic* **62** (1997), No. 4, 1297–1314.

346. S. Shelah, On Ciesielski's problems, *J. Appl. Anal.* **3** (1997), No. 2, 191–209.

347. A. Rosłanowski and S. Shelah, Norms on possibilities, II. More ccc ideals on 2ω, *J. Appl. Anal.* **3** (1997), No. 1, 103–127.

348. S. Fuchino and S. Shelah, Soukup, Lajos Sticks and clubs, *Ann. Pure Appl. Logic* **90** (1997), No. 1–3, 57–77.

349. S. Shelah and S. Thomas, The cofinality spectrum of the infinite symmetric group, *J. Symbolic Logic* **62** (1997), No. 3, 902–916.

350. S. Lifsches and S. Shelah, Peano arithmetic may not be interpretable in the monadic theory of linear orders, *J. Symbolic Logic* **62** (1997), No. 3, 848–872.

351. K. Liu and S. Shelah, Cofinalities of elementary substructures of structures on $\aleph_\omega$, *Israel J. Math.* **99** (1997), 189–205.

352. S. Shelah, σ-entangled linear orders and narrowness of products of Boolean algebra, *Fund. Math.* **153** (1997), No. 3, 199–275.

353. E. Rosen, S. Shelah and S. Weinstein, *k*-universal finite graphs, in *Logic and Random Structures* (New Brunswick, NJ, 1995), pp. 65–77, DIMACS Ser. Discrete Math. Theor. Comput. Sci., Vol. 33 (Amer. Math. Soc., 1997).

354. S. Shelah, Set theory without choice: not everything on cofinality is possible, *Arch. Math. Logic* **36** (1997), No. 2, 81–125.
355. P. C. Eklof, B. Huisgen-Zimmermann and S. Shelah, Torsion modules, lattices and p-points, *Bull. London Math. Soc.* **29** (1997), No. 5, 547–555.
356. M. Goldstern and S. Shelah, A partial order where all monotone maps are definable, *Fund. Math.* **152** (1997), No. 3, 255–265.
357. R. Jin and S. Shelah, Can a small forcing create Kurepa trees, *Ann. Pure Appl. Logic* **85** (1997), No. 1, 47–68.
358. S. Shelah,and J. Zapletal, Embeddings of Cohen algebras, *Adv. Math.* **126** (1997), No. 2, 93–118.
359. S. Shelah, Colouring and non-productivity of $\aleph_2$-cc, *Ann. Pure Appl. Logic* **84** (1997), No. 2, 153–174.
360. S. Shelah, Existence of almost free abelian groups and reflection of stationary set, *Math. Jpn.* **45** (1997), No. 1, 1–14.
361. S. Shelah, The pcf theorem revisited, in *The Mathematics of Paul Erdös, II, pp.* 420–459, Algorithms Combin., Vol. 14 (Springer, 1997).
362. S. Shelah, A finite partition theorem with double exponential bound, in *The Mathematics of Paul Erdös, II*, 240–246, Algorithms Combin., Vol. 14 (Springer, 1997).
363. J. T. Baldwin and S. Shelah, Randomness and semigenericity, *Trans. Amer. Math. Soc.* **349** (1997), No. 4, 1359–1376.
364. F.-V. Kuhlmann, S. Kuhlmann and S. Shelah, Exponentiation in power series fields, *Proc. Amer. Math. Soc.* **125** (1997), No. 11, 3177–3183.
365. A. W. Apter and S. Shelah, Menas' result is best possible, *Trans. Amer. Math. Soc.* **349** (1997), No. 5, 2007–2034.
366. M. Gitik and S. Shelah, Less saturated ideals, *Proc. Amer. Math. Soc.* **125** (1997), No. 5, 1523–1530.
367. A. W. Apter and S. Shelah, On the strong equality between supercompactness and strong compactness, *Trans. Amer. Math. Soc.* **349** (1997), No. 1, 103–128.
368. S. Shelah, Very weak zero one law for random graphs with order and random binary functions, *Random Structures Algorithms* **9** (1996), No. 4, 351–358.
369. T. Jech and S. Shelah, On countably closed complete Boolean algebras, *J. Symbolic Logic* **61** (1996), No. 4, 1380–1386.
370. M. C. Laskowski and S. Shelah, Forcing isomorphism, II, *J. Symbolic Logic* **61** (1996), No. 4, 1305–1320.
371. S. Shelah, If there is an exactly λ-free abelian group then there is an exactly λ-separable one in λ, *J. Symbolic Logic* **61** (1996), No. 4, 1261–1278.

372. S. Lifsches and S. Shelah, Uniformization, choice functions and well orders in the class of trees, *J. Symbolic Logic* **61** (1996), No. 4, 1206–1227.

373. S. Koppelberg and S. Shelah, Subalgebras of Cohen algebras need not be Cohe, in *Logic: From Foundations to Applications* (Staffordshire, 1993), pp. 261–275 (Oxford Univ. Press, 1996).

374. S. Shelah, Finite canonization, *Comment. Math. Univ. Carolin.* **37** (1996), No. 3, 445–456.

375. S. Fuchino, S. Koppelberg and S. Shelah, A game on partial orderings, *Proceedings of the International Conference on Set-theoretic Topology and its Applications* (Matsuyama, 1994), *Topology Appl.* **74** (1996), No. 1–3, 141–148.

376. S. Shelah and O. Kolman, Categoricity of theories in $L_{\kappa\omega}$, when κ is a measurable cardinal, I, *Fund. Math.* **151** (1996), No. 3, 209–240.

377. M. Magidor and S. Shelah, The tree property at successors of singular cardinals, *Arch. Math. Logic* **35** (1996), No. 5–6, 385–404.

378. S. Shelah, Large normal ideals concentrating on a fixed small cardinality, *Arch. Math. Logic* **35** (1996), No. 5–6, 341–347.

379. A. Rosłanowski and S. Shelah, Localizations of infinite subsets of ω, *Arch. Math. Logic* **35** (1996), No. 5–6, 315–339.

380. A. Rosłanowski and S. Shelah, More forcing notions imply diamond, *Arch. Math. Logic* **35** (1996), No. 5–6, 299–313.

381. U. Abraham and S. Shelah, Martin's axiom and Δ_{21} well-ordering of the reals, *Arch. Math. Logic* **35** (1996), No. 5–6, 287–298.

382. S. Shelah, Further cardinal arithmetic, *Israel J. Math.* **95** (1996), 61–114.

383. S. Shelah, In the random graph $G(n,p)$, $p=n_{-a}$: If ψ has probability $O(n_{-\varepsilon})$ for every $\varepsilon>0$ then it has probability $O(e-n_{\varepsilon})$ for some $\varepsilon>0$, *Ann. Pure Appl. Logic* **82** (1996), No. 1, 97–102.

384. R. Göbel and S. Shelah, G.C.H. implies existence of many rigid almost free abelian groups, in *Abelian Groups and Modules* (Colorado Springs, CO, 1995), pp, 253–271, Lecture Notes in Pure and Appl. Math., Vol. 182 (Dekker, 1996).

385. P. C. Eklof and S. Shelah, New nonfree Whitehead groups by coloring, in *Abelian Groups and Modules* (Colorado Springs, CO, 1995), pp. 15–22, Lecture Notes in Pure and Appl. Math., Vol. 182 (Dekkerk, 1996).

386. M. Gilchrist and S. Shelah, Identities on cardinals less than $\aleph_\omega$, *J. Symbolic Logic* **61** (1996), No. 3, 780–787.

387. S. Shelah, On Monk's questions, *Fund. Math.* **151** (1996), No. 1, 1–19.

388. S. Shelah, Universal in ($<\lambda$)-stable abelian group, *Math. Jpn.* **44** (1996), No. 1, 1–9.

389. S. Shelah, Toward classifying unstable theories, *Ann. Pure Appl. Logic* **80** (1996), No. 3, 229–255.
390. T. Jech and S. Shelah, A complete Boolean algebra that has no proper atomless complete subalgebra, *J. Algebra* **182** (1996), No. 3, 748–755.
391. S. Fuchino, S. Koppelberg and S. Shelah, Partial orderings with the weak Freese-Nation property, *Ann. Pure Appl. Logic* **80** (1996), No. 1, 35–54.
392. M. Džamonja and S. Shelah, Saturated filters at successors of singular, weak reflection and yet another weak club principle, *Ann. Pure Appl. Logic* **79** (1996), No. 3, 289–316.
393. S. Shelah, Was Sierpiński right?, III. Can continuum-c.c. times c.c.c. be continuum-c.c.?, Papers in honor of the Symposium on Logical Foundations of Computer Science, "Logic at St. Petersburg" (St. Petersburg, 1994), *Ann. Pure Appl. Logic* **78** (1996), No. 1–3, 259–269.
394. Y. Gurevich and S. Shelah, On finite rigid structures, *J. Symbolic Logic* **61** (1996), No. 2, 549–562.
395. T. Jech and S. Shelah, Possible PCF algebras, *J. Symbolic Logic* **61** (1996), No. 1, 313–317.
396. T. Bartoszyński, A. Rosłanowski and S. Shelah, Adding one random real, *J. Symbolic Logic* **61** (1996), No. 1, 80–90.
397. A. H. Mekler, S. Shelah and O. Spinas, The essentially free spectrum of a variety, *Israel J. Math.* **93** (1996), 1–8.
398. S. Shelah, On the very weak 0–1 law for random graphs with orders, *J. Logic Comput.* **6** (1996), No. 1, 137–159.
399. J. Saxl, S. Shelah and S. Thomas, Infinite products of finite simple groups, *Trans. Amer. Math. Soc.* **348** (1996), No. 11, 4611–4641.
400. P. Komjáth and S. Shelah, On Taylor's problem, *Acta Math. Hungar.* **70** (1996), No. 3, 217–225.
401. S. Shelah, Remarks on $\aleph_1$-CWH not CWH first countable spaces, in *Set Theory* (Boise, ID, 1992–1994), pp. 103–145, Contemp. Math., Vol. 192 (Amer. Math. Soc., 1996).
402. S. Shelah, On some problems in general topology, in *Set Theory* (Boise, ID, 1992–1994), pp. 91–101, Contemp. Math., Vol. 192 (Amer. Math. Soc., 1996).
403. J. Brendle and S. Shelah, Evasion and prediction, II, *J. London Math. Soc.* (2) **53** (1996), No. 1, 19–27.
404. S. Shelah and O. Spinas, Gross spaces, *Trans. Amer. Math. Soc.* **348** (1996), No. 10, 4257–4277.
405. P. Komjáth and S. Shelah, Coloring finite subsets of uncountable sets, *Proc. Amer. Math. Soc.* **124** (1996), No. 11, 3501–3505.

406. S. Ben-David and S. Shelah, The two-cardinals transfer property and resurrection of supercompactness, *Proc. Amer. Math. Soc.* **124** (1996), No. 9, 2827–2837.
407. S. Shelah, Possibly every real function is continuous on a non-meagre set, Đuro Kurepa memorial volume, *Publ. Inst. Math.* (*Beograd*) (*N.S.*) **57**(71) (1995), 47–60.
408. R. Göbel and S. Shelah, On the existence of rigid $\aleph_1$-free abelian groups of cardinality $\aleph_1$, in *Abelian Groups and Modules* (Padova, 1994), pp. 227–237, Math. Appl., Vol. 343 (Kluwer, 1995).
409. T. Łuczak and S. Shelah, Convergence in homogeneous random graphs, *Random Structures Algorithms* **6** (1995), No. 4, 371–391.
410. T. Hyttinen and S. Shelah, Constructing strongly equivalent nonisomorphic models for unsuperstable theories, Part B, *J. Symbolic Logic* **60** (1995), No. 4, 1260–1272.
411. M. Džamonja and S. Shelah, On squares, outside guessing of clubs and $I<f[\lambda]$, *Fund. Math.* **148** (1995), No. 2, 165–198.
412. M. Kojman and S. Shelah, Universal abelian groups, *Israel J. Math.* **92** (1995), No. 1–3, 113–124.
413. M. Kojman and S. Shelah, Homogeneous families and their automorphism groups, *J. London Math. Soc.* (2) **52** (1995), No. 2, 303–317.
414. J. Cummings and S. Shelah, Cardinal invariants above the continuum, *Ann. Pure Appl. Logic* **75** (1995), No. 3, 251–268.
415. J. Cummings, M. Džamonja and S. Shelah, A consistency result on weak reflection, *Fund. Math.* **148** (1995), No. 1, 91–100.
416. J. Cummings and S. Shelah, A model in which every Boolean algebra has many subalgebras, *J. Symbolic Logic* **60** (1995), No. 3, 992–1004.
417. S. Shelah and L. Soukup, Some remarks on a problem of J. D. Monk, *Period. Math. Hungar.* **30** (1995), No. 2, 155–163.
418. J. Baldwin and S. Shelah, Abstract classes with few models have "homogeneous-universal" models, *J. Symbolic Logic* **60** (1995), No. 1, 246–265.
419. M. Goldstern and S. Shelah, The bounded proper forcing axiom, *J. Symbolic Logic* **60** (1995), No. 1, 58–73.
420. S. Shelah and L. J. Stanley, The combinatorics of combinatorial coding by a real, *J. Symbolic Logic* **60** (1995), No. 1, 36–57.
421. S. Shelah and L. J. Stanley, A combinatorial forcing for coding the universe by a real when there are no sharps, *J. Symbolic Logic* **60** (1995), No. 1, 1–35.
422. S. Shelah, Every null-additive set is meager-additive, *Israel J. Math.* **89** (1995), No. 1–3, 357–376.

423. A. H. Mekler and S. Shelah, Almost free algebras, *Israel J. Math.* **89** (1995), No. 1–3, 237–259.
424. S. Koppelberg and S. Shelah, Densities of ultraproducts of Boolean algebras, *Canad. J. Math.* **47** (1995), No. 1, 132–145.
425. P. C. Eklof, M. Foreman and S. Shelah, On invariants for ω_1-separable groups, *Trans. Amer. Math. Soc.* **347** (1995), No. 11, 4385–4402.
426. P. Komjáth and S. Shelah, Universal graphs without large cliques, *J. Combin. Theory Ser. B* **63** (1995), No. 1, 125–135.
427. A. Kanamori and S. Shelah, Complete quotient Boolean algebras, *Trans. Amer. Math. Soc.* **347** (1995), No. 6, 1963–1979.
428. M. Goldstern, M. Repický, S. Shelah and O. Spinas, On tree ideals, *Proc. Amer. Math. Soc.* **123** (1995), No. 5, 1573–1581.
429. S. Shelah, *Cardinal Arithmetic*, Oxford Logic Guides, Vol. 29 (The Clarendon Press and Oxford University Press, 1994).
430. P. Komjáth and S. Shelah, A note on a set-mapping problem of Hajnal and Máté, *Period. Math. Hungar.* **28** (1994), No. 1, 39–42.
431. P.C. Eklof, A. H. Mekler and S. Shelah, Hereditarily separable groups and monochromatic uniformization, *Israel J. Math.* **88** (1994), No. 1–3, 213–235.
432. S. Shelah, How special are Cohen and random forcings, i.e. Boolean algebras of the family of subsets of reals modulo meagre or null, *Israel J. Math.* **88** (1994), No. 1–3, 159–174.
433. G. Melles and S. Shelah, Aut(M) has a large dense free subgroup for saturated M, *Bull. London Math. Soc.* **26** (1994), No. 4, 339–344.
434. R. Jin and S. Shelah, Essential Kurepa trees versus essential Jech–Kunen trees, *Ann. Pure Appl. Logic* **69** (1994), No. 1, 107–131.
435. S. Givant and S. Shelah, Universal theories categorical in power and κ-generated models, *Ann. Pure Appl. Logic* **69** (1994), No. 1, 27–51.
436. S. Shelah and J. Steprāns, Decomposing Baire class 1 functions into continuous functions, *Fund. Math.* **145** (1994), No. 2, 171 180.
437. S. Fuchino, S. Shelah and L. Soukup, On a theorem of Shapiro, *Math. Jpn.* **40** (1994), No. 2, 199–206.
438. T. Hyttinen and S. Shelah, Constructing strongly equivalent nonisomorphic models for unsuperstable theories, Part A, *J. Symbolic Logic* **59** (1994), No. 3, 984–996.
439. M. Magidor and S. Shelah, Bext$_2(G, T)$ can be nontrivial, even assuming GCH, in *Abelian Group Theory and Related Topics* (Oberwolfach, 1993), pp. 287–294, Contemp. Math., Vol. 171 (Amer. Math. Soc., 1994).

440. P. C. Eklof and S. Shelah, A combinatorial principle equivalent to the existence of non-free Whitehead groups, in *Abelian Group Theory and Related Topics* (Oberwolfach, 1993), pp. 79–98, Contemp. Math., Vol. 171 (Amer. Math. Soc., 1994).
441. G. Melles and S. Shelah, A saturated model of an unsuperstable theory of cardinality greater than its theory has the small index property, *Proc. London Math. Soc.* (3) **69** (1994), No. 3, 449–463.
442. S. Shelah, Cardinalities of topologies with small base, *Ann. Pure Appl. Logic* **68** (1994), No. 1, 95–113.
443. S. Shelah and J. Spencer, Random sparse unary predicates, *Random Structures Algorithms* **5** (1994), No. 3, 375–394.
444. A. Huck, F. Niedermeyer and S. Shelah, Large κ-preserving sets in infinite graphs, *J. Graph Theory* **18** (1994), No. 4, 413–426.
445. S. Shelah and L. Soukup, On the number of nonisomorphic subgraphs, *Israel J. Math.* **86** (1994), No. 1–3, 349–371.
446. S. Shelah and J. Steprāns, Somewhere trivial autohomeomorphisms, *J. London Math. Soc.* (2) **49** (1994), No. 3, 569–580.
447. H. Judah, A. Rosłanowski and S. Shelah, Examples for Souslin forcing, *Fund. Math.* **144** (1994), No. 1, 23–42.
448. S. Shelah and J. Steprāns, Erratum: "Maximal chains in $\omega\omega$ and ultra-powers of the integers" [*Arch. Math. Logic* **32** (1993), No. 5, 305–319], *Arch. Math. Logic* **33** (1994), No. 2, 167–168.
449. R. Jin and S. Shelah, The strength of the isomorphism property, *J. Symbolic Logic* **59** (1994), No. 1, 292–301.
450. L. Halbeisen and S. Shelah, Consequences of arithmetic for set theory, *J. Symbolic Logic* **59** (1994), No. 1, 30–40.
451. S. Shelah, Vive la différence, II, The Ax–Kochen isomorphism theorem, *Israel J. Math.* **85** (1994), No. 1–3, 351–390.
452. S. Shelah, The number of independent elements in the product of interval Boolean algebras, *Math. Jpn.* **39** (1994), No. 1, 1–5.
453. S. Shelah and J. Steprāns, Homogeneous almost disjoint families, *Algebra Universalis* **31** (1994), No. 2, 196–203.
454. R. Diestel, S. Shelah and J. Steprāns, Dominating functions and graphs, *J. London Math. Soc.* (2) **49** (1994), No. 1, 16–24.
455. M. Magidor and S. Shelah, When does almost free imply free? (For groups, transversals, etc.), *J. Amer. Math. Soc.* **7** (1994), No. 4, 769–830.
456. S. Shelah and J. Spencer, Can you feel the double jump?, *Proceedings of the Fifth International Seminar on Random Graphs and Probabilistic Methods in Combinatorics and Computer Science* (Poznań, 1991). *Random Structures Algorithms* **5** (1994), No. 1, 191–204.

457. H. Judah and S. Shelah, Killing Luzin and Sierpiński sets, *Proc. Amer. Math. Soc.* **120** (1994), No. 3, 917–920.
458. S. Shelah, On CH + $2^{\aleph_1} \to (\alpha)_2^2$ for $\alpha < \omega_2$, *Logic Colloquium '90* (Helsinki, 1990), pp. 281–289, Lecture Notes Logic, Vol. 2 (Springer, 1993).
459. W. W. Comfort, A. Kato and S. Shelah, Topological partition relations of the form $\omega^* \to (Y)_2^1$, in *Papers on General Topology and Applications* (Madison, WI, 1991), pp. 70–79, Ann. New York Acad. Sci., Vol. 704 (New York Acad. Sci., 1993).
460. P. Komjáth and S. Shelah, On uniformly antisymmetric functions, *Real Anal. Exchange* **19** (1993/94), No. 1, 218–225.
461. S. Shelah, Advances in cardinal arithmetic, in *Finite and Infinite Combinatorics in Sets and Logic* (Banff, AB, 1991), pp. 355–383, NATO Adv. Sci. Inst. Ser. C Math. Phys. Sci., Vol. 411 (Kluwer, 1993).
462. S. Shelah, H. Tuuri and J. Väänänen, On the number of automorphisms of uncountable models, *J. Symbolic Logic* **58** (1993), No. 4, 1402–1418.
463. M. Goldstern, H. Judah and S. Shelah, Strong measure zero sets without Cohen reals, *J. Symbolic Logic* **58** (1993), No. 4, 1323–1341.
464. J. T. Baldwin, M. C. Laskowski and S. Shelah, Forcing isomorphism, *J. Symbolic Logic* **58** (1993), No. 4, 1291–1301.
465. M. C. Laskowski and S. Shelah, On the existence of atomic models, *J. Symbolic Logic* **58** (1993), No. 4, 1189–1194.
466. R. Frankiewicz, S. Shelah and P. Zbierski, On closed *P*-sets with ccc in the space ω^*, *J. Symbolic Logic* **58** (1993), No. 4, 1171–1176.
467. S. Shelah, The universality spectrum: Consistency for more classes, in *Combinatorics, Paul Erdös is Eighty*, Vol. 1, pp. 403–420 (János Bolyai Math. Soc., 1993).
468. S. Shelah, More on cardinal arithmetic, *Arch. Math. Logic* **32** (1993), No. 6, 399–428.
469. H. Judah and S. Shelah, Baire property and axiom of choice, *Israel J. Math.* **84** (1993), No. 3, 435–450.
470. R. Bonnet and S. Shelah, On HCO spaces. An uncountable compact T_2 space, different from $\aleph_1+1$, which is homeomorphic to each of its uncountable closed subspaces, *Israel J. Math.* **84** (1993), No. 3, 289–332.
471. S. Shelah and R. Jin, A model in which there are Jech–Kunen trees but there are no Kurepa trees, *Israel J. Math.* **84** (1993), No. 1–2, 1–16.
472. S. Shelah, C. Laflamme and B. Hart, Models with second order properties, V. A general principle, *Ann. Pure Appl. Logic* **64** (1993), No. 2, 169–194.
473. S. Shelah, Number of open sets for a topology with a countable basis, *Israel J. Math.* **83** (1993), No. 3, 369–374.

474. I. Hodkinson and S. Shelah, A construction of many uncountable rings using SFP domains and Aronszajn trees, *Proc. London Math. Soc.* (3) **67** (1993), No. 3, 449–492.
475. A. Louveau, S. Shelah and B. Veličković, Borel partitions of infinite subtrees of a perfect tree, *Ann. Pure Appl. Logic* **63** (1993), No. 3, 271–281.
476. S. Shelah, The future of set theory, in *Set Theory of the Reals* (Ramat Gan, 1991), pp. 1–12, Israel Math. Conf. Proc., Vol. 6 (Bar-Ilan Univ., 1993).
477. S. Shelah and D. H. Fremlin, Pointwise compact and stable sets of measurable functions, *J. Symbolic Logic* **58** (1993), No. 2, 435–455.
478. T. Bartoszyński, H. Judah and S. Shelah, The Cichoń diagram, *J. Symbolic Logic* **58** (1993), No. 2, 401–423.
479. W. Hodges, I. Hodkinson, D. Lascar and S. Shelah, The small index property for ω-stable ω-categorical structures and for the random graph, *J. London Math. Soc.* (2) **48** (1993), No. 2, 204–218.
480. S. Shelah and L. Soukup, The existence of large ω_1-homogeneous but not ω-homogeneous permutation groups is consistent with ZFC + GCH, *J. London Math. Soc.* (2) **48** (1993), No. 2, 193–203.
481. T. Hyttinen, S. Shelah and H. Tuuri, Remarks on strong nonstructure theorems, *Notre Dame J. Formal Logic* **34** (1993), No. 2, 157–168.
482. A. H. Mekler and S. Shelah, Every coseparable group may be free, *Israel J. Math.* **81** (1993), No. 1–2, 161–178.
483. S. Shelah and L. Stanley, More consistency results in partition calculus, *Israel J. Math.* **81** (1993), No. 1–2, 97–110.
484. S. Shelah and J. Steprāns, Maximal chains in ${}_{\omega}\omega$ and ultrapowers of the integers, *Arch. Math. Logic* **32** (1993), No. 5, 305–319.
485. A. Mekler, R. Schipperus, S. Shelah and J. K. Truss, The random graph and automorphisms of the rational world, *Bull. London Math. Soc.* **5** (1993), No. 4, 343–346.
486. A. H. Mekler and S. Shelah, The canary tree, *Canad. Math. Bull.* **36** (1993), No. 2, 209–215.
487. H. Judah and S. Shelah, Δ_{13}-sets of reals, *J. Symbolic Logic* **58** (1993), No. 1, 72–80.
488. T. Jech and S. Shelah, Full reflection of stationary sets at regular cardinals, *Amer. J. Math.* **115** (1993), No. 2, 435–453.
489. J. E. Baumgartner, S. Shelah and S. Thomas, Maximal subgroups of infinite symmetric groups, *Notre Dame J. Formal Logic* **34** (1993), No. 1, 1–11.
490. M. Gitik and S. Shelah, More on simple forcing notions and forcings with ideals, Fourth Asian Logic Conference (Tokyo, 1990). *Ann. Pure Appl. Logic* **59** (1993), No. 3, 219–238.

491. P. C. Eklof and S. Shelah, Explicitly nonstandard uniserial modules, *J. Pure Appl. Algebra* **86** (1993), No. 1, 35–50.
492. A. H. Mekler and S. Shelah, Some compact logics — Results in ZFC, *Ann. of Math.* (2) **137** (1993), No. 2, 221–248.
493. D. Lascar and S. Shelah, Uncountable saturated structures have the small index property, *Bull. London Math. Soc.* **25** (1993), No. 2, 125–131.
494. M. Goldstern and S. Shelah, Many simple cardinal invariants, *Arch. Math. Logic* **32** (1993), No. 3, 203–221.
495. M. Kojman and S. Shelah, μ-complete Souslin trees on μ_+, *Arch. Math. Logic* **32** (1993), No. 3, 195–201.
496. P. Komjáth and S. Shelah, A consistent edge partition theorem for infinite graphs, *Acta Math. Hungar.* **61** (1993), No. 1–2, 115–120.
497. A. H. Mekler, E. Nelson and S. Shelah, A variety with solvable, but not uniformly solvable, word problem, *Proc. London Math. Soc.* (3) **66** (1993), No. 2, 225–256.
498. U. Abraham and S. Shelah, A Δ_{22} well-order of the reals and incompactness of $L(Q_{MM})$, *Ann. Pure Appl. Logic* **59** (1993), No. 1, 1–32.
499. P. C. Eklof, A. H. Mekler and S. Shelah, On coherent systems of projections for $\aleph_1$-separable groups, *Comm. Algebra* **21** (1993), No. 1, 343–353.
500. A. Mekler, S. Shelah and J. Väänänen, The Ehrenfeucht–Fraïssé-game of length ω_1, *Trans. Amer. Math. Soc.* **339** (1993), No. 2, 567–580.
501. P. C. Eklof and S. Shelah, On a conjecture regarding nonstandard uniserial modules. *Trans. Amer. Math. Soc.* **340** (1993), No. 1, 337–351.
502. H. Judah and S. Shelah, Adding dominating reals with the random algebra. *Proc. Amer. Math. Soc.* **119** (1993), No. 1, 267–273.
503. T. Bartoszyński, M. Goldstern, H. Judah and S. Shelah, All meager filters may be null, *Proc. Amer. Math. Soc.* **117** (1993), No. 2, 515–521.
504. M. R. Burke and S. Shelah, Linear liftings for noncomplete probability spaces, *Israel J. Math.* **79** (1992), No. 2–3, 289–296.
505. S. Shelah and L. J. Stanley, Coding and reshaping when there are no sharps, in *Set Theory of the Continuum* (Berkeley, CA, 1989), pp. 407–416, Math. Sci. Res. Inst. Publ., Vol. 26 (Springer, 1992).
506. S. Shelah, Vive la différence, I. Nonisomorphism of ultrapowers of countable models, in *Set Theory of the Continuum* (Berkeley, CA, 1989), pp. 357–405, Math. Sci. Res. Inst. Publ., Vol. 26 (Springer, 1992).
507. S. Shelah, Strong partition relations below the power set: Consistency; was Sierpiński right?, II, *Sets, Graphs and Numbers* (Budapest, 1991), pp. 637–668, Colloq. Math. Soc. János Bolyai, Vol. 60 (North-Holland, 1992).
508. P. C. Eklof, A. H. Mekler and S. Shelah, Uniformization and the diversity of Whitehead groups, *Israel J. Math.* **80** (1992), No. 3, 301–321.

509. S. Shelah and R. Jin, Planting Kurepa trees and killing Jech-Kunen trees in a model by using one inaccessible cardinal, *Fund. Math.* **141** (1992), No. 3, 287–296.
510. J. Brendle, H. Judah and S. Shelah, Combinatorial properties of Hechler forcing, *Ann. Pure Appl. Logic* **58** (1992), No. 3, 185–199.
511. M. Kojman and S. Shelah, Nonexistence of universal orders in many cardinals, *J. Symbolic Logic* **57** (1992), No. 3, 875–891.
512. T. Bartoszyński and S. Shelah, Closed measure zero sets, *Ann. Pure Appl. Logic* **58** (1992), No. 2, 93–110.
513. S. Shelah, CON($u > i$), *Arch. Math. Logic* **31** (1992), No. 6, 433–443.
514. I. Juhász and S. Shelah, On partitioning the triples of a topological space, *Proceedings of the Symposium on General Topology and Applications* (Oxford, 1989), *Topology Appl.* **44** (1992), No. 1–3, 203–208.
515. M. Kojman and S. Shelah, The universality spectrum of stable unsuperstable theories, *Ann. Pure Appl. Logic* **58** (1992), No. 1, 57–72.
516. S. Shelah, Factor = quotient, uncountable Boolean algebras, number of endomorphism and width, *Math. Jpn.* **37** (1992), No. 2, 385–400.
517. T. Bartoszyński and S. Shelah, Intersection of $<2\aleph_0$ ultrafilters may have measure zero, *Arch. Math. Logic* **31** (1992), No. 4, 221–226.
518. S. Shelah, The Hanf numbers of stationary logic, II. Comparison with other logics, *Notre Dame J. Formal Logic* **33** (1992), No. 1, 1–12.
519. S. Lifsches and S. Shelah, The monadic theory of $(\omega_2, <)$ may be complicated, *Arch. Math. Logic* **31** (1992), No. 3, 207–213.
520. H. Judah, A. W. Miller and S. Shelah, Sacks forcing, Laver forcing, and Martin's axiom, *Arch. Math. Logic* **31** (1992), No. 3, 145–161.
521. S. Shelah, Cardinal arithmetic for skeptics, *Bull. Amer. Math. Soc. (N.S.)* **26** (1992), No. 2, 197–210.
522. S. Shelah, On a problem in cylindric algebra, in *Algebraic Logic* (Budapest, 1988), pp. 645–664, Colloq. Math. Soc. János Bolyai, Vol. 54 (North-Holland, 1991).
523. S. Shelah, Strong negative partition relations below the continuum, *Acta Math. Hungar.* **58** (1991), No. 1–2, 95–100.
524. E. Hrushovski and S. Shelah, Stability and omitting types, *Israel J. Math.* **74** (1991), No. 2–3, 289–321.
525. S. Shelah, Multi-dimensionality, *Israel J. Math.* **74** (1991), No. 2–3, 281–288.
526. S. Shelah, Kaplansky test problem for *R*-modules, *Israel J. Math.* **74** (1991), No. 1, 91–127.
527. J. T. Baldwin and S. Shelah, The primal framework, II. Smoothness, *Ann. Pure Appl. Logic* **55** (1991), No. 1, 1–34.

528. W. Hodges and S. Shelah, There are reasonably nice logics, *J. Symbolic Logic* **56** (1991), No. 1, 300–322.
529. H. Judah and S. Shelah, Forcing minimal degree of constructibility, *J. Symbolic Logic* **56** (1991), No. 3, 769–782.
530. P. C. Eklof and S. Shelah, On Whitehead modules, *J. Algebra* **142** (1991), No. 2, 492–510.
531. S. Shelah, Reflecting stationary sets and successors of singular cardinals, *Arch. Math. Logic* **31** (1991), No. 1, 25–53.
532. T. Jech and S. Shelah, A partition theorem for pairs of finite sets, *J. Amer. Math. Soc.* **4** (1991), No. 4, 647–656.
533. D. Macpherson, A. H. Mekler and S. Shelah, The number of infinite substructures, *Math. Proc. Cambridge Philos. Soc.* **109** (1991), No. 1, 193–209.
534. T. Jech and S. Shelah, On a conjecture of Tarski on products of cardinals, *Proc. Amer. Math. Soc.* **112** (1991), No. 4, 1117–1124.
535. M. Goldstern, H. I. Judah and S. Shelah, Saturated families, *Proc. Amer. Math. Soc.* **111** (1991), No. 4, 1095–1104.
536. M. Goldstern, H. I. Judah and S. Shelah, A regular topological space having no closed subsets of cardinality $\aleph_2$, *Proc. Amer. Math. Soc.* **111** (1991), No. 4, 1151–1159.
537. H. Judah and S. Shelah, *Q*-sets, Sierpiński sets, and rapid filters, *Proc. Amer. Math. Soc.* **111** (1991), No. 3, 821–832.
538. S. Shelah and E. C. Milner, Graphs with no unfriendly partitions, in *A Tribute to Paul Erdös*, pp. 373–384 (Cambridge Univ. Press, 1990).
539. S. Shelah, Incompactness for chromatic numbers of graphs, in *A Tribute to Paul Erdös*, pp. 361–371 (Cambridge Univ. Press, 1990).
540. H. Judah, S. Shelah and W. H. Woodin, The Borel conjecture, *Ann. Pure Appl. Logic* **50** (1990), No. 3, 255–269.
541. S. Shelah, Classification Theory and the Number of Nonisomorphic Models, Second edn., Studies in Logic and the Foundations of Mathematics, Vol. 92 (North-Holland, 1990).
542. H. Judah and S. Shelah, Around random algebra, *Arch. Math. Logic* **30** (1990), No. 3, 129–138.
543. M. Goldstern and and S. Shelah, Ramsey ultrafilters and the reaping number — Con($r < u$), *Ann. Pure Appl. Logic* **49** (1990), No. 2, 121–142.
544. S. Shelah and H. Woodin, Large cardinals imply that every reasonably definable set of reals is Lebesgue measurable, *Israel J. Math.* **70** (1990), No. 3, 381–394.

545. S. Shelah, More on monadic logic, Part C, Monadically interpreting in stable unsuperstable T and the monadic theory of $\omega\lambda$, *Israel J. Math.* **70** (1990), No. 3, 353–364.
546. M. Kojman, M. A. Perles and S. Shelah, Sets in a Euclidean space which are not a countable union of convex subsets, *Israel J. Math.* **70** (1990), No. 3, 313–342.
547. M. A. Perles and and S. Shelah, A closed $(n+1)$-convex set in $\mathbf{R}_2$ is a union of n_6 convex sets, *Israel J. Math.* **70** (1990), No. 3, 305–312.
548. Y. Gurevich and and S. Shelah, Nondeterministic linear-time tasks may require substantially nonlinear deterministic time in the case of sublinear work space, *J. Assoc. Comput. Mach.* **37** (1990), No. 3, 674–687.
549. H. Judah and and S. Shelah, The Kunen–Miller chart (Lebesgue measure, the Baire property, Laver reals and preservation theorems for forcing), *J. Symbolic Logic* **55** (1990), No. 3, 909–927.
550. S. Shelah and B, Hart, Categoricity over P for first order T or categoricity for $\varphi \in L_{\omega_1\omega}$ can stop at $\aleph_k$ while holding for $\aleph_0, \ldots, \aleph_{k-1}$, *Israel J. Math.* **70** (1990), No. 2, 219–235.
551. S. Shelah, Products of regular cardinals and cardinal invariants of products of Boolean algebras, *Israel J. Math.* **70** (1990), No. 2, 129–187.
552. S. Shelah, Universal graphs without instances of CH: Revisited, *Israel J. Math.* **70** (1990), No. 1, 69–81.
553. T. Jech and S. Shelah, Full reflection of stationary sets below $\aleph_\omega$, *J. Symbolic Logic* **55** (1990), No. 2, 822–830.
554. S. Shelah and M. Makkai, Categoricity of theories in $L_{\kappa\omega}$, with κ a compact cardinal, *Ann. Pure Appl. Logic* **47** (1990), No. 1, 41–97.
555. J. T. Baldwin and S. Shelah, The primal framework, I, *Ann. Pure Appl. Logic* **46** (1990), No. 3, 235–264.
556. A. H. Mekler and S. Shelah, Determining abelian p-groups from their n-socles, *Comm. Algebra* **18** (1990), No. 2, 287–307.
557. S. Shelah, Notes on monadic logic. B. Complexity of linear orders in ZFC, *Israel J. Math.* **69** (1990), No. 1, 94–116.
558. S. Shelah, The theorems of Beth and Craig in abstract model theory, III. Δ-logics and infinitary logics, *Israel J. Math.* **69** (1990), No. 2, 193–213.
559. J.-P. Levinski, M. Magidor and S. Shelah, Chang's conjecture for $\aleph_\omega$, *Israel J. Math.* **69** (1990), No. 2, 161–172.
560. S. Shelah and C. Steinhorn, The nonaxiomatizability of $L(Q^2_{\aleph_1})$by finitely many schemata, *Notre Dame J. Formal Logic* **31** (1990), No. 1, 1–13.
561. S. Shelah, Strong negative partition above the continuum, *J. Symbolic Logic* **55** (1990), No. 1, 21–31.

562. P. C. Eklof, L. Fuchs and S. Shelah, Baer modules over domains, *Trans. Amer. Math. Soc.* **322** (1990), No. 2, 547–560.
563. A. H. Mekler and S. Shelah, $L_{\infty\omega}$-free algebras. *Algebra Universalis* **26** (1989), No. 3, 351–366.
564. E. Hrushovski and S. Shelah, A dichotomy theorem for regular types, Stability in model theory, II (Trento, 1987). *Ann. Pure Appl. Logic* **45** (1989), No. 2, 157–169.
565. T. Jech and S. Shelah, A note on canonical functions, *Israel J. Math.* **68** (1989), No. 3, 376–380.
566. S. Shelah, More on monadic logic. D, A note on addition of theories, *Israel J. Math.* **68** (1989), No. 3, 302–306.
567. A. Blass and S. Shelah, Near coherence of filters, III, A simplified consistency proof, *Notre Dame J. Formal Logic* **30** (1989), No. 4, 530–538.
568. M. Gitik and S. Shelah, Forcings with ideals and simple forcing notions, *Israel J. Math.* **68** (1989), No. 2, 129–160.
569. H. Judah and S. Shelah, MA(σ-centered): Cohen reals, strong measure zero sets and strongly meager sets, *Israel J. Math.* **68** (1989), No. 1, 1–17.
570. S. Shelah and S. Buechler, On the existence of regular types, *Ann. Pure Appl. Logic* **45** (1989), No. 3, 277–308.
571. S. Shelah, Consistency of positive partition theorems for graphs and models, in *Set Theory and Its Applications* (Toronto, ON, 1987), pp. 167–193, Lecture Notes in Math., Vol. 1401 (Springer, 1989).
572. Y. Gurevich and S. Shelah, Nearly linear time, in *Logic at Botik '89* (Pereslavl'-Zalesskiy, 1989), pp. 108–118, Lecture Notes in Comput. Sci., Vol. 363 (Springer, 1989).
573. A. H. Mekler and S. Shelah, The consistency strength of "every stationary set reflects", *Israel J. Math.* **67** (1989), No. 3, 353–366.
574. S. Shelah, Baire irresolvable spaces and lifting for a layered ideal, *Topology Appl.* **33** (1989), No. 3, 217–221.
575. S. Shelah, The number of pairwise non-elementarily-embeddable models, *J. Symbolic Logic* **54** (1989), No. 4, 1431–1455.
576. E. Dror Farjoun, K. Orr and S. Shelah, Bousfield localization as an algebraic closure of groups, *Israel J. Math.* **66** (1989), No. 1–3, 143–153.
577. Y. Gurevich and S. Shelah, Time polynomial in input or output, *J. Symbolic Logic* **54** (1989), No. 3, 1083–1088.
578. T. Bartoszyński, J. I. Ihoda and S. Shelah, The cofinality of cardinal invariants related to measure and category, *J. Symbolic Logic* **54** (1989), No. 3, 719–726.
579. I. Juhász and S. Shelah, $\pi(X) = \delta(X)$ for compact X, *Topology Appl.* **32** (1989), No. 3, 289–294.

580. A. Blass and S. Shelah, Ultrafilters with small generating sets, *Israel J. Math.* **65** (1989), No. 3, 259–271.
581. S. Shelah and J. Steprāns, Nontrivial homeomorphisms of $\beta\mathbf{N}\backslash\mathbf{N}$ without the continuum hypothesis, *Fund. Math.* **132** (1989), No. 2, 135–141.
582. S. Shelah, A consistent counterexample in the theory of collectionwise Hausdorff spaces, *Israel J. Math.* **65** (1989), No. 2, 219–224.
583. S. Shelah, "Gap 1" two-cardinal principles and the omitting types theorem for $L(Q)$, *Israel J. Math.* **65** (1989), No. 2, 133–152.
584. J. I. Ihoda and S. Shelah, Δ_{12}-sets of reals, *Ann. Pure Appl. Logic* **42** (1989), No. 3, 207–223.
585. A. H. Mekler and S. Shelah, Uniformization principles, *J. Symbolic Logic* **54** (1989), No. 2, 441–459.
586. Y. Gurevich and S. Shelah, On the strength of the interpretation method, *J. Symbolic Logic* **54** (1989), No. 2, 305–323.
587. M. Foreman, M. Magidor and S. Shelah, Correction to: "Martin's maximum, saturated ideals, and nonregular ultrafilters. I" [*Ann. of Math.* (2) **127** (1988), No. 1, 1–47], *Ann. of Math.* (2) **129** (1989), No. 3, 651.
588. S. Shelah and S. Thomas, Homogeneity of infinite permutation groups, *Arch. Math. Logic* **28** (1989), No. 2, 143–147.
589. M. Dugas and S. Shelah, E-transitive groups in L, in *Abelian Group Theory* (Perth, 1987), pp. 191–199, Contemp. Math., Vol. 87 (Amer. Math. Soc., 1989).
590. W. G. Fleissner and S. Shelah, Collectionwise Hausdorff: Incompactness at singulars, *Topology Appl.* **31** (1989), No. 2, 101–107.
591. R. Grossberg and S. Shelah, On the structure of $\mathrm{Ext}_p(G, \mathbf{Z})$, *J. Algebra* **121** (1989), No. 1, 117–128.
592. M. Gitik and S. Shelah, On certain indestructibility of strong cardinals and a question of Hajnal, *Arch. Math. Logic* 28 (1989), No. 1, 35–42.
593. S. Shelah and S. Thomas, Subgroups of small index in infinite symmetric groups, II, *J. Symbolic Logic* **54** (1989), No. 1, 95–99.
594. J. I. Ihoda and S. Shelah, Martin's axioms, measurability and equiconsistency results, *J. Symbolic Logic* **54** (1989), No. 1, 78–94.
595. T. Becker, L. Fuchs and S. Shelah, Whitehead modules over domains, *Forum Math.* **1** (1989), No. 1, 53–68.
596. L. Fuchs and S. Shelah, Kaplansky's problem on valuation rings, *Proc. Amer. Math. Soc.* **105** (1989), No. 1, 25–30.
597. S. Shelah, Classifying general classes. A plenary address presented at the *International Congress of Mathematicians* held in Berkeley, California, August 1986. Introduced by Ronald L. Graham. ICM Series (Amer. Math. Soc., 1988).

598. J. I. Ihoda and S. Shelah, Souslin forcing, *J. Symbolic Logic* **53** (1988), No. 4, 1188–1207.
599. A. H. Mekler and S. Shelah, Diamond and λ-systems, *Fund. Math.* **131** (1988), No. 1, 45–51.
600. S. Shelah, Notes on monadic logic, Part A, Monadic theory of the real line, *Israel J. Math.* **63** (1988), No. 3, 335–352.
601. S. Shelah, Decomposing topological spaces into two rigid homeomorphic subspaces, *Israel J. Math.* **63** (1988), No. 2, 183–211.
602. L. Harrington, D. Marker and S. Shelah, Borel orderings, *Trans. Amer. Math. Soc.* **310** (1988), No. 1, 293–302.
603. S. Shelah, Number of strongly $\aleph_\varepsilon$ saturated models — An addition, *Ann. Pure Appl. Logic* **40** (1988), No. 1, 89–91.
604. S. Shelah and L. Stanley, Weakly compact cardinals and nonspecial Aronszajn trees, *Proc. Amer. Math. Soc.* **104** (1988), No. 3, 887–897.
605. B. Biró and S. Shelah, Isomorphic but not lower base-isomorphic cylindric set algebras. *J. Symbolic Logic* **53** (1988), No. 3, 846–853.
606. P. Komjáth and S. Shelah, Forcing constructions for uncountably chromatic graphs, *J. Symbolic Logic* **53** (1988), No. 3, 696–707.
607. S. Shelah and J. Steprāns, A Banach space on which there are few operators, *Proc. Amer. Math. Soc.* **104** (1988), No. 1, 101–105.
608. S. Shelah, Was Sierpiński right?, I, *Israel J. Math.* **62** (1988), No. 3, 355–380.
609. I. Juhász, S. Shelah and L. Soukup, More on countably compact, locally countable spaces, *Israel J. Math.* **62** (1988), No. 3, 302–310.
610. S. Shelah, Successors of singulars, cofinalities of reduced products of cardinals and productivity of chain conditions, *Israel J. Math.* **62** (1988), No. 2, 213–256.
611. S. Shelah, Can the fundamental (homotopy) group of a space be the rationals?, *Proc. Amer. Math. Soc.* **103** (1988), No. 2, 627–632.
612. M. Foreman, M. Magidor and S. Shelah, Martin's maximum, saturated ideals and nonregular ultrafilters, II, *Ann. of Math.* (2) **127** (1988), No. 3, 521–545.
613. S. Shelah and S. Thomas, Implausible subgroups of infinite symmetric groups, *Bull. London Math. Soc.* **20** (1988), No. 4, 313–318.
614. S. Shelah, A graph which embeds all small graphs on any large set of vertices, *Ann. Pure Appl. Logic* **38** (1988), No. 2, 171–183.
615. S. Shelah and J. Steprāns, PFA implies all automorphisms are trivial, *Proc. Amer. Math. Soc.* **104** (1988), No. 4, 1220–1225.
616. S. Shelah, Some notes on iterated forcing with $2^{\aleph_0} > \aleph_2$, *Notre Dame J. Formal Logic* **29** (1988), No. 1, 1–17.

617. S. Shelah, Primitive recursive bounds for van der Waerden numbers, *J. Amer. Math. Soc.* **1** (1988), No. 3, 683–697.
618. J. Ihoda and S. Shelah, Q-sets do not necessarily have strong measure zero, *Proc. Amer. Math. Soc.* **102** (1988), No. 3, 681–683.
619. S. Shelah and J. Spencer, Zero-one laws for sparse random graphs, *J. Amer. Math. Soc.* **1** (1988), No. 1, 97–115.
620. M. Foreman, M. Magidor and S. Shelah, Martin's maximum, saturated ideals, and nonregular ultrafilters, I, *Ann. of Math.* (2) **127** (1988), No. 1, 1–47.
621. S. Shelah, On almost categorical theories, in *Classification Theory* (Chicago, IL, 1985), pp. 498–500, Lecture Notes in Math., Vol. 1292 (Springer, 1987).
622. S. Shelah, Classification of nonelementary classes, II, Abstract elementary classes, in *Classification Theory* (Chicago, IL, 1985), pp. 419–497, Lecture Notes in Math., Vol. 1292 (Springer, 1987).
623. S. Shelah, Universal classes, in *Classification Theory* (Chicago, IL, 1985), pp. 264–418, Lecture Notes in Math., Vol. 1292 (Springer, 1987).
624. Z. Chatzidakis, G. Cherlin, S. Shelah, G. Srour and C. Wood, Orthogonality of types in separably closed fields, in *Classification Theory* (Chicago, IL, 1985), pp. 72–88, Lecture Notes in Math., Vol. 1292 (Springer, 1987).
625. P. C. Eklof and S. Shelah, On groups A such that $A \oplus \mathbf{Z}_n \cong A$, in *Abelian Group Theory* (Oberwolfach, 1985), pp. 149–163 (Gordon and Breach, 1987).
626. A. H. Mekler and S. Shelah, When κ-free implies strongly κ-free, in *Abelian Group Theory* (Oberwolfach, 1985), pp. 137–148 (Gordon and Breach, 1987).
627. S. Shelah, Iterated forcing and normal ideals on ω_1, *Israel J. Math.* **60** (1987), No. 3, 345–380.
628. P. C. Eklof and S. Shelah, A calculation of injective dimension over valuation domains, *Rend. Sem. Mat. Univ. Padova* **78** (1987), 279–284.
629. S. Shelah, Taxonomy of universal and other classes, in *Proceedings of the International Congress of Mathematicians*, Vol. 1 (Berkeley, Calif., 1986), pp. 154–162 (Amer. Math. Soc., 1987).
630. S. Shelah, On reconstructing separable reduced p-groups with a given socle, *Israel J. Math.* **60** (1987), No. 2, 146–166.
631. S. Shelah, More on powers of singular cardinals, *Israel J. Math.* **59** (1987), No. 3, 299–326.
632. P. C. Eklof, A. H. Mekler and S. Shelah, On strongly nonreflexive groups, *Israel J. Math.* **59** (1987), No. 3, 283–298.
633. S. Shelah, On the number of strongly $\aleph_\varepsilon$-saturated models of power λ, *Ann. Pure Appl. Logic* **36** (1987), No. 3, 279–287.

634. K. J. Compton, C. W. Henson and S. Shelah, Nonconvergence, undecidability, and intractability in asymptotic problems, *Ann. Pure Appl. Logic* **36** (1987), No. 3, 207–224.
635. S. Shelah, Uncountable groups have many nonconjugate subgroups, *Ann. Pure Appl. Logic* **36** (1987), No. 2, 153–206.
636. S. Shelah and L. Stanley, A theorem and some consistency results in partition calculus, *Ann. Pure Appl. Logic* **36** (1987), No. 2, 119–152.
637. S. Shelah, Existence of many $L_{\infty,\lambda}$-equivalent, nonisomorphic models of T of power λ, *Ann. Pure Appl. Logic* **34** (1987), No. 3, 291–310.
638. M. Dugas, T. H. Fay and S. Shelah, Singly cogenerated annihilator classes, *J. Algebra* **109** (1987), No. 1, 127–137.
639. S. Shelah, Semiproper forcing axiom implies Martin maximum but not PFA+, *J. Symbolic Logic* **52** (1987), No. 2, 360–367.
640. Y. Gurevich and S. Shelah, Expected computation time for Hamiltonian path problem, *SIAM J. Comput.* **16** (1987), No. 3, 486–502.
641. Z. Drezner and S. Shelah, On the complexity of the Elzinga–Hearn algorithm for the 1-center problem, *Math. Oper. Res.* **12** (1987), No. 2, 255–261.
642. S. Shelah and J. Steprāns, Extraspecial p-groups, *Ann. Pure Appl. Logic* **34** (1987), No. 1, 87–97.
643. U. Abraham, S. Shelah and R. M. Solovay, Squares with diamonds and Souslin trees with special squares, *Fund. Math.* **127** (1987), No. 2, 133–162.
644. M. Droste and S. Shelah, On the universality of systems of words in permutation groups, *Pacific J. Math.* **127** (1987), No. 2, 321–328.
645. A. Blass and S. Shelah, There may be simple $P_{\aleph_1}$- and $P_{\aleph_2}$-points and the Rudin–Keisler ordering may be downward directed, *Ann. Pure Appl. Logic* **33** (1987), No. 3, 213–243.
646. J. E. Baumgartner and S. Shelah, Remarks on superatomic Boolean algebras, *Ann. Pure Appl. Logic* **33** (1987), No. 2, 109–129.
647. M. Rubin and S. Shelah, Combinatorial problems on trees: Partitions, Δ-systems and large free subtrees, *Ann. Pure Appl. Logic* **33** (1987), No. 1, 43–81.
648. A. Hajnal, A. Kanamori and S. Shelah, Regressive partition relations for infinite cardinals, *Trans. Amer. Math. Soc.* **299** (1987), No. 1, 145–154.
649. W. C. Holland, A. H. Mekler and S. Shelah, Total orders whose carried groups satisfy no laws, in *Algebra and Order* (Luminy-Marseille, 1984), pp. 29–33, Res. Exp. Math., Vol. 14 (Heldermann, 1986).
650. R. Göbel and S. Shelah, Modules over arbitrary domains, II, *Fund. Math.* **126** (1986), No. 3, 217–243.

651. S. Shelah and L. J. Stanley, *S*-forcing, IIa, Adding diamonds and more applications: Coding sets, Arhangel'skiĭ's problem and $\mathcal{L}[Q^{<\omega_1}, Q_2^1]$. With an appendix by John P. Burgess, *Israel J. Math.* **56** (1986), No. 1, 1–65.
652. S. Shelah, The spectrum problem, III, Universal theories, *Israel J. Math.* **55** (1986), No. 2, 229–256.
653. Y. Gurevich and S. Shelah, Fixed-point extensions of first-order logic, *Ann. Pure Appl. Logic* **32** (1986), No. 3, 265–280.
654. S. Ben-David and S. Shelah, Nonspecial Aronszajn trees on $\aleph_{\omega+1}$, *Israel J. Math.* **53** (1986), No. 1, 93–96.
655. L. Denenberg, Y. Gurevich and S. Shelah, Definability by constant-depth polynomial-size circuits, *Inform. and Control* **70** (1986), No. 2–3, 216–240.
656. I. Juhász and S. Shelah, How large can a hereditarily separable or hereditarily Lindelöf space be?, *Israel J. Math.* **53** (1986), No. 3, 355–364.
657. S. Shelah and L. J. Stanley, Corrigendum to: "Generalized Martin's axiom and Souslin's hypothesis for higher cardinals" [*Israel J. Math.* **43** (1982), No. 3, 225–236], *Israel J. Math.* **53** (1986), No. 3, 304–314.
658. S. Shelah, Remarks on squares, in *Around Classification Theory of Models*, pp. 276–279, Lecture Notes in Math., Vol. 1182 (Springer, 1986).
659. S. Shelah, On decomposable sentences for finite models, in *Around Classification Theory of Models*, pp. 272–275, Lecture Notes in Math., Vol. 1182 (Springer, 1986).
660. S. Shelah, On countable theories with models — Homogeneous models only, in *Around Classification Theory of Models*, pp. 269–271, Lecture Notes in Math., Vol. 1182 (Springer, 1986).
661. S. Shelah, A note on κ-freeness of abelian groups, in *Around Classification Theory of Models*, pp. 260–268, Lecture Notes in Math., Vol. 1182 (Springer, 1986).
662. S. Shelah, On normal ideals and Boolean algebras, in *Around Classification Theory of Models*, pp. 247–259, Lecture Notes in Math., Vol. 1182 (Springer, 1986).
663. S. Shelah, More on stationary coding, in *Around Classification Theory of Models*, pp. 224–246, Lecture Notes in Math., Vol. 1182 (Springer, 1986).
664. S. Shelah, Monadic logic: Hanf numbers, in *Around Classification Theory of Models*, pp. 203–223, Lecture Notes in Math., Vol. 1182 (Springer, 1986).
665. S. Shelah, The existence of coding sets, in *Around Classification Theory of Models*, pp. 188–202, Lecture Notes in Math., Vol. 1182 (Springer, 1986).
666. S. Shelah, Remarks on the numbers of ideals of Boolean algebra and open sets of a topology, in *Around Classification Theory of Models*, pp. 151–187, Lecture Notes in Math., Vol. 1182 (Springer, 1986).

667. S. Shelah, Nonstandard uniserial module over a uniserial domain exists, in *Around Classification Theory of Models*, pp. 135–150, Lecture Notes in Math., Vol. 1182 (Springer, 1986).
668. S. Shelah, On the no(*M*) for *M* of singular power, in *Around Classification Theory of Models*, pp. 120–134, Lecture Notes in Math., Vol. 1182 (Springer, 1986).
669. S. Shelah, Existence of endo-rigid Boolean algebras, in *Around Classification Theory of Models*, pp. 91–119, Lecture Notes in Math., Vol. 1182 (Springer, 1986).
670. S. Shelah, Classification over a predicate, II, in *Around Classification Theory of Models*, pp. 47–90, Lecture Notes in Math., Vol. 1182 (Springer, 1986).
671. S. Shelah, Classifying generalized quantifiers, in *Around Classification Theory of Models*, pp. 1–46, Lecture Notes in Math., Vol. 1182 (Springer, 1986).
672. S. Shelah, *Around Classification Theory of Models*, Lecture Notes in Mathematics, Vol. 1182 (Springer-Verlag, 1986).
673. S. Shelah and S. Todorčević, A note on small Baire spaces, *Canad. J. Math.* **38** (1986), No. 3, 659–665.
674. S. Shelah, On power of singular cardinals, *Notre Dame J. Formal Logic* **27** (1986), No. 2, 263–299.
675. S. Shelah and A. Soifer, Countable $\aleph_0$-indecomposable mixed abelian groups of finite torsion-free rank, *J. Algebra* **100** (1986), No. 2, 421–429.
676. R. Grossberg and S. Shelah, On the number of nonisomorphic models of an infinitary theory which has the infinitary order property, I, *J. Symbolic Logic* **51** (1986), No. 2, 302–322.
677. R. Grossberg and S. Shelah, A nonstructure theorem for an infinitary theory which has the unsuperstability property, *Illinois J. Math.* **30** (1986), No. 2, 364–390.
678. S. Shelah and A. Soifer, Two problems on $\aleph_0$-indecomposable abelian groups, *J. Algebra* **99** (1986), No. 2, 359–369.
679. S. Ben-David and S. Shelah, Souslin trees and successors of singular cardinals, *Ann. Pure Appl. Logic* **30** (1986), No. 3, 207–217.
680. A. Hajnal, I. Juhász and S. Shelah, Splitting strongly almost disjoint families, *Trans. Amer. Math. Soc.* **295** (1986), No. 1, 369–387.
681. U. Abraham and S. Shelah, On the intersection of closed unbounded sets, *J. Symbolic Logic* **51** (1986), No. 1, 180–189.
682. M. Foreman, M. Magidor and S. Shelah, 0# and some forcing principles, *J. Symbolic Logic* **51** (1986), No. 1, 39–46.

683. W. Hodges and S. Shelah, Naturality and definability, I, *J. London Math. Soc.* (2) **33** (1986), No. 1, 1–12.

684. S. Shelah and M. Kaufmann, The Hanf number of stationary logic, *Notre Dame J. Formal Logic* **27** (1986), No. 1, 111–123.

685. A. H. Mekler and S. Shelah, Stationary logic and its friends, II, *Notre Dame J. Formal Logic* **27** (1986), No. 1, 39–50.

686. S. Shelah and C. Steinhorn, On the nonaxiomatizability of some logics by finitely many schemas, *Notre Dame J. Formal Logic* **27** (1986), No. 1, 1–11.

687. A. H. Mekler and S. Shelah, The solution to Crawley's problem, *Pacific J. Math.* **121** (1986), No. 1, 133–134.

688. A. H. Mekler and S. Shelah, ω-elongations and Crawley's problem, *Pacific J. Math.* **121** (1986), No. 1, 121–132.

689. S. Shelah, On measure and category, *Israel J. Math.* **52** (1985), No. 1–2, 110–114.

690. S. Shelah, Remarks in abstract model theory, *Ann. Pure Appl. Logic* **29** (1985), No. 3, 255–288.

691. Y. Gurevich and S. Shelah, The decision problem for branching time logic, *J. Symbolic Logic* **50** (1985), No. 3, 668–681.

692. S. Shelah, Uncountable constructions for B.A., e.c. groups and Banach spaces, *Israel J. Math.* **51** (1985), No. 4, 273–297.

693. M. Droste and S. Shelah, A construction of all normal subgroup lattices of 2-transitive automorphism groups of linearly ordered sets, *Israel J. Math.* **51** (1985), No. 3, 223–261.

694. U. Abraham, M. Rubin and S. Shelah, On the consistency of some partition theorems for continuous colorings, and the structure of $\aleph_1$-dense real order types, *Ann. Pure Appl. Logic* **29** (1985), No. 2, 123–206.

695. A. Pillay and S. Shelah, Classification theory over a predicate, I, *Notre Dame J. Formal Logic* **26** (1985), No. 4, 361–376.

696. Y. Gurevich and S. Shelah, To the decision problem for branching time logic, in *Foundations of Logic and Linguistics* (Salzburg, 1983), pp. 181–198 (Plenum, 1985).

697. J. T. Baldwin and S. Shelah, Second-order quantifiers and the complexity of theories, *Notre Dame J. Formal Logic* **26** (1985), No. 3, 229–303.

698. S. Shelah, Incompactness in regular cardinals, *Notre Dame J. Formal Logic* **26** (1985), No. 3, 195–228.

699. G. Sageev and S. Shelah, On the structure of $\mathrm{Ext}(A, \mathbf{Z})$ in ZFC+, *J. Symbolic Logic* **50** (1985), No. 2, 302–315.

700. M. Kaufmann and S. Shelah, On random models of finite power and monadic logic, *Discrete Math.* **54** (1985), No. 3, 285–293.

701. S. Shelah, More on the weak diamond, *Ann. Pure Appl. Logic* **28** (1985), No. 3, 315–318.
702. U. Abraham and S. Shelah, Isomorphism types of Aronszajn trees, *Israel J. Math.* **50** (1985), No. 1–2, 75–113.
703. W. C. Holland; A. H. Mekler and S. Shelah, Lawless order, *Order* **1** (1985), No. 4, 383–397.
704. L. Harrington and S. Shelah, Some exact equiconsistency results in set theory, *Notre Dame J. Formal Logic* **26** (1985), No. 2, 178–188.
705. A. H. Mekler and S. Shelah, Stationary logic and its friends, I, *Notre Dame J. Formal Logic* **26** (1985), No. 2, 129–138.
706. R. Göbel and S. Shelah, Semirigid classes of cotorsion-free abelian groups, *J. Algebra* **93** (1985), No. 1, 136–150.
707. S. Shelah, Monadic logic and Löwenheim numbers, *Ann. Pure Appl. Logic* **28** (1985), No. 2, 203–216.
708. S. Shelah, Classification of first order theories which have a structure theorem, *Bull. Amer. Math. Soc. (N.S.)* **12** (1985), No. 2, 227–232.
709. R. Bonnet and S. Shelah, Narrow Boolean algebras, *Ann. Pure Appl. Logic* **28** (1985), No. 1, 1–12.
710. R. Göbel and S. Shelah, Modules over arbitrary domains, *Math. Z.* **188** (1985), No. 3, 325–337.
711. S. Shelah, On the possible number no(M) = the number of nonisomorphic models $L_{\infty,\lambda}$-equivalent to M of power λ, for λ singular, *Notre Dame J. Formal Logic* **26** (1985), No. 1, 36–50.
712. S. Shelah, A combinatorial theorem and endomorphism rings of abelian groups, II, *Abelian Groups and Modules* (Udine, 1984), pp. 37–86, CISM Courses and Lectures, Vol. 287 (Springer, 1984).
713. S. Shelah, L. Harrington and M. Makkai, A proof of Vaught's conjecture for ω-stable theories, *Israel J. Math.* **49** (1984), No. 1–3, 259–280.
714. S. Shelah, A combinatorial principle and endomorphism rings, I, On p-groups, *Israel J. Math.* **49** (1984), No. 1–3, 239–257.
715. A. H. Lachlan and S. Shelah, Stable structures homogeneous for a finite binary language, *Israel J. Math.* **49** (1984), No. 1–3, 155–180.
716. Y. Gurevich and S. Shelah, The monadic theory and the "next world", *Israel J. Math.* **49** (1984), No. 1–3, 55–68.
717. P. C. Eklof, A. H. Mekler and S. Shelah, Almost disjoint abelian groups, *Israel J. Math.* **49** (1984), No. 1–3, 34–54.
718. R. Aharoni, C. St. J. A. Nash-Williams and S. Shelah, Marriage in infinite societies, in *Progress in Graph Theory* (Waterloo, Ont., 1982), pp. 71–79 (Academic Press, 1984).

719. W. D. Goldfarb, Y. Gurevich and S. Shelah, A decidable subclass of the minimal Gödel class with identity, *J. Symbolic Logic* **49** (1984), No. 4, 1253–1261.

720. S. Shelah and H. Woodin, Forcing the failure of CH by adding a real, *J. Symbolic Logic* **49** (1984), No. 4, 1185–1189.

721. S. Shelah, More on proper forcing, *J. Symbolic Logic* **49** (1984), No. 4, 1034–1038.

722. S. Shelah, Diamonds, uniformization, *J. Symbolic Logic* **49** (1984), No. 4, 1022–1033.

723. M. Gitik and S. Shelah, On the ▯-condition, *Israel J. Math.* **48** (1984), No. 2–3, 148–158.

724. S. Shelah, Can you take Solovay's inaccessible away?, *Israel J. Math.* **48** (1984), No. 1, 1–47.

725. M. Kaufmann and S. Shelah, A nonconservativity result on global choice, *Ann. Pure Appl. Logic* **27** (1984), No. 3, 209–214.

726. S. Shelah, On cardinal invariants of the continuum, in *Axiomatic Set Theory* (Boulder, Colo., 1983), pp. 183–207, Contemp. Math., Vol. 31 (Amer. Math. Soc., 1984).

727. S. Shelah, On logical sentences in PA, in *Logic Colloquium '82* (Florence, 1982), pp. 145–160, Stud. Logic Found. Math., Vol. 112 (North-Holland, 1984).

728. M. Magidor, S. Shelah and J. Stavi, Countably decomposable admissible sets, *Ann. Pure Appl. Logic* **26** (1984), No. 3, 287–361.

729. R. Aharoni, C. St. J. A. Nash-Williams and S. Shelah, Another form of a criterion for the existence of transversals, *J. London Math. Soc.* (2) **29** (1984), No. 2, 193–203.

730. S. Shelah, On universal graphs without instances of CH, *Ann. Pure Appl. Logic* **26** (1984), No. 1, 75–87.

731. D. Giorgetta and S. Shelah, Existentially closed structures in the power of the continuum, *Ann. Pure Appl. Logic* **26** (1984), No. 2, 123–148.

732. S. Shelah, On co-κ-Souslin relations, *Israel J. Math.* **47** (1984), No. 2–3, 139–153.

733. S. Shelah, A pair of nonisomorphic $\equiv_{\infty\lambda}$ models of power λ for λ singular with $\lambda^{\omega} = \lambda$, *Notre Dame J. Formal Logic* **25** (1984), No. 2, 97–104.

734. S. Shelah, The singular cardinals problem: independence results, in *Surveys in Set Theory*, pp. 116–134, London Math. Soc. Lecture Note Ser., Vol. 87, (Cambridge Univ. Press, 1983).

735. M. Magidor, S. Shelah and J. Stavi, On the standard part of nonstandard models of set theory, *J. Symbolic Logic* **48** (1983), No. 1, 33–38.

736. S. Shelah, Classification theory for nonelementary classes, I. The number of uncountable models of $\psi \in L_{\omega_1,\omega}$, Part A, *Israel J. Math.* **46** (1983), No. 3, 212–240.
737. J. A. Makowsky and S. Shelah, Positive results in abstract model theory: A theory of compact logics, *Ann. Pure Appl. Logic* **25** (1983), No. 3, 263–299.
738. S. Shelah, Classification theory for nonelementary classes, I. The number of uncountable models of $\psi \in L_{\omega_1,\omega}$, Part B, *Israel J. Math.* **46** (1983), No. 4, 241–273.
739. Y. Gurevich and S. Shelah, Random models and the Gödel case of the decision problem, *J. Symbolic Logic* **48** (1983), No. 4, 1120–1124 (1984).
740. Y. Gurevich and S. Shelah, Rabin's uniformization problem, *J. Symbolic Logic* **48** (1983), No. 4, 1105–1119 (1984).
741. Y. Gurevich, M. Magidor and S. Shelah, The monadic theory of ω_2, *J. Symbolic Logic* **48** (1983), No. 2, 387–398.
742. S. Shelah, On the number of nonconjugate subgroups, *Algebra Universalis* **16** (1983), No. 2, 131–146.
743. J. T. Baldwin and S. Shelah, The structure of saturated free algebras, *Algebra Universalis* **17** (1983), No. 2, 191–199.
744. S. Shelah, Models with second order properties, IV, A general method and eliminating diamonds, *Ann. Pure Appl. Logic* **25** (1983), No. 2, 183–212.
745. S. Shelah, Constructions of many complicated uncountable structures and Boolean algebras, *Israel J. Math.* **45** (1983), No. 2–3, 100–146.
746. Y. Gurevich and S. Shelah, Interpreting second-order logic in the monadic theory of order, *J. Symbolic Logic* **48** (1983), No. 3, 816–828.
747. U. Abraham and S. Shelah, Forcing closed unbounded sets, *J. Symbolic Logic* **48** (1983), No. 3, 643–657.
748. M. Rubin and S. Shelah, On the expressibility hierarchy of Magidor-Malitz quantifiers, *J. Symbolic Logic* **48** (1983), No. 3, 542–557.
749. S. Shelah, Lifting problem of the measure algebra, *Israel J. Math.* **45** (1983), No. 1, 90–96.
750. R. Grossberg and S. Shelah, On universal locally finite groups, *Israel J. Math.* **44** (1983), No. 4, 289–302.
751. M. Jarden and S. Shelah, Pseudo-algebraically closed fields over rational function fields, *Proc. Amer. Math. Soc.* **87** (1983), No. 2, 223–228.
752. Sy D. Friedman and S. Shelah, Tall α-recursive structures, *Proc. Amer. Math. Soc.* **88** (1983), No. 4, 672–678.
753. R. Aharoni, C. St. J. A. Nash-Williams and S. Shelah, A general criterion for the existence of transversals, *Proc. London Math. Soc.* (3) **47** (1983), No. 1, 43–68.

754. S. Shelah, The spectrum problem, II, Totally transcendental and infinite depth, *Israel J. Math.* **43** (1982), No. 4, 357–364.
755. S. Shelah, The spectrum problem, I, $\aleph_\varepsilon$-saturated models, the main gap, *Israel J. Math.* **43** (1982), No. 4, 324–356.
756. S. Shelah and L. Stanley, Generalized Martin's axiom and Souslin's hypothesis for higher cardinals, *Israel J. Math.* **43** (1982), No. 3, 225–236.
757. S. Shelah and L. Stanley, *S*-forcing, I, A "black-box" theorem for morasses, with applications to super-Souslin trees, *Israel J. Math.* **43** (1982), No. 3, 185–224.
758. S. Shelah, *Proper Forcing*, Lecture Notes in Mathematics, Vol. 940 (Springer, 1982).
759. S. Shelah, On the number of nonisomorphic models in $L_{\infty,\kappa}$ when κ is weakly compact, *Notre Dame J. Formal Logic* **23** (1982), No. 1, 21–26.
760. D. Lehmann and S. Shelah, Reasoning with time and chance, *Inform. and Control* **53** (1982), No. 3, 165–198.
761. S. Shelah and B. Weiss, Measurable recurrence and quasi-invariant measures, *Israel J. Math.* **43** (1982), No. 2, 154–160.
762. L. Harrington and S. Shelah, Counting equivalence classes for co-κ-Souslin equivalence relations, in *Logic Colloquium '80* (Prague, 1980), pp. 147–152, Stud. Logic Foundations Math., Vol. 108 (North-Holland, 1982).
763. Y. Gurevich and S. Shelah, Monadic theory of order and topology in ZFC, *Ann. Math. Logic* **23** (1982), No. 2–3, 179–198 (1983).
764. S. Shelah, Better quasi-orders for uncountable cardinals, *Israel J. Math.* **42** (1982), No. 3, 177–226.
765. L. Harrington and S. Shelah, The undecidability of the recursively enumerable degrees, *Bull. Amer. Math. Soc. (N.S.)* **6** (1982), No. 1, 79–80.
766. U. Avraham and S. Shelah, Forcing with stable posets, *J. Symbolic Logic* **47** (1982), No. 1, 37–42.
767. S. Shelah, Canonization theorems and applications, *J. Symbolic Logic* **46** (1981), No. 2, 345–353.
768. S. Shelah, On Fleissner's diamond, *Notre Dame J. Formal Logic* **22** (1981), No. 1, 29–35.
769. S. Shelah, Iterated forcing and changing cofinalities, *Israel J. Math.* **40** (1981), No. 1, 1–32.
770. J. A. Makowsky and S. Shelah, The theorems of Beth and Craig in abstract model theory, II, Compact logics, *Arch. Math. Logik Grundlag.* **21** (1981), No. 1–2, 13–35.
771. S. Shelah, On endo-rigid, strongly $\aleph_1$-free abelian groups in $\aleph_1$, *Israel J. Math.* **40** (1981), No. 3–4, 291–295 (1982).

772. G. Sageev and S. Shelah, Weak compactness and the structure of Ext(A, $\mathbf{Z}$), in *Abelian Group Theory* (Oberwolfach, 1981), pp. 87–92, Lecture Notes in Math., Vol. 874 (Springer, 1981).
773. S. Shelah, On uncountable Boolean algebras with no uncountable pairwise comparable or incomparable sets of elements, *Notre Dame J. Formal Logic* **22** (1981), No. 4, 301–308.
774. S. Shelah, On saturation for a predicate, *Notre Dame J. Formal Logic* **22** (1981), No. 3, 239–248.
775. S. Shelah, Models with second order properties, III, Omitting types for $L(Q)$, *Arch. Math. Logik Grundlag.* **21** (1981), No. 1–2, 1–11.
776. S. Shelah, Free limits of forcing and more on Aronszajn trees, *Israel J. Math.* **38** (1981), No. 4, 315–334.
777. S. Shelah, $\aleph_\omega$ may have a strong partition relation, *Israel J. Math.* **38** (1981), No. 4, 283–288.
778. S. Shelah, The consistency of $\mathrm{Ext}(G,\mathbf{Z})=\mathbf{Q}$, *Israel J. Math.* **39** (1981), No. 1–2, 74–82.
779. S. Shelah, On the number of nonisomorphic models of cardinality λ $L_{\infty\lambda}$-equivalent to a fixed model, *Notre Dame J. Formal Logic* **22** (1981), No. 1, 5–10.
780. W. Hodges and S. Shelah, Infinite games and reduced products, in *Ann. Math. Logic* **20** (1981), No. 1, 77–108.
781. R. Laver and S. Shelah, The $\aleph_2$-Souslin hypothesis, *Trans. Amer. Math. Soc.* **264** (1981), No. 2, 411–417.
782. A. M. W. Glass, Y. Gurevich, W. C. Holland and S. Shelah, Rigid homogeneous chains, *Math. Proc. Cambridge Philos. Soc.* **89** (1981), No. 1, 7–17.
783. U. Avraham and S. Shelah, Martin's axiom does not imply that every two $\aleph_1$-dense sets of reals are isomorphic, *Israel J. Math.* **38** (1981), No. 1–2, 161–176.
784. S. Shelah, Remarks on Boolean algebras, *Algebra Universalis* **11** (1980), No. 1, 77–89.
785. S. Shelah, Whitehead groups may not be free, even assuming CH, II, *Israel J. Math.* **35** (1980), No. 4, 257–285.
786. S. Shelah, Simple unstable theories, *Ann. Math. Logic* **19** (1980), No. 3, 177–203.
787. G. Cherlin and S. Shelah, Superstable fields and groups, *Ann. Math. Logic* **18** (1980), No. 3, 227–270.
788. S. Shelah, A note on cardinal exponentiation, *J. Symbolic Logic* **45** (1980), No. 1, 56–66.

789. S. Shelah, Independence results, *J. Symbolic Logic* **45** (1980), No. 3, 563–573.
790. S. Shelah, Independence of strong partition relation for small cardinals, and the free-subset problem, *J. Symbolic Logic* **45** (1980), No. 3, 505–509.
791. S. Shelah, On a problem of Kurosh, Jónsson groups, and applications, in *Word Problems, II, Conf. on Decision Problems in Algebra* (Oxford, 1976), pp. 373–394, Stud. Logic Foundations Math., Vol. 95 (North-Holland, 1980).
792. M. Rubin and S. Shelah, On the elementary equivalence of automorphism groups of Boolean algebras, downward Skolem–Löwenheim theorems and compactness of related quantifiers, *J. Symbolic Logic* **45** (1980), No. 2, 265–283.
793. S. Shelah, Boolean algebras with few endomorphisms, *Proc. Amer. Math. Soc.* **74** (1979), No. 1, 135–142.
794. S. Shelah, On uncountable abelian groups, *Israel J. Math.* **32**(1979), No. 4, 311–330.
795. S. Shelah, On successors of singular cardinals, in *Logic Colloquium '78* (Mons, 1978), pp. 357–380, Stud. Logic Foundations Math., Vol. 97 (North-Holland, 1979).
796. D. H. Fremlin and S. Shelah, On partitions of the real line, *Israel J. Math.* **32** (1979), No. 4, 299–304.
797. S. Shelah, Weakly compact cardinals: A combinatorial proof, *J. Symbolic Logic* **44** (1979), No. 4, 559–562.
798. K. J. Devlin and S. Shelah, A note on the normal Moore space conjecture, *Canad. J. Math.* **31** (1979), No. 2, 241–251.
799. J. A. Makowsky and S. Shelah, The theorems of Beth and Craig in abstract model theory, I, The abstract setting, *Trans. Amer. Math. Soc.* **256** (1979), 215–239.
800. Y. Gurevich and S. Shelah, Modest theory of short chains, II, *J. Symbolic Logic* **44** (1979), No. 4, 491–502.
801. K. J. Devlin and S. Shelah, Souslin properties and tree topologies, *Proc. London Math. Soc.* (3) **39** (1979), No. 2, 237–252.
802. S. Shelah, On uniqueness of prime models, *J. Symbolic Logic* **44** (1979), No. 2, 215–220.
803. S. Shelah, Hanf number of omitting type for simple first-order theories, *J. Symbolic Logic* **44** (1979), No. 3, 319–324.
804. S. Shelah and M. Ziegler, Algebraically closed groups of large cardinality, *J. Symbolic Logic* **44** (1979), No. 4, 522–532.

805. S. Shelah, *Classification Theory and the Number of Nonisomorphic Models*, Studies in Logic and the Foundations of Mathematics, Vol. 92 (North-Holland, 1978).
806. S. Shelah and M. E. Rudin, Unordered types of ultrafilters, *Proceedings of the 1978 Topology Conference* (Univ. Oklahoma, Norman, Okla., 1978), I, *Topology Proc.* **3** (1978), No. 1, 199–204 (1979).
807. S. Shelah, A Banach space with few operators, *Israel J. Math.* **30** (1978), No. 1–2, 181–191.
808. S. Shelah, Appendix to: "Models with second-order properties, II, Trees with no undefined branches" [*Ann. Math. Logic* **14** (1978), No. 1, 73–87]. *Ann. Math. Logic* **14** (1978), 223–226.
809. S. Shelah, End extensions and numbers of countable models, *J. Symbolic Logic* **43** (1978), No. 3, 550–562.
810. S. Shelah, Models with second-order properties, II, Trees with no undefined branches, *Ann. Math. Logic* **14** (1978), No. 1, 73–87.
811. S. Shelah, Models with second-order properties, I, Boolean algebras with no definable automorphisms, *Ann. Math. Logic* **14** (1978), No. 1, 57–72.
812. S. Shelah and J. Stern, The Hanf number of the first order theory of Banach spaces, *Trans. Amer. Math. Soc.* **244** (1978), 147–171.
813. S. Shelah, A weak generalization of MA to higher cardinals, *Israel J. Math.* **30** (1978), No. 4, 297–306.
814. U. Avraham; K. J. Devlin and S. Shelah, The consistency with CH of some consequences of Martin's axiom plus $2_{\aleph_0} > \aleph_1$, *Israel J. Math.* **31** (1978), No. 1, 19–33.
815. S. Shelah, Jonsson algebras in successor cardinals, *Israel J. Math.* **30** (1978), No. 1–2, 57–64.
816. H. L. Hiller, M. Huber and S. Shelah, The structure of Ext(A, **Z**) and $V = L$, *Math. Z.* **162** (1978), No. 1, 39–50.
817. S. Shelah, On the number of minimal models, *J. Symbolic Logic* **43** (1978), No. 3, 475–480.
818. K. J. Devlin and S. Shelah, A weak version of $\Diamond$ which follows from $2_{\aleph_0} < 2_{\aleph_1}$, *Israel J. Math.* **29** (1978), No. 2–3, 239–247.
819. S. Shelah, Remarks on λ-collectionwise Hausdorff spaces, *Proceedings of the 1977 Topology Conference* (Louisiana State Univ., Baton Rouge, La., 1977), II, *Topology Proc.* **2** (1977), No. 2, 583–592.
820. S. Shelah, Existentially-closed groups in $\aleph_1$ with special properties, *Bull. Soc. Math. Grèce (N.S.)* **18** (1977), No. 1, 17–27.
821. S. Shelah, Decidability of a portion of the predicate calculus, *Israel J. Math.* **28** (1977), No. 1–2, 32–44.

822. S. Shelah, Remarks on cardinal invariants in topology, *General Topology and Appl.* 7 (1977), No. 3, 251–259.
823. W. Hodges, A. H. Lachlan and S. Shelah, Possible orderings of an indiscernible sequence, *Bull. London Math. Soc.* **9** (1977), No. 2, 212–215.
824. S. Shelah, Whitehead groups may be not free, even assuming CH, I, *Israel J. Math.* **28** (1977), No. 3, 193–204.
825. A. Litman and S. Shelah, Models with few isomorphic expansions, *Israel J. Math.* **28** (1977), No. 4, 331–338.
826. H. L. Hiller and S. Shelah, Singular cohomology in *L*, *Israel J. Math.* **26** (1977), No. 3–4, 313–319.
827. S. Shelah, Interpreting set theory in the endomorphism semi-group of a free algebra or in a category, *Ann. Sci. Univ. Clermont* No. 60 Math. No. **13** (1976), 1–29.
828. S. Shelah, Refuting Ehrenfeucht conjecture on rigid models, *Israel J. Math.* **25** (1976), No. 3–4, 273–286.
829. R. Amit and S. Shelah, The complete finitely axiomatized theories of order are dense, *Israel J. Math.* **23** (1976), No. 3–4, 200–208.
830. J. A. Makowsky; S. Shelah and J. Stavi, *D*-logics and generalized quantifiers, *Ann. Math. Logic* **10** (1976), No. 2, 155–192.
831. A. Macintyre and S. Shelah, Uncountable universal locally finite groups, *J. Algebra* **43** (1976), No. 1, 168–175.
832. S. Shelah, A two-cardinal theorem and a combinatorial theorem, *Proc. Amer. Math. Soc.* **62** (1976), No. 1, 134–136.
833. S. Shelah, Decomposing uncountable squares to countably many chains, *J. Combinatorial Theory Ser. A* **21** (1976), No. 1, 110–114.
834. S. Shelah, The lazy model-theoretician's guide to stability, *Comptes Rendus de la Semaine d'Étude en Théorie des Modèles* (Inst. Math., Univ. Catholique Louvain, Louvain-la-Neuve, 1975). *Logique et Analyse (N.S.)* **18** (1975), No. 71–72, 241–308.
835. S. Shelah, The monadic theory of order, *Ann. of Math.* (2) **102** (1975), No. 3, 379–419.
836. S. Shelah, Colouring without triangles and partition relation, *Israel J. Math.* **20** (1975), 1–12.
837. S. Shelah, Why there are many nonisomorphic models for unsuperstable theories, in *Proceedings of the International Congress of Mathematicians* (Vancouver, B. C., 1974), Vol. 1, pp. 259–263 (Canad. Math. Congress, 1975).
838. S. Shelah, Existence of rigid-like families of abelian *p*-groups, in *Model Theory and Algebra* (*A Memorial Tribute to Abraham Robinson*), pp. 384–402, Lecture Notes in Math., Vol. 498 (Springer, 1975).

839. S. Shelah, Notes on partition calculus, in *Infinite and Finite Sets* (Colloq., Keszthely, 1973; Dedicated to P. Erdös on his 60th Birthday), Vol. III, pp. 1257–1276. Colloq. Math. Soc. Janos Bolyai, Vol. 10 (North-Holland, 1975).
840. S. Shelah, A compactness theorem for singular cardinals, free algebras, Whitehead problem and transversals, *Israel J. Math.* **21** (1975), No. 4, 319–349.
841. S. Shelah, Categoricity in $\aleph_1$ of sentences in $L_{\omega_1,\omega}(Q)$, *Israel J. Math.* **20** (1975), No. 2, 127–148.
842. E. C. Milner and S. Shelah, Some theorems on transversals, in *Infinite and Finite Sets* (Colloq., Keszthely, 1973; Dedicated to P. Erdös on his 60th Birthday), Vol III, pp. 1115–1126. Colloq. Math. Soc. Janos Bolyai, Vol. 10, (North Holland, 1975).
843. S. Shelah, Generalized quantifiers and compact logic, *Trans. Amer. Math. Soc.* **204** (1975), 342–364.
844. S. Shelah, Graphs with prescribed asymmetry and minimal number of edges, in *Infinite and Finite Sets* (Colloq., Keszthely, 1973; Dedicated to P. Erdös on his 60th Birthday), Vol. III, pp. 1241–1256. Colloq. Math. Soc. Janos Bolyai, Vol. 10 (North-Holland, 1975).
845. S. Shelah, A two-cardinal theorem, *Proc. Amer. Math. Soc.* **48**(1975), 207–213.
846. E. C. Milner and S. Shelah, Sufficiency conditions for the existence of transversals, *Canad. J. Math.* **26** (1974), 948–961.
847. S. Shelah, Categoricity of uncountable theories, in *Proceedings of the Tarski Symposium* (Univ. of California, Berkeley, Calif., 1971), pp. 187–203, Proc. Sympos. Pure Math., Vol. XXV (Amer. Math. Soc., 1974).
848. S. Shelah, The Hanf number of omitting complete types, *Pacific J. Math.* **50** (1974), 163–168.
849. R. McKenzie and S. Shelah, The cardinals of simple models for universal theories, in *Proceedings of the Tarski Symposium* (Univ. California, Berkeley, Calif., 1971), pp. 53–74, Proc. Sympos. Pure Math., Vol. XXV (Amer. Math. Soc., 1974).
850. P. Erdös, A. Hajnal and S. Shelah, On some general properties of chromatic numbers, in *Topics in Topology* (Proc. Colloq., Keszthely, 1972), pp. 243–255, Colloq. Math. Soc. Janos Bolyai, Vol. 8 (North-Holland, 1974).
851. S. Shelah, Infinite abelian groups, Whitehead problem and some constructions, *Israel J. Math.* **18** (1974), 243–256.
852. S. Shelah, A substitute for Hall's theorem for families with infinite sets, *J. Combinatorial Theory Ser. A* **16** (1974), 199–208.

853. S. Shelah, First order theory of permutation groups, *Israel J. Math.* **14** (1973), 149–162, errata, *ibid.* **15** (1973), 437–441.
854. S. Shelah, Weak definability in infinitary languages, *J. Symbolic Logic* **38** (1973), 399–404.
855. S. Shelah, Differentially closed fields, *Israel J. Math.* **16** (1973), 314–328.
856. S. Shelah, There are just four second-order quantifiers, *Israel J. Math.* **15** (1973), 282–300.
857. F. Galvin and S. Shelah, Some counterexamples in the partition calculus, *J. Combinatorial Theory Ser. A* **15** (1973), 167–174.
858. S. Shelah, Notes on combinatorial set theory, *Israel J. Math.* **14** (1973), 262–277.
859. G. Moran and S. Shelah, Size direction games over the real line, III, *Israel J. Math.* **14** (1973), 442–449.
860. S. Shelah, On models with power-like orderings, *J. Symbolic Logic* **37** (1972), 247–267.
861. P. Erdös and S. Shelah, On problems of Moser and Hanson, in *Graph Theory and Applications* (Proc. Conf., Western Michigan Univ., Kalamazoo, Mich., 1972; Dedicated to the memory of J. W. T. Youngs), pp. 75–79. Lecture Notes in Math., Vol. 303 (Springer, 1972).
862. P. Erdös and S. Shelah, Separability properties of almost-disjoint families of sets, *Israel J. Math.* **12** (1972), 207–214.
863. J. H. Schmerl and S. Shelah, On power-like models for hyperinaccessible cardinals, *J. Symbolic Logic* **37** (1972), 531–537.
864. S. Shelah, Uniqueness and characterization of prime models over sets for totally transcendental first-order theories, *J. Symbolic Logic* **37** (1972), 107–113.
865. S. Shelah, A combinatorial problem, stability and order for models and theories in infinitary languages, *Pacific J. Math.* **41** (1972), 247–261.
866. S. Shelah, For what filters is every reduced product saturated?, *Israel J. Math.* **12** (1972), 23–31.
867. S. Shelah, A note on model complete models and generic models, *Proc. Amer. Math. Soc.* **34** (1972), 509–514.
868. S. Shelah, Saturation of ultrapowers and Keisler's order, *Ann. Math. Logic* **4** (1972), 75–114.
869. S. Shelah, Stability, the f.c.p., and superstability, model theoretic properties of formulas in first order theory, *Ann. Math. Logic* **3** (1971), No. 3, 271–362.
870. S. Shelah, Two cardinal compactness, *Israel J. Math.* **9** (1971), 193–198.
871. S. Shelah, Every two elementarily equivalent models have isomorphic ultrapowers, *Israel J. Math.* **10** (1971), 224–233.

872. S. Shelah, On the number of non-almost isomorphic models of T in a power, *Pacific J. Math.* **36** (1971), 811–818.
873. S. Shelah, The number of non-isomorphic models of an unstable first-order theory, *Israel J. Math.* **9** (1971), 473–487.
874. S. Shelah, On the cardinality of ultraproduct of finite sets, *J. Symbolic Logic* **35** (1970), 83–84.
875. S. Shelah, Finite diagrams stable in power, *Ann. Math. Logic* **2** (1970/1971), No. 1, 69–118.
876. S. Shelah, Remark to "local definability theory" of Reyes, *Ann. Math. Logic* **2** (1970/1971), No. 4, 441–447.
877. S. Shelah, On theories T categorical in T, *J. Symbolic Logic* **35** (1970), 73–82.
878. S. Shelah, A note on Hanf numbers, *Pacific J. Math.* **34** (1970), 541–545.
879. S. Shelah, On languages with non-homogeneous strings of quantifiers, *Israel J. Math.* **8** (1970), 75–79.
880. S. Shelah, Stable theories, *Israel J. Math.* **7** (1969), 187–202.
881. S. Shelah, Note on a min-max problem of Leo Moser, *J. Combinatorial Theory* **6** (1969), 298–300.

INTERVIEW WITH SAHARON SHELAH

MOSHE KLEIN

In August 2000 we participated in the conference **Mathematical Challenges of the 21th Century** *on the 100th anniversary of the famous conference in Paris, in which Hilbert exposed 23 open problems in Mathematics. Do you believe the history of Mathematics to be a part of Mathematics?*

I'd rather say that the history of Mathematics is part and parcel of History. The history of Mathematics is a very interesting and worthwhile intellectual area, but it does not contribute directly to mathematical research. In several cases, historians wrote on Mathematics and were criticized for not understanding it, even though it was ancient Mathematics. This is not surprising since according to the traditional education of historians, you cannot expect them to know mathematics properly. On the other hand, historians would say that we mathematicians understand Mathematics but not History. There have been some interesting phenomena, such as the simultaneous discovery of differential and integral calculus by Newton and Leibniz, so one could assume that the time was ripe for it. But perhaps we should not exaggerate and suppose that history is predetermined and the only question is which individual will carry out the task. There are counter examples, such as the general method discovered by Archimedes to calculate areas and volumes by considering their center of gravity. Archimedes did not succeed in proving his method, perhaps because of lack of appropriate symbols and tools, but he used it in solving various questions to which he subsequently gave *rigorous* proofs. He described the method in a separate book, however. It fell into oblivion, and both Newton and Leibniz arrived to a similar point. Around 1900, a copy of Archimedes' book was found.

About 500 million years ago, only monocellular organisms lived upon the face of Earth and then, along with the so-called "Cambrium explosion", polycellular creatures commenced to emerge. But when we say 'along with' we mean plus-minus some millions of years, so several

Date: July 2001.

We would like to thank `shlhetal` for their help.

Moshe Klein is the founder of Gan Adam project for mathematical education in kindergarten. He is also a mathematician in BrianPOP company. His email address is gan_adam@netvision.net.il *thousand years of the History of Mathematics is not such a long time in terms of the Earth's age.*

You can regard different questions with different scales. Maybe Mathematics will develop differently in the flow of time. One can propose such plausible possibilities, but as humans are not perfect, and in order to have a good understanding, we should give predictions and verify them. If we have no such feed-backs we may tend to give nonsense opinions. For example, to the question: *Could intelligent beings develop another Mathematics or Physics?*, we probably will get convincing answers only if we meet different cultures. I feel that Mathematics, of all things, would be much easier to understand in a cross–cultural perspective than other cultural elements. One could try, as an exercise, to develop another kind of Mathematics, but the question is where to start? Should one give up common logic?

There have been different cultures on Earth and each one developed Mathematics to some extent. We have documents from roughly 5000 years. As far as we know, the development of Mathematics with precise demonstrations appeared first in the Greek culture.

What made you devote yourself to Mathematics?

While I was still in primary school, I knew I wanted to become a scientist, but as Mathematics was taught, it did not attract me especially. I was very interested, on the other hand, in Physics and Biology and I read popular scientific books. But when I reached the ninth grade I began studying geometry and my eyes opened to that beauty - a system of demonstrations and theorems based on a very small number of axioms which impressed me and captivated me. On the other hand laboratories were not to my taste and vice versa. So by the age of 15 I knew my desire to be a mathematician. Later I read Abraham Halevy Fraenkel's (1891-1965) book 'An Introduction to Mathematics'. I have seen many popular books on Mathematics in English, but I think Fraenkel's book is better, and indeed he help coment my choice of Mathematics.

As a Mathematician, what do you do in your office all day?

Since I have succeeded in demonstrating a substantial number of theorems, I have also a lot of work completing and correcting the demos. As I write, I have a secretary typing (I did have a lot of troubles concerning this) and I have to proof-read a lot. I write and make corrections, send to the typist, get it back and revise it again and again.

A great amount of time is used to verify what I wrote. If it is not accurate or utterly wrong, I ask myself what went wrong. I tell myself: there must be a hole somewhere, so I try to fill it. Or perhaps there is a wrong way of looking at things or a mistake of understanding. Therefore one must correct or change or even throw everything and start all over again, or leave the whole matter. Many times what I wrote first was right, but the following steps were not, therefore one should check everything cautiously. Sometimes, what seems to be a tiny inaccuracy leads to the conclusion that the method is inadequate. I have a primeval picture of my goal. Let us assume that I have heard of a problem and it seems alike to problems that I know how to resolve, provided we change some elements. It often happens that, having thought of a problem without solving it, I get a new idea. But if you only think or even talk but not sit down and write, you do not see all the defaults in your original idea. Writing does not provide a 100% assurance, but it forces you to be precise. I write something and then I get stuck and I ask myself perhaps it might work in another direction? As if you were pulling the blanket to one part and then another part is exposed. You should see that all parts are integrated into some kind of completeness.

Indeed, sometimes you are happy in a moment of discovery. But then you find out, while checking up, that you were wrong. There has been a joy of discovery, but that is not enough, for you should write and check all the details. My office is full of drafts, which turned to be nonsense.

When you make a mathematical discovery, do you think it exists also outside the human mind?

It is difficult for Man to describe the world without him. We can hardly imagine our death. Otherwise probably we would not be able to concentrate. Therefore we can also hardly imagine that there exists something outside of us. But under axioms and logical rules, it seems that Mathematics is absolute. Plato's kind of argument according to which Mathematics is an idea which we discover will always be controversial, but to questions such as: is a given solution to a mathematical problem the right one, the answer is clear–cut. True, we can have only evidence and not an absolute proof of the existence of Mathematics outside of man, because we only act in a human reality. Yet, if one day we meet the little green creatures and speak with them, we might understand a lot more. Most mathematicians, in order to work, must at least pretend they are Platonists. If I felt that what I do is only playing games with symbols without any other meaning, I would probably not do Mathematics.

In the thirties, Kurt Gödel (1906-1974) demonstrated that in every complicated enough mathematical field there is a problem which cannot be solved by following the axioms of that field. What did that discovery do to 20th century Mathematics?

I suppose that most mathematicians who were doing research in some particular field and were aware of Gödel's theorem thought it is interesting but with no direct concern with the problems which really bothered them. Some people were utterly shocked and lost interest in some part of Mathematics, but truth is important even when it is unpleasant. During the conference *Mathematical Challenges of the 21th Century* I presented the question whether we can find a parallel method to demonstrate independence in Number Theory, as Paul Cohen did in Set Theory. Hilbert's first question in Paris was one that Cantor had asked too and we will return to it later. It has been answered, but usually if you give a correct answer to a good question, seven other questions emerge instead.

You mean for example the demonstration that Goldbach's or Riemann's hypotheses are independent from the number theory axioms?

If that can be proved in Mathematics, it will be an astounding achievement. One could add more subtle distinctions of the hypotheses, but we will not give them in detail here.

> **Goldbach's hypothesis:** *Any even number can be described as a sum of two primes. For example, 17+13=30, 13+11=24.*

As yet, no proof has been found, nor has been found a counter-example.

> **Riemann's hypothesis:** *The roots of ζ-function are in the line of complex numbers whose real part is 1/2.*

$$\zeta(x) = \sum_{n=1}^{\infty}(1/n^x) \qquad \{\text{real}(x) > 1\}.$$

In 1976 it has been proved by the computer that any map can be coloured with 4 colours so that no country bears the same colour. Thus, a 100 years conjecture has been demonstrated. Do you think that in another 100 years too, mathematicians will search for axioms, definitions and theorems? Since computers may do it better?

Many mathematicians were quite perplex and did not know how to treat the 4-colour problem. When you think about a mathematical problem, you do not just sit there expecting a voice to tell you if the theorem is true or not. You want to understand the problem in depth. Take for example Fermat's conjecture, whose understanding led to a profound understanding in Algebraic Number Theory.

For the 4 color conjecture, the computer helped us to prove it by dividing to great number of cases. We feel, perhaps, like Alice in "Through the Looking Glass". When the red queen asks Alice if she can sum up, Alice answers that she does. Then, the queen asks her swiftly how much is $1 + 1 + 1 + 1 + 1 + 1 + 1 + 1 + \ldots$, and she is not able to answer even though she can sum up.

There may exist in some future an artificial intelligence which will keep Man in natural reserves, as we do with chimps. We may wonder if there is any point to have people making intellectual activity in that time. There may also exist, however, a situation in which people's brain and computers are not separated from each other. And perhaps then we will be discovered by little green creatures who will legislate to preserve mankind?

We may ask, can we obtain a plausible answer to each question, which we find interesting?

Once upon a time people used to beat with their hands, then they learnt to throw stones, then to shoot guns. But there will always be a limit to the range they can attain. Likewise, by using computers we can check more cases, but there will always be a limit. Gödel's incompleteness theorem ensures us that there will always be new questions that we cannot solve in a given level of axioms and computing power. It may be, however, that our Mathematics is but the opening stage to another Mathematics, and then another kind of problems, of a higher level, will emerge, which is akin to our situation as human beings.

Is it possible that the special combination of Logic and Paradoxes, as it is expressed in Lewis Carroll's book, become in the future the basis to a new form of Mathematics?

Archimedes created a new Mathematics but he had no direct heirs. Maybe the lack of adequate symbols contributed to the delay in evolution. Of course, the situation of culture and society had an influence too. If you look at mathematical formulae and write them in conventional language, the result will be much less clear. I admire Carroll's writing, but it includes no mathematical breakthrough. He did not try to prove anything new but presented in an amusing way some paradoxes, which is not, in itself, a new mathematical content. On the other hand, if books in Mathematics were written like Lewis Carroll used to write, surely more people would do Mathematics.

The Indian mathematician Srinivasa Ramanujan (1887-1920) was undoubtedly an extraordinary phenomenon in Mathematics. We know for example that he discovered 4000 formulae without being able to prove them. How did he manage to do it, in your opinion?

I have read Ramanujan's biography, which was not written by a mathematician, and I do not think there is any logical problem about his achievement. Many people have non-formalized ways of demonstration, and so it is not clear when they are valid. Certainly, there are cases in which the results obtained are wrong, yet in the framework of the questions they asked it did work out well. Ramanujan had probably some sort of fundamental understanding, which allowed him to work. One should also recall that he studied in India, which was similar to studying in Great Britain some 40 years earlier, namely Newton's methods without the precision of epsilon and delta. What mattered was to get a right result, less important was the method. There were also some theorems about the distribution of primes which Hardy showed him to be wrong. Nevertheless, he was one of the most talented individuals in the history of Mathematics. It would be quite interesting to try and reconstruct the way he thought.

Anyway, a physicist also checks his discoveries in the light of reality. For example, if the solution of an equation leads to the conclusion that during the Big Bang, there existed not enough matter, then he will know that there is some mistake, because the solution should correspond to reality to a rather high degree, so Physics has control criteria. It was true in number theory in Fermat's day, and to some extent in Ramanujan's research, but when you get to theorems for which there exists no direct feed-back then clearly without precision you will go nowhere.

Is there a part of your field of research which you can describe?

I am interested in the first problem that Hilbert exposed in 1900, though it has been solved by Gödel in the late forties (one half) and the other half by Paul Cohen in the sixties. If we managed to get Cantor out of his tomb, he would most certainly not be happy with the solutions to his questions, because they are meta-mathematic.

But we would better begin by explaining the discoveries made by Cantor and others. His view was, as a matter of fact, very ancient and simple. We know that primitive peoples do not have large numbers, certainly not beyond 40. So how do they trade? Very simple: they trade one for one. Two sets of sheep are equivalent if there is a one to one correspondence between them. I shall trade twenty sheep for ten gems in ten steps: Each time I trade two sheep for a gem. Likewise, if someone is the depositary of a herd of sheep, in order to check that all of them were given back, they would put for each sheep a stone in a jar, and then seal it. When the sheep got back, they would open the jar and check the correspondence between sheep and stone.

And back to Cantor. He said: let us forget about the finiteness of the set and use the same test of a one-to-one correspondence to see if two sets are equivalent. In this case, for example, the set of positive natural numbers is equivalent – or in terms of set theory it has the same cardinality, or the same number of elements – to the set of even integers, since you can trade any natural number for a number two times greater.

$$\begin{array}{cccccccccccc} 1 & 2 & 3 & 4 & 5 & 6 & 7 & 8 & 9 & 10 & \dots \\ | & | & | & | & | & | & | & | & | & | & \\ 2 & 4 & 6 & 8 & 10 & 12 & 14 & 16 & 18 & 20 & \dots \end{array}$$

Similarly, we define an order. The number of elements in set a A is smaller than or equal to the number of elements in a set B if there is a correspondence from A into B, that is there is one-to-one correspondence between the elements of A and the elements of some subset of B. But if the number of elements of A is smaller or equal to the number of elements of B and vice versa, then those sets are equivalent, however this require a proof. It is also possible to show that the set of positive rationals, namely the numbers which are quotients of two positive integers, is equivalent to the set of natural numbers, since you can arrange the (positive) rationals in one sequence: first you count all the numbers in which the sum of the denominator and the numerator is 1, then those where the sum is 2, etc. If you obtain a number which is listed already – you omit it.

It was very surprising to discover that the number of points on the line is equivalent to the infinite number of points on the plane, the idea is that you can split the decimal representation of a number into two numbers according to the figures in the even and in the odd positions:

$$0.x_1x_2x_3x_4x_5x_6\dots \quad \longleftrightarrow \quad (0.x_1x_3x_5\dots) : (0.x_2x_4x_6\dots).$$

It would have been sensible then to think that all infinite sets are equivalent. But it does not turn out to be the case since Cantor has shown, for example, that you cannot establish a one-to-one correspondence between the natural numbers and the real numbers (index by the natural numbers) . For any infinite sequence of real numbers, there are real numbers that do not appear in the sequence, and therefore there are strictly more real numbers than natural numbers. For example, any number b such as its nth digit is different from the nth digit of a_n, will not be in the sequence.

$$\begin{array}{lcllllll} a_1 & = & 0. & \mathbf{x_1^1} & x_2^1 & x_3^1 & x_4^1 & \dots \\ a_2 & = & 0. & x_1^2 & \mathbf{x_2^2} & x_3^2 & x_4^2 & \dots \\ a_3 & = & 0. & x_1^3 & x_2^3 & \mathbf{x_3^3} & x_4^3 & \dots \\ a_4 & = & 0. & x_1^4 & x_2^4 & x_3^4 & \mathbf{x_4^4} & \dots \\ & & & & \dots & & & \end{array}$$

Since it turned out that there are several levels of infinite, it became necessary to give them names, and Cantor began to use the $\aleph$ symbol. He denotes by $\aleph_0$ the cardinality of the set of natural numbers. It turned out that every infinity has an immediately greater one, a successor. So $\aleph_1$ is the successor of $\aleph_0$, $\aleph_2$ is the successor of $\aleph_1$ etc. The limit of $\aleph_n$'s is called $\aleph_\omega$.

$$\aleph_0, \aleph_1, \aleph_2, \aleph_3, \aleph_4, \dots, \aleph_n, \dots, \aleph_\omega.$$

The first number which has $\aleph_4$ numbers below it is symbolized by $\aleph_{\omega_4}$. Later Zermelo proved that any two sets can be compared.

What happens to the arithmetic operations with infinite numbers ?

You can define naturally a sum of infinite cardinals of two disjoint sets. The number of elements of the union is the sum of elements in both sets. There is also a natural definition to multiplication: From the sets A, B we can define the set C of pairs (a, b), where a is from the set A and b is from the set B. Now, the number of elements of the set C is the number of elements of A multiplied by the number of elements of B. (well, we have to prove that the operation is well defined i.e does not depend on A and B just on the number of their elements)

What happens to division or subtraction ?

There is no definition to that because you can not cancel. For example: any infinite number plus one is equal to itself, that is: $x+1 = x$. If the rule of cancellation was correct, we would obtain $0 = 1$. For finite sets we get the normal operations. It is not enough to define, we have to prove that the operations are well defined.

Does the arithmetic of infinite numbers have rules like the finitary ones?

They look wonderful! Any schoolboy would prefer them to the normal rules. From the beginning of the 20th century we know that the sum of two infinite numbers, or two numbers of which at least one is infinite, is the greater of the two, similarly for the product when both are not zero.

What about the exponent?

One can define a multiplication of possibly infinitely many numbers and especially exponentiation. For instance, 2^x is the number of subsets of a set A with x elements. For example: the set $\{1, 2, 3\}$ has 8 subsets as follows: $\{\}, \{1\}, \{2\}, \{3\}, \{1, 2\}, \{2, 3\}, \{1, 3\}, \{1, 2, 3\}$.

Are the operations concerning exponents as simple, too?

No, because Cantor showed that 2^x is always greater than x. We have then two operations which increase x: the successor number of x and 2^x. Our life would be simple if there were just one such operation, namely if both operations were equal. You can concentrate on the most simple case for the natural numbers, which is Cantor's Continuum Hypothesis and the first problem exposed by Hilbert in Paris. The question was: is any infinite set of real numbers either equivalent to the set of natural numbers or to the set of reals?

A way to express this question is: Does

$$\prod_n 2^n = \aleph_1 \ ?$$

At the end of the thirties Gödel showed that you cannot contradict the Continuum Hypothesis. In the sixties, Paul Cohen showed that you cannot prove the Continuum Hypothesis. It means that the Continuum Hypothesis does not depend on the axioms of set theory which are called ZFC, after Zermelo and Fraenkel. So you can assume it is right and you can assume it is wrong.

It reminds us of the axiom of parallels, which has been proved not to depend on the axioms of geometry. Don't you have a strange feeling when you deal with such questions, do you touch infinity?

I am very interested in it, and it is nice that one can say something on infinity after all. A lot has been done by Solovay, Easton, Galvin, Hajnal, Magidor, Woodin, Gitik and mitchal in the world in different directions. I'll tell one more thing. I tried to show, in a book published in the last decade, that you can prove many things on infinity on the basis of the ZFC set theory if you just ask the right questions.

To simplify, let us look at the simplest case left, the product

$$\aleph_0 \times \aleph_1 \times \aleph_2 \times \ldots \times \aleph_n \times \ldots = (\aleph_\omega)^{\aleph_0}.$$

This number is equal or greater to the number of reals, $2^{\aleph_0}$. Let us suppose that the number of reals is equivalent to $\aleph_n$ for some natural n. How big can the product be? Of course, there are infinite number of infinities smaller than the product, but not too much, the number of smaller numbers is smaller than $\aleph_{\omega_4}$:

$$\prod_n \aleph_n \leq \aleph_{\omega_4} + 2^{\aleph_0}.$$

It sounds rather amazing, why 4 in a formula on infinities? Is it linked to the problem of colouring a map with 4 colours or perhaps to the 4 of the polynomial degree, which can be solved as Evariste Galois (1811-1832) discovered?

Many mathematicians have been surprised to see a link between 4 and infinite. Perhaps it is due only to my limitations, and in the future we will see that there is a lower limit, like $\aleph_{\omega_1}$. Anyway, it is one of the questions I have asked during my lecture "Logical dreams" in the conference *Mathematical Challenges of the 21*th *Century* in U.C.L.A.

Proceedings of the International Congress of Mathematicians
Berkeley, California, USA, 1986

Taxonomy of Universal and Other Classes

SAHARON SHELAH

1. The problem. I was attracted to mathematics by its generality, its ability to give information where apparently total chaos prevails, rather than by its ability to give much concrete and exact information where we a priori know a great deal. So, not surprisingly, the following represents a theme which has been central in my mathematical interests since starting my thesis. (We give "universal classes" as an example, as the definition is "logic-free" (see Definition).)

1.1. *The first problem: The taxonomy = classification problem for universal classes.* Find the main dividing lines in the family of universal classes; each line is significant in the sense that all classes in one side "enjoy" common properties witnessing "simplicity," "analyzability," and those of the other side have common properties witnessing complications, unanalyzability; we define

1.2. *Universal classes.* (1) Examples are the class of groups, the class of rings, any variety, and the class of locally finite groups.

(2) Generally, let τ denote a vocabulary = set of function symbols and predicates (= relation symbols) each with an assigned arity, $(n(F), n(R))$; a τ-structure M is a nonempty set $|M|$ (its universe), and interpretations of any function symbol $F \in \tau$ and relation symbol $R \in \tau$ is an $n(F)$-place function from $|M|$ to $|M|$, and $n(R)$-place relation on $|M|$, respectively.

(3) A universal class K is a class of τ-structures for some $\tau = \tau(K)$ such that $M \in K$ *iff* every finitely generated substructure of M belongs to K.

I also love, in mathematics, that there is no argument (at least usually) about whether or not one solves a problem. It suffices to find the correct solution; being untalented in convincing people is no serious hindrance. Hence I like problems that are precise, preferably with a yes/no answer, and I believe that usually the way to treat more elusive problems is by choosing the right test question. So we shall specify our problem below.

1.3. *The second problem: The structure/nonstructure problem.* (1) Describe for some (e.g., universal classes) K a structure theory (see below) and prove for the other classes (in the family) nonstructure theorems; that is, demonstrate

This research was partially supported by the United States–Israel Binational Science Foundation and the National Science Foundation.

the impossibility of a structure theory, construction of many and/or complicated structures in K.

(2) A *structure theory* for K is a theory that gives, for each $M \in K$, a complete set of invariants; i.e., each invariant should depend only on the isomorphism type of M, and if M_1, M_2 have the same invariants, then they are isomorphic.

Of course, we do not want the invariants to be too complicated (e.g., the isomorphism type of M) (and, preferably, derivation of the invariant from the structure and vice versa are explicit constructions). We shall not deal with this, but it would not change the theory much. See **[Sh86]** for more.

What objects should we use as invariants for structures of cardinality λ? In the prototype of structure theorems, the celebrated Steinitz theorem, for algebraically closed fields of a fixed characteristic, a cardinality (= transcendence dimension) is used. Certainly if K_i has a structure theory for $i \in I$, then so does $\sum_{i\in I} K_i \stackrel{\mathrm{df}}{=} \{\sum_{i\in I} M_i : M_i \in K_i \text{ for } i \in I\}$ (with any reasonable definition of $\sum_i M_i$, such that $\sum_{i\in I} K_i$ is a class of the right kind). Also, if K has a structure theory so does $\sum K \stackrel{\mathrm{df}}{=} \{\sum_{j\in J} M_j$: for some set J, with $M_j \in K\}$. So we have to admit λ-values of kind (α, χ) as invariant for structures of cardinality λ, where

1.4. *Definition.* (1) A λ-value of kind (α, χ) (λ, χ cardinals, α an ordinal) is defined by induction on α:

$\alpha = 0$. A λ-value of kind (α, χ) is a cardinal $\leq \lambda$.

$\alpha = \beta + 1$. A λ-value of kind (α, χ) is a λ-value of kind (β, χ) *or* sequence of length χ, each entry a function from $\{x : x \text{ a } \lambda\text{-value of kind } (\beta, \chi)\}$ into the set of cardinals $\leq \lambda$.

α limit. A λ-value of kind (α, χ) is a λ-value of kind (β, χ) for some $\beta < \alpha$.

(2) We say an invariant of kind $(\beta, 1)$ is of depth β.

(3) An invariant of kind (α, χ), for K, is a function which gives, for each $M \in K$ of cardinality λ, a λ-value of kind (χ, α), and which depends on M only up to isomorphism.

Now if we do not bound α, even for $\chi = 1$, any structure of cardinality λ can be coded up to isomorphism; also note that for any χ, α there is a β such that every λ-value of kind (χ, α) can be coded naturally by a λ-value of depth β (i.e., kind $(\beta, 1)$). So we are led to

1.5. *First thesis*: A class K has a structure theory iff for some β there are invariants of depth β for K which determine each $M \in K$ up to isomorphism.

I think that the part of the thesis that says that such invariants give a structure theory is very strong. First of all

1.6. *Claim.* If a class K has a structure theory according to Thesis 1.5, by an invariant of kind (α, χ), then for every cardinal $\aleph_\gamma$, $I(\aleph_\gamma, K) \leq \beth_\alpha(|\gamma| + \chi)$ (see below), so if the Generalized Continuum Hypothesis (= G.C.H.) holds, then $\beth(\aleph_\gamma, K) \leq (\chi + |\gamma|)^{+\beta}$ where

1.7. *Definition.* (1) $|\gamma|$ is the cardinality of the ordinal γ.

(2) $I(\aleph_\gamma, K)$ is the number of structures from K of cardinality $\aleph_\gamma$, up to isomorphism.

(3) The G.C.H. says that $2^{\aleph_\gamma} = \aleph_{\gamma+1}$ for every γ.

(4) $(\aleph_\alpha)^{+\beta} = \aleph_{\alpha+\beta}$, $\beth_\alpha(\lambda)$ is defined by induction on α as $\aleph_0 + \sum_{\beta<\alpha} 2^{\beth_\beta(\lambda)}$. (Note that $\beth_\alpha(\chi + |\gamma|)$ may be $> 2^{\aleph_\gamma}$ even for $\alpha = 2$, $\chi = 1$.)

Because knowing the number of members of K in one cardinality is a natural and important problem, $I(\lambda, K)$ is an important function; if its values are small this signifies that the class K is simple. For me, the really important thing is Thesis 1.8 below (and not the detailed computation as $I(\lambda, K)$).

1.8. *Second thesis: The (first) main dividing line = the main gap.* For a "nice" family of classes (like the family of universal classes), the dividing line "is there α such that for every $\gamma, I(\aleph_\gamma, K) \leq \beth_\alpha(|\gamma|)$" is a good one, i.e.,

(a) It coincides with "having a structure theory according to 1.5";

(b) Every class in the "complicated side" has strong evidence for nonstructure: a jump in the lower bound for $I(\lambda, K)$—i.e., among the possible functions $I(\lambda, K)$ there is a large gap:

(1) $I(\lambda, K) = 2^\lambda$ for large enough λ which is the maximum value when $\lambda \geq |\tau(K)|$, or at least

(2) $I(\lambda, K) > \lambda$ for large enough class of cardinals λ. We much prefer that the nonstructure proof be carried out in ZFC alone, but note that in order to show that we cannot prove a structure theorem, the consistency of nonstructure is enough.

(c) We believe that if we succeed in solving (b), we will have developed extended taxonomy having many tools to deal with, and we will know much on each side of those dividing lines; i.e., we suggest 1.8(b) as the test question.

Note that we do not claim that only dividing lines are interesting: the class of rings is a very interesting subclass of the class of structures with two-place functions, although its complement is not interesting at all. On the other hand, dividing lines, in addition to having intrinsic interest, help in proving theorems by cases.

Implicit is

1.9. *Thesis.* (d) In understanding a class you should look at a large enough cardinality in order to iron out singularities.

Note that, e.g., theories having unique countable model can have many complicated uncountable ones. Note that a priori the answer to the following question is not clear.

1.10. *Question.* Is there a reasonably general family of classes for which Thesis 1.8 can be confirmed?

We shall return to this question later.

2. Background. Why do we speak on universal classes? Now in model theory the primary family of classes is the family $\{\mathrm{Mod}(T): T$ a countable first-order theory$\}$, where

2.1. *Definition.* $\mathrm{Mod}(T)$ is the class of models of T (and we write T instead of $\mathrm{Mod}(T)$).

Of course, more complicated classes—uncountable theories, theories in infinitary logics or ones with generalized quantifiers on the one hand, and universal theories and even varieties (= equational theories)—are also interesting. I think this approach is right, but here there is no need to justify it.

Let us return to what is for me prehistory (you can ignore notions unknown to you). Generally, around 1960, research in mathematical logic became deeper and more complicated mathematically. At that time, the aim of much of the research in model theory was advancement toward the solution of the Los Conjecture, which was as follows.

LOS CONJECTURE. *If T is (first-order) countable, then if T is categorical in λ for some $\lambda > \aleph_0$ (i.e., $I(\lambda, T) = 1$), then this holds for every $\lambda > \aleph_0$.*

Certainly, in wanting to know something about $I(\lambda, T)$, this was a very good problem to start with: it was foolish to consider far-reaching conjectures like 1.8(b), considering the knowledge available. Those investigations culminated in the positive solution of the Los Conjecture by Morley in [**Mo**], which used many of the tools developed previously—in particular, the works of Vaught and Ehrenfeucht-Moslowski. Morley's theorem is considered by many (including myself) to be one of the main achievements of mathematical logic during the sixties.

For quite some time, little happened. This certainly has its reasons—among them, that Morley and Keisler, at least, thought that the (model) theory of first-order theories was finished, or essentially finished (they told me so in 1969). However, since then the field has increased in popularity in model theory, as witnessed by the research book [**Sh78**], and numerous articles, as well as the (largely) expository books Pillay [**Pi**], Lascar [**La**], Poizat [**Po**], and Baldwin [**Ba**], and lately some conferences dedicated to it. On early history see [**Sh74**]; this article overlaps with [**Sh85**] (which speaks on countable first-order T) and Baldwin, introduction to [**Ba1**] (which deals generally with the theory).

The papers of most researchers in the field, however, reflect a very different outlook—a "fine structure" one—wanting to know much (or everything) about what we already have considerable knowledge about or investigating families of classes which have some structure theory and/or tendency to be (relatively) more concrete. Illustrious examples are the Baldwin-Lachlan theorem (on first-order countable T, categorical in $\aleph_1$ but not $\aleph_0$) and the works of Zilber and Cherlin, Harrington and Lachlan (on T categorical in $\aleph_0$, $\aleph_1$, or totally transcendental T categorical on $\aleph_0$). I hope Lachlan and Peretyatkin, in this volume, will do justice to some of this.

Note that there is no real conflict: solutions of Problem 1.1 give, and are intended to give, instances for fine structure investigation with considerable tools to start with. Here I shall continue to present my personal outlook. When I started in 1967, I was interested in Problem 1.1, introducing the stability and superstability as dividing lines, and in [**Sh71**, p. 283, (13)] (also in [**Sh74**]) I conjectured what the functions $I(\lambda, T)$ for λ large enough, T countable first-order, should be like; in particular, the "main gap" of 1.7(b) was of interest.

The first class for which this was confirmed (with a minor correction) was for the family of $\{M\colon M$ an $\aleph_\varepsilon$-saturated model of $T\}$: T first-order (announced in [**Sh74**], see [**Sh83**; **Sh83a**; **Sh77**, Chapter X, Example 3.3]). but I felt this was cheating, as I invented $\aleph_\varepsilon$-saturativity. The second one was for universal (first-order) theories T (see [**Sh86**]); i.e., the set of models of T, T consisting of formulas of the form $\forall x_1, \dots, x_n \ \bigvee_i \bigwedge_j \phi_{i,j}$, $\phi_{i,j}$ atomic or negation of atomic. We will give more details on this result.

2.2. THEOREM. *For every universal (first-order) T, exactly one of the following occurs:*

(A) *For every $\lambda > |T|$, $I(\lambda, T) = 2^\lambda$, and there are other signs of complicatedness (see* [**Sh85**]*).*

(B) *Still $I(\lambda, T) = 2^\lambda$ for every $\lambda \geq |T| + \aleph_1$, but*

$(*)$ *for every model M of T of cardinality λ, there is $\langle M_\eta\colon \eta \in I\rangle$ such that:*

(i) *I is a set of finite sequences of ordinals $< \lambda$, nonempty, closed under initial segments.*

(ii) *M_η is a submodel of M of cardinality $\leq |T|$, and if ν is an inital segment of η then M_ν is a submodel of M_λ.*

(iii) *M is freely generated by $\bigcup_{\eta \in I} M_\eta$, which means:*

(α) *the closure of $\bigcup_{\eta\in I} |M_\eta|$ (union of universes) under the functions of M is $|M|$;*

(β) *if $\eta \in I$ has length $n+1$, $\nu = \eta \restriction n$, then for every finite sequence $\bar c$ from M_η : $\oplus$ for every finite set Φ of quantifier-free formulas with parameters from $\bigcup\{|M_\rho|\colon \rho \in I,\ \eta$ not an initial segment of $\rho\}$ which $\bar c$ satisfies, there is a finite sequence $\bar c'$ from M_ν satisfying all formulas in Φ.*

(C) *The condition $(*)$ from* (B) *holds, moreover, for some ordinal* $\mathrm{Dp}(T)$, *which is countable if T is countable, and of cardinality $\leq 2^{|T|}$ generally; for every M there is $\langle M_\eta\colon \eta \in I\rangle$ as in $(*)$ with I of depth* $\langle \mathrm{Dp}(T)$: *i.e., there is a function $d\colon I \to \alpha$ such that if ν is a proper initial segment of η then $d(\nu) > d(\eta)$.*

So for α large enough, $I(\aleph_\alpha, T) \leq I_{\mathrm{Dp}(T)}(|\alpha|)$ (in fact, we get very detailed information on $I(\lambda, T)$).

Note that the first dividing line is between (A)+(B) and (C). But a second dividing line between (A) and (B)+(C) is worthwhile; see [**Sh85**] on this, but we shall not deal with it here.

This looks to be a reasonable answer to Question 1.10—a general family where we have a solution; but, having a model theoretic background, I was not satisfied until the solution for countable first-order T (see [**Sh85**, **Sh87**]). Though I thought 1.8 was the point, I felt it was my duty not to avoid the relevant problem which was the legacy of the previous generation—the Morley Conjecture. Having thus answered Question 1.10 fully, we have

2.3. *Question.* Is the theory a theory on first-order classes, or is there really a collection of such theories which may even need a general framework?

There were some advances, some changing of the family of classes, some changing of the questions. In Baldwin-Shelah [**BaSh8**] (see also Shelah [**Sh86a**, **Sh85a**]), the complexity of class $K = \text{Mod}(T)$, T first-order, in the monadic logic (finitary or infinitary) was investigated (relying on [**Sh78**]).

Quite complete classification (in the relevant sense) was obtained. The results explain why the quite large body of works on monadic logic (e.g., works of Buchi, Rabin, Shelah, and Gurevich (see [**Gu**])) concentrate on linear orders and trees, and discover some neglected cases; also, more general quantifiers were dealt with. This also indicates that the classification in [**Sh78**] is relevant to a reasonably wide spectrum of problems not thought of in the first place. On classifying all quantifiers see [**Sh86x**].

We may work with classes K of two sorted structures and ask how much a $M \in K$ can be described up to isomorphism over the first sort. The prototype of such problems is the structure of a vector space V over a field F, letting both vary; so one cardinal invariant, the dimension, suffices.

So nonstructure will mean that for each (or at least arbitrarily large) cardinal λ, there is a structure M_0 so that for many and complicated $N \in K$, N restricted to the first sort is M_0. It seems that for $K = \text{Mod}(T)$, T countable first-order, there is a complete answer *provided that* we accept indepenent results for the nonstructure side. We say "it seems," as some parts are not worked out, others need considerable expansion; see [**PiSh8**], [**Sh86**], and the notes [**Sh85b**]. However, that is enough to show that there is such a theory.

The situation is similar for universal classes [**Sh87a**].

Grossberg and Hart [**GH**] have been proving the main gap, etc., for the family of excellent classes. Excellent classes were introduced in [**Sh83c**] (in the following context: if $\psi \in L_{\omega_1,\omega}$ has an uncountable model and, for no $n > 0$, has many nonisomorphic models (essentially $2^{\aleph_n}$), then $\text{Mod}(\psi)$ is the union of few excellent classes).

3. Outside interactions. Of course, the theory answered almost all relevant problems of the model theorist of the sixties and some which do not a priori look connected (like investigating Keisler order on first-order theories).

Note that the conclusion applies also to universal algebra. In particular, Theorem 2.2 gives the possible function $I(\lambda, K)$ for K a variety, except that we do not know if all values of the parameters really appear (mainly whether the depth of the theory can be infinite; also there are some problems for small cardinals). It seemed that though they have interest in the problem, universal algebraists have not learned the material we mention.

We have been interested in this theory for its own sake; and applications were sought in order to convince the "heathen." However, we sincerely believe that it should help in investigating specific classes.

Note, however, that there is an asymmetry between the structure and nonstructure side. You can deal with a structure theory for a specific class without having any formal definition of what a structure theory is. But we need one (or some alternatives) for showing that no structure theory exists.

Note that having a theory as we desire makes it natural and profitable to investigate for specific classes where they are on this classification. This line of research starts with Macintyre [**Mc**], dealing with first-order theories of fields (via [**Mo**]). For field and division rings this was continued in Cherlin [**Ch**] and Cherlin and Shelah [**ChSh**], thus giving the following

Conclusion. If T is a first-order theory, Mod(T) a class of infinite fields (or division rings), not all algebraically closed fields, then T is not superstable; hence Mod(T) has many and complicated models and modules.

There is extensive literature on first-order theories of rings, modules and groups. A very successful case is the theory T^p_{dcf} of differentially closed fields. Robinson, relying on Seidenberg [**Se**], proved that T^0_{dcf} is first-order. Blum [**B**] gave concrete axioms for it, and proved it totally transcendental. She deduced (relying on Morley's work) that over every differential field F of characteristic 0, there is a prime differentially closed field over F (extending F) (i.e., one embeddable in every differentially closed field extending F). We deduce, by a theorem in classification theory (see [**Sh78**, Chapter IV, §4]), the uniqueness of the prime differentially closed field over F. Wood [**W**], relying again on algebraic work, proved T^p_{dcf} is first-order, but not totally transcendental. Independently, Shelah [**Sh**] and Wood [**W1**] proved the existence of prime differentially closed fields over any differential field. Shelah [**Sh73**] proved that the theory is stable; so by [**Sh78**, Chapter IV, §5], the prime differentially closed field above is unique. But by [**Sh73**], T^p_{dcf} ($p > 0$) is not superstable, hence there is no structure theory for Mod(T^p_{dcf}).

The existing theory on Mod(ψ), ψ a sentence in infinitary logic ($L_{\omega_1,\omega}$), is used in Mekler and Shelah [**MSh**] to prove (when $V = L$) that for every variety either $L_{\infty,\omega}$-freeness implies freeness *or* there are λ-free not free ones for every λ (continuing work of Eklof and Mekler [**EM**]).

In Grossberg and Shelah [**GSh**] a problem of Fuchs and Salce (see [**FS**]) on the possibility of a structure theory of torsion divisible modules over a uniserial ring is answered (using a general theorem proved there). This theorem can also be used to deduce directly an older result from [**Sh74a**] (solving a problem of Fuchs [**F**]) that there are many complicated separable reduced abelian p-groups in every $\lambda > \aleph_0$.

Quite naturally, in many cases the theorems have not applied directly; rather, the proofs or the method apply. We have tried to adapt the theorems to general use in [**Sh83b**], the applications there being constructions of Boolean algebras which are complicated in various ways (e.g., have no automorphisms or one-to-one endomorphisms, are complete and/or satisfy the CCC).

Another attempt to adapt the theorems for applications is [**Sh85c**], which has been of use in several instances in representing rings as endomorphism groups of abelian groups, in works of Corner, Gobel, and the author.

We generally believe that the method should be useful in constructing structures in specific classes which are "complicated," e.g., have no "nontrivial" automorphism or endomorphism or are indecomposable, etc.

REFERENCES

[**B**] L. Blum, *Generalized algebraic structures, a model theoretic approach*, P.D. Thesis, M.I.T., Cambridge, Mass., 1968.

[**Ba**] J. Baldwin, Springer Verlag.

[**Ba1**] J. Baldwin (ed.), Introduction, USA-Israel, Proc. Conference on Classification Theory (Chicago, December 1985), Lecture Notes in Math., Springer-Verlag, Berlin and New York (to appear).

[**Ba2**] J. Baldwin, *Definable second order quantifiers*, Model-Theoretic Logics (J. Barwise and S. Feferman, eds), Springer-Verlag, Berlin and New York, 1985, pp. 445–478.

[**BaSh**] J. Baldwin and S. Shelah, *Classification of theories by second order quantifiers*, Proc. 1980/Jerusalem Model Theory Year, Notre Dame J. Formal Logic **26** (1985), 229–303.

[**Ch**] G. Cherlin, *Superstable division rings*, Logic Colloquium 77 (Proc. Conf., Wrocław, 1977), North-Holland, Amsterdam, 1978, pp. 99–112.

[**ChSh**] G. Cherlin and S. Shelah, *Superstable fields and groups*, Ann. of Math. Logic **18** (1980), 227–280.

[**EM**] P. Eklof and A. Mekler, *Categoricity results for $L_{\infty\kappa}$-free algebras*, Ann. Pure Appl. Logic (to appear).

[**F**] L. Fuchs, *Infinite Abelian groups*, Academic Press, 1970, 1973.

[**FS**] L. Fuchs and L. Salce, *Modules over valuation domains*, Lecture Notes in Pure Appl. Math., vol. 97, Marcel Dekker, New York, 1985.

[**Gu**] Y. Gurevich, *Monadic second order theories*, Model-Theoretic Logics (J. Barwise and S. Feferman, eds.), Springer-Verlag, Berlin and New York, 1985, pp. 479–506.

[**GSh**] R. Grossberg and S. Shelah, *A nonstructure theorem for an infinitary theory which has the unsuperstability property*, Illinois J. Math. **30** (1986), 364–390.

[**K**] E. R. Kolchin, *Differential algebra and algebraic groups*, Academic Press, New York, 1973.

[**La**] D. Lascar, *Introduction to stability*.

[**Mc**] A. Macintyre, *On ω_1-categorical theories of fields*, Fund. Math. **71** (1971), 1–25.

[**Mo**] M. D. Morley, *Categoricity in power*, Trans. Amer. Math. Soc. **114** (1965), 514–538.

[**MSh**] A. Mekler and S. Shelah, *For which varieties $L_{\infty,\omega}$-freeness implies freeness and excellent classes*, in preparation.

[**Pl**] A. Pillay, *An introduction to stability theory*, Clarendon Press, Oxford, 1983.

[**PlSh**] A. Pillay and S. Shelah, *Classification over a predicate*. I, Notre Dame J. Formal Logic **26** (1985), 361–376.

[**Po**] B. Poizat, *Cours de théorie des modèles*, nur al-mantiq wal-ma'rifah, 1985.

[**Ro**] A. Robinson, *On the concept of a differentially closed field*, Bulletin Research Council of Israel, Section F (later: Israel J. Math) **8F** (1959), 113–128.

[**Se**] A. Seidenberg, *An elimination theory for differential fields*, Univ. Calif. Publ. Math. (N.S.) **3** (1956), 31–65.

[**Sh71**] S. Shelah, *Stability, the f.c.p. and superstability, model theoretic properties of formulas in first order theory*, Ann. of Math. Logic **3** (1971), 271–362.

[**Sh73**] ——, *Differentially closed fields*, Israel J. Math. **16** (1973), 314–328.

[**Sh74**] ——, *Categoricity of uncountable theories*, Proc. Sympos. Pure Math., vol. 25, Amer. Math. Soc., Providence, R.I., 1974, pp. 187–204.

[**Sh74a**] ——, *Infinite abelian groups, Whitehead problem and some constructions*, Israel J. Math. **18** (1974), 243–256.

[**Sh78**] ——, *Classification theory and the number of nonisomorphic models*, North-Holland, Amsterdam, 1978.

[**Sh83**] ——, *The spectrum problem*. I, *$\aleph_\varepsilon$-saturated models the main gap*, Israel J. Math. **43** (1982), 324–356.

[**Sh83a**] ——, *The spectrum problem. II, Totally transcendental theories and the infinite depth case*, Israel J. Math. **43** (1982), 357–364.

[**Sh83b**] ——, *Construction of many complicated uncountable structures and Boolean algebras*, Israel J. Math. **45** (1983), 100–146.

[**Sh83c**] ——, *Classification theory for non-elementary classes. I, The number of uncountable models, models of* $\psi \in L_{\omega_1,\omega}$, Israel J. Math. **46** (1983), 2-12-273.

[**Sh85**] ——, *Classification of first order theories which have a structure theory*, Bull. Amer. Math. Soc. (N.S.) **12** (1985), 227–232.

[**Sh85a**] ——, *Monadic Logic: Lowenheim numbers*, Ann. Pure Appl. Logic **28** (1985), 203–216.

[**Sh85b**] ——, *Classification over a predicate*, Notes from Lectures in Simon Fraser University, Summer 1985.

[**Sh85c**] ——, *A combinatorial principle and endomorphism rings of abelian groups*. II, Proc. of the Conference on Abelian Groups Indine 4/1984, CISM courses and Lecture ∞, No. 287, International Center for Mechanical Sciences, *Abelian Groups and Modules*, R. Gobel, C. Metelli, A. Orsatti, and L. Salce, eds., 1985, pp. 37–86.

[**Sh86**] ——, *Spectrum problem. III, Universal theories*, Israel J. Math. **55** (1986), 229–252.

[**Sh86a**] ——, *Monadic logic: Hanf numbers*, Around Classification Theory, Lecture Notes in Math., vol. 1182, Springer-Verlag, Berlin and New York, 1986, pp. 203–223.

[**Sh86b**] ——, *Classification over a predicate*. II, Around Classification Theory, Lecture Notes in Math., vol. 1182, Springer-Verlag, Berlin and New York, 1986, pp. 47–90.

[**Sh86c**] ——, *Classifying generalized quantifiers*, Around Classification Theory, Lecture Notes in Math., vol. 1182, Springer-Verlag, Berlin and New York, 1986, pp. 1–46.

[**Sh87**] ——, *Classification theory; completed for countable theories*, North-Holland, Amsterdam (to appear).

[**Sh87a**] ——, *Universal classes*, Proc. of the USA-Israel Sympos. on Classification Theory, Chicago 12/85, Springer-Verlag.

[**W**] C. Wood, *The model theory of differential fields of characteristic* $p \neq 0$, Proc. Amer. Math. Soc. **40** (1973), 577–584.

[**W1**] ——, *Prime model extensions for differential fields of characteristic* $p \neq 0$, J. Symbolic Logic **39** (1974), 469–477.

THE HEBREW UNIVERSITY, JERUSALEM, ISRAEL

RUTGERS UNIVERSITY, NEW BRUNSWICK, NEW JERSEY 08903, USA

BULLETIN (New Series) OF THE
AMERICAN MATHEMATICAL SOCIETY
Volume 26, Number 2, April 1992

CARDINAL ARITHMETIC FOR SKEPTICS

SAHARON SHELAH

When modern set theory is applied to conventional mathematical problems, it has a disconcerting tendency to produce independence results rather than theorems in the usual sense. The resulting preoccupation with "consistency" rather than "truth" may be felt to give the subject an air of unreality. Even elementary questions about the basic arithmetical operations of exponentiation in the context of infinite cardinalities, like the value of $2^{\aleph_0}$, cannot be settled on the basis of the usual axioms of set theory (ZFC).

Although much can be said in favor of such independence results, rather than undertaking to challenge such prejudices, we have a more modest goal; we wish to point out an area of contemporary set theory in which theorems are abundant, although the conventional wisdom views the subject as dominated by independence results, namely, cardinal arithmetic.

To see the subject in this light it will be necessary to carry out a substantial shift in our point of view. To make a very rough analogy with another generalization of ordinary arithmetic, the natural response to the loss of unique factorization caused by moving from $\mathbb{Z}$ to other rings of algebraic integers is to compensate by changing the definitions, rescuing the theorems. Similarly, after shifting the emphasis in cardinal arithmetic from the usual notion of exponentiation to a somewhat more subtle variant, a substantial body of results is uncovered that leads to new theorems in cardinal arithmetic and has applications in other areas as well. The first shift is from cardinal exponentiation to the more general notion of an infinite product of infinite cardinals; the second shift is from cardinality to cofinality; and the final shift is from true cofinality to potential cofinality (pcf). The first shift is quite minor and will be explained in §1. The main shift in viewpoint will be presented in §4 after a review of basics in §1, a brief look at history in §2, and some personal history in §3. The main results on pcf are presented in §5. Applications to cardinal arithmetic are described in §6. The limitations on independence proofs are discussed in §7, and in §8 we discuss the status of two axioms that arise in the new setting. Applications to other areas are found in §9.

The following result is a typical application of the theory.

Theorem A. *If* $2^{\aleph_n} < \aleph_\omega$ *for all* n *then* $2^{\aleph_\omega} < \aleph_{\omega_4}$.

The subscript 4 occurring here is admittedly very strange. Our thesis is that the theorem cannot really be understood in the framework of conventional cardinal arithmetic, but that it makes excellent sense as a theorem on pcf. Another way of putting the matter is that the theory of cardinal arithmetic involves two quite different aspects, one of which is totally independent of the usual axioms

Received by the editors October 1, 1990 and, in revised form, May 27, 1991.
1991 *Mathematics Subject Classification.* Primary 03E10; Secondary 03-03, 03E35.
Research partially supported by the BSF; Pub. no. 400A.

of set theory, while the other is quite amenable to investigation on the basis of ZFC. Since the usual approach to cardinal arithmetic mixes these two aspects, the independence results mask the theorems and the occasional result that survives this interference then looks quite surprising.

Of course, the most severe skeptics will even deny the mathematical content of Cantor's theorem ($2^{\aleph_0} > \aleph_0$). To these we have nothing to say at all, beyond a reasonable request that they refrain from using the countable additivity of Lebesgue measure.

Most of the results reported here were obtained in the past three years and are expected to appear in a projected volume to be published by Oxford University Press.

1. Basics

The reader is assumed to be comfortable with the axiom system ZFC or an equivalent formulation of axiomatic set theory including the Axiom of Choice, though knowing naive set theory is enough for reading almost everything . In particular we have the notions of ordinal, cardinal, and cardinality $|A|$ of a set A, the identification of ordinals with sets of ordinals, of cardinals with "initial" ordinals, and hence also with sets of ordinals. The cofinality $\operatorname{cof}\alpha$ of an ordinal α is $\min\{|A| : A$ is an unbounded subset of $\alpha\}$; we call an infinite cardinal λ regular if $\operatorname{cof}\lambda = \lambda$, otherwise we call it singular. (Note $\operatorname{cof}\aleph_{\alpha+1} = \aleph_{\alpha+1}$, $\operatorname{cof}\aleph_0 = \aleph_0$, $\operatorname{cf}\aleph_\omega = \aleph_0$, and for limit ordinal δ, $\operatorname{cof}\aleph_\delta = \operatorname{cf}\delta$). We write $\alpha, \beta, \gamma, \delta$ for ordinals, with δ typically a limit ordinal, and κ, λ, μ for cardinals. $\aleph_\alpha$ is the αth infinite cardinal . The first cardinal above λ is denoted λ^+ $(\aleph_\alpha^+ = \aleph_{\alpha+1})$.

The product of a set of cardinals is the cardinality of their cartesian product (with each cardinal thought of as a set of ordinals). Exponentiation is treated as a special case of the infinite product. We recall that for any ordinal α, $\operatorname{cof}\alpha$ is a regular cardinal, and that $\aleph_\alpha$ is singular iff α is a limit ordinal with $\operatorname{cof}\alpha < \aleph_\alpha$. A cardinal λ is a limit cardinal [strong limit cardinal] if $\mu < \lambda$ implies $\mu^+ < \lambda [2^\mu < \lambda]$. Throughout our discussion the single most interesting limit cardinal will be $\aleph_\omega$, as was already illustrated in the introduction.

As has been known from time immemorial, addition and multiplication of two cardinals trivializes when at least one of them is infinite, so the theory of cardinal arithmetic begins with cardinal exponentiation, and more generally with infinite products. In particular, the function $\lambda \mapsto 2^\lambda$ satisfies the following two classical laws (the first being entirely trivial):

$$\text{If } \mu \leq \lambda \text{ then } 2^\mu \leq 2^\lambda. \tag{1}$$

$$\operatorname{cof} 2^\lambda > \lambda. \tag{2}$$

The most basic problem in (conventional) cardinal arithmetic would be whether this function obeys other laws, and for this one quickly comes to consider the behavior of $2^{\aleph_\delta}$ for δ a limit ordinal (as the historical discussion in §2 explains). In this case one has various relations of the type

$$2^{\aleph_\delta} = \prod_{\alpha<\delta} 2^{\aleph_\alpha}. \tag{3a}$$

In particular, if $\aleph_\delta$ is a strong limit cardinal – that is, if $2^{\aleph_\alpha} < \aleph_\delta$ for all $\alpha < \delta$ – we have

$$2^{\aleph_\delta} = \aleph_\delta^{\operatorname{cof}\delta}. \tag{3b}$$

One of the difficulties in the study of cardinal arithmetic is the preoccupation with 2^λ, or more generally with exponentiation of a small base to a large exponent; the reverse situation is considerably more manageable, and a preoccupation with strong limit cardinals is merely an attempt to trade in one problem for the other via a relation of type (3b). Maybe this preoccupation is a vestige of the Generalized Continuum Hypothesis.

We return to the first shift in viewpoint discussed in the introduction. If $\aleph_\omega$ is a strong limit cardinal then $2^{\aleph_\omega} = \aleph_\omega{}^{\aleph_0}$, by (3b). In the sample Theorem A given in the introduction, we explicitly included the hypothesis that $\aleph_\omega$ is a strong limit cardinal. In fact, the theorem is cleaner without it:

Theorem A′. $\aleph_\omega{}^{\aleph_0} < \max\{\aleph_{\omega_4}, (2^{\aleph_0})^+\}$.

This is clearly a more "robust" statement than the original formulation, in that fewer extraneous considerations are involved. Unfortunately the meaning of the statement will still depend on the value of $2^{\aleph_0}$, the dependence being trivial if $2^{\aleph_0} > \aleph_\omega$, and we will have to work harder, beginning in §4, to eliminate this feature.

The following ideas belong to modern cardinal arithmetic, though absent from the classical theory. There is a natural topology on the class of all ordinals, in which a limit ordinal α is a limit point of any set X of ordinals for which $X \cap \alpha$ is unbounded below α. If α is an ordinal of uncountable cofinality, $\mathscr{F}_\alpha$ will denote the filter generated by the closed unbounded subsets of α. This filter is closed under countable intersections.

A *filter* on a set I is a collection of subsets of I closed upward (with respect to inclusion) and closed under intersection. We require filters to be *nontrivial*, that is, the empty set may not be in the filter.

If $\{X_i : i \in I\}$ is an indexed family of sets and $\mathscr{F}$ is any filter on I, then the *reduced product* $\prod_i X_i/\mathscr{F}$ is the set of functions $f \in \prod_i X_i$ modulo the equivalence relation $=_\mathscr{F}$ defined by $f =_\mathscr{F} g$ iff $\{i : f(i) = g(i)\} \in \mathscr{F}$. Similarly, if the X_i are linearly ordered, then $\prod_i X_i$ is partially ordered by pointwise comparison and $\prod_i X_i/\mathscr{F}$ is partially ordered by pointwise comparison modulo $\mathscr{F}$.

A maximal (nontrivial) filter is also called an ultrafilter and the corresponding reduced products are called ultraproducts. For our purposes, the main point is that an ultraproduct of linearly ordered sets is again linearly ordered and not just partially ordered.

We will touch lightly on issues connected with large cardinals and inner models. The reader who is unfamiliar with these subjects may ignore these remarks. On the other hand the following comments may be sufficient by way of background. There are various axioms concerning the existence of "large" cardinals (bearing names like: strongly inaccessible, measurable, supercompact, huge), which are easily seen to be unprovable on the basis of ZFC; even their consistency is known to be unprovable on the basis of ZFC, though these axioms are generally thought to be consistent. Some consistency results have been obtained only on the basis of the assumed consistency of large cardinal axioms, and the application of "inner models theory" is a method to prove that such consistency results require large cardinals. As our main concern is with provable theorems, this takes us rather far afield, but there is a constant interaction between the search for new theorems and the limitations imposed by independence results, which one cannot ignore in practice. In any case, when we refer to large cardinals, it is understood in that sense. When we wish to refer to cardinals that are large in a less problematic sense (bigger than some particular cardinal) we will refer to them as "moderately large."

2. History

Until 1974, the classical monotonicity and cofinality restrictions (1), (2) given above exhausted the known properties of the function $(\lambda \mapsto 2^\lambda)$. Gödel proved the consistency (with ZFC) of the generalized continuum hypothesis: $2^{\aleph_\alpha} = \aleph_{\alpha+1}$. In 1963 Cohen introduced the notion of forcing, setting off the wave of independence results that continues to this day, and used it to prove that $2^{\aleph_0}$ can be any cardinal of uncountable cofinality. Easton adapted this method to show that for any appropriate function $f(\lambda)$ satisfying the monotonicity and cofinality restrictions, it is consistent to assume $2^\lambda = f(\lambda)$ for all *regular* λ. In Easton's constructions, for λ singular 2^λ will always be the least value consistent with (1), (2), and the values of 2^μ for μ regular. For example, if Easton made $\aleph_\omega$ a strong limit cardinal, then he made $2^{\aleph_\omega} = \aleph_{\omega+1}$.

Thus the belief grew that cardinal arithmetic lay outside the realm of mathematical investigation, and to complete the picture it apparently remained only to modify Easton's approach to incorporate singular cardinals. Under large cardinal hypotheses, progress was made first for large singular cardinals, and then by Magidor in 1973 for $\aleph_\omega$, proving for example that one could have $\aleph_\omega$ a strong limit and $2^{\aleph_\omega} = \aleph_{\omega+2}$.

So it came as a great surprise in 1974 when Silver produced a new theorem of cardinal arithmetic: if $2^{\aleph_\alpha} = \aleph_{\alpha+1}$ for all countable α, then $2^{\aleph_{\omega_1}} = \aleph_{\omega_1+1}$. At this point we will leave the later independence results aside (Magidor, Foreman, Woodin, Gitik, Cummings, and the present author), as well as the complementary work on inner models and consistency strength (Jensen, Devlin, Dodd, Mitchell, Gitik), and earlier works (Scott, Solovay and Magidor) and concentrate on theorems provable in ZFC. The next section concentrates on singular cardinals of uncountable cofinality; one can skip over this development and continue in §4.

3. $\lambda^{\aleph_1}$

In what follows, $\aleph_1$ can be replaced by any uncountable regular cardinal, but not (in this section) by ω.

Answering a question of Silver, Galvin and Hajnal proved

Theorem [GH]. *Suppose that δ is a limit ordinal of cofinality $\aleph_1$, and ${\aleph_\alpha}^{\aleph_1} < \aleph_\delta$ for $\alpha < \delta$. Then ${\aleph_\delta}^{\aleph_1} < \aleph_{(2^{|\delta|})^+}$.*

More precisely, if we define $f : \delta \to \delta$ by ${\aleph_\alpha}^{\aleph_1} = \aleph_{f(\alpha)}$, the theorem states that ${\aleph_\delta}^{\aleph_1} < \aleph_{\|f\|}$, where $\|f\|$ is defined inductively as

$$\sup\{\|g\| + 1 : g < f \bmod \mathscr{F}_\delta\}.$$

(For notation, cf. §1.) This definition turns out to make sense because the filter $\mathscr{F}_\delta$ is closed under countable intersections. Following Silver, this line was developed in parallel and subsequent works by Baumgartner, Galvin and Hajnal, Jech and Prikry, Magidor, and the author.

When I became interested in the subject, I saw a great deal of activity and suspected I had come into the game too late; shortly thereafter I seemed to be the only one still interested in getting theorems in ZFC. I believed that the following thesis would be fruitful.

Thesis. *For λ of cofinality $\aleph_1$ satisfying*

$$(*) \qquad \mu^{\aleph_1} < \lambda \text{ for all } \mu < \lambda$$

if λ is small in any sense then $\lambda^{\aleph_1}$ is small in a related sense.

In practice this means one writes $\lambda = F_1(\alpha)$ for F_1 some natural function, and proves a bound $\lambda^{\aleph_1} < F_2(\alpha)$ where $F_2(\alpha) = F_1((|\alpha|^{\aleph_1})^+)$ or some quite similar function.

For example,

Theorem [Sh 111]. *If $\lambda > 2^{2^{\aleph_1}}$ satisfies $(*)$ and is below the first regular uncountable limit cardinal, then so is $\lambda^{\aleph_1}$. We also get results in this vain for $F_1(\alpha) =$ the αth fix point (i.e., cardinal λ equal to $\aleph_\lambda$).*

This uses a rank function similar to Galvin/Hajnal's, with respect to more (normal) filters, which we show is well defined unless Jensen's work trivializes the problem.

We got a similar bound in [Sh 256] when λ is the first fixed point of order ω and cofinality $\aleph_1$ in the enumeration of the cardinals, solving a problem raised by Hajnal following [Sh 111]. There the ranks were with respect to more complicated objects than normal filters, and in [Sh 333] similar bounds are obtained for functions defined inductively. We also prove that if the problem is not trivial then if we collapse $2^{2^{\aleph_1}}$ there is an ultrapower of the old universe in which for all regular $\lambda > 2^{2^{\aleph_1}}$ there is a λ-like element in the ultrapower (in [Sh 111] this was done for each λ seperately). In [Sh 386] $\lambda > 2^{2^{\aleph_1}}$ was replaced in the theorem above by $\lambda > 2^{\aleph_1}$. A posteriori the line [Sc, So, Si, GH, Sh 111, Sh 256, Sh 386] is quite straight. The rest follows a different line.

4. Possible cofinalities

Although Cohen and Easton showed us that powers of regular cardinals are easily manipulated, we learned from inner model theory that this is not the case for powers of singular cardinals. Similarly, manipulating λ^κ for $\lambda > 2^\kappa$ is much harder than manipulating 2^κ; the same applies to products of relatively few moderately large cardinals. There were indications [Sh b, Chapter XIII, §§5, 6] that cofinalities are at work behind the scenes. At a certain point we began to feel that we could split off the independence results from the hard core of truth by shifting the focus.

Let $\mathbf{a} = (\lambda_i)_{i \in I}$ be an indexed set of regular cardinals with each λ_i greater than $|I|$. Where cardinal arithmetic is concerned with the cardinality of $\prod \mathbf{a} = \prod_i \lambda_i$, we will be concerned with the following cardinal invariants involving more of the structure of the product.

4.1. **Definition.** 1. A cardinal λ is a *possible cofinality* of $\prod \mathbf{a}$ if there is an ultrafilter $\mathcal{F}$ on $\mathbf{a}$ for which the cofinality of $\prod \mathbf{a}/\mathcal{F}$ is λ. (Recall that these ultraproducts are linearly ordered: §1.)

2. $\operatorname{pcf} \mathbf{a}$ is the set of all possible cofinalities of $\prod \mathbf{a}$.

We first used pcf in [Sh 68] in a more structural context, to construct Jońsson algebras (algebras of a given cardinality with no proper subalgebra of the same cardinality). In [Sh b, Chapter XIII, §§5,6] we obtained results under the more restrictive hypothesis $\lambda_i > 2^{|\mathbf{a}|}$ bearing on cardinal arithmetic. An instance of the main theorem there is the following (note that [GH] did not give information on cardinals with countable cofinalities, Theorem A is a significant improvement though formally they are incomparable).

4.2. **Theorem B$'$.** *If* $2^{\aleph_0} < \aleph_\omega$ *then* $\aleph_\omega^{\aleph_0} < \aleph_{(2^{\aleph_0})^+}$.

Since there are $2^{2^{|I|}}$ ultrafilters on I, $\operatorname{pcf} \mathbf{a}$ could be quite large a priori; this would restrict heavily its applications to cardinal arithmetic. Fortunately there are various uniformities present that lead to a useful structure theory for pcf.

4.3. **Main Theorem.** *Let* $\mathbf{a} = (\lambda_i)_{i \in I}$ *be an indexed set of distinct regular cardinals, with each* $\lambda \in \mathbf{a}$ *greater than* $|\mathbf{a}|$. *Then*:

1. $\operatorname{pcf} \mathbf{a}$ *contains at most* $2^{|\mathbf{a}|}$ *cardinals*;
2. $\operatorname{pcf} \mathbf{a}$ *has a largest element* $\max \operatorname{pcf} \mathbf{a}$;
3. $\operatorname{cof} \prod \mathbf{a} = \max \operatorname{pcf} \mathbf{a}$ (*see remark below*);
4. *For each* $\lambda \in \operatorname{pcf} \mathbf{a}$ *there is a subset* $\mathbf{b}_\lambda$ *of* $\mathbf{a}$ *such that*
   a. $\lambda = \max \operatorname{pcf} \mathbf{b}_\lambda$, *and*
   b. $\lambda \notin \operatorname{pcf}(\mathbf{a} - \mathbf{b}_\lambda)$;
5. *If* $\mathcal{J}_\lambda$ *is the ideal on* I *generated by the sets* $\mathbf{b}_\mu$ *for* $\mu < \lambda$, *then for each* $\lambda \in \operatorname{pcf} \mathbf{a}$ *there are functions* f_i^λ $(i < \lambda)$ *such that*
   a. *for* $i < j$ *we have* $f_i^\lambda < f_j^\lambda \bmod \mathcal{J}^\lambda$;
   b. *for any* $f \in \prod \mathbf{a}$ *and* $\lambda \in \operatorname{pcf} \mathbf{a}$ *there is some* $i < \lambda$ *such that* $f < f_i^\lambda \bmod \langle \mathcal{J}_\lambda, (\mathbf{a} - \mathbf{b}_\lambda) \rangle$.

Remarks. The meaning of clause (3) is that for $\lambda = \max \operatorname{pcf} \mathbf{a}$, there is a subset P of $\prod \mathbf{a}$ of cardinality λ that is cofinal in the sense that every function in $\prod \mathbf{a}$ is dominated pointwise by a function in P. For example, the cofinality of $\omega \times \omega_1$ is ω_1. This does not mean that there is a pointwise nondecreasing sequence of length ω_1 that dominates every element of $\omega \times \omega_1$; there is no such sequence. When a product actually contains a cofinal pointwise nondecreasing sequence, we say its cofinality is *true*, and we write tcf for the cofinality when it is true.

By clause (5) $\prod \mathbf{b}_\lambda / \mathcal{J}_\lambda$ has true cofinality λ.

The following structural principle is also of great practical importance. Note that we do not know whether $|\operatorname{pcf} \mathbf{a}| \le |\mathbf{a}|$, the following still says that $\operatorname{pcf} \mathbf{a}$ is "small," "local," has small density character.

4.4. **Localization Theorem.** *Let* $\mathbf{a}$ *be a set of* κ *distinct regular cardinals with* $\lambda > \kappa$ *for all* $\lambda \in \mathbf{a}$; *and suppose* $\mathbf{b} \subseteq \operatorname{pcf} \mathbf{a}$ *with* $\lambda > |\mathbf{b}|$ *for all* λ *in* $\mathbf{b}$. *If* $\mu \in \operatorname{pcf} \mathbf{b}$ *then* $\mu \in \operatorname{pcf} \mathbf{a}$, *and for some* $\mathbf{c} \subseteq \mathbf{b}$ *of cardinality at most* κ *we have* $\lambda \in \operatorname{pcf} \mathbf{c}$.

Thus pcf defines a canonical closure operation on sets of regular cardinals with some good properties.

5. Pseudopowers

For $\operatorname{cof}\lambda \leq \kappa < \lambda$ we define the *pseudopower* $\mathrm{pp}_\kappa(\lambda)$ as follows.

5.1. **Definition.** 1. $\mathrm{pp}_\kappa(\lambda)$ is the supremum of the cofinalities of the ultraproducts $\prod \mathbf{a}/\mathcal{F}$ associated with a set of at most κ regular cardinals below λ and an ultrafilter $\mathcal{F}$ on $\mathbf{a}$ containing no bounded set bounded below λ.

2. $\mathrm{pp}(\lambda)$ is $\mathrm{pp}_{\operatorname{cof}\lambda}(\lambda)$.

If we have a model of set theory with a very interesting (read: bizarre) cardinal arithmetic, say $2^{\aleph_\omega} = \aleph_{\omega+2}$, and we adjust $2^{\aleph_0}$ by Cohen's method, putting in $\aleph_{\omega+3}$ Cohen reals, there is then no nontrivial operation of cardinal arithmetic that will yield the result $\aleph_{\omega+2}$. Not that the original phenomena have been erased; they are simply drowned out by static. The operation $\mathrm{pp}_\kappa(\lambda)$ is more robust. It will be convenient to call λ κ-inaccessible if $\mu^\kappa < \lambda$ for $\mu < \lambda$.

5.2. **Theorem.** 1. $\mathrm{pp}_\kappa(\lambda) \leq \lambda^\kappa$;

2. *If* $\operatorname{cof}\lambda \leq \kappa < \lambda < \aleph_\lambda$ *and* λ *is* κ*-inaccessible* (*i.e.,* $(\forall \mu < \lambda)[\mu^\kappa < \lambda]$), *then* $\mathrm{pp}_\kappa(\lambda) = \lambda^\kappa$.

In this sense, $\mathrm{pp}_\kappa(\lambda)$ is an antistatic device. If λ is κ-inaccessible in one model of set theory and this condition is then destroyed in an extension in a reasonably subdued manner (i.e., when we make 2^κ large), $\mathrm{pp}_\kappa(\lambda)$ continues to reflect the earlier value of λ^κ. On another level, one can prove results about this operation by induction on cardinals, which is not possible with the less robust notions. (The restriction $\lambda < \aleph_\lambda$ is certainly convenient, but we will discuss its removal below.)

Let $\mathrm{PP}_\kappa(\lambda)$ be the set of cofinalities whose supremum was taken to get $\mathrm{pp}_\kappa(\lambda)$. This turns out to have the simplest possible structure.

5.3. **Convexity Theorem.** *If* $\kappa \in [\operatorname{cof}\lambda, \lambda)$ *then* $\mathrm{PP}_\kappa(\lambda)$ *is an interval in the set of regular cardinals with minimum element* λ^+.

We can also give a cleaner description of the way a cardinal enters $\mathrm{PP}_\kappa(\lambda)$ in some cases.

5.4. **First Representation Theorem.** *Suppose* $\aleph_0 < \operatorname{cof}\lambda < \kappa < \lambda$ *and for all* $\mu \in (\kappa, \lambda)$; *if* $\operatorname{cof}\mu \leq \operatorname{cof}\lambda$ *then* $\mathrm{pp}(\mu) < \lambda$. *Let* $\mathcal{F}_0$ *be the filter on* $\operatorname{cof}\lambda$ *generated by the complements of the bounded sets. Then for any regular cardinal* $\lambda^* \in \mathrm{PP}(\lambda)$, *there is an increasing sequence* $(\lambda_i)_{i<\operatorname{cof}\lambda}$ *with limit* λ *such that the product* $\prod \lambda_i/\mathcal{F}_0$ *has true cofinality* λ^*.

In the case $\lambda^* = \lambda^+$ there is a better representation.

5.5. **Second Representation Theorem.** *If* λ *is singular of uncountable cofinality,* $\mathcal{F}_0$ *is the filter generated by the complements of the bounded subsets of* $\operatorname{cof}\lambda$, *and* $(\lambda_i)_{i<\operatorname{cof}\lambda}$ *is increasing, continuous and cofinal in* λ, *then there is a closed unbounded subset* C *of* $\operatorname{cof}\lambda$ *such that* $\prod_C \lambda_i^+/\langle \mathcal{F}_0, \operatorname{cof}\lambda - C\rangle$ *has true cofinality* λ^+.

Discussion. Why do we suggest $PP_\kappa(\lambda)$ as a replacement to λ^κ? Maybe we had better reconsider what we are doing. λ^κ is a measure of the size of the family of subset λ of cardinality $\leq \kappa$, i.e., its cardinality. Remember that when

$\lambda \le \kappa$ the independence results do not leave us much to say. When $\kappa < \lambda$ we shall present various natural measures of this set below. It is useful to prove that various such measures are equal, as we can then look at them as various characterizations of the same number each useful in suitable circumstances. We report reasonable success in this direction below.

5.6. **Definition.** For $\kappa \in [\operatorname{cof}\lambda, \lambda)$ let the κ-covering number for λ, $\operatorname{cov}(\lambda, \kappa)$, be the minimal cardinality of a family of subsets of λ, each of size less than λ, such that every subset of λ of size κ is contained in one of the specified sets. Note that, as is well known, for $\kappa < \lambda$, $\lambda^\kappa = 2^\kappa +$ "the κ-covering of λ."

5.7. **Theorem.** *If* $\kappa \in [\operatorname{cof}\lambda, \lambda)$ *and* $\lambda < \aleph_\lambda$ *then* $\operatorname{pp}_\kappa(\lambda)$ *is the* κ*-covering number for* λ.

To remove the restriction that $\lambda < \aleph_\lambda$, define the *weak* κ-covering number for λ as the minimal cardinality of a family of subsets of λ, each of size less than λ, such that every subset of λ of size κ is contained in the union of countably many of the specified sets.

5.8. **Theorem.** 1. *For* κ *uncountable, the weak* κ*-covering number of* λ *is the supremum of the true cofinalities of the reduced products* $\prod \mathbf{a}/\mathcal{F}$ *with* $\mathbf{a}$ *a set of at most* κ *regular cardinals below* λ *and* $\mathcal{F}$ *a filter on* $\mathbf{a}$ *that contains the complement of every subset of* $\mathbf{a}$ *bounded below* λ *and is closed under countable intersections.*

2. *If this cardinal is regular, the indicated supremum is attained.*

Thus our problems are connected with the case of cofinality ω. We give relevant partial information in §8.

We mention two more invariants that we can now prove coincide.

5.9. **Theorem.** *For* $\aleph_0 < \kappa \le \lambda$ *with* κ *regular the following two cardinals coincide:*

1. *The minimal cardinality of a family of subsets of* λ, *each of size less than* κ, *such that any subset of* λ *of cardinality less than* κ *is contained in one of the specified sets.*

2. *The minimal cardinality of a stationary subset of the family of subsets of* λ *of size less than* κ. *(Stationarity means that for every algebra with set of elements* λ *and countably many operations, there is a subalgebra* B, $|B| < \kappa$, $B \cap \kappa$ *is a ordinal and* $B \in S$.*)*

6. Cardinal arithmetic revisited

If $\kappa = \aleph_\alpha$, it will be convenient to call α the *index* of κ; otherwise our results tend to live entirely in the land of subscripts. In [Sh b, Chapter XIII, §§5,6] we showed

$$\text{The index of } \aleph_\delta{}^{\operatorname{cof}\delta} \text{ is less than } (|\delta|^{\operatorname{cof}\delta})^+.$$

In particular, for $\delta = \omega$, the corresponding index is below $(2^{\aleph_0})^+$ (construed as an ordinal). If $\aleph_\omega$ is itself below $2^{\aleph_0}$ then this contains no information and (our usual theme) inessential modifications of the universe can always make this happen. On the other hand known independence results show that when $\aleph_\omega$ is a strong limit, the index of $(\aleph_\omega)^{\aleph_0}$ can be any countable successor ordinal, so in principle one could hope to prove that the true bound is ω_1 (which we doubt). Our strongest result is

6.1. **Theorem.** $\text{pp}(\aleph_\omega) < \aleph_{\omega_4}$.

6.2. **Corollary.** $\aleph_\omega^{\aleph_0} < \max(\aleph_{\omega_4}, (2^{\aleph_0})^+)$.

The more general formulation is

6.3. **Theorem.** *If* δ *is a limit ordinal,* $\delta < \aleph_{\alpha+\delta}$ *then* $\text{pp}(\aleph_{\alpha+\delta}) < \aleph_{\alpha+|\delta|^{+4}}$.

For the proof, one looks carefully at the closure operation induced on ω_4 by the pcf structure on $\aleph_{\omega_4}$ (passing to indices) under the assumption that $\aleph_{\omega_4}$ is $\leq \text{pp}(\aleph_\omega)$. There are conflicts between the main theorem, the localization theorem and the second representation theorem on the one hand and combinatorics of closed unbounded sets inside ω_4 on the other hand, which yield a contradiction. Ultimately, the pcf structure cannot exist on ω_4 because no such closure operation exists. On ω_3, there are two questions: does such a closure operation exist and can it be given by pcf?

Structure of the proof. Assuming (toward a contradiction) that $\aleph_{\omega_4+1}$ belongs to $\text{pcf}(\aleph_\omega)$, for $X \subseteq \omega_4$ define $\text{cl}\, X = \{i < \omega_4 : \aleph_{i+1} \in \text{pcf}(\aleph_{j+1})_{j \in X}\}$. By the Convexity Theorem $\text{cl}(\omega) = \omega_4$ and by the Localization Theorem, if $i \in \text{cl}\, X$ then for some countable $Y \subseteq X$ we have $i \in \text{cl}\, Y$. By the Second Representation Theorem, if $\delta < \omega_4$ is a limit ordinal of uncountable cofinality then δ is the maximal element of $\text{cl}\, C$ for some closed unbounded subset of δ as well as any smaller closed unbounded set. These three properties of the closure operation (alone) eventually lead to a contradiction. □

This may indicate that it is interesting to investigate, e.g., the set $\{\aleph_{\alpha+1} : \aleph_{\alpha+1} \leq \text{pp}\aleph_\omega\} = \text{pcf}\{\aleph_n : n\}$ with the relation "$\lambda \in \text{pcf}\, \mathbf{b}$" and even the sequence $\langle \mathbf{b}_\lambda : \lambda \in \text{pcf}\{\aleph_n : n\}\rangle$ from 4.3 (for which there are theorems saying "we can choose nicely").

7. For true believers

Naturally our results also give a great deal of information regarding the types of forcing that are applicable to certain open questions. Any instance of $\lambda \in \text{pcf}\, \mathbf{a}$ normally results from some normal ultrafilter "at the time when the large cardinals were present." Some of this information is in the canonical spot in [Sh-b, Chapter XIII, §§5,6], and more is in [Sh 282], for example, if $\text{pp}(\aleph_{\omega_1})$ is greater than $\aleph_{\omega_2}$ then there are many ordinals $\delta < \omega_2$ of cofinality ω for which $\text{pp}(\aleph_\delta)$ is above $\aleph_{\delta+\omega_1}$.

These considerations also shed some light on the problem of resurrection of supercompactness. By our 9.6(1)

7.1. **Corollary.** *If in* V *we have* $\text{cof}(\lambda) < \kappa < \lambda$ *and* $\text{pp}(\lambda) > \lambda^+$, *then there is no universe extending* V *in which* λ^+ *remains a cardinal while* κ *becomes supercompact or even compact.*

Gitik has proved, for example, that if $\text{cof}\,\lambda = \omega$, $\mu^{\aleph_0} < \lambda$ for all $\mu < \lambda$, and $\text{pp}(\lambda) > \lambda^+$, then in Mitchell's inner model $o(\lambda) = \lambda^{++}$. By his previous independence results this settles the consistency strength of $2^{\aleph_0} < \aleph_{\omega+1} < \aleph_\omega^{\aleph_0}$ (using [Sh b]), though it did not settle the consistency strength of the full singular cardinal hypothesis, as there are fixed points. For this purpose we have proved

7.2. **Lemma.** *If* λ *is singular of cofinality* $\aleph_0$, *and if* $\text{pp}(\mu) < \lambda$ *for all singular cardinals* $\mu < \lambda$ *of cofinality* $\aleph_0$ *and* $pp(\mu) = \mu^+$ *for every large enough* $\mu < \lambda$

of cofinality $\aleph_1$, *then there is a family of countable subsets of* λ *containing* $\mathrm{pp}(\lambda)$ *sets, so that every countable subset of* λ *is contained in one of the specified sets. Thus if* $\mu^{\aleph_0} < \lambda$ *for all* $\mu < \lambda$, *we find* $\mathrm{pp}(\lambda) = \lambda^{\aleph_0}$.

Together this yields an equiconsistency result. However Gitik points out that this result is intrinsically global. For a localized result, we have

7.3. **Lemma.** *If* λ *is singular of cofinality* $\aleph_0$, *and if* $\mu^{\aleph_0} < \lambda$ *for* $\mu < \lambda$ *and* $\mathrm{pp}(\lambda) > \lambda^{\aleph_0}$, *then* $\mathrm{pp}(\lambda)$ *is quite large; there are at least* $\aleph_1$ *fixed points in the interval from* λ^+ *to* $\mathrm{pp}(\lambda)$.

This still does not settle the behavior of singular cardinals of cofinality $\aleph_0$. The following suggests there are few exceptional values.

7.4. **Lemma.** *If* κ *is a regular uncountable cardinal and* $(\lambda_i)_{i\leq\kappa}$ *is an increasing continuous sequence of cardinals with* $\lambda_i^\kappa < \lambda_\kappa$ *for* $i < \kappa$, *then for some closed unbounded subset* C *of* κ *we have* $\mathrm{pp}(\lambda_i) = \lambda_i^\kappa$ *on* $C \cup \{\kappa\}$.

This has application to the construction of "black boxes."

8. Some hypotheses

8.1. **The Strong Hypothesis.** *For all singular cardinals* λ, $\mathrm{pp}(\lambda) = \lambda^+$.

This is a replacement for GCH as well as "$0^\#$ does not exist" in some cases. It is weaker than either and is hard to change by forcing, but is consistent with large cardinals (and indeed holds above any compact cardinal [So] while having a more combinatorial character than "$\neg 0^\#$". Of course this will not give, e.g., a square sequence on the successor of a singular cardinal.)

8.2. **Lemma.** *The strong hypothesis implies that for any singular cardinal* λ *and any* $\kappa < \lambda$, $\lambda^\kappa \leq \lambda^+ + 2^\kappa$, *there are* λ^+ *subsets of* λ, *each of cardinality* κ, *such that every subset of* λ *of cardinality* κ *is contained in one of the specified sets; and a Jońsson algebra exists in every successor cardinal.*

8.3. **The Weak Hypothesis.** *For any singular cardinal* λ, *there are at most countably many singular cardinals* $\mu < \lambda$ *with* $\mathrm{pp}(\mu) \geq \lambda$.

In my opinion, it is a major problem to determine whether this follows from ZFC and is the real problem behind the determination of the true bound on $\mathrm{pp}(\aleph_\omega)$.

8.4. **Lemma.** *The weak hypothesis implies:*

- $\mathrm{pp}(\aleph_\omega) < \aleph_{\omega_1}$, *and more generally if* $\delta < \aleph_\delta$ *then* $\mathrm{pp}(\aleph_\delta) < \aleph_{|\delta|^+}$.
- $\mathrm{pcf}\,\mathbf{a}$ *has cardinality at most* $|\mathbf{a}|$.
- $\mathrm{pp}(\lambda)$ *has cofinality at least* λ^+ *for* λ *singular.*

If we strengthen the weak hypothesis by replacing "countable" by finite, but only for μ of cofinality $\aleph_0$, then Gitik has proved that the stronger version does not follow from ZFC.

9. Other applications

If the list of applications below does not contain "familiar faces" the reader will not lose by skipping to the concluding remarks.

We turn now to results involving more structural information, bearing on almost free abelian groups, partition problems, failure of preservation of chain conditions in Boolean algebras under products, existence of Jońsson algebras, existence of entangled linear orders, equivalently narrow Boolean algebras, and the existence of $L_{\infty,\lambda}$-equivalent nonisomorphic models.

9.1. **Model theory.** *If λ has cofinality greater than $\aleph_1$ then there are two $L_{\infty,\lambda}$-equivalent nonisomorphic models of cardinality λ.*

This was known for λ regular or $\lambda = \lambda^{\aleph_0}$, and for strong limit λ of uncountable cofinality, but fails for cofinality $\aleph_0$. There is still a small gap.

9.2. **Jońsson Algebras.** *There is a Jońsson algebra of cardinality $\aleph_{\omega+1}$.*

A Jońsson algebra is an algebra in a finite or countable language that has no proper subalgebra of the same cardinality.

This was known previously under the hypothesis $2^{\aleph_0} \le \aleph_{\omega+1}$. We can now show that if λ is singular and there is no Jońsson algebra of cardinality λ^+, then λ is quite large, for example, λ is a limit of weakly inaccessible cardinals that do not admit Jońsson algebras (and there is a Jońsson algebra of inaccessible cardinality μ if μ is not Mahlo, or even not $(\mu \times \omega)$-Mahlo).

9.3. **Chain conditions.** *For any cardinal $\lambda > \aleph_1$, there is a boolean algebra that satisfies the λ^+-chain condition while its square does not; hence its Stone-Čech compactification has cellularity λ, but its square has cellularity greater than λ*

Previous References: [T1, Sh 282] for some singular cases, [Sh 280] for regular above $2^{\aleph_0}$, [Sh 327] for regular $\lambda \ge \aleph_2$.

This really comes from coloring theorems. We give a sample of the latter.

9.4. **Coloring Theorem.** *For $\lambda > \aleph_1$ there is a binary symmetric function c: $(\lambda^+)^2 \to \operatorname{cof}\lambda$ such that for any sequence $(w_i)_{i<\lambda}$ of pairwise disjoint finite subsets of λ^+ and any $\gamma < \operatorname{cof}\lambda$, there is a pair $i < j < \lambda$ with $c[w_i \times w_j] = \{\gamma\}$.*

9.5. **Narrow Boolean algebras.** 1. *If λ is singular and less than $2^{\aleph_0}$, then there is a λ^+-narrow Boolean order algebra of cardinality λ^+ (see below). The same holds if $\kappa^{+3} < \operatorname{cof}\lambda < \lambda < 2^\kappa$.*

2. *The class $\{\lambda$: there is in λ^+ a λ^+-narrow Boolean algebra$\}$ is not bounded (really $(\lambda, \aleph_{\lambda^{+3}+1}]$ is not disjoint to it).*

9.5A. **Definitions.** 1. The boolean order algebra associated to a linear order L is the boolean algebra of subsets of L generated by the closed-open intervals.

2. A boolean algebra is λ-narrow if it has no set of pairwise incomparable elements of size λ (a, b are incomparable if $a \not\le b$ and $b \not\le a$).

3. A linear order L is *entangled* if for every n, for all choices of distinct $x_{m,i}$ for $m \le n$, $i < |L|$, and for any subset w of $\{1, \dots, n\}$, there are $i < j$ so that for all $m \le n$ we have $x_{m,i} < x_{m,j}$ iff $m \in w$.

For λ regular, the boolean order algebra associated to the linear order L of cardinality λ is λ-narrow if and only if I is entangled. For background see [ARSh 153, BoSh 210, T, Sh 345].

The bound λ^{+3} in the theorem above is connected to §6. Note that in addition we can use a subdivision into various cases, bearing in mind that by [BSh 212, T] there is such an algebra of cardinality $\operatorname{cof} 2^\lambda$ if there is a linear order of cardinality 2^λ and density λ, and usually the absence of such an order enables us to use our current method.

9.6. **Almost disjoint sets, almost free Abelian groups.** 1. *If λ is singular and* $\mathrm{pp}(\lambda) > \lambda^+$, *then there is a family of λ^+ subsets of λ, each of cardinality* $\mathrm{cof}\,\lambda$, *such that any λ of them admit an injective choice function (called transversal). Consequently if* $\mathrm{cof}\,\lambda = \aleph_0$ *then there is a λ^+-free but not free abelian group.*

2. *The following are equivalent for $\lambda > \mu \geq 2^{\aleph_0}$.*

a. *There are λ subsets of μ of size $\aleph_1$ such that any two have finite intersection.*

b. *For some n, there are regular cardinals $\lambda_{i,m}$ for $i < \omega_1$ and $m \leq n$, such that for every infinite subset X of ω_1, $\mu \leq \max \mathrm{pcf}(\lambda_{i,m} : i \in X, m \leq n)$.*

3. *If λ is regular, $2^{<\lambda} < 2^\lambda$, and there is no linear order of cardinality 2^λ and density λ, then in every regular $\mu \in [2^{<\lambda}, 2^\lambda]$ there is an entangled order.*

There are other results on almost disjoint sets and λ-free abelian groups and a topological question of Gretlis, Hajnal, and Szentmiclossy.

Concluding Remarks. The following is not surprising in view of Theorem 6.3, and part of the argument is similar but requires, at least from the author, considerably more work.

Theorem. *If $\delta < \aleph_{\omega_4}$ has cofinality $\aleph_0$, then* $\mathrm{pp}(\aleph_\delta) < \aleph_{\omega_4}$, *and hence the cofinality of the partially ordered set $\langle S_{\leq\aleph_0}(\aleph_{\omega_4}); \subseteq\rangle$ is $\aleph_{\omega_4}$ (where $S_{\aleph_0}(\lambda) = \{a : a \subseteq \lambda, |a| = \aleph_0\}$).*

Also the state of Tarski conjecture can be clarified, by Jech and Shelah [JeSh 385] and, e.g., if $\mathrm{pp}(\aleph_{\omega_1}) > \aleph_{\omega_2}$ then there is a Kurepa tree on ω_1.

We conclude with some words of the author. Reflecting on the above it seems that whereas once we knew a considerable amount about the uncountable cofinality case and nothing about the countable cofinality case (I mean theorems and not consistency results or consistency strength results), now the situation has reversed. This is not accurate—the results of Galvin and Hajnal and those discussed in §3 above, are not superceded by the later results. There does not seem to be a generalization of ranks to countable cofinality, nor have we suggested so far anything on fixed points or limits of inaccessibles below the contiuum. Recently we have succeed in getting some results on this topic, and they will be reported elsewhere [Sh 420]. See more in Gitik Shelah [GiSh 412] (mainly if λ is real valued measurable $\{2^\sigma : \sigma < \lambda\}$ is finite), [Sh 413] (more on coloring and Jonsson algebras), [Sh 430] (on $\mathrm{cf}(\mathscr{S}_{<\aleph_1}(\lambda), \subseteq) \leq \lambda$ for λ real valued measurable, and again of existence of trees with κ nodes and any regular $\lambda \in [\kappa, 2^\kappa]$ κ-branches and improvements of 7.3 and on the smallest values needed for canonization theorems).

References

[ARSh 153] U. Abraham, M. Rubin, and S. Shelah, *On the consistency of some partition theorems for continuous colorings and the structure of $\aleph_1$-dense real order types*, Ann. Pure Appl. Logic **29** (1985), 123–206.

[B] J. Baumgartner, *Almost disjoint sets, the dense set problem, and the partition calculus*, Ann. Pure Appl. Logic **9** (1975), 401–439.

[BP] J. Baumgartner and K. Prikry, *On a theorem of Silver*, Discrete Math. **14** (1976), 17–22.

[BoSh 210] R. Bonnet and S. Shelah, *Narrow boolean algebras*, Ann. Pure Appl. Logic **28** (1985), 1-12.

[C] P. Cohen, *Set theory and the continuum hypothesis*, Benjamin, NY, 1966.

[Cu] J. Cummings, *Consistency results in cardinal arithmetic*, Ph. D. thesis, Cal-Tech, Pasadena, CA.

[CW] J. Cummings and H. Woodin, preprint.

[DeJ] K. Devlin and R. B. Jensen, *Marginalia to a theorem of Silver*, Proceedings of the Logic Colloquium, Kiel 1974 (G. H. Müller, A. Oberschelp, and K. Potthoff, eds.), Lecture Notes in Math., vol. 499, Springer-Verlag, Berlin, 1975, pp. 115–142.

[DoJ] A. Dodd and R. B. Jensen, *The core model*, Annals Math. Logic (Ann. Pure Appl. Logic) **20** (1981), 43–75.

[FW] M. Foreman and H. Woodin, *G.C.H can fail everywhere*, Ann. of Math. (2) **133** (1991), 1–35.

[GH] F. Galvin and A. Hajnal, *Inequalities for cardinal powers*, Ann. of Math. (2) **101** (1975), 491–498.

[GHS] J. Gertlis, A. Hajnal, and Z. Szentmiklossy, *On the cardinality of certain Hausdorff spaces*, Frolic Memorial Volume, accepted.

[Gi] M. Gitik, *The negation of SCH from* $o(\kappa)^{++}$, Ann. Pure Appl. Logic **43** (1989), 209–234.

[Gi1] ——, *The strength of the failure of the singular cardinal hypothesis*, preprint.

[GM] M. Gitik and M. Magidor, *The singular cardinal hypothesis revisited* (in preparation).

[GiSh 344] M. Gitik and S. Shelah, *On certain indestructibility of stray cardinals and a question of Hajnal*, Arch. Math Logic **28** (1989), 35–42.

[GiSh 412] M. Gitik and S. Shelah, *More on ideals with simple forcing notions*, Ann. Pure Appl. Logic (to appear).

[JP] T. Jech and K. Prikry, *Ideals over uncountable sets: applications of almost disjoint sets and generic ultrapowers*, Mem. Amer. Math. Soc. **214** (1979).

[JeSh 385] T. Jech and S. Shelah, *On a conjecture of Tarski on products of cardinals*, Proc. Amer. Math. Soc. **112** (1991), 1117–1124.

[Mg] M. Magidor, *On the singular cardinals problem* I, Israel J. Math. **28** (1977), 1–31.

[Mg1] ——, *On the singular cardinals problem* II, Ann. of Math. (2) **106** (1977), 517–547.

[Mg2] ——, *Chang's conjecture and powers of singular cardinals*, J. Symbolic Logic **42** (1977), 272–276.

[M] W. Mitchell, *The core model for sequences of measures* I, Math. Proc. Cambridge Philos. Soc. **95** (1984), 229–260.

[Pr] K. Prikry, *Changing measurable to accessible cardinals*, Rozprawy Mat. **LXVIII** (1970), 1–52.

[Sh 68] S. Shelah, *Jońsson algebras in successor cardinals*, Israel J. Math. **39** (1978), 475–480.

[Sh 71] ——, *A note on cardinal exponentiation*, J. Symbolic Logic **45** (1980), 56–66.

[Sh 111] ——, *On powers of singular cardinals*, Notre Dame J. Formal Logic **27** (1986), 263–299.

[Sh b] ——, *Proper Forcing*, Lecture Notes in Math., vol. 940, Springer-Verlag, Berlin, 1982.

[Sh 137] ——, *The singular cardinal problem: independence results*, Proceedings of a Symposium on Set Theory, Cambridge 1978 (A. Mathias, ed.), London Math Soc. Lecture Notes Ser., vol. 87, Cambridge Univ. Press, Cambridge and New York, 1983, pp. 116–134.

[Sh 161] ——, *Incompleteness in regular cardinals*, Notre Dame J. Formal Logic **26** (1985), 195–228.

[Sh 256] ——, *More on powers of singular cardinals*, Israel J. Math. **59** (1987), 263–299.

[Sh 280] ——, *Strong negative partition relations above the continuum*, J. Symbolic Logic **55** (1990), 21–31.

[Sh 282] ——, *Successors of singulars, productivity of chain conditions, and cofinalities of reduced products of cardinals*, Israel J. Math. **60** (1987), 146–166.

[Sh 327] ——, *Strong negative partition relations below the continuum*, Acta Math. Hungar. in press.

[Sh 410] ——, *More on cardinal arithmetic*, Arch. Math. Logic (to appear).

[Sh 413] ——, *More Jonsson algebras and coloring*, preprint.

[Sh 420] ——, *Advances in cardinal arithmetic*, to appear.

[Sh 430] ——, *Further cardinal arithmetic*, preprint.

[Sh-g] ——, *Cardinal arithmetic*, OUP, (to appear).

[ShW 159] S. Shelah and H. Woodin, *Forcing the failure of the CH*, J. Symbolic Logic **49** (1984), 1185–1189.

[Sc] D. Scott, *Measurable cardinals and constructible sets*, Bull. Acad. Pol. Sci. Ser. Math. Astron. Phys. **9** (1961), 521–524.

[Si] J. Silver, *On the singular cardinal problem*, Proceedings ICM, Vancouver 1974, vol. I, pp. 265–268,

[So] R. Solovay, *Strongly compact cardinals and the GCH*, Proceedings of the Tarski Symposium, Berkeley 1971, Proc. Sympos. Pure Math., vol. xxi, Amer. Math. Soc., Providence, RI, 1974, pp. 365–372.

[T] S. Todorčevič, *Remarks on chain conditions in products*, Compositio Math. **56** (1985), 295–302.

[T1] ——, *Remarks on cellularity in products*, Compositio Math. **57** (1986), 357–372.

[W] H. Woodin, *The Collected Unwritten Works of H. Woodin.*

Mathematics Department, The Hebrew University, Jerusalem, Israel

Current address: Mathematics Department, Hill Center, Busch Campus, Rutgers–The State University, New Brunswick, New Jersey 08903

Curriculum Vitae

Stephen Smale

Date of Birth: July 15, 1930, Flint, Michigan, USA

Education:

1957	Ph.D., University of Michigan
1953	M.S., University of Michigan
1952	B.S., University of Michigan

Employment History:

2009–2014	Distinguished University Professor, City University of Hong Kong, Hong Kong
2002–	Professor, Toyota Technological Institute at Chicago, USA
1995–2001	Distinguished University Professor, City University of Hong Kong
1995–	Distinguished University Professor, City University of Hong Kong
1995–	Professor of Mathematics in the Graduate School, University of California, Berkeley, USA
1994	Professor of Mathematics (and Economics) Emeritus, University of California, Berkeley, USA
1994	Visiting Professor, Instituto de Matematica Pura e Aplicada, Rio de Janeiro, Brazil
1988	Visiting Professor, Instituto de Matematica Pura e Aplicada, Rio de Janeiro, Brazil
1987	Visiting Professor, Columbia University, USA (Fall)
1987	Visiting Scientist, IBM Corporation, Yorktown Heights, USA (Fall)
1979–1980	Research Professor, Miller Institute for Basic Research in Science, Berkeley, USA
1976–	Professor without stipend, Department of Economics, University of California, Berkeley, USA
1976	Visiting Professor, Instituto de Matematica Pura e Aplicada, Rio de Janeiro, Brazil
1976	Visiting Professor, Institut des Hautes Études Scientifiques, France (May, June)
1974	Visiting Professor, Yale University, USA (Fall)
1972–1973	Visiting Professor, University of Paris, Orsay, France (Fall)
1972–1973	Visiting Member, Institut des Hautes Études Scientifiques, France (Fall)

1969–1970	Visiting Member, Institut des Hautes Études Scientifiques, France (Fall)
1967–1968	Research Professor, Miller Institute for Basic Research in Science, Berkeley, USA
1966	Member, Institute for Advanced Study, Princeton, USA (Fall)
1964–94	Professor of Mathematics, University of California, Berkeley, USA
1961–64	Professor of Mathematics, Columbia University, USA
1960–61	Associate Professor of Mathematics, University of California, Berkeley, USA
1962	Visiting Professor, College de France, Paris, France (Spring)
1960–1962	Alfred P. Sloan Research Fellow
1958–1960	Member, Institute for Advanced Study, Princeton, USA
1956–1968	Instructor, University of Chicago, USA

Honors (Selected):

2007	Wolf Prize
2005	Jurgen Moser Prize of SIAM, Dynamics Group
2004	Honorary Doctorate of University of Genoa
1999	Doctor Honoris Causa of Rostov State University, Rostov, Russia
1998	Honorary Member of the London Mathematical Society
1997	Honorary Member of the Moscow Mathematical Society
1997	Honorary Doctor of Science, City University of Hong Kong
1997	Doctor Honoris Causa, Universite Pierre et Marie Curie, Paris
1996	National Medal of Science, Washington, D. C.
1996	Honorary Doctor of Science, University of Michigan
1994	Class of the Grand Cross of the Brasilian National Order of Scientific Merit

List of Publications

1. S. Smale, L. Rosasco, J. Bouvrie, A. Caponnetto and T. Poggio, Mathematics of the neural response, *Found. Comput. Math.* **10** (2010), No. 1, 67–91.
2. S. Smale and D.-X. Zhou, Geometry on probability spaces, *Constr. Approx.* **30** (2009), No. 3, 311–323.
3. S. Smale and D.-X. Zhou, Online learning with Markov sampling, *Anal. Appl.* **7** (2009), No. 1, 87–113.
4. P. Niyogi, S. Smale and S. Weinberger, Finding the homology of submanifolds with high confidence from random samples, *Disc. Comput. Geom.* **39** (2008), No. 1–3, 419–441.
5. A. Caponnetto and S. Smale, Risk bounds for random regression graphs, *Found. Comput. Math.* **7** (2007), No. 4, 495–528.
6. S. Smale and D.-X. Zhou, Learning theory estimates via integral operators and their approximations, *Constr. Approx.* **26** (2007), No. 2, 153–172.
7. F. Cucker and S. Smale, Emergent behavior in flocks, *IEEE Trans. Automat. Control* **52** (2007), No. 5, 852–862.
8. F. Cucker and S. Smale, On the mathematics of emergence, *Jpn. J. Math.* **2** (2007), No. 1, 197–227.
9. S. Smale and Y. Yao, Online learning algorithms, *Found. Comput. Math.* **6** (2006), No. 2, 145–170.
10. S. Smale and D.-X. Zhou, Shannon sampling, II. Connections to learning theory, *Appl. Comput. Harmon. Anal.* **19** (2005), No. 3, 285–302.
11. S. Smale, On problems of computational complexity, in *Surveys in Modern Mathematics*, pp. 255–259, London Math. Soc. Lecture Note Ser., Vol. 321 (Cambridge Univ. Press, 2005).
12. M. W. Hirsch, S. Smale and R. L. Devaney, *Differential Equations, Dynamical Systems, and An Introduction to Chaos*, Second edition, Pure and Applied Mathematics (Amsterdam), Vol. 60 (Elsevier/Academic Press, 2004).
13. F. Cucker, S. Smale and D.-X. Zhou, Modeling language evolution, *Found. Comput. Math.* **4** (2004), No. 3, 315–343.
14. S. Smale and D.-X. Zhou, Shannon sampling and function reconstruction from point values, *Bull. Amer. Math. Soc. (N.S.)* **41** (2004), No. 3, 279–305 (electronic).
15. T. Poggio and S. Smale, The mathematics of learning: Dealing with data. *Notices Amer. Math. Soc.* **50** (2003), No. 5, 537–544.

16. S. Smale and D.-X. Zhou, Estimating the approximation error in learning theory, *Anal. Appl.* **1** (2003), No. 1, 17–41.
17. F. Cucker and S. Smale, Best choices for regularization parameters in learning theory: On the bias-variance problem, *Found. Comput. Math.* **2** (2002), No. 4, 413–428.
18. F. Cucker and S. Smale, On the mathematical foundations of learning, *Bull. Amer. Math. Soc.* (*N.S.*) **39** (2002), No. 1, 1–49 (electronic).
19. S. Smale, Mathematical problems for the next century (in Spanish); translated from *Math. Intelligencer* **20** (1998), No. 2, 7–15 by M. J. Alcón. *Gac. R. Soc. Mat. Esp.* **3** (2000), No. 3, 413–434.
20. S. Smale, *The Collected Papers of Stephen Smale*, Vols. 1–3. Edited by F. Cucker and R. Wong (Singapore Univ. Press and World Scientific, 2000).
21. S. Smale, Problèmes mathématiques pour le prochain siècle (in French), [Mathematical problems for the next century] English original appeared in [*Math. Intelligencer* **20** (1998), No. 2, 7–15], *Gaz. Math.* No. 83 (2000), 11–27.
22. S. Smale, Mathematical problems for the next century, in *Mathematics: Frontiers and Perspectives*, pp. 271–294 (Amer. Math. Soc., 2000).
23. F. Cucker and S. Smale, Complexity estimates depending on condition and round-off error, *J. ACM* **46** (1999), No. 1, 113–184.
24. F. Cucker, P. Koiran and S. Smale, A polynomial time algorithm for Diophantine equations in one variable, *J. Symbolic Comput.* **27** (1999), No. 1, 21–29.
25. F. Cucker and S. Smale, Complexity estimates depending on condition and round-off error, in *Algorithms — ESA '98* (Venice), pp. 115–126, Lecture Notes in Comput. Sci., Vol. 1461 (Springer, 1998).
26. J.-P. Dedieu and S. Smale, Some lower bounds for the complexity of continuation methods, *J. Complexity* **14** (1998), No. 4, 454–465.
27. S. Smale, The work of Curtis T. McMullen, Proceedings of the International Congress of Mathematicians, Vol. I (Berlin, 1998). *Doc. Math.* 1998, Extra Vol. I, 127–132 (electronic).
28. S. Smale, The work of [Fields medalist] Curtis T. McMullen, *Mitt. Dtsch. Math.-Ver.* 1998, No. 3, 48–51.
29. S. Smale, Mathematical problems for the next century, *Math. Intelligencer* **20** (1998), No. 2, 7–15.
30. S. Smale, Finding a horseshoe on the beaches of Rio, *Math. Intelligencer* **20** (1998), No. 1, 39–44.
31. L. Blum, F. Cucker, M. Shub and S. Smale, in *Complexity and Real Computation*. With a foreword by Richard M. Karp (Springer-Verlag, 1998).

32. S. Smale, A survey of some recent developments in differential topology, in *Fields Medallists' Lectures*, pp. 142–155 (World Scientific Ser. 20th Century Math., Vol. 5 (World Scientific, 1997).
33. S. Smale, Complexity theory and numerical analysis, in *Acta Numerica, 1997*, pp. 523–551, Acta Numer., Vol. 6 (Cambridge Univ. Press, 1997).
34. L. Blum, F. Cucker, M. Shub and S. Smale, Algebraic settings for the problem "P ≠ NP?", in *The Mathematics of Numerical Analysis* (Park City, UT, 1995), pp. 125–144, Lectures in Appl. Math., Vol. 32 (Amer. Math. Soc., 1996).
35. L. Blum, F. Cucker, M. Shub and S. Smale, Complexity and real computation: A manifesto, *Internat. J. Bifur. Chaos Appl. Sci. Engrg.* **6** (1996), No. 1, 3–26.
36. M. Shub and S. Smale, Complexity of Bezout's theorem, IV. Probability of success, extensions, *SIAM J. Numer. Anal.* **33** (1996), No. 1, 128–148.
37. M. Shub and S. Smale, On the intractability of Hilbert's Nullstellensatz and an algebraic version of "NP ≠ P?", A celebration of John F. Nash, Jr., *Duke Math. J.* **81** (1995), No. 1, 47–54.
38. M. Shub and S. Smale, Complexity of Bezout's theorem, V. Polynomial time, selected papers of the *Workshop on Continuous Algorithms and Complexity* (Barcelona, 1993), *Theor. Comput. Sci.* **133** (1994), No. 1, 141–164.
39. F. Cucker, M. Shub and S. Smale, Separation of complexity classes in Koiran's weak model, selected papers of the *Workshop on Continuous Algorithms and Complexity* (Barcelona, 1993), *Theor. Comput. Sci.* **133** (1994), No. 1, 3–14.
40. S. Smale, Some early recollection of Raoul Bott, in *Raoul Bott: Collected Papers*, Vol. 1, p. 39 (Birkhäuser, 1994).
41. L. Blum and S. Smale, The Gödel incompleteness theorem and decidability over a ring, in *From Topology to Computation: Proceedings of the Smalefest* (Berkeley, CA, 1990), pp. 321–339 (Springer, 1993).
42. S. Smale, Professional biography, bibliography, and graduate students, in *From Topology to Computation: Proceedings of the Smalefest* (Berkeley, CA, 1990), pp. 53–63 (Springer, 1993).
43. S. Smale, On the steps of Moscow University, in *From Topology to Computation: Proceedings of the Smalefest* (Berkeley, CA, 1990), pp. 41–52 (Springer, 1993).
44. S. Smale, The story of the higher dimensional Poincaré conjecture (what actually happened on the beaches of Rio), in *From Topology to Computation: Proceedings of the Smalefest* (Berkeley, CA, 1990), pp. 27–40 (Springer, 1993).

45. S. Smale, On how I got started in dynamical systems, 1959–1962. *From Topology to Computation: Proceedings of the Smalefest* (Berkeley, CA, 1990), pp. 22–26 (Springer, 1993).
46. S. Smale, Some autobiographical notes, in *From Topology to Computation: Proceedings of the Smalefest* (Berkeley, CA, 1990), pp, 3–21 (Springer, 1993).
47. M. Shub and S. Smale, Complexity of Bezout's theorem, II. Volumes and probabilities, in *Computational Algebraic Geometry* (Nice, 1992), pp. 267–285, Progr. Math., Vol. 109 (Birkhäuser, 1993).
48. M. Shub and S. Smale, Complexity of Bezout's theorem, III. Condition number and packing. Festschrift for Joseph F. Traub, Part I, *J. Complexity* **9** (1993), No. 1, 4–14.
49. M. Shub and S. Smale, Complexity of Bézout's theorem, I. Geometric aspects, *J. Amer. Math. Soc.* **6** (1993), No. 2, 459–501.
50. A. Connes, G. Faltings, V. Jones, S. Smale, R. Thom and J. Wagensberg, Round-table discussion, in *Mathematical Research Today and Tomorrow* (Barcelona, 1991), pp. 87–108, Lecture Notes in Math., Vol. 1525 (Springer, 1992).
51. S. Smale, Theory of computation, in *Mathematical Research Today and Tomorrow* (Barcelona, 1991), pp. 59–69, Lecture Notes in Math., Vol. 1525 (Springer, 1992).
52. S. Smale, The story of the higher dimensional Poincaré conjecture (what actually happened on the beaches of Rio) (in Czech), Translated from the English by Oldřich Kowalski, *Pokroky Mat. Fyz. Astronom.* **36** (1991), No. 1, 38–49.
53. S. Smale, Dynamics retrospective: Great problems, attempts that failed, Nonlinear science: The next decade (Los Alamos, NM, 1990), *Phys. D* **51** (1991), No. 1–3, 267–273.
54. S. Smale, Some remarks on the foundations of numerical analysis, in *Chaos* (Woods Hole, MA, 1989), pp. 107–120 (Amer. Inst. Phys., 1990).
55. L. Blum, M. Shub and S. Smale, On a theory of computation over the real numbers: NP completeness, recursive functions and universal machines [Bull. Amer. Math. Soc. (N.S.) **21** (1989), No. 1, 1–46], in *Workshop on Dynamical Systems* (Trieste, 1988), pp. 23–52, Pitman Res. Notes Math. Ser., Vol. 221 (Longman, 1990).
56. S. Smale, Some remarks on the foundations of numerical analysis, *SIAM Rev.* **32** (1990), No. 2, 211–220.
57. S. Smale, The story of the higher dimensional Poincaré conjecture (what actually happened on the beaches of Rio), *Math. Intelligencer* **12** (1990), No. 2, 44–51.

58. S. Smale, The story of the higher dimensional Poincaré conjecture (what actually happened on the beaches of Rio de Janeiro), A joint AMS-MAA invited address presented in Phoenix, Arizona, January 1989. AMS-MAA Joint Lecture Series (Amer. Math. Soc., 1989).
59. S. Smale, Newton's contribution and the computer revolution, *Math. Medley* **17** (1989), No. 2, 51–57.
60. L. Blum, M. Shub and S. Smale, On a theory of computation and complexity over the real numbers: NP-completeness, recursive functions and universal machines, *Bull. Amer. Math. Soc. (N.S.)* **21** (1989), No. 1, 1–46.
61. S. Smale, Complexity aspects of numerical analysis. A plenary address presented at the International Congress of Mathematicians held in Berkeley, California, August 1986. Introduced by J. F. Traub. ICM Series (Amer. Math. Soc., 1988).
62. S. Smale, Algorithms for solving equations, *Proceedings of the International Congress of Mathematicians*, Vol. 1, 2 (Berkeley, Calif., 1986), 172–195 (Amer. Math. Soc., 1987).
63. S. Smale, On the topology of algorithms, I, *J. Complexity* **3** (1987), No. 2, 81–89.
64. M. Shub and S. Smale, On the existence of generally convergent algorithms, *J. Complexity* **2** (1986), No. 1, 2–11.
65. S. Smale, Newton's method estimates from data at one point, in *The Merging of Disciplines: New Directions in Pure, Applied, and Computational Mathematics* (Laramie, Wyo., 1985), pp. 185–196 (Springer, 1986).
66. M. Shub and S. Smale, Computational complexity: On the geometry of polynomials and a theory of cost, II, *SIAM J. Comput.* **15** (1986), No. 1, 145–161.
67. M. Shub and S. Smale, Computational complexity. On the geometry of polynomials and a theory of cost, I. *Ann. Sci. École Norm. Sup.* (4) **18** (1985), No. 1, 107–142.
68. S. Smale, On the efficiency of algorithms of analysis, *Bull. Amer. Math. Soc. (N.S.)* **13** (1985), No. 2, 87–121.
69. M. Shub and S. Smale, On the average cost of solving polynomial equations, in *Geometric Dynamics* (Rio de Janeiro, 1981), pp. 719–724, Lecture Notes in Math., Vol. 1007 (Springer, 1983).
70. M. Shub and S. Smale, On a theory of cost for equation solving, in *Homotopy Methods and Global Convergence* (Porto Cervo, 1981), pp. 263–265, NATO Conf. Ser. II: Systems Sci., Vol. 13 (Plenum, 1983).
71. S. Smale, On the average number of steps of the simplex method of linear programming, *Math. Programming* **27** (1983), No. 3, 241–262.

72. S. Smale, The problem of the average speed of the simplex method, in *Mathematical Programming: The State of the Art* (Bonn, 1982), pp. 530–539 (Springer, 1983).
73. S. Smale, The fundamental theorem of algebra and complexity theory, *Bull. Amer. Math. Soc.* (*N.S.*) **4** (1981), No. 1, 1–36.
74. S. Smale, Smooth solutions of the heat and wave equations, *Comment. Math. Helv.* **55** (1980), No. 1, 1–12.
75. S. Smale, The mathematics of time, in *Essays on Dynamical Systems, Economic Processes, and Related Topics* (Springer-Verlag, 1980).
76. S. Smale, The prisoner's dilemma and dynamical systems associated to noncooperative games, *Econometrica* **48** (1980), No. 7, 1617–1634.
77. S. Smale, On comparative statics and bifurcation in economic equilibrium theory, in *Bifurcation Theory and Applications in Scientific Disciplines* (Papers, Conf., New York, 1977), pp. 545–548, Ann. New York Acad. Sci., Vol. 316 (New York Acad. Sci., 1979).
78. M. W. Hirsch and S. Smale, On algorithms for solving $f(x)=0$, *Comm. Pure Appl. Math.* **32** (1979), No. 3, 281–313.
79. S. Smale, Marston Morse (1892–1977), *Math. Intelligencer* **1** (1978/79), No. 1, 33–34.
80. S. Smale, Dynamical systems and turbulence, in *Turbulence Seminar* (Univ. Calif., Berkeley, Calif., 1976/1977), pp. 48–70, Lecture Notes in Math., Vol. 615 (Springer, 1977).
81. S. Smale, Convergent process of price adjustment and global Newton methods, in *Frontiers of Quantitative Economics*, Vol. IIIA (*Invited papers, Econometric Soc., Third World Congress, Toronto, Ont., 1975*), pp. 191–205. Contributions to Economic Analysis, Vol. 105 (North-Holland, 1977).
82. S. Smale, Exchange processes with price adjustment. *J. Math. Econom.* **3** (1976), No. 3, 211–226.
83. S. Smale, Global analysis and economics, VI. Geometric analysis of Pareto optima and price equilibria under classical hypotheses, *J. Math. Econom.* **3** (1976), No. 1, 1–14.
84. S. Smale and R. F. Williams, The qualitative analysis of a difference equation of population growth, *J. Math. Biol.* **3** (1976), No. 1, 1–4.
85. S. Smale, A convergent process of price adjustment and global Newton methods, *J. Math. Econom.* **3** (1976), No. 2, 107–120.
86. S. Smale, On the differential equations of species in competition, *J. Math. Biol.* **3** (1976), No. 1, 5–7.
87. S. Smale, Some dynamical questions in mathematical economics, *Colloques International du Centre National de la Recherche Scientifique*, No. 259 (Centre National de la Recherche Scientifique, 1976), pp. 95–97.

88. S. Smale, Sufficient conditions for an optimum, in *Dynamical Systems — Warwick 1974* (Proc. Sympos. Appl. Topology and Dynamical Systems, Univ. Warwick, Coventry, 1973/1974; presented to E. C. Zeeman on his fiftieth birthday), pp. 287–292, Lecture Notes in Math., Vol. 468 (Springer, 1975).
89. S. Smale, Global analysis and economics. Pareto optimum and a generalization of Morse theory, Mathematical methods of the social sciences. *Synthese* **31** (1975), No. 2, 345–358.
90. S. Smale, Optimizing several functions, in *Manifolds–Tokyo 1973* (Proc. Internat. Conf., Tokyo, 1973), pp. 69–75 (Univ. Tokyo Press, 1975).
91. M. W. Hirsch and S. Smale, Differential equations, dynamical systems, and linear algebra, Pure and Applied Mathematics, Vol. 60 (Academic Press, 1974).
92. S. Smale, A mathematical model of two cells via Turing's equation, in *Some Mathematical Questions in Biology*, V (Proc. Seventh Sympos., Mathematical Biology, Mexico City, 1973), pp. 15–26. Lectures on Math. in the Life Sciences, Vol. 6 (Amer. Math. Soc., 1974).
93. S. Smale, Global analysis and economics, V. Pareto theory with constraints, *J. Math. Econom.* **1** (1974), No. 3, 213–221.
94. S. Smale, Global analysis and economics, IV. Finiteness and stability of equilibria with general consumption sets and production, *J. Math. Econom.* **1** (1974), No. 2, 119–127.
95. S. Smale, Glodal analysis and economics, III. Pareto optima and price equilibria, *J. Math. Econom.* **1** (1974), No. 2, 107–117.
96. S. Smale, Global analysis and economics, IIA. Extension of a theorem of Debreu, *J. Math. Econom.* **1** (1974), No. 1, 1–14.
97. S. Smale, Global analysis and economics, I. Pareto optimum and a generalization of Morse theory, *Dynamical Systems* (Proc. Sympos., Univ. Bahia, Salvador, 1971), pp. 531–544 (Academic Press, 1973).
98. S. Smale, Stability and isotopy in discrete dynamical systems, in *Dynamical Systems* (Proc. Sympos., Univ. Bahia, Salvador, 1971), pp. 527–530 (Academic Press, 1973).
99. S. Smeĭl, Global analysis and economics, I. Pareto optimum and a generalization of Morse theory (in Russian), Translated from [*Dynamical Systems* (Proc. Sympos., Univ. Bahia, Salvador, 1971), pp. 531–544, Academic Press, 1973] by N. N. Brušlinskaja, *Usp. Mat. Nauk* **27** (1972), No. 3 (165), 177–187.
100. S. Smeĭl, Topology and mechanics (in Russian), Translated from the English [Invent. Math. **10** (1970), No. 4, 305–331, *ibid.* **11** (1970), No. 1, 45–64] by S. B. Katok, *Usp. Mat. Nauk* **27** (1972), No. 2 (164), 77–133.

101. S. Smale, On the mathematical foundations of electrical circuit theory, *J. Differential Geom.* **7** (1972), 193–210.
102. S. Shub and S. Smale, Beyond hyperbolicity, *Ann. of Math.* (2) **96** (1972), 587–591.
103. S. Smale, Problems on the nature of relative equilibria in celestial mechanics. *1971 Manifolds–Amsterdam 1970* (*Proc. Nuffic Summer School*) pp. 194–198, Lecture Notes in Mathematics, Vol. 197 (Springer, 1970).
104. S. Smale, Topology and mechanics, II. The planar *n*-body problem, *Invent. Math.* **11** (1970), 45–64.
105. S. Smale, Topology and mechanics, I, *Invent. Math.* **10** (1970), 305–331.
106. R. Abraham and S. Smale, Nongenericity of Ω-stability, in 1970 *Global Analysis* (Proc. Sympos. Pure Math., Vol. XIV, Berkeley, Calif., 1968), pp. 5–8 (Amer. Math. Soc., 1970).
107. S. Smale, The Ω-stability theorem, in 1970 *Global Analysis* (Proc. Sympos. Pure Math., Vol. XIV, Berkeley, Calif., 1968), pp. 289–297 (Amer. Math. Soc., 1970).
108. J. Palis and S. Smale, Structural stability theorems, in 1970 *Global Analysis* (Proc. Sympos. Pure Math., Vol. XIV, Berkeley, Calif., 1968), pp. 223–231 (Amer. Math. Soc., 1970).
109. S. Smale, Notes on differentiable dynamical systems, in 1970 *Global Analysis* (Proc. Sympos. Pure Math., Vol. XIV, Berkeley, Calif., 1968), pp. 277–287 (Amer. Math. Soc, 1970).
110. S. Smeĭl, Differentiable dynamical systems (in Russian), *Usp. Mat. Nauk* **25** (1970), No. 1(151), 113–185.
111. S. Smale, Global stability questions in dynamical systems, in 1969 *Lectures in Modern Analysis and Applications, I*, pp. 150–158 (Springer, 1969).
112. S. Smale, What is global analysis?, *Amer. Math. Monthly* **76** (1969), 4–9.
113. S. Smale, Differentiable Dynamical Systems, 1968 *Proc. Internat. Congr. Math.* (Moscow, 1966), p. 139 (Mir, 1968).
114. S. Smale, Differentiable dynamical systems, *Bull. Amer. Math. Soc.* **73** (1967), 747–817.
115. S. Smale, Corrigendum: "On the Morse index theorem", *J. Math. Mech.* **16** (1967), 1069–1070.
116. S. Smale, Structurally stable systems are not dense, *Amer. J. Math.* **88** (1966), 491–496.
117. S. Smale, An infinite dimensional version of Sard's theorem, *Amer. J. Math.* **87** (1965), 861–866.
118. S. Smale, On the Morse index theorem, *J. Math. Mech.* **14** (1965), 1049–1055.

119. S. Smale, Diffeomorphisms with many periodic points, in 1965 *Differential and Combinatorial Topology* (A Symposium in Honor of Marston Morse), pp. 63–80 (Princeton Univ. Press, 1965).
120. S. Smale, On the calculus of variations, in 1964 *Differential Analysis, Bombay Colloq., 1964*, pp. 187–189 (Oxford Univ. Press, 1964).
121. S. Smale, Morse theory and a non-linear generalization of the Dirichlet problem, *Ann. of Math.* (2) **80** (1964), 382–396.
122. S. Smeĭl, A survey of some recent results in differential topology (in Russian), *Usp. Mat. Nauk* **19** (1964), No. 1 (115), 125–138.
123. R. S. Palais and S. Smale, A generalized Morse theory, *Bull. Amer. Math. Soc.* **70** (1964), 165–172.
124. S. Smale, Dynamical systems and the topological conjugacy problem for diffeomorphisms, in 1963 *Proc. Internat. Congr. Mathematicians* (Stockholm, 1962), pp. 490–496 (Inst. Mittag-Leffler, 1963).
125. S. Smale, Stable manifolds for differential equations and diffeomorphisms, in 1963 *Topologia Differenziale* (Centro Internaz. Mat. Estivo, 1deg Ciclo, Urbino, 1962), Lezione 4 (Edizioni Cremonese, 1963).
126. S. Smale, Stable manifolds for differential equations and diffeomorphisms, *Ann. Scuola Norm. Sup. Pisa* (3) **17** (1963), 97–116.
127. S. Smale, A structurally stable differentiable homeomorphism with an infinite number of periodic points, in 1963 *Qualitative Methods in the Theory of Non-linear Oscillations* (Proc. Internat. Sympos. Non-linear Vibrations, Vol. II, 1961), pp. 365–366 (Akad. Nauk Ukrain., 1963).
128. S. Smale, A survey of some recent developments in differential topology, *Bull. Amer. Math. Soc.* **69** (1963), 131–145.
129. S. Smale, On the structure of manifolds, *Amer. J. Math.* **84** (1962), 387–399.
130. S. Smale, On the structure of 5-manifolds, *Ann. of Math.* (2) **75** (1962), 38–46.
131. S. Smale, Generalized Poincaré's conjecture in dimensions greater than four, *Ann. of Math.* (2) **74** (1961), 391–406.
132. S. Smale, On gradient dynamical systems, *Ann. of Math.* (2) **74** (1961), 199–206.
133. S. Smale, Differentiable and combinatorial structures on manifolds, *Ann. of Math.* (2) **74** (1961), 498–502.
134. S. Smale, On dynamical systems, *Bol. Soc. Mat. Mexicana* (2) **5** (1960), 195–198.
135. S. Smale, The generalized Poincaré conjecture in higher dimensions, *Bull. Amer. Math. Soc.* **66** (1960), 373–375.

136. S. Smale, Morse inequalities for a dynamical system, *Bull. Amer. Math. Soc.* **66** (1960), 43–49.
137. S. Smale, Diffeomorphisms of the 2-sphere, *Proc. Amer. Math. Soc.* **10** (1959), 621–626.
138. M. W. Hirsch and S. Smale, On involutions of the 3-sphere, *Amer. J. Math.* **81** (1959), 893–900.
139. S. Smale, The classification of immersions of spheres in Euclidean spaces, *Ann. of Math.* (2) **69** (1959), 327–344.
140. R. K. Lashof and S. Smale, Self-intersections of immersed manifolds, *J. Math. Mech.* **8** (1959), 143–157.
141. S. Smale, A classification of immersions of the two-sphere, *Trans. Amer. Math. Soc.* **90** (1958), 281–290.
142. R. Lashof and S. Smale, On the immersion of manifolds in euclidean space, *Ann. of Math.* (2) **68** (1958), 562–583.
143. S. Smale, Regular curves on Riemannian manifolds, *Trans. Amer. Math. Soc.* **87** (1958), 492–512.
144. S. Smale, A Vietoris mapping theorem for homotopy, *Proc. Amer. Math. Soc.* **8** (1957), 604–610.
145. S. Smale, A note on open maps, *Proc. Amer. Math. Soc.* **8** (1957), 391–393.

Mathematics: Frontiers and Perspectives 2000

Mathematical Problems for the Next Century

Steve Smale

Introduction

V. I. Arnold, on behalf of the International Mathematical Union, has written to a number of mathematicians with a suggestion that they describe some great problems for the next century. This report is my response.

Arnold's invitation is inspired in part by Hilbert's list of 1900 (see e.g., (Browder, 1976)) and I have used that list to help design this essay.

I have listed 18 problems, chosen with these criteria:

1. Simple statement. Also preferably mathematically precise, and best even with a yes or no answer.
2. Personal acquaintance with the problem. I have not found it easy.
3. A belief that the question, its solution, partial results or even attempts at its solution are likely to have great importance for mathematics and its development in the next century.

Some of these problems are well known. In fact, included are what I believe to be the three greatest open problems of mathematics: the Riemann Hypothesis, Poincaré Conjecture, and "Does P=NP?" Besides the Riemann Hypothesis, one below is on Hilbert's list (Hilbert's 16th Problem). There is a certain overlap with my earlier paper "Dynamics retrospective, great problems, attempts that failed" (Smale, 1991).

Let us begin.

Problem 1: The Riemann Hypothesis

Are those zeros of the Riemann zeta function, defined by analytic continuation from

$$\zeta(s) = \sum_{n=1}^{\infty} \frac{1}{n^s}, \qquad Re(s) > 1$$

Lecture given on the occasion of Arnold's 60th birthday at the Fields Institute, Toronto, June 1997. Original version appeared in the *Mathematical Intelligencer*, Vol. 20, (Spring 1998), pp. 7-15.

which are in the critical strip $0 \leq Re(s) \leq 1$, *all on the line* $Re(s) = \frac{1}{2}$?

This was problem #8 on Hilbert's list. There are many fine books on the zeta function and the Riemann hypothesis which are easy to locate. I leave the matter at this.

Problem 2: The Poincaré Conjecture

Suppose that a compact connected 3-dimensional manifold has the property that every circle in it can be deformed to a point. Then must it be homeomorphic to the 3-sphere?

The *n-sphere* is the space

$$\{x \in \mathbb{R}^{n+1} \mid \|x\| = 1\}\,, \qquad \|x\|^2 = \sum_{i=1}^{n+1} x_i^2\,.$$

A compact n-dimensional manifold can be thought of as a closed bounded n-dimensional surface (differentiable and nonsingular) in some Euclidean space.

The n-dimensional Poincaré conjecture asserts that a compact n-dimensional manifold M having the property that every map $f : S^k \to M$, $k < n$ (or equivalently, $k \leq n/2$) can be deformed to a point, must be homeomorphic to S^n.

Henri Poincaré studied these problems in his pioneering papers in topology. Poincaré in 1900 (see Poincaré, 1953, pp 338–370) announced a proof of the general n-dimensional case. Subsequently (in 1904) he found a counterexample to his first version of the statement (Poincaré 1953, pp 435–498). In the second paper he limits himself to $n = 3$ and states the 3-dimensional case as the problem above (not actually as a "conjecture").

My own relationship with this problem is described in the story (Smale, 1990a). There I wrote

> I first heard of the Poincaré conjecture in 1955 in Ann Arbor at the time I was writing a thesis on a problem of topology. Just a short time later, I felt that I had found a proof (3 dimensions). Hans Samelson was in his office, and very excitedly I sketched my ideas to him. ... After leaving the office, I realized that my "proof" hadn't used any hypothesis on the 3-manifold.

In 1960, "on the beaches of Rio", I gave an affirmative answer to the n-dimensional Poincaré conjecture for $n > 4$. In 1982, Mike Freedman gave an affirmative answer for $n = 4$. (Note: for $n > 4$, I proved the stronger result that M was the smooth union of two balls, $M = D^n \cup D^n$; that result is unproved for $n = 4$, today.)

For background on these matters, besides the above references, see (Smale, 1963).

Many other mathematicians after Poincaré have claimed proofs of the 3-dimensional case. See (Taubes, 1987) for an account of some of these attempts.

A reason that Poincaré's conjecture is fundamental in the history of mathematics is that it helped give focus to a manifold as an object of study in its own right. In this way, Poincaré influenced much of 20th century mathematics with its attention to geometric objects including eventually algebraic varieties, Riemannian manifolds, etc.

I hold the conviction that there is a comparable phenomenon today in the notion of a "polynomial time algorithm". Algorithms are becoming worthy of analysis in their own right, not merely as a means to solve other problems. Thus I am suggesting that, as the study of the set of solutions of an equation (e.g., a manifold) played such an important role in 20th century mathematics, the study of finding the solutions (e.g. an algorithm) may play an equally important role in the next century.

Problem 3: Does P=NP?

I sometimes consider this problem as a gift to mathematics from computer science. It may be useful to put it into a form which looks more like traditional mathematics.

Towards this end, first consider the Hilbert Nullstellensatz over the complex numbers. Thus let $f_1, \ldots, f_k$ be complex polynomials in n variables; we are asked to decide if they have a common zero $\zeta \in \mathbb{C}^n$. The Nullstellensatz asserts that this is not the case if and only if there are complex polynomials $g_1, \ldots, g_k$ in n variables satisfying

$$\sum_1^k g_i f_i = 1 \qquad (1)$$

as an identity of polynomials.

The effective Nullstellensatz as established by Brownawell (1987) and others, states that in (1), the degrees of the g_i may be assumed to satisfy

$$\deg g_i \leq \max(3, D)^n \,, \qquad D = \max \ \deg \ f_i.$$

With this degree bound the decidability problem becomes one of linear algebra. Given the coefficients of the f_i one can check if (1) has a solution whose unknowns are the coefficients of the g_i. Thus one has an algorithm to decide the Nullstellensatz. The number of arithmetic steps required grows exponentially in the number of coefficients of the f_i (the input size).

Conjecture (over $\mathbb{C}$). *There is no polynomial time algorithm for deciding the Hilbert nullstellensatz over $\mathbb{C}$.*

A polynomial time algorithm is one in which the number of arithmetic steps is bounded by a polynomial in the number of coefficients of the f_i, or in other words, is polynomially bounded.

To make mathematical sense of this conjecture, one has need of a formal definition of algorithm. In this context, the traditional definition of Turing machine makes no sense. In (Blum-Shub-Smale, 1989) a satisfactory definition is proposed, and the associated theory is exposed in (Blum-Cucker-Shub-Smale (or BCSS), 1997).

Very briefly, a machine over $\mathbb{C}$ has as inputs a finite string (... x_{-1}, x_0, x_1, ...) of complex numbers and the same for states and outputs. Computations on states include arithmetic operations and shifts on the string. Finally, a branch operation on "$x_1 = 0$?" is provided.

The size of an input is the number of elements in the input string. The time of a computation is the number of machine operations used in the passage from input to output. Thus a polynomial time algorithm over $\mathbb{C}$ is well-defined.

Note that all that has been said about the machines and the conjecture use only the structure of $\mathbb{C}$ as a field and hence both the machines and conjecture make sense over any field. In particular if the field is $\mathbb{Z}_2$ of two elements, we have the Turing machines.

Consider the decision problem: Given (as input) k polynomials in n variables with coefficients in $\mathbb{Z}_2$. *Decide if there is a common zero* $\zeta \in (\mathbb{Z}_2)^n$?

> **Conjecture.** *There is no polynomial time algorithm over* $\mathbb{Z}_2$ *deciding this problem.*

This is a reformulation of the classic conjecture $P \neq NP$.

In the above we have bypassed the basic ideas and theorems related to NP-completeness. For the classic case of Cook and Karp, see (Garey-Johnson, 1979), and for the theory over an arbitrary field, see BCSS.

It is useful for some of the problems below (7,9,17,18) to have the definition of a machine over the real numbers $\mathbb{R}$. In fact, the only change in the definition over $\mathbb{C}$, is to take the branch operation to be "$x_1 \leq 0$?"

REMARK. In his foreward to BCSS, Dick Karp writes that he is inclined to think that the complexity question over $\mathbb{C}$ above and the classical P versus NP question are very different and need to be attacked independently. On the other hand just after our book went to press, I noticed that there was a strong connection between the old and the new theories. BPP denotes the set of problems that can be solved in polynomial time (classically) using randomization. It is in a certain practical sense, almost the same as the class P. By using a short argument, reduction of polynomial systems modulo a random prime, supported by a useful conversation with Manuel Blum, I saw that, not $NP \subset BPP$ (classic) implies P $\neq$ NP over $\mathbb{C}$, that is, the conjecture over $\mathbb{C}$ above. This result is also implicit in (Cucker et al, 1995).

Problem 4: Integer zeros of a polynomial of one variable

Let us start by defining a diophantine invariant τ motivated by complexity theory. A *program* for a polynomial $f \in \mathbb{Z}[t]$ of one variable with integer coefficients is the object $(1, t, u_1, \ldots, u_k)$ where $u_k = f$, and for all ℓ, $u_\ell = u_i \circ u_j$, $i, j < \ell$, and $\circ$ is $+$ or $-$ or $\times$. Here $u_0 = t$, $u_{-1} = 1$. Then $\tau(f)$ is the minimum of k over all such programs.

Is the number of distinct integer zeros of f, polynomially bounded by $\tau(f)$? In other words, is

$$Z(f) \leq a\tau(f)^c \qquad \text{all } f \in \mathbb{Z}[t]?$$

Here $Z(f)$ is the number of distinct integer zeros of f with a, c universal constants.

From earlier results of Strassen, communicated via Schönhage, Shub, and Bürgisser, it follows that the exponent c has to be at least 2.

Mike Shub and I discovered this problem in our complexity studies. We proved that an affirmative answer implied the intractibility of the Nullstellensatz as a decision problem over $\mathbb{C}$ and thus $P \neq NP$ over $\mathbb{C}$. See (Shub-Smale, 1995) and also BCSS.

Since the degree of f is less than or equal to 2^τ, $\tau = \tau(f)$, there are no more than 2^τ zeros altogether.

For Chebyshev polynomials, the number of distinct real zeros grows exponentially with τ.

Many of the classic diophantine problems are in two or more variables. This problem asks for an estimate in just one variable, and nevertheless seems not so easy.

Here is a related problem. A *program* for an integer m is the object $(1, m_1, \ldots, m_\ell)$ where $m_\ell = m$, $m_0 = 1$, $m_q = m_i \circ m_j$, $i, j < q$ and $\circ = +, -$ or $\times$. Then let $\tau(m)$ be the minimum of ℓ, over all such programs. Thus $\tau(m)$ represents the shortest way to build up an integer m starting from 1 using plus, minus, and times.

PROBLEM. Is there a constant c such that $\tau(k!) \leq (\log k)^c$ for all integers k? One might expect this to be false, so that $k!$ is "hard to compute" (see Shub-Smale, 1995).

Problem 5: Height bounds for diophantine curves

Can one decide if a diophantine equation $f(x, y) = 0$ (input $f \in \mathbb{Z}[u, v]$) has an integer solution, (x, y), in time 2^{s^c} where c is a universal constant? That is, can the problem be decided in exponential time?

Here $s = s(f)$ is the size of f defined by

$$s(f) = \sum_{|\alpha| \leq d} max(\log |a_\alpha|, 1), \qquad f(x, y) = \sum_{|\alpha| \leq d} a_\alpha x^{\alpha_1} y^{\alpha_2}, \qquad \alpha = (\alpha_1, \alpha_2)$$

and $|\alpha| = \alpha_1 + \alpha_2$, $\alpha_i \geq 0$.

The Turing model of computation is supposed, so "time" is the number of Turing operations.

This problem is essentially posed in (Cucker-Koiran-Smale, 1997), but it is a version of a well-known problem in number theory. The size $s(f)$ is a version of the "height" of f. Our problem is likely to be very difficult since it is not even known if one can decide this diophantine problem at all, let alone in exponential time. The solution of Hilbert's tenth problem by Matiyasevich (see Matiyasevich, 1993), using work of Davis, Robinson, Putnam shows the undecidability if the number of variables (27 is sufficient if not 20 or even 11) is not restricted. A remaining important unsolved problem in this connection is:

> Can one decide if there is a rational number solution to a given diophantine equation (any number of variables)?

The computer science notion of NP is relevant to our main problem above. A problem in NP is seen to be solvable in exponential time. The simple standard argument is given by noting that the test, evaluation of a polynomial, is done in polynomial time. Thus one might well ask the stronger question; is the two variable diophantine problem in NP?

To simplify the discussion we will now assume that the genus of the curve (defined by) f is positive. The genus of a nonsingular curve is the number of "handles" in the homogenized curve of complex zeros. For a singular curve one can define the genus by taking an appropriate associated nonsingular curve.

Consider the following hypothesis.

> Height bound hypothesis: If the curve f, of positive genus, has any integer solution, then it has a solution (a, b) satisfying the estimate; $\log \max(|a|, |b|)$ is polynomially bounded by $s(f)$.

For curves of positive genus, the height bound hypothesis implies that such a class of diophantine equations is in NP, and hence answers our main problem affirmatively.

One may ask how the height bound hypothesis relates to older conjectures in number theory. Thus let the *strong height bound hypothesis* be the strengthening of the height bound hypothesis to include all integer solutions. The Lang-Stark conjecture (see Lang, 1991), on certain curves of genus one, is implied by the strong height bound hypothesis, if one replaces our polynomially bounded by linearly bounded.

We don't claim real evidence for the height bound hypothesis. However there is in the background everywhere in this section, the Siegel theorem that there are only finitely many integer points on any diophantine curve of positive genus. A great challenge is to make Siegel's theorem effective; the height bound hypothesis is a version of such a goal.

For the case of genus one, and genus zero in case of a finite number of zeros, one does have the effectiveness results of Baker, and Baker-Coates, (see Baker, 1979), but even here they are somewhat weaker than the estimate of the strong height bound hypothesis.

The height bound hypothesis is false without the condition on the genus. This follows from the fact that the smallest integer solution of the "non-Pellian" equation $x^2 - dy^2 = -1$ can be very large relative to a family of d (see Lagarias 1980). (Mazur (1994) points out a similar phenomenon for Pell's equation, conditional on the Gauss class number conjecture). Lagarias (1979) proves none the less that this equation (i.e., the feasibility) is in NP and moreover that the set of general binary quadratic equations is in NP. Manders-Adleman (1978) have shown some NP-completeness results on these problems.

There is also the more difficult version of all these problems when one uses the sparse representation of f, i.e., in the definition of $s(f)$ above delete the terms in which $a_\alpha = 0$. Even the one variable case while true is not immediate. See (Cucker-Koiran-Smale, 1997), and (Lenstra, 1997).

Problem 6: Finiteness of the number of relative equilibria in celestial mechanics

Is the number of relative equilibria finite, in the n-body problem of celestial mechanics, for any choice of positive real numbers $m_1, \ldots, m_n$ as the masses?

The problem is in Wintner's book (1941) on celestial mechanics. A relative equilibrium is a solution to Newton's equations which is induced by a plane rotation.

For the 3-body problem there are five relative equilibria: three found by Lagrange, two by Euler. There are "the Trojans" in the solar system, which correspond to the Lagrange relative equilibria. For 4-bodies the finiteness is unknown.

In (Smale, 1970), I interpreted the relative equilibria as critical points of a function induced by the potential of the planar n-body problem. More precisely the relative equilibria correspond to the critical points of

$$\hat{V} : (S - \Delta)/SO(2) \to \mathbb{R} \tag{2}$$

where $S = \{x \in (\mathbb{R}^2)^n \mid \sum m_i x_i = 0,\ \frac{1}{2}\sum m_i \|x_i\|^2 = 1\}$, $\Delta = \{x \in S \mid x_i = x_j \text{ some } i \neq j\}$. The rotation group $SO(2)$ acts on $S - \Delta$ and $\hat{V}$ is induced on the quotient from the potential function

$$V(x) = \sum_{i<j} \frac{m_i m_j}{\|x_i - x_j\|}\,.$$

Note that $V : S \to \mathbb{R}$ is invariant under the rotation group $SO(2)$ and that the quotient space $S/SO(2)$ is homeomorphic to complex projective space of dimension $n - 2$.

Thus the question has the equivalent form:

For any choice of $m_1, \ldots, m_n$, does $\hat{V}$ of (2) have a finite number of critical points?

Mike Shub (1970) has shown that the set of critical points is compact.

Say that $(m_1, \ldots, m_n)$ is critical if the corresponding $\hat{V}$ has a degenerate critical point. I asked in (Browder, 1976):

> What is the nature of the set of critical masses in the n-dimensional spaces of masses? Does it have measure zero? or finite Betti numbers?

Palmore (1976) has shown that it is empty in case $n = 3$ and not empty in case $n = 4$. Kuz'mina, Moeckel, Xia, Albouy, and McCord, (see e.g., McCord, 1996), have further results on these problems.

G. D. Birkhoff asked the question: what is the topology of the constant angular momentum submanifolds of the n body problem? In (Smale, 1970) I solved this problem for the case of n bodies in the plane. See also (Easton, 1971). The case for n bodies in 3-dimensional space remains open: one obstacle is the solution of Wintner's problem above. However, recently, McCord-Meyer-Wang (1998), solved Birkhoff's problem for the 3 body problem in 3-space. See this paper also for a good historical and mathematical background for many of these things.

Further background may be found in (Abraham-Marsden, 1978).

Problem 7: Distribution of points on the 2-sphere

Let $V_N(x) = \sum_{1 \le i < j \le N} \log \frac{1}{\|x_i - x_j\|}$ where $x = (x_1, \ldots, x_N)$, the x_i are distinct points on the 2-sphere $S^2 \subset \mathbb{R}^3$, and $\|x_i - x_j\|$ is the distance in $\mathbb{R}^3$. Denote $\min_x V_N(x)$ by V_N.

Can one find $(x_1, \ldots, x_N)$ *such that*

$$V_N(x) - V_N \le c \log N\,, \qquad c \text{ a universal constant.} \tag{3}$$

For a precise version one could ask for a real number algorithm in the sense of BCSS which on input N produces as output distinct $x_1, \ldots, x_N$ on the 2-sphere satisfying (3) with halting time polynomial in N.

This problem emerged from complexity theory, jointly with Mike Shub (see Shub-Smale, 1993). It is motivated by finding a good starting polynomial for a homotopy algorithm for realizing the Fundamental Theorem of Algebra.

An $(x_1, \ldots, x_N) = x$ such that $V_N(x) = V_N$ is called an N-tuple of elliptic Fekete points (see Tsuji, 1959).

The function V_N as a function of N satisfies

$$V_N = -\tfrac{1}{4} \log \left(\tfrac{4}{e}\right) N^2 - \tfrac{1}{4} N \log N + O(N)\,.$$

It is natural also to consider the functions

$$V_N(x, s) = \sum_{i<j} \frac{1}{\|x_i - x_j\|^s}, \qquad V_N(s) = \min_x V_N(x, s)\,,$$

x as before and $0 < s < 2$. The original $V_N(x)$, V_N correspond in a natural way to $s = 0$, and for $s = 1$, $V_N(x, 1)$ is the Coulomb potential, and $V_N(1)$ corresponds to an equilibrium position of N electrons constrained to lie on the two-sphere. There are similar problems for various s. One might equally well consider higher-dimensional spheres.

I had asked Ed Saff for some help in dealing with the main problem above. Subsequently, he and his colleagues produced a number of fine papers dealing with the subject and its ramifications. See (Kuijlaars-Saff, 1997) and (Saff-Kuijlaars, 1997) for background and further references. In Rakhmanov-Saff-Zhou (1994), one can find an algorithm where numerical evidence is provided to support (3) for $N \leq 12,000$, with $c = 114$.

Another way of looking at our main problem here is to optimize the function

$$W_N(x) = (\exp V_N(x))^{-1} = \prod_{i<j} \|x_i - x_j\| \ .$$

However, as was written in (Shub-Smale, 1993), "... this may not be so easy since there are saddle points of index N (on a great circle in S^2, evenly spaced N points, $x_1, \ldots, x_N$). Also the various symmetries that W_N possesses will confuse the picture."

Problem 8: Introduction of dynamics into economic theory

The following problem is not one of pure mathematics, but lies on the interface of economics and mathematics. It has been solved only in quite limited situations.

Extend the mathematical model of general equilibrium theory to include price adjustments.

There is a (static) theory of equilibrium prices in economics starting with Walras and firmly grounded in the work of Arrow and Debreu (see Debreu, 1959). For the simplest case of one market this amounts to the equation "supply equals demand" and a natural dynamics is easily found (Samuelson, 1971). For several markets, the situation is complex.

There is a function called the excess demand, $Z(p) = D(p) - S(p)$ from the space of prices to the space of commodities. Both the demand D and supply S are defined by aggregation over the individual agents. Economics justifies conditions on individual behavior which lead to axioms on Z. These axioms for the excess demand map $Z : \mathbb{R}^{\ell}_{+} \to \mathbb{R}^{\ell}$ are:

1. $Z(\lambda p) = Z(p)$, all $p = (p_1, \ldots, p_\ell)$, $p_i \geq 0$, $\lambda \in \mathbb{R}$, $\lambda > 0$.
2. $\sum_{i=1}^{\ell} p_i Z_i(p) = 0$, Walras' Law (the total value is zero).
3. $Z_i(p) > 0$ if $p_i = 0$ (positive demand for a free good).

By (1), (2), (3), Z may be interpreted as a vector field on the intersection of the $(\ell - 1)$-sphere with the positive orthant, pointing inward on the boundary. The existence of an equilibrium price vector p^* follows from Hopf's theorem, so that $Z(p^*) = 0$, and "supply equals demand".

Problem 8 asks for a dynamical model, whose states are price vectors (perhaps enlarged to include other economic variables). This theory should be compatible with the existing equilibrium theory. A most desirable feature is to have the time development of prices determined by the individual actions of economic agents.

I worked on this problem for several years, feeling that it was the main problem of economic theory (Smale, 1976). See also (Smale, 1981a) for background.

Problem 9: The linear programming problem

Is there a polynomial time algorithm over the real numbers which decides the feasibility of the linear system of inequalities $Ax \geq b$?

The algorithm requested by this problem is one given by a real number machine in the sense of BCSS (see also Problem 3). The system $Ax \geq b$ has as input an $m \times n$ real matrix A and a vector $b \in \mathbb{R}^m$ and the problem asks, is there some $x \in \mathbb{R}^n$ with $\sum_{j=1}^n a_{ij}x_j \geq b_i$ for all $i = 1, \ldots, m$? Time is measured by the number of arithmetic operations. This problem is in BCSS, page 275.

This is a decision version of the optimization problem of linear programming. Given A, b as above and $c \in \mathbb{R}^n$ decide if

$$\max_{x \in \mathbb{R}^n} c \cdot x \quad \text{subject to} \quad Ax \geq b$$

exists, and if so, output such an x.

The famous simplex method of Dantzig provides an algorithm for both problems (over $\mathbb{R}$) but Klee and Minty showed that it was exponentially slow in the worst case. On the other hand, Borgwardt, and I, each approach with subsequent important support from Haimovich, showed that it was polynomial time on the average. For all of these things, see (Schrijver, 1986).

In terms of the Turing model of computation, using rational numbers, $\mathbb{Q}$, and cost measured by "bits", there is a parallel development. Starting with ideas of Yudin-Nemirovsky, Khachian found a polynomial time algorithm (the ellipsoid method) for the linear programming problem. Subsequently Karmarkar with his "interior point method" found a practical algorithm for this problem, which he showed ran in polynomial time in the Turing model. For all these things one can see (Grötschel-Lovász-Schrijver, 1993) as well as (Schrijver, 1986).

Closer to the main problem above over $\mathbb{R}$ is a similar problem asking for a "strongly polynomial algorithm" for solving these linear programming problems. This is an algorithm over $\mathbb{Q}$ which is polynomial time in the sense of BCSS and moreover is polynomial time in the Turing sense. Partial results are due to Megiddo and especially Tardos (see (Grötschel-Lovász-Schrijver, 1993)).

For the problem over $\mathbb{R}$ there are also references (Barvinok-Vershik, 1993) and (Traub-Woźniakowski, 1982).

It is my belief that this problem number nine is the main unsolved problem of linear programming theory. The use of real number algorithms plays a natural role in a subject where the most important measure of cost is the number of arithmetic operations. See Problem 3; see also Chapter 1 of BCSS where the use of our real number model in the theory of computation is argued.

The ability of real number machines to deal with round-off error is investigated in (Cucker-Smale, 1997).

Problem 10: The Closing Lemma

Let p be a nonwandering point of a diffeomorphism $S : M \to M$ of a compact manifold. Can S be arbitrarily well approximated with derivatives of order r (C^r approximation) for each r, by $T : M \to M$ so that p is a periodic point of T?

A nonwandering point $p \in M$ is one with the property that for each neighborhood U of p there is a $k \in \mathbb{Z}$ such that $S^k U \cap U \neq \emptyset$. Here S^k is the kth iterate of S. Moreover p is a periodic point of period m if $T^m(p) = p$.

This is the discrete form of the famous "closing lemma" which in the C^1 case has been solved affirmatively by Charles Pugh (1967).

There is an easy C^0 approximation with the desired property. Peixoto observed that this argument failed for C^1 approximations correcting a mistake of René Thom (René told me that this was his biggest mistake).

Pugh-Robinson (1983) proved the closing lemma with C^1 approximations for the Hamiltonian version. Peixoto gave an affirmative answer with C^r approximations, any r, for the circle. Recently the closing lemma has been given additional importance by the work of Hayashi (1997); see also (Wen-Xia, 1997).

Problem 11: Is one-dimensional dynamics generally hyperbolic?

Can a complex polynomial T be approximated by one of the same degree with the property that every critical point tends to a periodic sink under iteration?

This is unsolved even for polynomials of degree 2. Here a polynomial map $T : \mathbb{C} \to \mathbb{C}$ ($\mathbb{C}$ the complex numbers) is considered a discrete dynamical system by iteration. So if $z \in \mathbb{C}$, its orbit in time, $z = z_0, z_1, z_2, \ldots$ is defined by $z_i = T(z_{i-1})$ and i may be interpreted as time (discrete). A fixed point w of T, ($T(w) = w$) is a *sink* if the derivative $T'(w)$ of T at w has absolute value less than 1. A periodic sink of T of period p is a sink for T^p. A critical point of T is just a point where the derivative of T is zero.

While the problem is now made precise it is useful to see it in the framework of hyperbolic dynamics from the 1960's.

A fixed point x of a diffeomorphism $T : M \to M$ is *hyperbolic* if the derivative $DT(x)$ of T at x (as a linear automorphism of the tangent space) has no eigenvalue of absolute value 1. If x is a periodic point of period p, then x is hyperbolic if it is a hyperbolic fixed point of T^p. The notion of hyperbolic extends naturally to Ω, the closure of the set of nonwandering points (see Problem 10).

A dynamical system $T \in \text{Diff}(M)$ is called *hyperbolic* (or satisfies Axiom A) if the periodic points are dense in Ω and Ω is hyperbolic (see Smale, 1967 or 1980). We assume also a no cycle condition. The work of many people, especially Ricardo Mañé, has identified hyperbolic dynamics with a strong notion of the stability of the dynamics called structural stability. There is even the beginning of a structure theory for this class of dynamics.

While hyperbolic systems constitute a large set of dynamics, an even larger set, including applied chaotic dynamics, lies beyond. The concept of hyperbolicity extends from the invertible dynamics to the case of our problem above, polynomial maps from $\mathbb{C}$ to $\mathbb{C}$. Classical complex variable theory permits recasting the problem to an equivalent one:

Can a polynomial map $T : \mathbb{C} \to \mathbb{C}$ be approximated by one which is hyperbolic?

The theory of complex one-dimensional dynamics was begun by Fatou and Julia towards the beginning of this century. In the 1960's I asked my thesis student John Guckenheimer to look at this literature and try to solve the above problem (among other things). His thesis (see Chern-Smale, 1970) contains the affirmative answer, but with a gap in the proof. Now the problem stands open as the fundamental problem of one-dimensional dynamics.

Complex 1-dimensional dynamics has become a fluorishing subject and includes important contributions of Douady-Hubbard, Sullivan, Yoccoz, McMullen, among many others. See (McMullen, 1994).

There is a parallel field of real 1-dimensional dynamics of a smooth map $T : I \to I$, $I = [0, 1]$.

PROBLEM. Can a smooth map $T : [0, 1] \to [0, 1]$ be C^r approximated by one which is hyperbolic, for all $r > 1$?

About the time of Guckenheimer's thesis, I asked Ziggy Nitecki to study this problem. My earlier negligence was compounded in not catching the mistake in Nitecki's thesis (see Chern-Smale, 1970), which purported to give an affirmative proof.

Subsequently, Jakobson (1971) answered the problem for C^1 approximations, but the general case remains open. See de Melo-van Strien (1993) for background.

More recently there is the work of Lyubich (1997), and of Graczyk-Swiatek (1997), which gives a positive solution for the real case when T is a quadratic map.

Let me remark on the mistakes in the published theses of my students, Guckenheimer and Nitecki mentioned above, Thom's mistake referred to in Problem 10 and Poincaré's mistake in the Poincaré conjecture of Problem 2. Mistakes happen frequently in published mathematics; I certainly have made my share. Especially in the early development of a subject, one is likely to err. Here oftentimes not only are the main concepts confused, but even the definitions are ambiguous. Thus there is the need to heed well the establishment of good foundations for a new subject. These considerations are a motivating factor in my efforts to help create a more solid foundation for numerical analysis (as in BCSS or Smale, 1990b).

Let me point out a story about Poincaré as told by Diacu-Holmes (1996) or Barrow-Green (1997). When Poincaré discovered a mistake in his theory of celestial mechanics, he had the copies of Acta Mathematica in which his article appeared, destroyed; but at the same time he discovered a phenomenon in dynamics which is now called "chaos". About 60 years later, oblivious of these events, I foolishly hypothesized (see Smale, 1998) that chaos did not exist in dynamics!

Problem 12: Centralizers of diffeomorphisms

Can a diffeomorphism of a compact manifold M onto itself be C^r approximated, all $r \geq 1$, by one $T : M \to M$ which commutes with only its iterates?

Thus the centralizer of T in the group of diffeomorphisms, Diff(M) should be $\{T^k \mid k \in \mathbb{Z}\}$.

I had started thinking about the centralizer in (Smale, 1963), but it was after Nancy Kopell's thesis with me (see Chern-Smale, 1970), answering the question affirmatively in case dim $M = 1$, that I proposed this problem (Smale, 1967). Today it remains unsolved even for the 2-sphere.

One may also ask if the set of diffeomorphisms of M with trivial centralizer is dense and *open* in Diff(M) with the C^r topology.

The main work on these problems has been done by Palis-Yoccoz (1989), with almost complete answers in the case of hyperbolic dynamics (see Problem 11) for any manifold.

I wrote in (Smale, 1991), "I find this problem interesting in that it gives some focus in the dark realm, beyond hyperbolicity, where even the problems are hard to pose clearly."

Problem 13: Hilbert's 16th Problem

Consider the differential equation in $\mathbb{R}^2$

$$\frac{dx}{dt} = P(x, y) , \qquad \frac{dy}{dt} = Q(x, y) \tag{4}$$

where P and Q are polynomials. Is there a bound K on the number of limit cycles of the form $K \leq d^q$ where d is the maximum of the degrees of P and Q, and q is a universal constant?

This is a modern version of the second half of Hilbert's sixteenth problem. Except for the Riemann hypothesis, it seems to be the most elusive of Hilbert's problems.

In fact, since a paper of Petrovskii and Landis (1957) purporting to give a positive solution, the progress seems to be backwards. Earlier, Dulac

(1923) claimed that the system (4) always has a finite number of limit cycles. After a gap in Petrovskii-Landis was found (see Petrovskii-Landis, 1959), Ilyashenko (1985) found an error in Dulac's paper. Moreover Shi Songling (1982) found a counter-example to the specific bounds of Petrovskii-Landis for the case $d = 2$. Subsequently, two long works have appeared, independently, giving proofs of Dulac's assertion (Écalle, 1992) and (Ilyashenko, 1991). These two papers have yet to be thoroughly digested by the mathematical community.

Thus one has the finiteness, but no bounds. We will consider a special class where the finiteness is simple, but the bounds remain unproved.

The following corresponds to Lienard's equation (see e.g., (Hirsch-Smale, 1974))

$$\frac{dx}{dt} = y - f(x)\,, \qquad \frac{dy}{dt} = -x \tag{5}$$

where $f(x)$ is a real polynomial with leading term x^{2k+1} and satisfying $f(0) = 0$.

If $f(x) = x^3 - x$ then (5) is van der Pol's equation with one limit cycle. More generally, it can be easily shown that all the solutions of (5) circle around the unique equilibrium at (0,0) in a clockwise direction. By following these curves, one defines a "Poincaré section", $T : \mathbb{R}^+ \to \mathbb{R}^+$ where $\mathbb{R}^+$ is the positive y axis. The limit cycles of (5) are precisely the fixed points of T. In various talks I raised the question of estimating the number of these fixed points (via some new kind of fixed point theorem?). In response, Linz, de Melo and Pugh (1977) found examples with k different limit cycles and conjectured this number k for the upper bound. Still no upper bound of the form $(\deg f)^q$ has been found. Since T is analytic, it follows that (5) has a finite number of limit cycles for each f.

Moreover, I spoke on what I called the "Pugh problem" having learned of it from Charles Pugh. Consider the 1-variable differential equation

$$\frac{dx}{dt} = x^d + h_1(t)x^{d-1} + h_2(t)x^{d-2} + \cdots\,, \qquad 0 \le t \le 1 \tag{6}$$

where the h_i are C^∞ functions. Pugh had asked for bounds on $K(d)$ where $K(d)$ is the number of solutions satisfying the boundary condition $x(0) = x(1)$. Subsequently Curt McMullen has given an answer, sketched as follows.

For $d = 0, 1, 2$, the corresponding map is a translation, affine, Möbius respectively. For $d > 2$ there can be arbitrarily many fixed points. To see this, notice that the closure of the space of mappings is a group. There are no closed subgroups lying between Möbius and the whole group of diffeomorphisms. More explicitly, if we start with polynomials of degree > 2 and take the Lie algebra they generate, it is dense among continuous vector fields on the real line (in the topology of uniform convergence on compact sets.)

For more background see (Browder, 1976), (Ilyashenko-Yakovenko, 1995), (Lloyd and Lynch, 1988), and (Smale, 1991).

Problem 14: Lorenz attractor

Is the dynamics of the ordinary differential equations of Lorenz (1963), that of the geometric Lorenz attractor of Williams, Guckenheimer and Yorke?

The Lorenz equations are:

$$\begin{aligned} \dot{x} &= -10x + 10y \\ \dot{y} &= 28x - y - xz \\ \dot{z} &= -\tfrac{8}{3}z + xy. \end{aligned}$$

Lorenz analysed by computer these equations to find that most solutions tended to a certain attracting set, and in so doing, he produced an important early example of "chaos". However mathematical proofs were lacking. This numerical work inspired the rigorous mathematical development of a geometrically defined ordinary differential equation which seems to have the same behavior (Yorke, Williams (1979), Guckenheimer-Williams (1979)). This geometric attractor has been analysed in detail and proved to be chaotic. Problem 14 asks if the dynamics of the original equations is the same as that of the geometric model. The most complete positive answer would be to describe a homeomorphism of $\mathbb{R}^3$ to $\mathbb{R}^3$ which would take solutions of the Lorenz equations to solutions of the geometric attractor. Actually the geometric Lorenz attractor is a two parameter family of dynamical systems and we are speaking of a member of this family.

An answer to this problem would be a step in establishing foundations for the field of applied chaotic dynamical systems. Up to the present time, in the equations of engineering and physics, chaos has only been established in a weaker sense, that of proving the existence of horseshoes (e.g., Melnikov, Marsden and Holmes; see Guckenheimer and Holmes (1990)).

One problem is a paradoxical situation which occurs from the accumulation of round-off error. While using the machine to study solutions of chaotic differential equations, the round-off error increases exponentially in time, by a fundamental property of chaos! The shadowing lemma of Anosov-Bowen has been extended to help deal with that paradox by Hammel-Yorke-Grebogi, Coomes-Kocak-Palmer (1996) and others.

Geometric, structurally stable, chaotic attractors in dynamics are in my paper (Smale (1967)). But these did not arise from any physical system.

Some partial results on problem 14 are due to Rychlik and Robinson, see Robinson (1989).

Another related problem is: can one decide if a given dynamical system is chaotic? Is there an algorithm which, with coefficients of a dynamical system as input, outputs yes if the dynamics is chaotic and no otherwise. For precision one could say that a dynamics is chaotic exactly when it contains a horseshoe.

Consider first the Turing machine point of view. Matiaysevich (1993) has described a polynomial over $\mathbb{Z}$, $P(u, x_1, ... x_n)$, $n = 27$, with the property that it can not be decided on input u in $\mathbb{Z}$ whether there is a zero $(x_1, ...x_n)$ in $\mathbb{Z}^n$. Richardson and Costa-Doria (1991) made a study of the function:

$$F(u, x) = P(u, x_1, ..., x_n) + \sum (\sin \pi x_i)^2, \quad u \in \mathbb{Z}, \; x \in \mathbb{R}^n.$$

Observe that one cannot decide the question: on input u, does this function have a zero in $\mathbb{R}^n$. If one can't decide the existence of a zero, one can hardly expect to decide the existence of chaos. In fact Costa-Doria prove just that, chaos is undecidable.

However one could well object in a number of ways; this analysis is far from the standard and useful models of dynamics. For one thing, Turing machines would seem to be a poor idealization of algorithms used in this subject as argued in our manifesto (see Chapter one of BCSS).

Another approach is the way L. Blum and I (1993) dealt with the question of Penrose (1991), "Is the Mandelbrot set decidable?". We made the question precise by putting it into the framework of machines over $\mathbb{R}$ and then answered it in the negative.

The Mandelbrot set is a certain set of complex numbers which might be interpreted as having a high level of chaos (in fact having a one dimensional chaotic set) for a corresponding dynamics. Thus my result with Blum asserts that one can't decide if a given dynamics has a high level of chaos. It suggests an alternate route to Costa-Doria in questions of generally deciding if a given dynamics is chaotic.

There are many nuances here, related to properties of approximation, random algorithms, polynomial time, etc. See Problem 18 for further thoughts.

Problem 15: Navier-Stokes equations

Do the Navier-Stokes equations on a 3-dimensional domain Ω in $\mathbb{R}^3$ have a unique smooth solution for all time?

This is perhaps the most celebrated problem in partial differential equations. Let us be a little more precise. The Navier-Stokes equations may be written

$$\frac{\partial u}{\partial t} + (u \cdot \nabla)u - \nu \Delta u + \text{grad } p = 0, \qquad \text{div } u = 0$$

where a C^∞ map $u : \mathbb{R}_+ \times \Omega \to \mathbb{R}^3$ and $p : \Omega \to \mathbb{R}$ are to be found satisfying these equations, u prescribed at $t = 0$ and on the boundary $\partial\Omega$. Here $\mathbb{R}_+ = [0, \infty)$, and $u \cdot \nabla$ is the operator $\sum_1^3 u_i \frac{\partial}{\partial x_i}$, and ν is a positive constant. See e.g., (Chorin-Marsden, 1993) for details.

Many mathematicians have contributed toward the understanding of this problem. An affirmative answer has been given in dimension 2 and in dimension 3 for t in a small interval $[0, T]$. See (Temam, 1979) for background.

The solution of this problem might well be a fundamental step toward the very big problem of understanding turbulence. For example it could help realize the ideas of Ruelle-Takens (1971), which put the notion of a chaotic attractor into a model of turbulence. See also (Chorin-Marsden-Smale, 1977).

In (Smale, 1991) I asked if the solutions of the 2-dimensional Navier-Stokes equation with a forcing term on a torus must converge to an equilibrium as time tends to infinity. Babin-Vishik (1983) had given some evidence to the contrary. Subsequently Liu (1992) provided examples to show convergence to a more complicated attractor.

Problem 16: The Jacobian Conjecture

Suppose $f : \mathbb{C}^n \to \mathbb{C}^n$ is a polynomial map with the property that the derivative at each point is nonsingular. Then must f be one to one?

Here $\mathbb{C}^n$ is an n-dimensional complex Cartesian space, $f(z) = (f_1(z), \ldots, f_n(z))$ and each f_i is a polynomial in n variables. The derivative of f at z, $Df(z) : \mathbb{C}^n \to \mathbb{C}^n$ may be thought of as the matrix of partial derivatives and the nonsingularity condition as Det $Df(z) \neq 0$.

If f is indeed injective then it is surjective and has an inverse which is a polynomial map.

The problem goes back to the 1930's and one can see the excellent surveys (Bass-Connell-Wright, 1982) and (van den Essen, 1997) for the importance, background, and related results.

Problem 17: Solving polynomial equations

Can a zero of n complex polynomial equations in n unknowns be found approximately, on the average, in polynomial time with a uniform algorithm?

More broadly, what are the features that distinguish the tractable from the intractable in the realm of solving polynomial systems of equations?

The final theorem in the five paper series, jointly done with Mike Shub, *Bez* $I - V$ (see Shub-Smale, 1994) is exactly the italicized result without "uniform".

We review the definitions. Consider $f : \mathbb{C}^n \to \mathbb{C}^n$, $f(z) = (f_1(z), \ldots, f_n(z))$, $z = (z_1, \ldots, z_n)$ where each f_i is a polynomial in n variables of degree d_i.

It is reasonable to make the f_i into homogeneous polynomials by adding a new variable z_0, work in the corresponding projective space, and then translate the algorithm and results back to the initial affine problem.

Approximately can be defined intrinsically using Newton's method and is necessary because of the classic results of Abel and Galois. Time is measured by the number of arithmetic operations and comparisons, using real machines (as in Problem 3) if one wants to be formal.

A probability measure must be put on the space of all such f, for each $d = (d_1, \ldots, d_n)$ and the time of an algorithm is averaged over the space of f. Is there such an algorithm where the average time is bounded by a polynomial in the number of coefficients of f (the input size)?

In (Shub-Smale, 1994) it is proved that this can be done, but the algorithm is different for each n-tuple d. A uniform algorithm is one that is independent of d (d is part of the input).

Certainly finding zeros of polynomials and polynomial systems is one of the oldest and most central problems of mathematics. Our problem asks if, under some conditions specified in the problem, it can be solved systematically by computers. If there is no polynomial time way of doing it, then no computer will ever succeed.

Morever, as developed in BCSS, the problem of zeros of polynomials plays a universal role. The Hilbert Nullstellensatz (as a decision problem) is NP-complete over any field (see Problem 3). In this form, no polynomial time algorithm seems likely to be found.

Similar, more difficult, problems may be raised for real polynomial systems (and even with inequalities).

Problem 18: Limits of intelligence

What are the limits of intelligence, both artificial and human?

Penrose (1991) attempts to show some limitations of artificial intelligence. Involved in his argumentation is the interesting question, "Is the Mandelbrot set decidable?" (see problem 14) and implications of the Gödel incompleteness theorem.

However a broader study is called for, one which involves deeper models of the brain, and of the computer, in a search of what artificial and human intelligence have in common, and how they differ.

This project requires the development of a mathematical model of intelligence, with variations to take into account the differences between kinds of intelligence.

It is useful to realize that there can be no unique model. Even in physics which is more clearly defined, one has classical mechanics, quantum mechanics, and relativity theory, each yielding its own insights and understandings and each with its own limitations. Models are idealizations with drastic simplifications which capture main truths.

An important part of intelligent activity is problem solving. For this one has a traditional model, the Turing machine, as well as a newer machine which processes real numbers (see BCSS), referred to previously in problem 3. The Turing machine has been accepted as a reasonable model for the digital computer. We have argued for the alternative real number machine as a more appropriate model for the digital computer's use in scientific computation and in situations where arithmetic operations dominate (the Manifesto as reprinted as Chapter 1 of BCSS). Such mathematical models for human intelligence are less developed.

There is one example of a general problem that comes to the forefront; that is the problem of equation solving for polynomial systems, over some field of numbers. The real numbers with inequalities are an important special case of this problem. Artificial intelligence has encountered it in its study of robotics. Moreover, over any field, equation solving possesses a universality in a formal mathematical sense in the theory of NP completeness.

One might ask, is there a form of intelligence that can solve general systems of polynomial equations. This problem is anticipated by the previous problems 3 and 17.

The use of the Turing machine versus its real counterpart is a manifestation of the age old conflict between the discrete and the continuous. I believe that the real number machine is the more important of the two for understanding the problem solving limitations of humans. Physical laws underpin biological proccesses and even such discrete activity as the firing of neurons has associated differential equations. Differential equations and equilibria are pervasive in the physical and biological worlds. Even discrete quantum levels are best understood in terms of the eigenvalues of the Schrodinger equation.

Venturing further in this direction, I would be skeptical about the use of Godel's incompleteness theorem (as in Penrose, 1991) for arguing the limitations of any kind of intelligence.

But real number computations and algorithms which work only in exact arithmetic can offer only limited understanding. Models which process approximate inputs and which permit round-off computations are called for. In the context of real number machines one can see Cucker-Smale (1997) in this respect. Moreover randomness in the input and in the processing itself would seem to be an important ingredient in our search for models of intelligence.

Complexity of computation must be considered in attempting to fathom the limits of intelligence. Any worthy model has to deal with this issue, and in the most drastic idealization this comes down to the requirement of polynomial time.

Finally problem solving as exemplified by Turing and real number machines is only part of the story of intelligence. Continual interaction with the environment must be incorporated into a good model. Learning is a part of human intelligent activity. The corresponding mathematics is suggested by the theory of repeated games, neural nets and genetic algorithms.

Addenda

Here we add a few problems that don't seem important enough to merit a place on our main list, but it would still be nice to solve them.

Add.1 Mean Value Problem

Given a complex polynomial f and a complex number z, is there a critical point θ of f (i.e., $f'(\theta) = 0$) such that

$$\frac{|f(z) - f(\theta)|}{|z - \theta|} \leq c|f'(z)|, \quad c = 1?$$

This was posed in (Smale, 1981b) where it was proved for $c = 4$. The constant c has to be at least $(d-1)/d$ from the example $f(z) = z^d - dz$. Tischler (1989) has some partial results.

Add.2 Is the three-sphere a minimal set?

Can a C^∞ vector field be found on the three-sphere so that every solution curve is dense?

I raised this problem in (Browder, 1976).

Add.3 Is an Anosov diffeomorphism of a compact manifold topologically the same as the Lie group model of John Franks?

This problem is in Franks' article in (Chern-Smale, 1970) where everything is made precise. Briefly an Anosov diffeomorphism is one where the tangent bunndle has a global invariant splitting into contracting and expanding subbundles (global hyperbolicity) and the nonwandering set is dense (see problems 10 and 11). It is not even known if the universal covering manifold must be Euclidean space. See also (Smale, 1967, 1980) for background.

References

Abraham,R. and Marsden,J., (1978). *Foundations of Mechanics.* Addison-Wesley Publishing Co., Reading, Mass.

Babin,A.V. and Vishik,M.I., (1983). Attractors of partial differential evolution equations and their dimension. *Russian Math. Survey* **38**, 151–213.

Baker, A., (1979). *Transcendental Number Theory*, Cambridge University Press, Cambridge, UK.

Barrow-Green, J., (1997). *Poincaré and the Three Body Problem*, American Math Society, Providence, RI.

Barvinok,A. and Vershik,A., (1993). Polynomial-time, computable approximation of families of semi-algebraic sets and combinatorial complexity. *Amer. Math. Soc. Trans.* **155**, 1–17.

Bass,H., Connell,E., and Wright,D., (1982). The Jacobian conjecture: reduction on degree and formal expansion of the inverse. *Bull. Amer. Math. Soc.* **7**, 287–330.

BCSS: Blum,L., Cucker,F., Shub,M., and Smale,S., (1997). *Complexity and Real Computation*, Springer-Verlag, New York.

Blum,L., Shub,M. and Smale,S., (1989). On a theory of computation and complexity over the real numbers: NP-completness, recursive functions and universal machines. *Bulletin of the Amer. Math. Soc.* **21**, 1–46.

Blum,L. and Smale,S., (1993). The Gödel incompleteness theorem and decidability over a ring. Pages 321–339 in M.Hirsch, J.Marsden and M.Shub (editors), *From Topology to Computation: Proceedings of the Smalefest*, Springer-Verlag, New York.

Browder,F. ed., (1976). *Mathematical Developments Arising from Hilbert Problems*, American Math Society, Providence, RI.

Brownawell,W., (1987). Bounds for the degrees in the Nullstellensatz. *Annals of Math.* **126**, 577–591.

Chern,S. and Smale,S., eds. (1970). *Proceedings of the Symposium on Pure Mathematics*, vol. **XIV**, American Math Society, Providence, RI.

Chorin,A., Marsden,J., (1993). *A Mathematical Introduction to Fluid Mechanics*, 3rd edition, Springer-Verlag, New York.

Chorin,A., Marsden,J. and Smale,S., (1977). Turbulence Seminar, Berkeley 1976–77, *Lecture Notes in Math.* **615**, Springer-Verlag, New York.

Coomes, B., Kocak, H., and Palmer, K., (1996). Shadowing in discrete dynamical systems. in *Six Lectures in Dynamical Systems, (eds. Aulbach, B. and Colonius, F.)* World Scientific, Singapore.
Costa N. da and Doria, F., (1991). Undecidability and Incompleteness in Classical Mechanics. *Internat. Jour. Theoretical Physics* **30** 1041–1073.
Cucker, F., M. Karpinski, M., Koiran, P., Lickteig, T., amd Werther, K., (1995). On real Turing machines that toss coins. In *27th annual ACM Symp. on the Theory of computing* pp 335-342.
Cucker,F., Koiran,P., and Smale,S., (1999). A polynomial time algorithm for Diophantine equations in one variable. *Jour. of Symbolic Computation* **27**, no. 1, 21–29.
Cucker, F. and Smale, S. (1997). Complexity estimates depending on condition and round-off error. *J. ACM* **46**, 1999, no. 1, 113–184.
Debreu,G., (1959). *Theory of Value*, Wiley, New York.
Diacu, F. and Holmes, P., (1996). *Celestial Encounters*, Princeton University Press, Princeton, N. J.
Dulac,H., (1923). Sur les cycles limites. *Bull. Soc. Math. France* **51**, 45–188.
Easton, R., (1971). Some topology of the 3 body problem. *Jour. of Diff. Equations.* **10**, 371-377.
Écalle,J., (1992). *Introduction aux Fonctions Analysables et Preuve Constructive de la Conjecture de Dulac.* Hermann, Paris.
van den Essen,A., (1997) Polynomial automorphisms and the Jacobian conjecture. in *"Algèbre non commutative, groupes quantiques et invariants, septième contact Franco-Belge, Reims, Juin 1995".* eds. J. ALev and G. Cauchon, "Société mathématique de France", Paris.
Freedman,M. (1982). The topology of 4-manifolds. *J. Diff. Geom.* **17**, 357–454.
Garey,M. and Johnson,D., (1979). *Computers and Intractability.* Freeman, San Francisco.
Graczyk,J. and Swiatek,G., (1997). Generic hyperbolicity in the logistic family. *Annals of Math.* **146**, 1-52.
Grötschel,M., Lovász,L., and Schrijver,A., (1993). *Geometric Algorithms and Combinatorial Optimization*, Springer-Verlag, New York.
Guckenheimer,J. and Holmes,P. (1990). *Nonlinear Oscillations, Dynamical Systems and Bifurcations of Vector Fields*, third printing, Springer-Verlag, New York.
Guckenheimer,J. and Williams,R.F. (1979). Structual stability of Lorenz attractors. *Publ. Math. IHES* **50**, 59–72.
Hayashi,S., (1997). Connecting invariant manifolds and the solution of the C^1 stability conjecture and Ω-stability conjecture for flows. *Annals of Math.* **145**, 81–137.
Hirsch,M. and Smale,S., (1974). *Differential Equations, Dynamical Systems, and Linear Algebra*, Academic Press, New York.
Ilyashenko,Yu., (1985). Dulac's memoir "On limit cycles" and related problems of the local theory of differential equations. *Russian Math. Surveys* VHO, 1–49.
Ilyashenko,Yu., (1991). *Finiteness Theorems for Limit Cycles*, American Math Society, Providence, RI.
Ilyashenko,Yu. and Yakovenko,S., (1995). Concerning the Hilbert 16th problem. *AMS Translations*, series 2, vol. **165**, AMS, Providence, RI.
Jakobson,M., (1971). On smooth mappings of the circle onto itself. *Math. USSR Sb.* **14**, 161–185.
Kuijlaars,A.B.J. and Saff,E.B., (1998). Asymptotics for minimal discrete energy on the sphere. *Trans. Amer. Math. Soc.* **350**, no. 2, 523–538.
Lagarias, J., (1979). Succinct certificates for the solvability of binary quadratic diophantine equations, *Proc. 20th IEEE Symposium on Foundations of Computer Science*, Proc IEEE, 47-54.

Lagarias, J., (1980). On the computational complexity of determining the solvability or unsolvability of the equation $X^2 - DY^2 = -1$. *Trans. Amer. Math. Soc.* **260**, 485-508.
Lang,S., (1991). *Number Theory III*, vol. **60** of *Encyclopaedia of Mathematical Sciences*, Springer-Verlag, New York.
Lenstra, H., (1997). Finding small degree factors of lacunary polynomials, preprint.
Linz,A., de Melo,W., and Pugh,C., (1977), in Geometry and Topology, *Lecture Notes in Math.* **597**, Springer-Verlag, New York.
Liu,V., (1992). An example of instability for the Navier-Stokes equations on the 2-dimensional torus. *Commun. PDE* **17**, 1995–2012.
Lloyd,N.G. and Lynch,S., (1988). Small amplitude limit cycles of certain Lienard systems. *Proceedings Roy. Soc. London* **418**, 199–208.
Lorenz,E., (1963). Deterministic non-periodic flow. *J. Atmosph. Sci.* **20**, 130–141.
Lyubich,M., (1997), Dynamics of Quadratic Polynomials. 1-2, *Acta Mathematica* **178**, 185–297.
Manders,K.L. and Adleman,L., (1978). NP-complete decision problems for binary quadratics. *J. Comput. System Sci.* **16**, 168–184.
Matiyasevich, Y., (1993). *Hilbert's Tenth Problem.* The MIT Press, Cambridge, Mass.
Mazur, B., (1994). Questions of decidability and undecidability in number theory. *The Journal of Symbolic Logic* **59**, 353–371.
McCord, C., (1996). Planar central configuration estimates in the n-body problem. *Ergodic Theory Dynamical Systems* **5**, 1059–1070.
McCord, C., Meyer, K., and Wang, Q., (1998) *The Integral Manifolds of the Three Body Problem. Memoirs, Amer. Math. Soc.* **132**, no. 628, viii+90pp., Providence, R.I.
McMullen,C., (1994). Frontiers in complex dynamics. *Bull. Amer. Math. Soc.* **31**, 155–172.
de Melo,W. and van Strien,S., (1993). *One-Dimensional Dynamics.* Springer-Verlag, New York.
Palis,J. and Yoccoz,J.C., (1989). (1) Rigidity of centralizers of diffeomorphisms. *Ann. Scient. Ecole Normale Sup.* **22**, 81–98; (2) Centralizer of Anosov diffeomorphisms. *Ann. Scient. Ecole Normale Sup.* **22**, 99–108.
Palmore,J., (1976). Measure of degenerate relative equilibria, I. *Annals of Math.* **104**, 421–429.
Petrovskii,I.G. and Landis,E.M., (1957). On the number of limit cycles of the equation $dy/dx = P(x,y)/Q(x,y)$, where P and Q are polynomials. *Mat. Sb. N.S.* **43** (85), 149–168 (in Russian), and (1960) *Amer. Math. Soc. Transl.* (2) **14**, 181–200.
Petrovskii,I.G. and Landis,E.M., (1959). Corrections to the articles "On the number of limit cycles of the equation $dy/dx = P(x,y)/Q(x,y)$, where P and Q are polynomials". *Mat. Sb. N.S.* **48** (90), 255–263 (in Russian)
Penrose,R. (1991). *The Emperor's New Mind.* Penguin Books, New York.
Poincaré,H., (1953). *Oeuvres*, VI. Gauthier-Villars, Paris. Deuxième Complément à L'Analysis Situs.
Peixoto,M., (1962). Structural stability on two-dimensional manifolds. *Topology* **1**, 101–120.
Pugh,C., (1967). An improved closing lemma and a general density theorem. *Amer. J. Math.* **89**, 1010–1022.
Pugh,C. and Robinson,C., (1983). The C^1 closing lemma including Hamiltonians. *Ergod. Theory Dynam. Systems* **3**, 261–313.
Rakhmanov,E.A., Saff,E.B. and Zhou,Y.M., (1994). Minimal discrete energy on the sphere. *Math. Res. Lett.* **1**, 647–662.
Robinson,C., (1989). Homoclinic bifurcation to a transitive attractor of Lorenz type. *Nonlinearity* **2**, 495–518.

Ruelle,D. and Takens,F., (1971). On the nature of turbulence. *Commun. Math. Phys.* **20**, 167–192.

Samuelson,P., (1971). *Foundations of Economic Analysis*, Atheneum, New York.

Saff,E. and Kuijlaars,A., (1997). Distributing many points on a sphere. *Math Intelligencer* **10**, 5–11.

Schrijver,A., (1986). *Theory of Linear and Integer Programming*. John Wiley & Sons, New York.

Shi,S., (1982). On limit cycles of plane quadratic systems. *Sci. Sin.* **25**, 41–50.

Shub,M., (1970). Appendix to Smale's paper: Diagonals and relative equilibria in manifolds, Amsterdam, 1970. *Lecture Notes in Math.* **197**, Springer-Verlag, New York.

Shub,M. and Smale,S., (1995). On the intractibility of Hilbert's Nullstellensatz and an algebraic version of "P=NP", *Duke Math. J.* **81**, 47–54.

Shub,M. and Smale,S., (1993). Complexity of Bezout's theorem, III: condition number and packing. *J. of Complexity* **9**, 4–14.

Shub,M. and Smale,S., (1994). Complexity of Bezout's theorem, V: polynomial time. *Theoret. Comp. Sci.* **133**, 141–164.

Smale,S., (1963). Dynamical systems and the topological conjugacy problem for diffeomorphisms, pages 490–496 in: *Proceedings of the International Congress of Mathematicians*, Inst. Mittag-Leffler, Sweden, 1962. (V.Stenström, ed.)

Smale,S., (1963). A survey of some recent developments in differential topology. *Bull. Amer. Math. Soc.* **69**, 131–146.

Smale,S., (1967). Differentiable dynamical systems. *Bull. Amer. Math. Soc.* **73**, 747–817.

Smale,S., (1970). Topology and mechanics, I and II. *Invent. Math.* **10**, 305–331 and *Invent. Math.* **11**, 45–64.

Smale,S., (1976). Dynamics in general equilibrium theory. *Amer. Economic Review* **66**, 288–294.

Smale,S., (1980). *Mathematics of Time*, Springer-Verlag, New York.

Smale,S., (1981a). Global analysis and economics, pages 331–370 in *Handbook of Mathematical Economics* **1**, editors K.J.Arrow and M.D.Intrilligator. North-Holland, Amsterdam.

Smale, S. (1981b). The fundamental theorem of algebra and complexity theory. *Bulletin of the Amer. Math. Soc.* *4*, 1–36.

Smale,S., (1990a). The story of the higher dimensional Poincaré conjecture. *Mathematical Intelligencer* **12**, no. 2, 40–51. Also in M.Hirsch, J.Marsden and M.Shub, editors, *From Topology to Computation: Proceedings of the Smalefest*, 281–301 (1992).

Smale, S., (1990b). Some remarks on the foundations of numerical analysis. *SIAM review 32*, 211–220.

Smale,S., (1991). Dynamics retrospective: great problems, attempts that failed. *Physica D* **51**, 267–273.

Smale, S., (1998). Finding a horseshoe on the beaches of Rio, *The Mathematical Intelligencer* **20**, 39–44.

Taubes,G., (July 1987). What happens when Hubris meets Nemesis? *Discover*.

Temam,R., (1979). *Navier-Stokes Equations*, revised edition. North-Holland, Amsterdam.

Tischler, D., (1989). Critical points and values of complex polynomials. *Jour. of Complexity* **5**, 438–456.

Traub,J. and Woźniakowski,H., (1982). Complexity of linear programming. *Oper. Res. Letts.* **1**, 59–62.

Tsuji,M., (1959). *Potential Theory in Modern Function Theory*. Maruzen Co., Ltd., Tokyo.

Wen,L. and Xia,Z., (1997). A simpler proof of the C^1 connecting lemma. To appear, Trans. of the Amer. Math. Soc.
Williams,R., (1979). The structure of Lorenz attractors. *Publ. IHES* **50**, 101–152.
Wintner,A., (1941). *The Analytical Foundations of Celestial Mechanics.* Princeton University Press, Princeton, NJ.

Department of Mathematics, City University of Hong Kong, Kowloon, Hong Kong
E-mail address: `masmale@math.cityu.edu.hk`

The Story of the Higher Dimensional Poincaré Conjecture (What Actually Happened on the Beaches of Rio)*

Steve Smale

This paper is dedicated to the memory of Allen Shields.

Although these pages tell mainly a personal story, let us start with a description of the "n-dimensional Poincaré Conjecture." It asserts:

A compact n-dimensional manifold M^n that has the homotopy type of the n-dimensional sphere

$$S^n = \{x \in \mathbf{R}^{n+1} \mid \|x\| = 1\}$$

is homeomorphic to S^n.

A "compact n-dimensional manifold" could be taken as a closed and bounded n-dimensional surface (differentiable and non-singular) in some Euclidean space.

The homotopy condition could be alternatively defined by saying there is a continuous map $f{:}M^n \to S^n$ inducing an isomorphism on the homotopy groups; or that every continuous map $g{:}S^k \to M^n$, $k < n$ (or just $k \leq n/2$) can be deformed to a point. One could equivalently demand that M^n be simply connected and have the homology groups of S^n.

Henri Poincaré studied this problem in his pioneering papers in topology. In [13], 1900, he announced a proof of the general n-dimensional case. A counter-example to his method is exhibited in a subsequent paper [14] 1904, where he limits himself this time to 3 dimensions. In this paper he states his famous problem, but not as a conjecture. The traditional description of the problem as "Poincaré's Conjecture" is inaccurate in this respect.

Many other mathematicians after Poincaré have claimed proofs of the 3-dimensional case. See, for example, [18] for a popular account of some of these attempts. On the other hand, there has been a solidly developing body of theorems and techniques of topology since Poincaré.

In 1960 I showed that the assertion is true for all $n > 4$, and this is an account of that discovery. The story here is complemented by two articles [15] and [16], but the overlap is minimal.

I first heard of the Poincaré conjecture in 1955 in Ann Arbor at the time I was writing a thesis on a problem of topology. Just a short time later, I felt that I had found a proof (3 dimensions). Hans Samelson was

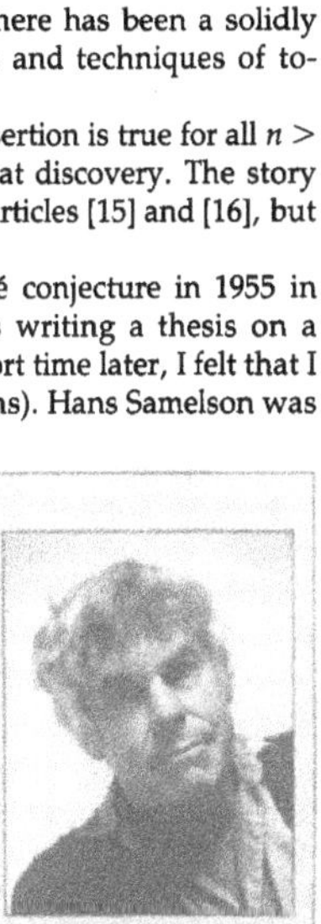

Steve Smale

Steve Smale is a member of the National Academy of Sciences (USA) and of the American Academy of Arts and Sciences. He has been awarded the Veblen Prize for Geometry (American Mathematical Society, 1965), the Fields Medal (International Mathematical Union, 1966), and the Chauvenet Prize (Mathematical Association of America, 1988).

Steve Smale, a correspondent for the *Mathematical Intelligencer*, will celebrate his 60th birthday this year. The *Mathematical Intelligencer* wishes him a happy birthday.

* This article is an expanded version of a talk given at the 1989 annual joint meeting of the American Mathematical Society, and the Mathematical Association of America.

in his office, and very excitedly I sketched my ideas to him. First triangulate the 3-manifold and remove one 3-dimensional simplex. It is sufficient to show the remaining manifold is homeomorphic to a 3-simplex. Then remove one 3-simplex at a time. This process doesn't change the homeomorphism type and finally one is left with a single 3-simplex. Q.E.D. Samelson didn't say much. After leaving his office, I realized that my "proof" hadn't used any hypothesis on the 3-manifold.

Less than 5 years later in Rio de Janeiro, I found a counterexample to the 3-dimensional Poincaré Conjecture. The "proof" used an invariant of the Leningrad mathematician Rohlin, and I wrote out the mathematics in detail. It would complement nicely the proof I had just found that for dimensions larger than 4, the result was true. Luckily, in reviewing the counterexample, I noticed a fatal mistake.

Let us go back in time. I was born into the "Golden Age of Topology." Today it is easy to forget how much topology in that era dominated the frontiers of mathematics. It has been said that half of all the Sloan Postdoctoral Fellowships awarded in those years went to topologists. Today it would be hard to conceive of such a lopsided distribution. Topology of that time, in fact, had revolutionizing effects in algebra (*K*-theory, algebraic geometry) and analysis (dynamical systems, the global study of partial differential equations).

In 1954 Thom's cobordism paper was published. That theory was used by Hirzebruch to prove his "signature theorem" (as part of his development of Riemann-Roch). In turn, already by 1956, Milnor used the signature theorem to prove the existence of exotic 7-dimensional spheres. I followed these results closely. Also as a student I learned from Raoul Bott about Serre's use of spectral sequences and Morse theory to find information on the homotopy groups of spheres. A little later, Bott himself was proving his periodicity theorems, also with the use of Morse Theory.

I received my doctorate in Ann Arbor in 1956 with Bott. My first encounter with the mathematical world at large occurred that summer with the Mexico City meeting in Algebraic Topology. I had never been to any conference before. And this conference was a historic event in mathematics by any standard. I have not seen that concentration of creative mathematics matched. My wife Clara and I took a bus from Ann Arbor to Mexico City, originally knowing only Bott and Samelson. Before leaving Mexico, I had met most of the stars of topology. There I also met two graduate students from the University of Chicago, Moe Hirsch and Elon Lima, who were to become part of our story.

That fall I took up my first regular position as an instructor in the college at the University of Chicago, primarily teaching set theory to humanities students. I had good relations with the mathematics department and went to the lectures of visiting professor René Thom (whom I had met in Mexico City) on transversality theory. I also pursued work in topology showing that one could "turn a sphere inside out."

Steve Smale with his son Nat, Chicago, 1958.

Chicago was a leading mathematics center at that time, before Weil, Chern, and a number of other important mathematicians left. An important part of the environment was created by the younger mathematicians, especially Moe Hirsch, Elon Lima, Dick Lashof, Dick Palais, and Shlomo Sternberg. I was lucky to be there during that period.

A two-year National Science Foundation (NSF) postdoctoral fellowship enabled me to go to the Institute for Advanced Study in the fall of 1958. Topology was very active in Princeton then. I shared an office with Moe Hirsch and we attended Milnor's crowded lectures on characteristic classes and Borel's seminar on transformation groups. I frequently encountered Deane Montgomery, Marston Morse, and Hassler Whitney, played go (with handicap) with Ralph Fox and met his students Lee Neuwirth and John Stallings.

Moreover, in the summer of 1958, Lima introduced me to Mauricio Peixoto, who sparked my interest in structural stability. That interest led to an invitation to spend the last six months of my NSF in Rio de Janeiro at I.M.P.A. (Instituto de Matematica, Pura e Aplicada).

Thus, at the beginning of January 1960, with Clara and our children, Laura and Nat, I arrived in Rio to meet our Brazilian friends. We arrived in Brazil just after a coup had been attempted by an Air Force colonel. He fled the country to take refuge in Argentina, and we were able to rent his apartment! It was an 11-room luxurious place, and we also hired his two maids. The U.S. dollar went a long way in those days.

Dynamics and Manifold Decomposition

Consider a manifold M^n, with some Riemannian metric, and some function $f{:}M^n \to \mathbf{R}$. Construct the dynamical system defined by the differential equation $dx/dt = -\text{grad } f$ on M.

If p is a non-degenerate critical point of f, then the set $W^s(p)$ of all points tending to p under the dynamics as $t \to \infty$ is an imbedded cell; the same applies to the set $W^u(p)$ of points tending to p as $t \to -\infty$. In the 2-dimensional picture, both $W^s(p)$ and $W^u(p)$ are 1-cells (or arcs).

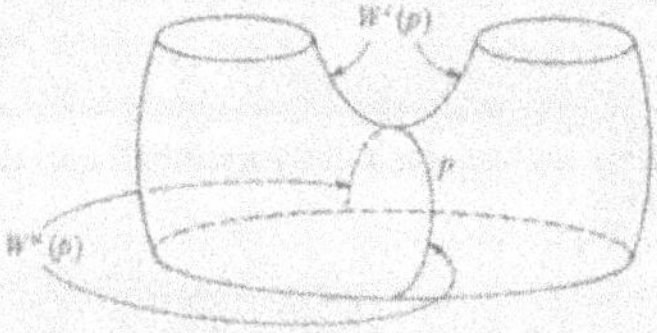

Suppose now that f has only non-degenerate critical points. We may assume that $W^s(p) \pitchfork W^u(q)$ for all critical points p,q of f; or, one could say that "the stable and unstable manifolds of grad f meet transversally." Under this hypothesis, the $W^s(p)$ give a nice decomposition of M. The boundary of a cell is a union of lower dimensional cells. René Thom had earlier considered such a decomposition, but without our transversality hypothesis so that the boundary of a cell could be in higher dimensional cells.

Next, from this dynamical system I constructed what I called handlebodies.

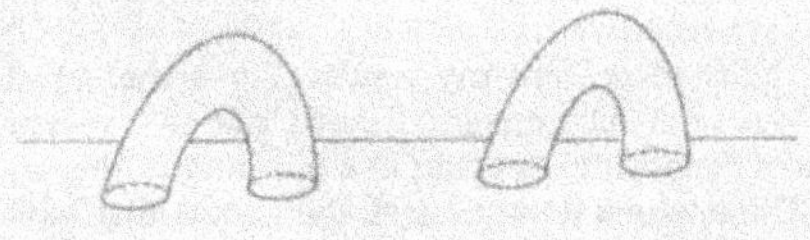

This is a diagram of a 1-handlebody, which consists of thickened 1-cells attached to a 3-disk. Then one can add all the 2-handles at one time, etc., until the manifold is given a filtrated structure by these handles.

This gives a starting point for their elimination, a pair at a time, using finally the homotopy and dimension hypotheses on M.

Eventually, one obtains the following description of our original manifold, which was to be proved.

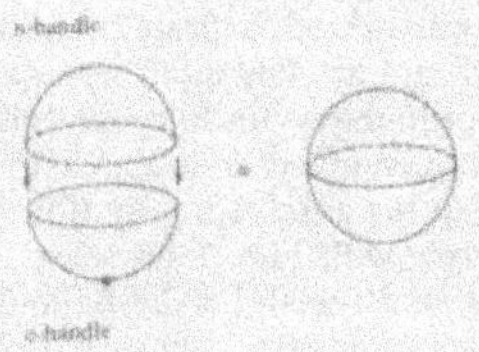

Most of our immediate neighbors were in the U.S. or Brazilian military. We could sit in our highly elevated gardened patio and look across to the hill of the *favela* (Babylonia) where *Black Orpheus* was filmed. In the hot, humid evenings preceding *carnaval*, we would watch hundreds of the favela dwellers winding down their path to dance the samba in the streets. Sometimes I would join their wild dancing, which paraded for many miles.

Just one block in the opposite direction from the hill lay the famous beaches of Copacabana (the Leme end). I would spend the mornings on that wide, beautiful, sandy beach sometimes swimming, or, depending on the height of the waves, body surfing. Also I took a pen and pad of paper and would work on mathematics.

Afternoons I would spend at I.M.P.A. discussing differential equations with Peixoto and topology with Lima. At that time I.M.P.A. was located in a small old building on a busy street. The next time I was to visit Rio, I.M.P.A. was in a much bigger building in a much busier street. I.M.P.A. has now moved to an enormous modern palace surrounded by jungle in the suburbs of Rio.

Returning to the story, my mathematical attention was at first directed towards dynamical systems and I constructed the "horseshoe" [15]. As I continued working in gradient dynamical systems, I noticed how the dynamics led to a new way of decomposing a manifold simply into cells. The possibilities of using this decomposition to attack the Poincaré conjecture soon developed, and before long all my work focused on that problem.

With apparent success in dimensions greater than 4, I reviewed my proof carefully; then I went through the details with Lima. Gaining confidence, I wrote Hirsch in Princeton and sent off a research announcement to Sammy Eilenberg. The box titled "Dynamics and Manifold Decomposition" contains a mathematical description of what was happening.

I was already planning to leave Rio for three weeks in Europe during June of that year, 1960. There was the famous *Arbeitstagung*, an annual mathematics event organized by Hirzebruch. This was to be followed by a topology conference in Zürich to which I had been invited. The two meetings provided a good opportunity to present my results. For a change of pace and with his consent, I have put two recent letters of Stallings to Zeeman in a box later in this article. These letters describe well the events that took place in Europe. While my memory in general is consistent with Stallings', I don't believe Hirsch helped me as Stallings conjectured. I do recall spending some relaxed days in St. Moritz with Moe Hirsch and Raoul Bott, after a more dramatic and traumatic week in Bonn.

My best-known work was done on the beaches of Rio de Janeiro.

Sometimes I have become upset at what I feel are inaccuracies in historical accounts of the discovery of the higher dimensional Poincaré conjecture. For example, Andy Gleason wrote in 1964 [1]: ". . . It was a great surprise, therefore, when Stallings in 1960 (4) proved that the generalized Poincaré conjecture is true for dimensions 7 and up. His result was extended by Zeeman (10) shortly thereafter to cover dimensions 5 and 6." (I wasn't mentioned in his article.)

Paul Halmos in his autobiography ([8], page 398) writes of my anger with him. I am sorry that I was angry with Paul, and I wish I could be more relaxed about this subject in general. My recent correspondence with Jack Milnor illustrates the issue.

Dear Jack,

I am writing about your article on the work of M. Freedman in connection with his winning the Fields Medal (*Proc. of the Int. Congress* 1986). There you say in discussing the "n-dimensional Poincaré Hypothesis"

> The cases $n = 1,2$ were known in the nineteenth century, while the cases $n \geq 5$ were proved by Smale, and independently by Stallings and Zeeman and by Wallace, in 1960–61.

The word independently seems inconsistent with the history of that discovery.

Stallings in his preprint "The topology of high-dimensional piecewise-linear manifolds" writes

> When I heard that Smale had proved the Generalized Poincaré Conjecture for manifolds of dimension 5 or more, I began to look for a proof of this fact myself.

I also recall Zeeman having written accurately of these events and giving a similar picture.

After my announcement appeared, Wallace wrote me (Sept. 29, 1960) asking me for details of my proof and telling me where he was blocked in his attempts. I sent him my preprints of the proof which he acknowledged in October (I still have his letters).

I made a mistake in the first draft of my proof, which I easily repaired; but that doesn't affect these issues.

Certainly Stallings and Zeeman, and Wallace have done fine work on this subject. Yet I do wish that mathematicians were more aware of the facts that I have just described. I am sending copies of this to Stallings, Zeeman and Wallace, and to a few other mathematicians.

Best regards,
Steve Smale

Milnor wrote back (as mildly modified and expanded by him):

February 27, 1988

Dear Steve,

I am very sorry that my attempt at sketching the history of the Poincaré conjecture was inaccurate. This was partly a result of not doing my homework properly, but more a matter of expressing myself very badly. What I should have said of course is that the Stallings proof, completed by Zeeman for dimensions 5 and 6, was *logically* independent of your proof; but it is certainly true that yours came first. Similarly, I should have said that Wallace's proof, for dimensions strictly greater than five, definitely came later, and was not really different from your proof.

It is worth noting that the Smale (or Wallace) proof, at the cost of the stronger hypothesis of differentiability, gives a much stronger conclusion. Namely, it shows that the smooth n-manifold contains a smoothly embedded standard $(n - 1)$-sphere which cuts it up into two smoothly embedded standard closed n-balls. The Stallings-Zeeman proof starts with the weaker hypothesis that M is a combinatorial manifold, and obtains the weaker conclusion that M with a single point removed is combinatorially homeomorphic to the standard Euclidean space.

Sincerely,
John Milnor

I wrote him, thanking him, saying that I appreciated his letter.

March 10, 1988

John Stallings to

Dear Chris,

I got your letter to Milnor, which must be about something he said or wrote about Smale, you, and me in the 1960 era. I think I got some comment from Smale about this too, but I cleaned up my office between then and now, and I have no idea what Smale said; and I don't know what Milnor said. And probably I should just go back to sleep.

Your memory of this is much more specific than mine. I remember spring of 1960. I remember the Smale rumor, which I disbelieved until Papakyriakopoulos visited; I think Papa said something like, Eilenberg had looked over Smale's proof of the high-dimensional PC and thought it was OK. I vaguely remember your giving a talk; it was about a paper by Penrose, Whitehead, and Zeeman; and it had some sort of engulfing thing in it; and it mainly strikes me that you or somebody said that it was the only joint paper JHC Whitehead had been involved in, in which his name didn't come last in alphabetical order. I remember sitting up in my little office in the olden Maths. Institute at Oxford and staring at the blackboard, trying to think about how to cancel handles; for some reason this was how I thought Smale would go about it, but I don't think anybody told me that. I couldn't figure out how to get rid of the pesky 1-handles, because I know some awful presentations of the trivial group. Then it occurred to me that I could, so to speak, push all the trouble out to infinity, using the engulfing stuff plus what I think of as an idea due to Barry Mazur in the Morton Brown formulation.

Then I remember being driven to Bonn by Ioan James; there was some other American along too, maybe Bob Hunter or Dick Swan. Somewhere in the middle of Holland, Ioan got out of the car to piss at a farm; we Americans thought this was very British and were too scared to do that ourselves. In Bonn, both Smale and I talked. I remember that A. Borel complained in my talk that I was waving my hands too much on matters in the foundations of PL topology that hadn't been properly proved; and this complaint obsessed me for the next six or seven years. Of Smale's talk, I remember thinking that he hadn't really considered the possibility that there could be awful presentations of the trivial group; in other words, he was just going to cancel the 1-handles with some 2-handles without thinking hard about it; I thought this was a fatal error in Smale's method, and this made me secretly very gleeful. However, the next week, at Zürich, Smale had fixed that up; in fact, if I had memorized Whitehead's "Simplicial Spaces, Nuclei and m-Groups," I would have seen how to do this myself. My impression, for which I have no evidence at all, is that Moe Hirsch, who is probably a much better scholar than Smale (in other words, Smale is the Mad Genius and Hirsch is the Hard Worker), told Smale how to fix it up. The point was that you could add trivial pairs of 2- and 3-cells, which on the 2-skeleton is equivalent to adding trivial relators; this is the extra, delicate type of "Tietze transformation" which you need in order to manipulate the 2-cells algebraically to get them to cancel the 1-cells. If the dimension is high enough, this manipulation can be done geometrically. I guess it is in the delicate part of this geometry that the question of whether Smale did it in dimension 5 or not comes up.

After that, I remember enjoying your hospitality, changing at Bletchley, and doing a lot of nice geometrical mathematics during the summer of 1960.

I do not really feel the reason why people are so interested in what exactly happened then. After all, it is not a big nationalistic thing like Newton versus Leibniz. We have all done more mathematics since then; and what I am really interested in is what I am doing and learning about now. Smale is a colleague who really means a lot to Berkeley and to me.

After the Zürich conference, I returned to join my family in Rio and shortly thereafter took up my position in Berkeley.

During the next year I wrote several papers extending the results, culminating with a paper proving the "h-cobordism theorem" in June 1961. A mathematical survey of all these matters, with references, is given in [17].

Because of Serge Lang and an irresistible offer from Columbia, we sold our house and left Berkeley in the summer of 1961. After three years of studying various aspects of global analysis, we returned from New York to Berkeley because of the prospect of better working conditions.

That was the fall of 1964, and the Free Speech Movement (FSM) caught my attention. After the big sit-in, I helped obtain the release from jail of mathematics graduate students David Frank and Mike Shub.

The Vietnam war was drastically escalated early in 1965. I felt that the U.S. heavy bombing of Vietnam

Christopher Zeeman

There is a funny thing about time. Not only does existence bifurcate infinitely towards the future, but also towards the past. In other words, if I manage to cross the street and not get flattened by a truck, I think, well, in an alternate universe I was flattened by the truck; I'm just not on that path towards the future, but another me was there. Now, I had the opportunity to receive a visit a few months ago from a mathematician I had last seen in 1977, namely Passi. I remember his appearance in 1977 very clearly. Now he looks 10 or 11 years older, but other than that very similar to how he used to look; *except* that nowadays he is wearing a *different nose!* So I think there were several Passis ten years ago; my existence passed by one of them then, but somehow a slightly different one came out of a different past to visit Berkeley recently.—And so I take these reminiscences of what happened 28 years ago with a little sense of amusement, and I presume that we are all coming together out of slightly different pasts.

Yours,
John R. Stallings

March 14, 1988

Dear Chris,

On March 10, I sent you a description of my memories of PL days in 1960. Now, instead of trying to be purely factual, I want to add some things about my impressions and my philosophy. I plan to send a copy of this to Smale and to Milnor.

I was a graduate student at Princeton, 1956–59; not as well-prepared as some of the other students, and with an idea that doing math was much more important than learning it. I learned something of the background of these high-dimensional topology results that I have been discussing. The methods of JHC Whitehead, MHA Newman, V Guggenheim, etc., were taught in Fox's classes. Milnor had had his construction of exotic structures on S^7 and was giving classes on differential topology. So the time was ripe for something to happen.

Most of the graduate students seemed to be trying to show off to each other about their vast knowledge and intellectual sharpness; sheaves and spectral sequences were big topics of conversation. Then came Mazur's sphere-embedding theorem. Suddenly (in winter or spring of 57–58), Barry Mazur reappeared, after a long stretch of god-knows-what in Paris, with his theorem. It was a very easy and immediately understandable proof of something that had seemed so complex that no one dared conjecture it. The result was completed and polished up by others later on, but the Genius was Mazur's; this Genius consisted in the audacity and simplicity of Callow Youth, mathematical mode.

In mathematics there are plenty of people who are quick and quite enough people who have assimilated detailed volumes of knowledge. I consider the Mazur-like phenomenon to be far more attractive and important. After someone like Mazur makes his incursion into new territory, the mathematical masses follow, licking up drops of the leader's sweat, computing lists and tables and obstructions and sheaves and spectral sequences. This gives employment to many.

I did my thesis on Grushko's Theorem, on a level that seemed at the time mundane (in my old age, I have learned to appreciate this thesis much more). But what I admired and envied, and who I hoped to imitate—Barry Mazur.

Here is the story I make up about Smale's work on the high-dimensional PC. Smale had already been working in high-dimensional differential topology; he was just a few steps behind Milnor in the incursion into this subject. Then the PC idea came to him "on the beaches of Rio."

I imagine that if I had been on the beaches of Rio, the idea wouldn't have come to me so easily, be-

was indefensible and threatened world peace. My involvement in the protest increased, and included organizing teach-ins and militant troop train demonstrations. I became cochairman with Jerry Rubin of the VDC, or Vietnam Day Committee. (Our headquarters near campus was later destroyed by a bomb.)

Already during the fall of 1965, I was becoming disillusioned with the VDC and returned to proving theorems. The subsequent part of this story is described in [16]. In Moscow, August 1966, I criticized the U.S. in Vietnam (and Russia as well) to return home under a storm of criticism. The University of California stopped payment of my NSF summer salary.

Dan Greenberg [2] gives a full account of what happened next.

> Upon being informed of the agitation surrounding him, and the withholding of his check, Smale sent to Connick an account of his summer researches—an account which

Continued on page 51

cause, as I pointed out before, I perceived a big problem in getting rid of the 1-handles. I imagine Smale did not know enough to be worried by this, and so he plowed into this big result. I do think there was a little bug in his proof; but this was a non-fatal bug. The 1-handle problem had been, in a somewhat different context, dealt with by JHC Whitehead 20 years earlier. I really think that Smale's Big Theorem was a matter of putting together a new audacity in high dimensions with techniques from olden times due to Whitehead and Whitney, and with his own particular developments. Perhaps the finishing-up of this theorem, and the pushing onward (to computations and obstructions and tables) were done by others. In particular, I think that Milnor had a great deal to do with creating the fundamental techniques of differential topology and expounding them clearly; and that the biggest fruit of this was to make the proof of Smale's Theorem quite solid.

Then there was the engulfing-theory proof of the high-dimensional PC. I'll call this "my theorem." In fact, my theorem and Smale's theorem, although they have a similar sound, are logically distinct. Smale's theorem was that if a differentiable manifold is a homotopy-sphere, then its underlying PL-structure is that of a sphere. My theorem was that if a PL-manifold is a homotopy-sphere, then its underlying topological structure is that of a sphere.

There were several things which paved the way for my theorem. I was familiar with Mazur's agrument (I think I was the first person he convinced; and I had come up with a funny fact, when I was an undergraduate, that a group in which you could define an infinite product was the trivial group, a trivial fact). And I was familiar with PL topology from Fox's courses. The mental block against proving something about high-dimensional homotopy-spheres had been reduced by hearing about the fact that Smale's alleged proof was being favorably received. And then, as you pointed out, I had the Penrose-Whitehead-Zeeman trick at my fingertips.

Once I had my theorem, it seems to me that its proof was much simpler and more obvious than Smale's proof of his theorem. Although you and I and others wrote up a lot about the foundations of PL topology, the truth is that there is nothing deep and interesting in this (comparable, say, to Sard's Theorem in differential topology); all the stuff about general position is somehow not too impressive or hard to believe, in spite of the fact that the details are painful. What was interesting about the books on PL topology were the end results, such as the unknotting theorems and h- and s-cobordism theorems; most of these results have both engulfing versions and handle versions. And they have, of course, later on given birth to new generations of mathematical theories.

Now, both my theorem and Smale's are valid from dimension 5 on up. They are easier to prove in dimensions 7 and higher. The intuition that they work in dimension 5 was there from the start. It was mainly a question of formulating the right lemmas and making plausible arguments. With reference to my theorem, your idea of "piping away the highest-dimensional singularities" was a good formulation.

(It occurs to me that I am saying something here that would warm the cockles of the hearts of Thurston and Gromov. The crimes of Thurston and Gromov, which consist of asserting things that are true if interpreted correctly, without giving really good proofs, thus claiming for themselves whole regions of mathematics and all the theorems therein, depriving the hard workers of well-earned credit—something like these crimes was indulged in by both Smale and me in 1960, because we knew some of these theorems were true, and said so, at least in private, without really knowing how to prove them, or perhaps without wanting to indulge in the labor this would involve.)

As I see it, Smale did his work on his own mostly, with perhaps some very minor collaboration with Hirsch. You and I did a considerable amount of collaboration in the summer of 1960, both in polishing the arguments down to dimension 5 and in using similar arguments for various unknotting results. Eventually, we published our own papers separately, although there were little pieces of joint work here and there. If we were really concerned about who thought of what first, we should have had a secretary taking down all our conversations. Neither of us was interested in producing joint papers. But that collaboration was stimulating and helped us produce enough good theorems for both of us.

Finally, one last word: After working on this subject for a few years, I lost interest in it. There were these big, easy theorems. They were done. Full stop. If someone else wants to lecture about them, they are welcome to do so. If someone wants to use these theorems to compute and classify, they have my minor blessing. But it all sounds tiring and boring to me.

Cheers,
John R. Stallings

Continued from page 49

will probably be a classic document in the literature of science and government. He quickly established that he had satisfied the requirement of 2 months of research for 2 months of salary.

Connick was Vice-Chancellor of Academic Affairs at Berkeley. Greenberg's "classic document" surprised me, but my letter did contain my most famous quote. As recounted by Greenberg, I wrote to Connick:

> However, during the remainder of this time I was also doing mathematics, e.g., in campgrounds, hotel rooms, or on a steamship. On the *S.S. France*, for example, I discussed problems with top mathematicians and worked on mathematics in the lounge of the boat. (My best-known work was done on the beaches of Rio de Janeiro, 1960!)
>
> I would like to repeat that I resent your stopping of my NSF support money for superficial technicalities. The reason goes back to my being issued a subpoena by the House Un-American Activities Committee and the subsequent congressional and newspaper attacks on me.

Before long, I did receive the NSF funds and things quieted down. However, within a year, a much bigger explosion took place when the NSF returned to me my new proposal amidst Congressional pressures.

The articles [3–7] in *Science* by Dan Greenberg chronicle these events. See also [11] and [12].

Eventually as the situation was settling down once more, my statement about the beaches of Rio surfaced. This time it was the science adviser to President Johnson, Donald Hornig [9], who wrote in *Science:*

> This blithe spirit leads mathematicians to seriously propose that the common man who pays the taxes ought to feel that mathematical creation should be supported with public funds on the beaches of Rio de Janeiro or in the Aegean Islands.

(I also had visited the Greek islands in August 1966 but, of course, not with NSF money.)

I was very happy at the response of the Council and President C. Morrey of the American Mathematical Society. Morrey's letter [10] in *Science* started:

> The Council of the American Mathematical Society at its meeting on 28 August asked me to forward the following comments to *Science:*
>
> Many mathematicians were dismayed and shocked by the excerpts of the speech by Donald Hornig, the Presidential Science Adviser (19 July, p. 248). His . . . comments about mathematics and mathematicians are . . . uncalled for. Implicit in Hornig's remarks about vacations on the beaches of Rio or the Aegean Islands was a thinly veiled attack on Stephen Smale. The allegations against Smale were adequately disproved by Daniel S. Greenberg in his articles in *Science* on the Smale-NSF controversy.

At the same time a letter to the *Notices* wound up:

> . . . the policy of creating a major scandal involving the implied application of political criteria in grant administration, apparently for the purpose of placating and warding off the demagogic attack of a single Congressman, is not a policy that will cause anyone (and least of all Congress) to have any great respect for the principles and integrity of those who adopt it. The conspicuous public silence of the whole class of Federal science administrators with regard to the future effects of current draft policy and the Vietnam war upon American science has been positively deafening. In this context, Hornig's apparent attempt to turn the discussion of the current crisis in Federal support for basic science into a hunt for scapegoats in the form of "mathematicians on the beaches" would be ludicrous if it were not so destructive.
>
> Hyman Bass — E. R. Kolchin
> F. E. Browder — S. Lang
> William Browder — M. Loève
> S. S. Chern — R. S. Palais
> Robert A. Herrmann — M. H. Protter
> I. N. Herstein — G. Washnitzer

It was especially gratifying to see such support from my friends.

Let me end, as I did in Phoenix: "Thanks very much for listening to my story."

References

1. Gleason, A., Evolution of an active mathematical theory, *Science* 145 (July 31, 1964) 451–457.
2. Greenberg, Dan, The Smale case: NSF and Berkeley pass through a case of jitters, *Science* 154 (Oct. 7, 1966) 130–133.
3. ———, Smale and NSF: A new dispute erupts, *Science* 157 (Sept. 15, 1967) 1285.
4. ———, Handler statements on Smale case, *Science* 157 (Sept. 22, 1967) 1411.
5. ———, The Smale case: Tracing the path that led to NSF's decision, *Science* 157 (Sept. 29, 1967) 1536–1539.
6. ———, Smale: NSF shifts position, *Science* 158 (Oct. 6, 1967) 98.
7. ———, Smale: NSF's records do not support the charges, *Science* 158 (Nov. 3, 1967) 618–619.
8. Halmos, P., *I Want to be a Mathematician, an Automathography*. New York: Springer-Verlag (1985).
9. Hornig, D., A point of view, *Science* 161 (July 19, 1968) 248.
10. Morrey, C., Letter to the editor, *Science* 162 (Nov. 1, 1968) 514–515.
11. ———, The case of Stephen Smale, *Notices of the American Math. Soc.* 14 (Oct. 1967) 778–782
12. ———, The case of Stephen Smale: Conclusion, *Notices of the American Math. Soc.* 15 (Jan. 1968) 49–52 and 16 (Feb. 1968) 297 (by Serge Lang).
13. Poincaré, H., *Oeuvres*, VI, Gauthier-Villars, Paris 1953, Deuxième Complément à L' Analysis Situs, 338–370.
14. ———, Cinquième Complément à L'Analysis Situs, 435–498.
15. Smale, S., On How I Got Started in Dynamical Systems, *The Mathematics of Time*. New York: Springer-Verlag (1980).
16. ———, On the Steps of Moscow University, *Math. Intelligencer* 6 no. 2 (1984) 21–27.
17. ———, A survey of some recent developments in differential topology, *Bull. Amer. Math. Soc.* 69 (1963) 131–145.
18. Taubes, G., What happens when Hubris meets Nemesis, *Discover*, July 1987.

Department of Mathematics
University of California
Berkeley, CA 94720 USA

Symposium on the Current State
and Prospects of Mathematics

Barcelona, June 1991

Theory of Computation

by

Stephen Smale

Fields Medal 1966

for his work in differential topology, where he proved the generalized Poincaré conjecture in dimension $n \geq 5$: Every closed n-dimensional manifold homotopy equivalent to the n-dimensional sphere is homeomorphic to it. He introduced the method of handle-bodies to solve this and related problems.

Abstract: It could be said that the modern theory of computation began with Alan Turing in the 1930's. After a period of steady development, work in complexity, specially that of Steve Cook and Richard Karp around 1970, gave a deeper tie of the Turing framework to the practice of the machine. I will discuss an expansion of the above to a theory of computation and complexity over the real numbers (joint work with L. Blum and M. Shub).

60

Theory of Computation

The reason for a theory of computation, for me in particular, comes from an attempt to understand algorithms in a more systematic way. The notion of algorithm is very old in mathematics; it goes back a couple of thousand years. Mathematicians have talked about algorithms for a long time, but it was not until Gödel that they tried to formalize the notion of algorithm. In Gödel's incompleteness theorem one saw for the first time the limitations of computations or the need to study more clearly what could be done.

To do so, one has to establish more explicitly what an algorithm is, and I think that this became clearer in the way that Turing interpreted Gödel. So let us stop for a moment and look more closely at Turing and his achievements. I think it is fair to say that he laid down the first theory of computation. Perhaps I will be more specific later about what Turing's notion of computation was.

We can take as set of inputs the integers $\mathbf{Z}$, so in the Turing abstraction the input is some integer; perhaps not every integer is allowed but only those that were eventually called the *halting set* Ω_M of the machine M. This is the domain of computation of the machine M. Given an integer in Ω_M, we feed it to the machine M and obtain as output another integer.

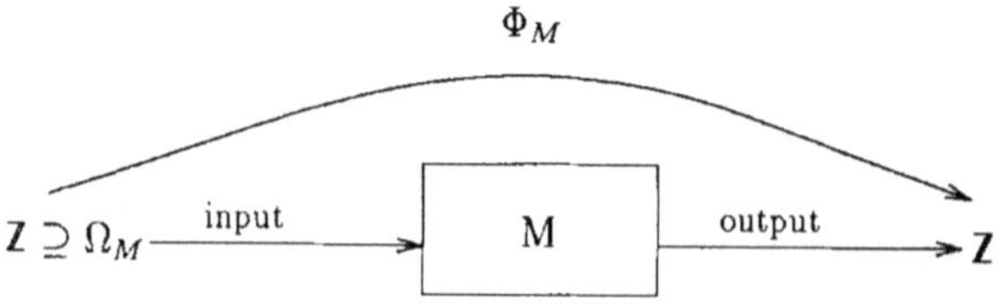

There is some kind of mechanism here, described by Turing, which I will later on formalize in my own way. Turing gave different versions of the input set; for instance, finite sequences of zeroes and ones.

Thus Gödel's incompleteness theorem can be stated in the following way.

THEOREM. *There is some set $S \subset \mathbf{Z}$ which is definable in terms of a finite number of polynomial conditions and is not decidable.*

This is Gödel's incompleteness theorem as formulated by Turing. *Not decidable* means that there is no computable function over $\mathbf{Z}$ which is 1 on S and 0 out of S. In other words, S is *decidable* if its characteristic function is computable by some machine. Thus, Gödel's incompleteness theorem asserts that there exists a set S which is very definable mathematically, yet is not decidable.

This is in some sense the beginning of the theory of computation, which shows the limits of decidability. Eventually, from this evolved a theory for present-day computers. It is from this formulation that it evolved into one of the foundations of computer science. Even a very refined theory of computer science is developed from this: This is complexity theory, which I could say today lies in the center of theoretical computer science, specially after the work of Cook [3] and Karp [7]. They made use of the notion of speed of computation; now the question is not whether a set is decidable, but whether it is decidable in a time that can be affordable by present-day machines, or whether it is a "crackable" problem.

The fundamental question is

$$P \neq NP?$$

This is a very famous conjecture and it is the most important new problem in mathematics in the last half of this century; it is only 20 years old. To me it is the most beautiful new problem in mathematics. Very hard to solve, a very fine notion coming from this theory of Cook and Karp.

So we have a very active subject in this area but there is something that is missing. I have talked about the need for a notion of definable algorithm, yet the algorithms mathematicians have used for a couple of thousand years at least do not fit into this framework. The algorithms we are talking about have to do with real numbers, and specially since the time of Newton they have had to do with differential equations, nonlinear systems, etc. We see the notion of the real numbers $\mathbb{R}$ is central.

Newton's method to me is a paradigm of a great classical algorithm like the procedure of the Greeks for finding square roots, and it does not fit naturally into this framework, because the framework is quite discrete and to fit Newton's method into it requires destroying geometric concepts. One can do this in a very cumbersome way —I find this a very destructive way— to deal with the algorithms of continuous mathematics with Turing machines.

Indeed, some work has now been done to adapt the Turing machine framework to deal with real numbers. Let me mention two such attempts. One of them is recursive analysis, which initially was worked out by Ker-I-Ko and Harvey Friedman [5] and the main name connected with it is Marian Pour-El. She worked with Richards [8] in developing a kind of real number analysis based on Turing machines. There has been very extensive work on this which deals with partial differential equations, and the way to deal with real numbers in this context is to consider a real number s defined by its decimal expansion

$$s = 1.2378\ldots .$$

A real number is computable in this sense if there exists a Turing machine which says that the first digit is 1, the second 2, the third 3, and so on, with the decimal point in the appropriate place. So a computable real number is given by a Turing machine. These work with computable real numbers and eventually provide a very successful theory. Similarly there is the notion of *interval arithmetic* from R. E. Moore (see [1]). In some ways it is close to the work of Pour-El but in a quite different direction. Thus, the foundations are probably being laid for a theory of computation over the real numbers.

Now, continuing from a very different point of view, I will devise a notion of computation taking the real numbers as something axiomatically given. So, a real number to me is something not given by a decimal expansion; a real number is just like a real number that we use in geometry or analysis, something which is given by its properties and not by its decimal expansion. Eventually, I will talk about a notion of computability over the real numbers which takes this point of view. There one thinks of inputing a real number not as its decimal expansion but as an abstract entity in its own right.

Some mathematicians and computer scientists have trouble with the idea that a machine takes as input an arbitrary real number. I wrote a paper [13] on precisely this point, saying that here one idealizes, as in physics Newton idealized the atomistic universe —making it a continuum— in order to use differential equations. One can idealize the machine itself by conceiving it as allowing an arbitrary real number as input, but I am not going to argue about this point today.

In a preliminary phase, I was concerned with the problem of root finding for polynomials for many years. In that process I faced the kind of objects known as *tame machines.* It is not a theory of computation, but just a preliminary.

We take as input now the coefficients $\{a_0, a_1, \ldots, a_d\}$ of a complex polynomial f of degree d, and we think of it over the real numbers, i.e., each a_i is given by its real and imaginary parts. Thus, we think of this as the input and we describe the computation in the language of flowcharts. Then comes a box describing the computations. We replace a by $g(a)$, where g is a rational function. This vector of numbers $[a_0, a_1, \ldots, a_d]$ can be considered as a state and in this step this state is transformed by a rational function. Then we can put down and answer the question of whether some coordinate of the state is less than or equal to 0. Depending on the outcome of this comparison, we continue along the corresponding branch.

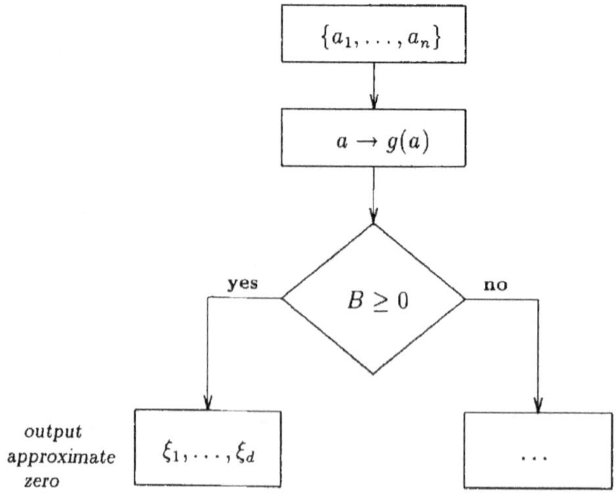

So we go down the tree in this way, and eventually we may output the approximate zeroes $\{\xi_1, \ldots, \xi_d\}$ of f up to some ε.

In fact this is a good way to express an algorithm for solving this equation. We next ask: How about these nodes which branch? To what extent are they necessary?

The topological complexity of a problem in general is the minimum number of these branch nodes for any machine which solves the problem.

I am not being completely precise about what resolving a problem means. One can imagine this example as a prototype of the general situation of a machine that solves problems. In particular, for this problem of finding the zeroes of polynomials we have the notion of its *topological complexity*. And the theorem [12] is as follows:

THEOREM. *The topological complexity of the root finding problem is greater than or equal to* $\log d$.

So the topological complexity increases with the degree, and the proof of this theorem actually is not so easy; it uses the cohomology of the braid group worked by Fuchs in Russia [6]. Subsequently, Vasiliev [14] extended this bound to $\simeq d$. So the answer eventually emerged that the topological complexity grows linearly with d and this is a sharp bound.

One notes that the Turing machine framework could never deal with this way of looking at all possible algorithms, even in this limited class. There is no useful way of thinking about it in terms of Turing machines, whereas using this kind of tree we were able to give necessary conditions on all algorithms, what we call *lower bound theorems*. There is some early work dealing with this kind of tree, but this is the first time we have obtained topological complexity results using algebraic topology.

Then, shortly after this, we did a joint work with Lenore Blum and Mike Shub [2] and developed this into a complete theory of computation over the reals by allowing loops. We certainly increased the computational power of tame machines by allowing loops to give a notion of computation in general. This situation is reflected in the next picture.

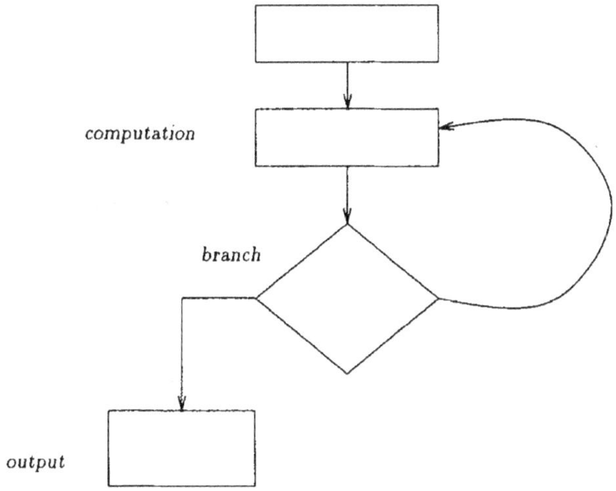

So here we have also

- an input space $\mathbf{R}^l$,
- an output space $\mathbf{R}^k$,
- and also a state space $\mathcal{S} = \mathbf{R}^j$ for things happening inside the machine.

If l, k, j are finite, this essentially defines a machine. We can take here an oriented graph where the nodes are computation nodes given by rational functions, branch nodes given by inequalities, and input and output defined inside accordingly. At each computation node there is a single output, a single branch going out of the node. A decision node has two. Remember that the number of input branches is arbitrary except for the input node where nothing comes in and there is a branch going out, and an output node, where nothing comes out.

This gives a theory of computation for finitely dimensional input and output spaces motivated directly by the flowcharts used in scientific computation. Yet the full theory will have to allow $l, k, j = \infty$ and we will have to have a little more technical process to access far out coordinates, but this is the idea.

This model gives an algebraic flavour to the process of computation. We defined this not only over the real numbers but over any ordered ring, eventually any ring. In particular, if we take the ring to be $\mathbf{Z}$, the input space to be a subset of $\mathbf{Z}$, the output space again $\mathbf{Z}$, and the state space $\mathbf{Z}^\infty$, we obtain Turing theory, and so this extends the Turing theory of computation.

We can now say that a Turing computable function is one which is given on some Ω of the machine, a domain of inputs, by following the flow of the machine and doing what is said at each node.

$$\{admissible\ inputs\} = \Omega_M \xrightarrow{\Phi_M} \mathbf{R}^k$$

And this essentially is a complete picture of what we mean by *computable function.* A function Φ_M defined by a machine going from the admissible inputs or the halting set of the machine to the output set.

And it is precisely equivalent —or practically so— to the notion of *Turing computable* in the case when the ring is the ring of integers.

We have developed for this model notions of computability itself; we have, for instance, shown the existence of universal machines. We have a complexity theory and the problem "$P \neq NP$?" is also defined over these rings; for example, the theory for $\mathbf{R}$ or $\mathbf{C}$ possesses universal or NP-complete problems just as in the case of Cook and Karp.

An NP-complete problem over $\mathbf{C}$ (a machine over $\mathbf{C}$ is like one over $\mathbf{R}$ except that the branch nodes just ask "$\neq 0$?") is the following:

Does a system of quadratic polynomials have a zero?

The idea of the reduction is to have more polynomials than variables. So it is an open question whether there is a machine that can decide in polynomial time if there is such a zero. All this is written very carefully in our paper.

In Barcelona, Felipe Cucker [4] gave an analog for the real numbers of the arithmetical hierarchy of classical recursion theory. There have been developments in different directions in the theory of computation of these machines from the point of view both of complexity theory and of computability.

There has also been a lot of controversy and criticism. Let me deal with one main point, making some comments on two sharp critiques by Pour-El and Moore. Our theory of computation is very different from their two theories. In a way this is more or less the basis of their criticism. It has to do with the branching

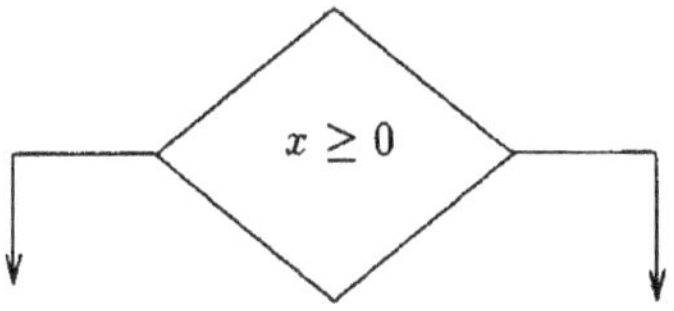

We branch according to whether one of the coordinates of the state space is greater than or equal to 0. This is in some respects one of the most controversial elements in the kind of machines we have, because the question is that an actual machine cannot do this. Given a number, for example

$$0.0000\ldots00\ldots,$$

it may or may not have a one after that eventually. If it never has a one, and we input it to an actual machine, we can never decide this question. If it does have a one, we wait long enough and we can decide it.

So we have a problem here when branching at ≥ 0, or equivalently at $= 0$, and this is the focus of attack of both Pour-El and Moore. Let me give an example here.

Both of their theories of computation lead to a notion of computable function which is continuous. Every computable function here is continuous. Even in a strong sense: They have to be constructively continuous.

Now the clue to this lies in the philosophy of thinking about the real numbers as abstractly given, and choosing the idealization of the right machines. For example, in scientific computation this is the kind of computation carried out traditionally by algorithms like Newton's method. One does test if something is ≥ 0, then do this, if not do something else.

Moreover, the need for these branchings is given by our earlier results on topological complexity. Topological complexity states that if one wants to find zeroes of polynomials then one has to branch, and the number of branchings in the machine is given approximately by the degree. Even to approximately solve the fundamental theorem of algebra one needs to branch. So I would imply that these two theories of computation do not lead even to an approximate solution of the fundamental theorem of algebra.

Here I would refer to a letter I received from Moore a year ago. I do not intend to dwell here on my opinion that numerical analysis and scientifical computing have weak foundations. Moore is the main developer of interval arithmetic and he wrote that "There are foundations for scientifical computation. More than 2000 papers and dozens of books. I invite you to read all of Aberth's book" [1]; it is a book that Moore even sent to me. He said "It will open your eyes to a whole new world." So I opened the book —actually a few weeks ago— and read on page 34 of the book (called *Precise Numerical Analysis*) "The problem of deciding whether two computable real numbers are equal is therefore a computational problem one should avoid." But problem 3.1 on this book reads: "Given two numbers a, b decide whether $a = b$." Later, on page 62, Aberth says: "Solve the problem 6.1: Find k decimals for the real and imaginary parts of the zeroes of a polynomial of positive degree." But the answer to this solvable problem —the fundamental theorem of algebra— needs to pass through d versions of this single problem that "one should avoid."

Marian Pour-El very kindly sent me a review that she has given of our paper in the Journal of Symbolic Logic [9], in which she says that it is a very good, highly developed theory of computability over the reals. In the review she confirms that in her theory she can only produce continuous functions and so she cannot solve the fundamental theorem of algebra, not even approximately.

What I will now do is to pass on to something which relates to this problem of NP-completeness if only a little indirectly. This is work done jointly with Mike Shub in the last few months. It is an example of an algorithm which fits into our framework. But it is a simple algorithm, so the fact that it is an algorithm in our strict sense is secondary. It is the problem of the complexity analysis of Bézout's theorem. Let me say a little bit about what this is. The situation we look at is as follows: We have a polynomial system

$$f : \mathbf{C}^n \to \mathbf{C}^n$$

of n polynomials in n variables of degrees $d_1, \ldots, d_n$ respectively. One wants to find an algorithm and analize its speed for solving the equation

$$f(z) = 0\,;$$

not to produce a solution but to analize how much time it takes. The idea is to make complexity analysis on this.

The work done so far on polynomial equation solving can be summarized by dividing it into two parts: one is Newton's method as the basic algorithm —it is essentially the method used by the Greeks for finding square roots— and the other method is elimination theory, a very algebraic method; it works over arbitrary fields. In the first one we are using some kind of norm or metric, so it is metric-oriented. It is the method of choice of numerical analysts. The second is probably the method that would be chosen by a computer scientist. My own inclination is to the first side. Numerical analysts have a better focus on the problem. They do not have a complexity theory or any kind of foundation, but they have a better instinct about how to solve this problem.

In any case, what we use is some global version of Newton's method to solve Bézout's theorem. That is, we use Newton's method to follow a path in the space of polynomials.

All these are homogeneized. It is more elegant to think entirely in terms of homogeneous coordinates and projections

$$\mathcal{P}_{(d)} = \{f : \mathbf{C}^n \to \mathbf{C}^n\}$$

and

$$\mathcal{H}_{(d)} = \{f : \mathbf{C}^{n+1} \to \mathbf{C}^n \text{ homogeneous of degree } d\}\,,$$

with

$$\mathcal{P}_{(d)} \equiv \mathcal{H}_{(d)}\,.$$

Thus we are going to work in projective space, and the main thing we will analize is a projective version of Newton's method which is due to M. Shub [11]. The previous work has been done mostly in one variable on this problem, using Newton's method to solve this; I spent many years doing that. Jim Renegar [10] has some extension to n variables.

What we want to do here is give a very conceptual process. All the ideas are lying there; we can try to understand the best, most elegant ways of looking at algorithms to solve this. In this way perhaps eventually we will be able to see more clearly the problem "$P \neq NP$?" over the real numbers or the complexes. I hope it will eventually shed some light on the practical problem "$P \neq NP$?"

Given the space $\mathcal{H}_{(d)}$ we consider a function f_0 for which we know the zeroes. The zeroes of f_0 could be given by a set of intersections in a grid so we get a set of equally spaced zeroes

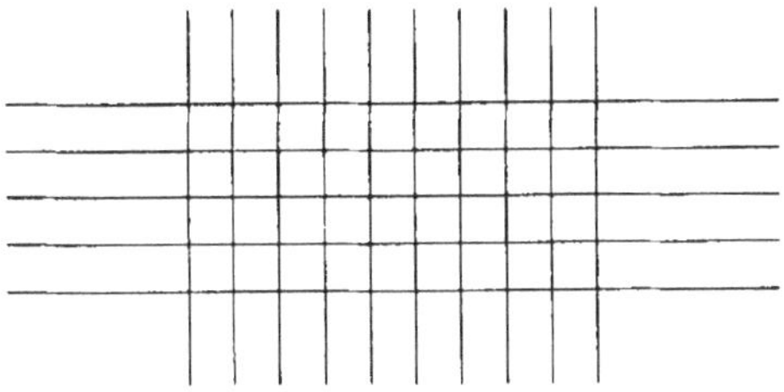

This could be the initial element in $\mathcal{H}_{(d)}$ for which we know the answer, and we simply homotope that back

$$f_t = tf + (1-t)f_0,$$

where t goes from 0 to 1. Let us denote by $\mathcal{F}$ the curve f_t in the space $\mathcal{H}_{(d)}$.

We try to trace the zeroes, which we know for f_0, to give the answer for f_1. This seems to be a very good method that has been used for the last decade or two. It embodies some kind of global Newton's method. The idea is to consider some sequence t_i and apply Newton's method to the function $f_{t_{i+1}}$, starting from some approximation X_i of the solutions for f_{t_i}

$$N_{f_{t_{i+1}}}(X_i) = X_{i+1}$$

where, here, $N_f(X)$ stands for applying Newton's method to solve f starting with the initial guess X. In the projective space $\mathbf{P}_n(\mathbf{R})$ we can see the paths given by the solutions X_t of f_t and the algorithm provides a sequence of points X_i following this path very closely.

So, what kind of results can we expect here? What kind of things may we prove? Here is the main theorem. The question is how many iterative steps are necessary, how many t_i, in such a way that we can follow this path very closely, and our result is

THEOREM. *The number of steps is bounded above by*

$$\frac{\hat{\alpha} D^{3/2}}{\rho^2}$$

where $\hat{\alpha}$ is a universal constant (given by a set of equations which can be solved itself by Newton's method) which is approximately $1/16$, *D is* $\max\{d_1, \ldots, d_n\}$ *and ρ is the distance from the arc joining f_0 and f_1 to the discriminant variety.*

It should be recalled that the discriminant variety is the subset of $\mathcal{H}_{(d)}$ of all singular polynomial systems. It is the variety of polynomial systems which are degenerate at some zero. And this is an algebraic variety that we shall call Σ. The theorem says that what is crucial are not the coefficients of f. They do not even enter. In fact, not even the dimension comes directly here; this is even dimension-free. But what is crucial here is the distance ρ between $\mathcal{F}$ and the discriminant variety Σ. This is the crucial factor —the only factor— in estimating the complexity for finding the zeroes of a polynomial system. Now we have to make a little caveat here because we are not finding the zeroes of every polynomial system. There may be a continuum of zeroes and then we cannot do this. So we have to put some kind of condition, let us say to solve $f + \varepsilon$ where ε is a small polynomial. This is the thing we solve. We cannot find the solution of arbitrary polynomial systems; there may be a continuum of solutions, but for some deformation we can find the zeroes in a very exact sense.

Now, the great problem to me is: To what extent is the term $D^{3/2}$ necessary?

While we have no proof of this, we suspect that the D itself could be eliminated from the formula. For a polynomial in one variable, this is the fundamental theorem of algebra, and we show that we can take off the $3/2$ to get D. This is what we have done in the last months; the proof is in handwritten form. Since last week we believe that we can eliminate the D in the one variable case, but this uses the theory of Schlicht functions, which is only available for one variable. There is no theory of Bieberbach conjecture for more than one variable. If it is true, if it is D-free, if we can do this, then one can find for example one zero of a polynomial in one variable in a universal number of steps, say one hundred.

References

[1] O. Aberth, *Precise Numerical Analysis*, Brown Publishers, Dubuque, Iowa, 1988.

[2] L. Blum, M. Shub and S. Smale, On a theory of computation and complexity over the real numbers: *NP*-completeness, recursive functions and universal machines, *Bull. Amer. Math. Soc. (N.S.)* **21** (1989), no. 1, 1–46.

[3] S. A. Cook, The complexity of theorem-proving procedures, Proceedings 3rd ACM STOC (1983), 80–86.

[4] F. Cucker, The arithmetical hierarchy over the reals, to appear in *J. Logic Comput.*

[5] H. Friedman and K. Ko, Computational complexity of real functions, *Theoret. Comput. Sci.* **20** (1986), 323–352.

[6] D. Fuchs, Cohomologies of the braid group mod 2, *Functional Anal. Appl.* **4** (1970), 143–151.

[7] R. Karp, Reducibility among combinatorial problems, in *Complexity of Computer Computations*, R. Miller and J. Thatcher (eds.), Plenum Press, New York, 1972, 85–104.

[8] M. B. Pour-El and I. Richards, Computability and noncomputability in classical analysis, *Trans. Amer. Math. Soc.* **275** (1983), 539–560.

[9] M. Pour-El, Review of [2], to appear in *J. Symbolic Logic.*

[10] J. Renegar, On the efficiency of Newton's method in approximating all the zeroes of a system of complex polynomials, *Math. Oper. Res.* **12** (1987), 121–148.

[11] M. Shub, Some remarks on Bézout's theorem and complexity theory, to appear in *Proceedings of the Smalefest*, M. Hirsch, J. Marsden and M. Shub (eds.).

[12] S. Smale, On the topology of algorithms I, *J. Complexity* **3** (1987), 81–89.

[13] S. Smale, Some remarks on the foundations of numerical analysis, *SIAM Rev.* **32** (1990), no. 2, 211–220.

[14] V. Vasiliev, Cohomology of the braid group and the complexity of algorithms, to appear in *Proceedings of the Smalefest*, M. Hirsch, J. Marsden and M. Shub (eds.).

Stephen Smale
Mathematics Department
University of California
Berkeley, California 94720
USA

Transcribed from the videotape of the talk by Felipe Cucker, Francesc Rosselló and Álvaro Vinacua; revised by the author.

MORSE INEQUALITIES FOR A DYNAMICAL SYSTEM

BY STEPHEN SMALE[1]

Communicated by S. Eilenberg, October 24, 1959

1. Introduction. We consider dynamical systems (X, M), where X is a C^∞ vector field on a C^∞ closed manifold M satisfying the following conditions.

(1) There are a finite number of singular points of X, say $\beta_1, \cdots, \beta_k$, each of simple type. This means that at each β_i, the matrix of first partial derivatives of X in local coordinates has eigenvalues with real part nonzero.

(2) There are a finite number of closed orbits (i.e., integral curves) of X, say $\beta_{k+1}, \cdots, \beta_m$, each of simple type. This means that no characteristic exponent (see, e.g., [2]) of β_i, $i>k$, has absolute value 1.

(3) The limit points of all the orbits of X as $t\to\pm\infty$ lie on the β_i. In other words, denote by ϕ_t the 1-parameter group of transformations generated by X (as we do throughout this paper). Let

$$\alpha(y) = \lim_{t\to-\infty} \text{set}\, \phi_t(y), \qquad \omega(y) = \lim_{t\to\infty} \text{set}\, \phi_t(y), \qquad y \in M.$$

Then for each y, $\alpha(y)$ and $\omega(y)$ are contained in the union of the β_i.

(4) The stable and unstable manifolds of the β_i (see §2 for the definition) have normal intersection with each other. More precisely for each i let W_i be the unstable manifold and W_i^* the stable manifold of β_i and for $x\in W_i$ (or W_i^*) let W_{ix} (or W_{ix}^*) be the tangent space of W_i (or W_i^*) at x. Then for each i, j if $x\in W_i\cap W_j^*$,

$$\dim W_i + \dim W_j^* - n = \dim (W_{ix} \cap W_{jx}^*).$$

See [5] for example for more details.

(5) If β_i is a closed orbit there is no $y\in M$ with $\alpha(y)=\omega(y)=\beta_i$.

First we remark that systems satisfying (1)–(5) may be very important because of the following possibilities.

(A) It seems at least plausible that systems satisfying (1)–(5) form an open dense set in the space (with the C^1 topology) of all vector fields on M.

(B) It seems likely that conditions (1)–(5) are necessary and sufficient for X to be structurally stable in the sense of Andronov and Pontrjagin [1]. See also [6].

(A) and (B) have been proved for the case M is a 2-disk, [3] and [9].

---

[1] Supported by a National Science Foundation Postdoctoral Fellowship.

We expect to have more to say about this subject at another time. It is true that conditions (1)–(5) are independent.

With (X, M) as above let $\sigma_i = \dim W_i$. Then if $i \leqq k$, σ_i is the number of eigenvalues associated to β_i with real part positive. Let a_q be the number of β_i, $i \leqq k$ with $\sigma_i = q$. If $i > k$, σ_i is one more than the number of characteristic exponents of β_i with absolute value greater than one. Let b_q be the number of β_i, $i > k$, with $\sigma_i = q$.

The main goal of this paper is to prove

THEOREM 1.1. *Let (X, M) be a system satisfying* (1)–(5), *K any field, R_q the rank of $H^q(M, K)$, and $M_q = a_q + b_q + b_{q+1}$. Then M_q and R_q satisfy the Morse relations*

$$M_0 \geqq R_0,$$
$$M_1 - M_0 \geqq R_1 - R_0,$$
$$M_2 - M_1 + M_0 \geqq R_2 - R_1 + R_0,$$
$$\cdots\cdots\cdots\cdots\cdots$$
$$\sum_{k=0}^{n} (-1)^k M_k = (-1)^n \chi$$

where $\dim M = n$ *and χ is the Euler characteristic of M with respect to K.*

Theorem 1.1 contains the Theorem of Èl'sgol'c [4] which excludes closed orbits. It also contains Reeb's theorem [11] which excludes singular points. However, both Èl'sgol'c and Reeb made the highly restrictive assumption[2] that no orbit joined saddle points (i.e., β_i, $i \leqq k$ with $\sigma_i \neq 0, n$) or saddle type closed orbits (i.e., β_i, $i > k$ with $\sigma_i \neq 1, n$).

Also it follows from the following theorem which we prove elsewhere that Theorem 1.1 includes the classical theorem of Morse [8] for a function f on M with nondegenerate critical points.

THEOREM 1.2. *If $X = \text{grad} f$, f a C^∞ function on M with nondegenerate critical points, then X can be C^1 approximated by a C^∞ field Y on M such that (Y, M) satisfies* (1)–(5) *with no closed orbits.*

2. **Construction of the stable and unstable manifolds.** 2.1. Suppose β is a singular point of simple type of the C^∞ system (X, M). Let k be the number of eigenvalues associated to β with real part positive. Then (e.g., [2, p. 330]) there is a k dimensional C^∞ submanifold W of M passing through β such that if $x \in W$ then $\alpha(x) = \beta$. If $k = 0$, let

[2] Reeb has asked me to note that his footnote 3, 2nd paragraph, of [11, p. 62] (that this assumption is unnecessary) is incorrect

$W=\beta$. Then W is tangent at β to the linear subspace of the tangent space M_β of M at β defined by these k eigenvalues [2, p. 333]. W is called the *unstable manifold* of X at β. Let R^k denote Euclidean k-space considered as a vector space. We will show that W is the image of a continuous 1-1 onto map $f: R^k \to W$, with $f(0)=\beta$, and f is C^∞ with Jacobian of rank k except at 0. Consider the new system X^* obtained by reversing the direction of each vector of X on M. Then β is a simple singularity of X^* and the above applies to yield the unstable $(n-k)$-dimensional manifold W^* of X^* at β. Call W^* the *stable manifold* of X at β. Note W and W^* have normal intersection at β.

2.2. Suppose β is a closed orbit of (X, M) of simple type. Let $k-1$ be the number of characteristic exponents of β with absolute value greater than one. Then ([7] or [13]) there is a k-dimensional C^∞ submanifold W of M passing through β such that if $x \in W$ then $\alpha(x)=\beta$. If $k=1$ let $W=\beta$. Also W is tangent at each point y of β to the linear subspace of M_y defined by these $k-1$ characteristic exponents and the tangent vector of β at y. Call W the *unstable manifold* of X at β. We will show there is a continuous 1-1 onto map $f: R^{k-1} \times S^1 \to W$, with $f(0 \times S^1)=\beta$, and except along $0 \times S^1$ is C^∞ with Jacobian of rank k. Similarly to 2.1, one defines the *stable manifold* W^* of X at β whose dimension in this case is $n-k+1$.

We now construct the map f of 2.1.

There exists[3] a differentiably imbedded $(k-1)$-sphere K in W, which is everywhere transversal to X. Let S_0 be the unit sphere of R^k and $h: S_0 \to K$ be a diffeomorphism. (A diffeomorphism is C^∞ homeomorphism with a differentiable inverse.) Let ψ_t be the 1-parameter group of transformations of R^k generated by the vector field $Y(x)=x$ on R^k. For $x \in R^k$, $x \neq 0$, let $t(x)$ be the unique t such that $x/\|x\| = \psi_{t(x)}(x) \in S_0$. Then let $f(0)=\beta$ and $f(x)=\phi_{-t(x)} h \psi_{t(x)}(x)$. It is easy to check that $f: R^k \to W$ thus defined has the desired properties.

To construct the map $f: R^{k-1} \times S^1 \to W$ of 2.2, first let Y be the vector field $(x, 1)$ on $R^{k-1} \times S^1$. Then if ψ_t is the 1-parameter group of transformations generated by Y we have $\psi_t(x, 0) = (xe^t, t \bmod 2\pi)$. Let $R^{k-1} = R^{k-1} \times 0 \subset R^{k-1} \times S^1$, and C be the unit ball in R^{k-1}, $\partial C = S_0$. Define $q: R^{k-1} \to R^{k-1}$ by $q(x) = xe^{2\pi}$ and let $q^i S_0 = S_i$ for each integer i.

Let Q be a surface of section (i.e., transversal to X, see [6]) locally about a point of β in W, diffeomorphic to a $(k-1)$-cell. Then [6] the orbits of X define a diffeomorphism $h: Q \to Q$ in a neighborhood of $\beta \cap Q$ leaving $\beta \cap Q$ fixed. There is[3] a closed k-cell B differentiably

[3] By Liaponov theory for example.

imbedded in Q, $\partial B = F_0$ such that $h^{-1}(F_0)$ is contained in the interior of B. Let $h^i(F_0) = F_i$, $i \leqq 0$.

Let f be an orientation preserving diffeomorphism of a neighborhood V_0 of S_0 in R^{k-1} into a neighborhood of F_0 in Q. Then extend f to a neighborhood of $\bigcup_{i \leqq 0} S_i$ in R^{k-1} into a neighborhood of $\bigcup_{i \leqq 0} F_i$ in Q by the formula

$$f(x) = h^{-i} f q^i(x), \qquad x \in \text{nbd. } V_i \text{ of } S_i. \tag{2.3}$$

This makes sense for an appropriate choice of the V_i's. Now consider the closed region U in R^{k-1} bounded by S_0 and S_{-1}. We have defined f in a neighborhood of the boundary ∂U of U. After restricting f to a smaller neighborhood of ∂U, f can be extended to a diffeomorphism of all of U into the region of Q bounded by F_0 and F_{-1}. This fact follows from arguments which are now standard in differential topology. We won't include them here. Then as in 2.3 we can extend f to a map of all of C into B which is a diffeomorphism except at $f(0) = \beta \cap Q$.

Next define f on $P = \{\psi_t(x) \mid x \in C,\ t < 0\}$ by the following: Let $\tau(x, \theta)$ be the smallest positive number such that $\phi_{\tau(x,\theta)} f \psi_{-\theta}(x, \theta)$ has θ as its second coordinate in a fixed product structure $Q \times \beta, (x, \theta) \in P$. Then let $f(x, \theta) = \phi_{\tau(x,\theta)} f \psi_{-\theta}(x, \theta)$. Define f: $0 \times S^1 \to \beta$ by $f(0 \times \theta) = \theta$.

Consider now the surface of section $S_{-1} \times S^1 = A$ in $R^{k-1} \times S^1$ and its image under f. Restrict f to the closure of the bounded component K of A. Finally extend f to all of $R^{k-1} \times S^1$ as follows. For $y \in R^{k-1} \times S^1 - K$ let $t(y)$ be the unique t such that $\psi_{t(y)}(y) \in A$. Then let $f(y) = \phi_{-t(y)} f \psi_{t(y)}(y)$. After a change of parameter near A, f will have our desired properties.

3. **Implications of** (1)–(5). Assume throughout this section that (X, M) is given as in §1. If β_i is a singular point then f_i: $R^k \to W_i$ is as in 2.1. If β_i is a closed orbit then f_i: $R^{k-1} \times S^1 \to W_i$ is as in 2.2.

LEMMA 3.1. *If* $x \in M$, $\alpha(x) = \beta_i$, $\omega(x) = \beta_j$, *then* $\dim W_i \geqq \dim W_j$ *and equality can occur only if* β_j *is a closed orbit.*

PROOF. Clearly $x \in W_i \cap W_j^*$ and by (4) we have that $\dim W_i + \dim W_j^* - n \geqq 1$. But $\dim W_j^* = n - \dim W_j$ if β_j is a singular point and $\dim W_j^* = n - \dim W_j + 1$ if β_j is a closed orbit. Then 3.1 follows.

See [12] for the following.

LEMMA 3.2. *Suppose* $W_i \cap W_j^* \neq \emptyset$ *and* $x \in W_j$. *Then there exists a cell neighborhood* H *of* x *in* W_j *such that given* $\delta > 0$, *there is a* $y \in W_i$ *with* $d(x, y) < \delta$ *and if* $\dim W_i = \dim W_j$, *there is a subcell* K *of* W_i *such that* H *and* K *are within* δ *in a* C^1 *metric.*

Define $\partial W_j = \{\lim_{k\to\infty} f_j(x_k) \mid x_k$ any sequence in R^k with no lps.$\}$. Then let $\partial^2 W_j = \partial(\partial W_j)$, etc. Note Cl $W_i = W_i \cup \partial W_i$.

LEMMA 3.3. *If* $W_i \cap W_j^* \neq \emptyset$, $\partial W_i \supset W_j$.

This follows from 3.2.

LEMMA 3.4. *Suppose* dim W_i = dim W_k = dim W_j. *If* $W_i \cap W_k^* \neq \emptyset$ *and* $W_k \cap W_j^* \neq \emptyset$ *then* $W_i \cap W_j^* \neq \emptyset$.

PROOF. Let $x \in W_k \cap W_j^*$; apply 3.2 using the fact that $W_i \cap W_k^* \neq \emptyset$. Since W_k and W_j^* have normal intersection at x, it follows from 3.2 that $W_i \cap W_j^* \neq \emptyset$.

LEMMA 3.5. *Suppose* $W_{i_k} \cap W_{i_{k+1}}^* \neq \emptyset$, $k = 1, \cdots, m$. *Then* $W_{i_k} \neq W_{i_j}$ *if* $j \neq k$.

PROOF. First note by 3.1, dim $W_{i_{k+1}} \leqq$ dim W_{i_k} and equality occurs only if $\beta_{i_{k+1}}$ is a closed orbit. This implies we can restrict ourselves to the case of the lemma where all the W_{i_k}'s are of the same dimension. Then if $W_{i_k} = W_{i_j}$, $k \neq j$, 3.4 implies that $W_{i_k} \cap W_{i_j}^* \neq \emptyset$. This contradicts condition (5).

LEMMA 3.6. *If* $\partial W_\gamma \cap W_\delta \neq \emptyset$, *then there is a sequence* $W_{i_1}, \cdots, W_{i_m}$ *such that* $W_{i_k} \cap W_{i_{k+1}}^* \neq \emptyset$, $W_\gamma = W_{i_1}$, *and* $W_\delta = W_{i_m}$.

PROOF. Let $\alpha(W_\delta^*) = \lim_{t\to-\infty} W_\delta^*$. Then it follows that Cl $W_\gamma \cap \alpha(W_\delta^*) \neq \emptyset$. Let $\beta_j \in$ Cl $W_\gamma \cap \alpha(W_\delta^*)$. Then $W_j \cap W_\delta^* \neq \emptyset$. If $j \neq \gamma$, similarly let $\beta_k \in$ Cl $W_\gamma \cap \alpha(W_j^*)$. Induction and 3.5 yield 3.6.

LEMMA 3.7. *If* $\partial W_i \cap W_j \neq \emptyset$, *then* $\partial W_i \supset W_j$ *and either* dim W^i > dim W_j *or* dim W_i = dim W_j, $W_i \cap W_j^* \neq \emptyset$, *and* β_j *is a closed orbit*

This follows from 3.6, 3.5, 3.3, 3.1, and 3.4.

LEMMA 3.8. *Each* W_i *is an imbedded* R^p *or* $R^{p-1} \times S^1$.

This follows from §2, 3.7 and (5).

LEMMA 3.9. $\partial^k W_i \neq \emptyset$, *any* i *for large enough* k.

If not there is a W_j such that $\partial^m W_j \cap W_j \neq \emptyset$. By 3.7 then $\partial W_j \cap W_j \neq \emptyset$, contradicting 3.8.

4. **On Morse theory.** A version of one of the standard theorems of Morse theory is stated in this section. The proof is a short well-known argument using the exact cohomology sequence of a pair and for example can be found in [10].

THEOREM 4.1. *Let* M *be an* n*-dimensional topological space with closed subspace* L^p *for each integer* p *such that* $L^p \supset L^{p-1}$, *and there* *exist integers a, b with* $L^a = \emptyset$ *and* $L^b = M$. *Using any fixed cohomology theory and coefficient field, assume dimension* $H^q(L^p, L^{p-1})$ *is finite for each p and q. Let* $B_q = \dim H^q(M)$ *and* $M_q = \sum_{r=a}^{b} \dim H^q(L^r, L^{r-1})$. *Then* M_q *and* B_q *satisfy the Morse relations*

$$M_0 \geqq B_0,$$
$$M_1 - M_0 \geqq B_1 - B_0,$$
$$M_2 - M_1 + M_0 \geqq B_2 - B_1 + B_0,$$
$$\cdots\cdots\cdots\cdots$$
$$\sum_{k=0}^{n} (-1)^k M_k = \sum_{k=0}^{n} (-1)^k B_k.$$

5. **Proof of the main theorem.** Define $K^p = \bigcup_{\dim W_i \leq p} W_i$. Thom in [14] considers subspaces related to K^p to prove the classical Morse inequalities. By 3.7 it follows that the K^p are closed sets. However, examples show that K^p has the following bad property. It may be that for $W_i \subset K^p$, ∂W_i is not contained in K^{p-1}. To avoid this we define a new structure on M.

We define by induction a sequence of closed subsets L_i of M with $L_i \supset L_{i-1}$ and $L_i = M$ for large enough i. Define $L_0 = \emptyset$, and if L_{i-1} has been defined, let L_i be the union of all the W_j whose boundary lies in L_{i-1}. It is immediate that $L_i \supset L_{i-1}$, that L_i is closed in M and that $L_i - L_{i-1}$ consists of a disjoint union of W_j. It follows from 3.9 that there is an integer b such that $L_b = M$.

One can construct an example to show that the L_i need not be locally connected and that for $W_0 \subset L_i - L_{i-1}$, $W_1 \subset L_{i-1}$, $\dim W_0 = \dim W_1$.

Lemma 5.1. *Using Čech theory if* M_q *is as in* 1.1 *then*

$$M_q = \sum_{i=0}^{b} \dim H^q(L_i, L_{i-1}).$$

Proof. As noted previously $L_i - L_{i-1}$ consists of a disjoint union of the W_j and as i ranges from 0 to b all the W_j are obtained. Denoting cohomology with compact carriers by K_K^q, since $H_K^q(P-Q) = H^q(P, Q)$ for Čech theory, we have

$$\sum_{i=0}^{b} \dim H^q(L_i, L_{i-1}) = \sum_{\text{all } W_j} \dim H_K^q(W_j).$$

Using 3.8 and Poincaré duality $H_K^q(W_j) = H_{\dim W_j - q}(W_j)$. Furthermore $\dim H_0(W_j) = 1$ for all j, $\dim H_1(W_j) = 1$ if β_j is a closed orbit and $\dim H_p(W_j) = 0$ otherwise. The lemma follows.

Theorem 1.1 follows from 4.1 and 5.1.

6. **An analogue of the main theorem.** Suppose instead of a vector field X on M, we just have given a C^∞ diffeomorphism h on M which satisfies certain conditions analogous to (1)–(5).

(1′) There are a finite number of periodic points (i.e., $x \in M$ such that $h^p(x) = x$ for some integer p) of h of simple type (i.e., the differential of h at p has no eigenvalue of absolute value 1).

(3′) The limit points of all the orbits of h (i.e., $\{h^p(x) \mid \text{all integers } p\} =$ orbit of x) are periodic points.

(4′) The "stable" and "unstable" manifolds of the periodic points have normal intersection.

The previous theory extends to cover this case. In particular if M_q is the number of periodic points with q eigenvalues having absolute value greater than one, the Morse relations of 1.1 hold.

One can ask the corresponding questions of (A) and (B) of §1 for the above situation.

References

1. A. A. Andronov and L. S. Pontrjagin, *Systèmes grossiers*, Dokl. Akad. Nauk vol. 14 (1937) pp. 247–250.

2. E. A. Coddington and N. Levinson, *Theory of ordinary differential equations*, New York, McGraw-Hill, 1955.

3. H. De Baggis, *Dynamical systems with stable structure*, Contributions to the Theory of Nonlinear Oscillations, Princeton University Press, vol. 2, 1952, pp. 37–59 (Annals of Mathematics Studies, no. 29).

4. L. E. Èl'sgol'c, *An estimate for the number of singular points of a dynamical system defined on a manifold*, Amer. Math. Soc. Translations, no. 68, 1952. Translated from Mat. Sb. (N.S.) vol. 26 (68) (1950) pp. 215–223.

5. R. K. Lashof and S. Smale, *Self-intersections of immersed manifolds*, J. Math. Mech. vol. 8 (1959) pp. 143–158.

6. S. Lefschetz, *Differential equations: geometric theory*, New York, Interscience Publishers, 1957.

7. D. C. Lewis, *Invariant manifolds near an invariant point of unstable type*, Amer. J. Math. vol. 60 (1938) pp. 577–587.

8. M. Morse, *Calculus of variations in the large*, Amer. Math. Soc. Colloquium Publications, vol. 18, 1934.

9. M. Peixoto, *On structural stability*, Ann. of Math. vol. 69 (1959) pp. 199–222.

10. E. Pitcher, *Inequalities of critical point theory*, Bull. Amer. Math. Soc. vol. 64 (1958) pp. 1–30.

11. G. Reeb, *Sur certaines propriétés topologiques des trajectoires des systèmes dynamiques*, Acad. Roy. Belg. Cl. Sci. Mém. Coll. in 8° 27 no. 9 (1952).

12. S. Smale, *On structural stability*, to appear.

13. S. Sternberg, *Local contractions and a theorem of Poincaré*, Amer. J. Math. vol. 79 (1957) pp. 809–724.

14. R. Thom, *Sur une partition en cellules associée à une fonction sur une variété*, C. R. Acad. Sci. Paris vol. 228 (1949) pp. 973–975.

Institute for Advanced Study

DYNAMICAL SYSTEMS AND THE TOPOLOGICAL CONJUGACY PROBLEM FOR DIFFEOMORPHISMS

By S. SMALE

For simplicity we consider an ordinary *differential equation* (or a dynamical system) to be a C^r vector field on a C^r manifold which generates a 1-parameter group of diffeomorphisms. An *equivalence* (or topological equivalence) between two differential equations is a homeomorphism preserving sensed trajectories (or orbits). The qualitative problem of differential equations is to obtain information on equivalence classes of differential equations on a given manifold.

This motivates us to consider the topological conjugacy problem for diffeomorphisms (a diffeomorphism is a differentiable homeomorphism with a differentiable inverse). More precisely, we say two diffeomorphisms $T_1, T_2: M \to M$ (say C^r diffeomorphisms of a C^r manifold, r always positive) are topologically conjugate if there exists a homeomorphism $h: M \to M$ such that $hT_2 = T_1 h$. The problem then is to study the topological conjugacy classes of diffeomorphisms of a given manifold.

There are several reasons for studying the latter problem, the most important being the following. It appears that usually a qualitative problem in differential equations has an analogue in the conjugacy problem. This analogue is a little simpler than the original, and if solved, its solution seems to give a way of doing the original problem. In any case, everything said in what follows on the conjugacy problem can be translated into statements about differential equations. At the end of our survey we indicate how this can be done.

It should also be noted that the above problems may be viewed as special cases in the study of a non-compact Lie Group G acting differentiably on a manifold, corresponding to $G = R$ and $G = Z$.

As enunciated in [7] for differential equations, the main conjugacy problem as we see it is the following. Given compact M, let $\mathcal{D}_M$ be the space of C^r diffeomorphisms of M in the C^r topology (diffeomorphisms are C^r close if they are pointwise close together with their first r derivatives). Then one seeks an open dense subset $\mathcal{C}$ of $\mathcal{D}_M$, somehow amenable to classification (say by numerical and algebraic invariants).

A fruitful notion relative to this problem is that of a structurally stable diffeomorphism (the analogous definition for a differential equation was given by Andronov and Pontrjagin in 1937; see [3]). The h in the definition of topologically conjugate is called an equivalence between T_1 and T_2, and if h is pointwise within ε of the identity (in some fixed metric on M) it is called an ε-equivalence. Then a diffeomorphism $T_1: M \to M$, compact M, is ***structurally stable*** if given $\varepsilon > 0$, there exists $\delta > 0$ such that if $d_{c^1}(T_1, T_2) < \delta$ for some diffeomorphism $T_2: M \to M$, then T_1 and T_2 are ε-equivalent. Here d_{c^1} is a C^1 metric on $\mathcal{D}_M$.

The problem of structural stability (for diffeomorphisms) is: given M, are the structurally stable diffeomorphisms dense in $\mathcal{D}_M$? If dim $M>1$, this is an open and difficult problem. In fact it is not known if there exists even one structurally stable diffeomorphism on a given manifold.

In any case it is clear that the periodic points (i.e. points $x\in M$ such that $T^m x=x$, $m\neq 0$, T^m the composition $T:M\to M$ with itself m times) and associated global stable and unstable manifolds will play a basic role in the topological conjugacy problem. So at this point we give the "stable manifold theorem". For more details and history, see [10].

Let $T:M\to M$ be a diffeomorphism with fixed point $p\in M$. The derivative of T at p is a linear automorphism of the tangent space M_p of M at p. The point p will be called an *elementary* fixed point of T if this derivative has no eigenvalue of absolute value one.

(A) STABLE MANIFOLD THEOREM. *Let p be an elementary fixed point of a C^∞ diffeomorphism $T:M\to M$ and E_1 the (eigen) subspace of M_p corresponding to the eigenvalues of the derivative of T at p of absolute value less than one. Then there is a C^∞ map $R:E_1\to M$ which is an immersion (i.e. with Jacobian of rank$=$dim E_1 everywhere), 1–1, and has the property $TR=RT_1$ for some contraction $T_1:E_1\to E_1$. Also $R(p)=p$ and the derivative of R at p is the inclusion of E_1 into M_p.*

By a *contraction* we mean a diffeomorphism T_1 of E_1 onto itself such that there is a differentiably imbedded disk D in E_1 with $T_1 D\subset$ interior D, $\bigcap_{i>0} T_1^i\, D=$origin of E_1, $\bigcup_{i<0} T_1^i\, D=E_1$. The map $R:E_1\to M$ or sometimes the image of R is called the *stable manifold* of p, or of T at p. The *unstable manifold* of T at p is the stable manifold of T^{-1} at p.

A point $p\in M$ will be called an *elementary* periodic point of $T:M\to M$ if p is an elementary fixed point of T^m for some m. An (*elementary*) *periodic orbit* is the finite set $\bigcup_{i\in Z} T^i\, p$ where p is an (elementary) periodic point. If $T^m\, p=p, m$ the minimal positive integer with this property, then m is called the least period of p. The definition of the stable manifold of an elementary periodic orbit is as follows. Let $\varphi:E_1\to M$ be the stable manifold of T^m at p where m is the least period of p,p in our periodic orbit. Then $R:E_1^i\to M$ is defined by $R=T^i\varphi$ where $0\leq i<m$ and E_1^i is a copy of E_1. Thus the stable manifold of a periodic orbit is a 1–1 immersion of the disjoint union of m copies of a Euclidean space. The unstable manifold of a periodic orbit is the stable manifold of the periodic orbit relative to T^{-1}.

To give some direction to our survey we list some axioms a diffeomorphism of a manifold might happen to satisfy. Fixing a compact manifold M, the space of C^r diffeomorphisms of M with the C^r topology is denoted as usual by $\mathcal{D}_M$. We will define T to be in a subspace $\mathcal{C}_M$ of $\mathcal{D}_M$ if and only if T has the following properties, i.e. T satisfies Axioms 1, 2′, 3.

Axiom 1. Every periodic point of T is elementary.

Axiom 1 implies that T has only countably many periodic points.

Axiom 2. T has the normal intersection property.

This means the following: if β_1,β_2 are elementary periodic orbits of T, let W_1^u, W_2^s be the unstable manifold of β_1, stable manifold of β_2 respectively. Then if $x\in W_1^u\cap W_2^s$ the tangent spaces of W_2^u and W_1^u at x span the tangent space of M at x.

The following is a strengthening of Axiom 2.

Axiom 2'. Let $\beta_1, \beta_2, W_1^u, W_2^s$ be as above. Then there exists a neighborhood $\mathcal{U}$ of β_1 with the following property. Each component of $\mathcal{U} \cap W_2^s$ is a cell which has a non-empty transversal intersection with W_1^u. The same is true with W_1^u, W_2^s replaced by W_1^s, W_2^u respectively.

Axiom 3. (*a*) Let Ω be the closure of the set of periodic points of T in M. Then for every $x \in M$, limit $m \to \pm \infty$ $T_x^m \subset \Omega$.

(*b*) The union of the stable manifolds of all the periodic orbits of T is a dense subset of M. The same is true of the unstable manifolds.

We pose two questions: (*a*) Is $\mathcal{C}_M$ open, dense, in $\mathcal{D}_M$? (*b*) Is $T \in \mathcal{C}_M$ a necessary and sufficient condition that T be structurally stable?

Although very possibly, in the final picture, $\mathcal{C}_M$ will not be the structurally stable diffeomorphisms, it seems that to date it is the best guess for such ((compare [7] or [8]!).

One can study these problems from the following point of view:

I. *The approximation problem.* Approximate a given diffeomorphism by a diffeomorphism in $\mathcal{C}_M$, and

II. *The regularity problem.* Find regularity properties of elements of $\mathcal{C}_M$.

We first discuss I. One can approximate an arbitrary $T \in \mathcal{D}_M$ by T satisfying Axiom 1. In fact,

(B) Theorem. *Let $\mathcal{E}$ be the subspace of $\mathcal{D}_M$ consisting of T with every periodic point elementary. Then $\mathcal{E}$ is the countable intersection of open and dense sets in $\mathcal{D}_M$.*

For a proof see [10]. Independently, R. Abraham has shown that this follows from a general transversality theorem [1]. The paper of L. Markus [4] is also in the direction of Theorem C.

A similar situation holds for Axiom 2.

(C) Theorem. *Let $\mathcal{J}$ be the subspace of $\mathcal{E}$ (of the previous theorem) of diffeomorphisms with the normal intersection property. Then $\mathcal{J}$ is the countable intersection of open and dense sets in $\mathcal{D}_M$.*

See [10] for a proof.

Unfortunately, there is no similar theorem known for Axiom 2'.

In Axiom 3, parts (*a*) and (*b*) seem to be related in some fashion, but it is not clear how. Does (*a*) imply (*b*), or conversely? To approximate a given diffeomorphism by one satisfying 3*a* or 3*b* is a central problem, related to what is sometimes called the problem of the "closing Lemma". See Peixoto [5] for an account of this important problem.

A special case of our approximation problem is the following.

(D) *Problem.* Let $T: M \to M$ be a diffeomorphism of a compact manifold. Is there a C' approximation T' of T such that T' has a periodic point?

The answer is not known for the 2-dimensional torus.

In the discussion of the regularity problem stated above, we start with the case of diffeomorphisms satisfying a highly restrictive axiom in addition to Axioms 1, 2', 3.

Axiom 4. T has a finite number of periodic points.

The set of diffeomorphisms satisfying Axioms 1, 2', 3, 4 is denoted by $\mathcal{C}_M^4$.

It is important to note that in assuming Axiom 4, an open set of diffeomorphisms of $\mathcal{D}_M$ is lost (see e.g. [11] or below). On the other hand one can say some substantial things about elements of C_M^4.

(E) THEOREM. *Diffeomorphisms in C_M^4 satisfy a form of the Morse inequalities relating the periodic points. The union of the stable manifolds associated to the periodic points is all of M. The boundary of a stable manifold of dimension p is the union of stable manifolds of dimension $\leqslant p$.*

The proof is essentially contained in [8]. A task which seems important and yet tractable is to show that the $T \in \mathsf{C}_M^4$ are structurally stable. However this has not even been carried out under the additional assumption of Axiom 5.

Axiom 5. Let W_i^u, W_j^s be an unstable, stable manifold respectively of periodic orbits of T which have non-empty intersection. Then

$$\dim W_i^u + \dim W_j^s > \dim M.$$

We say that the set of T satisfying all of our axioms is C_M^5. The following can be proved along the lines of [8], [9].

(F) THEOREM. *On every manifold there exist non-empty open sets of $\mathcal{D}_M$ which are contained in C_M^5. If $T \in \mathsf{C}_M^5$, the boundary of a stable manifold of T is the union of lower dimensional stable manifolds. The components of the stable manifolds generate the homology of M in a natural way and the correspondence between the stable manifold of a periodic point and the unstable manifold induces Poincaré duality. The Morse inequalities in the previous theorem can be interpreted to include the usual ones.*

This theorem shows that C_M^5, and hence C_M^4, C_M are not empty for any M. Thus if it could be shown that every $T \in \mathsf{C}_M^5$ is structurally stable, we would have proved that there exist structurally stable diffeomorphisms on every compact manifold.

Next the question comes up as to the existence of elements of C_M which are not in C_M^4, i.e. those with an infinite number of periodic points. The following is in [11].

(G) THEOREM. *There exist open sets in $\mathcal{D}_M$ for M an aribitrary n-sphere, $n > 1$, with the following properties:* (1) *the diffeomorphisms are in C_M;* (2) *they are structurally stable;* (3) *the diffeomorphisms have an infinite number of periodic points* (*and minimal sets homeomorphic to a Cantor set*).

This theorem answers the question as to whether a structurally stable diffeomorphism (differential equation) can have an infinite number of periodic points (closed orbits). It seems that this theorem and its proof are quite important. It shows that one can cope successfully with difficult phenomena present in differential equations of dimension greater than two and not present in two-dimensional differential equations.

In the examples of Theorem G, one has present homoclinic points. A *homoclinic point* associated to a periodic orbit β of a diffeomorphism is a point of intersection of the stable and unstable manifolds associated to β. Homoclinic points were first discovered by Poincaré [6] in the restricted 3-body problem, and studied by Birkhoff [2]. Merely the existence of a homoclinic point implies considerable complications. At the end of his three volumes on celestial mechanics [6, p. 389], referring to homoclinic points, Poincaré wrote: "On sera frappé de la complexité de cette figure, que je ne cherche même pas à tracer. Rien n'est plus propre à nous donner une idée de la complication du problème des trois corps et en général de tous les problèmes de Dynamique ....."

The methods used in proving Theorem G not only completely describe the homoclinic situation in those examples, but can be applied to give some understanding of arbitrary homoclinic points as well.

A different example which exhibits stable manifolds and homoclinic points has a simple description. In the plane E^2 let T_0 be a linear automorphism given by a 2×2 matrix with integer entries, determinant ± 1, and an eigenvalue greater than one in absolute value. Then T_0 induces a diffeomorphism T of the torus. The origin of E^2 projects into a fixed point p of T and the eigendirections of T_0 project into the stable and unstable manifolds of p. One easily sees that the homoclinic points associated to p are dense in the torus.

(H) THEOREM. *The T described above is a structurally stable diffeomorphism of the torus.*

After a meeting in September 1961 in Kiev on non-linear oscillations (where I announced Theorem G in dimension 2 [14]). I visited Moscow and spoke with mathematicians D. V. Anosov, V. I. Arnold, and Y. G. Sinai, among others. There I conjectured Theorem H and that geodesic flows on compact Riemannian manifolds of negative curvature were also structurally stable. Since then I and (as I have learned at this Congress), Arnold and Sinai have independently proved Theorem H, the latter proof having the advantage of being published [13]. In addition, Anosov has very recently proved (as I have learned also at this Congress) a beautiful theorem which settles the above conjecture for geodesic flows affirmatively and gives the n-dimensional generalization of Theorem H [12].

It seems likely that if $T \in \mathfrak{C}_M$ and $T \in \mathfrak{C}_M^4$, then T has homoclinic points. In the same vein one can ask if Axiom 5 implies Axiom 4 (certainly the converse is false).

An elementary periodic point will be called *elliptic* if all the associated eigenvalues have absolute value less than one, or all greater. It seems resonable to expect that if $T \in \mathfrak{C}_M$, then T will have only a finite number of elliptic points. It would be nice to have a proof of this. Also if $T \in \mathfrak{C}_M$ where M is the 2-sphere, must T have at least one elliptic point?

We indicate briefly how the previous discussion goes over into the analogous situation for differential equations. The definition of a structurally stable differential equation is exactly the same as for diffeomorphisms. The problem of structural stability for differential equations asks if the structurally stable differential equations on a compact manifold M are dense in the Banach space $\mathfrak{B}_M$, C^r norm, of all C^r differential equations on M. This has been answered in the affirmative if dim $M \leqslant 2$ by M. Peixoto, see Theorem I, below. One constructs stable and unstable manifolds for differential equations associated to each singular point and each closed orbit of a general type (corresponding to elementary periodic points). For details see [10]. The analogues of the previous axioms can be stated for this case. The union of the singular points and closed orbits of the differential equation replace the periodic points of the diffeomorphism. Let $\mathfrak{B}_M^*$ be the subspace of $\mathfrak{B}_M$ of differential equations satisfying the analogues of axioms 1, 2′, 3. The elements of $\mathcal{B}_M^*$ will have only a finite number of singular points but may have an infinite number of closed orbits if dim $M > 2$.

For the 2-dimensional case there is the following important theorem of Peixoto [5].

(I) THEOREM. *Let M^2 be a compact 2-manifold. Then (a) X on M is structurally stable if and only if $X \in \mathcal{B}_M^*$; (b) $\mathcal{B}_M^*$ is open and dense in $\mathcal{B}_M$.*

Also the analogues of the previous theorems are all valid for the differential equations case.

We comment on the problem of the existence of a first integral of a differential equation. A *first integral* of X on M in a C^r function $f: M \to R$ such that f is constant on each orbit but not on any open set. Problem E has the following analogue.

(E*) *Problem.* Can every non-singular vector field on a compact manifold be C' approximated by one with a closed orbit?

If Problem E* has an affirmative solution it follows that:

(J) The subset of $X \in \mathcal{B}_M$ with an elementary closed orbit is open and dense in $\mathcal{B}_M$.

Here *elementary* corresponds to elementary periodic point and the precise definition is in [10]. Putting J together with the analogue of Theorem B, as has been essentially observed by R. Thom, we obtain:

(J*) There exists a subset of $X \in \mathcal{B}_M$ with no first integral which is open and dense in $\mathcal{B}_M$.

A relation between the topological conjugacy problem for diffeomorphisms and the equivalence problem of differential equations is given by cross-sections, used by Poincaré and Birkhoff (e.g. see [2] or [10]). In a different direction, since a differential equation generates a 1-parameter group, one may ask under what conditions, is a diffeomorphism imbeddable in a flow? The following is not difficult.

(K) THEOREM. *Let $T \in C_M$. If T can be imbeeded in a flow, then T satisfies Axioms 4, 5, every periodic point is fixed and in the neighborhood of every fixed point, T is imbeddable in a flow.*

It would be interesting to know if the converse is true.

Certainly the main problems stated here are very difficult. On the other hand, it seems quite possible to us that this field may develop rapidly and already as indicated here, there have been some initial steps in this direction.

REFERENCES

[1]. ABRAHAM, R., Transversality of manifolds of mappings. (To appear.)

[2]. BIRKHOFF, G. D., *Collected Mathematical Papers*. New York, 1950.

[3]. LEFSCHETZ, L., *Differential Equations: Geometric Theory*. Interscience Publishers, New York, 1957.

[4]. MARCUS, L., Structurally stable differential systems. *Ann. Math.*, 73 (1961), 1–19.

[5]. PEIXOTO, M., Structural stability on 2-dimensional manifolds. *Topology*, 2 (1962), 101–121.

[6]. Poincaré, H., *Les Méthodes Nouvelles de la Mécanique Céleste*, Vols. I–III. Gauthier-Villars, Paris, 1899. (Reprinted New York, 1957.)
[7]. Smale, S., On dynamical systems. *Boletin de la Sociedad Matematica*, 1960, 195–198.
[8]. —— Morse inequalities for a dynamical system. *Bull. Amer. Math. Soc.*, 66 (1960), 43–49.
[9]. —— On gradient dynamical systems. *Ann. Math.*, 74 (1961), 199–206.
[10]. —— Stable manifolds for differential equations and diffeomorphisms. (To appear.)
[11]. —— Diffeomorphisms with many periodic points. (To appear.)
[12]. Anosov, D. V., Structural stability and ergodicity of geodesic flows on compact, Riemannian manifolds of negative curvature. (To appear.)
[13]. Arnold, V. I. & Sinai, Y. G., *Dokl. Akad. Nauk S.S.S.R.*, 144, 4 (1962), 695.
[14]. Smale, S., *Report on the Symposium on Non-linear Oscillations*. Kiev Math. Institute, 1961.

Colloques Internationaux du C.N.R.S.
N° 259. – Systèmes dynamiques et modèles économiques

SOME DYNAMICAL QUESTIONS IN MATHEMATICAL ECONOMICS

Steve SMALE
Université de Californie - Berkeley

RESUME

Cette courte note met à jour mon article de l'American Economic Review [3]. Un thème particulier de cet article est développé ici, à savoir le lien entre la nature des biens et la notion d'équilibre à retenir.

Let me start by posing what I like to call "the fundamental problem of equilibrium theory" : *how is economic equilibrium attained* ? A dual question more commonly raised is : *why is economic equilibrium stable ?* Behind these questions lie the problem of modeling economic processes and introducing dynamics into equilibrium theory. A successful attack here would give greater validity to equilibrium theory. It may be however that a resolution of this fundamental problem will require a recasting of the foundations of equilibrium theory. One might well keep in mind some historical perspective from physics, making an analogy between Walrasian equilibrium theory and Newtonian mechanics.

How did Relativity Theory respect classical mechanics ? For one thing Einstein worked from a very deep understanding of the Newtonian theory. Another point to remember is that while Relativity Theory lies in contradiction to Newtonian theory, even after Einstein, classical mechanics remains central to physics. I can well imagine that a revolution in economic theory could take place over the question of dynamics, which would both restructure the foundations of Walras and leave the classical theory playing a central role.

In the direction of attacking this fundamental problem, it seems to me important to idealize economic goods into two extreme classes. One one side are the durable goods, and on the other, the perishable, renewable goods with especially labor as an example. To each of these two classes of goods, one can let correspond two basic branches of equilibrium theory. As an illustration, Debreu's "Theory of Value" [1] has two substantive chapters,

Chapter 5 on the existence of (Walras) equilibria and Chapter 6 on "the fundamental theorem of welfare economics". I believe that one can associate the durable goods most naturally to the models in welfare economics and the renewable goods to the Walras equilibrium theory.

To see these things, it is useful to explicate the conditions for equilibrium. Assume classical (differentiable version) hypotheses on preferences for example as in [2]. Let there be l commodities, m agents in a pure exchange economy. Let price systems be denoted by $p = (p_1, \ldots, p_l)$ each $p_i \geqslant 0$, with $\Sigma p_i^2 = 1$. The endowment e_i of the ith agent will be a vector in $R = \{(e^1, \ldots, e^l) \mid e^j \geqslant 0\}$ and an allocation will be an m-tuple

$$x = (x_1, \ldots, x_m),$$

with each x_i in R^l. The preference of agent i is supposed to be represented by a utility function $u_i : R_+^l \to R$.

A pair (x, p) consisting of an allocation and a price system is a Walras equilibrium if these equations are satisfied:

(1) The gradient, grad $u_i(x_i)$ equals λp for some $\lambda > 0$, each $i = 1, \ldots, m$. This is a necessary condition for x_i to maximize satisfaction for agent i.

(2) $\Sigma x_i = \Sigma e_i$. This is a total resource condition on the allocation x. In other terms, x is attainable or even "supply equals demand".

(3) $p \cdot x_i = p \cdot e_i$, $i = 1, \ldots, m$. These dot products give the values and this is a budget condition.

The first two equations by themselves describe the kind of equilibrium used in welfare economics (e.g. Debreu's Chapter VI).

Returning to problem of dynamics, observe that if one is trading a non-tatonment situation with durable goods (or stocks), then the endowment allocation, after some trades will lose its effect and therefore play no role in any equilibrium attained (see [4]). Thus the notion of equilibrium which is relevant is not that of Walras but that of welfare economics.

While in the durable goods market, a commodity vector x_i in R_+^l is interpreted as a stock of goods, in renewable goods models, a point in commodity space is more naturally interpreted as a rate (or flow) of endowments or consumptions.

Thus in a model where the endowment of goods is being renewed continually, the endowments e_i should play a role in the equilibrium attained and therefore, a Walras equilibrium defined by the full set of equations (1) - (3) is most reasonable.

Perhaps the non-tatonment theory initiated by Hahn, Negishi, Uzawa has developed to handle the dynamics of durable goods of pure exchange in principle. On the other hand, clearly there is no satisfactory model for dynamics of renewable goods and Walras equilibria.

REFERENCES

[1] DEBREU G. – Theory of Value, New York 1959.

[2] SMALE S. – "Global Analysis and Economics VI", Jour. *Math. Econ.*, 3, (1976), 1-4.

[3] SMALE S. – "Dynamics in General Equilibrium Theory". *Amer. Econ. Rev.*, 66, (1976), 288-294.

[4] SMALE S. – "Exchange Processes with price adjustment", (to appear, Journ. Math. Econ.).

DISCUSSION

The discussant, Egbert Dierker, points out that, in his opinion, not only the case of perishable but also that of durable goods exhibits a Walrasian character, since the distribution of initial endowments is important for the final outcome. Smale answers that it is useful to study the pure laboratory cases first. Gabszewicz points out that the use derived from a durable good can be treated as a flow. Smale answers that the market for minerals or for houses cannot naturally be described in terms of flows. Bliss asks whether Smale distinction is appropriate. Smale answers that it is essential to know how to deal with labor, a purely perishable good. Production may then play the role of bringing both, perishable and durable goods, together into one model. The problem, however, is how to put stocks and flows into the same model. Harsanyi remarks that the distinction between durable and perishable goods is not that between a tatonnement and a non-tatonnement situation. The distinction rather lies in the fact that resale is possible in the first case but not in the latter.

Fuchs supports Smale's distinction and remarks that the case of perishable goods can be treated by the theory of temporary equilibria. The discussant points that in most models of price formation expectations about future prices play a major role. He asks to what extent agents anticipate price changes in Smale's model. Smale answers that his model had to be altered if individual expectations of price variations are to be taken explicitly into account.

Gabszewicz further remarks that the question of how to connect a given state with a Pareto optimum in the Malinvaud-Drèze-de la Vallée Poussin model is closely related to Smale's treatment of the durable goods case. Champsaur and Cornet relate Smale's process to Malinvaud's. But the latter process does not converge in finite time. Smale remarks that it is important how equations are defined near equilibria. Fuchs points out that an interest of Smale's model is that the Pareto set is reached in finite time, so one may minimize the time necessary to reach the Pareto set from the initial state. Smale answers that he originally considered processes responsive to time cost.

Guesnerie asks whether it is possible to reach any individually rational Pareto optimum in the durable goods case. Smale answers that it is likely that one can reach a subset of the Pareto surface of full dimension. A result of this kind has been shown by Schecter in a similar model without price adjustment. Kirman explains that in Smale's process an individual may continuously make losses because expectation about price changes are neglected. Smale says that the conditions characterizing his process require an essential change if one wants to handle this problem. Last Selten remarks that a behavioral point of view may be more appropriate than the requirement of full maximization. In order to relate a theory to experiments is should be put in a discrete framework. Smale answers that is should be possible reformulate his theory in a discrete set-up.

Curriculum Vitae

John T. Tate

Date of Birth: March 13, 1925, Minneapolis, Minnesota, USA

Education:

B.Sc. Harvard University
1950 Ph.D. Princeton University

Employment History:

1954–1989 Lecturer, Harvard University
1990–2009 Lecturer, University of Texas

Awards:

1956 Cole Prize
1995 Steele Prize
2002/03 Wolf Prize
2010 Abel Prize

List of Publications

1. E. Artin and J. Tate, Class Field Theory, Reprinted with corrections from the 1967 original (AMS Chelsea Publishing, 2009).
2. B. Mazur, W. Stein and J. Tate, Computation of p-adic heights and log convergence, *Doc. Math.* **2006**, Extra Vol., 577–614 (electronic).
3. M. Artin, F. Rodriguez-Villegas and J. Tate, On the Jacobians of plane cubics, *Adv. Math.* **198** (2005), No. 1, 366–382.
4. J. Tate, Refining Gross's conjecture on the values of abelian L-functions, in *Stark's Conjectures*: *Recent Work and New Directions*, pp. 189–192, Contemp. Math., Vol. 358 (Amer. Math. Soc., 2004).
5. J. Tate, The millennium prize problems. A lecture by John Tate. Springer VideoMATH. CMI Millennium Meeting Collection (Springer, 2002).
6. J. Tate, On a conjecture of Finotti, *Bull. Braz. Math. Soc.* (*N.S.*) **33** (2002), No. 2, 225–229.
7. J. Tate, Galois cohomology, *Arithmetic Algebraic Geometry* (Park City, UT, 1999), pp. 465–479, IAS/Park City Math. Ser., 9 (Amer. Math. Soc., 2001).
8. N. M. Katz and J. Tate, Bernard Dwork (1923–1998), *Notices Amer. Math. Soc.* **46** (1999), No. 3, 338–343.
9. J. Tate, Finite flat group schemes, in *Modular Forms and Fermat's Last Theorem* (Boston, MA, 1995), pp. 121–154 (Springer, 1997).
10. J. Tate, The work of David Mumford, in *Fields Medallists' Lectures*, pp. 219–223, World Scientific Ser. 20th Century Math., Vol. 5 (World Scientific, 1997).
11. J. Tate and J. F. Voloch, Linear forms in p-adic roots of unity, *Internat. Math. Res. Notices* **1996**, No. 12, 589–601.
12. J. Tate and M. van den Bergh, Homological properties of Sklyanin algebras, *Invent. Math.* **124** (1996), No. 1–3, 619–647.
13. J. Tate, On the conjectures of Birch and Swinnerton-Dyer and a geometric analog, in *Séminaire Bourbaki*, Vol. 9, Exp. No. 306, pp. 415–440 (Soc. Math. France, 1995).
14. J. Tate, WC-groups over p-adic fields, in *Séminaire Bourbaki*, Vol. 4, Exp. No. 156, pp. 265–277 (Soc. Math. France, 1995).
15. J. Tate, A review of non-Archimedean elliptic functions, in *Elliptic Curves, Modular Forms, & Fermat's Last Theorem* (Hong Kong, 1993), pp. 162–184, Ser. Number Theory, I (Int. Press, 1995).
16. M. Artin, W. Schelter and J. Tate, The centers of 3-dimensional Sklyanin algebras. *Barsotti Symposium in Algebraic Geometry* (Abano Terme, 1991), pp. 1–10, Perspect. Math., Vol. 15 (Academic Press, 1994).

17. J. Tate, The non-existence of certain Galois extensions of Qunramified outside 2, in *Arithmetic Geometry* (Tempe, AZ, 1993), pp. 153–156, Contemp. Math., Vol. 174 (Amer. Math. Soc., 1994).
18. S. P. Smith and J. Tate, The center of the 3-dimensional and 4-dimensional Sklyanin algebras. Proceedings of Conference on Algebraic Geometry and Ring Theory in honor of Michael Artin, Part I (Antwerp, 1992). *K-Theory* **8** (1994), No. 1, 19–63.
19. J. Tate, Conjectures on algebraic cycles in ℓ-adic cohomology, in *Motives* (Seattle, WA, 1991), pp. 71–83, Proc. Sympos. Pure Math., Vol. 55, Part 1 (Amer. Math. Soc., 1994).
20. J. H. Silverman and J. Tate, Rational points on elliptic curves, Undergraduate Texts in Mathematics (Springer, 1992).
21. M. Artin, J. Tate and M. van den Bergh, Modules over regular algebras of dimension 3, *Invent. Math.* **106** (1991), No. 2, 335–388.
22. M. Artin, W. Schelter and J. Tate, Quantum deformations of GL_n, *Comm. Pure Appl. Math.* **44** (1991), No. 8–9, 879–895.
23. B. Mazur and J. Tate, The p-adic sigma function, *Duke Math. J.* **62** (1991), No. 3, 663–688.
24. M. Artin, J. Tate and M. van den Bergh, Some algebras associated to automorphisms of elliptic curves, *The Grothendieck Festschrift*, Vol. I, 33–85, Progr. Math., Vol. 86 (Birkhäuser, 1990).
25. E. Artin and J. Tate, Class Field Theory, Second edition (Addison-Wesley, 1990).
26. B. Gross and J. Tate, Commentary on algebra, in *A Century of Mathematics in America*, Part II, pp. 335–336, Hist. Math., Vol. 2 (Amer. Math. Soc., 1989).
27. B. Mazur and J. Tate, Refined conjectures of the "Birch and Swinnerton–Dyer type", *Duke Math. J.* **54** (1987), No. 2, 711–750.
28. B. Mazur, J. Tate and J. Teitelbaum, On p-adic analogues of the conjectures of Birch and Swinnerton–Dyer, *Invent. Math.* **84** (1986), No. 1, 1–48.
29. J. Tate, Les conjectures de Stark sur les fonctions L d'Artin en $s = 0$ (in French) [Stark's conjectures on Artin L-functions at $s = 0$], Lecture notes edited by Dominique Bernardi and Norbert Schappacher. Progress in Mathematics, Vol. 47 (Birkhäuser, 1984).
30. J. Tate, Variation of the canonical height of a point depending on a parameter, *Amer. J. Math.* **105** (1983), No. 1, 287–294.
31. B. Mazur and J. Tate, Canonical height pairings via biextensions, in *Arithmetic and Geometry*, Vol. I, pp. 195–237, Progr. Math., Vol. 35 (Birkhäuser, 1983).

32. S. Rosset and J. Tate, A reciprocity law for K_2-traces, *Comment. Math. Helv.* **58** (1983), No. 1, 38–47.
33. J. Tate, On Stark's conjectures on the behavior of $L(s, \chi)$ at $s = 0$, *J. Fac. Sci. Univ. Tokyo Sect. IA Math.* **28** (1981), No. 3, 963–978.
34. J. Tate, Brumer–Stark–Stickelberger (in French), *Seminar on Number Theory*, 1980–1981 (Talence, 1980–1981), Exp. No. 24, Univ. Bordeaux I, 1981.
35. J. Tate, Number theoretic background, in *Automorphic Forms, Representations and L-functions* (Proc. Sympos. Pure Math., Oregon State Univ., Corvallis, Ore., 1977), Part 2, pp. 3–26, Proc. Sympos. Pure Math., Vol. XXXIII (Amer. Math. Soc., 1979).
36. P. Cartier and J. Tate, A simple proof of the main theorem of elimination theory in algebraic geometry, *Enseign. Math.* (2) **24** (1978), No. 3–4, 311–317.
37. D. Mumford and J. Tate, Fields medals, IV. An instinct for the key idea, *Science* **202** (1978), No. 4369, 737–739.
38. J. Tate, On the torsion in K_2 of fields, *Algebraic Number Theory* (Kyoto Internat. Sympos., Res. Inst. Math. Sci., Univ. Kyoto, Kyoto, 1976), pp. 243–261 (Japan Soc. Promotion Sci., 1977).
39. J. Tate, Local Constants. Prepared in collaboration with C. J. Bushnell and M. J. Taylor, in *Algebraic Number Fields: L-functions and Galois Properties* (Proc. Sympos., Univ. Durham, Durham, 1975), pp. 89–131 (Academic Press, 1977).
40. J. Tate, Problem 9: The general reciprocity law, in *Mathematical Developments Arising from Hilbert Problems* (Proc. Sympos. Pure Math., Northern Illinois Univ., De Kalb, Ill., 1974), pp. 311–322. Proc. Sympos. Pure Math., Vol. XXVIII (Amer. Math. Soc., 1976).
41. J. Tate, Relations between K_2 and Galois cohomology, *Invent. Math.* **36** (1976), 257–274.
42. J. Tate, The work of David Mumford, *Proceedings of the International Congress of Mathematicians* (Vancouver, B. C., 1974), Vol. 1, pp. 11–15. Canad. Math. Congress, Montreal, Que., 1975.
43. J. Tate, F. J. Almgren, Jr. and H. Montgomery, The 1974 Fields medals, (Bulgarian), translated from the English (*Science* **186** (1974), No. 4158, 39–40, *ibid.* **186** (1974), No. 4159, 130–131). *Fiz.-Mat. Spis. Akad. Nauk.* **18**(51) (1975), No. 1, 68–72.
44. J. Tate, Algorithm for determining the type of a singular fiber in an elliptic pencil, in *Modular Functions of One Variable*, IV (Proc. Internat. Summer School, Univ. Antwerp, Antwerp, 1972), pp. 33–52, Lecture Notes in Math., Vol. 476 (Springer, 1975).

45. J. Tate, The arithmetic of elliptic curves, *Invent. Math.* **23** (1974), 179–206.
46. J. Tate, The 1974 Fields medals, I. An algebraic geometer, *Science* **186** (1974), No. 4158, 39–40.
47. H. Bass and J. Tate, The Milnor ring of a global field, *Algebraic K-theory, II: "Classical" algebraic K-theory and connections with arithmetic* (Proc. Conf., Seattle, Wash., Battelle Memorial Inst., 1972), pp. 349–446, Lecture Notes in Math., Vol. 342 (Springer, 1973).
48. J. Tate, Letter from Tate to Iwasawa on a relation between K_2 and Galois cohomology, *Algebraic K-theory, II: "Classical" algebraic K-theory and connections with arithmetic* (Proc. Conf., Seattle Res. Center, Battelle Memorial Inst., 1972), pp. 524–527, Lecture Notes in Math., Vol. 342 (Springer, 1973).
49. B. Mazur and J. Tate, Points of order 13 on elliptic curves, *Invent. Math.* **22** (1973/74), 41–49.
50. J. Tate, Symbols in arithmetic, *Actes du Congrès International des Mathématiciens* (Nice, 1970), Tome 1, pp. 201–211 (Gauthier-Villars, 1971).
51. J. Tate, Rigid analytic spaces, *Invent. Math.* **12** (1971), 257–289.
52. J. Tate and F. Oort, Group schemes of prime order, *Ann. Sci. École Norm. Sup.* (4) **3** (1970), 1–21.
53. Dž. Tèĭt, Rigid analytic spaces (in Russian), *Mathematics: Periodical Collection of Translations of Foreign Articles*, Vol. 13, No. 3, pp. 3–37 (Mir, 1969).
54. J.-P. Serre and J. Tate, Good reduction of abelian varieties, *Ann. of Math.* (2) **88** (1968), 492–517.
55. J. Tate, Residues of differentials on curves, *Ann. Sci. École Norm. Sup.* (4) **1** (1968), 149–159.
56. E. Artin and J. Tate, *Class Field Theory* (W. A. Benjamin, Inc., 1968).
57. J. Tèĭt and I. R. Šafarevič, The rank of elliptic curves (in Russian), *Dokl. Akad. Nauk SSSR* **175** (1967), 770–773.
58. J. Tate, *p-divisible groups*, in 1967 *Proc. Conf. Local Fields* (Driebergen, 1966), pp. 158–183 (Springer, 1967).
59. J. Tate, Global class field theory, in 1967 *Algebraic Number Theory* (Proc. Instructional Conf., Brighton, 1965), pp. 162–203 (Thompson, 1967).
60. J. Tate, Fourier analysis in number fields, and Hecke's zeta-functions, in 1967 *Algebraic Number Theory* (Proc. Instructional Conf., Brighton, 1965), pp. 305–347 (Thompson, 1966)..
61. J. Lubin and J. Tate, Formal moduli for one-parameter formal Lie groups, *Bull. Soc. Math. France* **94** (1966), 49–59.
62. J. Tate, The cohomology groups of tori in finite Galois extensions of number fields, *Nagoya Math. J.* **27** (1966), 709–719.

63. J. Tate, Endomorphisms of abelian varieties over finite fields, *Invent. Math.* **2** (1966), 134–144.
64. J. Tate, Multiplication complexe formelle dans les corps locaux (in French), 1966 *Les Tendances Géom. en Algèbre et Théorie des Nombres*, pp. 257–258 (Éditions du Centre National de la Recherche Scientifique, 1966).
65. J. Tate, Algebraic cycles and poles of zeta functions, 1965 *Arithmetical Algebraic Geometry* (Proc. Conf. Purdue Univ., 1963), pp. 93–110 (Harper & Row, 1966).
66. Dž. Tèĭt, Algebraic classes of cohomologies (in Russian), *Usp. Mat. Nauk* **20** (1965), No. 6 (126), 27–40.
67. J. Lubin and J. Tate, Formal complex multiplication in local fields, *Ann. of Math.* (2) **81** (1965), 380–387.
68. J. Tate, Nilpotent quotient groups, *Topology* **3** (1964), suppl. 1, 109–111.
69. S. Sen and J. Tate, Ramification groups of local fields, *J. Indian Math. Soc. (N.S.)* **27** (1963), 197–202.
70. J. Tate, Duality theorems in Galois cohomology over number fields, in 1963 *Proc. Internat. Congr. Mathematicians* (Stockholm, 1962), pp. 288–295 (Inst. Mittag-Leffler, 1963).
71. A. Fröhlich, J.-P. Serre and J. Tate, A different with an odd class, *J. Reine Angew. Math.* **209** (1962), 6–7.
72. J. Tate, Principal homogeneous spaces for Abelian varieties, *J. Reine Angew. Math.* **209** (1962), 98–99.
73. S. Lang and J. Tate, Principal homogeneous spaces over abelian varieties, *Amer. J. Math.* **80** (1958), 659–684.
74. J. Tate, *WC*-groups over p-adic fields. Séminaire Bourbaki; 10e année: 1957/1958. Textes des conférences, Exposés 152 à 168, 2e éd. corrigée, Exposé 156, 13 pp. *Secrétariat mathématique*, Paris 1958, 189 pp. (mimeographed).
75. A. Mattuck and J. Tate, On the inequality of Castelnuovo–Severi, *Abh. Math. Sem. Univ. Hamburg* **22** (1958), 295–299.
76. J. Tate, Homology of Noetherian rings and local rings, *Illinois J. Math.* **1** (1957), 14–27.
77. R. Brauer and J. Tate, On the characters of finite groups, *Ann. of Math.* (2) **62** (1955), 1–7.
78. Y. Kawada and J. Tate, On the Galois cohomology of unramified extensions of function fields in one variable, *Amer. J. Math.* **77** (1955), 197–217.
79. J. Tate, The higher dimensional cohomology groups of class field theory, *Ann. of Math.* (2) **56** (1952), 294–297.
80. S. Lang and J. Tate, On Chevalley's proof of Luroth's theorem, *Proc. Amer. Math. Soc.* **3** (1952), 621–624.

81. J. Tate, Genus change in inseparable extensions of function fields, *Proc. Amer. Math. Soc.* **3** (1952), 400–406.
82. E. Artin and J. T. Tate, A note on finite ring extensions, *J. Math. Soc. Japan* **3** (1951), 74–77.
83. J. Tate, On the relation between extremal points of convex sets and homomorphisms of algebras, *Comm. Pure Appl. Math.* **4** (1951), 31–32.

Interview with Abel Laureate John Tate

Martin Raussen and Christian Skau

John Tate is the recipient of the 2009 Abel Prize of the Norwegian Academy of Science and Letters. This interview took place on May 25, 2010, prior to the Abel Prize celebration in Oslo, and originally appeared in the September 2010 issue of the *Newsletter of the European Mathematical Society.*

Education

Raussen and Skau: *Professor Tate, you have been selected as this year's Abel Prize Laureate for your decisive and lasting impact on number theory. Before we start to ask you questions, we would like to congratulate you warmly on this achievement. You were born in 1925 in Minneapolis in the United States. Your father was a professor of physics at the University of Minnesota. We guess he had some influence on your attraction to the natural sciences and mathematics. Is that correct?*

Tate: It certainly is. He never pushed me in any way, but on a few occasions he simply explained something to me. I remember once he told me how one could estimate the height of a bridge over a river with a stopwatch, by dropping a rock, explaining that the height in feet is approximately sixteen times the square of the number of seconds it takes until the rock hits the water. Another time he explained Cartesian coordinates and how one could graph an equation and, in particular, how the solution to two simultaneous linear equations is the point where two lines meet. Very rarely, but beautifully, he just explained something to me. He did not have to explain negative numbers—I learned about them from the temperature in the Minnesota winters.

But I have always, in any case, been interested in puzzles and trying to find the answers to questions. My father had several puzzle books. I liked reading them and trying to solve the puzzles. I enjoyed thinking about them, even though I did not often find a solution.

Raussen and Skau: *Are there other persons that have had an influence on your choice of fields of interest during your youth?*

Tate: No. I think my interest is more innate. My father certainly helped, but I think I would have done something like physics or mathematics anyway.

Raussen and Skau: *You started to study physics at Harvard University. This was probably during the Second World War?*

Tate: I was in my last year of secondary school in December 1941 when Pearl Harbor was bombed. Because of the war Harvard began holding classes in the summer, and I started there the following June. A year later I volunteered for a Naval Officer Training Program in order to avoid being drafted into the army. A group of us was later sent to M.I.T. to learn meteorology, but by the time we finished that training and Midshipman School it was VE day.[1] Our campaign in the Pacific had been so successful that more meteorologists were not needed, and I was sent to do minesweeping research. I was in the Navy for three years and never aboard a ship! It was frustrating.

Raussen and Skau: *Study conditions in those times must have been quite different from conditions today. Did you have classes regularly?*

Tate: Yes, for the first year, except that it was accelerated. But then in the Navy I had specific classes to attend, along with a few others of my choice I could manage to squeeze in. It was a good program, but it was not the normal one.

It was not the normal college social life either, with parties and such. We had to be in bed or in a

Martin Raussen is associate professor of mathematics at Aalborg University, Denmark. His email address is raussen@math.aau.dk.

Christian Skau is professor of mathematics at the Norwegian University of Science and Technology, Trondheim, Norway. His email address is csk@math.ntnu.no.

This is a slightly edited version of an interview taken on the morning preceding the prize ceremony: May 25, 2010, at Oslo.

[1] *Victory in Europe day: May 8, 1945.*

study hall by ten and were roused at 6:30 A.M. by a recording of reveille, to start the day with calisthenics and running.

Raussen and Skau: *Then you graduated in 1946 and went to Princeton?*

Tate: Yes, that's true. Harvard had a very generous policy of giving credit for military activities that might qualify—for instance, some of my navy training. This and the wartime acceleration enabled me to finish the work for my undergraduate degree in 1945. On my discharge in 1946, I went straight from the Navy to graduate school in Princeton.

Raussen and Skau: *When you went to Princeton University, it was still with the intention to become a physicist?*

Tate: That's correct. Although my degree from Harvard was in mathematics, I entered Princeton graduate school in physics. It was rather silly, and I have told the story many times: I had read the book *Men of Mathematics* by Eric Temple Bell. That book was about the lives of the greatest mathematicians in history, people like Abel. I knew I wasn't in their league and I thought that unless I was, I wouldn't really be able to do much in mathematics. I didn't realize that a less talented person could still contribute effectively. Since my father was a physicist, that field seemed more human and accessible to me, and I thought that was a safer way to go, where I might contribute more. But after one term it became obvious that my interest was really in mathematics. A deeper interest, which should have been clear anyway, but I just was too afraid and thought I never would be able to do much research if I went into mathematics.

Raussen and Skau: *Were you particularly interested in number theory from the very beginning?*

Tate: Yes. Since I was a teenager I had an interest in number theory. Fortunately, I came across a good number theory book by L. E. Dickson, so I knew a little number theory. Also I had been reading Bell's histories of people like Gauss. I liked number theory. It's natural, in a way, because many wonderful problems and theorems in number theory can be explained to any interested high-school student. Number theory is easier to get into in that sense. But of course it depends on one's intuition and taste also.

Raussen and Skau: *Many important questions are easy to explain, but answers are often very tough to find.*

Tate: Yes. In number theory that is certainly true, but finding good questions is also an important part of the game.

Teachers and Fellows

Raussen and Skau: *When you started your career at Princeton you very quickly met Emil Artin, who became your supervisor. Emil Artin was born in Austria and became a professor in mathematics at the University of Hamburg. He had to leave*

Photo by Eirik F. Baardsen/ Abelprisen/Norwegian Academy of Sciences.

Abel interview, from left to right: Martin Raussen, Christian Skau, and John Tate.

Germany in 1937 and came to the United States. Can you tell us more about his background? Why did he leave his chair, and how did he adjust when he came to the States?

Tate: His wife was half Jewish, and he eventually lost his position in Germany. The family left in '37, but at that time there weren't so many open jobs in the United States. He took a position at the University of Notre Dame in spite of unpleasant memories of discipline at a Catholic school he had attended in his youth. After a year or two he accepted an offer from Indiana University and stayed there until 1946. He and his wife enjoyed Bloomington, Indiana, very much. He told me it wasn't even clear that he would have accepted Princeton's offer in 1946 except that President H. B. Wells of Indiana University, an educational visionary, was on a world tour, and somehow Indiana didn't respond very well to Princeton's offer. Artin went to Princeton the same year I did.

Raussen and Skau: *Artin had apparently a very special personality. First of all, he was an eminent number theorist, but also a very intriguing person; a special character. Could you please tell us a bit more about him?*

Tate: I think he would have made a great actor. His lectures were polished: He would finish at the right moment and march off the scene. A very lively individual with many interests: music, astronomy, chemistry, history.... He loved to teach. I had a feeling that he loved to teach anybody anything. Being his student was a wonderful experience; I couldn't have had a better start to my mathematical career. It was a remarkable accident. My favorite theorem, which I had first learned from Bell's book, was Gauss's law of quadratic reciprocity, and there, entirely by chance, I found myself at the same university as the man who had discovered *the* ultimate law of reciprocity. It was just amazing.

Raussen and Skau: *What a coincidence!*

Tate: Yes, it really was.

Raussen and Skau: *You wrote your thesis with Artin, and we will certainly come back to it. After that you organized a seminar together with Artin on class field theory. Could you comment on this seminar: What was the framework and how did it develop?*

Tate: During his first two years in Princeton, Artin gave seminars in algebraic number theory, followed by class field theory. I did not attend the former, but one of the first things I heard about Artin concerned an incident in it. A young British student, Douglas Northcott, who had been captured when the Japanese trapped the British army in Singapore and barely survived in the Japanese prison camp, was in Princeton on a Commonwealth Fellowship after the war. Though his thesis was in analysis under G. H. Hardy, he attended Artin's seminar, and when one of the first speakers mentioned the characteristic of a field, Northcott raised his hand and asked what that meant. His question begot laughter from several students, whereupon Artin delivered a short lecture on the fact that one could be a fine mathematician without knowing what the characteristic of a field was. And, indeed, it turned out that Northcott was the most gifted student in that seminar.

But I'm not answering your question. I attended the second year, in which class field theory was treated, with Chevalley's nonanalytic proof of the second inequality, but not much cohomology. This was the seminar at the end of which Wang discovered that both published proofs of Grunwald's theorem, and in fact the theorem itself, were not correct at the prime 2.

At about that time, Gerhard Hochschild and Tadasi Nakayama were introducing cohomological methods in class field theory and used them to prove the main theorems, including the existence of the global fundamental class which A. Weil had recently discovered. In 1951-52 Artin and I ran another seminar giving a complete treatment of class field theory incorporating these new ideas. That is the seminar you are asking about. Serge Lang took notes, and thanks to his efforts they were eventually published, first as informal mimeographed notes and, in 1968, commercially, under the title *Class Field Theory*. A new edition (2008) is available from AMS-Chelsea.

Raussen and Skau: *Serge Lang was also a student of Emil Artin and became a famous number theorist. He is probably best known as author of many textbooks; almost every graduate student in mathematics has read a textbook by Serge Lang. He is also quite known for his intense temper, and he got into a lot of arguments with people. What can you tell us about Serge Lang? What are your impressions?*

Tate: He was indeed a memorable person. The memories of Lang in the May 2006 issue of the *Notices of the AMS*, written by about twenty of his many friends, give a good picture of him. He started Princeton graduate school in philosophy, a year after I started in physics, but he, too, soon switched to math. He was a bit younger than I and had served a year and a half in the U.S. Army in Europe after the war, where he had a clerical position in which he learned to type at incredible speed, an ability which served him well in his later book writing.

He had many interests and talents. I think his undergraduate degree from Caltech was in physics. He knew a lot of history and he played the piano brilliantly.

He didn't have the volatile personality you refer to until he got his degree. It seemed to me that he changed. It was almost a discontinuity; as soon he got his Ph.D. he became more authoritative and asserted himself more.

It has been noted that there are many mathematical notions linked to my name. I think that's largely due to Lang's drive to make information accessible. He wrote voluminously. I didn't write easily and didn't get around to publishing; I was always interested in thinking about the next problem. To promote access, Serge published some of my stuff and, in reference, called things "Tate this" and "Tate that" in a way I would not have done had I been the author.

Throughout his life, Serge addressed great energy to disseminating information; to sharing where he felt it was important. We remained friends over the years.

Research Contributions

Raussen and Skau: *This brings us to the next topic: Your Ph.D. thesis from 1950, when you were twenty-five years old. It has been extensively cited in the literature under the sobriquet "Tate's thesis". Several mathematicians have described your thesis as unsurpassable in conciseness and lucidity and as representing a watershed in the study of number fields. Could you tell us what was so novel and fruitful in your thesis?*

Tate: Well, first of all, it was not a new result, except perhaps for some local aspects. The big global theorem had been proved around 1920 by the great German mathematician Erich Hecke, namely the fact that all L-functions of number fields, abelian L-functions, generalizations of Dirichlet's L-functions, have an analytic continuation throughout the plane with a functional equation of the expected type. In the course of proving it Hecke saw that his proof even applied to a new kind of L-function, the so-called L-functions with Grössencharacter. Artin suggested to me that one might prove Hecke's theorem using abstract harmonic analysis on what is now called the adele ring, treating all places of the field equally, instead of using classical Fourier analysis at the archimedian places and finite Fourier analysis with congruences at the p-adic places as Hecke had done. I think I did a good job —it might even have been lucid and concise!—but in a way it was just a wonderful exercise to carry out this idea. And it was also in the air. So often there is a time in mathematics for something to be done. My thesis is an example. Iwasawa would have done it had I not.

Raussen and Skau: *What do you think of the fact that, after your thesis, all places of number fields are treated on an equal footing in analytic number theory, whereas the situation is very different in the classical study of zeta functions; in fact, gamma factors are very different from nonarchimedean local factors.*

Tate: Of course there is a big difference between archimedean and nonarchimedean places, in particular as regards the local factors, but that is no reason to discriminate. Treating them equally, using adeles and ideles, is the simplest way to proceed, bringing the local-global relationship into clear focus.

Raussen and Skau: *The title of your thesis was* Fourier Analysis in Number Fields and Hecke's Zeta-Functions. *Atle Selberg said in an interview five years ago that he preferred—and was most inspired by—Erich Hecke's approach to algebraic number theory, modular forms and L-functions. Do you share that sentiment?*

Tate: Hecke and Artin were both at Hamburg University for a long time before Artin left. I think Artin came to number theory more from an algebraic side, whereas Hecke and Selberg came more from an analytic side. Their basic intuition was more analytic and Artin's was more algebraic. Mine was also more algebraic, so the more I learned of Hecke's work, the more I appreciated it, but somehow I did not instinctively follow him, especially as to modular forms. I didn't know much about them when I was young.

I have told the story before, but it is ironic that being at the same university, Artin had discovered a new type of L-series and Hecke, in trying to figure out what kind of modular forms of weight one there were, said they should correspond to some kind of L-function. The L-functions Hecke sought were among those that Artin had defined, but they never made contact—it took almost forty years until this connection was guessed and ten more before it was proved, by Langlands. Hecke was older than Artin by about ten years, but I think the main reason they did not make contact was their difference in mathematical taste. Moral: Be open to all approaches to a subject.

Raussen and Skau: *You mentioned that Serge Lang had named several concepts after you, but there are lots of further concepts and conjectures bearing your name. Just to mention a few: Tate module, Tate curve, Tate cohomology group, Shafarevich-Tate group, Tate conjecture, Sato-Tate conjecture, etc. Good definitions and fruitful concepts, as well as good problems, are perhaps as important as theorems in mathematics. You excel in all these categories. Did all or most of these concepts grow out of your thesis?*

Tate: No, I wouldn't say that. In fact, I would say that almost none of them grew out of my thesis. Some of them, like the Tate curve, grew out of my interest in p-adic fields, which were also very central in my thesis, but they didn't grow out of my thesis. They came from different directions. The Tate cohomology came from my understanding the cohomology of class field theory in the seminar that we discussed. The Shafarevich-Tate group came from applying that cohomology to elliptic curves and abelian varieties. In general, my conjectures came from an optimistic outlook, generalizing from special cases.

Although concepts, definitions, and conjectures are certainly important, the bottom line is to prove a theorem. But you do have to know what to prove, or what to try to prove.

Raussen and Skau: *In the introduction to your delightful book* Rational Points on Elliptic Curves *that you coauthored with your earlier Ph.D. student Joseph Silverman, you say, citing Serge Lang, that it is possible to write endlessly on elliptic curves. Can you comment on why the theory of elliptic curves is so rich and how it interacts and makes contact with so many different branches of mathematics?*

Tate: For one thing, they are very concrete objects. An elliptic curve is described by a cubic polynomial in two variables, so they are very easy to experiment with. On the other hand, elliptic curves illustrate very deep notions. They are the first nontrivial examples of abelian varieties. An elliptic curve is an abelian variety of dimension one, so you can get into this more advanced subject very easily by thinking about elliptic curves. On the other hand, they are algebraic curves. They are curves of genus one, the first example of a curve which isn't birationally equivalent to a projective line. The analytic and algebraic relations which occur in the theory of elliptic curves and elliptic functions are beautiful and unbelievably fascinating. The modularity theorem stating that every elliptic curve over the rational field can be found in the Jacobian variety of the curve which parametrizes elliptic curves with level structure its conductor is mind-boggling.

By the way, by my count about one quarter of Abel's published work is devoted to elliptic functions.

Raussen and Skau: *Among the Abel Prize laureates so far, you are probably the one whose contributions would have been closest to Abel's own interests. Could we challenge you to make a historical sweep, to put Abel's work in some perspective and to compare it to your research? In modern parlance, Abel studied the multiplication-by-n map for elliptic equal parts and studied the algebraic equations that* *arose. He studied also complex multiplication and showed that, in this case, it gave rise to a commutative Galois group. These are very central concepts and observations, aren't they?*

Tate: Yes, absolutely, yes. Well, there's no comparison between Abel's work and mine. I am in awe of what I know of it. His understanding of algebraic equations, and of elliptic integrals and the more general, abelian integrals, at that time in history is just amazing. Even more for a person so isolated. I guess he could read works of Legendre and other great predecessors, but he went far beyond. I don't really know enough to say more. Abel was a great analyst and a great algebraist. His work contains the germs of many important modern developments.

Raussen and Skau: *Could you comment on how the concept of "good reduction" for an elliptic curve is so crucial, and how it arose?*

Tate: If one has an equation with integer coefficients, it is completely natural, at least since Gauss, to consider the equation mod p for a prime p, which is an equation over the finite field F_p with p elements.

If the original equation is the equation of an elliptic curve E over the rational number field, then the reduced equation may or may not define an elliptic curve over F_p. If it does, we say E has "good reduction at p". This happens for all but a finite set of "bad primes for E", those dividing the discriminant of E.

Raussen and Skau: *The Hasse principle in the study of Diophantine equations says, roughly speaking: If an equation has a solution in* p*-adic numbers, then it can be solved in the rational numbers. It does not hold in general. There is an example for this failure given by the Norwegian mathematician Ernst Selmer...*

Tate: Yes. The equation $3x^3 + 4y^3 + 5z^3 = 0$.

Raussen and Skau: *Exactly! The extent of the failure of the Hasse principle for curves of genus 1 is quantified by the Shafarevich-Tate group. The so-called Selmer groups are related groups, which are known to be finite, but as far as we know the Shafarevich-Tate group is not known to be finite. It is only a conjecture that it is always finite. What is the status concerning this conjecture?*

Tate: The conjecture that the Shafarevich group Sha is finite should be viewed as part of the conjecture of Birch and Swinnerton-Dyer. That conjecture, BSD for short, involves the L-function of the elliptic curve, which is a function of a complex variable s. Over the rational number field, $L(s)$ is known to be defined near $s=1$, thanks to the modularity theorem of A. Wiles, R. Taylor, et al. If $L(s)$ either does not vanish or has a simple zero at $s=1$, then Sha is finite and BSD is true, thanks to the joint work of B. Gross and D. Zagier on Heegner points and the work of Kolyvagin on Euler systems. So, by three big results which are the work of many people, we know a very special circumstance in which Sha is finite.

If $L(s)$ has a higher order zero at $s=1$, we know nothing, even over the field of rational numbers. Over an imaginary quadratic field we know nothing, period.

Raussen and Skau: *Do you think that this group is finite?*

Tate: Yes. I firmly believe the conjecture is correct. But who knows? The curves of higher rank, or whose L-functions have a higher order zero—BSD says the order of the zero is the rank of the curve—one knows nothing about.

Raussen and Skau: *What is the origin of the Tate conjecture?*

Tate: Early on I somehow had the idea that the special case about endomorphisms of abelian varieties over finite fields might be true. A bit later I realized that a generalization fit perfectly with the function field version of the Birch and Swinnerton-Dyer conjecture. Also it was true in various particular examples which I looked at and gave a heuristic reason for the Sato-Tate distribution. So it seemed a reasonable conjecture.

Raussen and Skau: *In the arithmetic theory of elliptic curves, there have been major breakthroughs like the Mordell-Weil theorem, Faltings' proof of the Mordell conjecture, using the known reduction to a case of the Tate conjecture. Then we have Wiles's breakthrough proving the Shimura-Taniyama-Weil conjecture. Do you hope the next big breakthrough will come with the Birch and Swinnerton-Dyer conjecture? Or the Tate conjecture, maybe?*

Tate: Who knows what the next big breakthrough will be, but certainly the Birch and Swinnerton-Dyer conjecture is a big challenge; and also the modularity, i.e., the Shimura-Taniyama-Weil idea, which is now seen as part of the Langlands program. If the number field is not totally real, we don't know much about either of these problems. There has been great progress in the last thirty years, but it is just the very beginning. Proving these things for all number fields and for all orders of vanishing, to say nothing of doing it for abelian varieties of higher dimension, will require much deeper insight than we have now.

Raussen and Skau: *Is there any particular work from your hand that you are most proud of, that you think is your most important contribution?*

Tate: I don't feel that any one of my results stands out as most important. I certainly enjoyed working out the proofs in my thesis. I enjoyed very much proving a very special case of the so-called Tate conjecture, the result about endomorphisms of abelian varieties over finite fields. It was great to be able to prove at least one nontrivial case and not have only a conjecture! That's a case that is useful in cryptography, especially elliptic curves over finite fields. Over number fields, even finitely generated fields, that case of my conjecture was proved by Faltings, building on work of Zarhin over function fields, as the first step in his proof of the Mordell conjecture. I enjoyed very much the paper which I dedicated to Jean-Pierre Serre on the K^2 groups of number fields. I also had fun with a paper on residues of differentials on curves giving a new definition of residue and a new proof that the sum of the residues is zero, even though I failed to see a more important aspect of the construction.

Applied Number Theory

Raussen and Skau: *Number theory stretches from the mysteries of the prime numbers to the way we save, transmit, and secure information on modern computers. Can you comment on the amazing fact that number theory, in particular the arithmetic of elliptic curves, has been put to use in practical applications?*

Tate: It certainly is amazing to me. When I first studied and worked on elliptic curves I had no idea that they ever would be of any practical use. I did not foresee that. It is the high-speed computers which made the applications possible, but of course many new ideas were needed also.

Raussen and Skau: *And now it's an industry: elliptic curves, cryptography, intelligence, and communication!*

Tate: It's quite remarkable. It often happens that things which are discovered just for their own interest and beauty later turn out to be useful in practical affairs.

Raussen and Skau: *We interviewed Jacques Tits a couple of years ago. His comment was that the Monster group, the biggest of all the sporadic simple groups, is so beautiful that it has to have some application in physics or whatever.*

Tate: That would be interesting!

Collaboration and Teaching

Raussen and Skau: *You have been one of the few non-French members of the Bourbaki group, the group of mathematicians that had the endeavor to put all existing mathematics into a rigid format. Can you explain what this was all about and how you got involved?*

Tate: I would not say it was about putting mathematics in a rigid format. I view Bourbaki as a modern Euclid. His aim was to write a coherent series of books which would contain the fundamental definitions and results of all mathematics as of mid-twentieth century. I think he succeeded pretty well, though the books are somewhat unbalanced—weak in classical analysis and heavy on Lie theory. Bourbaki did a very useful service for a large part of the mathematics community just by establishing some standard notations and conventions.

The presentation is axiomatic and severe, with no motivation except for the logic and beauty of the development itself. I was always a fan of Bourbaki. That I was invited to collaborate may have been at Serge Lang's suggestion, or perhaps Jean-Pierre Serre's also. As I mentioned, I am not a very prolific writer. I usually write a few pages and then tear them up and start over, so I never was able to contribute much to the writing. Perhaps I helped somewhat in the discussion of the material. The conferences were enjoyable, all over France, in the Alps, and even on Corsica. It was a lot of fun.

Raussen and Skau: *You mentioned Jean-Pierre Serre, who was the first Abel Prize laureate. He was one of the driving forces in the Bourbaki project after the Second World War. We were told that he was—as [was] Serge Lang—instrumental in getting some of your results published in the form of lecture notes and textbooks. Do you have an ongoing personal relation with Jean-Pierre Serre?*

Tate: Yes. I'm looking forward to meeting him next week when we will both be at Harvard for a conference in honor of Dick Gross on his sixtieth birthday. Gross was one of my Ph.D. students.

I think Serre was a perfect choice for the first Abel Prize laureate.

Raussen and Skau: *Another possible choice would have been Alexander Grothendieck. But he went into reclusion. Did you meet him while you were in Paris or maybe at Harvard?*

Tate: I met him in Paris. I had a wonderful year. Harvard had the enlightened policy of giving a tenure-track professor a year's sabbatical leave. I went to Paris for the academic year 1957–58, and it was a great experience. I met Serre, I met Grothendieck, and I was free from any duty. I could think and I could learn. Later, they both visited Harvard several times, so I saw them there too. It's great good fortune to be able to know such people.

Raussen and Skau: *Did you follow Grothendieck's program reconstructing the foundations of algebraic geometry closely?*

Tate: Well, yes, to the extent I could. I felt "ah, at last, we have a good foundation for algebraic geometry." It just seemed to me to be the right thing. Earlier I was always puzzled, do we have affine varieties, projective varieties? But it wasn't a category. Grothendieck's schemes, however, did form a category. And breaking away from a ground field to a ground ring, or even a ground scheme, so that the foundations could handle not only polynomial equations but also Diophantine equations and reduction mod p, was just what number theorists needed.

Raussen and Skau: *We have a question of a more general and philosophical nature: A great mathematician once mentioned that it is essential to possess a certain naiveté in order to be able to create something really new in mathematics. One can do impressive things requiring complicated techniques, but one rarely makes original discoveries without being a bit naive. In the same vein, André Weil claimed that breakthroughs in mathematics* *are typically not done by people with long experience and lots of knowledge. New ideas often come without that baggage. Do you agree?*

Tate: I think it's quite true. Most mathematicians do their best work when they are young and don't have a lot of baggage. They haven't worn grooves in their brains that they follow. Their brains are fresher, and certainly it's important to think for oneself rather than just learning what others have done. Of course, you have to build on what has been done before or else it's hopeless; you can't rediscover everything. But one should not be prejudiced by the past work. I agree with the point of view you describe.

Raussen and Skau: *Did you read the masters of number theory already early in your career?*

Tate: I've never been such a good reader. My instincts have been to err on the side of trying to be independent and trying to do things myself. But as I said, I was very fortunate to be in contact with brilliant people, and I learned very much from personal conversations. I never was a great reader of the classics. I enjoyed that more as I got older.

Raussen and Skau: *You have had some outstanding students who have made important contributions to mathematics. How did you attract these students in the first place, and how did you interact with them, both as students and later?*

Tate: I think we were all simply interested in the same kind of mathematics. You know, with such gifted students there is usually no problem: After getting to know them and their interests you suggest things to read and think about, then just hear about progress and problems, offering support and encouragement as they find their way.

Raussen and Skau: *Did you give them problems to work on or did they find the problems themselves?*

Tate: It varies. Several found their own problems. With others I made somewhat more specific suggestions. I urged Dick Gross to think about a problem which I had been trying unsuccessfully to solve, but very sensibly he wrote a thesis on a quite different subject of his own choosing. I was fortunate to have such able students. I continued to see many of them later, and many are good friends.

Raussen and Skau: *You have taught mathematics for more than sixty years, both at Harvard and at Austin, Texas. How much did you appreciate this aspect of your professional duties? Is there a particular way of teaching mathematics that you prefer?*

Tate: I always enjoyed teaching at all levels. Teaching a subject is one of the best ways to learn it thoroughly. A few times, I've been led to a good new idea in preparing a lecture for an advanced course. That was how I found my definition of Neron's height, for example.

Work Style

Raussen and Skau: *Would you consider yourself mainly a theory builder or a problem solver?*

Tate: I suppose I'm a theory builder or maybe a conjecture maker. I'm not a conjecture prover very much, but I don't know. It's true that I'm not good at solving problems. For example, I would never be good in the Math Olympiad. There speed counts and I am certainly not a speedy worker. That's one pleasant thing in mathematics: It doesn't matter how long it takes if the end result is a good theorem. Speed is an advantage, but it is not essential.

Raussen and Skau: *But you are persistent. You have the energy to stay with a problem.*

Tate: At least, I did at one time.

Raussen and Skau: *May we ask you a question that we, in various ways, have asked almost everybody in previous interviews: Look back on how you came up with new concepts or made a breakthrough in an area you had been working on for some time. Did that usually happen when you were concentrated and working intensely on the problem, or did it happen in a more relaxed situation? Do you have concrete examples?*

Tate: The first thing I did after my thesis was the determination of the higher-dimensional cohomology groups in class field theory. I had been working on that for several months, off and on. This was at the time of the seminar after my thesis at Princeton. One evening I went to a party and had a few drinks. I came home after midnight and thought I would think a little about the problem. About one or two in the morning I saw how to do it!

Raussen and Skau: *So this was a "Poincaré moment"?*

Tate: In a way. I think that, like him, I had put the work aside for a longer time when this happened. I remember what it was: I had been invited to give some talks at MIT on class field theory and I thought "what am I going to say?" So it was after a party, motivated by needing something to say at MIT, that this idea struck me. It was very fortunate.

But it varies. Sometimes I've had an idea after talking to someone and had the impression the person I was talking to had the idea and told me about it. The Ph.D. thesis of my student Jonathan Lubin was on what should be called the Lubin groups. They somehow have been called the Lubin-Tate groups. Incidentally, I think it's useful in math that theorems or ideas have two names so you can identify them. If I say Serre's theorem, my God, that doesn't say too much. But anyway, they are called Lubin-Tate groups, and it occurred to me, just out of the blue, that they might be useful in class field theory. And then we worked it out and indeed they were. One gets ideas in different ways, and it's a wonderful feeling for a few minutes, but then there is a letdown after you get used to the idea.

Raussen and Skau: *Group cohomology had been studied in various guises, long before the* *notion of group cohomology was formulated in the 1940s. You invented what is called Tate cohomology groups, which are widely used in class field theory, for instance. Could you elaborate?*

Tate: In connection with class field theory it suddenly dawned on me that if the group is finite—the operating group G—then one could view the homology theory of that group as negative dimensional cohomology. Usually the homology and the cohomology are defined in nonnegative dimensions, but suddenly it became clear to me that for a finite group you could glue the two theories together. The ith homology group can be viewed as a $(1-i)$th cohomology group and then you can glue these two sequences together so that the cohomology goes off to plus infinity and the homology goes off, with renumbering, to minus infinity, and you fiddle a little with the joining point and then you have one theory going from negative infinity to plus infinity.

Raussen and Skau: *Was this a flash of insight?*

Tate: Perhaps. There was a clue from the finite cyclic case, where there is periodicity; a periodicity of length two. For example, H^3 is isomorphic to H^1, the H^5 is isomorphic to H^3, etc., and it's obvious that you could go on to infinity in both directions. Somehow it occurred to me that one could do that for an arbitrary finite group. I don't remember exactly how it happened in my head.

The Roles of Mathematics

Raussen and Skau: *Can we speculate a little about the future development of mathematics? When the Clay Millennium Prizes for solving outstanding problems in mathematics were established back in the year 2000, you presented three of these problems to the mathematical public. Not necessarily restricting to those, would you venture a guess about new trends in mathematics: the twenty-first century compared to the twentieth century? Are there trends that are entirely new? What developments can we expect in mathematics and particularly in your own field, number theory?*

Tate: We certainly have plenty of problems to work on. One big difference in mathematics generally is the advent of high-speed computers. Even for pure math, that will increase the experimental possibilities enormously. It has been said that number theory is an experimental science, and until recently that meant experimenting by looking at examples by hand and discovering patterns that way. Now we have a zillionfold more powerful way to do that, which may very well lead to new ideas even in pure math, but certainly also for applications.

Mathematics somehow swings between the development of new abstract theories and the application of these to more concrete problems and from concrete problems to theories needed to solve them. The pendulum swings. When I was young better foundations were being developed, things were becoming more functorial, if you will, and a very abstract point of view led to much progress. But then the pendulum swung the other way to more concrete things in the 1970s and 1980s. There were modular forms and the Langlands program, the proof of the Mordell conjecture and of Fermat's last theorem. In the first half of my career, theoretical physics and mathematics were not so close. There was the time when the development of mathematics went in the abstract direction, and the physicists were stuck. But now in the last thirty years they have come together. It is hard to tell whether string theory is math or physics. And noncommutative geometry has both sides.

Who knows what the future will be? I don't think I can contribute much in answering that question. Maybe a younger person would have a better idea.

Raussen and Skau: *Are you just as interested in mathematics now as you were when you were young?*

Tate: Well, not as intensely. I'm certainly still very much interested, but I don't have the energy to really go so deeply into things.

Raussen and Skau: *But you try to follow what is happening in your field?*

Tate: Yes, I try. I'm in awe of what people are doing today.

Raussen and Skau: *Your teacher Emil Artin, when asked about whether mathematics was a science, would rather say: "No. It's an art." On the other hand, mathematics is connected to the natural sciences, to computing and so on. Perhaps it has become more important in other fields than ever; the mutual interaction between science and engineering on one side and mathematics on the other has become more visible. Is mathematics an art, is it rather to be applied in science, or is it both?*

Tate: It's both, for heaven's sake! I think Artin simply was trying to make a point that there certainly is an artistic aspect to mathematics. It's just beautiful. Unfortunately it's only beautiful to the initiated, to the people who do it. It can't really be understood or appreciated much on a popular level the way music can. You don't have to be a composer to enjoy music, but in mathematics you do. That's a really big drawback of the profession. A nonmathematician has to make a big effort to appreciate our work; it's almost impossible.

Yes, it's both. Mathematics is an art, but there are stricter rules than in other arts. Theorems must be proved as well as formulated; words must have precise meanings. The happy thing is that mathematics does have applications which enable us to earn a good living doing what we would do even if we weren't paid for it. We are paid mainly to teach the useful stuff.

Public Awareness of Mathematics

Raussen and Skau: *Have you tried to popularize mathematics yourself?*

Tate: When I was young I tried to share my enthusiasm with friends, but I soon realized that's almost impossible.

Raussen and Skau: *We all feel the difficulty of communicating with the general audience. This interview is one of the rare occasions providing public attention for mathematics![2] Do you have any ideas about how mathematicians can make themselves and what they do more well known? How can we increase the esteem of mathematics among the general public and among politicians?*

Tate: Well, I think prizes like this do some good in that respect. And the Clay Prizes likewise. They give publicity to mathematics. At least people are aware. I think the appreciation of science in general and mathematics in particular varies with the country. What fraction of the people in Norway would you say have an idea about Abel?

Raussen and Skau: *Almost everyone in Norway knows about Abel, but they do not know anything about Lie. And not necessarily anything about Abel's work, either. They may know about the quintic.*

Tate: I see. And how about Sylow?

Raussen and Skau: *He is not known either. Abel's portrait has appeared on stamps and also on bills, but neither Lie's nor Sylow's.*

Tate: I think in Japan, people are more aware. I once was in Japan and eating alone. A Japanese couple came and wanted to practice their English. They asked me what I did. I said I was a mathematician but could not get the idea across until I said: "Like Hironaka". Wow! It's as though in America I'd said "Like Babe Ruth", or Michael Jordan, or Tiger Woods. Perhaps Hironaka's name is, like Abel's, the only one known, but in America I don't think any mathematician's name would get any response.

Private Interests

Raussen and Skau: *Our last question: What other interests do you have in life? What are you occupied with when you are not thinking about mathematics? Certainly that happens once in a while, as well?*

Tate: I'm certainly not a Renaissance man. I don't have wide knowledge or interests. I have enjoyed very much the outdoors, hiking, and also sports. Basketball was my favorite sport. I played on the Southeast Methodist church team as a teenager and we won the Minneapolis church league championship one year. There were several of us who went to church three out of four Sundays during a certain period in the winter, in order to play on the team. In the Navy I coached a team from the minesweeping research base which beat Coca-Cola for the Panama City league championship. Anyway, I have enjoyed sports and the outdoors.

I like to read a reasonable amount and I enjoy music, but I don't have a really deep or serious hobby. I think I'm more concentrated in mathematics than many people. My feeling is that to do some mathematics I just have to concentrate. I don't have the kind of mind that absorbs things very easily.

Raussen and Skau: *We would like to thank you very much for this interview; as well as on behalf of the Norwegian, Danish, and European mathematical societies. Thank you very much!*

Tate: Well, thank you for not asking more difficult questions! I have enjoyed talking with you.

---

[2] *The interview was broadcast on Norwegian television; cf.* `http://www.abelprisen.no/en/multimedia/2010/`.

Pure and Applied Mathematics Quarterly
Volume 5, Number 4
(*Special Issue: In honor of
John Tate, Part 1 of 2*)
1429—1433, 2009

The Mathematical Circle Around Tate In The Late Fifties and Early Sixties

Stephen S. Shatz

Abstract: A discussion from the personal and historical point of view of the graduate mathematics students and the mathematical activity around John Tate at Harvard in the period: 1958-1962.
Keywords: History, graduate mathematics, algebraic geometry, number theory, John T. Tate, Jr.

After three years as an instructor at Princeton (1950-1953) and one year as an assistant professor at Columbia ('53-'54), John Tate came to Harvard in the autumn of 1954. He was already known among mathematicians for his share of the famous Artin-Tate Seminar, for his thesis (yet unpublished) and for his complete clarification of the cohomological structure of class field theory (for which he would be awarded the AMS Cole Prize in Number Theory–1956). I met him that autumn, but I was still only a sophomore in college and still a physics student then. Nonetheless, there was about him already an aura that attracted many people including myself though I would have been hard pressed then to say why. My switch to mathematics came from some inner compulsion as well as being befriended by Richard Brauer and Lars Ahlfors. Some people have a great deal of luck!

Bernard Dwork also came to Harvard that autumn and, while he was officially a student of E. Artin and thus Tate's mathematical brother, he really considered

Received November 30, 2006.

himself a Tate student (the first in what would be a long line). In any case, Tate and Dwork were already very close and had many animated discussions in a little seminar room where we sometimes had tea. I was extremely fortunate to make friends with Dwork and learn a great deal from him while still an undergraduate; as I said, some people have a great deal of luck. At the time, Dwork must have been formulating his deep studies of p-adic analysis which led to his proof of the rationality of the Weil Zeta function of a variety over a finite field. Many times I caught suggestive words as I passed by this little room: root number, functional equation, Frobenius, Euler product, etc. It is impossible to indicate the excitement that radiated from the two of them in this little room and I was not alone in receiving it.

One must remember that this was also the time that Jean-Pierre Serre began his continued visits to Harvard, sometimes for a year, more normally for one of the two semesters. Serre and Tate had become fast friends and the atmosphere was quite charged when they were both around–though it was hard to get to see Tate then as he was very busy with Serre.

At Harvard, Tate's first student was Edward Assmus. He graduated in 1958 and so I do not know about his participation in what may have been a small circle around Tate in the middle fifties. He did not participate in the growing circle of slightly younger students that was forming around Tate about the middle of 1958. The younger students who formed the group around Tate were some of his own and some working with other advisors–principally Zariski, though at least one of Mackey's students was involved. Here are the names of those students, listed in order of their degree dates and with their advisors also listed: Michael Artin (1960-Zariski), Leonard Evens (1960-Tate), Calvin Moore (1960-Mackey), James Cohn (1961-Tate), Andrew Ogg (1961-Tate), Stephen Shatz (1962-Tate), Jonathan Lubin (1963-Tate), Judith Obermayer (1963-Tate), Stephen Lichtenbaum (1964-Tate), J. Michael Schlessinger (1964-Tate), John McCabe (1967-Tate). Occasionally, we had the participation of Heisuke Hironaka (1960-Zariski) and David Mumford (1961-Zariski) as well as Warren Wong (1959-Brauer) and Morton Harris (1960-Brauer). Students from MIT were also occasionally involved, but the principal participants were from Harvard. As one can imagine, this was a lively group; there were even "secret seminars" in which one could let one's hair down and confess ignorance or expose the struggling we all were going through to master hard material. The seminars were secret only in that we took care not to invite faculty!

The year 1957-58 found Tate in Paris where he interacted with Serre and doubtless with Grothendieck. I recall a letter from Tate to Evens from approximately this time in which the Atiyah-Hirzebruch Spectral Sequence was discussed. Evens, then becoming an expert on the cohomology ring of a finite group, was to prove that this ring, with cup product as its multiplication, is noetherian. However, the point of my recollection is that already the nascent group around Tate was functioning even though the leader was away in Paris.

The year 1959-60 was a banner year both for Harvard's Mathematics Department and for the Tate circle. During that time Akizuki, Grothendieck and Nagata were in residence the entire year. Actually, due to some visa problems, Grothendieck arrived about one and a half months late. And so, Tate was the lecturer in the graduate course that Grothendieck had announced: Theory of Sheaves. What an audience was in attendance! Of course, there was the usual complement of graduate students, but the first row was filled with dignitaries: Akizuki, Nagata, Zariski, Lang (when he was in town), Tate (when he had stopped filling in for Grothendieck). We were going to learn scheme theory from Grothendieck in a seminar and, for that, we really had to know sheaves. In addition, there was another course with a powerhouse audience: Bott had arrived from Michigan and gave an introductory graduate course on Algebraic Topology, but everybody–simply everybody–had to hear Bott on Algebraic Topology. It was during this year, too, that Tate and Bott formed their fast friendship.

Actually, the year '59-'60 began in the summer when the Tate circle held its first summer seminar. These seminars ran every summer after that and usually discussed unpublished work of Tate (of which there was plenty!) frequently from typescripts by Lang. During the school year, the seminar continued and in 1959-60 it was dedicated to the theory of profinite groups, their cohomology groups, and Galois Cohomology. Here, though we didn't realize it, Tate was developing the *étale* cohomology of $Spec(k)$–where k is a field and this was very important as a test case for Grothendieck who was starting his drive toward the Weil Conjectures.

In particular, we saw the theory of cohomological dimension and the application of these methods and techniques to local class field theory. The following summer we concentrated on duality, in particular for abelian varieties over local fields. There were many rough spots for us as neophytes because we had to absorb the idea of $Pic(A)$ as an Ext, the Weil pairing, isogenies, and the connection of Cartier Duality (which we didn't really know) with all this stuff. I cannot communicate the excitement we felt, especially as it was mixed with the mystery engendered by our poor understanding.

During the academic year, Tate was also active with the circle (almost as a whole) and with its individual members. In particular, when several of us were taking Zariski's course on algebraic curves (Zariski always called this course: "Algebraic Functions of One Variable") and listening in on his seminar on abelian varieties, we needed Tate to explain to us the geometry (even the topology) behind all the algebra we saw from Zariski. This he did in his usual direct fashion and we were privileged to learn from two masters at once.

In the year '60-'61, the seminar was again very exciting. For, we studied elliptic curves and their arithmetic; in particular, Tate developed–before our eyes–the theory of what is now known as the "Tate Curve." There was the mysterious business of good reduction and various types of bad reduction. The structure of the formal group began to emerge as we studied the work of E. Lutz (in our secret seminar we looked at Mattuck's generalization) and we started our introduction to arithmetic algebraic geometry.

Many thesis topics came out of these seminars. Tate had the policy that, where possible, one should pick one's own thesis topic. He would comment on suitability and his approval meant you were more or less on your own–he had a minimal help policy, as well. This was, of course, all to the good. It was also tough on the student and served to inculcate the independence needed to pursue one's own ideas.

In the last year that I can speak about from personal knowledge, namely '61-'62, Grothendieck again visited. His course/seminar that year was a fair proportion of what is now the complete EGA IV. He went very fast, wrote practically nothing on the board (we were in effect taking dictation–a favorite subject of French students!) and left us to figure it out on our own later. Many hours were spent with Tate deciphering this material. Tate also lectured on his own way of seeing the material in EGA III on formal functions and the basic finiteness theorems for proper morphisms. The preparatory notes for the seminar on duality, run by Robin Hartshorne later, also seem to be derived from this visit. In addition, there was the well-attended seminar offered by Mike Artin on "Grothendieck Topologies"; so, we had plenty to do besides our own work.

I am certain the circle around Tate maintained its excitement and hectic pace after I graduated–it's hard to see how those two attributes could *increase.* What is clear is that the experience of mathematics at high level, in vast quantities, among interested and good students marked all of us for life. No mathematician in his role as passer on of mathematical culture and knowledge could have given us more. This is the heritage we took from Tate.

Stephen S. Shatz
Department of Mathematics
University of Pennsylvania
209 S. 33rd Street
Philadelphia, PA 19104-6395 U.S.A.
E-mail: sss@sas.upenn.edu

Pure and Applied Mathematics Quarterly
Volume 6, Number 1
(*Special Issue: In honor of
John Tate, Part 2 of 2*)
1—20, 2010

Recent Progress and Open Problems in Function Field Arithmetic — The Influence of John Tate's Work

Dinesh S. Thakur*

Abstract: The goal of this paper is to give a quick survey of some important recent results and open problems in the area of function field arithmetic, which studies geometric analogs of arithmetic questions. We will sketch related developments and try to trace the multiple influences of works of John Tate in this context.
keywords: Tate, function field, Drinfeld modules, elliptic curves, zeta values.

This paper is dedicated to my teacher John Tate. I am glad and honored to be invited to contribute to this special volume celebrating his 80th birthday. The goal of this paper is to give a quick survey of some important recent results and open problems in the area of function field arithmetic, which studies geometric analogs of arithmetic questions. We will sketch related developments and try to trace the multiple influences of works of John Tate in this context. We mainly, but not fully, limit ourselves to topics where these are clearly visible. Also, we focus mainly on results simple to state, and leave variants or generalizations to the references. General references for background on recent results in function field arithmetic are [Ros02, Gos96, G^+92, G^+97, Tha04].

Received November 13, 2006.
*Supported in part by NSF grant.

Tate's fundamental contributions to the foundations of function field studies (sometimes in the context of global fields and sometimes over more general base fields) and of non-archimedean studies have been very influential in the development of number theory. Let us quickly recall some of them:

- Work with Artin on the function field basics and on the foundations of global class field theory,
- Galois cohomology of global fields,
- Study of local constants in global context,
- Study of Milnor ring in global context,
- Work on theorems of Luroth, Castelnuovo-Severi,
- Work on genus change in inseparable extensions of function fields,
- Novel treatment of residues and Riemann-Roch,
- Tate conjectures and function field analog of Birch and Swinnerton-Dyer conjectures,
- Rigid analysis in non-archimedean setting,
- Lubin-Tate's treatment of local class field theory,
- Tate elliptic curves,
- Tate-Shafarevich elliptic curves giving arbitrary large Mordell-Weil rank over global function fields,
- Tate's formulation of the Stark conjectures in general global setting and his proof of Stark-Stickelberger in the function field case,
- Conjectures of Mazur-Tate-Teitelbaum about p-adic Birch and Swinnerton-Dyer and Mazur-Tate refined conjectures in various global settings.

I have omitted specific references, as they can be easily found out from his bibliography obtainable, e.g., from MathSciNet.

It is hard to trace the influence of Tate's work on other works accurately, as the basic objects, ideas and theorems that he introduced have permeated (and been generalized by others in well-developed standard theories) throughout mathematics as the following common terms show: Tate modules, Tate elliptic curves, Tate uniformization, Tate's q, Lubin-Tate theory, Honda-Tate theory, Sato-Tate conjecture, p-divisible or Barsotti-Tate groups, Hodge-Tate decomposition, Tate cohomology, Poitou-Tate duality, Serre-Tate parameters, Tate cycles, Tate conjectures, Mumford-Tate group, Mumford-Tate conjectures, Tate twists, Tate motives, Tate spaces, Tate ring, Tate algebra, Tate residue, Cassels-Tate pairing,

Neron-Tate heights, Shafarevich-Tate groups, Koszul-Tate resolution (complex, derivation) etc.

1. Class field theory and Langlands conjectures

We have already mentioned important contributions of Tate in the 1950's to global class field theory and its cohomological aspects.

In the 1930's, Carlitz developed what is now called a Carlitz module and related analogs for $\mathbb{F}_q[t]$ of the exponential, $2\pi i$, roots of unity and cyclotomic theory. His contributions were forgotten to a large extent, probably because he wrote a huge number of papers and because this theory was developed in a series of papers with uninformative titles such as 'A class of polynomials', 'A set of polynomials', 'Some properties of polynomials' etc. Carlitz's student David Hayes attended Tate's lectures at Harvard giving an exposition of explicit local class field theory by the Lubin-Tate approach of formal groups and noticed a similarity with the explicit cyclotomic approach that he learned from Carlitz. When he pointed this out, Tate encouraged Hayes to develop it further.

As Hayes writes in [Hay], while Artin, Weil and others were treating all places of function fields on equal footing, Carlitz developed his theory by singling out a distinguished place at infinity in analogy with the distinguished (archimedean this time) place at infinity for the field of rational numbers. But doing this, he missed some abelian extensions (those wildly ramified at infinity). Hayes [Hay74] then developed the full explicit class field theory for the rational function field and started considering the higher genus case.

Let K be a function field of one variable over $\mathbb{F}_q$ of characteristic p, ∞ be a place of K, A be the ring of elements of K integral outside ∞, K_∞ be the completion of K, and C_∞ be the completion of an algebraic closure of K_∞. We then consider K, A, K_∞, C_∞ as analogs of $\mathbb{Q}$, $\mathbb{Z}$, $\mathbb{R}$, $\mathbb{C}$ respectively, with ∞ corresponding to the archimedean place of $\mathbb{Q}$. For a finite prime $\wp$ of A, we can consider $A_\wp$ as an analog of $\mathbb{Q}_p$. For more on these analogies, see [Tha04, Sec. 1.1].

We start with the simplest case: $K = \mathbb{F}_q(t)$, $A = \mathbb{F}_q[t]$, so that $K_\infty = \mathbb{F}_q((1/t))$.

Let us define $C_a(u) \in A[u]$, for $a \in A$, by

$$C_1(u) = u,\ C_t(u) = tu + u^q,\ C_{t^n}(u) = C_t(C_{t^{n-1}}(u)),$$

and asking $C_a(u)$ to be $\mathbb{F}_q$-linear as a function of a. Then the $C_a(u)$ are analogs of the cyclotomic polynomials $u^{|n|} - 1$. Thus their roots, ζ_a's say, give analogs of roots of unity ζ_n's. In this case,

$$e(z) = \sum_{n=0}^{\infty} \frac{z^{q^n}}{(t^{q^n} - t)(t^{q^n} - t^q) \cdots (t^{q^n} - t^{q^{n-1}})}$$

gives an analog of the exponential function e^z and satisfies $e(az) = C_a(e(z))$ analogous to $e^{nz} = (e^z)^n$. There is $\tilde{\pi} \in C_\infty$, transcendental over K, with $e(z) = 0$ if and only if $z = \tilde{\pi}a$, with $a \in A$. This thus gives analog of $2\pi i$, because the kernel of the usual exponential is $2\pi i\mathbb{Z}$. (Note that $\tilde{\pi}$ is well-defined up to multiplication by an element of $A^* = \mathbb{F}_q^*$, just as $2\pi i$ is well-defined up to multiplication by an element of $\mathbb{Z}^* = \{\pm 1\}$). Further, $e(\tilde{\pi}a_1/a)$ represent roots of C_a, for $a_1 \in A$, just as $e^{2\pi ik/n}$ represent roots of the n-th cyclotomic polynomial.

Further, we have the following analog [Hay74, Hay] of cyclotomic theory (explicit class field theory) and Kronecker-Weber theorem over $\mathbb{Q}$:

Theorem 1. *With the notation as above, if a is a non-constant polynomial in A, then $K(e(\tilde{\pi}/a))$ is an abelian extension of K, with Galois group $(A/(a))^*$, unramified outside ∞ and the primes of A dividing a. The Frobenius at prime $\wp$ of A acts on the generator by applying $C_\wp$.*

The maximal abelian extension of K can be obtained as the compositum of all such extensions together with those obtained by doing the same procedure with $\mathbb{F}_q[1/t]$ replacing $A = \mathbb{F}_q[t]$.

For a proof, more details and generalization to any K, see [Gos96, Tha04, Hay79, Hay85, Hay].

Around the same time, Drinfeld, again unaware of Carlitz work, but using ideas of Lubin-Tate and taking a clue from Deligne's work relating Galois representations to modular forms, developed [Dri74], what are now called Drinfeld modules, in the setting of any function field K and any place at infinity (considered as an analog of $\mathbb{Q}$ or a totally imaginary field with unique archimedean infinity, where we have explicit class field theory classically). He developed an attack not only on class field theory (which corresponded to the case of rank one), but also on Gl_n Langlands conjecture analogs. (Deligne observed that the modularity theorem in the function field case, now connecting elliptic curves over function fields to the modular curves of Drinfeld modules, follows by Drinfeld's work combined with previous works. See [Tha04, Pa. 215] for a sketch and note, in particular, that one step is Zarhin's proof of Tate's isogeny conjectures for abelian varieties over function fields.) Lafforgue [Laf02] finally succeeded in proving these, building on this work of Drinfeld and of many others.

Theorem 2. *Let K be a function field of characteristic p. Fix a prime $l \neq p$ and an integer $r \geq 1$. Let A_K denote the adele ring of K and G_K be the absolute Galois group of K.*

Then there is an explicit bijection $\pi \to \sigma(\pi)$ between (i) the set of irreducible cuspidal automorphic representations π of $Gl_r(A_K)$ which have central character of finite order, and (ii) the set of of irreducible representations $\sigma : G_K \to Gl_r(\overline{\mathbb{Q}_l})$ unramified outside a finite set of primes and with determinant of finite order, with the property that Frobenius and Hecke eigenvalues (or equivalently the local L factors) of π and $\sigma(\pi)$ match.

In addition to influence of Lubin-Tate work, Drinfeld's work also used ideas of rigid analysis, Tate curves, Tate uniformization, Honda-Tate theory, and Tate modules. (Note that Tate modules of Drinfeld modules give $Gl_n(A_\wp)$ representations in contrast to $Gl_2(\mathbb{Q}_l)$ representations given by elliptic curves over function fields). Tate's results on local-global analysis of L-functions, and local constants also play an important role in the Langlands program.

Gaitsgory and de Jong, Böckle-Khare proved (see [Kha] for details and references) the function field analog of Serre's conjectures that a continuous, absolutely irreducible, n-dimensional representation over a finite field of characteristic $l \neq p$, of the fundamental group of a geometrically irreducible smooth curve over a finite field of characteristic p, is automorphic. Because of Lafforgue's theorem this reduces to appropriate lifting theorems.

On the other hand, over $\mathbb{Q}$, works of Hida, Ribet and especially of Wiles, Taylor and others settling the modularity conjecture of Shimura-Taniyama-Weil, opened up powerful methods of attack in the area of the original Langlands' and Serre's conjectures. These works together with those of Böckle, Dieulefait, Khare, Kisin, Taylor, Ramakrishna, Winterberger and many others have helped to settle most of the original Serre conjectures. The level one case due to Khare [Kha06] has just been published and the earlier manuscripts of Khare and Winterberger as well as those announcing the odd conductor case are available on the archives. For general account and references, we refer to Khare's excellent survey paper [Kha].

In the beautiful inductive method used for this proof, Tate's 1973 proof [Tat94b] of $p = 2$ case of Serre's conjectures forms the base case. Also, $p = 3$ case done by Serre by similar discriminant bound methods and other works using this method of Tate are heavily used. Other influences of Tate's foundational works are seen [Kha] in uses of Barsotti-Tate representations, Hodge-Tate theory, Poitou-Tate exact sequences. Schoof's extensions of Fontaine's famous result that there is no abelian variety over $\mathbb{Z}$ played an important role in Khare's work. See [Ogg66] for Tate's earlier observation and elementary proof of the one dimensional case of elliptic curves.

2. Zeta functions: arithmetic of special values

Consider $A = \mathbb{F}_q[t]$ case again. Let $A+$ and $\mathbb{Z}+$ denote the set of monic polynomials and of positive integers respectively. Consider Carlitz zeta values

$$\zeta(s) = \sum_{a \in A+} \frac{1}{a^s} \in K_\infty, \quad s \in \mathbb{Z}+.$$

While $\zeta(1)$ makes sense, and $\zeta(sp) = \zeta(s)^p$, in contrast to the Riemann zeta function situation, Carlitz proved (analog of Euler's result on Riemann zeta values) that the values of ζ at 'even' positive integers s are rational multiples of $\tilde{\pi}^s$, where 'even' now means a multiple of $q-1$, which is a cardinality of $A^* = \mathbb{F}_q^*$, analogous to $\mathbb{Z}^* = \{\pm 1\}$. In particular, they are transcendental, as $\tilde{\pi}$ is transcendental.

The nature of Riemann zeta values at odd positive integers is still a mystery. On the other hand, we have

Theorem 3. *If s is a positive integer, then $\zeta(s)$ is transcendental. If, further, s is 'odd', i.e., not divisible by $q-1$, then $\zeta(s)/\tilde{\pi}^s$ as well as $\zeta_\wp(s)$ are transcendental, where $\zeta_\wp$ is the interpolation due to Goss of ζ at prime $\wp$ of A.*

How was this proved? Greg Anderson developed higher dimensional generalization 't-motives' of Drinfeld modules, put them in a framework analogous to motives and constructed $C^{\otimes s}$, the s-th tensor power of the Carlitz motive (corresponding to the Carlitz module above). This Carlitz-Tate motive is analog of the Tate twist $\mathbb{Z}(s) = \mathbb{Z}(1)^{\otimes s}$ and has corresponding exponential and logarithm attached to it. In [AT90], we constructed an algebraic point (torsion if and only if s is 'even') on $C^{\otimes s}$ and expressed $\zeta(s)$ in terms of its logarithm. An analog of the Hermite-Lindemann theorem on transcendence of logarithm values (and its $\wp$-adic incarnation) was then proved [Yu91] by Jing Yu, which implied the theorem above.

In [Tha04] multizeta values were defined by

$$\zeta(s_1, \cdots, s_k) = \sum_{|n_1|>\cdots>|n_k|} \frac{1}{n_1^{s_1} \cdots n_k^{s_k}}$$

(where $n_i \in A+$) and their properties, relations between them and interpolations were studied.

Theorem 4. [APT] *The multizeta values defined above are periods of explicit iterated extension of $C^{\otimes s_i}$'s and thus transcendental.*

The relations satisfied by them are under investigation. These relations are not quite the classical sum-shuffle and integral-shuffle relations and involve subtle 'digit phenomena'.

Further in [AT], the analogs of Ihara power series, Deligne-Soule cocycles and higher circular units are constructed. Ihara power series occurs in Grothendieck-Ihara program of the study of absolute Galois group over $\mathbb{Q}$ by realizing it in the automorphism group of the algebraic fundamental group of the projective line minus three points. When we consider the more manageable nilpotent quotient of the fundamental group, the mixed motive structure of iterated extensions of Tate twists shows up, with multizeta values occurring in the DeRham-Betti aspects and Ihara power series and Deligne-Soule cocycles occurring in the etale aspects of the meta-abelian simplification. We have these ingredients, as we have the mixed Tate-motives theory in function fields, but we have no good understanding of the fundamental group background yet!

David Goss (see [Gos96]) interpolated the zeta values at positive integers above to a much larger continuous space containing, in particular, the negative integers. (Cohomological aspects of L-functions in this situation have been developed by Taguchi, Wan [TW96], Pink and Böckle [Böc02]). The special values at positive as well as negative integers are related in a way analogous to the Herbrand-Ribet theorem to class groups of cyclotomic extensions, with the values at positive integers linking to the class groups of integral closures of $\mathbb{F}_q[t]$ in the cyclotomic extensions, while the values at the negative integers linking to the full Pic^0 of the cyclotomic extensions. But still no functional equation connecting the two sets of values is understood in contrast to the Riemann zeta function functional equation whose nice analysis was done in Tate's thesis. We do have good analogs [Tha04] of Gamma functions, but do not understand them as a factor at infinity for zeta as in Tate's thesis.

Classically, a simple analysis of poles of gamma factors occurring in the functional equations of general zeta functions leads to determination of orders of vanishing of zeta functions at negative integers. In function fields, the story is not even conjecturally fully understood. There is some interesting systematic 'extra vanishing' phenomena of zeta at $-s$ depending on digit combinatorics of s and Weierstrass gaps [Tha04, Sec. 5.4] and [Gos].

3. Zeta functions: Stark conjectures

Tate generalized Brumer and Stark conjectures (giving ideal class annihilators and abelian extensions via analytic processes in the ground field) to global settings, considered refined conjectures and systematically worked out many basic functorialities and results on them.

For a finite abelian extension L of a global field K with Galois group G and a non-empty set T of places of K containing at least all those ramified in L, by character/Fourier theory, there is a unique $\theta = \theta_{T,L} \in \mathbb{C}[G]$ such that

$$\psi(\theta) = L_T(0, \overline{\psi})$$

for all complex valued characters ψ of G (extended linearly to the group algebra), where the L function

$$L_T(s, \psi) := \prod_{\wp \notin T} (1 - \psi(\mathrm{Frob}_\wp)\mathrm{Norm}(\wp)^{-s})^{-1}, Re(s) > 1$$

is a rational function in q^{-s} and is finite at 0. Let μ be the number of roots of unity in L and $\omega := \mu\theta$.

Theorem 5. *(Tate and Deligne* [Tat84]*) Let K be a function field and P any prime divisor of L. We have*

(1) $\omega \in \mathbb{Z}[G]$,

(2) If $|T| \geq 2$, *then* P^ω *is a principal divisor* (ℓ) *of some* $\ell \in L$,

(3) If $T = \{v\}$, *then* P^ω *is* $(\ell) + nv_L$, *for some* $\alpha \in L$ *and* $n \in \mathbb{Z}$ *and where* v_L *denotes the simple sum of places of* L *above* v.

(4) If $\lambda^\mu = \ell$, *then* $L(\lambda)$ *is abelian over* K.

Let us give a quick sketch of Tate's nice proof of (2) for a geometric extension of a function field (an analog of the Stickelberger theorem on ideal class annihilators): The class group of a curve is the group of $\mathbb{F}_q$-rational points i.e., the part of the $\overline{\mathbb{F}_q}$-points of its Jacobian where the Frobenius F acts as the identity, whereas its L-function is essentially the characteristic function of the Frobenius on the corresponding component of the Jacobian or rather the Tate module. Hence it kills the component when $t = F$ by the Cayley-Hamilton theorem. Hence ω which is just a linear combination of L-values at $s = 0$ (i.e. $t = 1$) kills the class group. (See [Tha04] and references there for more details.) Using one-motives in place of Jacobian, Deligne generalized from a geometric extension to any extension and proved the theorem above.

These ideas inspired the proof [GS85] of the Herbrand-Ribet analog (using the zeta values at negative integers) that was mentioned in the last section

Tate then suggested to Hayes that his Drinfeld module work might imply another proof of this theorem giving λ (as in the Theorem) explicitly (after the reduction technology developed in [Tat84] is used). This was proved in [Hay85]

In 1987, Mazur and Tate [MT87] developed a refinement of L-function conjectures in the case of elliptic curves. They ran a seminar at Harvard working out these ideas in various situations, where Gross and Hayes worked out (and Hayes proved [Hay88]) the refined p-adic abelian Stark conjectures in the number field and the function field case respectively. See works of Ki Seng Tan and Joongul Lee for some results on the refined elliptic curves case.

In a very interesting recent work, Greg Anderson has a refinement of the Stark conjectures [And06] (see also his recent preprints for proofs using Lang-Serre geometric class field theory and Drinfeld's shtukas) which involves two rather than one variable algebraic functions. This generalizes his earlier work [And] in the $\mathbb{F}_q[t]$ case. For a long time, examples suggested existence of a generalization, but a good formulation was not at all clear. Anderson generalizes to higher genus (see also [And94]) by using an adelic harmonic analysis formulation in the language and framework of Tate's thesis.

4. Diophantine geometry

Tate's foundational work developing tools of group schemes, Tate modules, p-divisible groups, Hodge-Tate decompositions, Neron-Tate heights, p-adic sigma functions, Tate curves (with their place in moduli studies) etc. has played an important role in Diophantine geometry over global fields. Works of Serre, Mazur, Faltings, Fontaine and others made heavy use of these tools.

Tate's conjectures and his results on them have also been very influential. For a nice presentation of the precise statements and of the connections between Tate conjectures relating algebraic cycles on varieties over fields finitely generated over the prime fields, galois invariant cohomology classes, order of poles of their zeta functions, Tate conjectures on isogenies of abelian varieties, Tate's analogs of conjectures of Birch and Swinnerton-Dyer (BSD) in the function field case, as well as for the summary of evidence, see [Tat94a].

Tate proved the Tate conjecture for abelian varieties over finite fields basically via a Shafarevich type finiteness result (trivial over finite fields). Faltings reversed the direction to deduce the Shafarevich finiteness conjecture from his Tate's conjecture proof, which in turn was modelled on Tate's original proof together with Zarhin's ideas which proved Tate conjectures for function fields. Faltings then deduced Mordell's finiteness conjecture from the Shafarevich conjecture using ideas of Pashin in Parshin's proof of the Mordell conjecture over function fields.

In many respects, Drinfeld modules (of rank two, for best analogies) are analogs of elliptic curves. Their higher dimensional generalizations, such as t-motives or A-motives, are analogs of abelian varieties (or of motive representing their H_1). Drinfeld modules also have 'characteristic' which can be generic (analog of characteristic zero) or finite prime $\wp \in A$, and we can consider Drinfeld modules in finite field, local or global setting. Thus many theorems or conjectures about abelian varieties have two different analogs in function fields: in terms of abelian varieties over function fields or in terms of t-motives.

In addition to proofs of Tate, Zarhin, Faltings respectively, over finite fields, function fields, number fields respectively of Tate conjectures mentioned, let us look at the function field situation in this t-motive realm:

For Drinfeld modules over finite fields, an analog of the Tate isogeny theorem was proved [Dri77] by Drinfeld. For Drinfeld modules of generic characteristic (in fact, in a much more general setting), the analog of the Tate conjecture/Faltings theorem was established by Tamagawa and Taguchi [Tam94, Tag95, Tag96]. This was done by a method (quite different from Faltings in the classical case) inspired by the previous important work of Anderson [And93], where he proves an analog of Tate conjectures for 'formal t-modules', by approximating solutions of '$\wp$-adic linear Frobenius equations'. Taguchi [Tag93, Tag91] also proved the semisimplicity of the Galois representation on the Tate module, for both finite and generic characteristic Drinfeld modules. Taguchi [Tag99] proved that a given L-isogeny class of Drinfeld A-module contains only finitely many L-isomorphism classes, for L a finite extension of K. Oliver Watson has recently proved an analog of the Tate conjecture in the equi-characteristic case, in his 2003 University of Pennsylvania thesis.

Classically, there is a well-known theorem of Serre on the image of the Galois representation (with the general conjecture for abelian varieties being that of Mumford-Tate) obtained from the torsion of elliptic curves. Pink [Pin97] showed that if ρ has no more endomorphisms than A, then for a finite set S of places $\wp \neq \infty$, the image of $\mathrm{Gal}(K^{sep}/K)$ in $\prod_{\wp \in S} GL_n(A_\wp)$ for the corresponding representation for rank n Drinfeld modules is open. (Note that this is weaker than the Serre type adelic version, but much stronger (unlike the classical case) than the case of one prime $\wp$, because we are dealing with all huge pro-p groups here, even though the primes $\wp$ change. So the simple classical argument combining p-adic and l-adic information to go from the result for one place to the result for finitely many places does not work). This has been generalized and improved in recent works of Pink, Traulsen and Gardeyn.

Let us mention that while over number fields, the Tate isogeny conjecture and the Shafarevich finiteness conjecture follow from each other for abelian varieties, it is easy to exhibit [Tha04, Sec. 6.1] infinitely many non-isomorphic rank 2 Drinfeld modules, with good reduction everywhere (so not only the support of the discriminant is bounded, but the discriminant is one). So with many analogies, there are a few important contrasts as well.

Finally, let us mention that the analog for Drinfeld modules of the conjecture of Tate and Voloch [TV96] (predicting existence of a lower bound for the p-adic distance of torsion of a semi-abelian variety over C_p from a closed subvariety not containing it) has been recently announced in a preprint by D. Ghioca.

5. Diophantine Geometry: Elliptic curves over function fields

Tate's Haverford lecture notes and Inventiones survey on elliptic curves beautifully and systematically laid out the subject area and became the necessary first reading for anybody interested in learning the subject.

In a very influential Bourbaki seminar [Tat95] of 1966, Tate explained (joint work with Michael Artin) the homological machinery behind the geometric case of BSD conjectures and proved

Theorem 6. *For an elliptic curve E over a function field K, (1) its analytic rank (i.e., the order of vanishing of its L function at $s = 1$) is at least its arithmetic rank (i.e. the rank of the Mordell-Weil group of its rational points); (2) These two ranks are equal if and only if the Shafarevich-Tate group Ш is finite if and only if the ℓ-primary part of Ш is finite for one prime ℓ. Further, if these equivalent conditions hold, the full BSD conjecture, giving the conjectural formula for the leading term of the L function at $s = 1$, holds.*

As stated, this uses the later refinements due to Milne and others taking care of the p-part, allowing $\ell = p$ and $p = 2$ etc. These were not handled in [Tat95], essentially because Tate used etale cohomology and crystalline cohomology machinery needed for taking care of p was developed soon thereafter. For a generalization to abelian varieties, see [Sch82].

Iwasawa used analogies between function fields and number fields to develop Iwasawa theory and carried over Hasse-Weil's geometric understanding and cohomological machinery related to zeta functions to the p-adic case in number fields. Transferring Tate's ideas to this realm led to the development of results on the p-adic BSD analog by Perrin-Riou [PR83], using the Iwasawa theoretic machinery. More recently, Bertiloni and Darmon [BD01] developed another analog, this time dealing with p-adic L function defined analytically and using rigid analysis and p-adic uniformization ideas, initiated by Tate!

It suffices for our purposes below to say that in the function field case, Tate proved (see [Tat94a]) the rank equality and thus the full BSD conjecture for E which is iso-trivial or E whose corresponding elliptic surface over $\mathbb{F}_q$ is dominated by a product of curves, in particular Fermat surfaces.

Now we describe 1967 work of Shafarevich-Tate and a recent nice advance by my colleague Douglas Ulmer:

It is not known whether the (arithmetic or analytic) ranks of $E(\mathbb{Q})$ are bounded as E runs through elliptic curves E over $\mathbb{Q}$, the usual bet seems to be that they are not. The largest rank known today is around 28.

In a famous paper, Shafarevich and Tate [TS67] showed that over a fixed $\mathbb{F}_p(t)$, there do exist elliptic curves of arbitrarily large arithmetic rank: They considered certain hyperelliptic quotient C over $\mathbb{F}_p$ of the Fermat curve of degree p^n+1. Its zeta function can be calculated in terms of Gauss sums following Weil, and the Gauss sums were made more explicit by them for this degree. This allowed them to show that the Jacobian of C has a supersingular elliptic curve E_0 as an isogeny factor to a high multiplicity m over $\mathbb{F}_p$. (Note that this multiplicity can be calculated by Honda-Tate theory). If E is the constant curve over $\mathbb{F}_p(t)$ based on E_0, over quadratic extension $F=\mathbb{F}_p(C)$ it has rank m, as $E(F)/\text{torsion} = \text{Hom}_{\mathbb{F}_p}(J(C), E_0)$. Thus the quadratic twist of E by F has a large arithmetic rank over $\mathbb{F}_p(t)$.

But the curves they exhibited are isotrivial (j-invariant is in $\mathbb{F}_q$). Thus the situation is not readily comparable to the number field case. Shioda exhibited non-isotrivial elliptic curves with arbitrarily large rank over $\overline{\mathbb{F}_q}(t)$. Ulmer [Ulm02] showed that, in fact, they have arbitrarily large arithmetic rank over $\mathbb{F}_q(t)$ and proved (for a variant of equations working for all p) the following

Theorem 7. *For non-isotrivial E over $K=\mathbb{F}_p(t)$ given by $y^2+xy=x^3-t^{p^n+1}$, BSD holds and its rank is at least $(p^n-1)/2n$. Further, the degree of the conductor of E is p^n+2 or p^n+4 depending on whether p^n+1 is divisible by 6 or not.*

Here is a sketch showing influence of Tate's ideas and results: Following Shioda, Ulmer gets a dominant rational map from Fermat surface of degree p^n+1 to the surface over $\mathbb{F}_p$ corresponding to E, which gives the BSD conjecture for E

by results of Tate mentioned above. By detailed analysis of the Fermat quotient isomorphic to this surface, he expresses its zeta function (related to the L-function of E) in terms of the zeta function of the Fermat surface. Using the Shafarevich-Tate calculation mentioned above, he calculates the analytic rank, which is now also the arithmetic rank. Finally, the conductor calculation is done by Tate's algorithm for minimal models!

Ulmer further showed that these curves (as well as Shafarevich-Tate isotrivial curves) asymptotically attain the upper bound for the rank in terms of the size of the conductor of the elliptic curve. This gives stronger support to the corresponding number fields conjecture predicting the existence of a family of elliptic curves E with conductors N tending to infinity, with rank at least $c \log(N)/\log(\log(N))$ for some $c > 0$, coming from the random matrix theory analogies compared to a competing conjecture coming from probabilistic models. (In the number field case, it was shown earlier [KM00] using Heegner points that the conjecture obtained from the first conjecture by replacing elliptic curves by abelian varieties is true). Darmon produced more examples of high rank curves, by an alternate method of analysis of zeta functions, but as pointed out in [Ulm04], the calculation of root numbers he needs requires knowing the local representation of decomposition groups on the Tate module at places of bad reduction and ultimately boils down to analyzing super-singular Gauss sums using Shafarevich-Tate!

Trying to answer a question of Ellenberg about behavior of ranks as you go up a certain tower of extensions, Ulmer realized that the examples above are not really special and proved

Theorem 8. *Given any non-isotrivial elliptic curve E over $\mathbb{F}_q(t)$, there exists a finite extension $\mathbb{F}_r(u)$ of $\mathbb{F}_q(t)$ such that E has unbounded analytic ranks in the tower $\mathbb{F}_r(u^{1/d})$.*

(See [Ulm] for this, many more related interesting results and generalizations to abelian varieties.)

As $\mathbb{F}_r(u^{1/d})$ is isomorphic to $\mathbb{F}_r(x)$, E thus gives a family of curves with arbitrarily large rank over this base. Note that like isotriviality, this property of a rational function field that it contains a copy of itself in several different ways, by replacing t by a rational function of t, is again not readily comparable to the number field case!

6. Diophantine Geometry: Diophantine approximation

We just mention that in the related field of Diophantine approximation of algebraic elements over function fields, the naive analog of Roth's theorem that 'diophantine approximation exponent of algebraic real numbers is two' fails. For connections with deformation theory and many interesting recent results and open questions about distribution of Diophantine approximation exponents of algebraic elements and about their continued fraction expansions, we refer to [Tha04] and [Tha].

7. Transcendence of special values and periods

Let us begin with a transcendence result on the period of the Tate elliptic curve:

Let p be a prime number, and k be an algebraic closure of $\mathbb{F}_p$. Let q_0 be a variable and consider $a_4, a_6 \in k[[q_0]]$ defined by

$$a_4 := \sum_{n\geq 1} \frac{-5n^3 q_0^n}{1-q_0^n}, \qquad a_6 := \sum_{n\geq 1} \frac{-(7n^5+5n^3)q_0^n}{12(1-q_0^n)}.$$

Theorem 9. *The period q_0 of the Tate elliptic curve $y^2 + xy = x^3 + a_4 x + a_6$ over $K := k(a_4, a_6)$ is transcendental over K.*

This function field analog of Mahler-Manin conjecture was proved by Voloch [Vol96], by approximating q_0 by algebraic quantities and getting a contradiction by analyzing the Galois action using Igusa's theorem.

Soon afterwards, the original conjecture was proved. See [Wal97] for the history and an account of the proof. Nice application is that '$\log_p q_0$' appearing in the p-adic Birch and Swinnerton-Dyer conjectures of Mazur, Tate and Teitelbaum (Theorem of Stevens/Greenberg) does not then vanish, so that the order of vanishing of the L-functions is exactly as predicted in the conjectures.

Voloch explained his theorem and proof in a seminar at the University of Arizona and I could give another proof [Tha96] (and yet another [AT99] with Allouche) using the automata criterion for algebraicity of Christol, which says that $\sum f_n x^n$ is algebraic over $\mathbb{F}_q(x)$, if and only if $f_n \in \mathbb{F}_q$ is produced by a q-automaton, if and only if there are only finitely many subsequences of the form $f_{q^k n + r}$ with $0 \leq r < q^k$. (See [Tha04, Cha. 11] for definitions and details).

This result on the transcendence of the Tate period was the start of my work on applications of automata theory to naturally occurring quantities in the setting of function field arithmetic, by giving refined transcendence theorems for them using results in computer science, formal language theory and logic. In my thesis, special values of arithmetic gamma for $\mathbb{F}_q[t]$ defined by Carlitz and Goss were calculated and related to the periods of Drinfeld modules and an analog of the Chowla-Selberg formula was proved. Thus applying the results of Jing Yu on the transcendence of periods, knowledge of transcendence of gamma values at fractions was parallel in $\mathbb{Z}$ and $\mathbb{F}_q[t]$ case. The automata techniques (and help from automata experts Allouche, Mendes-France, Yao) allowed me to get a complete result showing ([Tha04, Cha. 11] for details) that all the monomials in gamma values at fractions that I had not shown to be algebraic in my thesis are, in fact, transcendental!

For the $\wp$-adic interpolation due to Goss of the gamma, Yao and myself only managed partial results [Tha04, Cha. 11], and the general case is still open.

There is another gamma function (geometric case)

$$\Gamma(x) = \frac{1}{x} \prod_{a \in A+} (1 + \frac{x}{a})^{-1}$$

for which there is an even more satisfying result. I had proved the functional equations for this gamma function giving algebraicity of some explicit monomials, and proved transcendence of a few (all, if $q = 2$) values at proper fractions in K by connecting them to the periods of Drinfeld modules.

By Γ-monomial we will mean an element of the subgroup of C_∞^* generated by $\tilde{\pi}$ and the values of Γ at proper fractions in K.

Theorem 10. [ABP04] *A set of Γ-monomials is $\overline{K}$-linearly dependent exactly when some pair of Γ-monomials is. Pairwise $\overline{K}$-linear dependence is decided by an explicit combinatorial criterion and exactly those monomials mentioned above are the only algebraic ones.*

In particular, the transcendence degree of the field extension of $\overline{K}$ generated by $\tilde{\pi}$ and gamma values at proper fractions with denominators dividing $a \in A$ over $\overline{K}$ is $1 + (1 - 1/(q-1))|(A/a)^*|$.

(We remark here that there is [Tha04, Sec. 4.12] a unified Galois-theoretic description of the 'explicit algebraic monomials' mentioned above, in both classical and function field cases, and that in the classical as well as in the geometric gamma case (but not in the arithmetic case) they follow from reflection and multiplication formulas.)

This was done by expressing the gamma monomials as periods of appropriate t-motives (these motives are defined using the same two variable functions mentioned in the Stark conjecture section) and by the powerful transcendence criterion of [ABP04] that all period relations come from motivic relations. This was further developed by Papanikolas [Pap] who proved the analog of the Grothendieck conjecture that the transcendence degree of the field generated by the periods of a motive is the dimension of its motivic Galois group (or Mumford-Tate group if considering the analogy in the abelian varieties situation), i.e., the tannakian group of the tannakian category generated by its powers. Jing Yu and Chang [YC] have applied this to get all algebraic dependence relations between Carlitz zeta values and we should soon have a result about multizeta values. See [Tha] for a sketch.

There are still many open questions about the nature of values of $\wp$-adic interpolations, algebraic dependence relations for the Carlitz-Goss arithmetic gamma function mentioned above as well as their generalizations to arbitrary function fields, and for another gamma function developed in [Tha04, Sec.8.3, 8.7].

Acknowledgement: I am very grateful to Jean-Pierre Serre and Greg Anderson for their detailed suggestions for improvement.

References

[ABP04] G. Anderson, W. Brownawell, and M. Papanikolas. Determination of the algebraic relations among special Γ-values in positive characteristic. *Ann. of Math. (2)*, 160(1):237–313, 2004.

[And] G. Anderson. A two-dimensional analogue of Stickelberger's theorem. In *[G+92]*, pages 51–73.

[And93] G. Anderson. On Tate modules of formal t-modules. *Internat. Math. Res. Notices*, (2):41–52, 1993.

[And94] G. Anderson. Rank one elliptic A-modules and A-harmonic series. *Duke Math. J.*, 73(3):491–542, 1994.

[And06] G. Anderson. A two-variable refinement of the Stark conjecture in the function-field case. *Compos. Math.*, 142(3):563–615, 2006.

[APT] G. Anderson, M. Papanikolas, and D. Thakur. Multizeta values for $F_q[t]$, their period interpretation and transcendence properties. *Preprint.*

[AT] G. Anderson and D. Thakur. Ihara power series for $F_q[t]$. *Preprint.*

[AT90] G. Anderson and D. Thakur. Tensor powers of the Carlitz module and zeta values. *Ann. of Math. (2)*, 132(1):159–191, 1990.

[AT99] J-P. Allouche and D. Thakur. Automata and transcendence of the Tate period in finite characteristic. *Proc. Amer. Math. Soc.*, 127(5):1309–1312, 1999.

[BD01] M. Bertolini and H. Darmon. The p-adic L-functions of modular elliptic curves. In *Mathematics unlimited—2001 and beyond*, pages 109–170. Springer, Berlin, 2001.

[Böc02] G. Böckle. Global L-functions over function fields. *Math. Ann.*, 323(4):737–795, 2002.

[Dri74] V. Drinfel′d. Elliptic modules. *Mat. Sb. (N.S.)*, 94(136):594–627, 656, 1974.

[Dri77] V. Drinfel′d. Elliptic modules. II. *Mat. Sb. (N.S.)*, 102(144)(2):182–194, 325, 1977.

[G$^+$92] D. Goss et al., editors. *The arithmetic of function fields*, volume 2 of *Ohio State University Mathematical Research Institute Publications*, Berlin, 1992. Walter de Gruyter & Co.

[G$^+$97] E.-U. Gekeler et al., editors. *Drinfeld modules, modular schemes and applications*, River Edge, NJ, 1997. World Scientific Publishing Co. Inc.

[Gos] D. Goss. Zeros of L-series in characteristic p. *Preprint.*

[Gos96] D. Goss. *Basic structures of function field arithmetic.*). Springer-Verlag, Berlin, 1996.

[GS85] D. Goss and W. Sinnott. Class-groups of function fields. *Duke Math. J.*, 52(2):507–516, 1985.

[Hay] D. Hayes. A brief introduction to Drinfel′d modules. In *[G+92]*, pages 1–32.

[Hay74] D. Hayes. Explicit class field theory for rational function fields. *Trans. Amer. Math. Soc.*, 189:77–91, 1974.

[Hay79] D. Hayes. Explicit class field theory in global function fields. In *Studies in algebra and number theory*, pages 173–217. Academic Press, New York, 1979.

[Hay85] D. Hayes. Stickelberger elements in function fields. *Compositio Math.*, 55(2):209–239, 1985.

[Hay88] D. Hayes. The refined $\wp$-adic abelian Stark conjecture in function fields. *Invent. Math.*, 94(3):505–527, 1988.

[Kha] C. Khare. Serre's modularity conjecture: a survey of level one case. *To appear in Proc. of LMS Durham Conference.*

[Kha06] C. Khare. Serre's modularity conjecture: the level one case. *Duke Math. J.*, 134(3):557–589, 2006.

[KM00] E. Kowalski and P. Michel. A lower bound for the rank of $J_0(q)$. *Acta Arith.*, 94(4):303–343, 2000.

[Laf02] L. Lafforgue. Chtoucas de Drinfeld et correspondance de Langlands. *Invent. Math.*, 147(1):1–241, 2002.

[MT87] B. Mazur and J. Tate. Refined conjectures of the "Birch and Swinnerton-Dyer type". *Duke Math. J.*, 54(2):711–750, 1987.

[Ogg66] A. Ogg. Abelian curves of 2-power conductor. *Proc. Cambridge Philos. Soc.*, 62:143–148, 1966.

[Pap] M. Papanikolas. Tannakian duality for anderson-drinfeld motives and algebraic independence of carlitz logarithms. *Preprint.*

[Pin97] R. Pink. The Mumford-Tate conjecture for Drinfeld-modules. *Publ. Res. Inst. Math. Sci.*, 33(3):393–425, 1997.

[PR83] B. Perrin-Riou. Descente infinie et hauteur p-adique sur les courbes elliptiques à multiplication complexe. *Invent. Math.*, 70(3):369–398, 1982/83.

[Ros02] M. Rosen. *Number theory in function fields.* Springer-Verlag, New York, 2002.

[Sch82] P. Schneider. Zur Vermutung von Birch und Swinnerton-Dyer über globalen Funktionenkörpern. *Math. Ann.*, 260(4):495–510, 1982.

[Tag91] Y. Taguchi. Semisimplicity of the Galois representations attached to Drinfel'd modules over fields of "finite characteristics". *Duke Math. J.*, 62(3):593–599, 1991.

[Tag93] Y. Taguchi. Semi-simplicity of the Galois representations attached to Drinfel'd modules over fields of "infinite characteristics". *J. Number Theory*, 44(3):292–314, 1993.

[Tag95] Y. Taguchi. The Tate conjecture for t-motives. *Proc. Amer. Math. Soc.*, 123(11):3285–3287, 1995.

[Tag96] Y. Taguchi. On ϕ-modules. *J. Number Theory*, 60(1):124–141, 1996.

[Tag99] Y. Taguchi. Finiteness of an isogeny class of Drinfeld modules. Correction to a previous paper: "Ramifications arising from Drinfel'd modules" [in *the arithmetic of function fields (columbus, oh, 1991)*, 171–187, de Gruyter, Berlin, 1992; MR 94b:11049]. *J. Number Theory*, 74(2):337–348, 1999.

[Tam94] A. Tamagawa. Generalization of Anderson's t-motives and Tate conjecture. *Sūrikaisekikenkyūsho Kōkyūroku*, (884):154–159, 1994.

[Tat84] J. Tate. *Les conjectures de Stark sur les fonctions L d'Artin en $s = 0$*, volume 47 of *Progress in Mathematics.* Birkhäuser Boston Inc., Boston, MA, 1984. Lecture notes edited by Dominique Bernardi and Norbert Schappacher.

[Tat94a] J. Tate. Conjectures on algebraic cycles in l-adic cohomology. In *Motives (Seattle, WA, 1991)*, volume 55 of *Proc. Sympos. Pure Math.*, pages 71–83. Amer. Math. Soc., Providence, RI, 1994.

[Tat94b] J. Tate. The non-existence of certain Galois extensions of $\mathbf{Q}$ unramified outside 2. In *Arithmetic geometry (Tempe, AZ, 1993)*, volume 174 of *Contemp. Math.*, pages 153–156. Amer. Math. Soc., Providence, RI, 1994.

[Tat95] J. Tate. On the conjectures of Birch and Swinnerton-Dyer and a geometric analog. In *Séminaire Bourbaki, Vol. 9, Exp. 306 (1966)*, pages 415–440. Soc. Math. France, Paris, 1995.

[Tha] D. Thakur. Diophantine approximation and transcendence in finite characteristic. To appear in proceedings of International conference in Diophantine equations, DION 2005, TIFR, to be published by Springer.

[Tha96] D. Thakur. Automata-style proof of Voloch's result on transcendence. *J. Number Theory*, 58(1):60–63, 1996.

[Tha04] D. Thakur. *Function field arithmetic.* World Scientific Publishing Co. Inc., River Edge, NJ, 2004.

[TS67] J. Tate and I. Shafarevich. The rank of elliptic curves. *Dokl. Akad. Nauk SSSR*, 175:770–773, 1967.

[TV96] J. Tate and J-F. Voloch. Linear forms in p-adic roots of unity. *Internat. Math. Res. Notices*, (12):589–601, 1996.

[TW96] Y. Taguchi and D. Wan. L-functions of ϕ-sheaves and Drinfeld modules. *J. Amer. Math. Soc.*, 9(3):755–781, 1996.

[Ulm] D. Ulmer. L-functions with large analytic rank and abelian varieties with large algebraic rank over function fields. *To appear in Inven. Math.*

[Ulm02] D. Ulmer. Elliptic curves with large rank over function fields. *Ann. of Math. (2)*, 155(1):295–315, 2002.

[Ulm04] D. Ulmer. Elliptic curves and analogies between number fields and function fields. In *Heegner points and Rankin L-series*, volume 49 of *Math. Sci. Res. Inst. Publ.*, pages 285–315. Cambridge Univ. Press, Cambridge, 2004.

[Vol96] J-F. Voloch. Transcendence of elliptic modular functions in characteristic p. *J. Number Theory*, 58(1):55–59, 1996.

[Wal97] M. Waldschmidt. Sur la nature arithmétique des valeurs de fonctions modulaires. *Astérisque*, (245):Exp. No. 824, 3, 105–140, 1997. Séminaire Bourbaki, Vol. 1996/97.

[YC] J. Yu and C. Chang. Determination of algebraic relations among special zeta values in positive characteristic. *Preprint.*

[Yu91] J. Yu. Transcendence and special zeta values in characteristic p. *Ann. of Math. (2)*, 134(1):1–23, 1991.

Dinesh S. Thakur
University of Arizona, Tucson
E-mail: thakur@math.arizona.edu

DUALITY THEOREMS IN GALOIS COHOMOLOGY OVER NUMBER FIELDS

By JOHN TATE

1. Notation and terminology

Let X be a Dedekind ring with field of fractions k and let C be a commutative group scheme over X. Except in the special case $X=\mathbf{R}$ or $\mathbf{C}$ (real or complex field) we put, for all $r\in\mathbf{Z}$,

$$H^r(X,C)=\varinjlim_{K} H^r(G_{K/k}, C_Y),$$

the direct limit taken over all finite Galois extensions K of k in which the integral closure Y of X is unramified over X, where $G_{K/k}$ denotes the Galois group of such an extension, and where C_Y denotes the group of points of C with coordinates in Y. For example, if $X=k$, our notation coincides with that of [10]. For any X, the group $H^r(X,C)$ is the r-th cohomology group of the profinite group $G_X=\varprojlim_{K} G_{K/k}$ (fundamental group of Spec X) with coefficients in the G_X-module $\varinjlim C_Y$ of points of C with coordinates in the maximal unramified extension of X; a general discussion of the cohomology theory of profinite groups can be found in [5]. In the special case $X\approx\mathbf{R}$ or $\mathbf{C}$ we put

$$H^r(\mathbf{R},C)=\hat{H}^r(G_{\mathbf{C}/\mathbf{R}},C) \quad\text{and}\quad H^r(\mathbf{C},C)=\hat{H}^r(G_{\mathbf{C}/\mathbf{C}},C)=0,$$

where $\hat{H}$ denotes the complete cohomology sequence of the finite group G, in general non trivial in negative dimensions ([2], Ch. 12).

In our applications, X will be a ring associated with an algebraic number field, or with an algebraic function field in one variable over a finite constant field, and the group scheme C will be one of two special types, which we will denote by M and A, respectively. By M we shall always understand (the group scheme of relative dimension zero over X associated with) a finite G_X-module whose order, $|M|=\text{card } M$, is prime to the characteristics of the residue class fields of X. By A we shall denote an abelian scheme over X (i.e., an abelian variety defined over k having "non-degenerate reduction" at every prime of X).

Underlying our whole theory is the cohomology of the multiplicative group, $\mathbf{G}_m$, as determined by class field theory. For any M, we put $M'=\text{Hom}(M,\mathbf{G}_m)$. By our assumption that $|M|$ is invertible in X, we see that M' is a group scheme of the same type as M, namely the one associated with the G_X-module $M'=\text{Hom}(M,\mu)$ where μ denotes the group of roots of unity. Moreover we have $|M|=|M'|$, and $M\approx(M')'$. For any A we put $A'=\text{Ext}(A,\mathbf{G}_m)$, the dual abelian variety; then $A\approx(A')'$ by "biduality". Our aim is to discuss dualities between the cohomology of C and C' in both cases, $C=M$ and $C=A$. Notice that $\text{Ext}(M,\mathbf{G}_m)=0$ and $\text{Hom}(A,\mathbf{G}_m)=0$, so that in each case, C' denotes the only non-vanishing group in the sequence $\text{Ext}^r(C,\mathbf{G}_m)$. Thus our results are presumably special cases of a vastly more general hyperduality theorem for commutative algebraic groups envisaged by Grothendieck, involving all the $\mathrm{Ext}^r(C, \mathbf{G}_m)$ simultaneously.

Finally, for any locally compact abelian group H, we let H^* denote its Pontrjagin character group.

2. Local results

Let k be either $\mathbf{R}$ or $\mathbf{C}$ (archimedean cases), or be complete with respect to a discrete valuation with finite residue class field (non-archimedean case). By local class field theory we have a canonical injection $H^2(k, \mathbf{G}_m) \to \mathbf{Q}/\mathbf{Z}$ which associates with each element of the Brauer group of k its "invariant" which is a rational number (mod. 1). Hence the cup product with respect to the canonical pairing $M \times M' \to \mathbf{G}_m$ gives pairings $H^r(k, M) \times H^{2-r}(k, M') \to \mathbf{Q}/\mathbf{Z}$.

THEOREM 2.1. *For all M and r the group $H^r(k, M)$ is finite and the pairing just discussed yields an isomorphism $H^r(k, M) \approx H^{2-r}(k, M')^*$.*

In particular we have $H^r(k, M) = 0$ for $r > 2$ when k is non-archimedean. In fact, even more is true in that case, namely the group G_k has strict cohomological dimension 2 in the sense of [5]. This fact, together with the theorem, can be proved easily using results of Nakayama [11], by writing $M = F/R$, where F is a $\mathbf{Z}$-free G_k-module.

THEOREM 2.2. *If k is archimedean, then $H^r(k, M)$ is an elementary abelian 2-group whose order is independent of r. If k is non-archimedean and $\mathfrak{o}$ is the valuation ring in k, then the "Euler-characteristic" of M has the value*

$$\chi(k, M) = \frac{|H^0(k, M)|\,|H^2(k, M)|}{|H^1(k, M)|} = \frac{1}{(\mathfrak{o} : |M|\,\mathfrak{o})}.$$

The archimedean case is trivial. In the non-archimedean case one uses the multiplicativity of χ to reduce to various special types of simple M's, only one of which is difficult. In that one the determination of χ is essentially equivalent to a result of Iwasawa [7] on the structure of $K^*/(K^*)^p$ as $G_{K/k}$-module, for a certain type of extension K/k.

In case of an abelian variety A defined over k the relationship $A' = \mathrm{Ext}(A, \mathbf{G}_m)$ leads to a "derived cup product" pairing:

$$H^r(k, A) \times H^{1-r}(k, A') \to \mathbf{Q}/\mathbf{Z},$$

as explained in [16]. The group $H^0(k, A)$, which is the group of rational points of A in k (modulo its connected component in case $k = \mathbf{R}$ or $\mathbf{C}$) is compact and totally disconnected because A is complete and k locally compact. On the other hand, we view $H^1(k, A)$ as discrete and have then

THEOREM 2.3. *For all A and r, the pairing just mentioned yields an isomorphism $H^r(k, A) \approx H^{1-r}(k, A')^*$ (possibly provided we ignore the p-primary components of the groups in case k is of characteristic $p > 0$).*

In the archimedean case this theorem is due to Witt [17]. For k non-archimedean of characteristic 0, the case $r = 0$ and 1 is proved in [16]. One can simplify that proof and at the same time extend the result to all r (i.e., show $H^2(k,A)=0$ in the non-archimedean case), by applying theorems 2.1 and 2.2 to the kernel, M, of the isogeny $A \xrightarrow{m} A$. The same method works for k of positive characteristic, with the proviso in the theorem. Although that proviso is probably unnecessary, new methods will be required to remove it, possibly those of Shatz [15], where the analog of theorem 2.1 is proved for arbitrary finite commutative group schemes M over k.

Suppose now that k is non-archimedean with valuation ring $\mathfrak{o}$. Let M be a finite $G_{\mathfrak{o}}$-module such that $|M|$ is invertible in $\mathfrak{o}$.

THEOREM 2.4. *We have* $|H^0(\mathfrak{o},M)|=|H^1(\mathfrak{o},M)|$ *and* $H^r(\mathfrak{o},M)=0$ *for* $r>1$. *In the duality of Theorem* 2.1 *between* $H^1(k,M)$ *and* $H^1(k,M')$, *the subgroups* $H^1(\mathfrak{o},M)$ *and* $H^1(\mathfrak{o},M')$ *are the exact annihilators of each other.*

The first statements follow from the fact that $G_{\mathfrak{o}}$ is a free profinite group on one generator. The annihilation results from $H^2(\mathfrak{o},\mathbf{G}_m)=0$. The *exact* annihilation now follows by counting, using theorems 2.1 and 2.2.

THEOREM 2.5. *If A is an abelian scheme over* $\mathfrak{o}$, *we have* $H^1(\mathfrak{o},A)=0$.

To prove this one has only to combine results of Lang [8] and Greenberg [6].

3. Global results

Let k be a finite extension of $\mathbf{Q}$ (case (N)), or a function field in one variable over a finite field (case (F)). Let S be a non-empty (possibly infinite) set of prime divisors of k, including the archimedean ones in case (N), and let k_S denote the ring of elements in k which are integers at all primes P *not* in S. For example, if S is the set of all prime divisors of k, then $k_S=k$. Let M be a finite module for the Galois group of the maximal extension of k unramified outside S, and such that $|M|k_S=k_S$. For each prime P *in* S, let k_P denote the completion of k at P. The localization maps $H^r(k_S,M)\to H^r(k_P,M)$ taken all together yield a map

$$H^r(k_S, M) \xrightarrow{\alpha_r} \underline{\prod_{P\in S}} H^r(k_P, M),$$

where the symbol $\underline{\prod}$ denotes the (compact) *direct product* for $r=0$, the (locally compact) *restricted direct product relative to the subgroups* $H^1(\mathfrak{o}_p, M)$ for $r=1$, and the (discrete) *direct sum* for $r\geqslant 2$. By Theorems 2.1 and 2.4, our local dualities yield isomorphisms

$$\underline{\prod_{P\in S}} H^r(k_P,M) \approx \Big[\underline{\prod_{P\in S}} H^{2-r}(k_P, M')\Big]^*.$$

Thus by duality we obtain maps

$$\underline{\prod_{P\in S}} H^r(k_P,M) \xrightarrow{\beta_r} H^{2-r}(k_P,M')^*,$$

namely $\beta_r=(\alpha'_{2-r})^*$, where α' is to M' as α is to M.

Let $\mathrm{Ker}^r(k_S,M)$ denote the kernel of α_r, that is, the group of elements in

$H^r(k_S,M)$ which are zero locally at all primes $P \in S$. There is a canonical pairing

$$(*) \qquad \mathrm{Ker}^2(k_S,M) \times \mathrm{Ker}^1(k_S,M') \to \mathbf{Q}/\mathbf{Z}$$

defined as follows: we represent the cohomology classes to be paired by a 2-cocycle f and a 1-cocycle f'. Then for each $P \in S$ we have, over k_P, a 1-cochain g_P and a 0-cochain g'_P such that $f = \delta g_P$ and $f' = \delta g'_P$. Also, since $H^3(k_S, \mathbf{G}_m)$ has no non-zero elements of order dividing $|M|$, there is, over k_S, a 2-cochain h with coefficients in $\mathbf{G}_m$, such that $f \cup f' = \delta h$. Then, over k_P, we have $\delta(g_P \cup f') = \delta h = \delta(f \cup g'_P)$ and $\delta(g_P \cup g'_P) = f \cup g'_P - g_P \cup f'$, so that for each P the cochains $(g_P \cup f') - h'_P$ and $(f \cup g'_P) - h$ are cocycles representing the same class, say x_P, in $H^2(k_P, \mathbf{G}_m)$. We pair our original elements to the sum (over $P \in S$) of the invariants of these x_P; it is easy to see that the result is independent of the choices involved.

THEOREM 3.1. (*a*) *The pairing* (*) *just discussed is a perfect duality of finite groups.*

(*b*) α_0 *is injective*, β_2 *is surjective, and for* $r = 0,1,2$ *we have* $\mathrm{Im}\ \alpha_r = \mathrm{Ker}\ \beta_r$.

(*c*) α_r *is bijective for* $r \geqslant 3$.

Notice that these statements imply, and are, in turn, summarized by, the existence of an exact sequence:

$$\begin{array}{c}
0 \to H^0(k_S, M \xrightarrow{\alpha_0} \prod_{P \in S} H^0(k'_P, M) \xrightarrow{\beta_0} H^2(k_S, M')^* \to H^1(k_S, M) \\
\searrow \alpha_1 \\
\prod_{P \in S} H^1(k_P, M) \\
\swarrow \beta_1 \\
0 \leftarrow H^0(k_S, M')^* \xleftarrow{\beta_2} \prod_{P \in S} H^2(k_P, M) \xleftarrow{\alpha_2} H^2(k_S, M) \leftarrow H^1(k_S, M')^*
\end{array}$$

together with isomorphisms

$$H^r(k_S, M) \simeq \prod_{P \text{ real}} H^r(k_P, M) \quad \text{for} \quad r \geqslant 3,$$

where the unlabeled arrows in the exact sequence require the non-degeneracy of the pairing (*) for their definition.

I understand that a large part of Theorem 3.1 has been obtained independently by Poitou, and I suspect that the theorem is closely related to results of Shafaryevitch on the extension problem to which he alluded in his talk at this Congress.

If $M = \mu_m$, the group of mth roots of unity, then Theorem 3.1 summarizes well-known statements in class field theory. For general M, all statements of the theorem except case $r = 1$ of (*b*) can be proved by considering the pairing $M \times \mathrm{Hom}(M, C) \to C$, where C is the S-idele-class group of the maximal extension of k unramified outside S; denoting by G the Galois group of that extension, one shows that the resulting pairing $H^2(k_S, M) \times \mathrm{Hom}_G(M, C) \to \mathbf{Q}/\mathbf{Z}$ is non-degenerate, except that in case (N), there is a kernel on the right-hand side, namely the norm from K to k of $\mathrm{Hom}_{\mathbf{Z}}(M, D_K)$, where D_K is the connected component of the idele class group of a sufficiently large finite extension K of k. For finite S all groups involved are finite and the case $r = 1$ of (*b*) then follows by counting, using Theorem 2.2 and a method of Ogg [12]. The passage to infinite S is not difficult. As a by-product of the proof one finds that the group G has strict cohomological dimension 2 for all primes l such that $lk_S = k_S$, except of course if $l=2$ and k is not totally imaginary.

Let A be an abelian scheme over k_S and let m be a natural number such that $mk_S = k_S$. For $X = k_S$ or $X = k_P$, we put:

$$H^r(X, A; m) = \varprojlim_n [\mathrm{Coker}\,(H^r(X, A) \xrightarrow{m^n} H^r(X, A))] \quad \text{for} \quad r \leqslant 0,$$

and $$H^r(X, A; m) = \varinjlim_n [\mathrm{Ker}\,(H^r(X, A) \xrightarrow{m^n} H^r(X, A))] \quad \text{for} \quad r \geqslant 1.$$

The localization maps give homorphisms

$$H^r(k_S, A; m) \xrightarrow{\alpha_r} \prod_{P \in S} H^r(k_P, A; m),$$

where now $\prod$ denotes the (compact) *direct product* for $r=0$, and the (discrete) *direct sum* for $r \geqslant 1$. By Theorem 2.3 we have isomorphisms

$$\prod_{P \in S} H^r(k_P, A; m) \approx \Big[\prod_{P \in S} H^{1-r}(k_P, A'; m)\Big]^*$$

and consequently by duality we have maps $\beta_r = (\alpha'_{1-r})^*$:

$$\prod_{P \in S} H^r(k_P, A; m) \xrightarrow{\beta_r} H^{1-r}(k_S, A'; m)^*.$$

Let $\mathrm{Ker}^r(k_S, A; m)$ denote the kernel of α_r. For $r \geqslant 1$ and fixed m and A, this group is independent of S, by Theorem 2.5. Hence, $\mathrm{Ker}^1(k_S, A, m)$ is the m-primary component of the group of everywhere locally trivial principal homogeneous spaces for A over k. As is well known (and follows for example from [10], Theorem 5) this group is an extension of a finite group by a divisible group of "finite rank". There is canonical pairing

$$(^{**}) \qquad \mathrm{Ker}^1(k_S, A; m) \times \mathrm{Ker}^1(k_S, A'; m) \rightarrow \mathbf{Q}/\mathbf{Z}$$

which can be defined either by a method using finite modules of m-primary division points and quite analogous to the definition of the pairing (*) above, or else by generalizing the method used by Cassels [3] in case dim $A=1$, the generalization involving the "reciprocity law" of Lang [9].

THEOREM 3.2. *The pairing (**) annihilates only the divisible part of* $\mathrm{Ker}^1(k, A; m)$, *nothing more.*

In case dim $A=1$ this theorem is due to Cassels [3], and his methods suffice for the case of general A, once one has Theorem 3.1 at one's disposal. Cassels' proof of skew symmetry in dimension 1 gives in the general case:

THEOREM 3.3. *If E is a divisor on A rational over k, and $\varphi_E : A \rightarrow A'$ the corresponding homomorphism, defined by $\varphi_E(a) = Cl(E_a - E)$, then for any $\alpha \in \mathrm{Ker}^1(k_S, A; m)$ the elements α and $\varphi_E(\alpha)$ annihilate each other in the pairing (**).*

There is also a canonical pairing

$$(***)\qquad \mathrm{Ker}^0(k_S, A; m)\times \mathrm{Ker}^2(k_S, A', m)\to \mathbf{Q}/\mathbf{Z}$$

and we have

THEOREM 3.4. *The map α_2 is surjective and its kernel is the divisible part of $H^2(k_S,A;\, m)$. Moreover, the following statements are equivalent:*

(i) $\mathrm{Ker}^1(k_S,A;\, m)$ *is finite (i.e. its divisible part is* 0).

(ii) *$\mathrm{Im}\ \alpha_0=\mathrm{Ker}\ \beta_0$, and the pairing (***) gives a perfect duality between the compact group* Ker^0 *and the discrete group* Ker^2.

Thus *if* these equivalent conditions (i) and (ii) are satisfied, we have an exact sequence

$$\begin{array}{ccccccc}
 & & 0\to \prod\limits_{P\ \mathrm{real}} H^2(k_P, A'; m)^* & \xrightarrow{\alpha_2^*} & H^2(k_S, A'; m)^* \to H^0(k_S, A; m) & & \\
 & & & & & \downarrow{\scriptstyle \alpha_0} & \\
\prod\limits_{P\in S} H^1(k_P, A, m) & \xleftarrow{\alpha_1} & H^1(k_S, A; m) \leftarrow H^1(k_S, A'; m)^* & \xleftarrow{\beta_0} & \prod\limits_{P\in S} H^0(k_P, A; m) & & \\
 & {\scriptstyle \beta_1}\downarrow & & & & & \\
 & & H^0(k_S, A'; m)^* \to H^2(k_S, A; m) & \xrightarrow{\alpha_2} & \prod\limits_{P\ \mathrm{real}} H^2(k_P, A; m)\to 0 & &
\end{array}$$

quite analogous to (3.1), but with the appropriate shift of dimensions by 1.

4. Conjectures

In view of Theorem 3.4, one would have to be more pessimistic than I not to make the following

CONJECTURE 4.1. $\mathrm{Ker}^1(k_S,A;\, m)$ *is finite.*

There is some numerical evidence for this. For example, Selmer [13] has shown $\mathrm{Ker}^1(\mathbf{Q},A,3)$ is finite for all but a few elliptic curves A of the form $x^3+y^3=cz^3$ with $0\leqslant c\leqslant 500$. A proof of 4.1 which yielded an a priori estimate for the order of $\mathrm{Ker}^1(k_S, A;\, m)$ would yield an effective procedure for computing the rank of, and finding generators for, the group of rational points on an abelian variety. In general, conjecture 4.1, is in the nature of an existence theorem for rational points of infinite order on abelian varieties.

Another conjecture in the same direction is that of Birch and Swynnerton-Dyer, discussed by Cassels in his talk at this Congress, to the effect that the rank of the group of rational points on an abelian variety A of dimension 1 is determined by the order of the pole of the zeta-function of A at $s=1$. In case (F), i.e., if k is a function field over a finite field k_0 with q elements, the two conjectures are conjecturally equivalent. Namely, let Y be the complete non-singular model of k/k_0, and X the unique complete non-singular model of $k(A)/k_0$ which is minimal with respect to the morphism $X\to Y$. Let $\bar{X}=X\times \bar{k}_0$ be the variety obtained by extending the finite ground field to its algebraic closure, and let $\varphi:\bar{X}\to\bar{X}$ be the Frobenius morphism of $\bar{X}$ relative to k_0. Combining the result of Ogg [12] and Shafaryevitch [14] with recent results of M. Artin [1] on the Grothendieck cohomology of algebraic surfaces, one sees that conjecture 4.1 is equivalent in case (F) to

Conjecture 4.2. *The operator $\varphi - q$ annihilates exactly that part of $H^2(\bar{X}, \mathbf{Z}_m)$ which is "algebraic" and rational over k_0, and no more.*

Clearly, 4.2 makes sense for any complete non-singular surface X over a finite field, not only for a pencil of elliptic curves. So generalized, conjecture 4.2 is equivalent, modulo Weil's well-known conjectures, to the following function theoretic analog of the conjecture of Birch and Swynnerton-Dyer.

Conjecture 4.3. *Let X be a complete non-singular algebraic surface defined over a finite field k_0. Then the order of the pole of the zeta-function of X at the point $s=1$ is equal to the number of algebraically independent divisors on X rational over k_0, i.e., to the k_0-picard number of X.*

Mumford has called my attention to the following interpretation of 4.3 in the special case when X is the product of two curves, one of which is elliptic.

Conjecture 4.3′. *Let E and E' be two complete non-singular curves defined over a finite field k_0, and suppose E is of genus 1. Then there exists a non-constant rational map $E' \to E$ defined over k_0 if and only if the zeta-function of E divides that of E'.*

In particular, if E and E' have genus 1, then they are isogenous over k_0 if and only if they have the same zeta-function. This beautiful statement has been proved by Birch and Swynnerton-Dyer and (independently) by Mumford, using results of Deuring [4] on the lifting to characteristic 0 of the Frobenius automorphism.

REFERENCES

[1]. Artin, M., *Grothendieck topologies.* Mimeographed notes, Harvard, 1962.

[2]. Cartan-Eilenberg, *Homological Algebra.* Princeton Univ. Press, 1956.

[3]. Cassels, J. W. S., Arithmetic on curves of genus 1 (IV). Proof of the Hauptvermutung, *J. reine angew. Math.*, 211 (1962), 95–112.

[4]. Deuring, M., Die Typen der Multiplikatorenringe elliptischer Funktionenkörper. *Abh. Math. Sem. Hamburg*, 14 (1941), 197–272.

[5]. Douady, A., Cohomologie des groupes compacts totalement discontinus. *Séminaire Bourbaki*, 12, 1959–60, exposé 189.

[6]. Greenberg, M., Schemata over local rings, *Ann. Math.*, 73, 1961, 624–648.

[7]. Iwasawa, K., On Galois groups of local fields. *Trans. Amer. Math. Soc.*, 80 (1955), 448–469.

[8]. Lang, S., Algebraic groups over finite fields. *Amer. J. Math.*, 78 (1956), 555–563.

[9]. Lang, S., *Abelian Varieties.* Interscience Publishers, New York, 1959.

[10]. Lang, S. & Tate, J., Principal homogeneous spaces over abelian varieties, *Amer. J. Math.*, 80 (1958), 659–684.

[11]. Nakayama, T., Cohomology of class field theory and tensor products of modules I. *Ann. Math.*, 65 (1957), 255–267.

[12]. Ogg, A., Cohomology of Abelian varieties over function fields, *Ann. Math.*, 76 (1962), 185–212.

[13]. Selmer, E. S., The diophantine equation $ax^3+by^3+cz^3=0$. *Acta Math.*, 85 (1951), 203–362.

[14]. Shafaryevitch, I.R., Главные однородные пространства, определенные над полем функций. *Труды Математиуеского института имени В. А. Стеклова*, Том LXIV.

[15]. SHATZ, S., *Cohomology of Artinian group schemes over local fields*. Thesis, Harvard, June, 1962.

[16]. TATE, J., W. C.-groups over *P*-adic fields, *Séminaire Bourbaki*, 1957–58, exposé 156.

[17]. WITT, E., Zerlegung reeller algebraischer Funktionen in Quadrate, Schiefkörper über reellen Funktionenkörper, *J. reine angew. Math.*, 171 (1934), 4–11.

Actes, Congrès intern. math., 1970. Tome 1, p. 201 à 211.

SYMBOLS IN ARITHMETIC

by John TATE

§ 1. Symbols and K_2.

Let F be a commutative field, $F^{\cdot}$ its multiplicative group, and let C be a commutative group. A *symbol on F with values in C* is a function

$$(,): F^{\cdot} \times F^{\cdot} \to C$$

satisfying the two identities

(1) $$(aa', b) = (a, b)(a', b) \quad \text{and} \quad (a, bb') = (a, b)(a, b')$$

(2) $$(a, 1 - a) = 1.$$

Replacing a by a^{-1} in (2) and then using (1) and (2) gives

(3) $$(a, -a) = 1, \quad \text{and hence} \quad (a, a) = (a, -1).$$

Replacing a by ab in (3) and expanding using (1) and (3) gives then

(4) $$(a, b) = (b, a)^{-1}.$$

Example 1. — Suppose v is a discrete valuation of F, with residue field k_v. Then

(5) $$(a, b)_v = \text{residue class of } (-1)^{v(a)v(b)} \frac{a^{v(b)}}{b^{v(a)}}$$

is a symbol on F with values in $k_v^{\cdot}$, called the *tame symbol at v*; Cf. e. g. [12, Ch. III, no. 4].

Let F_s denote a separable algebraic closure of F, and let $G_F = \text{Gal}(F_s/F)$. If X is a topological G_F-module we shall write $H^r(F, X)$ for the r-th cohomology group of the complex $C^*(G_F, X)$ of continuous standard cochains on G_F with values in X. When X is discrete, these groups are the usual Galois cohomology groups; cf. e. g. [13].

Example 2. — Let m be a natural number not divisible by the characteristic of F, and let μ_m be the group of m-th roots of unity in F_s. The exact sequence

(6) $$0 \to \mu_m \to F_s^{\cdot} \xrightarrow{m} F_s^{\cdot} \to 0$$

gives rise to a homomorphism

(7) $$\delta_m : F^{\cdot} \to H^1(F, \mu_m),$$

Putting

(8) $$(a, b)_m = \delta_m a . \delta_m b,$$

we obtain a symbol on F with values in $H^2(F, \mu_m \otimes \mu_m)$ which was discussed briefly in [16]. Its main property is this

(9) $$\{a, b\} \in (K_2F)^m \Leftrightarrow (a, b)_m = 1 \Leftrightarrow b \in N_{E_a/F}E_a^{\cdot},$$

where E_a is the F-algebra $F[X]/(X^m - a)$, and the notations $\{\ ,\ \}$ and K_2 are as explained below.

When $\mu_m \subset F$ we have

$$H^2(F, \mu_m \otimes \mu_m) = H^2(F, \mu_m) \otimes \mu_m = (\mathrm{Br}\ F)_m \otimes \mu_m,$$

where $(\mathrm{Br}\ F)_m$ is the group of elements of order dividing m in the Brauer group of F. In this case the symbol $(a, b)_m$ is well known; cf. e. g. [11, Ch. XIV].

EXAMPLE 3. — The formula

$$(a, b)_{\mathrm{diff}} = \frac{da}{a} \wedge \frac{db}{b}.$$

defines a symbol on F with values in the group $\Omega^2_{F/\mathbb{Z}}$.

Steinberg [15] showed that for $n \geq 3$, the group $H_2(SL_n(F), \mathbb{Z})$ was generated by the values $\{a, b\}_n$ of a certain canonical symbol on F with values in that group, except if F has 2, 3 or 4 elements, in which case the same result holds for $n \geq 5$. Matsumoto [7], showed that Steinberg's symbols $\{\ ,\ \}_n$ are universal. It follows that the group $K_2F = H_2(SL_\infty(F), \mathbb{Z})$ discussed by Swan at this Congress is the target group of a universal symbol $\{\ ,\ \}$; for each abelian group C, the map $f \mapsto f(\{\ ,\ \})$ is a bijection between Hom (K_2F, C) and the group of symbols on F with values in C. In other words, K_2F is presented as an abelian group by the generators $\{a, b\}$, for a and b in F, and the relations $\{1\}$ and $\{2\}$ obtained by replacing parentheses by curly brackets in (1) and (2) above. This is the " computation " of K_2F referred to yesterday by Swan.

The notions of *dimension* of a vector space and *determinant* of a linear transformation give rise to isomorphisms

$$K_0F \approx \mathbb{Z} \qquad \text{and} \qquad K_1F \approx F^{\cdot}.$$

Milnor [8], [9] interprets $\{a, b\}$ as the product of a and b under a " multiplication "

$$K_1F \times K_1F \to K_2F.$$

In general one might ask whether the K-theories discussed by Swan are furnished with products

$$K_iR \times K_jR \to K_{i+j}(R \underset{\mathbb{Z}}{\otimes} R')$$

which for a commutative ring R lead to a graded ring structure on $K_*R \doteq \Sigma_n K_nR$ via the homomorphism $R \otimes R \to R$.

Suppose E/F is a field extension of finite degree. Then there is [8] a transfer homomorphism

(10) $$\mathrm{Tr}_{E/F} : K_2E \to K_2F,$$

whose importance was first emphasized by Bass, such that

$$\mathrm{Tr}_{E/F}\{a, b\}_E = \{a, N_{E/F}b\}_F \qquad \text{for} \qquad a \in F^{\cdot},\ b \in E^{\cdot}.$$

In general, if $R \subset R'$ are commutative rings such that R' is a projective R module of finite rank, one would expect a K_*R-linear map $\mathrm{Tr}: K_*R' \to K_*R$.

Suppose F is the fraction field of a Dedekind ring R. The maximal ideals of R correspond to certain discrete valuations v of F; for each such v, let $\lambda_{v,\mathrm{tame}}: K_2F \to k_v^{\cdot}$ be the homomorphism corresponding to the tame symbol (5) at v, and let

$$\lambda_R : K_2F \to \coprod_v k_v^{\cdot} \tag{11}$$

be the homomorphism whose components are the $\lambda_{v,\mathrm{tame}}$. (We can write direct sum $\coprod_v$ instead of product, because for each fixed pair of elements a, b in $F^{\cdot}$ we have $(a, b)_v = 1$ for almost all v). Bass [1] has shown that the cokernel of λ_R is canonically isomorphic to SK_1R, and that if the set of maximal ideals of R is countable then the kernel of λ_R is the image of K_2R, and even of $H_2(SL_nR, \mathbb{Z})$ for $n \geq 3$.

§ 2. **Results of C. Moore.**

Suppose now that F is a global arithmetic field, i. e. a number field of finite degree, or a function field in one variable over a finite constant field k. For each place v of F, let F_v denote the completion of F at v, and let μ_v denote the group of roots of unity in F_v. For v non-complex, the group μ_v is of finite order $m_v = |\mu_v|$ and the Hilbert m_v-*norm residue symbol* $\left(\frac{a, b}{v}\right) = (a, b)_{m_v}$ is a symbol on F with values in $\mu_v = (\mathrm{Br}\, F_v)_m \otimes \mu_v$ (cf. Example 2 above). Calvin Moore [10] has shown that this symbol is a universal for *continuous* symbols on F_v with values in locally compact abelian groups (see also his talk at this Congress). Thus μ_v should be viewed as a *topological* K_2 of the locally compact field F_v.

Lichtenbaum has raised the question whether μ_v is the ordinary non-topological K_2F_v' where F_v' is the field of algebraic numbers in F_v. This is true at least for $F_v = \mathbb{R}$.

For each place v let

$$\lambda_v : K_2F_v \to \mu_v \tag{12}$$

be the homomorphism corresponding to the symbol $\left(\frac{a, b}{v}\right)$. For non-archimedean v we have

$$\lambda_{v,\mathrm{tame}} = \text{residue of } \lambda_v^{g_v}, \tag{13}$$

where g_v is the power of the residue characteristic p_v dividing m_v. Note that $g_v = 1$ for all v in the function field case, and for almost all v in the number field case (e. g. those v such that $p_v - 1 > [F_v : \mathbb{Q}_{p_v}]$). The kernel of λ_v is divisible. It is uniquely divisible if $F = \mathbb{R}$ or $\mathbb{Q}_2$; I wonder whether it is always so.

The maps λ_v give a homomorphism

$$K_2F \xrightarrow{\lambda} \coprod_{v\,\text{non-complex}} \mu_v . \tag{14}$$

Calvin Moore *(loc. cit.)* has shown that the cokernel of λ is the group μ_F of roots of unity in F. The reciprocity law

$$\prod_v \left(\frac{a, b}{v}\right)^{m_v/m} = \prod_v \left(\frac{a, b}{v}\right)_m = 1$$

for $a, b \in F$ and $m = |\mu_F|$ gives a map of coker λ onto μ_F. The new thing is that this map is injective, and it is really new; in classical class field theory, one never considered simultaneously norm residue symbols whose orders m_v vary with v!

§ 3. Finiteness of Ker λ.

The main subject of this talk is the study of Ker λ which has taken place over the past two years. If Ker $\lambda = 0$, then there are no " exotic " symbols on F, every global symbol is expressible in terms of the local ones $\left(\frac{a, b}{v}\right)$. On the other hand, if Ker λ is large, then there are many exotic symbols. In this section we discuss work which has limited the size of Ker λ.

Two years ago, Bass and I showed that Ker λ is finitely generated and is finite and prime to the characteristic in the function field case, by the following method. For each finite set of places S, let $U_S \subset F^{\cdot}$ be the group of S-units and let $K_2^S F$ be the subgroup of $K_2 F$ generated by the elements $\{a, b\}$ for $a, b \in U_S$. If a_i are generators for U_S, then $\{a_i, a_j\}$ generate $K_2^S F$. Hence the groups $K_2^S F$ are finitely generated. Bass and I proved that if S is a sufficiently large initial segment of the places, relative to an ordering by increasing norms, then the sequence

$$(15) \qquad 0 \to K_2^S F \hookrightarrow K_2 F \xrightarrow{\lambda_{R_S}} \coprod_{v \notin S} k_v^{\cdot} \to 0$$

is exact, where R_S is the ring of S-integers in F. The exactness of (15) follows by induction over the set of places from the exactness of the sequences

$$(16) \qquad 0 \to K_2^{S'} F \to K_2^{S''} F \xrightarrow{\lambda_{v,\text{tame}}} k_v^{\cdot} \to 0$$

where S' and $S'' = S \cup \{v\}$ are successive initial segments containing S. It is proving the exactness of (16) that is difficult, or at least tedious.

For the rational field $F = \mathbb{Q}$ the exactness of (16) follows easily from the Euclidean algorithm. The argument is essentially that used by Gauss in his first proof of the quadratic reciprocity law in the Disquisitiones, by an induction over the primes. Gauss was in fact classifying symbols on $\mathbb{Q}$ with values in a group of order 2. His methods give the isomorphism

$$(17) \qquad K_2 \mathbb{Q} \approx (\pm 1) \times \coprod_p \mathbb{F}_p^{\cdot},$$

the direct sum taken over all odd primes p, and $\mathbb{F}_p$ denoting the prime field with p elements; cf. [8, § 11] for details.

Since Ker λ is contained in Ker λ_{R_S}, it follows from (15) that Ker λ is finitely generated. In the function field case one can show that Ker λ is divisible by the characteristic p of F and is therefore finite and prime to p. By (14) it is enough to prove this divisibility by p for K_2F itself, and this holds for any field E of characteristic $p > 0$ such that $[E: E^p] = p$. Indeed, let $a, b \in E^{\cdot}$. Let $\alpha, \beta \in E' = E^{1/p}$ such that $\alpha^p = a$ and $\beta^p = b$. Then

$$\{ a, b \} = \{ a, N_{E'/E}\beta \} = \mathrm{Tr}_{E'/E} \{ a, \beta \} = \mathrm{Tr}_{E'/E} \{ \alpha, \beta \}^p.$$

Hence $K_2E = (K_2E)^p$.

In the number field case, the result of Bass mentioned at the end of § 1, for R the ring of integers in F, gave another proof of the finite generation of Ker λ and showed that Ker λ is finite if $H_2(SL_nR, \mathbb{R}) = 0$ for some n. This vanishing of H_2 for n sufficiently large and therewith the finiteness of Ker λ has recently been proved by H. Garland [4], using results from differential geometry and analysis on symmetric spaces.

§ 4. **The structure of** Ker λ.

By the method of Bass-Tate discussed above, one can effectively construct generators for Ker λ, but the set of defining relations is infinite. Roughly speaking, there is one relation, induced by $\{ a, 1 - a \} = 1$, for each element $a \in F$, and the method does not give a procedure for deciding which finite subsets of these relations suffice to define Ker λ. Thus, by taking the generators and some set A of relations one obtains a group X_A of which Ker λ is a quotient, and if A is large, there is a good chance that $X_A =$ Ker λ, although the precedure gives no way to prove this, unless $X_A = 0$.

Indeed for some F one can show Ker $\lambda = 0$ by this method. This is true, for example, for imaginary quadratic fields F of small discriminant d; certainly for $|d| \leq 11$, and almost certainly for $|d| \leq 23$ (there is a theoretical bound on the norm of primes to be considered which increases like $|d|^{3/2}$ and consequently the amount of computation required to achieve absolute certainty increases rapidly with $|d|$).

After the imaginary quadratic fields, our next experiment was with the five function fields F of genus 1 over the field with 2 elements (for genus 0 one has Ker $\lambda = 0$; cf. [9]). For these fields we have $h = 1, 2, 3, 4$ and 5, where h is the number of divisor classes of degree 0. After many mistakes and considerable effort we found groups X_A of orders $h' = 5, 7, 9, 11$ and 13, respectively, for these fields. If α and α' are the characteristic roots of the Frobenius endomorphism π, then

$$h = (1 - \alpha)(1 - \bar{\alpha}) \quad \text{and} \quad \alpha \bar{\alpha} = 2,$$

hence

$$h' = 3 + 2h = (1 - 2\alpha)(1 - 2\bar{\alpha}). \tag{18}$$

A theoretical explanation for this experimental result is as follows.

Suppose F is a function field whose constant field k has q elements. Let $\bar{k}$ be an algebraic closure of k, let $F_\infty = F\bar{k}$, let $G = \mathrm{Gal}\,(F_\infty/F)$, and let $\mu = (\bar{k})^{\cdot}$ be the group of roots of unity in F_∞. Let D and C be the groups of divisors and divisor classes of F_∞. Tensoring the exact sequence

$$0 \to \mu \to F_\infty^{\cdot} \to D \to C \to 0 \tag{19}$$

with μ gives an exact sequence

$$0 \to \mathrm{Tor}\,(\mu, C) \to \mu \otimes F_\infty^{\cdot} \to \mu \otimes D \to \mu \to 0 \tag{20}$$

Taking invariants under G we obtain

$$0 \to \operatorname{Tor}(\mu, C)^G \to (\mu \otimes F^{\cdot}_{\infty})^G \xrightarrow{\lambda'} (\mu \otimes D)^G, \tag{21}$$

still exact. Now consider the homomorphism

$$\mu \otimes F^{\cdot}_{\infty} \to K_2 F_{\infty} \tag{22}$$

for which $\zeta \otimes f \mapsto \{\zeta, f\}$, and the homomorphism

$$K_2 F \to (K_2 F_{\infty})^G \tag{23}$$

induced by the inclusion $F \subset F_{\infty}$. Suppose (22) and (23) are isomorphisms. Then

$$K_2 F \approx (\mu \otimes F^{\cdot}_{\infty})^G \tag{24}$$

On the other hand, it is easy to see that

$$\coprod_v \mu_v \approx (\mu \otimes D)^G \tag{25}$$

and that the map λ' in (21) becomes identified with λ via the isomorphisms (24) and (25). Hence we would have

$$\operatorname{Ker} \lambda \approx (\operatorname{Tor}(\mu, C))^G \tag{26}$$

This last group is the kernel of $1 - \sigma$ acting on $\operatorname{Tor}(\mu, C)$, where $\sigma \in G$ is the Frobenius automorphism, and this kernel is non-canonically isomorphic to the kernel of $1 - q\sigma$ acting on C, or, what is the same, the kernel of $1 - q\pi$ acting on the Jacobian variety of F_{∞}, where π is the Frobenius endomorphism. The order of this kernel is

$$\deg(1 - q\pi) = \prod_{i=1}^{2g} (1 - q\alpha_i) = (q - 1)(q^2 - 1)\zeta_F(-1), \tag{27}$$

where the α_i are the characteristic roots of π and where ζ_F is the zeta function of F. Since $|\operatorname{Coker} \lambda| = q - 1$ (by Moore) this would give the formula

$$\frac{|\operatorname{Ker} \lambda|}{|\operatorname{Coker} \lambda|} = (q^2 - 1)\zeta_F(-1). \tag{28}$$

THEOREM 1. — *For a function field F the maps* (22) *and* (23) *are bijective, and consequently we have isomorphisms* (24) *and* (26), *and formula* (28) *holds.*

At the time of the Congress in Nice, only the 2-primary part of this theorem was proved, via methods of Birch [2]. Shortly afterwards, the theorem was proved in general, using the cohomological methods described in § 5 below, which were inspired by a suggestion of Lichtenbaum.

When Theorem 1 was first conjectured two years earlier, it was natural to seek an analog for number fields. A formula like (28) could not make much sense in general, because the zeta function of a number field F has a zero of order r_2 at $s = -1$, where r_2 is the number of complex places of F. But Birch suggested that some such formula might very well hold in the totally real case ($r_2 = 0$), and he and Atkin quickly produced numerical evidence to the effect that, for real quadratic F of discriminant $d < 50$, a prime like 5 or 7 occuring in the numerator of $\zeta_F(-1)$ always divides the order of X_A.

Using the Atlas computer, Atkin was able to take very large sets A of relations, so this evidence was quite convincing. Further experimentation suggested that the conjecture should be

$$(29) \qquad |\operatorname{Ker} \lambda_R| = \pm\, w_F^{(2)} \zeta_F(-1),$$

where R is the ring of integers of the totally real field F and where $w_F^{(2)}$ is a certain integer, the simplest description of which was recently suggested by Lichtenbaum, namely, for $r > 0$, $w_F^{(r)}$ denotes the largest integer m such that Gal $(\overline{F}/F)$ acts trivially on the r-fold tensor product $\mu_m \otimes \ldots \otimes \mu_m$, where μ_m is the group of m-th roots of unity in an algebraic closure $\overline{F}$ of F. In the function field case we have $w^{(r)} = q^r - 1$. Since λ_R is surjective, formulas (28) and (29) are two special cases of one formula.

Formula (29) predicts various non-trivial divisibility properties of $\zeta_F(-1)$ which have been proven by various people. An especially good example is due to Serre [14 (3.7)] (our $w^{(2)}$ is Serre's w).

For a given number field F, one can show that, for prime numbers l outside a certain finite set Σ_F, the order of Ker $\lambda/(l \operatorname{Ker} \lambda)$ divides the order of $(C_l \otimes \mu_l)^{G_l}$, where C_l is the ideal class group of $F(\mu_l)$ and $G_l = \operatorname{Gal}(F(\mu_l)/F)$. Using methods of Leopoldt-Kubota [6] and Iwasawa, Brumer showed that if F is totally real and abelian over $\mathbb{Q}$ and l outside another finite set Σ'_F, then $(C_l \otimes \mu_l)^{G_l} = 0$ if l does not divide $\zeta_F(-1)$. This gave added evidence for (29), and it also proved the finiteness of Ker λ for totally real abelian fields before Garland proved the finiteness for all number fields.

More recently, John Coates has produced evidence for a number field analog of (26) in certain cases. Here one must treat the l-primary part of Ker λ separately for each prime l, and one replaces F_∞ by $F(\mu)$ where μ is the group of l^n-th roots of unity, all n. Coates' work shows there are some close relations between conjectures about Ker λ and conjectures in Iwasawa's theory of the $\mathbb{Z}_l$-extension $F_\infty/F(\mu_l)$ (cf. Iwasawa's talk at this Congress).

§ 5. Symbols in Galois cohomology.

Let m be a natural number not divisible by the characteristic of F, and let

$$(30) \qquad h_m : K_2F \to H^2(F, \mu_m \otimes \mu_m)$$

be the homomorphism corresponding to the symbol $(a, b)_m$ discussed in example 2 of § 1 above.

THEOREM 2. — *The map h_m is surjective, and its kernel is $(K_2F)^m$, i. e., h_m induces an isomorphism*

$$(31) \qquad K_2F/(K_2F)^m \approx H^2(F, \mu_m \otimes \mu_m)$$

A series of reduction steps reduces the proof of this theorem to the case in which m is a prime and $\mu_m \subset F$. In that case the surjectivity is well known. For injectivity one starts from the fact (9) that an element of the form $\{a, b\}$ is in $(K_2F)^m$ if and only if it is in Ker h_m. To treat an arbitrary element $\Pi_i \{a_i, b_i\}$, one uses the following two lemmas, which are true at least for arithmetic fields containing μ_m, m a prime.

LEMMA 1. — *Given elements $a_1, b_1, a_2, b_2, a_3, b_3 \in F^{\cdot}$, there exist elements c_1, c_2, c_3 and d in $F^{\cdot}$ such that $(a_i, b_i)_m = (c_i, d)_m$ for $i = 1, 2, 3$.*

This just means that any three elements of order m in the Brauer group of F have a common cyclic splitting field $F(d^{1/m})$ of degree m.

LEMMA 2. — *Given a, b, c, d in $F^{\cdot}$ such that $(a, b)_m = (c, d)_m$, there exist x, y in $F^{\cdot}$ such that*

$$(a, b)_m = (x, b)_m = (x, y)_m = (c, y)_m = (c, d)_m \tag{32}$$

To prove Lemma 2 one selects y such that for each place v an x_v exists such that (32) holds locally; then a standard lemma on norm residue symbols guarantees the existence of an x globally.

I do not know whether these lemmas, and consequently Theorem 2, hold for all fields, or for a wide class of fields, or only for very special fields. The situation for $m = 2$ seems a bit special. For $m = 2$, Lemma 2 holds for all fields, even with $y = d$, and Milnor [9] has interpreted $K_2F/(K_2F)^2$ in terms of the Witt ring for all fields F.

It can be shown that the map

$$H^2(F, \mu_m \otimes \mu_m) \to \coprod_v H^2(F_v, \mu_m \otimes \mu_m) \tag{33}$$

is injective if m is not divisible by 8, and in any case has kernel of order 1 or 2. Thus the common kernel of the h_m, which by Theorem 2 is the group $\bigcap_m (K_2F)^m$ of elements divisible by m in K_2F for all m, is of index 1 or 2 in Ker λ. For this reason I thought until recently that Galois cohomology could not help in the attempt to show Ker λ very non-trivial and to determine its structure. I could not have been more wrong; one has only to make a symbol with values in

$$H^2(F, \varprojlim_m (\mu_m \otimes \mu_m)) \quad \text{instead of in} \quad \varprojlim_m H^2(F, \mu_m \otimes \mu_m)$$

in order to get a cohomological symbol which promises to be universal!

It is better for this to fix a prime l different from the characteristic of F and to restrict m to be a power of l. Let

$$T = \varprojlim_n (\mu_{l^n}) \tag{34}$$

be the free $\mathbb{Z}_l$-module of rank one on which Galois acts according to its action on the l^n-th roots of unity. For each integer $r > 0$, let

$$0 \to T^{(r)} \to V^{(r)} \to W^{(r)} \to 0 \tag{35}$$

denote the exact sequence obtained by tensoring the exact sequence

$$0 \to \mathbb{Z}_l \to \mathbb{Q}_l \to \mathbb{Q}_l/\mathbb{Z}_l \to 0$$

r times over $\mathbb{Z}_l$ with T.

Recently S. Lichtenbaum suggested that the l-primary part of K_2F should be isomorphic to $H^1(F, W^{(2)})$ if $r_2 = 0$, and to a quotient of $H^1(F, W^{(2)})$ if $r_2 > 0$. Considering Lichtenbaum's idea together with the connecting homomorphisms

$$H^q(F, W^{(r)}) \to H^{q+1}(F, T^{(r)}) \tag{36}$$

associated with the sequence (35) suggests that there is a symbol with values in

$$H^2(F, T^{(2)}) = H^2(F, \varprojlim_n (\mu_{l^n} \otimes \mu_{l^n})) \tag{37}$$

analogous to the m-symbol (8). Such a symbol is easy to make; one simply replaces the δ_m in definition (8) by the analogous homomorphism $F^{\cdot} \to H^1(F, T)$. To show that the resulting pairing is a symbol one proves that a and $1 - a$ are paired to an element in the divisible part of $H^2(F, T^{(2)})$, and uses

PROPOSITION. — *The divisible part of* $H^q(F, T^{(r)})$ *is* 0 *for all* q *and* r.

Since $H^q(F, V^{(r)})$ is uniquely divisible and $H^q(F, W^{(r)})$ is a torsion group we get

COROLLARY. — *The homomorphism* (36) *induces an isomorphism*

$$H^q(F, W^{(r)})/H^q(F, W^{(r)})_{\text{div}} \xrightarrow{\sim} H^{q+1}(F, T^{(r)})_{\text{tors}} \tag{38}$$

Here X_{div} (resp. X_{tors}) denotes the largest divisible (resp. torsion) subgroup of the abelian group X.

Let

$$h : K_2F \to H^2(F, T^{(2)})$$

denote the homomorphism corresponding to the symbol just discussed. If $\mu_l \subset F$, there is an exact commutative diagram

$$\begin{array}{ccccccccc} 0 & \longrightarrow & \operatorname{Ker}\alpha & \longrightarrow & \mu_l \otimes F^{\cdot} & \xrightarrow{\alpha} & (K_2F)_l & & \\ & & \downarrow & & \downarrow \wr & & \downarrow h & & \\ 0 & \to & H^1(F, T^{(2)})/lH^1(F, T^{(2)}) & \to & H^1(F, \mu_l \otimes \mu_l) & \to & (H^2(F, T^{(2)}))_l & \to & 0 \end{array} \tag{39}$$

in which the bottom row comes from the exact sequence

$$0 \to T^{(2)} \xrightarrow{l} T^{(2)} \to \mu_l \otimes \mu_l \to 0. \tag{40}$$

The symbol X_l denotes the kernel of the map $X \xrightarrow{l} X$. The homomorphism α is characterized by $\alpha(\zeta \otimes a) = \{\zeta, a\}$ for $\zeta \in \mu_l$ and $a \in F^{\cdot}$. The middle vertical isomorphism is that obtained by tensoring (7) with μ_l (with $m = l$).

Suppose now that we do not known that Ker λ is finite, but only that it is finitely generated, say of rank ρ (so $\rho = 0 \Leftrightarrow$ Ker λ finite).

LEMMA 3. — *If* $\mu_l \subset F$, *then* $|\operatorname{Ker}\alpha| / |\operatorname{Coker}\alpha| = l^{1+r_2+\rho}$.

This is proved by considering the composed map $\lambda\alpha$, and using the following

COROLLARY OF THEOREM 2. — *The map* λ *induces an injection*

$$K_2F/(K_2F)^l \to \coprod_v \mu_v/\mu_v^l .$$

Combining Lemma 3 with diagram (39) gives

THEOREM 3. — *Suppose* $\mu_l \subset F$. *Then* $(H^1(F, T^{(2)}) : lH^1(F, T^{(2)})) \geq l^{1+r_2}$, *and equality holds if and only if* Ker λ *is finite*, α *is surjective, and* h *is injective on the* l*-primary part of* K_2F.

By the Corollary of the Proposition, we have

$$H^1(F, T^{(2)})_{\text{tors}} \approx H^0(F, W^{(2)}), \tag{41}$$

a cyclic group of order l^n for some $n \geq 1$. Hence Theorem 3 suggests the

MAIN CONJECTURE [1] — *The rank of the $\mathbb{Z}_l$-module $H^1(F, T^{(2)})$ is r_2, or, equivalently, the divisible part of $H^1(F, W^{(2)})$ is isomorphic to $(\mathbb{Q}_l/\mathbb{Z}_l)^{r_2}$.*

Let $F_\infty = \bigcup_n F(\mu_{l^n})$ and let $G = \text{Gal}(F_\infty/F)$. Let M be the maximal abelian pro-l extension of F_∞ which is unramified outside places dividing l, and let $X = \text{Gal}(M/F_\infty)$. Then the Main conjecture translates into $\text{Hom}_G(X, T^{(2)}) \approx \mathbb{Z}_l^{r_2}$, i. e. into a statement about the G-module X. This module has been intensively studied by Iwasawa [5] in the number field case. In the function field case, the module is described by the action of the Frobenius endomorphism on the Jacobian, and results of A. Weil show $\text{Hom}_G(X, T^{(2)}) = 0$, hence

THEOREM 4. — *The Main conjecture is true for function fields.*

A proof of the Main conjecture for number fields would give (via Theorem 3) a new proof of the finiteness of Ker λ, completely different from Garland's. In fact, a proof of the Main conjecture would justify the ideas of Lichtenbaum which inspired it and would reduce all questions about K_2F to questions about Galois cohomology, in view of

THEOREM 5. — *If the Main conjecture is true for the field $F(\mu_l)$, then h induces an isomorphism*

$$K_2F(l) \xrightarrow{\sim} H^2(F, T^{(2)})_{\text{tors}} \approx H^1(F, W^{(2)})/H^1(F, W^{(2)})_{\text{div}} \tag{42}$$

where $K_2F(l)$ denotes the l-primary part of K_2F.

In the function field case these cohomology groups are readily computed and one easily derives Theorem 1 from Theorems 4 and 5.

In the number field case such definitive results await a proof of the Main conjecture; nevertheless Theorem 5 seems to be an excellent guide as to what to expect, and it should at least suggest partial and special results which can be proven.

For $l = 2$ there is one general result which is slightly weaker than the Main conjecture but still can be used to prove the existence of plenty of exotic symbols, i. e. to show that Ker $\lambda/(\text{Ker }\lambda)^2$ can be very large. This is the fact that the map α in diagram (39) is surjective for $l = 2$. Indeed, methods suggested by Birch [2] lead to a proof of the following algebraic theorem which applies in particular to arithmetic fields.

---

[1] (Note added during the correction of proofs). It seems that this « Main Conjecture » has now been proved for number fields as well as function fields (cf. H. BASS, K_2 des Corps globaux (d'après TATE, GARLAND,...), *Séminaire Bourbaki*, No. 394, June, 1971). The proof uses Garland's finiteness theorem [4] (hence the remark in the text following Theorem 4 is a bit ridiculous), as well as Matsumoto's theorem [7], Moore's result on Coker λ [10], and a fundamental result of Iwasawa.

THEOREM 6. — *Suppose F is any field of characteristic* $\neq 2$ *for which Lemma* 1 *holds for* $m = 2$. *Then every element of order* 2 *in* K_2F *is of the form* $\{-1, a\}$.

I do not know whether the hypothesis about Lemma 1 is essential, nor do I have any idea how to prove a corresponding statement for $l \neq 2$.

Let me finish by emphasizing one question to which we have not even a conjectural answer at present, namely, what should be the analog of (29) in case $r_2 \neq 0$, e. g. for an imaginary quadratic field F? Lichtenbaum suggests that the right hand side should involve some $r_2 \times r_2$ determinant multiplied by the value at $s = -1$ of $(s+1)^{-r_2}\zeta_F(s)$, but which determinant? Is there a bilinear form on $H^2(F, T^{(2)})$?

REFERENCES

[1] H. BASS. — K_2 of global fields, *Lecture at A. M. S. meeting in Cambridge*, Mass. October 1969 (Tape recording and supplementary manual available at American Mathematical Society, Providence, Rhode Island).

[2] B. J. BIRCH. — K_2 of global fields, *Proc. Sympos. Pure Math.*, vol. 20, *Amer. Math. Soc.*, Providence, Rhode Island, 1970.

[3] CASSELS-FRÖHLICH. — *Algebraic Number Theory*, Academic Press, New York, 1967.

[4] H. GARLAND. — A finiteness theorem for K_2 of a number field (to appear in *Annals of Math.*).

[5] K. IWASAWA. — (His talk at Nice, in this volume).

[6] T. KUBOTA and H. W. LEOPOLDT. — Eine p-adische Theorie der Zetawerte, *Jour. f. reine. angew. Math.*, vol. 214-215 (1964), pp. 328-339.

[7] H. MATSUMOTO. — Sur les sous-groupes arithmétiques des groupes semi-simples déployés, *Ann. Sci. École Norm. Sup.* (4), 2 (1969), pp. 1-62.

[8] J. MILNOR. — Notes on algebraic K-theory (to appear in *Annals of Math. Studies*).

[9] —. — Algebraic K-theory and quadratic forms, *Inventiones math.*, 9 (1970), pp. 318-344.

[10] C. C. MOORE. — Group extensions of p-adic and adelic linear groups, *Inst. Hautes Études Sci. Publ. Math.*, No. 35 (1968), pp. 157-221.

[11] J.-P. SERRE. — *Corps locaux*, Hermann, Paris, 1962.

[12] —. — *Groupes algébriques et corps de classes*, Hermann, Paris (1959).

[13] —. — *Cohomologie Galoisienne*, troisième édition, Springer, *Lecture notes*, vol. 5 (1965).

[14] —. — Cohomologie des groupes discrets (to appear in *Annals of Math. Studies*).

[15] R. STEINBERG. — *Générateurs, relations et revêtements de groupes algébriques*, Coll. Théorie des Groupes (Bruxelles, 1962), Librairie Universitaire, Louvain et Gauthier-Villars, Paris (1962), pp. 113-127.

[16] J. TATE. — K_2 of global fields, *Lecture at A. M. S. meeting at Cambridge*, Mass., October 1969 (Tape recording and supplementary manual available at American Mathematical Society, Providence, Rhode Island).

Harvard University
Department of Mathematics,
2, Divinity Avenue
Cambridge, Mass. 02138
(U. S. A.)

Proceedings of Symposia in Pure Mathematics
Volume 55 (1994), Part 1

Conjectures on Algebraic Cycles in ℓ-adic Cohomology

JOHN TATE

1. Introduction and notation

Throughout, k is a field finitely generated over its prime field, $\bar{k}$ is a separable algebraic closure of k, the Galois group is $G = \mathrm{Gal}(\bar{k}/k)$, and ℓ is a prime different from the characteristic of k. We consider smooth projective equidimensional k-schemes $X, Y, \ldots$ and put $\overline{X} = X \times_k \bar{k}$. The geometric ℓ-adic cohomology of X will be denoted simply by $H^i(X) := H^i(\overline{X}_{\text{ét}}, \mathbb{Q}_\ell)$. For $0 \le j \le \dim X$, we put $V^j(X) := H^{2j}(X)(j)$, where (j) means the standard jth twist, and let $A^j(X) \subset V^j(X)$ denote the $\mathbb{Q}$-span of the image of the cycle class map $\mathscr{Z}^j(X) \to V^j(X)$. Thus $A^j(X)$ is (the isomorphic image of) the group of classes of algebraic cycles of codimension j on X, with coefficients in $\mathbb{Q}$, for ℓ-adic homological equivalence. Let $N^j(X) \subset A^j(X)$ denote the group of classes of cycles that are numerically equivalent to zero. In other words, put, for $j' = \dim X - j$,

$$N^j(X) := A^j(X) \cap A^{j'}(X)^{\perp} = \{a \in A^j(X) \mid \langle a, a' \rangle = 0 \quad \text{for all } a' \in A^{j'}(X)\} ,$$

where $\langle\ ,\ \rangle$, denotes the pairing via cup product

$$V^j(X) \times V^{j'}(X) \to V^{\dim X}(X) \overset{\mathrm{Tr}}{\to} \mathbb{Q}_\ell .$$

This pairing gives the Poincaré duality between V^j and $V^{j'}$ and has the property that $\langle a, a' \rangle \in \mathbb{Q}$ for $a \in A^j$ and $a' \in A^{j'}$, because $\langle a, a' \rangle = a \cdot a'$ is the total intersection multiplicity of the two cycles.

The group G acts continuously on $V(X)$, compatibly with the pairing, and fixes the elements of $A(X)$. We have a map

$$\mathbb{Q}_\ell \underset{\mathbb{Q}}{\otimes} A^j(X) \to V^j(X) \tag{1.1}$$

induced by inclusion.

Here are our notations for some optimistic conjectural statements:

1991 *Mathematics Subject Classification*. Primary 14C25, 14F20, 14G10, 14G15, 14K15.

$T^j(X)$: The map (1.1) is surjective, i.e., $\mathbb{Q}_\ell A^j(X) = V^j(X)^G$.

$I^j(X)$: The map (1.1) is injective.

$E^j(X)$: $N^j(X) = 0$, i.e., numerical equivalence is equal to ℓ-adic homological equivalence for algebraic cycles of codimension j on X with rational coefficients.

$SS^i(X)$: G acts semisimply on $H^i(X)$.

$S^j(X)$: The map $V^j(X)^G \to V^j(X)_G$ induced by identity is bijective.

Another statement, closely related to T^1, is now a theorem, proved by Zarhin in characteristic $p > 0$ and by Faltings in characteristic 0 [**34, 35, 8**; **10**, Chapter VI]:

$H(A, B)$: The map $\mathbb{Q}_\ell \otimes \mathrm{Hom}_k(A, B) \to \mathrm{Hom}_G(H^1(B), H^1(A))$ is bijective for the abelian varieties A and B.

(Memory aids: T is for Tate, I for injective, E for equality of equivalence relations, SS for semisimple, S for partially semisimple, and H for Homs.)

In the early 1960s I got the idea that T^j, especially T^1, might be true [**30**]. The origin of the idea for T^1 was my conviction [**29**] that the Shafarevitch group Ш of an abelian variety over a global field should be finite. If it, or at least its ℓ-primary part, were not finite, then the Galois cohomology of the abelian variety would be a mess, and the determination of the group of rational points by "descent" would be ineffective. Thinking about these things in the light of the new étale cohomology, with much help from M. Artin and D. Mumford, I was led to T^1, then to T^j. Especially helpful for T^1 was the consideration of the function field analog [**31**] of the conjecture of Birch and Swinnerton-Dyer and Artin's interpretation of the Shafarevitch group as the Brauer group (cf. [**21**] or [**12**]). Other clues to the reasonableness of T^1 were the logical equivalence

$$T^1(X \times Y) \Leftrightarrow T^1(X) + T^1(Y) + H(A, B), \tag{1.2}$$

where A (resp. B) is the Albanese (or Picard) variety of X (resp. Y), and the implication

$$H(A, A) \Rightarrow T^1(A) \tag{1.3}$$

for an abelian variety A over k. Indeed $H(A, B)$ seemed extremely plausible to me thirty years ago, especially after Mumford explained how it followed for elliptic curves over a finite field from results of Deuring, and Serre proved it for elliptic curves over a number field if the j invariant of one of them was not an integer. Later I was very happy to prove $H(A, B)$ in general for k finite (in [**32**], where the implications (1.2) and (1.3) are explained). A key idea for that proof was suggested by Lichtenbaum, and several conversations about it with Serre were of great help.

My idea that T^j might be true for $j > 1$ was based on the analogy with $i = 1$ and some very meager evidence from the case of Fermat surfaces [**30**].

The plan of this paper is as follows. In §2 we discuss various logical interrelationships between the statements T, E, I, and S which can be deduced from Poincaré duality and the hard Lefschetz theorem. In §3 we give a proof by Deligne that T implies that the Künneth components of an algebraic cycle on a product $X \times Y$ are algebraic. In §4 we discuss relations with arithmetic cohomology and recall the equivalence, for k finite, between $T^1(X)$ and the finiteness of the ℓ-primary part of $\mathrm{Br}(X)$. In §5 we discuss, with no claim to completeness, some cases in which the conjectures have been proven. Most of these cases concern divisors, and we include a proof of the birational invariance of T^1.

Two topics we have not discussed are the related conjectures about orders of poles of L-functions for infinite k **[30]**, and the relations between the ℓ-adic conjectures and the Hodge conjectures in case k is of characteristic 0.

I would like to thank P. Deligne for several helpful discussions during the preparation of this paper, and D. Ramakrishnan for his help with §5.

2. Some folklore

This section is a distillation of ideas more or less known to experts which were informally explained at the conference by N. Katz, W. Messing, and U. Jannsen. I thank them for their help, and also J. Milne from whom I have copied (2.6) below. For a striking result involving related ideas, see **[14]**.

In this section we fix an X and j and drop them from the notation, writing simply V instead of $V^j(X)$, T for $T^j(X)$, etc. We also denote objects in the complementary dimension by a "prime", writing V' for $V^{d-j}(X)$, T' for $T^{d-j}(X)$, etc., where $d = \dim X$. The representations V and V' are noncanonically isomorphic by the hard Lefschetz theorem **[6]** and are canonically dual by Poincaré duality. Thus the space V^G is of the same dimension as $(V')^G$, and V^G is dual to $(V')_G$. Moreover, the canonical maps

$$V^G \to V_G \quad \text{and} \quad (V')^G \to (V')_G \tag{2.1}$$

are dual to each other. Hence

(2.2) Lemma.

(a) *The four spaces* V^G, V_G, $(V')^G$, *and* $(V')_G$ *have the same dimension.*

(b) *If either one of the maps* (2.1) *is either injective or surjective, then both maps are bijective, i.e.,* S *and* S' *hold. In particular,* $S \Leftrightarrow S'$.

Now consider the following diagram in which A means $A(X)$, not an abelian variety, and $*$ means dual:

$$\begin{array}{ccccc} \mathbb{Q}_\ell \otimes A & \xrightarrow{b} & V^G = (V'_G)^* & \xrightarrow{c} & ((V')^G)^* \\ {\scriptstyle a}\downarrow & & & & \downarrow{\scriptstyle d} \\ \mathbb{Q}_\ell \otimes (A/N) & \overset{f}{\hookrightarrow} & Q_\ell \otimes \mathrm{Hom}_{\mathbb{Q}}(A'_j, \mathbb{Q}) & \overset{e}{\hookrightarrow} & (\mathbb{Q}_\ell \otimes A')^* . \end{array} \tag{2.3}$$

It is commutative, and the arrows e and f are injective. From the definitions and (2.2) we have

$$
\begin{aligned}
&E \Leftrightarrow a \text{ is injective},\\
&T \Leftrightarrow b \text{ is surjective},\\
&I \Leftrightarrow b \text{ is injective},\\
(2.4)\qquad &S \Leftrightarrow S' \Leftrightarrow c \text{ is bijective}, \Leftrightarrow c \text{ is surjective}, \Leftrightarrow c \text{ is injective},\\
&T' \Leftrightarrow d \text{ is injective},\\
&I' \Leftrightarrow d \text{ is surjective}.
\end{aligned}
$$

(2.5) **Lemma.** $\operatorname{Ker} b \subset \operatorname{Ker} a = \mathbb{Q}_\ell \otimes N$. *In particular,* $E \Rightarrow I$. *Moreover,* $\operatorname{Ker} b = \operatorname{Ker} a \Leftrightarrow E$.

Proof. We have $\operatorname{Ker} b \subset \operatorname{Ker} a$ because ef is injective. Moreover, $b(\operatorname{Ker} a) = \mathbb{Q}_\ell N$, so if $\operatorname{Ker} a = \operatorname{Ker} b$, then $\mathbb{Q}_\ell N = 0$, i.e., $N = 0$.

(2.6) **Proposition** [22, Proposition 8.4]. *The following implications hold:*

$$T + E \Rightarrow T' + S \Rightarrow E .$$

Proof. Assume $T + E$. Then a is injective and b is bijective; hence dc is injective. The injectivity of c implies both S and the bijectivity of c, which in turn gives the injectivity of d, i.e., T'. Assume $T' + S$. Then c and d are injective, so $\operatorname{Ker} b = \operatorname{Ker} a$ which implies E by (2.5).

(2.7) **Corollary.** *If T holds, then*

$$T' + S \Leftrightarrow E \quad \textit{and} \quad S \Rightarrow E' .$$

This is an immediate consequence of (2.6).

(2.8) **Proposition.**
(i) $\dim_{\mathbb{Q}}(A/N) \le \dim \mathbb{Q}_\ell A$, *with equality if and only if* E.
(ii) $\dim \mathbb{Q}_\ell A \le \dim V^G$, *with equality if and only if* T.
(iii) $\dim_{\mathbb{Q}}(A/N) \le \dim V^G$, *with equality if and only if* $T + E$.

Proof. (i) follows from (2.5) and the surjection

$$\mathbb{Q}_\ell A = (\mathbb{Q}_\ell \otimes A)/\operatorname{Ker} b \to (\mathbb{Q}_\ell \otimes A)/\operatorname{Ker} a = \mathbb{Q}_\ell \otimes A/N .$$

(ii) Obvious.
(iii) Follows from (i) and (ii).

(2.9) **Theorem.** *The following are equivalent:*
(a) $\dim_{\mathbb{Q}}(A/N) = \dim_{\mathbb{Q}_\ell} V^G$,
(b) $T + E$,
(c) $T + T' + S$,
(d) $T + T' + E + E' + I + I' + S + S'$.
Moreover, if k is finite then these statements are independent of the prime ℓ and are equivalent to

(e) *The order of the pole of the zeta function* $Z(X, t)$ *at* $t = q^{-j}$ *is equal to* $\dim_{\mathbf{Q}}(A/N)$, *the rank of the group of numerical equivalence classes of cycles of codimension* j *on* X.

PROOF. The equivalence of (a), (b), (c), and (d) follows immediately from the preceding discussion. Suppose k is finite. By [5], the order, m, of the pole in question is equal to the multiplicity of $t = 1$ as a root of $\det(1 - \varphi t, V)$, where $\varphi \in G$ is the Frobenius topological generator of G. Thus m is equal to the dimension of the kernel of the operator $(1-\varphi)^N$ on V for large N. Hence $m \geq \dim \operatorname{Ker}(1-\varphi) = \dim V^G$, and equality holds if and only if $\operatorname{Ker}(1-\varphi) = \operatorname{Ker}(1-\varphi)^2$, i.e., $\operatorname{Ker}(1-\varphi) \cap (1-\varphi)V = 0$, i.e., S. Combining this with (2.6) (iii) we find that $m = \dim_{\mathbf{Q}}(A/N) \Leftrightarrow T + E + S$. This shows the equivalence of (e) with the other four statements, and since (e) is independent of ℓ they all are.

For use in §5 we add here one more statement in the spirit of this section:

(2.10) PROPOSITION. *Suppose* $B \subset A$ *is a subspace such that* $B \cap N = 0$. *Then* $\mathbf{Q}_\ell \otimes B \approx \mathbf{Q}_\ell B$ *is a direct summand of the G-module* V.

PROOF. Since $B \cap N = 0$, there is a subspace $B' \subset A'$ such that B and B' are put in perfect duality by the intersection pairing. Then the orthogonal space to $\mathbf{Q}_\ell B'$ is a complementary submodule to $\mathbf{Q}_\ell B$ in V.

3. Künneth components

Almost 30 years ago, Grothendieck remarked that the conjecture T should imply that the Künneth components of an algebraic cycle on a product $X \times Y$ are algebraic. Not recalling how, I asked Deligne at the conference. This section gives his answer.

(3.1) THEOREM (Deligne). *Let* $d = \dim X$. *If* $T^d(X \times X)$ *is true, then the Künneth components of the (class of) the diagonal,* $\Delta \in A^d(X \times X) \subset V^d(X \times X)$, *are algebraic.*

PROOF. We identify $V^d(X \times X)$ with the space of degree 0 endomorphisms of $H^*(X)$. Then Δ is the identity map, $A^d(X \times X)$ is the $\mathbf{Q}$-algebra of algebraic endomorphisms, the Künneth decomposition is

$$\text{(3.2)} \qquad \operatorname{End}(H^*(X)) = \bigoplus_{i=0}^{2d} \operatorname{End}(H^i(X)) ,$$

and for each i, the ith Künneth component of Δ is the idempotent projection

$$\text{(3.3)} \qquad p_i \colon H^*(X) \longrightarrow H^i(X) \hookrightarrow H^*(X) .$$

For $u \in \operatorname{End}(H^*(X))$, let $u_i \colon H^i(X) \to H^i(X)$ denote the effect of u on H^i.

(3.4) Lemma. *For* $u \in A^d(X \times X)$ *the characteristic polynomial* $\det(1 - u_i t, H^i(X))$ *has coefficients in* $\mathbb{Q}$.

Proof. View X as the generic fiber of a smooth projective S-scheme $\mathscr{X}$, where $S =$ Spec R with R a regular integral domain of finite type over $\mathbb{Z}$ with fraction field k. It suffices to prove the statement for the special fiber $\mathscr{X}_s$ at a suitable closed point $s \in S$. We may therefore suppose k is finite. In that case the statement is Theorem 2.2 of [**15**], in which it is shown that the projection p_i is algebraic, given by a rational linear combination of powers of the Frobenius morphism. Hence up_i is algebraic and by an argument in [**16**], it follows that $\mathrm{Tr}(u_i)$ is rational by the Lefschetz formula namely

$$(-1)^i \mathrm{Tr}(u_i, H^i(X)) = \sum_{\nu=0}^{2d} (-1)^\nu \mathrm{Tr}(up_i, H^\nu(X)) = (up_i) \cdot \Delta \in \mathbb{Q}.$$

For the same reason, $\mathrm{Tr}(u_i^n) \in \mathbb{Q}$ for every n and the result follows.

To prove the theorem, note that p_i is fixed by G. Thus, if $T^d(X \times X)$ is true, then there exists $u \in A^d(X \times X)$ approximating p_i ℓ-adically arbitrarily closely. We can therefore choose u such that u_i and u_j for $j \neq i$ have no common eigenvalues, those of u_i being near 1 and the others near 0. By the lemma, there is a polynomial $P \in \mathbb{Q}[X]$ such that P takes the value 1 (with multiplicity if necessary) at the eigenvalues of u_i, and takes the value 0 at the others (with multiplicity if necessary.) Then $p_i = P(u) \in A^d(X \times X)$.

If we define motives in terms of algebraic cycles mod homological equivalence, then $\mathrm{Hom}(h(X), h(Y)) = A^d(X \times Y)$, where $d = \dim X$, and the statements $T^d(X \times Y)$ and $I^d(X \times Y)$ concern the surjectivity and injectivity of the map

$$\mathbb{Q}_\ell \otimes \mathrm{Hom}(h(X), h(Y)) \to \mathrm{Hom}_G(H(X), H(Y)).$$

In the spirit of this conference, we note that the general truth of T and/or I would imply the surjectivity and/or injectivity of the map

$$\mathbb{Q}_\ell \otimes \mathrm{Hom}(M, N) \to \mathrm{Hom}_G(M_\ell, N_\ell)$$

for all motives M, N. Here M_ℓ denotes the ℓ-adic realization of M.

The proof of Lemma (3.4) above shows that it applies not only to $u \in A^d(X \times X)$ that are algebraic, but also to $u \in V^d(X \times X)^G$ that are *almost algebraic*, in the sense that, with S as in the proof of (3.4) above, the reduction of u at s is algebraic for all closed points s in some open dense subset of S. This notion of almost algebraic class seems to be part of the folklore. It is mentioned explicitly in [**27**, 5.2] that the Künneth components of Δ are almost algebraic (by [**15**]). When Deligne told me the proof of (3.1) he remarked that if one defines motives in terms of almost algebraic classes then a motive M would have a grading (M^i) by weight and the traces would still be rational. Thus one might consider a weaker conjecture than T, namely, that V^G is spanned by almost algebraic classes.

In fact all three statements T, I, and E have "almost algebraic" analogues, in which A is replaced by the $\mathbb{Q}$-span of the almost algebraic classes and N by the subspace of A orthogonal to all almost algebraic classes in the complementary dimension. All of the results of §2 hold for the almost algebraic versions of T, I, and E, because the arguments there are based entirely on the following abstract situation: A group G, a field K (namely $\mathbb{Q}_\ell$), two isomorphic finite-dimensional K-representations V and V' of G which are canonically dual by a pairing $\langle\ ,\ \rangle$, a subfield $F \subset K$ (namely $F = \mathbb{Q}$), and F-subspaces $A \subset V$ and $A' \subset V'$ such that $\langle A, A'\rangle \subset F$.

4. Relation with arithmetic cohomology and the Brauer group

The cycle class map

$$\mathscr{Z}_i(X) \to V^i(X)^G = H^{2i}(\overline{X}_{\text{ét}}, \mathbb{Q}_\ell(i))^G$$

factors through the arithmetic cohomology group

$$V^i_{\text{arithm}}(X) := H^{2i}(X_{\text{ét}}, \mathbb{Q}_\ell(i)) ,$$

and the map $V^i_{\text{arithm}} \to (V^i)^G$ is an edge homomorphism at position $(0, 2i)$ in a spectral sequence

$$H(G, H_{\text{geom}}(X)) \Rightarrow H_{\text{arithm}}(X) .$$

Thus the conjecture $T^i(X)$ implies that the differentials

$$d_r^{0,2i} : E_r^{0,2i} \to E_r^{4,2i+1-r}$$

in that sequence are 0 for $r \geq 2$. For k finite this is easily seen to be true by consideration of Frobenius eigenvalues. In fact, it seems that the spectral sequence degenerates completely in all cases [3]. Thus, some obvious obstructions to the truth of $T^i(X)$ do vanish. On the other hand, for infinite k we do not have even a conjectural characterization of the $\mathbb{Q}_\ell$-span of the algebraic cycles in $V^i_{\text{arithm}}(X)$ (cf. [7]).

From now on in this section we assume k is finite. Then $G = \widehat{\mathbb{Z}}$ and for a finite G-module Λ,

$$H^0(G, \Lambda) = \Lambda^G , \quad H^1(G, \Lambda) = \Lambda_G , \qquad H^p(G, \Lambda) = 0 \quad \text{for } p \neq 0, 1 .$$

It follows that for coefficients in $(\mathbb{Z}/\ell^n\mathbb{Z})(i)$ the spectral sequence above becomes a collection of short exact sequences, one of which, on passage to the inverse limit over n, becomes

$$(4.1)\quad 0 \to H^{2i-1}(\overline{X}_{\text{ét}}, \mathbb{Z}_\ell(i))_G \to H^{2i}(X_{\text{ét}}, \mathbb{Z}_\ell(i)) \to H^{2i}(\overline{X}_{\text{ét}}, \mathbb{Z}_\ell(i))^G \to 0 .$$

By [5], the eigenvalues of the Frobenius acting on $H^i(j)$ are of absolute value $q^{j-\frac{i}{2}}$. Hence the left-hand group in (4.1) is finite, and tensoring with $\mathbb{Q}_\ell$ we find that $V^i_{\text{arithm}} = (V^i)^G$. Thus, for k finite, the conjecture $T^i(X)$ is simply that the classes of algebraic cycles of codimension i span the arithmetic cohomology $H^{2i}(X_{\text{ét}}, \mathbb{Q}_\ell(i))$ or equivalently, that they generate a $\mathbb{Z}_\ell$-submodule of finite index in $H^{2i}(X_{\text{ét}}, \mathbb{Z}_\ell(i))$.

For $i = 1$ this has a nice interpretation in terms of the Brauer group of X. The exact sequence

$$0 \to \mu_{\ell^n} \to \mathbb{G}_m \xrightarrow{\ell^n} \mathbb{G}_m \to 0$$

of sheaves on $X_{\text{ét}}$ gives, on taking cohomology, an exact sequence of finite groups

$$0 \to (\mathbb{Z}/\ell^n\mathbb{Z}) \otimes \operatorname{Pic} X \to H^2(X_{\text{ét}}, \mu_{\ell^n}) \to \operatorname{Hom}(\mathbb{Z}/\ell^n\mathbb{Z}, \operatorname{Br}(X)) \to 0$$

where $\operatorname{Br}(X)$ is the Brauer group of the scheme X (cf. **[20, 12]**.) Passing to the inverse limit we get

$$(4.2) \quad 0 \to \mathbb{Z}_\ell \otimes \operatorname{Pic} X \to H^2(X_{\text{ét}}, \mathbb{Z}_\ell(1)) \to \operatorname{Hom}(\mathbb{Q}_\ell/\mathbb{Z}_\ell, \operatorname{Br}(X)) \to 0.$$

The ℓ-primary part of $\operatorname{Br}(X)$ has the form $(\mathbb{Q}_\ell/\mathbb{Z}_\ell)^r \times$ finite, because $\operatorname{Hom}(\mathbb{Z}/\ell\mathbb{Z}, \operatorname{Br}(X))$ is finite, and the right-hand group in (4.2) is then isomorphic to $\mathbb{Z}_\ell^r$. Putting these considerations together with (2.9) we obtain

(4.3) Proposition. *For k finite, the following statements are equivalent:*
(i) $T^1(X)$.
(ii) *The ℓ-primary part of* $\operatorname{Br}(X)$ *is finite.*
(ii) $\mathbb{Z}_\ell \otimes \operatorname{Pic} X \to H^2(X_{\text{ét}}, \mathbb{Z}_\ell(1))$ *is bijective.*
(iv) *The order of pole of $Z(X, t)$ at $t = q^{-1}$ is equal to the rank of* $\operatorname{Pic} X$.

It is known that not only are these statements independent of $\ell \neq p = \operatorname{char} k$ (as is clear from (iv)), but they imply the finiteness of the whole group $\operatorname{Br}(X)$ and this finiteness is also implied by the finiteness of the p-primary part of $\operatorname{Br}(X)$ **[31]**, **[19]**, **[18]**, **[22**, 0.4].

5. Known cases

We now discuss some cases in which the conjecture T has been proven. Most of the results concern divisors and, accordingly, most of this section deals with T^1. At the end there is a brief discussion of T^i for $i > 1$.

We begin with some general properties of T^1. It is well known (cf., e.g., SGA 6 XIII, Theorem 4.6) that numerical equivalence implies τ-equivalence for divisors. For cycles with coefficients in $\mathbb{Q}$, τ-equivalence is the same as algebraic equivalence, and implies homological equivalence. Hence $E^1(X)$ holds for all X. From (2.7) we conclude

(5.1) Proposition. *Let $d = \dim X$. Then* $T^1(X) \Rightarrow T^{d-1}(X) + E^{d-1}(X) + S^1(X)$.

If we know T^1 for one type of variety X, it follows for many others.

(5.2) Theorem. (a) $T^1(X) + T^1(Y) \Leftrightarrow T^1(X \times Y)$.

(b) *The conjecture $T^1(X)$ is birationally invariant. More generally, if $X \to Y$ is a dominant rational map between varieties, then $T^1(X)$ implies $T^1(Y)$.*

PROOF.

(a) As already noted (1.2) this follows from the theorem $H(A, B)$ of Faltings and Zarhin.

(b) For k finite this follows easily from (4.3) and the birational invariance of the ℓ-primary part of the Brauer group of X. When I asked Deligne what could be said for arbitrary k, he pointed out that one could get around the use of Brauer group by considering $T^1(U)$ for an arbitrary open dense subscheme $U \subset X$ and proving $T^1(X) \Leftrightarrow T^1(U)$. To get from X to U one first removes a subscheme Z of codimension ≥ 2, and then removes smooth prime divisors D_i, their singularities having been removed by the first step. The first operation has no effect on H^2, and the effect of the second is calculated by the Gysin exact sequence

$$\bigoplus_{i=1}^{N} H^0(D_i)(-1) \to H^2(X) \to H^2(U) \to \bigoplus_{i=1}^{N} H^1(D_i)(-1) .$$

Twisting once and taking the part of weight ≤ 0 gives

$$\mathbb{Z}_\ell^N \to H^2(\overline{X}_{\text{ét}}, \mathbb{Z}_\ell)(1) \to H^2(\overline{U}_{\text{ét}}, \mathbb{Z}_\ell)(1) \to 0 ,$$

because the groups $H^1(D_i)$ are of weight ≥ 1 (cf. [6]). Tensoring with $\mathbb{Q}$ yields a short exact sequence

$$(5.3) \qquad 0 \to \mathbb{Q}_\ell B \to V^1(X) \to V^1(U) \to 0$$

where $B \subset A^1(X) \subset V^1(X)$ is the $\mathbb{Q}$-span of the classes of the divisors D_i. By (2.10) the sequence (5.3) of G-modules splits because, as noted above, $E^1(X)$ is true, i.e., $N^1(X) = 0$. Hence

$$0 \to \mathbb{Q}_\ell B \to V^1(X)^G \to V^1(U)^G \to 0$$

is exact, and $T^1(X) \Leftrightarrow T^1(U)$ follows. From this, the birational invariance is clear.

Let $f: X \to Y$ be a dominant rational map. Let $i: X' \hookrightarrow X$ be a linear section of the same dimension as Y with $X' \to Y$ dominant. Replacing X and Y by suitable open dense subvarieties of themselves we can assume that f is a morphism, X' is smooth, and $fi: X' \to Y$ is finite, say of degree n. Then the equation $n = (fi)_*(fi)^* = (f_* i_* i^*) f^*$ shows that f^* maps the situation for Y isomorphically onto a direct summand of the situation for X. Thus $T^1(X)$ implies $T^1(Y)$.

REMARK. The proof shows that, whether or not $T^1(X)$ holds, the quotient $V^1(U)^G/\mathbb{Q}_\ell A(U)$ is independent of the open dense U in X, so is birationally invariant. Thus $T^1(X)$ is equivalent to the vanishing of $V^1(U)^G$

for one, hence all, U's such that $A(U) = 0$, i.e., the U's obtained by removing from X divisors that generate the Néron-Severi group of X mod torsion.

A nice application of (5.2) is the following theorem, noted in [**28**].

(5.5) THEOREM. *The statement* T^1 *is true for every Fermat surface in* $\mathbb{P}^3$.

Indeed, the surface

$$a_0x_0^n + a_1x_1^n + a_2x_2^n + a_3x_3^n = 0$$

is dominated by the product of the two curves

$$a_0x_0^n + a_1x_1^n = y^n \quad \text{and} \quad a_2x_2^n + a_3x_3^n = z^n ,$$

and T^1 is trivially true for curves.

Of course much more generally, T^1 holds for every variety that is dominated by a product of curves, or a product of curves and an abelian variety, since (1.3) T^1 is true for abelian varieties.

Close to abelian varieties, in some sense, are $K3$ surfaces.

(5.6) THEOREM.

(a) T^1 *holds for all* $K3$ *surfaces* X *in characteristic* 0.

(b) *Over a finite field* k *of characteristic* $p \geq 5$, T^1 *holds for all nonsupersingular and all elliptic* $K3$ *surfaces.*

Statement (a) and its proof were told to me by D. Ramakrishnan. The proof is an easy exercise, given (1) the existence [**4**] of an abelian variety A and an absolute Hodge cycle on $X \times A$ inducing an injection $H^2(X) \hookrightarrow H^2(A)$; (2) the theorem of Faltings that T^1 is true for A; and (3) the theorem of Lefschetz that rational classes of type $(1, 1)$ are algebraic.

(b) See [**25**] for the nonsupersingular case, and [**1**] for the elliptic case.

(5.7) QUESTION. What about $K3$ surfaces over infinite fields of characteristic p?

In recent years there has been work on conjecture T^1 for various types of modular surfaces defined over number fields.

(5.8) THEOREM. T^1 *holds for*

(a) *Hilbert modular surfaces,*

(b) *many quaternionic Shimura surfaces,*

(c) *Picard modular surfaces, i.e., compactifications of congruence arithmeticquotients of the unit ball in* $\mathbb{C}^2$,

(d) *Siegel modular threefolds.*

(a) The references are [**13, 17, 23**]. In the first of these it is proved T^1 holds for $k/\mathbb{Q}$ abelian by showing that the Hirzebruch-Zagier cycles give enough in that case to fill out V^G, thereby proving T^1. Soon after, Klingenberg and, independently by quite different methods, Murty and Ramakrishnan were able to treat the case of arbitrary k, in which the existence of some more exotic cycle classes, not defined over $\mathbb{Q}^{ab}$, must be proved. Both teams show their existence only indirectly, via the Lefschetz $(1, 1)$-theorem. It is an open problem to find divisors representing these classes concretely. The intersection of such divisors with the "diagonal" modular curve might give interesting points on the modular curve. (The Heegner points occur in the intersection of Hirzebruch-Zagier cycles.)

(b) Such a surface X is constructed by means of a totally real number field F and a quaternion division algebra B over F which is split at exactly two real places $\sigma, \tau: F \to \mathbb{R}$, and is defined over an abelian extension k_0 of the field $F^\sigma F^\tau$. In a work **[24]**, $T^1(X)$ is proved for arbitrary $k \supset k_0$ in case $F^\sigma F^\tau / F^\sigma$ is solvable—in particular, in case $F/\mathbb{Q}$ is Galois. The method here consists in establishing period relations between X and certain modular surfaces where it can be shown directly that V^G is spanned by algebraic classes.

(c) See **[2]**. An important ingredient of the proof of Blasius-Rogawski is the proof of irreducibility of certain ℓ-adic representations occurring in $V(X)$. This generalizes the theorem of Ribet for elliptic modular forms. Their proof also uses p-adic Hodge theory.

(d) In **[33]** it is shown that for Siegel modular threefolds the whole of H^2 is algebraic, so T^1 holds.

What about T^i for $i > 1$? In the few cases I know in which it has been proved it is true because there are as many algebraic cycles as there are room for.

One such case is the old one **[30]** of the $2r$-dimensional Fermat hypersurface $\sum_{\nu=0}^{2r+1} x_\nu^{q+1} = 0$. Over the field with q^2 elements this equation can be viewed as $\sum \overline{x}_\nu x_\nu = 0$. This shows that there is a large group U (unitary over the finite field) of automorphisms. It turns out that the representation of U on V^r is the direct sum of the trivial representation and the irreducible representation of lowest degree > 1, forcing $\mathbb{Q}_\ell A^r = V^r$. For a more geometric proof, see **[28]**.

In characteristic 0 one does not need the dimension of $\mathbb{Q}_\ell A^r$ to be the whole $2r$th Betti number to conclude that T^r holds; it is enough to have $\dim \mathbb{Q}_\ell A^r = h^{r,r}$, because the p-adic Hodge theorem of Faltings gives for $\ell = p$ the inequality $\dim(V^r)^G \leq h^{r,r}$ as a corollary. Although this seems to be well known and indeed is used implicitly in (5.8)(b) and (c) above, I first learned it from Brent Gordon who attributes the argument to Faltings in a remark in **[11]**. Indeed, if we imbed $\overline{k}$ in the completion C_p of an algebraic closure of $\mathbb{Q}_p$, let k_p be the closure of k in C_p, put $X_p = X \times_k k_p$ and $G_p = \mathrm{Gal}(\overline{k}/\overline{k} \cap k_p)$, then the p-adic Hodge decomposition **[9]** is

$$V^r(X) \underset{\mathbb{Q}_p}{\otimes} C_p = \bigoplus_\nu H^{r-\nu}(X_p, \Omega^{r+\nu}_{X_p/k_p}) \otimes_{k_p} C_p(-\nu) .$$

Hence

$$(V_r(X) \underset{\mathbb{Q}_p}{\otimes} C_p)^{G_p} = H^r(X_p, \Omega^r_{X_p/k_p})$$

and this gives the required inequality because $(V^r)^G \otimes_{\mathbb{Q}_p} k_p$ is a subspace of the left-hand side.

Gordon [11] shows, for all r, that the equality $\dim \mathbb{Q}_\ell A^r = h^{r,r}$ holds, and hence T^r is true, for X a smooth compactification of the k-fold fiber product of the universal family $A \to M$ of elliptic curves with level N-structure.

We conclude with a challenge. In [26] some exotic Hodge classes on abelian fourfolds with complex multiplication by cube roots of unity are shown to be algebraic and the Hodge conjecture thereby proved in a non-trivial case. D. Ramakrishnan asks whether it is possible to prove T^2 for these fourfolds.

References

1. M. Artin and H. P. F. Swinnerton-Dyer, *The Shafarevich-Tate conjecture for pencils of elliptic curves on K3 surfaces*, Invent. Math. **20** (1973), 249–266.
2. D. Blasius and J. Rogawski, *Tate classes and arithmetic quotients of the two ball*, in Zeta Functions of Picard Modular Surfaces, Les publications CRM (R. P. Langlands and D. Ramakrishnan, eds.), Montreal, 1992, pp. 421–444.
3. P. Deligne, *Théorème de Lefschetz et critères de dégénérescence de suites spectrales*, Inst. Hautes Études Sci. Publ. Math. **35** (1968), 107–126.
4. ——, *La conjecture de Weil pour les surfaces K3*, Invent. Math. **15** (1972), 206–226.
5. ——, *La conjecture de Weil.* I, Inst. Hautes Études Sci. Publ. Math. **43** (1974) 273–308.
6. ——, *La conjecture de Weil.* II, Inst. Hautes Études Sci. Publ. Math. **52** (1981), 137–251.
7. ——, *A quoi servent les motifs?*, these Proceedings, vol. 1, pp. 143–161.
8. G. Faltings, *Endlichkeitssätze für abelsche Varietäten über Zahlkörpern*, Invent. Math. **73** (1983), 349–366.
9. ——, *p-adic Hodge theory*, J. Amer. Math. Soc. **1** (1988), 255–299.
10. G. Faltings and G. Wüstholz, *Rational Points*, Braunschweig, Viehweg, 1984.
11. B. Gordon, *Algebraic cycles and the Hodge structure of a Kuga fiber variety*, Trans. Amer. Math. Soc. **336**, no. 2 (1993), 933–947.
12. A. Grothendieck, *Le groupe de Brauer*, I. *Algèbres d'Azumaya et interprétations diverses*, II. *Théorie cohomologique*, III. *Exemples et compléments*, Dix Exposés sur la Cohomologie des Schémas, North-Holland, Amsterdam, 1968, pp. 46–188.
13. G. Harder, R. P. Langlands, and M. Rapoport, *Algebraische Zyklen auf Hilbert-Blumenthal-Flächen*, J. Reine Angew. Math. **366** (1986), 53–120.
14. U. Jannsen, *Motives, numerical equivalence, and semi-simplicity*, Invent. Math. **107** (1992), 447–452.
15. N. Katz and W. Messing, *Some consequences of the Riemann hypothesis for varieties over finite fields*, Invent. Math. **23** (1974), 73–77.
16. S. Kleiman, *Algebraic cycles and the Weil conjectures*, Dix Exposés sur la Cohomologie des Schémas, North-Holland, Amsterdam, 1968, pp. 359–386.
17. C. Klingenberg, *Die Tate-Vermutungen für Hilbert-Blumenthal-Flächen*, Invent. Math. **89** (1987), 291–317.
18. S. Lichtenbaum, *Zeta-functions of varieties over finite fields at $s = 1$*, Arithmetic and Geometry, Progr. Math., vol. 35, Birkhäuser, Boston, 1983, 173–194.
19. J. Milne, *On a conjecture of Artin and Tate*, Ann. of Math. (2) **102** (1975), 517–533.
20. ——, *Étale cohomology*, Princeton Math. Ser., vol. 33, Princeton Univ. Press, Princeton, NJ, 1980.
21. ——, *Comparison of the Brauer group with the Tate-Shafarevitch group*, J. Fac. Sci. Univ. Tokyo Sect. IA Math. **28** (1982), 735–743.

22. ——, *Values of zeta functions of varieties over finite fields*, Amer. J. Math. **108** (1986), 297–360.
23. V. K. Murty and D. Ramakrishnan, *Period relations and the Tate conjecture for Hilbert modular surfaces*, Invent. Math. **89** (1987), 319–345.
24. ——, *Cycles on quaternionic Shimura surfaces* (in preparation).
25. N. Nygaard and A. Ogus, *Tate's conjecture for $K3$ surfaces of finite height*, Ann. of Math. (2) **122** (1985), 461–507.
26. C. Schoen, *Hodge classes on self products of a variety with an automorphism*, Compositio Math. **65** (1988), 3–32.
27. J.-P. Serre, *Valeurs propres des endomorphismes de Frobenius*, Sém. Bourbaki, no. 446, 1974, Lecture Notes in Math., vol. 431, Springer-Verlag, 1974, pp. 190–204.
28. T. Shioda and T. Katsura, *On Fermat varieties*, Tôhoku J. Math. **31** (1979), 97–115.
29. J. Tate, *Duality theorems in Galois cohomology over number fields*, Proc. Internat. Congr. Math. (Amsterdam 1962), pp. 288–295.
30. ——, *Algebraic cycles and poles of zeta functions*, Arithmetic Algebraic Geometry, Harper and Row, New York, 1965, pp. 93–110.
31. ——, *On the conjecture of Birch and Swinnerton-Dyer and a geometric analog*, Sém. Bourbaki, no. 306, 1966, pp. 1–26.
32. ——, *Endomorphisms of abelian varieties over finite fields*, Invent. Math. **2** (1966), 134–144.
33. R. Weissauer, *Differentialformen zu Untergruppen der Siegelschen Modulgruppe zweiten Grades*, J. Reine Angew. Math. **391** (1988), 100–156.
34. J. G. Zarhin, *Isogenies of abelian varieties over fields of finite characteristics*, Math USSR-Sb. **24** (1974), 451–461.
35. ——, *A remark on endomorphisms of abelian varieties over function fields of finite characteristics*, Math. USSR-Izv. **8** (1974), 477–480.

THE UNIVERSITY OF TEXAS AT AUSTIN, AUSTIN, TEXAS
E-mail address: `tate@math.utexas.edu`

Curriculum Vitae

John G. Thompson

Date of Birth: October 13, 1932, Ottawa, Kansas, USA

Education:

1955 B.A. Yale University
1959 Ph.D. University of Chicago

Employment History:

1961–1962	Lecturer, Harvard University
1962–1968	University of Chicago
1968–1993	University of Cambridge
1993	University of Florida

Awards and Honors:

1965	Cole Prize
1970	Fields Medal
1971	Member of the National Academy of Sciences, USA
1979	Fellow of the Royal Society of London
1982	Senior Berwick Prize from the London Mathematical Society
1985	Sylvester Medal from the Royal Society
1992	Wolf Prize
1992	Poincaré Prize
2000	The National Medal of Science
2008	Abel Prize

List of Publications

1. J. G. Thompson, Unipotent elements, standard involutions, and the divisor matrix, *Comm. Algebra* **36** (2008), No. 9, 3363–3371.
2. J. G. Thompson, Composition factors of rational finite groups, *J. Algebra* **319** (2008), No. 2, 558–594.
3. L. Scott, R. Solomon, J. G. Thompson, J. Walter and E. Zelmanov, Walter Feit (1930–2004). *Notices Amer. Math. Soc.* **52** (2005), No. 7, 728–735.
4. J. G. Thompson and H. Völklein, Braid-abelian tuples in $\mathrm{Sp}_n(K)$, in *Aspects of Galois Theory* (Gainesville, FL, 1996), pp. 218–238, London Math. Soc. Lecture Note Ser., Vol. 256 (Cambridge Univ. Press, 1999).
5. J. G. Thompson and H. Völklein, Symplectic groups as Galois groups, *J. Group Theory* **1** (1998), No. 1, 1–58.
6. J. G. Thompson, Incidence matrices of finite projective planes and their eigenvalues, *J. Algebra* **191** (1997), No. 1, 265–278.
7. J. G. Thompson, Power maps and completions of free groups and of the modular group, *J. Algebra* **191** (1997), No. 1, 252–264.
8. G. R. Robinson and J. G. Thompson, On Brauer's $k(B)$-problem, *J. Algebra* **184** (1996), No. 3, 1143–1160.
9. J. G. Thompson, Some generalized characters, *J. Algebra* **179** (1996), No. 3, 889–893.
10. J. G. Thompson, Sylow 2-subgroups of simple groups, in *Séminaire Bourbaki*, Vol. 10, Exp. No. 345, pp. 543–545 (Soc. Math. France, 1995).
11. G. R. Robinson and J. G. Thompson, Sums of squares and the fields $\mathbf{Q}_{An}$, *J. Algebra* **174** (1995), No. 1, 225–228.
12. J. G. Thompson, 4-punctured spheres, *J. Algebra* **171** (1995), No. 2, 587–605.
13. J. G. Thompson, Note on $H(4)$, *Comm. Algebra* **22** (1994), No. 14, 5683–5687.
14. J. G. Thompson, Note on realizable sequences of partitions, *Comm. Algebra* **22** (1994), No. 14, 5679–5682.
15. J. G. Thompson, Algebraic integers all of whose algebraic conjugates have the same absolute value, in *Coding Theory, Design Theory, Group Theory* (Burlington, VT, 1990), pp. 107–110 (Wiley-Interscience, 1993).
16. J. G. Thompson, Discrete groups and Galois theory, in *Groups, Combinatorics & Geometry* (Durham, 1990), pp. 476–479, London Math. Soc. Lecture Note Ser., Vol. 165 (Cambridge Univ. Press, 1992).
17. J. G. Thompson, Galois groups, in *Groups — St. Andrews 1989*, Vol. 2, pp. 455–462, London Math. Soc. Lecture Note Ser., Vol. 160 (Cambridge Univ. Press, 1991).

18. J. G. Thompson, Rigidity, GL(n, q), and the braid group, Algebra, groups and geometry, *Bull. Soc. Math. Belg. Sér. A* **42** (1990), No. 3, 723–733.
19. J. G. Thompson, Groups of genus zero and certain rational functions, in *Groups — Canberra 1989*, pp. 185–190, Lecture Notes in Math., Vol. 1456, (Springer, 1990).
20. R. M. Guralnick and J. G. Thompson, Finite groups of genus zero, *J. Algebra* **131** (1990), No. 1, 303–341.
21. J. G. Thompson, Hecke operators and noncongruence subgroups. Including a letter from J.-P. Serre, *Group Theory* (Singapore, 1987), pp. 215–224 (de Gruyter, 1989).
22. J. G. Thompson, Fricke, free groups and SL_2, in *Group Theory* (Singapore, 1987), pp. 207–214 (de Gruyter, 1989).
23. J. G. Thompson, Archimedes and continued fractions, *Math. Medley* **15** (1987), No. 2, 67–75.
24. J. G. Thompson, Algebraic numbers associated to certain punctured spheres, *J. Algebra* **104** (1986), No. 1, 61–73.
25. J. G. Thompson, Some finite groups which appear as Gal L/K, where $K \subseteq Q(\mu_n)$, in *Group Theory*, Beijing, 1984, pp. 210–230, Lecture Notes in Math., Vol. 1185 (Springer, 1986).
26. J. G. Thompson, Primitive roots and rigidity, in *Proceedings of the Rutgers Group Theory Year, 1983–1984* (New Brunswick, N.J., 1983–1984), pp. 327–350 (Cambridge Univ. Press, 1985).
27. J. G. Thompson, Rational rigidity of $G_2(5)$, in *Proceedings of the Rutgers Group Theory Year, 1983–1984* (New Brunswick, N.J., 1983–1984), pp. 321–322 (Cambridge Univ. Press, 1985).
28. J. G. Thompson, PSL_3 and Galois groups over **Q**, in *Proceedings of the Rutgers Group Theory Year, 1983–1984* (New Brunswick, N.J., 1983–1984), pp. 309–319 (Cambridge Univ. Press, 1985).
29. J. G. Thompson, Finite nonsolvable groups, in *Group Theory*, pp. 1–12 (Academic Press, 1984).
30. J. G. Thompson, Some finite groups which appear as Gal L/K, where $K \subseteq Q(\mu_n)$, *J. Algebra* **89** (1984), No. 2, 437–499.
31. N. J. A. Sloane and J. G. Thompson, Cyclic self-dual codes, *IEEE Trans. Inform. Theory* **29** (1983), No. 3, 364–366.
32. V. Pless and J. G. Thompson, 17 does not divide the order of the group of a (72,36,16) doubly even code, *IEEE Trans. Inform. Theory* **28** (1982), No. 3, 537–541.
33. J. G. Thompson, Ovals in a projective plane of order 10, in *Combinatorics* (Swansea, 1981), pp. 187–190, London Math. Soc. Lecture Note Ser., Vol. 52 (Cambridge Univ. Press, 1981).

34. J. G. Thompson, Rational functions associated to presentations of finite groups, *J. Algebra* **71** (1981), No. 2, 481–489.
35. N. J. A. Sloane and J. G. Thompson, The nonexistence of a certain Steiner system $S(3,12,112)$, *J. Combin. Theory Ser. A* **30** (1981), No. 3, 209–236.
36. J. G. Thompson, Finite-dimensional representations of free products with an amalgamated subgroup, *J. Algebra* **69** (1981), No. 1, 146–149.
37. J. G. Thompson, Invariants of finite groups, *J. Algebra* **69** (1981), No. 1, 143–145.
38. R. P. Anstee, M. Hall, Jr. and J. G. Thompson, Planes of order 10 do not have a collineation of order 5, *J. Combin. Theory Ser. A* **29** (1980), No. 1, 39–58.
39. J. G. Thompson, A finiteness theorem for subgroups of PSL(2,**R**) which are commensurable with PSL(2,**Z**), in *The Santa Cruz Conference on Finite Groups* (Univ. California, Santa Cruz, Calif., 1979), pp. 533–555, Proc. Sympos. Pure Math., Vol. 37 (Amer. Math. Soc., 1980).
40. J. G. Thompson, Some numerology between the Fischer-Griess Monster and the elliptic modular function, *Bull. London Math. Soc.* **11** (1979), No. 3, 352–353.
41. J. G. Thompson, Finite groups and modular functions, *Bull. London Math. Soc.* **11** (1979), No. 3, 347–351.
42. J. G. Thompson, Uniqueness of the Fischer–Griess monster, *Bull. London Math. Soc.* **11** (1979), No. 3, 340–346.
43. J. G. Thompson, Remarks on finite groups, in *Proceedings of the 5th School of Algebra* (Rio de Janeiro, 1978), pp. 75–77 (Soc. Brasil. Mat., 1978).
44. J. G. Thompson, Toward a characterization of $F_2^*(q)$, III, *J. Algebra* **49** (1977), No. 1, 162–166.
45. J. G. Thompson, Finite groups and even lattices, *J. Algebra* **38** (1976), No. 2, 523–524.
46. J. G. Thompson, A conjugacy theorem for E_8, *J. Algebra* **38** (1976), No. 2, 525–530.
47. J. G. Thompson, Simple $3'$-groups, in *Symposia Mathematica*, Vol. XIII (Convegno di Gruppi e loro Rappresentazioni, INDAM, Rome, 1972), pp. 517–530 (Academic Press, 1974).
48. J. G. Thompson, Weighted averages associated to some codes, Collection of articles dedicated to the memory of Abraham Adrian Albert, *Scripta Math.* **29** (1973), No. 3–4, 449–452.
49. J. G. Thompson, Nonsolvable finite groups all of whose local subgroups are solvable, IV, V, VI, *Pacific J. Math.* **48** (1973), 511–592, *ibid.* **50** (1974), 215–297, *ibid.* **51** (1974), 573–630.

50. J. G. Thompson, Isomorphisms induced by automorphisms, Collection of articles dedicated to the memory of Hanna Neumann, I, *J. Austral. Math. Soc.* **16** (1973), 16–17.
51. F. J. MacWilliams, N. J. A. Sloane and J. G. Thompson, On the existence of a projective plane of order 10, *J. Combinatorial Theory Ser. A* **14** (1973), 66–78.
52. J. G. Thompson, Toward a characterization of $E_2^*(q)$, II, *J. Algebra* **20** (1972), 610–621.
53. F. J. MacWilliams, N. J. A. Sloane and J. G. Thompson, Good self dual codes exist, *Dis. Math.* **3** (1972), 153–162.
54. J. G. Thompson, Quadratic pairs, in *Actes du Congrès International des Mathématiciens* (Nice, 1970), Tome 1, pp. 375–376 (Gauthier-Villars, 1971).
55. J. G. Thompson, Nonsolvable finite groups all of whose local subgroups are solvable, III, *Pacific J. Math.* **39** (1971), 483–534.
56. Z. Janko and J. G. Thompson, On finite simple groups whose Sylow2-subgroups have no normal elementary subgroups of order 8, *Math. Z.* **113** (1970), 385–397.
57. J. G. Thompson, Nonsolvable finite groups all of whose local subgroups are solvable, II, *Pacific J. Math.* **33** (1970), 451–536.
58. J. G. Thompson, Bounds for orders of maximal subgroups, *J. Algebra* **14** (1970), 135–138.
59. J. G. Thompson, Normal p-complements and irreducible characters, *J. Algebra* **14** (1970), 129–134.
60. J. G. Thompson, A non-duality theorem for finite groups, *J. Algebra* **14** (1970), 1–4.
61. J. G. Thompson, A replacement theorem for p-groups and a conjecture, *J. Algebra* **13** (1969), 149–151.
62 J. G. Thompson, Envelopes and p-signalizers of finite groups, *Illinois J. Math.* **13** (1969), 87–90.
63. J. G. Thompson, Characterization of finite simple groups, *Proc. Internat. Congr. Math.* (Moscow, 1966), pp. 158–162 (Mir, 1968).
64. G. Glauberman and J. G. Thompson, Weakly closed direct factors of Sylow subgroups, *Pacific J. Math.* **26** (1968), 73–83.
65. J. G. Thompson, Nonsolvable finite groups all of whose local subgroups are solvable, *Bull. Amer. Math. Soc.* **74** (1968), 383–437.
66. J. G. Thompson, Toward a characterization of $E_2^*(q)$, *J. Algebra* **7** (1967), 406–414.
67. J. G. Thompson, Defect groups are Sylow intersections, *Math. Z.* **100** (1967), 146.

68. J. G. Thompson, On a question of L. J. Paige, *Math. Z.* **99** (1967), 26–27.
69. J. G. Thompson, Centralizers of elements in p-groups, *Math. Z.* **96** (1967), 292–293.
70. J. G. Thompson, Vertices and sources, *J. Algebra* **6** (1967), 1–6.
71. J. G. Thompson, An example of core-free quasinormal subgroups of p-groups, *Math. Z.* **96** (1967), 226–227.
72. Z. Janko and J. G. Thompson, On a class of finite simple groups of Ree, *J. Algebra* **4** (1966), 274–292.
73. J. G. Thompson, Hall subgroups of the symmetric groups, *J. Combinatorial Theory* **1** (1966), 271–279.
74. O. Rothaus and J. G. Thompson, A combinatorial problem in the symmetric group, *Pacific J. Math.* **18** (1966), 175–178.
75. J. G. Thompson, Factorizations of p-solvable groups, *Pacific J. Math.* **16** (1966), 371–372.
76. J. G. Thompson, Automorphisms of solvable groups, *J. Algebra* **1** (1964), 259–267.
77. J. G. Thompson, Fixed points of p-groups acting on p-groups, *Math. Z.* **86** (1964), 12–13.
78. J. G. Thompson, Normal p-complements for finite groups, *J. Algebra* **1** (1964), 43–46.
79. J. G. Thompson, 2-signalizers of finite groups, *Pacific J. Math.* **14** (1964), 363–364.
80. J. G. Thompson, Two results about finite groups, 1963 *Proc. Internat. Congr. Mathematicians* (Stockholm, 1962), pp. 296–300 (Inst. Mittag-Leffler, 1963).
81. W. Feit and J. G. Thompson, Solvability of groups of odd order, *Pacific J. Math.* **13** (1963), 775–1029.
82. J. Boen, O. Rothaus and J. Thompson, Further results on p-automorphic p-groups, *Pacific J. Math.* **12** (1962), 817–821.
83. W. Feit and J. G. Thompson, A solvability criterion for finite groups and some consequences, *Proc. Nat. Acad. Sci. U.S.A.* **48** (1962), 968–970.
84. W. Feit and J. G. Thompson, Finite groups which contain a self-centralizing subgroup of order 3, *Nagoya Math. J.* **21** (1962), 185–197.
85. W. Feit and J. G. Thompson, Groups which have a faithful representation of degree less than $(p - 1/2)$, *Pacific J. Math.* **11** (1961), 1257–1262.
86. W. Feit, M. Hall, Jr. and J. G. Thompson, Finite groups in which the centralizer of any non-identity element is nilpotent, *Math. Z.* **74** (1960), 1–17.
87. J. Thompson, Finite groups with normal p-complements, 1959 *Proc. Sympos. Pure Math.*, Vol. 1, pp. 1–3 (Amer. Math. Soc., 1959).

88. J. G. Thompson, Normal p-complements for finite groups, *Math. Z.* **72** (1959/1960), 332–354.
89. J. G. Thompson, A special class of non-solvable groups, *Math. Z.* **72** (1959/1960), 458–462.
90. D. R. Hughes and J. G. Thompson, The H-problem and the structure of H-groups, *Pacific J. Math.* **9** (1959), 1097–1101.
91. A. A. Albert and J. Thompson, Two-element generation of the projective unimodular group, *Illinois J. Math.* **3** (1959), 421–439.
92. J. Thompson, Finite groups with fixed-point-free automorphisms of prime order, *Proc. Nat. Acad. Sci. U.S.A.* **45** (1959), 578–581.
93. A. A. Albert and J. Thompson, Two element generation of the projective unimodular group, *Bull. Amer. Math. Soc.* **64** (1958), 92–93.
94. J. Thompson, A method for finding primes, *Amer. Math. Monthly* **60** (1953), 175.

Actes, Congrès intern. math., 1970. Tome 1, p. 15 à 16.

ON THE WORK OF JOHN THOMPSON

by R. BRAUER

It is an honor to be called upon to describe to you the brilliant work for which John Thompson has just been awarded the Fields medal. The pleasure is tempered by the feeling that he himself could do this job much better. But perhaps I can say some things he would never say since he is a modest person.

The central outstanding problem in the theory of finite groups today is that of determining the simple finite groups. One may say that this problem goes back to Galois. In any case, Camille Jordan must have been aware of it. Important classes of simple groups have been constructed as well as some individual types of such groups. French mathematicians, Galois, Jordan, Mathieu, Chevalley, have been the pioneers in this work. In recent years, mathematicians of many different countries have joined. However, the general problem is unsolved. We do not know at all how close we are to knowing all simple finite groups. I shall not discuss the present situation of the problem since this will be the topic of Feit's address at this congress. I may only say that up to the early 1960's, really nothing of real interest was known about general simple groups of finite order.

I shall now describe Thompson's contribution. The first paper I have to mention is a joint paper by Walter Feit and John Thompson and, of course, Feit's part in it should not be overlooked. Here, the authors proved a famous conjecture, to the effect that all non-cyclic finite simple groups have even order. I am not sure who was the first to observe this. Fifty years ago this was already referred to as a very old conjecture. While it was usually mentioned in courses on algebra, it is only fair to say that nobody ever did anything about it, simply because nobody had any idea how to get even started. It was not even clear that the whole problem made much sense. Was the role of the prime 2 simply a little accident; did 2 play an entirely exceptional role, or were there properties of other prime divisors of the group order which bore at least some resemblance to those of 2? It was only after the Feit-Thompson paper that one could be sure that the whole question has been a reasonable one.

Thompson's work which has now been honored by the Fields medal is a sequel to this first paper. In it, he determines the minimal simple finite groups, this is to say, the simple finite groups, whose proper subgroups are solvable. Actually, a more general problem is solved. It suffices to assume that only certain subgroups, the so-called local subgroups, are solvable. These are the normalizers of subgroups of prime power order larger than I.

These results are the first substantial results achieved concerning simple groups. A number of important corollaries show that one is now able to answer questions on finite groups which had been completely out of reach before. I mention one: a finite groups is solvable, if and only if every subgroup generated by two elements is solvable. You only have to try to prove this yourself if you want to see how deep the result lies.

Both investigations are very long and complicated and their logical structure is extremely intricate. Unfortunately, I cannot even give you a vague idea of the methods. Reading the papers, one reaches stages repeatedly that one feels caught in a hopeless situation, in an abyss from which there is no escape. Then, miraculously, a way out appears, an amazing turn, which saves us. A famous 19-th century mathematician once remarked that group theory could be done by people who did not know much else of mathematics. There may be some truth in this, but I think, this was not meant in a very nice way. However I believe it was overlooked that if you work in a field where you have few tools, you have to create your own tools. In order to reach positive achievements, mathematical imagination must replace knowledge from other fields.

There is other important work of Thompson in group theory which I cannot discuss here. His methods have already been used successfully by other mathematicians who have developed some of them further. In this way, Thompson has had a tremendons influence. Since he first appeared at the International Congress in Stockholm eight years ago, finite group theory simply is not the same any more.

Let me finish with a personal remark. One reaches a point in life where one wonders what one still expects of life, what one would still like to see happen. This applies to events in Mathematics too. I have passed the point I mentioned. I like to say that I would like to see the solution of the problem of the finite simple groups and the part I expect Thompson's work to play in it. Quite generally, I would like to see to what further heights Thompson's future work will take him. I feel I should also say the same about the three other Fields medallists.

Richard BRAUER
Harvard University
Department of Mathematics,
2 Divinity Avenue,
Cambridge, Mass. 02138
(U. S. A.)

John, G. THOMPSON
University of Cambridge
Department of Mathematics,
16 Mill Lane
Cambridge
(Grande-Bretagne)

CHARACTERIZATIONS OF FINITE SIMPLE GROUPS

JOHN G. THOMPSON

1. Introduction

In the last four years, several results about finite groups have been obtained. The methods of proof are not easy to master, though in large measure they bear a striking fidelity to the foundations laid at the turn of the century by Frobenius, Schur, Burnside and Sylow.

At the moment we have no idea how much further effort will be necessary to classify the finite simple groups. Considering the time which has elapsed since Mathieu discovered his groups, and considering that no one pretends to understand these groups, any optimism must be guarded.

However, it may also be said that recent techniques have led to results which ten years ago seemed impenetrable, and that the power of these techniques is not yet exhausted.

2. Characterizations

Let $\mathcal{S} = \{L_2(q), L_3(q), S_z(q), U_3(q), A_7, M_{11}\}$. Most, but not all, of the recent characterization theorems deal with $\mathcal{S}$. For example,

Theorem. *If G is a non abelian simple group and $G \notin \mathcal{S}$, then*

(1) *G contains a solvable subgroup $\neq 1$ with non solvable normalizer.*

(2) (Gorenstein-Walter) [5]. *Sylow 2-subgroups of G are not dihedral.*

(3) (Suzuki) [9]. *G contains an involution whose centralizer does not have a normal Sylow 2-subgroup.*

In 1963, I announced that (1) held with finitely many exceptions. The complete proof of (1) has not yet appeared. The proof is complicated and it will take several years to determine the extent to which the various arguments admit of useful generalization.

One of the pregnant and technical parts of the proof of (1) is given by

Theorem ES. *$E_2(3)$ and $S_4(3)$ are the only simple groups G such that*

(i) $2,3 \in \pi_4(G)$.

(ii) *If $p \in \{2,3\}$, G_p is a S_p-subgroup of G and $A \in \mathcal{S}cn_3(G_p)$, then $\mathsf{N}(A)$ contains only 1.*

(iii) *The normalizer of every non identity 3-subgroup of G is solvable.*

(iv) *The centralizer of every involution of G is solvable.*

(v) $2 \sim 3$.

Gorenstein [4] has substantially generalized a portion of this theorem by characterizing the groups $E_2(3^n)$.

To explain the meaning of the various statements of the theorem requires a bit of notation. If π is a set of primes, π' is the complementary set of primes. A π-signalizer of a group X is a subgroup A such that $|A|$ and $|X : \mathbf{N}(A)|$ are π'-numbers. Let $\mathcal{S}cn(X)$ be the set of self centralizing normal subgroups of X and let $m(X)$ be the minimal number of generators of X. Let
$\mathcal{S}cn_m(X) = \{A \mid A \in \mathcal{S}cn(X),\ m(A) \geqslant m\}$. Let

$\pi_1(X) = \{p \mid \text{a } S_p\text{-subgroup of } X \text{ is a cyclic group} \neq 1\}$.

$\pi_2(X) = \{p \mid \text{a } S_p\text{-subgroup } X_p \text{ of } X \text{ is non cyclic and } \mathcal{S}cn_3(X)_p = \varnothing\}$.

$\pi_3(X) = \{p \mid \text{if } X_p \text{ is a } S_p\text{-subgroup of } X, \text{ then } \mathcal{S}cn_3(X_p) \neq \varnothing$ and 1 is not the only p-signalizer of $G\}$.

$\pi_4(X) = \{p \mid \text{if } X_p \text{ is a } S_p\text{-subgroup of } X, \text{ then } \mathcal{S}cn_3(X_p) \neq \varnothing$ and 1 is the only p-signalizer of $X\}$.

If H is a subgroup of X, $\mathsf{M}(H) = \mathsf{M}_X(H)$ is the set of all subgroups K of X such that $K \cap H = 1$ and $H \subseteq \mathbf{N}(K)$. If X is a p-group, define $\mathcal{U}(X)$ as follows: if $\mathbf{Z}(X)$ is non cyclic, $\mathcal{U}(X)$ is the set of non cyclic subgroups of $\mathbf{Z}(X)$ of order p^2; if $\mathbf{Z}(X)$ is cyclic, $\mathcal{U}(X)$ is the set of non cyclic normal subgroups of X of order p^2. For general X, let $\mathcal{U}(p) = \mathcal{U}_X(p) = \bigcup \mathcal{U}(X_p)$, where X_p ranges over all the S_p-subgroups of X. If p is an odd prime, write $2 \sim p$ if and only if X has a solvable subgroup which contains a non cyclic abelian subgroup of order 8 and a non cyclic p-subgroup each element of which centralizes an element of $\mathcal{U}(p)$. These definitions explain the hypotheses of Theorem ES and serve to introduce some of the objects of current interest.

Another result deals with the groups $E_2^*(q)$ of Ree.

Theorem R. (Ward-Janko-Thompson) [10], [7]. *If G is a simple group with abelian S_2-subgroups and if G contains an involution i such that $\mathbf{C}(i) = \langle i \rangle \times L$, where $L \simeq L_2(q)$ and $q > 5$, then*

(a) *q is an odd power of 3.*

(b) $|G| = q^3(q-1)(q^3+1)$.

(c) *If P is a S_3-subgroup of G and N is its normalizer, then G is doubly transitive on the cosets of N in G and $G = N \cup NtP$ where t is an involution of G which inverts a $S_{3'}$-subgroup of N.*

We would like to conclude that $G \simeq E_2^*(q)$. This is still open and gives rise to the conjecture

(C_1) The character table of a group determines the Brauer characters.

If the hypothesis $q > 5$ is replaced by $q \geqslant 5$ in Theorem R, Janko [6] has shown that precisely one further group arises. This group is new and is another tantalizing reason for studying simple groups.

Janko's work disclosed a lamentable error in one of my announcements which vitiates some results of Sah [8]. Janko's work contains a lovely application of the results of Brauer for blocks of defect 1.

Brauer and Fong [1] have characterized M_{12} and Wong [11] has characterized A_8. The isomorphism $A_8 \simeq L_4(2)$ places A_8 in the family of Chevalley groups. We have yet to take our first steps toward the characterization of A_n for $n \geqslant 9$.

3. Corollaries

(I) The group G is solvable if and only if every pair of elements generates a solvable subgroup.

(II) G is non solvable if and only if there are non identity elements a, b, c of G of pairwise coprime orders with $abc = 1$.

(III) (Gallagher [2]) G is non solvable if and only if there is a non principal irreducible character of G whose restriction to each Sylow subgroup contains the principal character.

(IV) If G is non solvable of order $p^a q^b r^c$, then one of the following groups is a subquotient of G:

$$L_2(q), \quad q = 5,\ 7,\ 8,\ 17, \quad L_3(3).$$

We may mention the conjectures

(C_2) There are only finitely many simple non abelian groups of order $p^a q^b r^c$.

(C_3) If G is a non abelian simple group and $3 \nmid |G|$, then G is a Suzuki group.

4. Techniques

If N_1, N_2, N_3 are subgroups of a group X and if for every permutation σ of $\{1, 2, 3\}$, $N_{\sigma(1)} \subseteq N_{\sigma(2)} N_{\sigma(3)}$, then $N_1 N_2$ is a subgroup. This elementary observation has numerous applications. When stripped of their group theoretic significance, these applications sometimes depend on the fact that the proposition $(p \vee q) \wedge (q \vee r) \wedge (r \vee p)$ is equivalent to the proposition obtained by interchanging $\vee$ and $\wedge$. I can illustrate this symmetry rather easily. Suppose P is a S_p-subgroup of a group X, A_1, A_2, A_3 are weakly closed subgroups of P, and $N_i = \mathsf{N}(A_i)$. Suppose also that

(a) $\mathsf{N}(P)$ is a maximal subgroup of X.

(b) $X = N_1 N_2 = N_2 N_3 = N_3 N_1$.

Then at least 2 of A_1, A_2, A_3 are normal in X.

There are many variations of this theme, and taken together, they are quite helpful. The difficulty is in finding subgroups A_i of P. This requires some discussion.

In working on the problem so brilliantly solved by Shafarevich nd Golod, I was led to consider the following invariant of a p-group $^{\flat}$: $d = d(P) = \max\{m(A)\}$, where A ranges over all the abelian ubgroups of P. This invariant leads into thickets which I could not ›enetrate. However, my efforts were not without value, for I even- ually realized that the related group $\mathsf{J}(P) = \langle A \mid A' = 1,$ $\imath(A) = d(P)\rangle$ plays an exploitable role in the structure of p-solvable roups. In particular, if X is a p-solvable group, $\mathrm{O}_{p'}(X) = 1$ and $;L(2, p)$ is not a subquotient of X, then $X = N_1N_2$, $N_i = \mathrm{N}(A_i)$, $1_1 = \mathrm{Z}(X_p)$, $A_2 = \mathrm{J}(X_p)$, and where X_p is a S_p-subgroup of X.

When one considers the above result one is led to try to find . *uniformly normal* subgroup. This term requires some elaboration. ›uppose P is a p-group $\neq 1$. Let $\mathscr{S}(P)$ be the set of all p-solvable roups X such that $\mathrm{O}_{p'}(X) = 1$ and P is a S_p-subgroup of X. The ubgroup H of P is said to be uniformly normal provided $1 \neq H \lhd X$ or all X in $\mathscr{S}(P)$. It is a remarkable result of Glauberman [3] that f $p \geqslant 5$, $\mathrm{Z}(\mathrm{J}(P))$ is uniformly normal. This result has already led o a subtheory of finite groups with substantial ramifications. f $p \leqslant 3$, the "old" factorizations still appear indispensable, though nuch remains to be done.

A second technique involves transitivity theorems. As these have ›een discussed elsewhere, I need not elaborate.

The object of the factorizations and the transitivity theorems is :o obtain information about $\mathscr{M}^*(G)$. To define $\mathscr{M}^*(G)$, we let $\mathscr{S}ol(G)$ ›e the set of solvable subgroups of G. $\mathscr{S}ol(G)$ is partially ordered by nclusion; $\mathscr{M}\mathscr{S}(G)$ is the set of maximal elements of $\mathscr{S}ol(G)$ and $\mathscr{M}^*(G)$ is the set of elements of $\mathscr{S}ol(G)$ which are contained in just l element of $\mathscr{M}\mathscr{S}(G)$. Thus, there is a map $\mathsf{M}: \mathscr{M}^*(G) \to \mathscr{M}\mathscr{S}(G)$ lefined by $\mathsf{M}(H) =$ the unique element of $\mathscr{M}\mathscr{S}(G)$ which contains H. Several difficult results in the proof of (1) are of the shape $H \in \mathscr{M}^*(G)$.

A third technique has been introduced by Suzuki and has its roots in work of Zassenhaus. So far this technique has been used only for a limited class of doubly transitive groups. Suppose G is doubly tran- sitive on Ω and $\alpha \in \Omega$. Let $H = G_\alpha$ and suppose $H = QK$, where $Q \lhd H$, $Q \cap K = 1$, and where Q is regular and transitive on $\Omega - \alpha$. Suppose also that t is an involution of G which normalizes K. Then $G = H \cup HtQ$ and if $x \in Q - \{1\}$, $txt = h(x)\,tf(x)$, where $h(x) \in H$, $f(x) \in Q$. These equations are the structure equations for G and G is completely deretmined by H, the structure equations and the auto- morphism of K induced by t. The groups $L_2(q)$, $U_3(q)$, $S_z(q)$ and $E_2^*(q)$ satisfy these hypotheses.

Elegant and subtle arguments of Suzuki deal with the structure equations. In the hope of extending Suzuki's ideas to obtain a characte- rization of $E_2^*(q)$, I have studied the structure equations of the groups which appear in Theorem R. One of the difficulties is that the structure-equations for $E_2^*(q)$ have never been determined. If (C_1) is true the difficulties can probably be avoided.

Dept. of Mathematics,
University of Chicago, USA

REFERENCES

[1] Brauer R., Fong P., A characterization of the Mathieu group M_{12} to appear.
[2] Gallagher P. X., Group characters and Sylow subgroups, *J. London Math. Soc.*, **39** (1964), 720-722.
[3] Glauberman G., A characteristic subgroup of a p-stable group to appear.
[4] Gorenstein D., Finite simple groups and the family $G_2(3^n)$ to appear.
[5] Gorenstein D., Walter J., The characterization of finit groups with dihedral Sylow 2-subgroups, *J. of Alg.*, **2** (1965), 85-151 218-270, 354-393.
[6] Janko Z., A new finite simple group with Abelian Sylow 2-sub groups and its characterization, *J. of Alg.*, **3**, No. 2, 147-186.
[7] Janko Z., Thompson J., On a class of finite simple groups o Ree, *J. of Alg.*, to appear.
[8] Sah C. H., A glass of finite groups with Abelian 2-Sylow subgroups *Math. Zeit.*, **82** (1936), 335-346.
[9] Suzuki M., Finite groups in which the centralizer of any elemen of order 2 is 2-closed, *Ann. of Math.*, **82** (1965), 191-212.
[10] Ward H. N., On Ree's series of simple groups, *Transactions of th Amer. Math. Soc.*, **121**, No. 1 (1966), 62-89.
[11] Wong W., A characterization of the alternating group of degree 8 *Proc. London Math. Soc.*, **13** (1963), 359-383.

Reprinted with permission from the American Mathematical Society, *Bull. Amer. Math. Soc.* **74** (1968) 383–437.

NONSOLVABLE FINITE GROUPS ALL OF WHOSE LOCAL SUBGROUPS ARE SOLVABLE[1]

BY JOHN G. THOMPSON

TABLE OF CONTENTS

1. Introduction. The results of this paper grew from an attempt to classify the minimal simple groups. For obvious reasons, this paper is a natural successor to 0.[2] The structure of the proof showed that a larger class of groups could be mastered with some further effort. An easy corollary classifies the minimal simple groups.

In a broad way, this paper may be thought of as a successful translation of the theory of solvable groups to the theory of simple groups. By this is meant that a substantial structure is constructed which makes it possible to exploit properties of solvable groups to obtain delicate information about the structure and embedding of many solvable subgroups of the simple group under consideration. In this way, routine results about solvable groups acquire great power.

In somewhat more detail, the arguments go as follows, apart from numerous special cases which involve groups of small order: Let $\mathfrak{G}$ be a finite group. Let $\mathcal{S}o\ell(\mathfrak{G})$ be the set of all solvable subgroups of $\mathfrak{G}$. Then $\mathcal{S}o\ell(\mathfrak{G})$ is partially ordered by inclusion and we let $\mathfrak{M}\mathcal{S}(\mathfrak{G})$ be the set of maximal elements of $\mathcal{S}o\ell(\mathfrak{G})$. Let $\mathfrak{M}^*(\mathfrak{G})$ be the set of all elements of $\mathcal{S}o\ell(\mathfrak{G})$ which are contained in precisely one element of

[1] Research supported by a National Science Foundation Grant, GN-530, to the American Mathematical Society. The author also thanks the Sloan Foundation for its extended support.

[2] 0 refers to *Solvability of groups of odd order*, W. Feit and J. Thompson, Pacific J. Math. (3) **13** (1963), and Result X of 0 is here referred to as Result 0.X. Also, as in 0, (B) refers to Theorem B of [26].

$\mathfrak{MS}(\mathfrak{G})$, so that $\mathfrak{M}^*(\mathfrak{G}) \supseteq \mathfrak{MS}(\mathfrak{G})$. The theory of solvable groups makes it possible to prove statements of the sort $\mathfrak{H} \in \mathfrak{M}^*(\mathfrak{G})$, and most of the important technical results of this paper are of this type.

The characterizations of $E_2(3)$ and $S_4(3)$ which emerge are the result of detailed and careful study. These characterizations could be avoided in classifying the minimal simple groups, but the effort this requires is comparable to the characterizations themselves. Furthermore, these characterizations have an independent interest. They are prototypes for the translation referred to above.

A portion of an earlier version of this paper was read by E. C. Dade, whose comments have led to several improvements. Recent results of J. Alperin [1], [2], G. Glauberman [16], [17], [18], and P. Fong [14], have also eased the proofs somewhat. A recent result of C. Sims [33] is helpful.

It is somewhat anomalous that the character theory is not used in this paper. The reason for this is that the relevant character theory is in the literature [8], [9], [10], [11], [14], [15], [16], [18], [28], [35], [45]. This anomaly is in marked contrast with 0, where character theory was needed and was not readily available.

The work is flawed because as yet I have been unable to axiomatize the properties of solvable groups which are "really" needed. To carry out the axiomatization of the various parts of this paper will require several years further study. If this is done, the usual benefits will undoubtedly accrue: stronger theorems, shorter proofs.

This first paper sets the stage. §5 introduces many of the configurations which are relevant to the study of simple groups, and §6 deals with the notion of transitivity.

2. **Notation and definitions.** A *minimal simple group* is a simple group of composite order all of whose proper subgroups are solvable.

Following Alperin [2], the subgroup $\mathfrak{H}$ of the group $\mathfrak{G}$ is a *local subgroup* of $\mathfrak{G}$ if and only if, for some prime p, there is a nonidentity p-subgroup $\mathfrak{P}$ of $\mathfrak{G}$ such that $\mathfrak{H} = N(\mathfrak{P})$.

An *N-group* is a group all of whose local subgroups are solvable. Since every nonidentity solvable group contains a nonidentity characteristic p-subgroup for some prime p, it follows that N-groups are precisely those groups such that the normalizer of every nonidentity solvable subgroup is solvable.

An *involution* is a group element of order 2.

A noncyclic group of order 8 with exactly 1 involution is a *quaternion group*. A noncyclic 2-group with exactly 1 involution is a *generalized quaternion group*. A group which is generated by two distinct involutions is a *dihedral group*. A *four-group* is a dihedral group of order 4.

The techniques and results of 0 are used freely here. The terminology and notation of this paper extend that of 0.

Artin's notation [**4**] for simple groups is used. In addition, $Sz(q)$ is the group of order $q^2(q^2+1)(q-1)$ discovered by Suzuki [**37**], M_{11} is the Mathieu group of order 7920, and Σ_n, A_n denote the symmetric group and alternating group on n letters. The group of inner automorphisms of the group $\mathfrak{X}$ is $I(\mathfrak{X})$.

The number of conjugacy classes of involutions of $\mathfrak{X}$ is $i(\mathfrak{X})$.

If $\mathfrak{A}$, $\mathfrak{B}$ are permutation groups, $\mathfrak{A} \wr \mathfrak{B}$ is the wreath product of $\mathfrak{A}$ and $\mathfrak{B}$, and if $\mathfrak{A}$, $\mathfrak{B}$ have not been presented as permutation groups, $\mathfrak{A} \wr \mathfrak{B}$ is the wreath product of the regular representations of $\mathfrak{A}$, $\mathfrak{B}$. This is the regularity convention [**24**] and will be used on occasion.

Let $\mathfrak{S}=\mathfrak{K}/\mathfrak{L}$ be a section of the group $\mathfrak{X}$. There is thus a homomorphism of $\boldsymbol{N}(\mathfrak{K})\cap\boldsymbol{N}(\mathfrak{L})$ into Aut$(\mathfrak{S})$ induced by conjugation. The image of $\boldsymbol{N}(\mathfrak{K})\cap\boldsymbol{N}(\mathfrak{L})$ in Aut$(\mathfrak{S})$ is denoted $\boldsymbol{A}_{\mathfrak{X}}(\mathfrak{S})$. More generally, if $\mathfrak{M}$ is a subgroup of $\mathfrak{S}$, $\boldsymbol{A}_{\mathfrak{M}}(\mathfrak{S})$ denotes the image of $\mathfrak{M}\cap\boldsymbol{N}(\mathfrak{K})\cap\boldsymbol{N}(\mathfrak{L})$ in $\boldsymbol{A}_{\mathfrak{X}}(\mathfrak{S})$. If X is in $\boldsymbol{N}(\mathfrak{K})\cap\boldsymbol{N}(\mathfrak{L})$ and $S=\mathfrak{L}K$ is in $\mathfrak{S}$, then $[X, S]$ denotes $\mathfrak{L}[X, K]$. Similarly, if $\mathfrak{T}=\mathfrak{M}/\mathfrak{N}$ is a section of $\mathfrak{X}$, if $\mathfrak{M}$ normalizes both $\mathfrak{K}$ and $\mathfrak{L}$, and if $[\mathfrak{N}, \mathfrak{K}]\subseteq\mathfrak{L}$, then we will view $\mathfrak{T}$ as a group of operators of $\mathfrak{S}$, and we let $\boldsymbol{A}_{\mathfrak{T}}(\mathfrak{S})=\boldsymbol{A}_{\mathfrak{M}}(\mathfrak{S})$.

If $1=\mathfrak{P}_0\subseteq\mathfrak{N}_0\subseteq\mathfrak{P}_1\subseteq\mathfrak{N}_1\subseteq\cdots\subseteq\mathfrak{P}_r\subseteq\mathfrak{N}_r=\mathfrak{G}$ is the upper π-series for the π-solvable group $\mathfrak{G}$[3] defined via $\mathfrak{N}_n=\boldsymbol{O}_{\pi'}$ ($\mathfrak{G}$ mod $\mathfrak{P}_n$), $\mathfrak{P}_{n+1}=\boldsymbol{O}_\pi(\mathfrak{G}$ mod $\mathfrak{N}_n)$, $n=0, 1, \cdots$, we set $\boldsymbol{P}_\pi^n(\mathfrak{G})=\mathfrak{P}_n/\mathfrak{N}_{n-1}$, $n=1, \cdots, r$, and $\boldsymbol{Q}_\pi^n(\mathfrak{G})=\mathfrak{N}_n/\mathfrak{P}_n$, $n=0, \cdots, r$. Here $r=l_\pi(\mathfrak{G})$ is the π-length of $\mathfrak{G}$. As in 0, the major attention is focussed on $\boldsymbol{P}_p^1(\mathfrak{G})$, $\boldsymbol{Q}_p^1(\mathfrak{G})$ and $\boldsymbol{Q}_p^0(\mathfrak{G})(=\boldsymbol{O}_{p'}(\mathfrak{G}))$.

If $\mathfrak{A}$ is a group of operators of the group $\mathfrak{B}$ and $1=\boldsymbol{C}_{\mathfrak{B}}(\mathfrak{A})$, we say that $\mathfrak{A}$ has no fixed points on $\mathfrak{B}$.

DEFINITION 2.1. The group $\mathfrak{G}$ is *π-reduced* if and only if $\boldsymbol{O}_\pi(\mathfrak{G})=1$. The subgroup $\mathfrak{A}$ of the group $\mathfrak{G}$ is *π-reducible* if and only if $\boldsymbol{A}_{\mathfrak{G}}(\mathfrak{A})$ is π-reduced.

DEFINITION 2.2. $\boldsymbol{R}_\pi(\mathfrak{G})$ is the subgroup of $\mathfrak{G}$ generated by all the normal π-reducible π-subgroups of $\mathfrak{G}$.

DEFINITION 2.3. The subgroup $\mathfrak{A}$ of the group $\mathfrak{G}$ is a π-signalizer of $\mathfrak{G}$ if and only if $|\mathfrak{A}|$ and $|\mathfrak{G}: \boldsymbol{N}(\mathfrak{A})|$ are π'-numbers.

DEFINITION 2.4. A noncyclic p-group $\mathfrak{P}$ is of symplectic type if and only if every characteristic abelian subgroup of $\mathfrak{P}$ is cyclic.

[3] By a π-solvable group, we mean a group each of whose c.f. is either a p-group for some p in π, or a π'-group, that is, we adhere to the terminology of [**23**], not [**26**]. A π-separable group is one for which every c.f. is either a π-group or a π'-group.

REMARK. The groups of symplectic type are classified in [24]. This classification is of importance in this paper. If $\mathfrak{P}$ is a p-group of symplectic type, then $\mathfrak{P}$ is the central product of a cyclic group and an extra special group, or $p=2$ and $\mathfrak{P}$ is the central product of an extra special group and a group of maximal class, or $p=2$ and $\mathfrak{P}$ is of maximal class. The explicit nature of the groups of symplectic type will be used frequently.

DEFINITION 2.5. If $\mathfrak{P}$ is a p-group of symplectic type, the *width* of $\mathfrak{P}$ is the largest integer n such that $\mathfrak{P}$ contains an extra special subgroup $\mathfrak{H}$ of order p^{2n+1} such that $\mathfrak{P}=\mathfrak{H}\cdot \boldsymbol{C}(\mathfrak{H})$. If $\mathfrak{P}$ contains no such extra special subgroups, the width of $\mathfrak{P}$ is 0.

DEFINITION 2.6. If $\mathfrak{X}$ is a nilpotent group and $\mathfrak{A}$ is a characteristic abelian subgroup of $\mathfrak{X}$, $\mathcal{B}(\mathfrak{X};\mathfrak{A})$ is the set of all subgroups $\mathfrak{B}$ of $\mathfrak{X}$ such that

(a) $\mathfrak{B}$ char $\mathfrak{X}$.
(b) $\ker(\mathrm{Aut}(\mathfrak{X})\xrightarrow{\text{res}}\mathrm{Aut}(\mathfrak{B}))$ is an abelian $\pi(\mathfrak{X})$-group.
(c) $\mathfrak{A}\subseteq \boldsymbol{Z}(\mathfrak{B})$.
(d) $[\mathfrak{X},\mathfrak{B}]\subseteq \boldsymbol{Z}(\mathfrak{B})$.
(e) $\boldsymbol{D}(\mathfrak{B})\subseteq \boldsymbol{Z}(\mathfrak{B})$.
(f) $\boldsymbol{C}_{\mathfrak{X}}(\mathfrak{B})=\boldsymbol{Z}(\mathfrak{B})$.

We set $\mathcal{B}(\mathfrak{X})=\mathcal{B}(\mathfrak{X};1)$ and observe that $\mathcal{B}(\mathfrak{X})\supseteq\mathcal{B}(\mathfrak{X};\mathfrak{A})$ for every characteristic abelian subgroup $\mathfrak{A}$ of $\mathfrak{X}$.

If $\mathfrak{C}$: $\mathfrak{A}=\mathfrak{A}_0\supseteq\mathfrak{A}_1\supseteq\cdots\supseteq\mathfrak{A}_n=1$ is a chain, $\boldsymbol{A}(\mathfrak{C})$ denotes the stability group of $\mathfrak{C}$, that is, the group of all automorphisms α such that for $i=1, 2, \cdots, n$, α fixes each coset of $\mathfrak{A}_i$ in $\mathfrak{A}_{i-1}$. If $\mathfrak{A}$ is a section of $\mathfrak{X}$, set $\boldsymbol{A}_{\mathfrak{X}}(\mathfrak{C})=\boldsymbol{A}_{\mathfrak{X}}(\mathfrak{A})\cap\boldsymbol{A}(\mathfrak{C})$.

C denotes the field of complex numbers, F_q the field of q elements. If K is a field and $\mathfrak{G}$ is a group, $K\mathfrak{G}$ denotes the group algebra of $\mathfrak{G}$ over K.

Let K be a field of characteristic p. It is well known that the subgroup $\mathfrak{A}$ of the group $\mathfrak{G}$ is represented trivially on every irreducible $K\mathfrak{G}$-module if and only if $\mathfrak{A}$ lies in $\boldsymbol{O}_p(\mathfrak{G})$. Thus, if $\mathfrak{A}$ is a subgroup of $\mathfrak{G}$ which does not lie in $\boldsymbol{O}_p(\mathfrak{G})$, we may define $r_K(\mathfrak{A};\mathfrak{G})$ to be the smallest integer r such that $\mathfrak{G}$ has an r-dimensional irreducible representation over K which does not represent $\mathfrak{A}$ trivially. In particular, $r_K(\mathfrak{A};\mathfrak{G})$ is defined for all fields of characteristic 0, with the convention that $\boldsymbol{O}_0(\mathfrak{G})=1$. We also set $r_q(\mathfrak{A};\mathfrak{G})=r_{F_q}(\mathfrak{A};\mathfrak{G})$.

DEFINITION 2.7. $\mathcal{S}o\ell(\mathfrak{G})$ is the set of solvable subgroups of $\mathfrak{G}$, and $\mathcal{MS}(\mathfrak{G})$ is the set of maximal elements of $\mathcal{S}o\ell(\mathfrak{G})$ under inclusion. $\mathcal{M}^*(\mathfrak{G})$ is the set of all solvable subgroups of $\mathfrak{G}$ which are contained in precisely one element of $\mathcal{MS}(\mathfrak{G})$, and if $\mathfrak{H}\in\mathcal{M}^*(\mathfrak{G})$, $\boldsymbol{M}(\mathfrak{H})$ is the unique element of $\mathcal{MS}(\mathfrak{G})$ which contains $\mathfrak{H}$.

We define for each group $\mathfrak{G}$, the following sets of primes:

$\pi_1(\mathfrak{G}) = \{p \mid \text{A } S_p\text{-subgroup of } \mathfrak{G} \text{ is a nonidentity cyclic group}\}$.

$\pi_2(\mathfrak{G}) = \{p \mid$ (i) A S_p-subgroup $\mathfrak{P}$ of $\mathfrak{G}$ is noncyclic. (ii) $\mathcal{S}cn_3(\mathfrak{P}) = \varnothing\}$.

$\pi_3(\mathfrak{G}) = \{p \mid$ (i) If $\mathfrak{P}$ is a S_p-subgroup of $\mathfrak{G}$, then $\mathcal{S}cn_3(\mathfrak{P}) \neq \varnothing$. (ii) $\mathsf{N}_{\mathfrak{G}}(\mathfrak{P})$ contains a nonidentity subgroup$\}$.

$\pi_4(\mathfrak{G}) = \{p \mid$ (i) If $\mathfrak{P}$ is a S_p-subgroup of $\mathfrak{G}$, then $\mathcal{S}cn_3(\mathfrak{P}) \neq \varnothing$. (ii) $\mathsf{N}_{\mathfrak{G}}(\mathfrak{P})$ contains only $1\}$.

As proved in [**5**], if p is an odd prime in $\pi_2(\mathfrak{G})$, the structure of the S_p-subgroups of $\mathfrak{G}$ is known. Those 2-groups $\mathfrak{T}$ with $\mathcal{S}cn_3(\mathfrak{T}) = \varnothing$ are as yet undetermined, an awkward situation.[4]

If p and q are odd primes, we write $p \sim q$ if and only if $\mathcal{S}o\ell(\mathfrak{G})$ contains an element which contains elementary subgroups of order p^3 and q^3, otherwise $p \nsim q$. This definition conforms with 0. We wish to extend the relation in a useful fashion. This is difficult. We need the sets $\mathcal{U}(p)$ explicitly.

DEFINITION 2.8. Let $p \in \pi(\mathfrak{G})$ and $\mathfrak{P}$ be a S_p-subgroup of $\mathfrak{G}$. If every normal abelian subgroup of $\mathfrak{P}$ is cyclic, then $\mathcal{U}(\mathfrak{P}) = \varnothing$. If $\boldsymbol{Z}(\mathfrak{P})$ is noncyclic, then $\mathcal{U}(\mathfrak{P}) = \{\mathfrak{A} \mid \mathfrak{A} \subseteq \boldsymbol{Z}(\mathfrak{P}), \mathfrak{A} \text{ is of type } (p, p)\}$. If $\mathfrak{P}$ contains a noncyclic normal abelian subgroup and $\boldsymbol{Z}(\mathfrak{P})$ is cyclic, then $\mathcal{U}(\mathfrak{P}) = \{\mathfrak{A} \mid \mathfrak{A} \triangleleft \mathfrak{P}, \mathfrak{A} \text{ is of type } (p, p)\}$. $\mathcal{U}(p) = \bigcup \mathcal{U}(\mathfrak{P})$, $\mathfrak{P}$ ranging over all the S_p-subgroups of $\mathfrak{G}$. In case we wish to emphasize the dependence on $\mathfrak{G}$, we write $\mathcal{U}_{\mathfrak{G}}(p)$ for $\mathcal{U}(p)$.

DEFINITION 2.9. If p is odd, we set $\mathfrak{J}(p) = \{\mathfrak{A} \mid \mathfrak{A}$ is a p-subgroup of $\mathfrak{G}$ and $\mathfrak{A}$ contains a subgroup $\mathfrak{B}$ of type (p, p) such that for each B in $\mathfrak{B}$, $\boldsymbol{C}_{\mathfrak{G}}(B)$ contains an element of $\mathcal{U}(p)\}$ $\mathfrak{J}(2) = \{\mathfrak{A} \mid \mathfrak{A}$ is a 2-subgroup of $\mathfrak{G}$ and $\mathfrak{A}$ contains a noncyclic abelian subgroup of order $8\}$.

For a prime q, we write $q \sim 2$ and $2 \sim q$ if and only if there is an element of $\mathcal{S}o\ell(\mathfrak{G})$ which contains elements of $\mathfrak{J}(q)$ and $\mathfrak{J}(2)$.

DEFINITION 2.10. $\mathcal{E}(p) = \mathcal{E}_{\mathfrak{G}}(p)$ is the set of subgroups $\mathfrak{E}$ of $\mathfrak{G}$ of type (p, p) which centralize every element in $\mathsf{N}_{\mathfrak{G}}(\mathfrak{E}; p')$.

Let $p \in \pi_3(\mathfrak{G}) \cup \pi_4(\mathfrak{G})$ and let $\mathfrak{P}$ be a S_p-subgroup of $\mathfrak{G}$. The sets $\mathcal{A}_i(\mathfrak{P})$ are relevant. Here, as in 0,

$\mathcal{A}_1(\mathfrak{P}) = \{\mathfrak{A} \mid$ (i) $\mathfrak{A}$ is a subgroup of $\mathfrak{P}$. (ii) $\mathfrak{A}$ contains some element of $\mathcal{S}cn_3(\mathfrak{P})\}$.

$\mathcal{A}_{i+1}(\mathfrak{P}) = \{\mathfrak{A} \mid$ (i) $\mathfrak{A}$ is a subgroup of $\mathfrak{P}$. (ii) $\mathfrak{A}$ contains a subgroup $\mathfrak{B}$ of type (p, p) such that $\boldsymbol{C}_{\mathfrak{P}}(B) \in \mathcal{A}_i(\mathfrak{P})$ for all B in $\mathfrak{B}\}$, $i = 1, 2, 3$.

Let $\mathcal{A}(\mathfrak{P}) = \mathcal{A}_4(\mathfrak{P})$ and $\mathcal{A}_i(p) = \bigcup \mathcal{A}_i(\mathfrak{P})$, $\mathcal{A}(p) = \bigcup \mathcal{A}(\mathfrak{P})$, where in both unions, $\mathfrak{P}$ ranges over all the S_p-subgroups of $\mathfrak{G}$.

[4] The S_2-subgroups of Janko's simple group of order 604,800 are of this type.

If G, $H \in \mathfrak{G}$, we write $G \sim_{\mathfrak{G}} H$ if and only if G and H are $\mathfrak{G}$-conjugate, and similarly for subsets of $\mathfrak{G}$. If there is no danger of confusion, we write $G \sim H$. The negation of $\sim$ is $\nsim$. We are thus using the symbol $\sim$ in two senses, but since a prime is hardly to be confused with an element of a finite group, no confusion is likely. Following Brauer, if $\mathfrak{H}$ is a subgroup of $\mathfrak{G}$ and G, $H \in \mathfrak{H}$ satisfy $G \nsim_{\mathfrak{H}} H$ and $G \sim_{\mathfrak{G}} H$, we say that G and H are *fused* in $\mathfrak{G}$, or that a *fusion* of G and H occurs in $\mathfrak{G}$.

3. Statement of main theorem and corollaries.

MAIN THEOREM. *Each nonsolvable N-group is isomorphic to a group $\mathfrak{G}$ such that $I(\mathfrak{S}) \subseteq \mathfrak{G} \subseteq \mathrm{Aut}(\mathfrak{S})$, where $\mathfrak{S}$ is one of the following N-groups:*

(a) $L_2(q)$, $q > 3$.
(b) $Sz(q)$, $q = 2^{2n+1}$, $n \geq 1$.
(c) $L_3(3)$.
(d) M_{11}.
(e) A_7.
(f) $U_3(3)$.

COROLLARY 1. *Every minimal simple group is isomorphic to one of the following minimal simple groups:*

(a) $L_2(2^p)$, *p any prime.*
(b) $L_2(3^p)$, *p any odd prime.*
(c) $L_2(p)$, *p any prime exceeding* 3 *such that* $p^2 + 1 \equiv 0 \pmod 5$.
(d) $Sz(2^p)$, *p any odd prime.*
(e) $L_3(3)$.

COROLLARY 2. *A finite group is solvable if and only if every pair of its elements generates a solvable group.*

COROLLARY 3. *A finite group is solvable if and only if it does not contain three nonidentity elements A, B, C of pairwise coprime orders such that $ABC = 1$.*

COROLLARY 4. *If $\mathfrak{G}$ is a nonsolvable group with $|\pi(\mathfrak{G})| = 3$, then one of the following groups is involved in $\mathfrak{G}$: $L_2(4)$, $L_2(7)$, $L_2(8)$, $L_2(17)$, $L_3(3)$.*

COROLLARY 5. *If every c.f. of the finite group $\mathfrak{G}$ is an N-group and n is a divisor of $|\mathfrak{G}|$ such that there are exactly n elements in $\mathfrak{G}$ of order dividing n, they form a subgroup.*

COROLLARY 6. *If $\mathfrak{G}$ is a nonsolvable group, then $|\pi(\mathfrak{G})| \geq 3$.*

Corollary 6 of Burnside is well known. The interested reader may extract the relevant results from 0 and the present paper to give a new proof of Corollary 6. The other five corollaries are probably new. The possible existence of Corollary 3 was mentioned in [22]. Corollary 5 is a minuscule contribution to an old problem and sheds no light on it. Finally, we state a characterization theorem for $E_2(3)$ and $S_4(3)$.

THEOREM *ES*. *$E_2(3)$ and $S_4(3)$ are the only simple groups $\mathfrak{G}$ such that*
(i) $2, 3 \in \pi_4(\mathfrak{G})$.
(ii) If $p \in \{2, 3\}$, *$\mathfrak{G}_p$ is a S_p-subgroup of $\mathfrak{G}$ and $\mathfrak{A} \in \mathcal{S}\mathfrak{cn}_3(\mathfrak{G}_p)$, then $\mathsf{N}(\mathfrak{A})$ is trivial.*
(iii) *The normalizer of every nonidentity 3-subgroup of $\mathfrak{G}$ is solvable.*
(iv) *The centralizer of every involution of $\mathfrak{G}$ is solvable.*
(v) $2 \sim 3$, *that is, $\mathfrak{G}$ has a solvable subgroup containing*
(a) *a noncyclic abelian subgroup of order* 8,
(b) *an elementary subgroup of type* (3, 3) *each element of which centralizes a subgroup in $\mathfrak{U}(3)$.*

4. **Proofs of corollaries.** It is a consequence of results of Dickson [12] that the groups listed in (a), (b), (c), (e) of Corollary 1 are minimal simple groups. Suzuki [37] has shown that the groups in (d) are minimal simple groups. By Lemma 5.33, $U_3(3) \supset L_2(7)$, so Corollary 1 follows from the Main Theorem.

Corollary 2 is an almost trivial consequence of Corollary 1. Explicit proofs are available for all the groups listed in Corollary 1 [34].

In proving Corollary 3, it suffices to show that for each minimal simple group $\mathfrak{G}$, there are elements A, B, C of $\mathfrak{G}$ of pairwise coprime order with $ABC=1$. As the character tables of all the minimal simple groups have been determined [12], [37], Corollary 3 may be easily verified. We remark that if $\mathfrak{G} = Sz(q)$, we may choose A, B, C of orders $q-1$, $q-r+1$, $q+r+1$, where $2q=r^2$.

Corollary 4 is a consequence of elementary number theory and Corollary 1.

In proving Corollary 5 for $\mathfrak{G}$, an appeal to a result of Zemlin [46] entitles us to assume that $\mathfrak{G}$ is simple. Rust [31] has verified Corollary 5 for $L_2(q)$, $L_3(3)$, A_7 and $Sz(q)$. We omit the discussion of M_{11} and $U_3(3)$, which is not difficult.

BIBLIOGRAPHY

1. J. Alperin, *Centralizers of abelian normal subgroups of p-groups*, J. Algebra (2) **1**(1964), 110–113.

2. ———, *Sylow intersections and fusion*, J. Algebra **6** (1967), 222–241.

3. J. Alperin and D. Gorenstein, *Transfer and fusion in finite groups*, J. Algebra **6** (1967), 242–255.

4. E. Artin, *The orders of the classical simple groups*, Comm. Pure. Appl. Math. **8** (1955), 455–472.

5. N. Blackburn, *Generalizations of certain elementary theorems on p-groups*, Proc. London Math. Soc. (3) **11** (1961), 1–22.

6. ———, *On prime-power groups with 2 generators*, Proc. Cambridge Philos. Soc. (54) **3**(1958), 327–337.

7. R. Brauer, *On the structure of groups of finite order*, Proc. Internat. Congress Math., vol. I, 1954, Noordhoff, Groningen and North-Holland, Amsterdam, 1957, pp. 209–217.

8. ———, *On finite Desarguesian planes*. I, II, Math. Z. **90**(1965), 117–123 and **91**(1966), 124–151.

9. ———, *Some applications of the theory of blocks of characters of finite groups*. II, J. Algebra (1) **4**(1964), 307–334.

10. R. Brauer and M. Suzuki, *On finite groups of even order whose 2-Sylow subgroup is a quaternion group*, Proc. Nat. Acad. Sci. U.S.A. (12) **45** 1757–1759.

11. R. Brauer and P. Fong, *A characterization of the Mathieu group* M_{12}, Trans. Amer. Math. Soc. **122**(1966), 18–47.

12. L. E. Dickson, *Linear groups*, Dover, New York, 1958.

13. W. Feit, *A characterization of the simple groups* SL$(2, 2^a)$, Amer. J. Math. (2) **82**(1960), 281–300.

14. P. Fong, *Some Sylow subgroups of order* 32 *and a characterization of* $U(3, 3)$, J. Algebra **6**(1967), 65–76.

15. G. Frobenius and I. Schur, *Ueber die reelen Darstellungen der endlichen Gruppen*, Berliner Sitz. (1906), 186–208.

16. G. Glauberman, *Central elements of core-free groups*, J. Algebra **4**(1966), 403–420.

17. ———, *A characteristic subgroup of a p-stable group*, Canad. J. Math. (to appear).

18. ———, *A characterization of Suzuki groups*, Illinois J. Math. (to appear).

19. D. Gorenstein and J. Walter, *The characterization of finite groups with dihedral Sylow 2-groups*, J. Algebra (2) (1965), 85–151, 218–270, 354–393.

20. ———, *On the maximal subgroups of finite simple groups*, J. Algebra (2) **1**(1964), 168–213.

21. M. Hall, Jr., *The theory of groups*, Macmillan, New York, 1959.

22. P. Hall, *A characteristic property of soluble groups*, J. London Math. Soc. **12**(1937), 198–200.

23. ———, *Theorems like Sylow's*, Proc. London Math. Soc. (3), **6**(1956), 286–304.

24. ———, Lecture Notes (unpublished).

25. ———, *On a theorem of Frobenius*, Proc. London Math. Soc. **40**(1935), 468–501.

26. P. Hall and G. Higman, *The p-length of a p-soluble group and reduction theorems for Burnside's problem*, Proc. London Math. Soc. (3) **7**(1956), 1–42.

27. G. Higman, *Suzuki 2-groups*, Illinois J. Math. **7**(1963), 79–95.

28. N. Ito, *On a theorem of H. F. Blichfeldt*, Nagoya Math. J. **5**(1954), 75–77.

29. B. H. Neumann, *Groups with automorphisms that leave only the neutral element fixed*, Archiv. Math. **7** (1956), 1–5.

30. R. Ree, *A family of simple groups associated with the simple Lie algebra of type* (G_2), Amer. J. Math. **83**(1961), 432–462.

31. J. Rust, *On a conjecture of Frobenius*, Ph.D. Thesis, University of Chicago, 1966.

32. I. Schur, *Zur Theorie der vertauschbaren Matrizen*, J. Reine Angew. Math. **130** (1905), 66–76.

33. C. Sims, *Graphs and finite permutation groups*, (to appear).

34. R. Steinberg, *Generators for simple groups*, Canad. J. Math. **14** (1962), 277–283.

35. M. Suzuki, *Finite groups with nilpotent centralizers*, Trans. Amer. Math. Soc. **99**(1961), 425–470.

36. ———, *Finite groups of even order whose Sylow 2-subgroups are independent*, Ann. of Math (80) **1** (1964), 56–77.

37. ———, *On a class of doubly transitive groups*, Ann. of Math. (2) **75**(1962), 105–145.

38. ———, *Characterizations of linear groups*. III, Nagoya Math. J. **21** (1962), 159–183.

39. ———, *Finite groups in which the centralizer of any element of order 2 is 2-closed*, Ann. of Math. (2) **82** (1965), 191–212.

40. J. Tits, *Theorie des groupes—theoreme de Bruhat et sous-groupes paraboliques*, C. R. Acad. Sci. Paris **254** (1962), 2910–2912.

41. J. Thompson, *Fixed points of p-groups acting on p-groups*, Math. Z. **86** (1964), 12–13.

42. ———, *Normal p-complements for finite groups*, J. Algebra (1) **1** (1964), 43–46.

43. ———, *Factorizations of p-solvable groups*, Pacific J. Math. (16) **2** (1966), 371–372.

44. H. Wielandt, *Beziehungen zwischen den Fixpunktzahlen von Automorphismengruppen einer endlicher Gruppe*, Math. Z. **73** (1960), 146–158.

45. W. Wong, *On finite groups whose 2-Sylow subgroups have cyclic subgroups of index* 2, J. Austral. Math. Soc. **4** (1964), 90–112.

46. R. Zemlin, *On a conjecture arising from a theorem of Frobenius*, Ph.D. Thesis, Ohio State University, 1954 (unpublished).

JOURNAL OF ALGEBRA **89**, 437–499 (1984)

Some Finite Groups Which Appear as Gal L/K, Where $K \subseteq Q(\mu_n)$

JOHN G. THOMPSON

Department of Mathematics, University of Cambridge, Cambridge CB2 1SB, England

Communicated by Walter Feit

Received April 20, 1983

Contents. 0. Introduction. 1. Preliminary definitions and lemmas. 2. Preliminaries about discrete groups and more function theory. 3. The groups Δ. 4. The construction of $F(Y, Z)$ and the auxiliary polynomial. 5. The Puiseux elements for $F(Y, Z)$ and $F_{\infty}(Y, Z)$. 6. Recovering Γ and Δ from the Puiseux elements for $F \circ \rho$ and $F_{\infty} \circ \rho$, where $\rho \in \mathscr{G}$. 7. Extensions and the big group extension. 8. The action of $\mathscr{G}^*$ on G. 9. More lemmas. 10. Rigidity and Γ/Λ. 11. The construction of E. 12. Proof of the Corollary.

0. INTRODUCTION

For each natural number n, let

$$\mu_n = \{z \in C \mid z^n = 1\}, \tag{0.1}$$

$$\zeta_n = \exp 2\pi i/n, \tag{0.2}$$

$$U_n = \text{Gal } Q(\mu_n)/Q, \tag{0.3}$$

where U_n is identified with $(Z/nZ)^{\times}$. I denote by $\sigma(l)$ the automorphism of $Q(\mu_n)$ which maps ζ_n to ζ_n^l ($(l, n) = 1$), and I write automorphisms on the right, so that

$$\zeta_n \circ \sigma(l) = \zeta_n^l.$$

Suppose F is a finite group and $\alpha \in \text{Aut}(F)$. Then α induces an action on the set of conjugacy classes of F, also denoted by α, where, if C is a conjugacy class,

$$C \circ \alpha = \{x \circ \alpha \mid x \in C\}.$$

If $\bar{l} \in U_{|F|}$, then $\bar{l}$ induces an action on the set of conjugacy classes of F, also denoted by $\bar{l}$, where

$$C \circ \bar{l} = \{x^l \mid x \in C, l \in \bar{l}\}.$$

For each ordered set $(C_1,..., C_k)$ of conjugacy classes of F, let

$$A = A(C_1,..., C_k) = A_F(C_1,..., C_k) = \{(x_1,..., x_k) \mid x_i \in C_i, x_1 \cdots x_k = 1\}. \tag{0.4}$$

Denote by $\mathcal{O}\!\ell_k(F)$ the set of all non-empty sets of the form $A_F(C_1,..., C_k)$, as $C_1,..., C_k$ range over the conjugacy classes of F, and let $\mathcal{O}\!\ell(F)$ be the union of the sets $\mathcal{O}\!\ell_k(F)$, $k = 1, 2,...$. Then Aut(F) and $U_{|F|}$ act on $\mathcal{O}\!\ell(F)$.

$$A \circ \alpha = \{(x_1 \circ \alpha,..., x_k \circ \alpha) \mid x_i \in C_i, x_1 \cdots x_k = 1\}$$
$$= \{(y_1,..., y_k) \mid y_i \in C_i \circ \alpha, y_1 \cdots y_k = 1\},$$

whereas for $\bar{l} \in U_{|F|}$, we merely have

$$A \circ \bar{l} \underset{\text{def}}{=} A(C_1 \circ \bar{l},..., C_k \circ \bar{l}). \tag{0.5}$$

It can be verified that $(A \circ \bar{l})\, \bar{l}_0 = A \circ \overline{l l_0}$, $A \circ \bar{1} = A$, so (0.5) defines an action of $U_{|F|}$ on $\mathcal{O}\!\ell(F)$.

Definition 0.1.[1] $A = A_F(C_1,..., C_k)$ is *rigid* if and only if

1. $A \neq \varnothing$.
2. F permutes A transitively by conjugation.
3. For each $(x_1,..., x_k) \in A$, $F = \langle x_1,..., x_k \rangle$.

The applications of the function-theoretic results of this paper concern a finite group F such that

(H)
1. F is a non Abelian simple group.
2. For some integer $k \leqslant 6$, and conjugacy classes $C_1,..., C_k$ of F, $A_F(C_1,..., C_k)$ is rigid.

A finite group F satisfying (H) is said to be rigid, and if we wish to make explicit the relevant conjugacy classes, we say that F is rigid with respect to $(C_1,..., C_k)$. I have little idea just how restrictive this condition is, and for the applications I have in mind, $k = 3$. The demand that $k \leqslant 6$ is perhaps unnatural, but is forced on me by the function-theory and the discrete groups I have at my disposal, as will become clear. In any case, in order to state my

[1] Similar definitions to the definition of rigidity appear in the following papers: M. Fried, Fields of definition of function fields and Hurwitz families—Groups as galois groups, *Comm. Alg.* **5** (1977), 17–82; B. Matzat, "Zur Konstruktion von Zahl- und Funktionenkörpern mit Vorgegebener Galoisgruppe," Karlsruhe, 1980; G. Belyi, On Galois extensions of a maximal cyclotomic field, *Math. USSR Izv.* **14** (1980), 247–256.

results, it is necessary first to derive several consequences of (H), so for the next few paragraphs, I assume that (H) holds for F and $(C_1,\dots, C_k)$.

Pick $x_\nu \in C_\nu$, and let

$$d_\nu = o(x_\nu), \qquad D = \text{l.c.m.}\{d_1,\dots, d_k\}.$$

Set

$$U = U_D.$$

The first thing to pin down is

$$A \circ \bar{l} \text{ is rigid for all } \bar{l} \in U. \tag{0.6}$$

To see this, I invoke character theory, which says that

$$\operatorname{card} A(C_1,\dots, C_k) = \frac{|F|^{k-1}}{c_F(x_1)\cdot \cdots \cdot c_F(x_k)} \frac{\sum \chi^{(x_1)}\cdot \cdots \cdot \chi^{(x_k)}}{\chi(1)^{k-2}}, \tag{0.7}$$

where $x_i \in C_i$, $c_F(x_i)$ is the order of the centralizer of x_i in F, and the summation is extended over all the irreducible characters χ of F. Since A is rigid and F is a group whose centre is 1 (F being non Abelian and simple), we have

$$\operatorname{card} A(C_1,\dots, C_k) = |F|.$$

If l is prime to D, then $\chi(x_i) \circ \sigma(l) = \chi(x_i^l)$, for all χ and all i, so if we apply $\sigma(l)$ to (0.7), we learn that

$$\operatorname{card} A \circ \bar{l} = |F|, \qquad \text{for all} \quad \bar{l} \in U.$$

Thus, in order to show that $A \circ \bar{l}$ is rigid, it suffices to show that if $(y_1,\dots, y_k) \in A \circ \bar{l}$, then $F = \langle y_1,\dots, y_k\rangle$. Let $F_0 = \langle y_1,\dots, y_k\rangle$. Choose l' with $ll' \equiv 1 \pmod D$. Let $C_{i,0}$ be the conjugacy class of F_0 which contains y_i. By the character-theoretic formula already used, and applying it this time to F_0, we have

$$\operatorname{card} A_{F_0}(C_{1,0},\dots, C_{k,0}) = \operatorname{card} A_{F_0}(C_{1,0} \circ \bar{l}',\dots, C_{k,0} \circ \bar{l}').$$

In particular, $\operatorname{card} A_{F_0}(C_{1,0} \circ \bar{l}',\dots, C_{k,0} \circ \bar{l}') \neq 0$, and since $A_{F_0}(C_{1,0} \circ \bar{l}',\dots, C_{k,0} \circ \bar{l}') \subseteq A_F(C_1,\dots, C_k)$, we get $F = F_0$, and so we get (0.6).

Let

$$\mathscr{S} = \{A \circ \bar{l} \mid \bar{l} \in U\}, \tag{0.8}$$

and

$$V = V_{F,\mathscr{S}} = \{\bar{l} \in U \mid S \circ \bar{l} = S \text{ for all } S \in \mathscr{S}\}. \tag{0.9}$$

Since U is Abelian, $V=\{\bar{l}\in U\mid A\circ\bar{l}=A\}$, and so

$$\operatorname{Card}\mathscr{S}=|U:V|.$$

V is a subgroup of U, and is characterized as those $\bar{l}$ in U such that x_i^l and x_i are conjugate in F for all $i=1,\ldots,k$.

There is a second, possibly larger, subgroup of U associated to this situation, namely,

$$W=W_{F,\mathscr{S}}=\{\bar{l}\in U\mid S\circ\bar{l}=S\circ\alpha \text{ for some } \alpha=\alpha(S,\bar{l})\in\operatorname{Aut}(F), \text{ and all } S\in\mathscr{S}\}. \tag{0.10}$$

Examples abound in which $V\subset W$.

The automorphisms α of F which are candidates to appear in (0.10) give rise to yet another group, denoted by $\mathscr{A}$, where

$$\mathscr{A}=\mathscr{A}(\mathscr{S})=\{\alpha\in\operatorname{Aut}(F)\mid \text{for some } \bar{l}\in U, A\circ\alpha=A\circ\bar{l}\}. \tag{0.11}$$

Note that $\mathscr{A}$ is characterized as the set of those automorphisms α of F such that $\mathscr{S}\circ\alpha=\mathscr{S}$, since if $A\circ\alpha=A\circ\bar{l}$, then $S\circ\alpha=S\circ\bar{l}$ for all $S\in\mathscr{S}$. Clearly, $\mathscr{A}\supseteq\operatorname{Inn}(F)$, the group of inner automorphisms of F. If we consider the subset X of $W\times\mathscr{A}$ defined by

$$(\bar{l},\alpha)\in X\Leftrightarrow A\circ\bar{l}=A\circ\alpha,$$

we get that $(\bar{l},\alpha)\in X\Rightarrow \bar{l}V\times\alpha\operatorname{Inn}(F)\subseteq X$, and I argue that this relation on $W/V\times\mathscr{A}/\operatorname{Inn}(F)$ is an isomorphism of groups. I emphasize that we are in a very unusual situation here, since one does not run across rigidity so frequently. So perhaps at the risk of redundancy, I should prove just why it is that if $\alpha\in\mathscr{A}$ and $S\in\mathscr{S}$ satisfy $S\circ\alpha=S$, then α is inner. We have $S=A(C_1',\ldots,C_k')$, where $C_i'=C_i\circ\bar{l}$ for some $\bar{l}\in U$. Pick $(y_1,\ldots,y_k)\in S$. Then there are elements $f_1,\ldots,f_k$ in F such that $y_i\circ\alpha=f_i^{-1}y_if_i$, since $S\circ\alpha=S$. Since α is an automorphism of F, we have $(f_1^{-1}y_1f_1,\ldots,f_k^{-1}y_kf_k)\in S$. But S is rigid, so there is f in F such that $f_i^{-1}y_if_i=f^{-1}y_if$ for all i. Since $\langle y_1,\ldots,y_k\rangle=F$, we get $\alpha=i_f$, the inner automorphism of F determined by f. As we see, rigidity converts an a priori very weak notion of "locally inner" to a global inner automorphism, and this fact has important function-theoretic consequences. In particular, at this stage, we get our isomorphism

$$W/V\cong\mathscr{A}/\operatorname{Inn}(F). \tag{0.12}$$

Out of this intertwining of $\mathscr{A}$ and W will emerge a group, which I call $F^*=F^*(F,A)$, determined by a rigid $A=A_F(C_1,\ldots,C_k)$. As far as I can tell, it is not possible to construct this group F^* explicitly. It is built up heuristically from two pieces, a bottom, which is the direct product of $|U : W|$ copies of F, and a top, which is U/V. I proceed to describe F^* as best I can, leaving it to the statement of the main result of this paper to formalize what I have in mind, and to the function-theory to show that an F^* really exists, and has splendid properties.

Let

$$B = F^{U/W}$$

be the set of all maps from U/W to F. If $b_1, b_2 \in B$, then by definition, $(b_1 \cdot b_2)(\xi) = b_1(\xi) \cdot b_2(\xi)$, $\xi \in U/W$. This is the "base" group, and if $\xi \in U/W$, set

$$B_\xi = \{b \in B \mid b(\xi') = 1_F \text{ for all } \xi' \neq \xi\}.$$

Then $B_\xi \cong F$, and B is the direct product of the B_ξ for $\xi \in U/W$, whence, as abstract group, B is the direct product of t copies of F, where $t = |U : W| = \text{card}\, \mathscr{S}/|W : V|$.

I shall use an explicit isomorphism from B_ξ to F, the obvious one,

$$j_\xi : B_\xi \to F, \qquad b \circ j_\xi = b(\xi). \tag{0.13}$$

Since $b(\xi') = 1$ for all $\xi' \neq \xi$, $\xi' \in U/W$, an element b of B_ξ is uniquely specified by its value at ξ.

I need to bring all this into relation with subfields of $Q(\mu_D)$, so let $V = V_{F, \mathscr{S}}$, where V is given by (0.9). Let

$$K = \text{fixed field of } V \text{ on } Q(\mu_D).$$

Thus, $\text{Gal}\, K/Q = U/V$. It is now possible to state the main result of this paper.

THEOREM. *Assume* (H) *for* F *and* $(C_1, \ldots, C_k)$. *Define the integer* D *and subgroups* V, W *of* $U = U_D$, *and the group* B, *as above. Let* $K_V = K$ *be the fixed field of* V *on* $Q(\mu_D)$. *Let*

$$1 \to V \to U \xrightarrow{\text{res}} U/V \to 1$$

be the exact sequence of groups given by the inclusion map of V *in* U, *where* $\text{res}(\bar{l}) = \sigma(l)|_K$.

There is an exact sequence of groups

$$1 \to B \xrightarrow{i} F^* \xrightarrow{\pi} U/V \to 1$$

such that if

$$P = \{(\beta, \bar{l}) \mid \beta \in F^*, \bar{l} \in U, \beta \circ \pi = \text{res}(\bar{l})\},$$

then the following hold:

(i) *The centralizer of* $B \circ \iota$ *in* F^* *is* 1.

(ii) $\beta^{-1} B_\xi \circ \iota\beta = B_{\bar{l}\xi} \circ \iota$ *for* $(\beta, \bar{l}) \in P$, $\xi \in U/W$.

(iii) *If* $\bar{l} \in W$ *and* $(\beta, \bar{l}) \in P$, *then for each* $\xi \in U/W$, *there is an automorphism* $\gamma = \gamma(\xi, \beta)$ *of* B_ξ *such that* $\beta^{-1} b \circ \iota\ \beta = (b \circ \gamma) \circ \iota$ *for all* $b \in B_\xi$.

(iv) *Let* $\mathscr{A}_\xi = \{\gamma(\xi, \beta) \mid (\beta, \bar{l}) \in P, \bar{l} \in W\}$. *Then* $j_\xi^{-1} \mathscr{A}_\xi j_\xi = \mathscr{A}$, *where* $\mathscr{A}$ *is defined in* (0.11) *and* j_ξ *is given in* (0.13).

(v) *There is a Galois extension E of Q, such that*

$$\psi : F^* \cong \operatorname{Gal} E/Q,$$

and such that under this isomorphism, the fixed field of $B \circ \psi$ *is K. In addition, there is a commutative diagram:*

$$\begin{array}{ccc} F^* & \xrightarrow{\ \psi\ } & \operatorname{Gal} E/Q \\ {\scriptstyle \pi}\big\downarrow & & \big\downarrow{\scriptstyle \text{res}} \\ U/V & = & \operatorname{Gal} K/Q. \end{array}$$

(vi) *There is a Galois extension L of K such that* $\operatorname{Gal} L/K \cong F$.

Initially, this result seemed astonishing to me, but one becomes accustomed to it, and we shall see it unfold in a fashion which by now seems natural to me. It was tempting to write this paper by retaining only the conclusion $\operatorname{Gal} L/K \cong F$, but *I* believe it is better, perhaps even imperative, to keep the entire picture.

To unravel some consequences of this theorem, we must get our hands on some occurrences of (H). Then we can perhaps identify B, V, W, and see what happens.

Case 1. $F = F_1 =$ *the Fischer–Griess group.* Let $C_1 =$ the class of Fischer involutions. For x_1 in this class, $C_{F_1}(x_1)/\langle x_1 \rangle \cong F_2$, the baby monster. Let $C_2 =$ the class of elements of order 3 which are of Suzuki type. If x_2 is in this class, then $C_{F_1}(x_2)/\text{solv. gp.} \cong Sz$, a sporadic simple group. Let $C_3 =$ any class of elements of order 71 (there are two of them), and let $C_4 =$ all elements of order 29, a single class in F_1. Then $A(C_1, C_2, C_3)$ and $A(C_1, C_2, C_4)$ are rigid. Call these two cases, (2, 3, 71) and (2, 3, 29), respectively.

Case 1(a) (2, 3, 71). Here $D = 2.3.71$, $U \supseteq V = W = \{\bar{l}$ is a square modulo $71\}$. So $K = Q(\sqrt{-71})$, $F^* \cong F_1 \iota C_2$, and F_1 is of the shape $\operatorname{Gal} L(71)/Q(\sqrt{-71})$ for a suitable field $L(71)$.

Case 1(b) (2, 3, 29). Here $D = 2.3.29$, $U = V = W$, $K = Q$, and

$$F_1 \cong \mathrm{Gal}\, L(29)/Q, \qquad \text{for a suitable field } L(29).$$

Case 2. $F = PSL_2(2^n)$, $n \geqslant 2$. Let x_1, x_2, x_3 have orders 2, $2^n - 1$, $2^n + 1$, respectively. Let C_i be the conjugacy class containing x_i. Then $A(C_1, C_2, C_3)$ is rigid. Set $D = 2(2^{2n} - 1)$. Here $K = K_V = K_n \subseteq Q(\mu_{2^{2n}-1})$. More precisely, K_V is the compositum of the real subfield of $Q(\mu_{2^n-1})$ and the real subfield of $Q(\mu_{2^n+1})$. The image of W in $U_{2^n-1} \times U_{2^n+1}$ consists of the elements $(\varepsilon 2^i, \varepsilon' 2^i)$, where ε, $\varepsilon' = 1$ or -1, and so W/V is identified with $\mathrm{Gal}\, F_{2^n}/F_2$.

It is surely possible to obtain rigidity for many simple groups, but these are the only examples given in this paper. Putting these assertions together (proofs will come later), we get

COROLLARY. (i) $F_1 \cong \mathrm{Gal}\, L(71)/Q(\sqrt{-71}) \cong \mathrm{Gal}\, L(29)/Q$, *where* F_1 *is the Fischer–Griess group, and* $L(71)$, $L(29)$ *are suitable fields. If* L *is the smallest Galois extension of* Q *which contains* $L(71)$, *then* $\mathrm{Gal}\, L/Q \cong F_1 \wr C_2$ (*wreath product*).

(ii) *If* $n \geqslant 2$, *then* $L_2(2^n) \cong \mathrm{Gal}\, L_n/K_n$, *where*

$$K_n = Q(\zeta_{2^n-1} + \zeta_{2^n-1}^{-1}, \zeta_{2^n+1} + \zeta_{2^n+1}^{-1}),$$

and L_n *is a suitable field.*

1. PRELIMINARY DEFINITIONS AND LEMMAS

My notation may be considered idiosyncratic both to group theorists and to algebraic geometers, but will not, I hope, be an obstacle to understanding the flow of the argument. I omit proofs of elementary function-theoretic results, but try to provide references and keep a record of what I use.

I had difficulty in sorting out questions about local variables, so this first section provides some material for the novice, and further discussion will appear in later sections. What follows is a part of my self-education, and may be helpful to group theorists.

Let

$$C[[t]] = \left\{ \sum_{i=0}^{\infty} c_i t^i \mid c_i \in C \right\} \subseteq C((t)) = \left\{ \sum_{i=h}^{\infty} c_i t^i \mid c_i \in C \right\}, \tag{1.1}$$

$$C\langle t\rangle = \Big\{ f = \sum_{i=h}^{\infty} c_i t^i \in C((t)) \mid \text{for some positive } A = A(f), \text{ and some natural number } N = N(f), |c_i| < A^i \text{ for all } i > N \Big\}. \tag{1.2}$$

So $C\langle t\rangle \cap C[[t]] = C_0\langle t\rangle$ is the ring of formal power series in t which converge in some neighbourhood of 0; $0 < |t| < 1/A(f)$ is a punctured neighbourhood of 0 in which $f (\in C\langle t\rangle)$ converges [2]. Both $C((t))$ and $C\langle t\rangle$ are fields, the fields of fractions of $C[[t]]$, $C_0\langle t\rangle$, respectively.

If $f \in C((t))$ and $f \neq 0$, let

$$v_t(f) = \text{smallest } i \text{ such that } c_i \neq 0.$$

A local variable is an f in $C((t))$ with $v_t(f) = 1$. If

$$f = \sum c_i t^i, \qquad g = \sum d_i t^i \tag{1.3}$$

are elements of $C((t))$ and $v_t(g) \geqslant 1$, then

$$f \circ g \underset{\text{def}}{=} \sum c_i g^i \in C((t)).$$

Moreover, it is straightforward to check that

$$f, g \in C\langle t\rangle \quad \text{and} \quad v_t(g) \geqslant 1 \Rightarrow f \circ g \in C\langle t\rangle. \tag{1.4}$$

As a special case, if $f, g \in C\langle t\rangle$ and g is a local variable, then $f \circ g \in C\langle t\rangle$. If $c \in C^\times$, then $g = ct$ is a local variable, and in this case, I set $f \circ g = f_c$, or, more formally, $f \circ g = f \circ h(c)$, defining thereby an automorphism $h(c)$ of $C((t))$. If n is a natural number, and $g = t^n$, I denote $f \circ g$ by $f^{(n)}$, or, on occasion, by $f \circ E_n$, defining thereby an endomorphism of $C((t))$ whose kernel is 0. Explicitly, then, if f is given by (1.3), then

$$f_c = f \circ h(c) = \sum c_i \cdot c^i t^i, \qquad f^{(n)} = f \circ E_n = \sum c_i t^{ni}.$$

Suppose f is a local variable, and $f \in C\langle t\rangle$. Then in $C((t))$, we have $t = \sum d_i f^i$, and it is a handy fact that $|d_i| < B^i$ for some positive B, that is [2],

$$t \in C\langle u\rangle \Leftrightarrow u \in C\langle t\rangle \qquad (\text{if } v_t(u) = 1). \tag{1.5}$$

Let $\mathscr{G}$ be the group of all automorphisms of C. If $\rho \in \mathscr{G}$, then ρ acts on $C((t))$ via

$$f = \sum c_i t^i \mapsto f \circ \rho = \sum (c_i \circ \rho)\, t^i, \tag{1.6}$$

and such a map does not necessarily preserve $C\langle t\rangle$. However, we check that $(f \circ \rho) \circ \rho' = f \circ \rho\rho'$, $f \circ 1 = f$, so we have an action of $\mathscr{G}$ on $C((t))$. For reasons of convenience later in this paper, I choose to let $\mathscr{G}$ act on both C and $C((t))$ on the right. I wish to make use of $\mathscr{G}$, so I need to check that those elements of $C\langle t\rangle$ which appear in this paper behave properly under $\mathscr{G}$. I shall also make use of the trivial but very important relation

$$\rho^{-1}h(c)\rho = h(c\circ\rho) \qquad (c\in C^{\times}, \rho\in\mathscr{G}). \tag{1.7}$$

If f is not constant, and $f\in C((t))$, set

$$\mu(f) = \{c\in C^{\times}, f=f_c\}.$$

Then

$$\mu(f) = \mu_n \qquad \text{for some natural number } n,$$

which I denote by $n(f)$. So

$$n(f) \text{ is the largest integer } n \text{ such that } f\in C((t^n)). \tag{1.8}$$

Alas, more notation. The open disc at z with radius a is

$$D_a(z) = \{\zeta\in C \,\big|\, |\zeta - z| < a\} \qquad (z\in C, a>0).$$

The punctured disc is $\dot{D}_a(z)$, in which the centre of the disc is removed.

I shall use (1.5) later on to make "good" choices of local variables. As I see it, and I am only about a century late, the reason this replacement of t by u comes into the picture is that for $z_0\in C$, the natural local variable is $z-z_0$, which is used in power series expansions of functions which are analytic in a neighbourhood of z_0, whereas if α, β are algebraic functions related by $F(\alpha,\beta)=0$, the data provided by the polynomial practically force the local variable t to satisfy $t^n=\beta-c$ for some natural number n dictated to us by the polynomial. In no other way (that I have yet learned) can we capture uniformly the way α depends on β. A further point can be made. Since $C_0\langle t\rangle$ is an integral domain and since functions f, g which are analytic in a neighbourhood of 0 have power series expansions which represent them, it follows that

> If f, g are analytic in $D_a(0)$ and $f(z)\cdot g(z)=0$ in this domain, then either $f=0$ or $g=0$ in this disc. (1.9)

It is this fact which enables us to stitch coverings together.

The main object of study here is a polynomial

$$F(Y,Z) = F_0(Z)\,Y^N + F_1(Z)\,Y^{N-1} + \cdots + F_N(Z), \tag{1.10}$$

which is irreducible in $C[Y,Z]$ and of positive degrees N in Y and n in Z. Set

$$F_i(Z) = c_{i0} + c_{i1}Z + \cdots c_{in}Z^n. \tag{1.11}$$

My object is, given the finite group F with its generating elements whose product is 1, to construct from a judiciously chosen discrete subgroup Δ of $PSL(2, R)$ a pair of generating functions for the function field of the Riemann surface associated to Δ in such a way that the polynomial (1.10) relating these two functions has very restrained properties, so tight in fact that applying any automorphism ρ of C to the coefficients c_{ij} of $F(Y, Z)$ leads to another pair of functions on the very same Riemann surface. In other words, my game is to force $\mathscr{G}$ to act on the function field of the Riemann surface associated to Δ. This is in fact a well known procedure. The problem is that in this fashion, we can construct normal subgroups Δ_ρ of a discrete group Γ, one for each automorphism ρ of C in such a way that Γ/Δ_ρ is isomorphic to a fixed group, the Galois group of a splitting field of $F(Y, Z)$ over $C(Z)$. This group, to within isomorphism, depends only on $F(Y, Z)$ and not on ρ, a fact of elementary Galois theory [1, Theorem 10]. But what is needed is an equality $\Delta = \Delta_\rho$. Isomorphism of Γ/Δ and Γ/Δ_ρ is too weak. As I shall prove, the trick is Lemma 3.3. Furthermore, the big group Γ will be chosen so that $\Gamma\backslash\mathfrak{H}^*$ has genus 0, so that one member of my pair of generating functions for K_Δ will be a suitably chosen generator for K_Γ. As for Γ itself, it is chosen very carefully indeed, since the problems are not over even when we manage to force $\mathscr{G}$ to act on K_Δ.

Returning to (1.10) and (1.11), I wish to derive a property of the Puiseux series associated to F at the point $Z = 0$. (It is a feature of English that the word "series" is the same in both singular and plural. In this case, I mean "Reihen.") I assume that $F_0(0) \neq 0$. Then, setting $a_i = F_i(0)$, we get

$$F(Y, 0) = a_0 Y^N + \cdots + a_N = a_0 \prod_{\alpha \in C} (Y - \alpha)^{n(\alpha)}.$$

If $n(\alpha) \leqslant 1$ for all α, it is easy to show that there are distinct elements $f_1, \ldots, f_N \in C_0\langle t\rangle$ such that

$$F((f_i, t) = 0 \quad \text{in } C((t)), \qquad i = 1, \ldots, N.$$

The case $n(0) > 1$ is much trickier, but has long been subdued. Pick a small positive number a such that if $z \in \dot{D}_a(0)$, then $F(Y, z)$ has N distinct roots. For each z_0 in $\dot{D}_a(0)$, there are elements $f_1^*, \ldots, f_N^* \in C_0\langle t\rangle$, such that $\{f_1^*(0), \ldots, f_N^*(0)\}$ is the set of roots of $F(Y, z_0) = 0$, and such that

$$F(f_i^*, z_0 + t) = 0 \quad \text{in } C((t)), \qquad i = 1, \ldots, N.$$

Let $a(z_0)$ be a positive number such that $a(z_0) < |z_0|$, and in addition, such that $a(z_0)$ is smaller than the radius of convergence of each f_i^*, $i = 1, \ldots, N$. If $z \in D_{a(z_0)}(z_0)$, define

$$f_i(z) = f_i^*(z - z_0).$$

Then f_i is analytic in $D_{a(z_0)}(z_0)$, and $F(f_i(z), z) = 0$ for $z \in D_{a(z_0)}(z_0)$. Moreover, if $i \neq j$, then $f_i(z) - f_j(z)$ is never 0 for $z \in D_{a(z_0)}(z_0)$. Obviously, the functions $f_1, \ldots, f_N$ depend on z_0, and we may write $f_i = f_{i,z_0}$.

We cover the punctured disc $\dot{D}_a(0)$ with discs $D_{a(z_0)}(z_0)$, as described in the preceding paragraph. Fix z_0 and continue analytically the ordered N-tuple $(f_{1,z_0}, \ldots, f_{N,z_0})$ around the circle centered at 0 and passing through z_0. This produces for us a permutation π of $\{1, \ldots, N\}$, with the property that in following f_{i,z_0} around the circle once in a positive sense, we return to $f_{\pi(i),z_0}$. So π has cycles of lengths $r_1, \ldots, r_k$, where $r_1 + \cdots + r_k = N$. As we traverse the circle $|z| = |z_0|$ once in a positive sense, we get, with a suitable renumbering of the f_{i,z_0}, $f_{1,z_0} \to f_{2,z_0} \to \cdots \to f_{r_1,z_0} \to f_{1,z_0}$.

Consider now two new things: first, the circle $|w| = |z_0|^{1/r}$, where $r = r_1$, and second, a brand new polynomial $\tilde{F}(Y, Z) = F(Y, Z^r)$. Fix a w_0 with $w_0^r = z_0$, and build $\tilde{f}_1, \ldots, \tilde{f}_N \in C_0\langle t\rangle$ such that $\{\tilde{f}_1(0), \ldots, \tilde{f}_N(0)\}$ is the set of roots of $\tilde{F}(Y, w_0) = 0$, and such that $\tilde{F}(\tilde{f}_i, w_0 + t) = 0$ in $C((t))$. Define $g_i(w) = \tilde{f}_i(w - w_0)$, obtaining functions analytic in $D_{b(w_0)}(w_0)$ such that $\tilde{F}(g_i(w), w) = 0$ for $w \in D_{b(w_0)}(w_0)$. Since $w_0^r = z_0$, the sets

$$\{f_1(z_0), \ldots, f_N(z_0)\}, \qquad \{g_1(w_0), \ldots, g_N(w_0)\}$$

coincide and are of cardinal N, and we may assume that

$$f_i(z_0) = g_i(w_0), \qquad i = 1, \ldots, r.$$

At this point, we are only interested in the cycle of π which contains $f_1 = f_{1,z_0}$, so we neglect to keep track of the remaining f_i for $i > r$.

At w_0, we have $g_i(w_0) = f_i(z_0)$. Since

$$\tilde{F}(g_i(w), w) = 0, \qquad w \in D_{b(w_0)}(w_0),$$
$$F(f_i(z), z) = 0, \qquad z \in D_{a(z_0)}(z_0),$$

and since $\tilde{F}(g_i(w), w) = F(g_i(w), w^r) = 0$, it follows that if $b(w_0)$ is small enough, then

$$g_i(w) = f_i(w^r), \qquad i = 1, \ldots, r, \quad w \in D_{b(w_0)}(w_0).$$

Now we start at w_0 and traverse the circle $|w| = |z_0|^{1/r}$, from w_0 to $w_0\zeta_r$, and we learn that each g_i has a uniquely determined analytic continuation to a function, which I still denote by g_i, such that

$$g_i(w_0\zeta_r) = f_{i+1}(z_0), \qquad i \pmod r,$$

since as we go from w_0 to $w_0\zeta_r$, w^r goes once around the circle $|z| = |z_0|$, from z_0 back to z_0. Repeating this process by going from $w_0\zeta_r$ to $w_0\zeta_r^2$, etc., we see that

$$\{g_1(w_0), g_1(w_0\zeta_r), \ldots, g_1(w_0\zeta_r^{r-1})\} \qquad \text{and} \qquad \{f_1(z_0), \ldots, f_r(z_0)\}$$

coincide and have cardinal r. Furthermore, in traversing the circle $|w| = |w_0|$ once in the positive sense from w_0 to w_0, g_1 returns to g_1. Similarly for $g_2, ..., g_r$. The theorem on removable singularities [2] applies, and produces for us r elements of $C_0\langle t\rangle$, say $h_0, ..., h_{r-1}$ such that

$$F(h_i, t^r) = 0 \quad \text{in} \quad C((t)),$$

and such that for some w_0, the two sequences

$$(h_0(w_0), h_0(w_0\zeta_r), ..., h_0(w_0\zeta_r^{r-1})) \quad \text{and} \quad (f_1(z_0), ..., f_r(z_0))$$

coincide. This implies that

$$(n(h_0), r) = 1 \qquad (\text{see } (1.8)),$$

and that

$$(h_0)_{\zeta_r^k} = h_k, \qquad 0 \leqslant k \leqslant r-1,$$

renumbering the h's if necessary. This sketch, details of which can be found in [6] and elsewhere, suggests the following old-fashioned approach, which is my method. Sheaf theory can wait.

Definition. Let $F(Y, Z)$ be given by (1.10) and (1.11). A Puiseux element for F at a is a triple (a, r, f), where

$$a \in C, \qquad r \in N, \qquad f \in C((t)),$$

and

(i) $F_0(a) \neq 0$,

(ii) $F(f, a + t^r) = 0$ in $C((t))$,

(iii) $n(f)$ and r are relatively prime.

It is of great importance that we use $C((t))$ (or, at the very least, $C[[t]]$), and not $C_0\langle t\rangle$, in the definition of a Puiseux element, for otherwise, our automorphisms of C would be dead on the spot. However, the great thing is the following

Lemma 1.1. *Let $F(Y, Z)$ be given by* (1.10) *and* (1.11). *If $a \in C$ and $F_0(a) \neq 0$, then there are precisely N Puiseux elements for F at a. If they are $(a, r_1, f_1), ..., (a, r_N, f_N)$, then $f_i \in C_0\langle t\rangle$, for $i = 1, ..., N$.*

I have no reference for the result, and have obviously been reading the wrong stuff. Siegel gives a hint [6, p. 137]. In any case, the proof is not difficult, so even though the result is vital, I merely sketch a proof. Without difficulty, we can find N Puiseux elements for F at a, say (a, r_ν, f_ν), such that $f_\nu \in C_0\langle t\rangle$, and it is obvious that $\{f_1, ..., f_N\}$ has cardinal N, since if $i \neq j$, then f_i and f_j disagree at every point of $\dot{D}_a(0)$ for a suitably small positive a. Suppose there were an $(N+1)$st culprit (a, r, f). Let

$$R = \text{l.c.m.}\{r_1, ..., r_N, r\},$$

$$R = r_\nu d_\nu = rd.$$

We then set $\tilde{f}_\nu = f_\nu^{(d_\nu)}$, $\tilde{f} = f^{(d)}$, and set $\tilde{F}(Y, Z) = F(Y, a + Z^R)$. By construction, we have $\tilde{F}(\tilde{f}_\nu, t) = F(\tilde{f}_\nu, a + t^R) = F(f_\nu^{(d_\nu)}, a + (t^{r_\nu})^{(d_\nu)}) = F(f_\nu, a + t^{r_\nu}) \circ E_{d_\nu} = 0$, and it is fairly obvious that $(r_\nu, n(f_\nu)) = 1$ forces $\{\tilde{f}_1, ..., \tilde{f}_N\}$ to have cardinal N. Since $F(\tilde{f}, a + t^R) = 0$, and since $F(Y, a + t^R)$ has N distinct roots in $C((t))$ which already happen to lie in $C_0\langle t\rangle$, and since $C((t))$ is a field, we get $f^{(d)} = f_\nu^{(d_\nu)}$ for some ν. This puts f in $C_0\langle t\rangle$, after all, and forcing $f = f_\nu$, $r = r_\nu$, is arithmetic.

From the preceding discussion, it should be clear that the integers r_ν of Lemma 1.1 have a topological meaning, and I aim to exploit this fact.

If $F \in C[Y, Z]$, let

$$\mathscr{P}(a, F) = \text{the set of Puiseux elements for } F \text{ at } a. \tag{1.12}$$

2. Preliminaries about Discrete Groups and More Function Theory

Again, I need to recapitulate the notation and terminology, some of which has been used in the preceding section without explanation.

If G is a group and S is a set on which G acts, $G\backslash S$ denotes the set of orbits of G on S. The group $PSL_2(C)$ acts on the Riemann sphere $C \cup \{\infty\} = C^*$ via

$$(\gamma, z) \mapsto \gamma(z) = \frac{az+b}{cz+d}, \qquad \gamma = \begin{pmatrix} a & b \\ c & d \end{pmatrix}^{\cdot} \in PSL_2(C). \tag{2.1}$$

The symbol $\cdot$ indicates the image in $PSL_2(C)$ of an element of $SL(2, C)$ and will often be omitted.

As is well known, $PSL_2(R)$ has 3 orbits on C^*, which are

$$R \cup \{\infty\}, \qquad \mathfrak{H} = \{x + iy \mid x, y \in R, y > 0\}, \qquad \bar{\mathfrak{H}}.$$

If S is a set, then $PSL_2(R)$ acts on the set of all functions from C^* to S via

$$(f, \gamma) \to (f \circ \gamma)(z) \underset{\text{def}}{=} f(\gamma(z)). \tag{2.2}$$

From (2.1), (2.2), we deduce the usual (but important) rules:

$$\begin{aligned} &(\alpha\beta)(\tau) = \alpha(\beta(\tau)), \qquad (f \circ \alpha) \circ \beta = f \circ (\alpha\beta), \\ &\alpha, \beta \in PSL_2(R), \qquad \tau \in C^*, \qquad f \in S^{C^*}. \end{aligned} \tag{2.3}$$

We check that

$$\begin{aligned} &\text{if } c \in C^*, \alpha, \beta \in PSL_2(R) \text{ and } \alpha(c) = c, \text{ then} \\ &(\beta^{-1}\alpha\beta)(\beta^{-1}(c)) = \beta^{-1}(c); \\ &\text{if } f \in S^{C^*} \text{ and } f \circ \alpha = f, \text{ then } (f \circ \beta)\, \beta^{-1}\alpha\beta = f \circ \beta. \end{aligned} \tag{2.4}$$

The rules (2.3) are too valuable to be altered, but they imply that fixed point sets and conjugation are related as in (2.4).

If Γ acts on $\mathfrak{H}$ discontinuously (for $\Gamma \subseteq PSL_2(R)$) and if in addition Γ has a fundamental domain whose hyperbolic area is finite, this area is denoted by $v(\Gamma)$, and Γ is said to be a Fuchsian group of the first kind.

If $c \in \mathfrak{H} \cup R \cup \{\infty\}$, let

$$\Gamma_c = \{\gamma \in \Gamma \mid \gamma(c) = c\},$$

and let

$$\mathrm{Uni} = \{\gamma \in PSL_2(R) \mid \gamma \text{ is unipotent}\}.$$

In order to compactify the orbit space $\Gamma \backslash \mathfrak{H}$, we adjoin to $\mathfrak{H}$ the set

$$Cp_\Gamma = Cp \underset{\text{def}}{=} \{c \in R \cup \{\infty\}, \Gamma_c \cap \mathrm{Uni} \neq \{1\}\}. \tag{2.5}$$

The numbers of orbits of Γ on Cp is finite, is called the cusp number, and is denoted by $m(\Gamma)$. An orbit of Γ on Cp is called a cusp, and sometimes by abuse of language, an element of Cp is called a cusp. We set

$$\mathfrak{H}^* = \mathfrak{H} \cup Cp.$$

The topology on $\mathfrak{H}^*$ which is relevant here is given by a basis of open sets: at ∞ we take

$$N_a = \{\infty\} \cup \{\tau \in \mathfrak{G} \mid \mathrm{Im}(\tau) > a\},$$

and $\dot{N}_a$ is the "punctured" neighbourhood. For $x \in Cp \cap R$,

$$N_{x,r} = \{x\} \cup \text{interior of a circle of radius } r \text{ whose centre is at } x + ir,$$

and $\dot{N}_{x,r}$ is the "punctured" neighbourhood. For $\tau \in \mathfrak{H}$, the usual neighbourhoods suffice: $D_a(\tau)$ with $a < \mathrm{Im}(\tau)$.

Given this topology on $\mathfrak{H}^*$, $\Gamma\backslash\mathfrak{H}^*$ is endowed with the quotient topology. It is a fact of life that $\Gamma\backslash\mathfrak{H}^*$ is compact, which looks preposterous at first glance.

For $c \in Cp$, Γ_c is an infinite cyclic group of unipotent elements, and so Cp depends only on the commensurability class of Γ. Moreover, Γ_c has two generators u_c, u_c^{-1}, and these generators are not conjugate in $PSL_2(R)$. In the Fuchsian groups of the first kind considered in this paper, $\infty \in Cp$, and so Γ_∞ has a unique generator $u_\infty(\lambda) = (\begin{smallmatrix} 1 & \lambda \\ 0 & 1 \end{smallmatrix})$, with $\lambda > 0$. I choose notation so that u_c and $u_\infty(\lambda)$ are conjugate in $PSL_2(R)$. The element $u_\infty(1)$ is so important that I call it

$$u = u_\infty(1) = \begin{pmatrix} 1 & 1 \\ 0 & 1 \end{pmatrix}. \tag{2.6}$$

Let

$$El = El_\Gamma \underset{\text{def}}{=} \{c \in \mathfrak{H} \mid \Gamma_c \neq 1\}.$$

Then El is a union of finitely many orbits of Γ, and for $c \in El$, Γ_c is a finite cyclic group of order $e(c)$. By convention, and because it is reasonable, we put $e(c) = 1$, if $c \in \mathfrak{H}$, $c \in El$.

The orbit of $c \in \mathfrak{H}^*$ under Γ is denoted by $\Gamma(c)$, and for $c \in \mathfrak{H}$, we set $e(\Gamma(c)) = e(c)$.

Letting $g(\Gamma)$ denote the genus of $\Gamma\backslash\mathfrak{H}^*$ [6] and setting

$$\sigma(\Gamma) = 2(g(\Gamma) - 1) + m(\Gamma) + \sum_{c \in \Gamma\backslash El} (1 - e(c)^{-1}), \tag{2.7}$$

we have the wonderful fact that

$$2\pi\sigma(\Gamma) = v(\Gamma). \tag{2.8}$$

Much easier is

$$v(\Delta) = |\Gamma : \Delta| v(\Gamma) \qquad \text{if} \quad \Delta \subseteq \Gamma \text{ and } |\Gamma : \Delta| < \infty. \tag{2.9}$$

The field of meromorphic functions on $\Gamma\backslash\mathfrak{H}^*$ is denoted by K, or by K_Γ, and C is identified with the constant functions. I aim to avoid using the Riemann–Roch theorem, not because of aversion, but because in this paper, cruder arguments suffice. Each element of $K^\times$ has only finitely many zeroes, so has only finitely many poles, and the only elements of K which have no zeroes are the non-zero constant functions, that is, $C^\times$ [5].

3. The Groups Δ

I must deal with infinitely many such, but they will all be normal subgroups (and very restricted ones, too) of a group Γ such that the conditions (C) which follow, all hold. It is a tight squeeze, but such a Γ exists.

(Ci) Γ is a Fuchsian group of the first kind.

(Cii) Γ is torsion free.

(Ciii) $g(\Gamma)=0$.

(Civ) $m(\Gamma)\geqslant 6$, and $\Gamma_\infty \neq 1$.

(Cv) A local variable at infinity is $q=\exp 2\pi i\tau$.

(Cvi) If T is the unique generator for K whose pole is at ∞, and which in addition has no constant term, such that

$$T=q^{-1}+\sum_{j=1} c_j(T)\,q^j,$$

then

$$c_j(T)\in Q \qquad \text{for all } j.$$

(Cvii) If $m=m(\Gamma)$, and $Cp=\Gamma(\infty)\cup\Gamma(c_1')\cup\cdots\cup\Gamma(c_{m-1}')$, then $T(c_i')\in Q$ for $i=1,\dots,m-1$.

These conditions enable us to play a nice game. Define

$$\mathrm{Ram}_\Gamma=\mathrm{Ram}=\{T(c_i')\mid 1\leqslant i\leqslant m-1\}. \tag{3.1}$$

Thus, Ram is a subset of Q of cardinal $m-1\geqslant 5$.

LEMMA 3.1. *$\Gamma_0(12)$ satisfies (C). For this Γ we have*

(i) $m=6$,

(ii) $T=\eta_4^4\eta_6^2/\eta_2^2\eta_{12}^4$, *where*

$$\eta_d(\tau)=\eta(d\tau),$$

and

$$\eta(\tau)=e^{\pi i\tau/12}\prod_{n=1}^{\infty}(1-q^n),$$

(iii) $\mathrm{Ram}_{\Gamma_0(12)}=\{0,1,-1,3,-3\}$.

For a proof of this lemma, I refer the reader to [4, Chapter 3]. However, I remark that the normalizer of $\Gamma_0(12)$ in $PSL_2(R)$ is $h^{-1}\Gamma_0(3)^+h$, $h=\left(\begin{smallmatrix}2&0\\0&1\end{smallmatrix}\right)$, and that $\Gamma_0(3)^+$ is the free product of two cyclic groups of orders 2 and 6, respectively, while $\Gamma_0(3)^{+h}/\Gamma_0(12)$ is a dihedral group of order 12. This is part of the algebraic background for the wonderful rationality properties of $\Gamma_0(12)$.

For the remainder of this paper, Γ denotes a group satisfying (C).

For uniformity of notation, set $u_0 = u$ (see (2.6)), $c_0 = \infty$, and choose $c_i \in \Gamma(c_i')$ such that

$$\Gamma_{c_i} = \langle u_i \rangle,$$

and

$$u_0 \cdot u_1 \cdots u_{m-1} = 1. \tag{3.2}$$

Such a choice is always possible. As an example, we have for $\Gamma_0(12)$,

$$\begin{pmatrix} 1 & 1 \\ 0 & 1 \end{pmatrix}\begin{pmatrix} 1 & 0 \\ -12 & 1 \end{pmatrix}\begin{pmatrix} -5 & 1 \\ -36 & 7 \end{pmatrix}\begin{pmatrix} -11 & 3 \\ -48 & 13 \end{pmatrix}\begin{pmatrix} -11 & 4 \\ -36 & 13 \end{pmatrix}\begin{pmatrix} -5 & 3 \\ -12 & 7 \end{pmatrix} = 1,$$

and the respective points of Cp fixed by these elements are, in order, ∞, 0, 1/6, 1/4, 1/3, 1/2. Relation (3.2) defines Γ by generators and relations, and so Γ is a free group of rank $m-1$.

For each $i = 0, \ldots, m-1$, let

$$P_i = \{\gamma \in \Gamma \mid \gamma \neq 1, \gamma \text{ is conjugate to a power of } u_i\}. \tag{3.3}$$

(The mnemonics of this notation stem from: u = unipotent; P = parabolic.) Set

$$P = P_0 \cup P_1 \cup \cdots \cup P_{m-1}. \tag{3.4}$$

If $d_0, d_1, \ldots, d_{m-1}$ are natural numbers, let

$$\Pi(d_0, d_1, \ldots, d_{m-1}) = \Pi = \langle u_0^{d_0}, \ldots, u_{m-1}^{d_{m-1}} \rangle^{\Gamma} \tag{3.5}$$

be the smallest normal subgroup of Γ which contains

$$\{u_0^{d_0}, \ldots, u_{m-1}^{d_{m-1}}\}.$$

Let

$$\bar{\Gamma} = \Gamma/\Pi, \tag{3.6}$$

$$\bar{u}_i = \Pi u_i, \tag{3.7}$$

$$\bar{U} = \{\bar{u}_0, \ldots, \bar{u}_{m-1}\}, \tag{3.8}$$

$$\bar{X} = X\Pi/\Pi \qquad (X \text{ any subset of } \Gamma). \tag{3.9}$$

The bar convention is not entirely satisfactory, since it depends on the integers $d_0, d_1, \ldots, d_{m-1}$, but in this paper, no ambiguity should arise.

Since Γ is a free group and $\{u_1,..., u_{m-1}\}$ is a free basis, we get

If $\phi : \bar{U} \to G$, where G is a group, then ϕ extends to a homomorphism of $\bar{\Gamma}$ into G if and only if

$$\text{(i)} \quad (\bar{u}_i \circ \phi)^{d_i} = 1, \qquad i = 0,..., m-1,$$
$$\text{(ii)} \quad \bar{u}_0 \circ \phi \cdot \bar{u}_1 \circ \phi \cdots \bar{u}_{m-1} \circ \phi = 1. \tag{3.10}$$

If $\mathscr{F}$ is a family of groups, a homomorphism ϕ of $\bar{\Gamma}$ into a group X is said to be $\mathscr{F}$-admissible if and only if $\bar{\Gamma} \circ \phi$ is isomorphic to an element of $\mathscr{F}$, and we call ker ϕ an $\mathscr{F}$-admissible subgroup of $\bar{\Gamma}$. The residual core of $\mathscr{F}$ in $\bar{\Gamma}$ is by definition

$$\bigcap \bar{\Gamma}_0,$$

where $\bar{\Gamma}_0$ ranges over all the $\mathscr{F}$-admissible subgroups of $\bar{\Gamma}$. This intersection is a subgroup of $\bar{\Gamma}$, so is of the form Δ/Π for a unique subgroup Δ of Γ which contains Π, and we write

$$\Delta = \Delta(\mathscr{F}). \tag{3.11}$$

It is certain of these groups which I wish to study.

Let

$$\begin{aligned}\mathscr{E} = \mathscr{E}(d_0,..., d_{m-1}) = \{F \mid F &\text{ is a finite group} \neq 1,\\ F = \langle f_0,...,f_{m-1}\rangle, &\ f_i \text{ has order } d_i,\\ \text{and } f_0 \cdot \cdots \cdot f_{m-1} &= 1\}.\end{aligned} \tag{3.12}$$

I do not wish to consider the case $d_0 = d_1 = \cdots = d_{m-1} = 1$, which is why I restrict myself to $F \neq 1$.

Let

$$\begin{aligned}\mathscr{D} = \mathscr{D}_\Gamma = \mathscr{D}_\Gamma(d_0,..., d_{m-1})\\ = \{\Delta \subseteq \Gamma \mid \text{there is a non-empty finite subset } \mathscr{F} \text{ of}\\ \mathscr{E}(d_0,..., d_{m-1}) \text{ such that } \Delta = \Delta(\mathscr{F})\}.\end{aligned} \tag{3.13}$$

If $\Delta = \Delta(\mathscr{F})$, let

$$n(\mathscr{F}) = \max_{F \in \mathscr{F}} |F|.$$

Since $\bar{\Gamma}$ is finitely generated, it contains only finitely many subgroups of index $\leqslant n(\mathscr{F})$, and the index in $\bar{\Gamma}$ of every $\mathscr{F}$-admissible subgroup is at most $n(\mathscr{F})$. From this, we get

$$\Gamma/\Delta = G \text{ is a finite group.} \tag{3.14}$$

This notation is fixed, and so is

$$|G| = N. \tag{3.15}$$

In addition, I shall try to write elements of G as g, g', g_1, etc. with corresponding elements of Γ as γ, γ', γ_1, so that

$$g = \Delta\gamma, \qquad g' = \Delta\gamma', \ldots$$

The index in Γ of Δ tends to be ineffably large, but nonetheless it is the fields $K_\Delta = K$ which I propose to study.

LEMMA 3.2. *Suppose X is a finitely generated group and $\mathscr{F}$ is a finite set of finite groups. Set*

$$Y = \bigcap H,$$

where H ranges over the set $\mathscr{H}$ of normal subgroups of X such that X/H is isomorphic to an element of $\mathscr{F}$. If Z is a normal subgroup of X and ψ is an isomorphism from X/Y to X/Z, then $Y = Z$.

Proof. I provide a proof, since although one possibly exists in the literature, I have not located it. If X/Z has order 1, so does X/Y, and so $Y = X = Z$. We may assume that $X/Z \neq 1$. For each x in X with $x \notin Z$, let $\psi^{-1}(Zx) = Yw$. Then $Yw \neq Y$, as ψ^{-1} is an isomorphism. By definition of Y, there is an H in $\mathscr{H}$ such that $w \notin H$. Since H is one of the groups appearing in the intersection which defines Y, we have $H \supseteq Y$. Thus, $\psi(H/Y)$ is defined and is of the shape K/Z for some subgroup K of X which contains Z. Since $w \notin H$, so also $Yw \notin H/Y$, owing to $Y \subseteq H$. So $\psi(Yw) \notin K/Z$. Hence, we get $Yw = \psi^{-1}(Zx)$, $\psi(Yw) \notin K/Z$, and $\psi(Yw) = \psi(\psi^{-1}(Zx)) = Zx$. Hence, $x \notin K$. On the other hand, there is F in $\mathscr{F}$ such that $F \cong X/H$, and we find that

$$F \cong X/H \cong X/Y/H/Y \overset{\psi}{\cong} X/Z/K/Z \cong X/K.$$

Thus, $X/K \cong F$, so $K \in \mathscr{H}$, so $K \supseteq Y$. Set

$$Z^* = \bigcap K,$$

where the intersection is extended over those K in $\mathscr{H}$ such that $Z \subseteq K$. By what we have just shown, $x \notin Z^*$. Since x is an arbitrary element of X which is not contained in Z, we conclude that $Z^* = Z$. Since each K in the above intersection contains Y, we get $Z \supseteq Y$. Since $|X : Z| = |X : Y|$, we have $Z = Y$.

Lemma 3.3. *Suppose Δ, Δ_0 are normal subgroups of Γ, $\Delta \supseteq \Pi$, $\Delta_0 \supseteq \Pi$, and $\Delta \in \mathscr{D}$. If $\Gamma/\Delta \cong \Gamma/\Delta_0$, then $\Delta = \Delta_0$.*

Proof. We have $\bar{\Gamma}/\bar{\Delta} \cong \Gamma/\Delta \cong \Gamma/\Delta_0 \cong \bar{\Gamma}/\bar{\Delta}_0$, so we apply the preceding lemma with $\bar{\Gamma}$ in the role of X, $\bar{\Delta}$ in the role of Y, $\bar{\Delta}_0$ in the role of Z, and $\mathscr{F}$ the finite set of finite groups in $\mathscr{E}$ which define $\Delta = \Delta(\mathscr{F})$. The preceding lemma gives $\bar{\Delta} = \bar{\Delta}_0$. Since $\Delta \supseteq \Pi$ and $\Delta_0 \supseteq \Pi$, we have $\Delta = \Delta_0$.

It might be just as well to say a word about the logic of the above situation. The first order of priority is the choice of the integers $d_0, \ldots, d_{m-1}$. Once these are given, we can look for groups F which are generated by elements $f_0, \ldots, f_{m-1}$ whose precise orders are $d_0, \ldots, d_{m-1}$, respectively, and such that the product of the f_i's is 1. We choose our groups $F \in \mathscr{F}$ from among such finite groups. We are then guaranteed that there are homomorphisms of $\bar{\Gamma}$ onto F such that the image of each $\bar{u}_i$ has order precisely d_i. However, we allow the possibility, as we must, that there are other homomorphisms of $\bar{\Gamma}$ onto F such that the image of some (or many) $\bar{u}_i$ has an order other than d_i. But since $\bar{u}_i$ has order precisely d_i in $\bar{\Gamma}$, we know that the order of the image of $\bar{u}_i$ under any homomorphism of $\bar{\Gamma}$ anywhere is a divisor of d_i. This is the best we can hope for, and still get Lemma 3.3.

We fix some more notation by setting

$$U_i = \Delta u_i, \qquad 0 \leqslant i \leqslant m-1. \tag{3.16}$$

By what has just been said, we see that U_i has order d_i in G.

Lemma 3.4. *Suppose $\Delta \subseteq \Lambda \subseteq \Gamma$ and $i \in \{0, \ldots, m-1\}$. If $\Lambda \cap P_i = \Delta \cap P_i$, then Λ has precisely $(1/d_i)\,|\Gamma : \Lambda|$ orbits on $\Gamma(c_i)$.*

Proof. (See (3.3) for the definition of P_i.) Let $\Gamma = \Lambda\gamma_1\langle u_i\rangle \cup \Lambda\gamma_2\langle u_i\rangle \cup \cdots \cup \Lambda\gamma_h\langle u_i\rangle$ be the double coset decomposition of Γ with respect to the pair $(\Lambda, \langle u\rangle)$. Then

$$\Gamma(c_i) = \Lambda\gamma_1(c_i) \cup \cdots \cup \Lambda\gamma_h(c_i).$$

If $\Lambda\gamma(c_i) = \Lambda\gamma'(c_i)$, then $\gamma(c_i) = \lambda\gamma'(c_i)$ for some $\lambda \in \Lambda$, whence $\gamma^{-1}\lambda\gamma' \in \langle u_i\rangle$, so $\gamma' \in \Lambda\gamma\langle u_i\rangle$, and $\Lambda\gamma'\langle u_1\rangle = \Lambda\gamma\langle u_i\rangle$. Thus, the lemma demands that we show that the number of $(\Lambda, \langle u_i\rangle)$ double cosets is $|\Gamma : \Lambda|/d_i$. This in turn is equivalent to showing that $\Lambda\gamma\langle u_i\rangle$ contains precisely d_i cosets of Λ. Since $u_i^{d_i}$ is in the normal subgroup Δ, the number of cosets of Λ in $\Lambda\gamma\langle u_i\rangle$ is a divisor of d_i. If it is a proper divisor d of d_i, we get $\Lambda\gamma = \Lambda\gamma u_i^d$, so that $\gamma u_i^d \gamma^{-1} \in \Lambda$, against our hypothesis that $\Lambda \cap P_i = \Delta \cap P_i$. The proof is complete.

We need one more property of groups. If $F \in \mathscr{E}(d_0, \ldots, d_{m-1})$, let

$$\mathscr{H}_\Gamma(F) = \{\Lambda \mid \Lambda \lhd \Gamma, \Lambda \supseteq \Pi(d_0, \ldots, d_{m-1}), \Gamma/\Lambda \cong F\}. \tag{3.17}$$

It is important to understand the connection between $\mathscr{H}_\Gamma(F)$ and the structure of F. For this purpose, let

$$P_F(d_0,\dots,d_{m-1}) = \{(f_0,\dots,f_{m-1}) \mid f_i \in F, f_i^{d_i}=1,$$
$$f_0 \cdot \cdots \cdot f_{m-1} = 1, F = \langle f_0,\dots,f_{m-1}\rangle\}.$$

Clearly, $A=\mathrm{Aut}(F)$ acts on $P=P_F(d_0,\dots,d_{m-1})$. If $\alpha \in A$, and $(f_0,\dots,f_{m-1}) \in P$, then $(f_0 \circ \alpha,\dots,f_{m-1}\circ\alpha) \in P$. A moment's thought convinces that

$$\mathscr{H}_\Gamma(F) \qquad \text{and} \qquad A\backslash P_F(d_0,\dots,d_{m-1}) \tag{3.18}$$

are in one-one correspondence via

$$\Lambda \to \text{the } A\text{-orbit which contains}$$
$$((\Lambda u_0)\,\phi, (\Lambda u_1)\,\phi,\dots,(\Lambda u_{m-1})\,\phi), \tag{3.19}$$

where ϕ is any isomorphism of Γ/Λ onto F. This assertion follows from the observation that if ϕ_1, ϕ_2 are isomorphisms from Γ/Λ to F, then $\phi_1^{-1}\phi_2 \in \mathrm{Aut}(F)$, so that every isomorphism of Γ/Λ to F is of the shape $\phi \circ \alpha$, where ϕ is a fixed isomorphism, and $\alpha \in \mathrm{Aut}(F)$.

4. The Construction of $F(Y,Z)$ and the Auxiliary Polynomial

Here is the situation:

$$\Gamma \text{ satisfies } (C),$$

$\mathscr{F} = \{F, F',\dots,\}$ is a finite subset of $\mathscr{E}(d_0,\dots,d_{m-1})$,

$$\Delta = \Delta(\mathscr{F}), \qquad G=\Gamma/\Delta, \qquad |G|=N,$$
$$K=K_\Delta = \text{function field of } \Delta\backslash\mathfrak{H}^*. \tag{4.1}$$

Then $K \supseteq C(T)$, so for each x in K, there is an irreducible $P_x(Y,Z) \in C[Y,Z]$ such that

$$P_x(x,T)=0 \quad \text{in} \quad K. \tag{4.2}$$

We normalize notation so that the coefficient of the highest power of Y which appears in $P_x(Y,Z)$ is a monic polynomial in Z.

As already mentioned, I shall follow the convention that group elements of G are $g, g', \ldots$, and that γ, γ' are corresponding elements of Γ, so that $g = \Delta\gamma$, $g' = \Delta\gamma', \ldots$. I also use the notation of Section 3, in particular, (3.1). Set

$$\xi_\nu = T(c_\nu), \qquad 1 \leqslant \nu \leqslant m-1, \tag{4.3}$$

so that $\xi_\nu \in Q$.

We have an action of G on $\Delta \backslash \mathfrak{H}^*$:

$$G \times (\Delta \backslash \mathfrak{H}^*) \to \Delta \backslash \mathfrak{H}^*$$
$$(g, p) \mapsto g \circ p,$$

defined as follows:

There is $x \in \mathfrak{H}^*$ such that $p = \Delta(x)$, and there is $\gamma \in \Gamma$ with $g = \Delta\gamma = \gamma\Delta$. Define

$$g \circ p = \Delta(\gamma(x)).$$

This depends neither on our choice of x' in $\Delta(x)$ nor on our choice of γ' in $\Delta\gamma$, almost from first principles; and we find that

$$(gg') \circ p = g \circ (g' \circ p), \qquad 1 \circ p = p.$$

The covering map

$$\Delta \backslash \mathfrak{H}^* \to \Gamma \backslash \mathfrak{H}^*,$$
$$\Delta(x) \mapsto \Gamma(x)$$

has the following properties:

> Each of the m points $\Gamma(c_i)$ of $\Gamma \backslash \mathfrak{H}^*$ has N/d_i preimages, all other points have N preimages.

Here I am using Lemma 3.4 with $\Delta = \Lambda$.

If we view $\Delta \backslash \mathfrak{H}^*$ as a permutation set for G, then with precisely m exceptions, every orbit is a copy of the regular G-set. In the m exceptional cases, the ith orbit is permutation-isomorphic to the action of G on the space of cosets $g\langle U_i \rangle (g \in G)$, where U_i is given by (3.16).

We have an action of G on K, too:

$$K \times G \to K,$$
$$(f, g) \mapsto f \circ g \qquad (g = \Delta\gamma),$$

where for $p = \Delta(x) \in \Delta \backslash \mathfrak{H}^*$, we have

$$(f \circ g)(p) = f(\gamma(x)).$$

And things are so arranged that this definition depends neither on x' in $\Delta(x)$, nor on γ' in $\Delta\gamma$, and in addition, it satisfies the axioms for an action:

$$f \circ 1 = f, \qquad (f \circ g) \circ g' = f \circ (gg').$$

This action of G preserves the addition and multiplication in K, so G acts as automorphisms of the field K. The fixed field of G on K is the fixed field of Γ on K, which is, by construction, $C(T)$, and so here I am using the fact that G acts faithfully on $K = K_\Delta$. This is no deep matter, since it is easy to construct enough modular forms for Δ and to find a ratio of two such which lies in K and is moved by every non-identity element of G.

The action of G on K and $\Delta\backslash\mathfrak{H}^*$ are related by

$$(f \circ g)(p) = f(g \circ p), \qquad f \in K, \quad g \in G, \quad p \in \Delta\backslash\mathfrak{H}^*$$

which is just a version of (2.2).

It behooves me now to give some further discussion of the local function-theory. First, I treat $\mathfrak{H}$, which is much easier than Cp.

If $\tau_0 \in \mathfrak{H}$, and $f \in K$, then there is an $a = a(f, \tau_0) > 0$ such that in $\dot{D}_a(\tau_0)$, we have

$$f(\tau) = \sum_{n=h}^{\infty} a_n(f) \cdot (\tau - \tau_0)^n.$$

In particular,

$$T(\tau) - T(\tau_0) = \sum_{n=1}^{\infty} a_n(T - T(\tau_0)) \cdot (\tau - \tau_0)^n,$$

and since Γ is torsion free, while $T - T(\tau_0)$ is a generator for K_Γ, we have $a_1(T - T(\tau_0)) \neq 0$. We use $T - T(\tau_0)$ as a local variable at τ_0, and get, for each f in K,

$$f(\tau) = \sum_{n=h}^{\infty} b_n(f)(T(\tau) - T(\tau_0))^n,$$

with

$$\sum_{n=h}^{\infty} b_n(f)\, t^n \in C\langle t\rangle.$$

It is (1.4) and (1.5) which give the convergence.

This gives us an injection

$$i(\tau_0) : K \to C\langle t\rangle, \qquad T - T(\tau_0) \mapsto t \qquad (\tau_0 \in \mathfrak{H}).$$

Furthermore, if $\delta \in \Delta$, then $f \circ \delta = f$ for all f in K, and so

$$f(\delta(\tau)) = f(\tau) = \sum b_n(f)(T(\tau) - T(\tau_0))^n$$

$$= \sum b_n(f)(T(\delta(\tau)) - T(\delta(\tau_0)))^n,$$

whence the relation $i(\tau_0) = i(\delta(\tau_0))$. This is even better, and it gives us

$$i(\Delta(x)) : K \to C\langle t \rangle, \qquad i(\Delta(x)) = i(x), \qquad x \in \mathfrak{H}. \tag{4.4}$$

Now however, come the cusps.

I start with the single point ∞. We have, in a suitable $\dot{N}_a$,

$$\frac{1}{T} = \sum_{n=1}^{\infty} a_n(T^{-1})\, q^n, \qquad q = \exp 2\pi i \tau.$$

However, u_0 has order d_0 modulo Δ, and so the standard local variable at ∞ for Δ is $q_0 = \exp(2\pi i \tau / d_0)$. We do not take this however, but we exploit the relation $a_1(T^{-1}) = 1$, and take an element

$$t_\infty = q_0 + a_1 q_0^{1+d_0} + a_2 q_0^{1+2d_0} + \cdots \tag{4.5}$$

such that

$$t_\infty^{d_0} = \frac{1}{T}. \tag{4.5a}$$

Of course, we can and do verify that

$$t + \sum_{n=1}^{\infty} a_n t^{1+nd_0} \in C_0\langle t \rangle.$$

And we observe that

$$t_\infty \circ u_0 = \zeta_{d_0} t_\infty.$$

So we get an injection

$$i(\infty) : K \to C\langle t \rangle \qquad (T^{-1} \mapsto t^{d_0}). \tag{4.6}$$

Now suppose $x \in \Delta(\infty)$, $x \neq \infty$, say $x = \delta(\infty)$. What is $\delta(N_a)$? We check that $\delta(N_a) = N_{x,r}$, or equivalently, $N_a = \delta^{-1}(N_{x,r})$. This suggests that $t_\infty \circ \delta^{-1}$ is the good choice of local variable at x. And it is; we set

$$t_{\delta(\infty)} = t_\infty \circ \delta^{-1}.$$

Is this well-defined? If $\delta' \in \Delta$, and $\delta(\infty) = \delta'(\infty)$, then $\delta^{-1}\delta'(\infty) = \infty$, so $\delta^{-1}\delta' \in \langle u_0^{d_0} \rangle$, $\delta' = \delta u_0^{d_0 l}$, say. Then $t_\infty \circ \delta'^{-1} = t_\infty \circ u_0^{-d_0 l}\delta^{-1}$, and the pieces fit together, since t_∞ is fixed by $u_0^{d_0}$. So $t_{\delta(\infty)}$ is holomorphic in $N_{x,r}$, and we get another injection $i(\delta(\infty)) : K \to C\langle t\rangle$ $(t_{\delta(\infty)} \mapsto t)$. But $f \circ \delta = f$ for all $f \in K$, and we find that $i(\delta(\infty)) = i(\infty)$, so we get a well-defined injection

$$i(\Delta(\infty)) : K \to C\langle t\rangle, \qquad i(\Delta(\infty)) = i(\delta(\infty)), \qquad \delta \in \Delta. \tag{4.7}$$

If $\gamma_1,\dots,\gamma_h$ is a set of representatives for the $(\Delta, \langle u_0\rangle)$ double cosets, we set $t_{\gamma_j(\infty)} = t_\infty \circ \gamma_j^{-1}$, and use this particular choice of a local variable at $\gamma_j(\infty)$ to give us an injection

$$i(\gamma_j(\infty)) : K \to C\langle t\rangle, \qquad (t_{\gamma_j(\infty)} \to t),$$

and define, for $\delta \in \Delta$, $t_{\gamma_j\delta(\infty)} = t_\infty \delta^{-1}\gamma_j^{-1}$, verifying that this is well-defined and that

$$i(\gamma_j(\infty)) = i(\gamma_j\delta(\infty)) \qquad \text{for all} \quad \delta \in \Delta,$$

whence

$$i(\Delta\gamma_j(\infty)) : K \to C\langle t\rangle, \qquad i(\Delta\gamma_j(\infty)) = i(\gamma_j(\infty)),$$

is a well-defined injection of K into $C\langle t\rangle$. (Since $\Delta \lhd \Gamma$, $\gamma_j\Delta = \Delta\gamma_j$.)

Next, suppose $1 \leqslant v \leqslant m-1$. Let's take c_v and bust it. We have, for some $\gamma \in PSL_2(R)$, $\gamma u_v \gamma^{-1} = \left(\begin{smallmatrix}1 & \lambda\\ 0 & 1\end{smallmatrix}\right)$ $(\lambda > 0)$, $\gamma^{-1}(\infty) = c_v$, $\gamma^{-1}(N_a) = N_{c_v,r}$. So we are led to take

$$\exp(2\pi i\gamma(\tau)/\lambda) \qquad (\tau \in N_{c_v,r})$$

as local variable at c_v for Γ. But u_v has order d_v modulo Δ, so a local variable at c_v for Δ is $\exp 2\pi i\gamma(\tau)/\lambda d_v$. On the other hand,

$$T(\tau) - \xi_v = \sum_{n=1}^{\infty} a_{n,v}(T)\exp(2\pi i n\gamma(\tau)/\lambda) \qquad (\tau \in N_{c_v,r}),$$

with $a_{1,v}(T) \neq 0$. So there is

$$\sum_{n=1}^{\infty} b_{n,v}t^n \in C_0\langle t\rangle$$

such that

$$T(\tau) - \xi_v = \left\{\sum_{n=1}^{\infty} b_{n,v}\exp(2\pi i n\gamma(\tau)/\lambda d_v)\right\}^{d_v} \qquad (\tau \in N_{c_v,r}),$$

and we get a local variable at c_v such that

$$(t_{c_v}(\tau))^{d_v} = T(\tau) - \xi_v \qquad (\tau \in N_{c_v,r}).$$

The remaining points in $\Gamma(c_v)$ fall into line, and we get a well-defined injection

$$i(\Delta\gamma(c_v)) : K \to C\langle t\rangle \qquad (i(\Delta\gamma(c_v)) = i(\gamma(c_v))), \tag{4.8}$$

where, for each ν, the elements γ which appear in (4.8) form a set of representatives for the $(\Delta, \langle u_\nu \rangle)$ double cosets in Γ. The stabilizer in G of $\Delta(c_\nu)$ is $\langle U_\nu \rangle$, and we check that

$$t_{c_\nu} \circ U_\nu = \zeta_{d_\nu} t_{c_\nu} \qquad (U_\nu = \Delta u_\nu,\ \nu = 0,\ldots, m-1).$$

This gives us the commutative diagram

$$\begin{array}{ccc} K & \xrightarrow{i(\Delta(c_\nu))} & C\langle t \rangle \\ \big\downarrow{\scriptstyle \circ U_\nu} & & \big\downarrow{\scriptstyle \circ h(\xi_{d_\nu})} \\ K & \xrightarrow{i(\Delta(c_\nu))} & C\langle t \rangle \end{array}. \tag{4.9}$$

Note that since $T \circ \gamma = T$ for all $\gamma \in \Gamma$, we get

$$t_x \circ \gamma = \zeta t_{\gamma^{-1}x} \qquad \text{for all} \quad x \in \mathfrak{H}^*,\ \gamma \in \Gamma.$$

Here ζ is a d_νth root of unity if $x \in \Gamma(c_\nu)$, and $\zeta = 1$ if $x \in \mathfrak{H}$. I recapitulate:

$$\begin{aligned} &\text{If} \quad x \in \mathfrak{H}, \quad t_x = T - T(x); \\ &\text{if} \quad x \in \Gamma(\infty), \quad \text{then} \quad t_x^{d_0} = 1/T; \\ &\text{if} \quad x \in \Gamma(c_\nu) \quad \text{and} \quad 1 \leqslant \nu \leqslant m-1, \quad \text{then} \quad t_x^{d_\nu} = T - \xi_\nu. \end{aligned} \tag{4.10}$$

Moreover, $i(x) = i(\Delta(x)) : K \to C\langle t \rangle$ $(t_x \mapsto t)$ is a well-defined injection.

We can consolidate this information, and it is worthwhile to do so. Define

$$\begin{aligned} &\mathrm{ram} : \mathfrak{H}^* \to \{1, d_0,\ldots, d_{m-1}\}, \\ &\mathrm{ram}(a) = 1 \qquad \text{if} \quad a \in \mathfrak{H}, \\ &\mathrm{ram}(a) = d_\nu \qquad \text{if} \quad a \in \Gamma(c_\nu) \qquad (c_0 = \infty). \end{aligned}$$

Then

$$\begin{aligned} t_x^{\mathrm{ram}(x)} &= T - T(x), \qquad x \notin \Gamma(\infty), \\ &= 1/T, \qquad x \in \Gamma(\infty). \end{aligned} \tag{4.11}$$

We are almost ready to begin. But since I wish to make use of $\mathscr{G}$, I move everything to $C((t))$ by letting $i_x = i_{\Delta(x)}$ be defined to be the relevant composition:

$$K \xrightarrow{i(x)} C\langle t \rangle \hookrightarrow C((t)), \qquad i_x : K \to C((t)).$$

So as a subscript, x carries us all the way to $C((t))$, but within parentheses x stops at $C\langle t\rangle$. We could get by with something smaller than $C((t))$, namely, with the subfield of $C((t))$ consisting of those elements which are algebraic over $C[t]$, the usual polynomial ring. But this definition of i_x is acceptable, too. I have been overly explicit, especially for those who have been over this territory before, but with these matters out of the way, let the construction begin.

Let

$$\text{Int} = \{f \in K \mid \text{if } x \in \mathfrak{H}^* \text{ and } x \notin \Gamma(\infty), \text{ then } f \text{ is regular at } x\}. \tag{4.12}$$

LEMMA 4.1. Int *is stable under* G.

Proof. Pick $f \in \text{Int}$, $g \in G$, $p = \Delta(x) \in \Delta\backslash\mathfrak{H}^*$, $\Delta(x) \not\subseteq \Gamma(\infty)$. Then f is regular at $g \circ p$, so $f \circ g$ is regular at p. That does it.

LEMMA 4.2. *For each* $f \in K$, *there is* $p_f(T) \in C[T]$ *such that* $p_f(T) \neq 0$, *and such that* $p_f(T) \cdot f \in \text{Int}$.

Proof. Let $x_1, \ldots, x_h \in \mathfrak{H}^*$ such that the poles of f on $\Delta\backslash\mathfrak{H}^*$ are at $\Delta(x_1), \ldots, \Delta(x_h)$. Let t_i be the local variable at x_i, and set $r_i = v_{t_i}(f)$. Choose notation so that $x_1, \ldots, x_{h_0} \notin \Gamma(\infty)$, $x_{h_0+1}, \ldots, x_h \in \Gamma(\infty)$. For each i, let m_i be the smallest integer such that

$$r_i + (\text{ram } x_i) \cdot m_i \geqslant 0.$$

Then, with considerable room to spare,

$$\prod_{i=1}^{h_0} (T - T(x_i))^{m_i} \cdot f \in \text{Int},$$

and the lemma follows.

We use the following notation: If $f \in K$, let $Rt_f = \{f \circ g \mid g \in G\}$, $N_0(f) = \text{card } Rt_f$,

$$P_f(Y, Z) = \sum_{i=0}^{N_0(f)} F_{i,f}(Z) \cdot Y^{N_0(f)-i} \qquad \text{(see (4.2))},$$

$$F_{i,f}(Z) \in C[Z], \qquad F_{0,f}(Z) \text{ monic}.$$

LEMMA 4.3. *If* $f \in \text{Int}$, *then* $F_{0,f}(Z) = 1$.

Proof. Since $F_{0,f}(Z)$ is monic, we have

$$F_{0,f}(Z) = \prod_{\alpha \in C} (Z - \alpha)^{n(\alpha)},$$

so it suffices to show that $n(\alpha)=0$ for all α. Let

$$I_\alpha=\{i \mid 0 \leqslant i \leqslant N_0(f), F_{i,f}(\alpha)\neq 0\}.$$

By definition of P_f, we get $I_\alpha \neq \varnothing$. Suppose by way of contradiction that $0 \notin I_\alpha$. Pick $a \in \mathfrak{H}^*$ with $T(a)=\alpha$. Then $a \notin \Gamma(\infty)$ so by Lemma 4.1, each member of Rt_f is regular at a. The ith elementary symmetric function σ_i of Rt_f is

$$(-1)^i F_{i,f}(T)/F_{0,f}(T),$$

which, for $i \in I_\alpha$, is not regular at a. This contradiction completes the proof.

Let $\Gamma(\infty)=\Delta(x_1')\cup \cdots \cup \Delta(x_{h_0}')$, $h_0=|\Gamma:\Delta|/d_0$. Let t_i be the local variable at x_i', let $f \in K$, and let

$$r_i(f)=v_{t_i}(f).$$

Let h be the smallest integer $\geqslant 0$ such that

$$r_i+hd_0 \geqslant 0, \qquad 1 \leqslant i \leqslant h_0.$$

Then for each $f' \in Rt_f$, f'/T^h is regular on $\Gamma(\infty)$. Set

$$P_f^\infty(Y,Z)=Z^{M_f}\cdot P_f(Y/Z^h, 1/Z), \tag{4.13}$$

where M_f is the smallest integer M such that $Z^M P_f(Y/Z^h, 1/Z) \in C[Y,Z]$. Let

$$I_f(\infty)=\{i \mid \text{the coefficient of } Y^{N_0(f)-i} \text{ in } P_f^\infty(Y.\,Z) \text{ does not vanish at } 0\}.$$

LEMMA 4.4. *If*

$$P_f^\infty(Y,Z)=\sum_{i=0}^{N_0(f)} F_{i,f}^\infty(Z)\cdot Y^{N_0(f)-i},$$

then $F_{0,f}^\infty(0)\neq 0$.

Proof. Rt_f/T^h is the set of roots of $P_f^\infty(Y,1/T)=0$ in K, so the proof follows the lines of Lemma 4.3.

Next, let

$$\text{reg}_0=\{f \in K \mid g \in G \text{ and } f \circ g=f \Rightarrow g=1\}. \tag{4.14}$$

These are the primitive elements. Since G is a Galois group, $\text{reg}_0 \neq \varnothing$. Let

$$\text{Reg}=\text{reg}_0 \cap \text{Int}. \tag{4.15}$$

By Lemma 4.2, $\text{Reg} \neq \varnothing$, since for each $p(T) \in C(T)^\times$, $p(T)\cdot \text{reg}_0=\text{reg}_0$.

Pick $f \in \text{Reg}$, and set

$$F(Y, Z) = P_f(Y, Z), \qquad F_\infty(Y, Z) = P_f^\infty(Y, Z). \tag{4.16}$$

Set

$$Rt = \{f \circ g \mid g \in G\},$$

so that card $Rt = N$. Let

$$Rt_h = \{f \circ g/T^h \mid g \in G\}$$

be the set of roots of $F_\infty(Y, 1/T) = 0$ in K.

5. The Puiseux Elements For $F(Y, Z)$ and $F_\infty(Y, Z)$

LEMMA 5.1. (i) *If* $\tau \in \mathfrak{H}$ *and* $\tau_1 \in \Gamma(\tau)$, *then* $\{(T(\tau_1), 1, i_{\tau_1}(f \circ g)) \mid g \in G\}$ *is the set of Puiseux elements for* $F(Y, Z)$ *at* $a = T(\tau)$.

(ii) *If* $1 \leqslant \nu \leqslant m-1$ *and* $x_1 \in \Gamma(c_\nu)$, *then* $\{(T(x_1), d_\nu, i_{x_1}(f \circ g)) \mid g \in G\}$ *is the set of Puiseux elements for* $F(Y, Z)$ *at* ξ_ν.

(iii) *If* $x_1 \in \Gamma(\infty)$, *then* $\{(0, d_0, i_{x_1}(f \circ g/T^h)) \mid g \in G\}$ *is the set of Puiseux elements for* $F_\infty(Y, Z)$ *at* 0.

Proof. (i) Since T is constant on $\Gamma(\tau)$, we have $T(\tau_1) = T(\tau) = a$. Obviously, $(1, n(i_{\tau_1}(f \circ g))) = 1$, and equally obviously, we have N triples, as Rt has cardinal N, and i_{τ_1} is an injection, so it suffices to show that

$$F(i_{\tau_1}(f \circ g), a + t) = 0 \quad \text{in} \quad C((t)).$$

But the left hand side is

$$i_{\tau_1}(F(f \circ g, T)),$$

which is 0.

(ii) is proved in the same way. The condition $(d_\nu, n(i_{x_1}(f \circ g)) = 1$ follows from (4.9), together with $Rt \subseteq \text{Reg}$, and $o(U_\nu) = d_\nu$.

(iii) is left as an exercise.

Set

$$\mathscr{P}(a, F) = \{(a, 1, i_x(f \circ g)) \mid g \in G\}, \qquad T(x) = a \in \text{Tame} = \{T(\tau) \mid \tau \in \mathfrak{H}\}, \tag{5.1}$$

$$\mathscr{P}(\xi_\nu, F) = \{(\xi_\nu, d_\nu, i_{c_\nu}(f \circ g)) \mid g \in G\}, \tag{5.2}$$

$$\mathscr{P}(0, F_\infty) = \{(0, d_0, i_\infty(f \circ g/T^h)) \mid g \in G\}. \tag{5.3}$$

These are full sets of Puiseux elements, each of cardinal N.

6. Recovering Γ and Δ from the Puiseux Elements for $F \circ \rho$ and $F_\infty \circ \rho$, Where $\rho \in \mathscr{G}$

We are about to leave G, temporarily it turns out, so we write $G = \{g_1, ..., g_N\}$, and for $a \in$ Tame, and $T(b) = a$, where b is fixed, we set

$$g_{\mu,a} = i_b(f \circ g_\mu), \qquad 1 \leqslant \mu \leqslant N.$$

Similarly, set

$$g_{\mu,\xi_\nu} = i_{c_\nu}(f \circ g_\mu), \qquad 1 \leqslant \mu \leqslant N, \quad 1 \leqslant \nu \leqslant m-1,$$
$$g_{\mu,\infty} = i_\infty(f \circ g_\mu / T^h), \qquad 1 \leqslant \mu \leqslant N.$$

Let $\rho \in \mathscr{G}$. Then ρ acts on $C[Y,Z]$, via

$$f = \sum a_{ij} Y^i Z^j \to f \circ \rho = \sum (a_{ij} \circ \rho)\, Y^i Z^j,$$

and we have corresponding actions of ρ on $C((t))$ given in (1.6). We apply ρ to (5.1), (5.2), (5.3), obtaining another miracle:

$$\mathscr{P}(a \circ \rho, F \circ \rho) = \{(a \circ \rho, 1, g_{\mu,a} \circ \rho) \mid 1 \leqslant \mu \leqslant N\}, \qquad a \in \text{Tame}, \quad (6.1)$$
$$\mathscr{P}(\xi_\nu, F \circ \rho) = \{(\xi_\nu, d_\nu, g_{\mu,\xi_\nu} \circ \rho) \mid 1 \leqslant \mu \leqslant N\}, \quad (6.2)$$
$$\mathscr{P}(0, F_\infty \circ \rho) = \{(0, d_0, g_{\mu,\infty} \circ \rho) \mid 1 \leqslant \mu \leqslant N\}. \quad (6.3)$$

Of course, (6.1), (6.2), (6.3) are immediate. To obtain (6.1), we simply apply ρ to $F(g_{\mu,a}, a + t) = 0$. It works, and it is utterly transparent that $n(f) = n(f \circ \rho)$ for all non-constant $f \in C((t))$.

I aim to use (6.1), (6.2), (6.3) to reconstruct Γ and Δ. More precisely, I shall prove that

$$(F \circ \rho)(Y, T) = 0 \text{ has } N \text{ roots} \quad \text{in} \quad K = K_\Delta. \quad (6.4)$$

Before I begin, let me recast (6.1) a bit. As a ranges over Tame, so does $a \circ \rho$, since ρ stabilises Tame. So if $b \in$ Tame, there is $a \in$ Tame such that $a \circ \rho = b$, that is, $a = b \circ \rho^{-1}$. Set

$$h_{\mu,b} = g_{\mu,a} \circ \rho = g_{\mu,b \circ \rho^{-1}} \circ \rho, \quad (6.5)$$

and rewrite (6.1) as

$$\mathscr{P}(b, F \circ \rho) = \{(b, 1, h_{\mu,b}) \mid 1 \leqslant \mu \leqslant N\}, \qquad b \in \text{Tame}. \quad (6.6)$$

I emphasize that this is only a relabelling.

Let us begin the reconstruction and the proof of (6.4).

Pick $\tau_0 \in \mathfrak{H}$, and let $b = T(\tau_0)$. Then

$$T(\tau) - b = \sum_{n=1}^{\infty} a_n(T, \tau_0)(\tau - \tau_0)^n, \qquad \tau \in D_{a(\tau_0)}(\tau_0).$$

We also have, therefore, for each $\mu = 1,..., N$, replacing $a(\tau_0)$ by a possibly smaller value, still called $a(\tau_0)$ (recall: $h_{\mu,b} = \sum_{n=0}^{\infty} a_n(\mu, b)\, t^n \in C_0\langle t \rangle$, by Lemma 1.1),

$$\begin{aligned} h_{\mu,b}(T(\tau) - b) &= \sum_{n=0}^{\infty} a_n(\mu, b)(T(\tau) - b)^n \\ &= \sum_{n=0}^{\infty} a_n'(\mu, b)(\tau - \tau_0)^n, \end{aligned}$$

so we set

$$k_{\mu,\tau_0}(\tau) = \sum_{n=0}^{\infty} a_n'(\mu, b)(\tau - \tau_0)^n, \qquad 1 \leqslant \mu \leqslant N,$$

and conclude that for $\tau \in D_{a(\tau_0)}(\tau_0)$, we have

$$(F \circ \rho)(k_{\mu,\tau_0}(\tau), T(\tau)) = 0, \tag{6.7}$$

and k_{μ,τ_0} is analytic in $D_{a(\tau_0)}(\tau_0)$. Visibly, $\{k_{1,\tau_0},..., k_{N,\tau_0}\}$ has cardinal N, since $\{h_{1,b},..., h_{N,b}\}$ has cardinal N. So we set

$$l_{\tau_0} = \{k_{\mu,\tau_0} \mid 1 \leqslant \mu \leqslant N\}, \tag{6.8}$$

and are guaranteed that (6.7) holds in $D_{a(\tau_0)}(\tau_0)$. That is all there is to it.

Let's see what we get from (1.9) and the Monodromy theorem for simply-connected subsets of the complex plane [2].

Suppose $\tau_0, \tau_1 \in \mathfrak{H}$, and

$$D = D_{a(\tau_0)}(\tau_0) \cap D_{a(\tau_1)}(\tau_1) \neq \emptyset.$$

Pick $\mu_0 \in \{1, 2,..., N\}$. I argue that for each $\tau \in D$,

$$\prod_{\mu=1}^{N} (k_{\mu_0,\tau_0}(\tau) - k_{\mu,\tau_1}(\tau)) = 0. \tag{6.9}$$

Namely, $\{k_{\mu,\tau_1}(\tau) \mid 1 \leqslant \mu \leqslant N\}$ exhausts all the roots of the polynomial $(F \circ \rho)(Y, T(\tau)) = 0$, and $k_{\mu_0,\tau_0}(\tau)$ is one of them. By (1.9) we get a permutation $'$ of $1,..., N$ such that

$$k_{\mu,\tau_0}(\tau) = k_{\mu',\tau_1}(\tau) \qquad \text{for} \quad \tau \in D_{a(\tau_0)}(\tau_0) \cap D_{a(\tau_1)}(\tau_1).$$

By the Monodromy theorem, we get functions $k_1,\ldots,k_N$ on $\mathfrak{H}$, holomorphic there, such that

$$(F\circ\rho)(k_\mu(\tau), T(\tau))=0 \qquad \text{for all} \quad \tau\in\mathfrak{H} \text{ and } 1\leqslant\mu\leqslant N.$$

Let

$$\mathscr{L}=\{k_\mu \mid 1\leqslant\mu\leqslant N\}. \tag{6.10}$$

The small l in (6.8) has been graduated to a large one, and τ_0 has vanished, but we still use $\mathscr{L}$ for "local," since the cusps are a barrier. But before tackling Cp (put off the evil day!), let us see what sort of thing we have on $\mathfrak{H}$. I argue that

$$\mathscr{L}\circ\gamma=\mathscr{L} \qquad \text{for all} \quad \gamma\in\Gamma. \tag{6.11}$$

Pick $\tau\in\mathfrak{H}$, and μ_0, $1\leqslant\mu_0\leqslant N$. Then

$$(F\circ\rho)((k_{\mu_0}\circ\gamma)(\tau), T(\tau))=(F\circ\rho)(k_{\mu_0}(\gamma(\tau)), T(\gamma(\tau)))=0,$$

and so

$$\prod_{\mu=1}^{N}(k_{\mu_0}\circ\gamma-k_\mu)=0 \qquad \text{on} \quad \mathfrak{G}.$$

By (1.9), we get (6.11), which is quite a bundle. The relevance of the auxiliary polynomial becomes clearer.

To simplify notation a bit (it is always on the verge of overwhelming us), set

$$p_{\mu,\infty}=g_{\mu,\infty}\circ\rho, \qquad p_{\mu,\nu}=g_{\mu,\xi_\nu}\circ\rho, \qquad 1\leqslant\mu\leqslant N, \quad 1\leqslant\nu\leqslant m-1.$$

Fix ν, $1\leqslant\nu\leqslant m-1$, and choose $x\in\Gamma(c_\nu)$. Let

$$p_{\mu,\nu}=\sum_{n=0}^{\infty}a_n(\mu,\nu)\,t^n \qquad (\in C_0\langle t\rangle!).$$

Then in a suitable neighbourhood $\dot{N}_{x,r}$ of x, set

$$\tilde{p}_{\mu,\nu}(\tau)=\sum_{n=0}^{\infty}a_n(\mu,\nu)\cdot t_x(\tau)^n \qquad (t_x \text{ is the local variable at } x).$$

We get immediately that

$$F\circ\rho(\tilde{p}_{\mu,\nu}(\tau), T(\tau))=0, \qquad \tau\in\dot{N}_{x,r},$$

since $\xi_\nu+t_x(\tau)^{d_\nu}=\xi_\nu+(T(\tau)-\xi_\nu)=T(\tau)$.

For each μ_0, $1 \leqslant N$, I argue that

$$\prod_{\mu=1}^{N} (k_{\mu_0}(\tau) - \tilde{p}_{\mu,\nu}(\tau)) = 0, \qquad \text{for} \quad \tau \in \dot{N}_{x,r}. \tag{6.12}$$

This is obvious, since $\{\tilde{p}_{\mu,\nu}(\tau) \mid 1 \leqslant \mu \leqslant N\}$ is the set of all roots of $F(Y, T(\tau)) = 0$ and $k_{\mu_0}(\tau)$ is one of them. This is (6.12) for each $\mu_0 = 1,..., N$, whence by (1.9), there is a permutation $''$ such that

$$k_\mu(\tau) = \tilde{p}_{\mu'',\nu}(\tau), \qquad \tau \in \dot{N}_{x,r}.$$

The permutation $''$ depends possibly on x $(\in \Gamma(c_\nu))$, but never mind, we still see that k_μ is regular at x, which is enough.

It only remains to treat $\Gamma(\infty)$, and this, believe it or not, is safely left as an exercise; $F_\infty \circ \rho$, (4.13), (4.16) do the trick.

We still retain the notation $\mathscr{L}$ for our functions k_μ, but we now view them as living on $\mathfrak{H}^*$, meromorphic there. Furthermore, $\mathscr{L} \circ \gamma = \mathscr{L}$ for all $\gamma \in \Gamma$, since we already know that $k_\mu \circ \gamma - k_{\mu'}$ vanishes on $\mathfrak{H}$, and as $k_\mu \circ \gamma$ and $k_{\mu'}$ are meromorphic on $\mathfrak{H}^*$, $k_\mu \circ \gamma = k_{\mu'}$.

Let

$$\Delta_0 = \{\gamma \in \Gamma \mid k_\mu \circ \gamma = k_\mu \text{ for all } \mu = 1,..., N\},$$
$$\tilde{K} = C(T, k_1,..., k_N).$$

Then

$$\begin{aligned} \Gamma/\Delta_0 &\cong \text{Gal } \tilde{K}/C(T) \\ &\cong \text{Gal(a splitting field for } (F \circ \rho)(Y, Z) \text{ over } C(Z)), \\ &\cong \text{Gal(a splitting field for } F(Y, Z) \text{ over } C(Z)), \\ &\cong \text{Gal } K_\Delta/C(T), \\ &\cong G \cong \Gamma/\Delta, \end{aligned}$$

so

$$\Gamma/\Delta \cong \Gamma/\Delta_0. \tag{6.13}$$

We next show that $\Delta_0 \supseteq \Pi(d_0,..., d_{m-1})$, that is, we check that

$$u_\nu^{d_\nu} \in \Delta_0, \qquad 0 \leqslant \nu \leqslant m-1. \tag{6.14}$$

For $\nu = 0$, we work in $\dot{N}_a$. We have

$$t_\infty(\tau)^{d_0} = T(\tau)^{-1},$$

and we also have

$$\frac{k_\mu(\tau)}{T(\tau)^h} = \sum_{n=0}^{\infty} a_n(\mu)\, t_\infty(\tau)^n,$$

and by (4.5)

$$t_\infty(\tau) = \sum_{m=0}^{\infty} a_m \exp 2\pi i\tau(1+md_0)/d_0 \qquad (a_0 = 1).$$

By (1.4), this yields

$$\frac{k_\mu(\tau)}{T(\tau)^h} = \sum_{n=0}^{\infty} b_n(\mu) \exp(2\pi i n\tau/d_0) \qquad (\tau \in \dot{N}_a),$$

and so $k_\mu(\tau + d_0) = k_\mu(\tau)$. This is (6.14) for $v = 0$, and there is no problem at the other cusps either, as

$$x \in \Gamma(c_v) \Rightarrow t_x^{d_v} = T - \xi_v.$$

By Lemma 3.3, (6.14) gives us $\Delta = \Delta_0$. This completes the proof of (6.4).

7. Extensions and the Big Group Extension

Let $\rho \in \mathscr{G}$, $f \in \mathrm{Reg}$, and let $k_1, \ldots, k_N$ be the roots of $(P_f \circ \rho)(Y, T) = 0$ which lie in K. We have an action of ρ on $C(T)$ determined by

$$\alpha = \sum c_i T^i \to \sum (c_i \circ \rho)\, T^i = \alpha \circ \rho.$$

Since $f \in \mathrm{Reg}$, we also have

$$K = C(T)[f],$$

so if $\xi \in K$, then

$$\xi = \sum_{i=0}^{N-1} \xi_i f^i, \qquad \xi_i \in C(T).$$

Pick μ, $1 \leqslant \mu \leqslant N$, and define

$$\rho_\mu : K \to K, \qquad \xi \circ \rho_\mu = \sum_{i=0}^{N-1} (\xi_i \circ \rho)\, k_\mu^i.$$

Then ρ_μ is an automorphism of K, since

$$P_f(f, T) = 0 \qquad \text{and} \qquad P_f \circ \rho(k_\mu, T) = 0.$$

Also, $\operatorname{res}\rho_\mu = \rho_\mu|_{C(T)} = \rho$, $1 \leqslant \mu \leqslant N$, so that ρ_μ is an extension of ρ from $C(T)$ to K. Furthermore, since G permutes transitively and regularly $\{k_1,\dots,k_N\}$, we get

$$\{\rho_1,\dots,\rho_N\} = \rho_\mu G \qquad \text{for} \quad 1 \leqslant \mu \leqslant N. \tag{7.1}$$

On the other hand, if $g \in G$, then $\rho_\mu^{-1} g \rho_\mu$ is an automorphism of K which is the identity on $C(T)$, and since G exhausts all such automorphisms, we have

$$\rho_\mu^{-1} G \rho_\mu = G. \tag{7.2}$$

Let $\mathscr{G}^* = \langle G, \rho_1 \mid \rho_1$ is an extension of ρ from $C(T)$ to K, $\rho \in \mathscr{G}.\rangle$. Then for each $\rho \in \mathscr{G}$, $\mathscr{G}^*$ contains all the extensions of ρ to K. As far as I can tell, there is no way in which the extensions of ρ to K can be given globally in a consistent function-theoretic fashion. The best I can achieve is the following lemma.

LEMMA 7.1. *Assume that* $a \in \mathfrak{H}^*$ *and that* $T(a) \in Q \cup \{\infty\}$. *For each* $\rho \in \mathscr{G}$, *there is an extension of* ρ *to an automorphism* ρ_a *of* K *defined as follows: for each* $f \in \operatorname{Reg}$, $f \circ \rho_a$ *is the unique element* k *of* K *such that*

$$i_a(k) = i_a(f) \circ \rho.$$

Then $\mathscr{G}_a = \{\rho_a \mid \rho \in \mathscr{G}\}$ *is a group and* $ex_a : \mathscr{G} \to \mathscr{G}_a$, $\rho \mapsto \rho_a$ *is an isomorphism. There is a split exact sequence*

$$1 \to G \to \mathscr{G}^* \xrightarrow{\operatorname{res}_a} \mathscr{G}_a \to 1$$

where $G \to \mathscr{G}^*$ *is the inclusion map, and* res_a *is the composition of* res *and* ex_a.

Proof. First suppose that $T(a) = q \in Q$, and that $a \in \mathfrak{H}$. Then $(q, 1, i_a(f))$ is a Puiseux element for $P_f(Y, Z)$ at q, and $(q, 1, i_a(f) \circ \rho)$ is a Puiseux element for $P_f \circ \rho(Y, Z)$ at q. By (6.4), there is k in K such that $i_a(k) = i_a(f) \circ \rho$. So ρ_a exists in this case.

Next, suppose $T(a) = q \in Q$, and $a \in Cp$. Then $q = \xi_\nu$ for some ν, $1 \leqslant \nu \leqslant m-1$, and

$$(\xi_\nu, d_\nu, i_a(f)) \text{ is a Puiseux element for } F(Y, Z) \text{ at } \xi_\nu,$$

whence $(\xi_\nu, d_\nu, i_a(f) \circ \rho)$ is a Puiseux element for $F \circ \rho(Y, Z)$ at ξ_ν. Again, (6.4) produces a k in K such that $i_a(k) = i_a(f) \circ \rho$, and ρ_a exists. Finally, if $T(a) = \infty$, then $a \in \Gamma(\infty)$, and we use i_a and $F_\infty(Y, Z)$. So ρ_a exists, and the isomorphism ex_a is obvious.

Since $\{\rho_1,\dots,\rho_N\} = \rho_\mu G$ for each $\mu = 1,\dots, N$, and each $\rho \in \mathscr{G}$, and since $G \lhd \mathscr{G}^*$, it only remains to show that $G \cap \mathscr{G}_a = 1$. If $\rho_a = \alpha \in G \cap \mathscr{G}_a$, then

α is the identity on $C(T)$, whereas if $c \in C$, then $c \circ \alpha = c \circ \rho_a$ is mapped by i_a to $c \circ \rho(\in C(t)))$. Thus, $\rho = id$ on C, so $\rho = 1$, $\rho_a = 1$, and the extension splits.

8. The Action of $\mathscr{G}^*$ on G

Let

$$D = \text{l.c.m.}\{d_0, \dots, d_{m-1}\}.$$

We have an exact sequence

$$1 \to \mathscr{G}_D^* \to \mathscr{G}^* \xrightarrow{\sigma} \text{Gal } Q(\mu_D)/Q = (Z/DZ)^\times \to 1, \tag{8.1}$$

given by

$$\rho^* \circ \sigma = \text{res } \rho^*|_{Q(\mu_D)}, \qquad \mathscr{G}_D^* = \ker \sigma.$$

Lemma 8.1. *Suppose $\rho^* \in \mathscr{G}^*$ and $\rho^* \circ \sigma = \bar{l}$. Then for each $v = 0, \dots, m-1$, there are elements $g(v, \rho^*) \in G$ such that*

$$\rho^{*-1} U_v \rho^* = g(v, \rho^*)^{-1}\, U_v^l\, g(v, \rho^*).$$

Proof. Let $\rho = \text{res } \rho^*$, $P = \rho^* G$. Then $\text{res } \tilde{\rho} = \rho$ for all $\tilde{\rho} \in P$, and so $\tilde{\rho} \circ \sigma = \bar{l}$. For each $a \in \mathfrak{H}^*$ such that $T(a) \in Q \cup \{\infty\}$, $P \cap \mathscr{G}_a = \{\rho_a^*\}$ has cardinal one. Take $a = c_v$, $v = 0, \dots, m-1$ $(c_0 = \infty)$. Then for all $f \in K$, $i_{c_v}(f \circ \rho_a^*) = i_{c_v}(f) \circ \rho$. By (1.7), $\rho^{-1} h(\zeta_{d_v}) \rho = h(\zeta_{d_v}^l)$. By (4.9), we conclude that

$$\rho_a^{*-1} U_v \rho_a^* = U_v^l. \tag{8.2}$$

Let's look at this closely, just to be sure. We can give a proof by diagrams which is utterly transparent. Here is one picture:

$$(a = c_v) \qquad \begin{array}{ccc} K & \xrightarrow{i_a} & C((t)) \\ \downarrow{\scriptstyle \circ \rho_a^{*-1}} & & \downarrow{\scriptstyle \circ \rho^{-1}} \\ K & \xrightarrow{i_a} & C((t)) \\ \downarrow{\scriptstyle \circ U_v} & & \downarrow{\scriptstyle h(\zeta_{d_v})} \\ K & \xrightarrow{i_a} & C((t)) \\ \downarrow{\scriptstyle \circ \rho_a^*} & & \downarrow{\scriptstyle \circ \rho} \\ K & \xrightarrow{i_a} & C((t)). \end{array}$$

The top and bottom rectangles are commutative by Lemma 7.1, the middle rectangle is commutative by (4.9), so the whole shebang is commutative, whence, contracting, and using (1.7),

$$\begin{array}{ccc} K & \xrightarrow{i_a} & C((t)) \\ \downarrow{\scriptstyle \circ \rho_a^{*-1} U_{r'} \rho_a^*} & & \downarrow{\scriptstyle h(\zeta_{d_{r'}}^l)} \\ K & \xrightarrow{i_a} & C((t)) \end{array} \tag{8.3}$$

is commutative. On the other hand, we may choose $l \in \bar{l}$ such that l is a natural number, and we can concatenate l examples of (4.9) on top of one another:

$$\begin{array}{ccc} K & \xrightarrow{i_a} & C((t)) \\ \downarrow{\scriptstyle \circ U_{r'}} & & \downarrow{\scriptstyle \circ h(\zeta_{d_{r'}})} \\ K & \xrightarrow{i_a} & C((t)) \\ \downarrow & & \downarrow \\ & \cdots & \\ \downarrow & & \downarrow \\ K & \xrightarrow{i_a} & C((t)) \\ \downarrow{\scriptstyle \circ U_{r'}} & & \downarrow{\scriptstyle \circ h(\zeta_{d_{r'}})} \\ K & \xrightarrow{i_a} & C((t)) \end{array} \qquad (l \text{ rectangles}).$$

So we get the commutatively of

$$\begin{array}{ccc} K & \xrightarrow{i_a} & C((t)) \\ \downarrow{\scriptstyle U_{r'}^l} & & \downarrow{\scriptstyle h(\zeta_{d_{r'}}^l)} \\ K & \xrightarrow{i_a} & C((t)). \end{array}$$

In this rectangle, the vertical maps are automorphisms, so we can reverse the arrows, obtaining a commutative diagram

$$\begin{array}{ccc} K & \xrightarrow{i_a} & C((t)) \\ \uparrow{\scriptstyle U_{r'}^{-l}} & & \uparrow{\scriptstyle h(\zeta_{d_{r'}}^{-l})} \\ K & \xrightarrow{ia} & C((t)). \end{array}$$

(Or we could choose a natural number $l' \in -\bar{l}$ and concatenate (4.9).) Composing this with (8.3) gives the commutativity of

$$\begin{array}{ccc} K & \xrightarrow{\;i_a\;} & C((t)) \\ \Big\downarrow{\scriptstyle \rho_a^{*-1}U_\nu\rho_a^*\cdot U_\nu^{-l}} & & \Big\downarrow{\scriptstyle id} \\ K & \xrightarrow{\;i_a\;} & C((t)), \end{array}$$

so if we set $g_0 = \rho_a^{*-1}U_\nu \rho_a^* U_\nu^{-l}$, we get the commutativity of

$$\begin{array}{ccc} K & & \\ \Big\downarrow{\scriptstyle \circ g_0} & \searrow^{i_a} & C((t)). \\ & \nearrow_{i_a} & \\ K & & \end{array}$$

This means that for all $f \in K$, $i_a(f \circ g_0 - f) = 0$, whence $f \circ g_0 = f$, $g_0 = id$.

Since $\rho_{c_\nu}^* \in P$, we have $\rho_{c_\nu}^* = \rho^* \cdot g(\nu, \rho^*)^{-1}$ for some $g(\nu, \rho^*) \in G$. Thus

$$\rho_{c_\nu}^{*-1} U_\nu \rho_{c_\nu}^* = U_\nu^l = g(\nu, \rho^*) \cdot \rho^{*-1} U_\nu \rho^* g(\nu, \rho^*)^{-1},$$

and the lemma is proved.

Let me try explain what is going on here function-theoretically as best I can. We can choose a cusp, and then focus on c_ν. (Say $\nu \geqslant 1$.) We can then take our automorphism ρ_{c_ν} such that in a neighbourhood $N_{c_\nu, r}$ of c_ν, the action of ρ_{c_ν} on $f \in K$ is the "obvious" one:

$$\text{if} \quad f(\tau) = \sum a_n(f) \cdot t_{c_\nu}(\tau)^n \qquad (\tau \in \dot{N}_{c_\nu, r}),$$

then

$$(f \circ \rho_{c_\nu})(\tau) = \sum (a_n(f) \circ \rho) \cdot t_{c_\nu}(\tau)^n \qquad (\tau \in \dot{N}_{c_\nu, r'}).$$

We then ask ourselves: what is the behaviour of ρ_{c_ν} on the power series expansion of f in a neighbourhood $N_{c_\mu, s}$ for $\mu \neq \nu$? We cannot conclude that ρ_{c_ν} acts in such a mindless fashion as in (1.6), but only that for some $g = g(\mu, \rho_{c_\nu})$, $\rho_{c_\nu} \cdot g^{-1}$ acts in the mindless way. We do *not* obtain at c_μ the standard action of ρ given by (1.6); we have to contend with a "fudge factor" coming from the element $g = g(\mu, \rho_{c_\nu}) \in G$. The precise way in which G intervenes in this situation seems to me to be a fascinating problem.

9. More Lemmas

In order not to interrupt the flow of argument too much, I interpolate several easy lemmas and remarks here.

Lemma 9.1. *Suppose X is a finite group, F is a non-Abelian finite simple group and $H_1,\dots,H_t$ are distinct normal subgroups of X such that $H_1 \cap H_2 \cap \cdots \cap H_t = 1$ and $X/H_i \cong F$ for all i. Let*

$$H(i) = \bigcap_{j \neq i} H_j.$$

Then $H(i) \cong F$, and $H = H(1) \times H(2) \times \cdots \times H(t) = H_i \times H(i)$, $1 \leqslant i \leqslant t$.

Proof. By induction on t, the case $t = 1$ being trivial. Let $D = H(t)$, and let $\bar{X} = X/D$. Then $\bar{H}_i = H_i/D$ are $t-1$ distinct normal subgroups of X, with $\bar{X}/\bar{H}_i \cong F$, and $\bar{H}_1 \cap \bar{H}_2 \cap \cdots \cap \bar{H}_{t-1} = 1$, so by induction

$$\bar{X} = \overline{H(1;t)} \times \cdots \times \overline{H(t-1;t)} = \bar{H}_i \times \overline{H(i;t)}, \qquad 1 \leqslant i \leqslant t-1 \tag{9.1}$$

where

$$\overline{H(i;t)} = \bigcap_{\substack{j \neq i \\ 1 \leqslant j \leqslant t-1}} \bar{H}_j.$$

Since F is a non Abelian simple group, $\bar{H}_1,\dots,\bar{H}_{t-1}$ are the only maximal normal subgroups of $\bar{X}$.

Suppose $D = 1$. Then H_t, as a maximal normal subgroup of X, is forced to coincide with one of $H_1,\dots,H_{t-1}$. This is false, so

$$D = H(t) \neq 1, \qquad D \cap H_t = 1.$$

Since $D \lhd X$, while X/H_t is simple, we get

$$X = H_t \cdot D = H_t \times D = H_t \times H(t), \qquad \bar{X} \cong H_t. \tag{9.2}$$

By symmetry, $H(i) \cong F$, $1 \leqslant i \leqslant t$, and since (9.1) and (9.2) hold, it follows that $X = X_1 \times \cdots \times X_t$, where $X_i \cong F$, and $\{X_1,\dots,X_t\}$ is the set of all minimal normal subgroups of F. Then for each $i = 1,\dots,t$, $H(i) \in \{X_1,\dots,X_t\}$. Obviously, $H(1), H(2),\dots,H(t)$ are pairwise distinct, since for example, $H(1) \nsubseteq H_1$, $H(2) \subseteq H_1$, $H(3) \subseteq H_1,\dots,H(t) \subseteq H_1$. So $\{X_1,\dots,X_t\} = \{H(1), H(2),\dots,H(t)\}$. Also, by (9.2) and symmetry, $X = H(i) \times H_i$, and the proof is complete.

LEMMA 9.2. *Suppose* $X = F_1 \times \cdots \times F_t$, $F_i \cong F$, *a non-Abelian finite simple group. Suppose Y is a group of automorphisms of X, that* $Y_0 \lhd Y$, *and that Y permutes transitively* $\{F_1, \ldots, F_t\}$.

(i) *If* Y_0 *normalizes* F_1, *then* Y_0 *normalizes every* F_1.

(ii) *If* Y_0 *normalizes* F_1 *and induces only inner automorphisms of* F_1, *then* Y_0 *induces inner automorphisms of X.*

Proof. Choose $y_i \in Y$ such that $F_1 \circ y_i = F_i$. Then $F_i \circ y_i^{-1} Y_0 y_i = F_i$. But $Y_0 \lhd Y$, so (i) follows.

Suppose $\psi : Y_0 \to F_1$ such that $f_1 \circ y_0 = \psi(y_0)^{-1} f_1 \psi(y_0)$ for all $y_0 \in Y_0$, $f_1 \in F_1$. Then

$$\begin{aligned}(f_1 \circ y_i)\, y_i^{-1} y_0 y_i &\\ &= f_1 \circ y_0 y_i = (\psi(y_0)^{-1} f_1 \psi(y_0)) \circ y_i \\ &= (\psi(y_0) \circ y_i)^{-1} \cdot (f_1 \circ y_i) \cdot \psi(y_0) \circ y_i, \qquad \text{with} \quad \psi(y_0) \circ y_i \in F_i.\end{aligned}$$

Thus, $y_i^{-1} y_0 y_i$ induces an inner automorphism of F_i. But $Y_0 \lhd Y$, so Y_0 induces inner automorphisms of F_i for all i, and (ii) follows.

LEMMA 9.3. *Suppose* $F \subseteq K \subseteq L \subseteq M$ *is a tower of fields and* $[L:F] < \infty$, *that* $[K:F] = l$, $[L:K] = n$, *and that* K/F *and* L/K *are normal extensions. Suppose*

$$f(X) = a_0 + a_1 X + \cdots a_n X^n$$

and the following hold:

(i) $F(a_0, a_1, \ldots, a_n) = K$.

(ii) $a_n \in F^{\times}$.

(iii) f *is irreducible in* $K[X]$.

(iv) $\alpha \in L$ *and* $f(\alpha) = 0$.

Assume further that $\sigma_1, \ldots, \sigma_l$ *are automorphisms of M which fix F elementwise and that*

(v) $\{\sigma_{1|K}, \ldots, \sigma_{l|K}\} = \Sigma$

has cardinal l. Assume finally that

(vi) $\alpha \circ \sigma_i = \alpha \cdot k_i \neq 0$, *where* $k_i \in K$, $1 \leqslant i \leqslant l$.

Then L is normal over F, and

$$L = F[\alpha].$$

Proof. Let $\tau_i = \sigma_{i|K}$, so that $\Sigma = \{\tau_1, \ldots, \tau_l\}$ is the group of automorphisms of K which are 1 on F, that is,

$$\Sigma = \mathrm{Gal}\, K/F.$$

By (i) and (ii), we have, for $1 \leqslant i \leqslant l$,

$$f \circ \sigma_i(X) = a_0 \circ \sigma_i + (a_1 \circ \sigma_i) X + \cdots + (a_{n-1} \circ \sigma_i) X^{n-1} + a_n X^n,$$

and the polynomials $f \circ \sigma_1, \ldots, f \circ \sigma_l$ are pairwise distinct. Since $f(\alpha) = 0$, so also $(f \circ \sigma_i)(\alpha \circ \sigma_i) = 0$. On the other hand, we check that

$$f_i(X) = k_i^n f(X/k_i)$$

has the properties

$$f_i(\alpha k_i) = 0, \qquad f_i(X) = a_0 k_i^n + a_1 k_i^{n-1} X + \cdots + a_{n-1} k_i X^{n-1} + a_n X^n.$$

Thus, $f \circ \sigma_i$ and f_i have a root in common, and $\deg(f \circ \sigma_i - f_i) < n$, so

$$f \circ \sigma_i = f_i, \qquad 1 \leqslant i \leqslant l.$$

Let $\alpha = \alpha_1, \alpha_2, \ldots, \alpha_{n'}$ be all the roots of f in L. Since $[L:K] = n = \deg f$, since L is normal over K, and since f has a root in L, we have $n = n'$, and $L = K[\alpha]$. Since $f \circ \sigma_i = f_i$, the roots of $f \circ \sigma_i$ in L are $\alpha_1 k_i, \ldots, \alpha_n k_i$. (I do not assert that $\alpha_r \circ \sigma_i = \alpha_r \cdot k_i$.)

We set

$$g(X) = \prod_{i=1}^{l} f_i(X) = \prod_{i=1}^{l} (f \circ \sigma_i)(X).$$

Then $g(X)$ has degree $ln = [L:F]$, and

$$\{\alpha_j k_i \mid 1 \leqslant j \leqslant n, 1 \leqslant i \leqslant l\}$$

is a set of ln roots of $g(X)$ in L. Since $f_1, \ldots, f_l$ are pairwise relatively prime, and since f is irreducible over K, and since $f_i = f \circ \sigma_i$, it follows that $g(X)$ is irreducible over F. Since $g(\alpha) = 0$, we have $L = F[\alpha]$, and L/F is a normal extension.

Remark. The lemma is messy, but since the relation "F_1 is a normal extension of F_2" is not transitive, in order to *force* transitivity we must put in something, and the above ingredients suffice.

10. Rigidity and Γ/Λ

Before launching into the next, penultimate phase, let me set up some notation and terminology for the action $\mathscr{G}^*$ on G. Let

$$\mathscr{L}(G) = \{H \mid H \subseteq G, H \text{ is } \mathscr{G}^*\text{-invariant}\}.$$

That is, $\mathscr{L}(G)$ is the set of subgroups of G which are normal in $\mathscr{G}^*$. Since $G \subseteq \mathscr{G}^*$, $\mathscr{L}(G)$ is a subset of the set of normal subgroups of G, and $\mathscr{L}(G)$ contains all the characteristic subgroups of G. But $\mathscr{L}(G)$ is tremendously mysterious. In any case, if H, $K \in \mathscr{L}(G)$, and $H \subseteq K$, then there are subgroups Λ, Λ' of Γ, both of which contain Δ, such that $H = \Lambda/\Delta$, $K = \Lambda'/\Delta$, and the induced action of $\mathscr{G}^*$ on K/H gives us an action of $\mathscr{G}^*$ on Λ'/Λ. In particular, when $K = G$, we get an action of $\mathscr{G}^*$ on Γ/Λ, and the exact sequence

$$1 \to \Gamma/\Lambda \xrightarrow{i} \mathscr{G}^*/H \to \mathscr{G}_a \to 1 \tag{10.1}$$

splits. Here the map i is given by: $\Lambda\gamma \circ i = Hg$, $g = \Delta\gamma \in G$. The second map is equally obvious, since $G = \ker \operatorname{res}_a$, where res_a is given in Lemma 7.1.

We are about to meet such a Λ, and it will have the property that Λu_v has order d_v in Γ/Λ. This means, and under our circumstances is equivalent to the assertion, that

$$\Delta\backslash\mathfrak{H}^* \to \Lambda\backslash\mathfrak{H}^*,$$
$$\Delta(x) \mapsto \Lambda(x)$$

is unramified. This in turn implies that for each $x \in \mathfrak{H}^*$, any local variable for Λ at x is also a local variable for Δ at x, a very good thing for bookkeeping.

I now give myself (H) for F and $(C_0, \ldots, C_{k-1})$, where I now only demand that $k \leqslant m$. Of course, $\Gamma_0(12)$ is the only horse in the stable, but there may be others later, so I keep the parameter m free. Anyway, m is easier to handle than 6. The conjugacy classes are numbered from 0 to $k-1$, to conform with the notation already in place.

Choose $(x_0, \ldots, x_{k-1}) \in A = A_F(C_0, \ldots, C_{k-1})$, and for $k \leqslant v \leqslant m-1$, set $x_v = 1$, $C_v = \{1\}$, so that $(x_0, \ldots, x_{m-1}) \in A(C_0, \ldots, C_{m-1})$. Define

$$d_v = o(x_v), \qquad 0 \leqslant v \leqslant m-1,$$
$$D = \text{l.c.m.}\{d_0, \ldots, d_{m-1}\}, \tag{10.2}$$

and use (3.5) to define $\Pi(d_0, \ldots, d_{m-1})$. Take

$$\mathscr{F} = \{F\}, \qquad \Delta = \Delta(\mathscr{F}).$$

By Lemma 9.1, G is the direct product of a certain number of groups, each isomorphic to F, so G is characteristically simple. One big problem is to determine the normal subgroups of G which are $\mathscr{G}^*$-invariant. A yet more subtle problem is to determine those subgroups of G (they will not all be normal in G) which are invariant under $\mathscr{G}_a$, where $a \in \mathfrak{H}^*$ and $T(a) \in Q \cup \{\infty\}$. As I have tried to indicate, there is a mystery here.

Define

$$\phi : \{u_0, \ldots, u_{m-1}\} \to F,$$
$$u_\nu \circ \phi = x_\nu.$$

Since $u_0 \cdot u_1 \cdot \cdots \cdot u_{m-1} = 1$ defines Γ, and since $x_0 \cdots x_{m-1} = 1$, ϕ extends to a homomorphism of Γ into F, also denoted by ϕ. Since $(x_0, \ldots, x_{k-1}) \in A$, the definition of rigidity gives

$$\Gamma \circ \phi = F.$$

Set

$$\Lambda_1 = \ker \phi. \tag{10.3}$$

I plan to use $(u_0, \ldots, u_{m-1})$, ϕ, $(x_0, \ldots, x_{m-1})$ as pivots on which various maps will turn. By construction, $\Lambda_1 \supseteq \Pi$, and so $\Lambda_1 \supseteq \Delta$, and

$$H_1 = \Lambda_1 / \Delta \lhd G. \tag{10.4}$$

For each $\rho^* \in \mathscr{G}^*$, there is a uniquely determined subgroup Λ_{ρ^*} of Γ which contains Δ and satisfies

$$\rho^{*-1} H_1 \rho^* = \Lambda_{\rho^*} / \Delta. \tag{10.5}$$

Set

$$\Lambda = \bigcap_{\rho^* \in \mathscr{G}^*} \Lambda_{\rho^*}, \tag{10.6}$$

so that

$$\Lambda / \Delta = H = \bigcap_{\rho^* \in \mathscr{G}^*} \rho^{*-1} H_1 \rho^* \lhd \mathscr{G}^*. \tag{10.7}$$

By construction and Lemma 9.1, Γ/Λ is the direct product of a certain number of groups, each isomorphic to F. Also, by construction, $\mathscr{G}^*$ permutes transitively the maximal normal subgroups of Γ/Λ, and so permutes transitively the minimal normal subgroups of Γ/Λ, that is, the set of normal subgroups of Γ/Λ which are isomorphic to F.

The next lemma busts open this situation.

LEMMA 10.1. *$\mathscr{G}_D^*$ normalizes H_1 and induces inner automorphisms of Γ/Λ_1, where $\mathscr{G}_D^*$ is given in* (8.1).

Proof. Choose $\rho^* \in \mathscr{G}_D^*$. By Lemma 8.1, together with $\mathscr{G}_D^* = \ker \sigma$, there are elements $g(\nu, \rho^*) \in G$ such that

$$\rho^{*-1} U_\nu \rho^* = g(\nu, \rho^*)^{-1}\, U_\nu\, g(\nu, \rho^*), \qquad 0 \leqslant \nu \leqslant m-1.$$

Since $\ker \phi = \Lambda_1 \supseteq \Delta$, we have a factorization

$$\begin{array}{ccc} \Gamma & \xrightarrow{\ \phi\ } & F. \\ & {\scriptstyle\pi}\searrow \quad \nearrow{\scriptstyle\lambda} & \\ & G & \end{array} \tag{10.8}$$

Then $x_\nu = u_\nu \circ \phi = u_\nu \circ \pi\lambda = U_\nu \circ \lambda$. Choose $\gamma(\nu, \rho^*) \in \Gamma$ with $\gamma(\nu, \rho^*) \circ \pi = g(\nu, \rho^*)$. Set

$$f_\nu = \gamma(\nu, \rho^*) \circ \phi,$$

so that

$$\begin{aligned} \gamma(\nu, \rho^*)^{-1}\, u_\nu \gamma(\nu, \rho^*) \circ \phi &= f_\nu^{-1} x_\nu f_\nu \\ &= g(\nu, \rho^*)^{-1}\, U_\nu\, g(\nu, \rho^*) \circ \lambda \\ &= (\rho^{*-1} U_\nu \rho^*) \circ \lambda. \end{aligned} \tag{10.9}$$

This implies that

$$f_0^{-1} x_0 f_0 \cdot f_1^{-1} x_1 f_1 \cdot \cdots \cdot f_{m-1}^{-1} x_{m-1} f_{m-1} = 1,$$

since the left hand side of this purported equation is the image under λ of $\rho^{*-1} U_0 \rho^* \cdot \rho^{*-1} U_1 \rho^* \cdots \rho^{*-1} U_{m-1} \rho^* = 1_G$. Hence

$$(f_0^{-1} x_0 f_0, \ldots, f_{m-1}^{-1} x_{m-1} f_{m-1}) \in A(C_0, \ldots, C_{m-1}),$$

and so

$$(f_0^{-1} x_0 f_0, \ldots, f_{k-1}^{-1} x_{k-1} f_{k-1}) \in A(C_0, \ldots, C_{k-1}) = A.$$

Since A is rigid, there is f in F such that

$$f^{-1} x_\nu f = f_\nu^{-1} x_\nu f_\nu, \qquad 0 \leqslant \nu \leqslant k-1, \tag{10.10}$$

and the same equation holds for $\nu = k, \ldots, m-1$, as $x_\nu = 1$ for ν in this range. Since $\Gamma \circ \phi = F$, there is γ in Γ with $\gamma \circ \phi = f$. Set

$$g = \gamma \circ \pi, \qquad \text{so that} \quad g \circ \lambda = f. \tag{10.11}$$

We now get

$$\rho^{*-1}U_\nu\rho^* \circ \lambda = f^{-1}x_\nu f, \tag{10.12}$$

by (10.9) and (10.10). By (10.11), we also get

$$g^{-1}U_\nu g \circ \lambda = f^{-1}(U_\nu \circ \lambda)\cdot f = f^{-1}u_\nu \circ \phi \cdot f = f^{-1}x_\nu f. \tag{10.13}$$

So (10.12), (10.13) yield

$$\rho^{*-1}U_\nu\rho^* \cdot g^{-1}U_\nu^{-1}\cdot g \in \ker\lambda, \qquad 0 \leqslant \nu \leqslant m-1. \tag{10.14}$$

Now $\ker\phi = \Lambda_1 = \ker\pi\lambda$, and $\Lambda_1 \circ \tau = H_1$, whence $H_1 \circ \lambda = \Lambda_1 \circ \pi\lambda = \Lambda_1 \circ \phi = 1$, and so $H_1 \subseteq \ker\lambda$. Since G/H_1 is simple and $G \circ \lambda = \Gamma \circ \pi\lambda = \Gamma \circ \phi = F$, we get $H_1 = \ker\lambda$, and so (10.14) gives

$$\rho^{*-1}U_\nu\rho^* \in H_1 \cdot g^{-1}U_\nu g = g^{-1}H_1U_\nu g, \qquad 0 \leqslant \nu \leqslant m-1.$$

Since $G = \langle U_0,\dots, U_{m-1}\rangle$, we find that

$$\rho^{*-1}z\rho^* \in g^{-1}H_1 zg \qquad \text{for all} \quad z \in G. \tag{10.15}$$

Taking $z \in H_1$ gives

$$\rho^{*-1}H_1\rho^* \subseteq g^{-1}H_1 g = H_1,$$

and since G is a finite group, ρ^* normalizes H_1.

If $\zeta \in \Gamma$, then $z = \Delta\zeta \in G$, and (10.15) becomes $\rho^{*-1}\Delta\zeta\rho^* \subseteq \gamma^{-1}\Lambda_1\zeta\gamma$ ($g = \Delta\gamma$, $H_1 = \Lambda_1/\Delta$). Since $\Delta \subseteq \Lambda_1$, we get

$$\rho^{*-1}\Lambda_1\zeta\rho^* = \gamma^{-1}\Lambda_1\zeta\gamma \qquad \text{for all} \quad \zeta \in \Gamma.$$

Hence, ρ^* induces conjugation by $\Lambda_1\gamma$ on Γ/Λ_1, and the proof is complete.

Set

$$\mathcal{N} = \text{normalizer in } \mathcal{G}^* \text{ of } H_1,$$

$$W = \mathcal{N} \circ \sigma \qquad (\subseteq U = U_D).$$

Since $\mathcal{G}^*/\mathcal{G}_D^*$ is Abelian, we have, by the preceding lemma, $\mathcal{G}_D^* \subseteq \mathcal{N} \lhd \mathcal{G}^*$, and we have an exact sequence

$$1 \to \mathcal{G}_D^* \to \mathcal{N} \xrightarrow{\sigma} U \xrightarrow{q} U/W \to 1, \tag{10.16}$$

where, for future reference, I have labelled by q the quotient map $\bar{l} \mapsto \bar{l}W$ ($\bar{l} \in U$).

For each $\rho^* \in \mathcal{G}^*$, we have already defined Λ_{ρ^*}. We next define $\Lambda(\rho^*)$ to be the unique subgroup of Γ which contains Λ and satisfies

$$\Lambda(\rho^*)/\Lambda = \text{centralizer in } \Gamma/\Lambda \text{ of } \Lambda_{\rho^*}/\Lambda. \tag{10.17}$$

By Lemma 9.1, $\Lambda(\rho^*)/\Lambda \cong F$. Since $\mathcal{N}$ acts on Λ_1/Λ, it follows that $\mathcal{N}$ acts on $\Lambda(1)/\Lambda$, and by Lemma 9.2(i), $\mathcal{N}$ acts on $\Lambda(\rho^*)/\Lambda$ for all $\rho^* \in \mathcal{G}^*$. Furthermore, by Lemma 9.2(ii), $\mathcal{G}_D^*$ induces inner automorphisms of Γ/Λ. Since

$$\Lambda(\rho^*)/\Delta = \rho^{*-1}(\Lambda(1)/\Delta)\,\rho^*,$$

it follows that $\Lambda(\rho^*)$ depends only on the coset $\mathcal{N}\rho^*$. From (10.16) we get an exact sequence

$$1 \to \mathcal{N} \to \mathcal{G}^* \xrightarrow{\bar{\sigma}} U/W \to 1, \tag{10.18}$$

where $\bar{\sigma}$ is the composition of σ with the quotient map q. So if $\rho^* \in \mathcal{G}^*$ and $\xi = \rho^* \circ \bar{\sigma} \in U/W$, then

$$\Lambda_\xi = \Lambda_{\rho^*}, \qquad \Lambda(\xi) = \Lambda(\rho^*) \tag{10.19}$$

are well-defined subgroups of Γ depending only on ξ. Hence, if we set

$$L = \Gamma/\Lambda, \qquad L_\xi = \Lambda_\xi/\Lambda, \qquad L(\xi) = \Lambda(\xi)/\Lambda, \tag{10.20}$$

then

$$L = L(\xi_1) \times \cdots \times L(\xi_t),\ \{\xi_1, \dots, \xi_t\} = U/W \qquad (\xi_1 = 1_{U/W}),$$
$$L_{\xi_1} = \Lambda_1/\Lambda, \qquad L(\xi) \cong F, \qquad \text{all } \xi \in U/W.$$

Also, set

$$\mathcal{H}^* = \mathcal{G}^*/H \qquad \text{(see (10.7))}. \tag{10.22}$$

We have an action of $\mathcal{H}^*$ on L, given by

$$L \times \mathcal{H}^* \to L,$$
$$(y, \eta^*) \mapsto y \circ \eta^*,$$

where $y \circ \eta^* = \rho^{*-1} y \Delta \rho^* \cdot \Lambda$, if $\eta^* = H\rho^*$, $\rho^* \in \mathcal{G}^*$, and $y = \Lambda\gamma$, $g = \Delta\gamma \in G$. Setting

$$\mathcal{H}_D^* = \mathcal{G}_D^*/H \qquad (\text{note that } \mathcal{G}_D^* \supseteq G),$$

we get from (8.1) an exact sequence

$$1 \to \mathcal{H}_D^* \to \mathcal{H}^* \xrightarrow{\sigma} U_D \to 1. \tag{10.23}$$

Strictly speaking, the map σ above comes from the commutative triangle

$$\begin{array}{ccc} \mathscr{G}^* & \xrightarrow{\sigma} & U \\ & {\scriptstyle q}\searrow \quad \nearrow & \\ & \mathscr{H}^* & \end{array} \qquad (\rho^* \circ q = H\rho^*).$$

And I choose to call the unlabelled arrow σ, too. Explicitly, then,

$$H\rho^* \circ \sigma = (\operatorname{res} \rho^*)|_{Q(\mu_p)},$$

which is well-defined, since $H \subseteq \ker \operatorname{res}$, H being the identity on $C(T)$.

Let

$$N = \mathcal{N}/H.$$

For each $\xi \in U/W$, we thus get a map

$$\alpha(\xi) : N \to \operatorname{Aut} L(\xi),$$

given by

$$n \circ \alpha(\xi) = \text{the automorphism of } L(\xi) \text{ induced by } \rho^*,$$
$$\text{where} \quad n = H\rho^*, \ \rho^* \in \mathcal{N}.$$

Denote by $\tilde{\alpha}(\xi)$ the composition of $\alpha(\xi)$ and the quotient map $\operatorname{Aut} L(\xi) \to \operatorname{Aut} L(\xi)/\operatorname{Inn} L(\xi)$, and set

$$N(\xi) = \ker \tilde{\alpha}(\xi). \tag{10.24}$$

LEMMA 10.2. *$N(\xi)$ is independent of ξ.*

Proof. This is a consequence of Lemma 9.2. Namely, by Lemmas 10.1 and 9.2, $\mathscr{G}_D^*$ induces inner automorphisms of L, so in particular induces inner automorphisms of $L(\xi)$, a direct factor of L. So $\mathscr{H}_D^* \subseteq N(\xi)$, and as

$$\mathscr{H}^*/\mathscr{H}_D^* \cong \mathscr{G}^*/\mathscr{G}_D^* \cong U, \qquad \text{an Abelian group,}$$

we have $N(\xi) \lhd \mathscr{H}^*$, so $N(\xi)$ induces inner automorphisms of $L(\xi')$ for all $\xi' \in U/W$, by Lemma 9.2. Hence, $N(\xi) \subseteq N(\xi')$, so by symmetry, we get this lemma.

Set

$$N_1 = N(\xi) \qquad (\xi \in U/W),$$
$$V = N_1 \circ \sigma. \tag{10.25}$$

Then $V \subseteq W$, and I assert that

$$W/V \cong N \circ \alpha(\xi)/\operatorname{Inn} L(\xi), \tag{10.26}$$

where I am using the trivial fact that

$$\operatorname{Inn} L(\xi) \subseteq N \circ \alpha(\xi),$$

which holds since $G/H \subseteq N$, and visibly

$$G/H \circ \alpha(\xi) = \operatorname{Inn} L(\xi).$$

This remark implies that the right hand side of (10.26) is defined. Where does the isomorphism come from? If $n = H\rho^* \in N$, then $n \circ \sigma = x \in W$. Furthermore, by definition of $N(\xi) = N_1$, $\tilde{a}(\xi)$ induces an isomorphism of N/N_1 and $N \circ \alpha(\xi)/\operatorname{Inn} L(\xi)$. On the other hand, $N \circ \sigma = W$, $N_1 \circ \sigma = V$, and $\ker \sigma = \mathcal{H}_D^* \subseteq N_1$, so σ induces an isomorphism of N/N_1 and W/V. This is (10.26).

In order to lay bare (10.26) more thoroughly, further work is required. Let

$$\Lambda u_\nu = \prod_{\xi \in U/W} u_\nu(\xi), \qquad u_\nu(\xi) \in L(\xi), \quad 0 \leqslant \nu \leqslant m - 1.$$

We need to relate these components $u_\nu(\xi)$ to ϕ. Here is the picture we use:

$$1 \longrightarrow \Lambda_1 \longrightarrow \Gamma \xrightarrow{\quad \phi \quad} F \longrightarrow 1.$$

$$\begin{array}{c} {}_q \searrow \quad \nearrow {}_p \\ L \end{array}$$

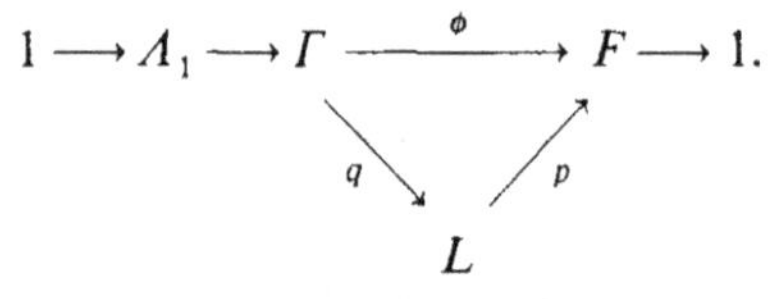

The triangle exists since $L = \Gamma/\Lambda$ with $\Lambda \subseteq \Lambda_1$. Now

$$\Lambda_1 \circ q = \prod_{\substack{\xi \in U/W \\ \xi \neq \xi_1}} L(\xi)$$

and so $u_\nu \circ \phi = x_\nu = u_\nu \circ qp = (\prod_{\xi \in U/W} u_\nu(\xi)) \circ p = u_\nu(\xi_1) \circ p$. This implies in particular that

$$o(u_\nu(\xi_1)) = d_\nu, \qquad 0 \leqslant \nu \leqslant m - 1.$$

On the other hand, if $\eta^* = H\rho^* \in \mathcal{H}^*$, and $\rho^* \circ \sigma = \bar{l}$, then

$$L(\xi) \circ \eta^* = L(\bar{l}\xi), \tag{10.27}$$

by definition of $L(\xi)$, and by (10.5) and (10.19). This is so, since $L(\xi)=$ centralizer in L of L_ξ, and $L_\xi \circ \eta^* = L_{\bar{l}\xi}$. So we pick $\rho^* \in \mathscr{G}^*$, set $\eta^* = H\rho^*$, and let

$$\rho_v^* \text{ be the unique element of } \rho^* G \cap \mathscr{G}_{c_v},$$

so that

$$\rho^* = \rho_v^* g(v, \rho^*)^{-1}, \qquad g(v, \rho^*) \in G.$$

Furthermore, by (8.2) (where ρ_v^* was written $\rho_{c_v}^*$),

$$\rho_v^{*-1} U_v \rho_v^* = U_v^l, \qquad l \in \rho^* \cdot \sigma.$$

This implies that, with $\eta_v^* = H\rho_v^*$,

$$(\Lambda u_v) \circ \eta_v^* = \prod_{\xi \in U/W} u_v(\xi)^l = \prod_{\xi \in U/W} (u_v(\xi) \circ \eta_v^*). \tag{10.28}$$

From (10.27), (10.28), we conclude that

$$u_v(\xi) \circ \eta_v^* \in L(\bar{l}\xi),$$

since $H \subseteq N_1$, and so we get the lovely, indisputable relation

$$u_v(\xi) \circ \eta_v^* = u_v(\bar{l}\xi)^l, \qquad \bar{l} = \eta_v^* \circ \sigma,$$
$$\eta_v^* = H\rho_v^*, \quad \rho_v^* \in \mathscr{G}_{c_v}. \tag{10.29}$$

This relation has various consequences. First,

$$o(u_v(\xi)) = d_v, \qquad \text{for} \quad \xi \in U/W, \text{ and } 0 \leqslant v \leqslant m-1. \tag{10.30}$$

This follows, since l is prime to D, so is prime to d_v, whence $o(u_v(\bar{l}\xi)^l) = o(u_v(\bar{l}\xi)) = o(u_v(\xi) \circ \eta_v^*) = o(u_v(\xi))$, and we may choose $\bar{l} \in \xi^{-1}$. Next, if $\eta^* = H\rho^* \in N$, then $\bar{l} \in W$, so $\bar{l}\xi = \xi$, and we get

$$C_v(\xi) \circ \eta^* = C_v(\xi) \circ \bar{l}, \qquad \bar{l} = \eta^* \circ \sigma, \qquad 0 \leqslant v \leqslant m-1, \tag{10.31}$$

where

$$C_v(\xi) \text{ is the conjugacy class of } L(\xi) \text{ which contains } u_v(\xi). \tag{10.32}$$

Setting

$$A_\xi = A_{L(\xi)}(C_0(\xi), \ldots, C_{m-1}(\xi)),$$

and

$$\mathscr{S}_\xi = \{A_\xi \circ \bar{l} \mid \bar{l} \in U_D\},$$

we can at last identify $N \circ \alpha(\xi)$.

LEMMA 10.3. $N \circ \alpha(\xi) = \{\theta \in \operatorname{Aut} L(\xi) \mid \mathscr{S}_\xi \circ \theta = \mathscr{S}_\xi\} = \mathscr{A}(\mathscr{S}_\xi)$.

Proof. By (10.31), we get that

$$N \circ \alpha(\xi) \subseteq \{\theta \in \operatorname{Aut} L(\xi) \mid \mathscr{S}_\xi \circ \theta = \mathscr{S}_\xi\}.$$

Conversely, however, suppose $\mathscr{S}_\xi \circ \theta = \mathscr{S}_\xi$ for some $\theta \in \operatorname{Aut} L(\xi)$. Then $A_\xi \circ \theta = A_\xi \circ \bar{l}$ for some $\bar{l} \in U_D$. Choose $\rho^* \in \mathscr{G}^*$ with $\rho^* \circ \sigma = \bar{l}$. Set $\eta^* = H\rho^*$. Then by (10.31)

$$C_\nu(\xi) \circ \eta^* = C_\nu(\xi) \circ \theta, \qquad 0 \leqslant \nu \leqslant m-1.$$

Hence, if we set

$$\beta = \eta^* \circ \alpha(\xi),$$

then $\theta\beta^{-1} = \delta$ is an automorphism of $L(\xi)$ such that $C_\nu(\xi) \circ \delta = C_\nu(\xi)$ for all $\nu = 0,\dots, m-1$. However, from (10.29), we conclude that $A(C_0(\xi),\dots, C_{m-1}(\xi))$ is rigid (more about this point in a moment), and $\delta \in \operatorname{Inn} L(\xi)$. Since $N \circ \alpha(\xi) \supseteq \operatorname{Inn} L(\xi)$, we get $\theta \in N \circ \alpha(\xi)$, and Lemma 10.3 is proved. This gives us complete control of W/V, which, to me, is the most unusual part of this paper.

The preceding results enable us to produce good isomorphisms from $L(\xi)$ to F. We use (10.8). Since $\ker \lambda = H_1 \supseteq H$, we can extend our triangle to the following picture:

$$\begin{array}{ccccc}
\Gamma & \xrightarrow{\ \phi\ } & F & & \\
{\scriptstyle\pi}\downarrow & \overset{\lambda}{\nearrow} & \uparrow{\scriptstyle\lambda'} & & \\
G & \xrightarrow{\ \pi'\ } & G/H & \xleftarrow{\ i\ } & L.
\end{array}$$

Now both triangles are commutative and i is an isomorphism.

Let's take $L(\xi_1)$ first. We have

$$L = L(\xi_1) \times L_{\xi_1}, \qquad L_{\xi_1} = \Lambda_1/\Lambda, \qquad \Lambda_1 = \ker \phi,$$
$$H_1 = \ker \lambda, \qquad H_1/H = \ker \lambda',$$

so we set

$$j'_{\xi_1} = i \circ \lambda'|_{L(\xi_1)}.$$

Since $L_{\xi_1} \circ i = H_1/H = \ker \lambda'$, we have

$$L(\xi_1) \circ j'_{\xi_1} = F,$$

and since F, $L(\xi_1)$ are simple, j'_{ξ_1} is an isomorphism. It visibly has the property that

$$u_\nu(\xi_1) \circ j'_{\xi_1} = x_\nu,$$

whence

$$A_{\xi_1}(C_0(\xi_1),\dots, C_{k-1}(\xi_1)) \circ j'_{\xi_1} = A_F(C_0,\dots, C_{k-1}).$$

Suppose now that $\xi \in U/W$, $\xi \neq \xi_1$. Choose $\bar{l} \in \xi^{-1}$, and then choose $\rho^* \in \mathscr{G}^*$ with $\rho^* \circ \sigma = \bar{l}$. Set $\eta^* = H\rho^* \in \mathscr{H}^*$. Then

$$L(\xi) \circ \eta^* = L(\bar{l}\xi) = L(\xi_1).$$

Then set

$$j'_\xi = \eta^* \circ j'_{\xi_1}|_{L(\xi)}.$$

We get $L(\xi) \circ j'_\xi = F$. Moreover, we can see what happens to $C_\nu(\xi) \circ j'_\xi$. Since $C_\nu(\xi)$ is stable under inner automorphisms, we may, for each ν, replace η^* by η^*_ν. That is, $C_\nu(\xi) \circ j'_\xi = C_\nu(\xi) \circ \eta^*_\nu \circ j'_{\xi_1}$, and $\eta^* \circ \sigma = \eta^*_\nu \circ \sigma = \bar{l}$. For this good choice, (10.29) gives $C_\nu(\xi) \circ \eta^*_\nu = C_\nu(\xi_1) \circ \bar{l}$, as $\bar{l}\xi = \xi_1$. So

$$C_\nu(\xi) \circ j'_\xi = (C_\nu(\xi_1) \circ \bar{l}) \circ j'_{\xi_1} = C_\nu \circ \bar{l}.$$

We conclude that $A(C_0(\xi),\dots, C_{k-1}(\xi)) \circ j'_\xi = A(C_0,\dots, C_{k-1}) \circ \bar{l}$, and thus by (0.6), $A(C_0(\xi),\dots, C_{k-1}(\xi))$ is rigid. Moreover,

$$\mathscr{S}_\xi \circ j'_\xi = \mathscr{S} \qquad \text{for all} \quad \xi \in U/W,$$

whence

$$(j'_\xi)^{-1} \mathscr{A}(\mathscr{S}_\xi) j'_\xi = \mathscr{A}(\mathscr{S}). \tag{10.33}$$

I am in awe that the function-theory has led to such things.

We are almost home.

Set

$$\mathscr{H}_\infty = \mathscr{G}_\infty H/H, \tag{10.34}$$

$$\mathscr{D} = \text{centralizer of } L \text{ in } \mathscr{H}^*, \tag{10.35}$$

$$\tilde{\mathscr{H}}_\infty = \mathscr{D} \cap \mathscr{H}_\infty, \tag{10.36}$$

$$F^* = \mathscr{H}^*/\mathscr{D}, \qquad B' = N_1/\mathscr{D}, \tag{10.37}$$

$$H^* = \mathscr{H}^*/\tilde{\mathscr{H}}_\infty, \tag{10.38}$$

$$E_0 = \text{fixed field of res } \tilde{\mathscr{H}}_\infty \text{ on } C. \tag{10.39}$$

Let

$$e_0 = [E_0 : Q]. \tag{10.40}$$

With this array, we can fit some pieces together. First,

$$N_1 = \mathscr{D} \times L \circ i. \tag{10.41}$$

In any case, by definition, N_1 is the largest subgroup of $\mathscr{H}^*$ which induces inner automorphisms of L, so $\mathscr{D} \subseteq N_1$, and $L \circ i = G/H \subseteq N_1$. Furthermore, the given action of $\mathscr{H}^*$ on L maps G/H isomorphically onto Inn L, and so $N_1 = \langle \mathscr{D}, L \circ i \rangle$. Since i is an isomorphism, $L \circ i$ has no centre, so $\mathscr{D} \cap (L \circ i) = 1$. Since $\mathscr{D}$ is the kernel of the action of $\mathscr{H}^*$ on L, we have $\mathscr{D} \lhd \mathscr{H}^*$. Since $G \lhd \mathscr{G}^*$, and $L \circ i = G/H$, we have $L \circ i \lhd \mathscr{G}/H = \mathscr{H}^*$. This is (10.41).

From (10.41), we get immediately that

$$\operatorname{res} N_1 = \operatorname{res} \mathscr{D}, \tag{10.42}$$

since $L \circ i = G/H \subseteq \ker \operatorname{res}$. And then we get

$$K_V \text{ is the fixed field of res } N_1 \text{ on } C. \tag{10.43}$$

To see this, we use $N_1 \supseteq \mathscr{H}_D^*$, a consequence of Lemma 10.2. By (8.1), $\mathscr{G}_D^* = \ker \sigma$ acts trivially on $Q(\mu_D)$. On the other hand, $\mathscr{G}_D^* \supseteq \ker \operatorname{res}$, and $\operatorname{res} \mathscr{G}_D^* \lhd \mathscr{G}$, with $|\mathscr{G} : \operatorname{res} \mathscr{G}_D^*| = [Q(\mu_D) : Q]$, again by (8.1). Since $\mathscr{G} = \operatorname{Aut} C$, it follows that $Q(\mu_D) =$ fixed field of res $\mathscr{G}_D^*$ on C. Since $N_1 \supseteq \mathscr{H}_D^*$, and $N_1 \circ \sigma = V$ (by (10.25)), we get (10.43).

These bits of information give us the important

$$E_0 \supseteq K_V. \tag{10.44}$$

For $\tilde{\mathscr{H}}_\infty \leqslant \mathscr{D}$, and so $\operatorname{res} \tilde{\mathscr{H}}_\infty \subseteq \operatorname{res} \mathscr{D} = \operatorname{res} N_1$, whence (10.44) follows from (10.43) and the definition of E_0.

11. The Construction of E

For each natural number h, let

$$\mathscr{L}_h = \{ f \in K_\Lambda \mid f \in \text{Int, for all } x \in \Gamma(\infty),\ v_{t_x}(f) \geqslant -hd_0, \\ \text{where } t_x \text{ is a local variable for } \Lambda \text{ at } x \}.$$

We can compute dim $\mathscr{L}_h$ from the Rieman–Roch theorem, but all we need is that dim $\mathscr{L}_h < \infty$, which is obvious.

Visibly $\mathscr{L}_h$ is stable under G. Slightly less visibly, $\mathscr{L}_h$ is stable under $\mathscr{G}^*$. In order to see this, we need to characterize $\mathscr{L}_h$ by polynomial relations. So, as already used implicitly,

$$\mathrm{Int} = \{f \in K_\Delta \mid P_f(Y, Z) \text{ is monic in } Y\}.$$

Lemma 4.3 provides half the proof. The other half is equally easy. As for $\mathscr{L}_h$, suppose $f \in K_\Lambda \cap \mathrm{Int}$. Then $f \in \mathscr{L}_h \Leftrightarrow f/T^h$ is regular at every point of $\Gamma(\infty) \Leftrightarrow f \circ g/T^h$ is regular at ∞ for all g in $G \Leftrightarrow \tilde{F}_0(0) \neq 0$, where

$$0 \neq \tilde{F}_{f,h}(Y, Z) = \sum_{i=0}^{N_0(f)} \tilde{F}_i(Z) \cdot Y^{N_0(f)-i} \in C[Y, Z],$$

and $\tilde{F}_{f,h}(f/T^h, 1/T) = 0$. ($N_0(f)$ is defined in Section 4.)

So the three conditions

(i) $f \in K_\Lambda$,

(ii) $f \in \mathrm{Int}$,

(iii) $f \circ g/T^h$ is regular at ∞, for all $g \in G$,

are all stable under $\mathscr{G}^*$, and so $\mathscr{G}^*$ acts on $\mathscr{L}_h$.

For each subfield F of C, let

$$\mathscr{L}_h(F) = \{f \in \mathscr{L}_h, i_\infty(f) \in F((t))\}.$$

LEMMA 11.1. *$\mathscr{L}_h(Q)$ contains a basis for $\mathscr{L}_h$.*

Proof. Let $\{f_1, ..., f_l\}$ be a basis for $\mathscr{L}_h$. Let

$$i_\infty(f_j) = \sum_{n \geqslant -hd_0} a_{jn} t^n, \qquad j = 1, ..., l.$$

This produces for us an $l \times \infty$ matrix $A = (a_{jn})$ which has rank l, so there are integers $k_1 < k_2 < \cdots < k_l$, with $-hd_0 \leqslant k_1$ such that $B = (b_{ij})$ is non-singular, where $b_{ij} = a_{ik_j}$. Replacing $\{f_1, ..., f_l\}$ by another basis if necessary, we may assume at the outset that $B = Id_l$. With this normalization, choose $\rho_\infty \in \mathscr{G}_\infty$ and let $\rho = \mathrm{res}\, \rho_\infty$. Then

$$i_\infty(f_j \circ \rho_\infty - f_j) = \sum_{n \geqslant -hd_0} c_{jn} t^n = \sum_{n \geqslant -hd_0} (a_{jn} \circ \rho - a_{jn})\, t^n,$$

and so $c_{jk_i} = 0$, $1 \leqslant i \leqslant l$. This means that $i_\infty(f_j \circ \rho_\infty - f_j)$ and $\{i_\infty(f_j) \mid 1 \leqslant j \leqslant l\}$ are linearly independent, whence $i_\infty(f_j \circ \rho_\infty - f_j) = 0$, so $f_j \circ \rho_\infty = f_j$. This holds for all $\rho_\infty \in \mathscr{G}_\infty$, and the lemma follows.

From this lemma, we get

$$\mathscr{L}_h(F) = \sum_{j=1}^{l} F f_j \qquad \text{for every subfield } F \text{ of } C.$$

Since $[K_\Lambda : C(T)] < \infty$, there is $x \in K_\Lambda$ with $K_\Lambda = C(T, x)$, and by Lemma 4.2, we may assume that $x \in \text{Int}$. Choose h so large that $x \in \mathscr{L}_h$. With such a choice of h, we are guaranteed that for each non-trivial subgroup X of G/H, say $X = K/H$, with $K = \Lambda'/\Delta$, $\Lambda' \supset \Lambda$, we have $K_{\Lambda'} \cap \mathscr{L}_h \subset \mathscr{L}_h$. If $\mathscr{L}_h(Q) \subseteq K_{\Lambda'}$, since $\mathscr{L}_h(Q)$ contains a basis for $\mathscr{L}_h$, while $K_\Lambda{}'$ is a vector space over C, we would get $\mathscr{L}_h \subseteq K_{\Lambda'}$, which is false. Hence, $\mathscr{L}_h(Q) \cap K_{\Lambda'} \subset \mathscr{L}_h(Q)$ for all $\Lambda' \supset \Lambda$. This implies that

$$\operatorname{Re} g_Q(\Lambda) = \{f \in \mathscr{L}_h(Q) \mid g_0 \in G/H \text{ and } f \circ g_0 = f \Rightarrow g_0 = 1\} \neq \varnothing.$$

Choose $f \in \operatorname{Re} g_Q(\Lambda)$, and let

$$F(Y, Z) = P_f(Y, Z).$$

For each $\rho \in \mathscr{G}$, $(F \circ \rho)(Y, T)$ and $F(Y, T)$ have a common root, namely, f. Since F is monic in Y, we get

$$F(Y, Z) \in Q[Y, Z].$$

Let

$$F(Y, Z) = Y^{N_0} + F_1(Z) \cdot Y^{N_0 - 1} + \cdots, \qquad N_0 = |G : H|.$$

Replacing f by $af + b$ where a, b are suitably chosen rational integers, we see that there is $f \in \operatorname{Re} g_Q(\Lambda)$ such that

(a) $F(Y, Z) = P_f(Y, Z) = Y^{N_0} + F_1(Z) \cdot Y^{N_0 - 1} + \cdots \in Z[Y, Z]$,

(b) $F_1(Z) \neq 0$.

I need another characterization of the field E_0 of (10.39).

LEMMA 11.2. *E_0 is the smallest subfield F_0 of C such that $i_\infty(f \circ g_0) \in F_0((t))$ for all $g_0 \in G/H$.*

Proof. Since $\tilde{\mathscr{H}}_\infty \lhd \mathscr{H}^*$, and since $i_\infty(f) \in Q((t)) \subseteq E_0((t))$, it follows that $i_\infty(f \circ g_0) \in E_0((t))$ for all $g_0 \in G/H$. Let F_0 be the smallest subfield of C such that $i_\infty(f \circ g_0) \in F_0((t))$ for all $g_0 \in G/H$, so that $F_0 \subseteq E_0$. If $\rho_\infty \in \mathscr{H}_\infty$, and $\operatorname{res} \rho_\infty = \rho$ is the identity on F_0, I argue that $\rho_\infty \in \tilde{\mathscr{H}}_\infty$, that is $\rho_\infty g_0 = g_0 \rho_\infty$ for all $g_0 \in G/H$. To see this, we use $i_\infty((f \circ g_0) \circ \rho_\infty) = i_\infty(f \circ g_0) \circ \rho$, and conclude that $i_\infty((f \circ g_0) \circ \rho_\infty) = i_\infty(f \circ g_0)$, whence $(f \circ g_0) \circ \rho_\infty = f \circ g_0$. On the other hand, $f \circ \rho_\infty = f$, and so $(f \circ \rho_\infty) \circ g_0 = f \circ g_0$, whence $f \circ (\rho_\infty g_0) = f \circ (g_0 \rho_\infty)$, so that $\rho_\infty g_0 \rho_\infty^{-1} g_0^{-1}$ fixes f. Since $f \in \operatorname{Re} g_Q(\Lambda)$, it follows that $\rho_\infty g_0 \rho_\infty^{-1} g_0^{-1} = 1$ in G/H, and $\rho_\infty \in \tilde{\mathscr{H}}_\infty$. Hence, the fixed field of $\operatorname{res} \tilde{\mathscr{H}}_\infty$ on C is contained in F_0, that is, $E_0 \subseteq F_0$, so $E_0 = F_0$, as required.

Since $\tilde{\mathscr{H}}_\infty \lhd \mathscr{H}_\infty$, we now see that E_0 is a normal extension of Q, and that

$$\text{Gal}\, E_0/Q \cong \mathscr{H}_\infty/\tilde{\mathscr{H}}_\infty,$$

and that $\rho_\infty \mapsto \text{res}\, \rho_\infty|_{E_0}$ induces the isomorphism.

Let $\mathscr{O}$ be the ring of integers of E_0. Choose $\kappa \in \mathscr{O}$ such that $E_0 = Q(\kappa)$. Set $f^* = \kappa \cdot f$, $H(Y, Z) = P_{f^*}(Y, Z)$. Then $P_{f^*}(Y, Z) = P_f(Y/\kappa, Z) \cdot \kappa^{N_0} = Y^{N_0} + \kappa \cdot F_1(Z) \cdot Y^{N_0 - 1} + \cdots \in \mathscr{O}[Y, Z]$, and in addition

$$\text{if} \quad \eta^* \in \mathscr{H}_\infty, \quad \eta^* = H\rho^*, \quad \rho = \text{res}\, \rho^*, \quad \text{and}$$
$$H(Y, Z) = H \circ \rho(Y, Z), \quad \text{then} \quad \eta^* \in \tilde{\mathscr{H}}_\infty. \tag{11.1}$$

Let $f^* = y_1, ..., y_{N_0}$ be all the roots of $H(Y, T) = 0$ in K_Λ. Since

$$y_1 \in \mathscr{L}_h(E_0) \quad \text{and} \quad \tilde{\mathscr{H}}_\infty \lhd \mathscr{H}^*,$$

it follows that

$$y_i \in \mathscr{L}_h(E_0) \quad \text{for all} \quad i = 1, ..., N_0. \tag{11.2}$$

I apply Lemma 9.3 to the tower

$$Q(T) \subseteq E_0(T) \subseteq E_0(T, y_1, ..., y_{N_0}) \subseteq K_\Lambda,$$

taking $f(X) = H(X, T)$, $\alpha = y_1$, $\{\sigma_1, ..., \sigma_{e_0}\}$ a set of coset representatives for $\tilde{\mathscr{H}}_\infty$ in $\mathscr{H}_\infty$. By that lemma, $Q(T, y_1) = E_0(T, y_1, ..., y_{N_0})$ is a normal extension of $Q(T)$. Moreover, setting

$$P(Y, Z) = \prod_{i=1}^{e_0} (H \circ \tau_i)(Y, Z) \qquad (\tau_i = \text{res}\, \sigma_i),$$

we have

$$\deg_X P(X, T) = N_0 e_0, \qquad P(Y, Z) \in Z[Y, Z],$$
$$P(y_1, T) = 0, \qquad P(X, T) \text{ is irreducible in } Q(T)[X].$$

Let $z_0, z_1, ..., z_{b-1}$ be all the roots of $P(X, T) = 0$ in K_Λ. Then $b = e_0 N_0$, and $Q(T, y_1)$ splits $P(X, T)$. Choose notation so that $z_0 = y_1$.

Since $Q(T, z_0)$ splits $P(X, T)$, and $P(y_1, T) = 0$, we have

$$z_i = \sum_{j=0}^{b-1} a_{ij} z_0^j, \qquad a_{ij} \in Q(T). \tag{11.3}$$

Furthermore, the unique automorphism ϕ_k of $Q(T, y_1)$ which is the identity on $Q(T)$ and sends z_0 to z_k, sends z_r to z_s, where $s = s(k, r)$ is uniquely determined by k and r. But

$$z_r = \sum_{j=0}^{b-1} a_{rj} z_0^j,$$

so

$$z_r \circ \phi_k = \sum_{j=0}^{b-1} a_{rj} z_k^j = z_s = \sum_{j=0}^{b-1} a_{sj} z_0^j$$

$$= \sum_{j=0}^{b-1} a_{rj} \left\{ \sum_{j'=0}^{b-1} a_{kj'} z_0^{j'} \right\}^j . \tag{11.4}$$

This tells us that the following congruences hold in $Q(Z)[Y]$:

$$P\left(\sum_{j=0}^{b-1} a_{ij}(Z)\, Y^j, Z \right) \equiv 0 \qquad (\text{mod } P(Y, Z)), \quad 0 \leqslant i \leqslant b-1, \tag{11.5}$$

by (11.3), and

$$\sum_{j=0}^{b-1} a_{sj}(Z)\, Y^j - \sum_{j=0}^{b-1} a_{rj}(Z) \left\{ \sum_{j'=0}^{b-1} a_{kj'}(Z)\, Y^{j'} \right\}^j$$

$$\equiv 0 \qquad (\text{mod } P(Y, Z)), \quad 0 \leqslant r, \quad k \leqslant b-1, \quad s = s(k, r), \tag{11.6}$$

by (11.4).

Pick $D(Z) \in Q[Z]$ such that

$$0 \neq D(Z)\, a_{ij}(Z) \in Q[Z] \qquad \text{for all} \quad i, j. \tag{11.7}$$

We are almost ready, but we need to identify our Galois group. It is the group H^* of (10.38), that is,

$$\text{Gal } Q(T, y_1)/Q(T) \cong H^*. \tag{11.8}$$

To see this, note that by construction, we have

$$Q(T, y_1) = E_0(T, y_1, \ldots, y_{N_0}).$$

Since $y_1, \ldots, y_{N_0} \in \mathscr{L}_h(E_0)$, we get that $\tilde{\mathscr{H}}_\infty$ acts trivially on $Q(T, y_1)$. Since $|\mathscr{H}^* : \tilde{\mathscr{H}}_\infty| = e_0 N_0$, we get our isomorphism (11.8).

We invoke Hilbert's irreducibility theorem [3], and choose a positive integer M such that

(a) $P(Y, M)$ is irreducible in $Q[Y]$.

(b) $D(M) \neq 0$.

(c) $F_1(M) \neq 0$. (11.9)

(d) M^{-1/d_0} is smaller than the radius of convergence of each $i_\infty(z_j)$, $0 \leqslant j \leqslant b-1$.

Choose $\tau = iy$, $y > 0$ such that $T(\tau) = M$, and such that y is maximal. Then $t_\infty(\tau) = M^{-1/d_0}$ is the unique positive number such that $t_\infty(\tau)^{d_0} = M^{-1}$, since

$$t_\infty = q_0 + a_1 q_0^{1+d_0} + \cdots, \qquad t_\infty^{d_0} = T^{-1},$$

and since (C.vi) of Section 3 leads immediately to $a_1, \ldots \in Q$. This shows that $t_\infty(\tau)$ is real. If $t_\infty(\tau) < 0$, then T would vanish at some $\tau' = iy'$, $y' > y$, and so T would assume the value M at $\tau'' = iy''$, $y'' > y' > y$, against maximality of y. This is not an important point, but I mention it in passing. Set

$$\alpha_j = i_\infty(z_j) \mid t = M^{-1/d_0}, \qquad 0 \leqslant j \leqslant b-1.$$

Let

$$E^* = Q(\alpha_0, \alpha_1, \ldots, \alpha_{b-1}).$$

I next define an action of H^* on $\{0, \ldots, b-1\}$, by the following rule: if $\eta \in H^*$ and $i \in \{0, \ldots, b-1\}$, then $i \circ \eta = j$ if and only if $z_i \circ \eta = z_j$. Then H^* permutes $\{0, \ldots, b-1\}$ transitively and regularly. Furthermore, by (11.5) and (11.6), for each $\eta \in H^*$, there is a unique automorphism $\eta \circ \psi$ of E^* such that

$$\alpha_i \circ (\eta \circ \psi) = \alpha_{i \circ \eta}. \tag{11.10}$$

Thus,

$$\psi : H^* \to \text{Gal } E^*/Q \text{ is an isomorphism.} \tag{11.11}$$

I next show that

(i) $E_0 \subseteq E^*$,

(ii) if $x \in E_0$, then $x \circ \eta = x \circ (\eta \circ \psi)$ for all $\eta \in H^*$. (11.12)

To see this, note that $\{y_1, \ldots, y_{N_0}\} \subseteq \{z_0, \ldots, z_{b-1}\}$, since $H(Y, Z)$ divides $P(Y, Z)$ in $E_0[Y, Z]$, so there is a subset I of $\{0, 1, \ldots, b-1\}$, such that

$$\sum_{i=1}^{N_0} y_i = \sum_{i \in I} z_i.$$

By construction, we have

$$\sum_{i=1}^{N_0} y_i = \kappa \cdot F_1(T),$$

with $0 \neq F_1(T) \in \mathbb{Z}[T]$.

So we have

$$\sum_{i\in I} z_i = \kappa \cdot F_1(T), \tag{11.13}$$

whence, at $\tau = iy$, we have

$$\sum_{i\in I} \alpha_i = \kappa \cdot F_1(M). \tag{11.14}$$

Apply η to (11.13), and get

$$\sum_{i\in I} z_i \circ \eta = (\kappa \cdot F_1(T)) \circ \eta = \kappa \circ \eta \cdot F_1(T),$$

and so

$$\sum_{i\in I} z_{i\circ\eta} = \kappa \circ \eta \cdot F_1(T),$$

whence, at $\tau = iy$, we get

$$\sum_{i\in I} \alpha_{i\circ\eta} = \kappa \circ \eta \cdot F_1(M). \tag{11.15}$$

Apply $\eta \circ \psi$ to (11.14), obtaining

$$\sum_{i\in I} \alpha_i \circ (\eta \circ \psi) = \kappa \circ (\eta \circ \psi) \cdot F_1(M). \tag{11.16}$$

Now (11.9), (11.10), (11.15), (11.16) yield

$$\kappa \circ \eta = \kappa \circ (\eta \circ \psi),$$

and (11.12) follows from $Q(\kappa) = E_0$.

Let

$$E = \text{fixed field of } (\mathscr{D}/\tilde{\mathscr{H}}_\infty) \circ \psi \qquad \text{on} \quad E^*. \tag{11.17}$$

Thus, ψ induces an isomorphism, still called ψ,

$$\psi : F^* \to \text{Gal}\, E/Q. \tag{11.18}$$

The next step is to check that

$$\text{the fixed field of } B' \circ \psi \text{ on } E \text{ is } K_V. \tag{11.19}$$

By (10.44) we have $K_V \subseteq E_0$. By (10.42), (10.43) and (11.12), we get $K_V \subseteq E$. Since $B' \cong B' \circ \psi$ is perfect, while $\text{Gal}\, K_V/Q$ is Abelian, we see that

$K_V \subseteq$ fixed field of $B' \circ \psi$ on E. Since $|F^* \circ \psi : B' \circ \psi| = |F^* : B'| = |U : V| = [K_V : Q]$, (11.19) follows.

Set

$$B'_\xi = L(\xi) \circ i',$$

where i' is the composition

$$L \to L \circ i \to N_1 \to N_1/\mathscr{D} = B' \qquad \text{(see (10.41) and (10.37))}.$$

Let i'_ξ be the inverse to $i'|_{L(\xi)}$, and set

$$j''_\xi = i'_\xi j'_\xi \to F.$$

It remains to find a good isomorphism of B' and the group $B = F^{U/W}$ of Section 0. This is staring at us. For $b' \in B'$, we

$$b' = \prod_{\xi \in U/W} b'(\xi), \qquad b'(\xi) \in L(\xi) \circ i'.$$

So we map b' to $f_{b'}$ via

$$f_{b'}(\xi) = b'(\xi) \circ j''_\xi.$$

This is an isomorphism, whose inverse gives us our map

$$\iota : B \to F^*,$$

which is the composition of the isomorphism $B \cong B'$ and the inclusion of B' in F^*.

Since $F^* = \mathscr{H}^*/\mathscr{D}$, we are able to use (10.23). We have

$$N_1 = \mathscr{D} \times L \circ i, \qquad N_1 \circ \sigma = \mathscr{D} \circ \sigma = V,$$

so (10.23) induces an exact sequence

$$1 \to B' \to F^* \xrightarrow{\pi} U/V \to 1, \qquad B' = N_1/\mathscr{D}. \tag{11.20}$$

Here π is defined explicitly, as follows:

$$\text{if } f^* = \mathscr{D}\eta^*, \qquad \eta^* \in \mathscr{H}^*, \qquad \text{then } f^* \circ \pi = (\eta^* \circ \sigma) \cdot V.$$

This is well defined, since by (10.25), $N_1 \circ \sigma = V$, and $N_1 = \mathscr{D} \cdot L \circ i$, with $L \circ i = G/H \subseteq \ker \sigma$, so that $\mathscr{D} \circ \sigma = V$.

We are now ready to prove the main theorem. We have an exact sequence which modifies (11.20) to

$$1 \to B \xrightarrow{\iota} F^* \xrightarrow{\pi} U/V \to 1,$$

since the map ι factors through the inclusion map of B' in F^*. We have $B \circ \iota = B' = \mathscr{D}$. $L \circ i/\mathscr{D} \cong L$, and the centralizer of B' in F^* is 1, by definition of $\mathscr{D}$, together with $L = [L, L]$. This is (i) of the main theorem; (ii) follows from (10.27) and the definition of π; (iii) and (iv) are properties of our isomorphisms j''_{ξ}, in particular, of (10.33). We have just constructed E and ψ so that (v) holds, and the commutativity of the diagram is a consequence of (11.12) and (10.44); (vi) is trivial.

12. Proof of the Corollary

I first treat $L = L_2(2^n)$, $n \geqslant 2$. Let x_1, x_2, x_3, C_1, C_2, C_3 be as in Section 0. We need a certain, limited, amount of information about the characters of L. The degrees of the irreducible characters are 1, $q-1$, q, $q+1$, so if ϕ is an irreducible character, and p is a prime which divides $\phi(1)$, then ϕ is of defect 0 for p, so vanishes on all p-singular classes. This implies that

$$\phi(x_1) \cdot \phi(x_2) \cdot \phi(x_3) = 0$$

for all irreducible characters of L except for the trivial character. Hence, by the character theoretic formula already used in Section 0,

$$\operatorname{card} A(C_1, C_2, C_3) = \frac{|L|^2}{c(x_1) \cdot c(x_2) \cdot c(x_3)} = |L|,$$

since $c(x_1) = q$, $c(x_2) = q-1$, $c(x_3) = q+1$. If $(x_1, x_2, x_3) \in A$, then $L_0 = \langle x_1, x_2, x_3 \rangle$ has order $2^a(q^2-1)$, with $a > 0$. Let T be a Sylow 2-subgroup of L. Then $L_0 \cdot T = L$, and so $L_0 \cap T = T_0$ has order $2^a > 1$. Since T is Abelian, it follows that

$$T_0 \subseteq \bigcap_{x \in T} L_0^x = \bigcap_{x \in L} L_0^x = D,$$

so D is a non-trivial normal subgroup of L. But L is simple, so $D = L = L_0$, and A is rigid.

We have $\operatorname{Aut} L/\operatorname{Inn} L \cong \operatorname{Gal} F_{2^n}/F_2$, and in $\Gamma L_2(2^n)$, there is an automorphism α such that

$$C_1 \circ \alpha = C_1 \circ \bar{l}, \qquad C_2 \circ \alpha = C_2 \circ \bar{l}, \qquad C_3 \circ \alpha = C_3 \circ \bar{l},$$

where $l = q^2 + 1$, so that

$$l \equiv 1 \pmod 2,$$

$$l \equiv 2 \pmod{q-1},$$

$$l \equiv 2 \pmod{q+1}.$$

Of course, α is the field automorphism of $L_2(2^n)$ induced by the Frobenius automorphism: $x \mapsto x^2$ $(x \in GF(2^n))$. Since $C_2 \circ \bar{l}' = C_2 \Leftrightarrow l' \equiv \pm 1 \pmod{q-1}$, and $C_3 \circ \bar{l}' = C_3 \Leftrightarrow l' \equiv \pm 1 \pmod{q+1}$, it follows that W is as described, and (ii) of the corollary follows.

Now for (i) of the corollary. Let me first worry about rigidity, and save V, W, $\mathscr{A}$ for the end

I invoke private communication and computer printout, both reluctantly, but I have no better arguments. In 1979, D. Hunt computed card $A(B_1, B_2, B_3)$ for all conjugacy classes of F_1 such that if $x_i \in B_i$, then $x_1^2 = x_2^3 = 1$. Setting C_1, C_2, C_3, C_4 be as in Section 0, he found that in the two given cases, card $A = |F_1|$.

Case (2, 3, 71). Pick $(x_1, x_2, x_3) \in A$, set $L = \langle x_1, x_2, x_3 \rangle$. To prove rigidity, it is necessary and sufficient to show that $L = F_1$. Let M be a maximal normal subgroup of L, and set $S = L/M$.

LEMMA. *If $X = \langle x, y, z \rangle$ is a finite group and $x^a = y^b = z^c = xyz = 1$, where a, b, c are pairwise relatively prime, then $X = [X, X] = \langle x \rangle^X$.*

Proof. That X is perfect is obvious. Set $Y = \langle x \rangle^X$, and let $\bar{y}$, $\bar{z}$ be the images of y, z in X/Y. Then $\bar{y}^b = \bar{z}^c = \bar{y}\bar{z} = 1$, whence $\bar{y} = \bar{z} = 1$, $Y = X$.

In our situation, the lemma gives us $71 \mid |S|$, and S has a class C of involutions such that $o(xy) \leqslant 6$ for all $x, y \in C$. Among sporadic groups, only F_1 has an order divisible by 71, so if S is sporadic, we get $M = 1$, $L = S = F_1$. S is certainly no alternating group, since $71 \mid |S|$, $67 \nmid |S|$. So if $S \neq F_1$, then S is a Chevalley group with the following properties:

(i) $|S|$ divides $|F_1|$.
(ii) $71 \mid |S|$.
(iii) $o(xy) \leqslant 6$ for all $x, y \in C$.

There are no such Chevalley groups.

Case (2, 3, 29). Choose $(x_1, x_2, x_4) \in A(C_1, C_2, C_4)$. Let $L = \langle x_1, x_2, x_4 \rangle$, M a maximal normal subgroup, $S = L/M$. Table I can be used here, and

TABLE I

q	$o(q)$ in $F_{71}^{\times}$	$o(q)$ in $F_{29}^{\times}$	$\nu_q(\|F_1\|)$
2	35	28	46
3	35	28	20
5	5	14	9
7	70	7	6

in the preceding case. If S is a sporadic group, then since $29 \mid |S|$, we have $S \cong F_1, F'_{24}, J_4$ or Ru. If S is an alternating group A_n, then $n \geqslant 29$, so there is an Abelian subgroup of S of order 3.13^2, and so there is an Abelian subgroup of L of that order. From the character table of F_1, we learn that F_1 has no Abelian subgroups of order 3.13^2. So S is not an alternating group. Conditions (i), (ii), (iii) above (with 29 replacing 71) eliminate the Chevalley groups.

So we must kill F'_{24}, J_4, and Ru. This was done by S. Norton, with customary dispatch. J_4 is excluded, since $37 \mid |J_4|$, $37 \nmid |F_1|$. As for Ru, it has two classes of involutions, neither of which satisfies (iii). More precisely, if D is any class of involutions of Ru, then $o(xy) = 7$ for suitably chosen $x, y \in D$.

If $S = L = F'_{24}$, then we appeal to the character table of that group. F_1 has an irreducible character χ of degree 47.59.71, and as $S \subseteq F_1$, we get $\chi|_S = \sum a_\phi \phi$, where the summation is over the irreducible characters of F'_{24} and the a_ϕ are non-negative integers. If $a_\phi \neq 0$, then $\phi(1) \leqslant \chi(1) = 196883$. There are just 3 irreducible characters of S with this property, and the values of these three characters on elements of orders 1, 17, 23, 29 are shown in Table II, where I also include the values of χ on these same elements. We have $\chi|_S = a_1\phi_1 + a_2\phi_2 + a_3\phi_3$, and comparison of the last three columns of the above table gives, by inspection,

$$a_1 = a_2 = 3, \qquad a_3 = 1.$$

But $3.1 + 3.8671 + 57477 = 83493 \neq 196883$. In fact, the difference $196883 - 83493$ is $113390 = 17.23.29.10$ (10, not 0).

So there remains the case $S \cong F'_{24}$, $M \neq 1$. Comparison of Table I with the fact that $C_{F_1}(x_4)$ has order 3.29 forces $M = \langle u \rangle$ to be of order 3. The reason M is forced to have odd order is that it is well known that $|E| \leqslant 2^{24}$ for every elementary Abelian 2-subgroup E of F_1, whereas the table reveals that 2 is a primitive root mod 29. So u has order 3 and from the structure of centralizers of elements of order 3 in F_1, we get that u is of Fischer type. Let $C'' =$ class of $x_4 \cdot u$, $C' =$ class of $x_2 \cdot u^{-1}$. We have $(x_1, x_2u^{-1}, x_4 \cdot u) \in$

TABLE II

	1	17	23	29
ϕ_1	1	1	1	1
ϕ_2	8671	1	0	0
ϕ_3	57477	0	0	−1
χ	196883	6	3	2

$A(C_1, C', C'')$. From Hunt's calculations, we learn that $C' =$ Thompson elements of order 3. Since every element of F_1 of order 3 is real, it follows that the elementary Abelian subgroup $\langle x_2, u\rangle$ of order 3^2 has at least 2 Suzuki elements, 2 Fischer elements, 2 Thompson elements. According to Norton, there are no such subgroups in F_1. More precisely, it can be shown that if X is any non cyclic subgroup of F_1 of order 3^2, then the number of Thompson elements in X is either 0 or 6.

As for V, W, $\mathscr{A}$, it is known that every automorphism of F_1 is inner, which forces $\mathscr{A} = \operatorname{Inn} F_1$, $W = V$. The character table for F_1 carries with it the information about V, and gives it as indicated in Section 0.

References

1. E. Artin, "Galois Theory," Notre Dame Mathematical Lectures, 1953.
2. L. Bieberbach, "Lehrbuch der Funktionentheorie," Vol. 1, Chelsea, New York, 1945.
3. D. Hilbert, "Gesammelte Abhandlungen," Chelsea, New York, 1965.
4. L. Queen, Ph.D. thesis, University of Cambridge, 1979
5. G. Shimura, "Introduction to the Arithmetic Theory of Automorphic Functions," Princeton Univ. Press, Princeton, N. J., 1971.
6. C. Siegel, "Vorlesungen über ausgewählte Kapitel der Funktionentheorie," Teil I, Göttingen, 1964.

www.ingramcontent.com/pod-product-compliance
Lightning Source LLC
Chambersburg PA
CBHW060940210326
41598CB00031B/4685